Neuroeconomics: Core Topics and Current Directions

David V. Smith • Patricia L. Lockwood •
Dominic S. Fareri
Editors

Neuroeconomics: Core Topics and Current Directions

Springer

Editors
David V. Smith
Department of Psychology and Neuroscience
Temple University
Philadelphia, PA, USA

Patricia L. Lockwood
Centre for Human Brain Health
University of Birmingham
Birmingham, UK

Dominic S. Fareri
Gordon F. Derner School of Psychology
Adelphi University
Garden City, NY, USA

ISBN 978-3-032-02924-9 ISBN 978-3-032-02925-6 (eBook)
https://doi.org/10.1007/978-3-032-02925-6

This Springer imprint is published by the registered company Springer Nature Switzerland AG
The registered company address is: Gewerbestrasse 11, 6330 Cham, Switzerland

Preface

Neuroeconomics has matured into a dynamic and methodologically sophisticated discipline dedicated to understanding the neural, cognitive, and computational mechanisms underlying decision-making. What began as an interdisciplinary effort integrating concepts from neuroscience, psychology, and economics has evolved into a cohesive research enterprise that investigates how organisms assign value, form preferences, and choose among competing options. At its core, neuroeconomics provides a robust framework for exploring how decisions are made under uncertainty, emotional influences, social contexts, or constraints of time and effort. This volume, *Neuroeconomics: Core Topics and Current Directions*, offers a structured overview of recent advances, reflecting both the conceptual maturity and ongoing evolution of the field.

This book is designed to serve as an essential resource for researchers, clinicians, practitioners, and graduate students. Rather than providing an exhaustive historical survey, the chapters focus explicitly on current directions in neuroeconomics, synthesizing recent empirical findings in concise and accessible reviews. A key aim is to foster dialogue not only among specialists within neuroeconomics but also with researchers from adjacent disciplines who wish to engage with contemporary work in the field. By highlighting current trends and new insights, we hope this volume can stimulate novel connections and interdisciplinary collaborations.

The initial chapters establish historical and methodological foundations, tracing neuroeconomics from its origins in reward processing and reinforcement learning. These contributions emphasize how computational models and neuroscientific techniques have reshaped traditional assumptions about value and choice. Interdisciplinary methods—including lesion studies, neuroimaging, and behavioral paradigms—are discussed as central tools, providing a common framework that unites research across the diverse topics covered in subsequent chapters.

Building upon these foundations, later chapters investigate how value is represented and updated within dynamic and uncertain environments. Research on learning and valuation elucidates the processes by which individuals acquire and revise beliefs through feedback, integrate probabilistic information, and encode risk and ambiguity across neural systems. Rather than assuming stable preferences, these chapters emphasize the dynamic nature of valuation and its responsiveness to contextual influences. Closely related are discussions of temporal discounting, effort-based choice, and self-control. These chapters clarify how people evaluate trade-offs between immediate and future outcomes, linking behavioral observations to underlying neural and cognitive processes of goal-directed behavior and impulse regulation.

Subsequent chapters explore the modulatory roles of affect and context on decision-making behavior. Here, authors examine how emotional states, environmental framing, and cognitive interpretations shape valuation and choice architecture. Whether through amplifying affective responses or subtly altering evaluations via contextual cues, these modulatory influences demonstrate the inherent flexibility of preferences. Social context emerges as particularly influential, profoundly shaping how decisions are evaluated and enacted. Chapters on social preferences and social decision-making highlight the role of neural circuits involved in interpersonal evaluations of fairness, cooperation, and reciprocity. These findings underscore the notion that decision-making rarely occurs in isolation; instead, it is deeply embedded within networks of relationships and shared experiences.

The book also addresses significant variability in decision-making processes across individuals. Chapters on individual differences explore how factors such as developmental stage, psychiatric conditions, hormonal influences, and sociocultural identities affect valuation processes and behavioral outcomes. Rather than treating such variability as extraneous noise, these contributions position it as essential for understanding how decision-making systems function across diverse populations. This emphasis challenges normative assumptions and advocates for more inclusive, representative models of neuroeconomic behavior.

Finally, chapters in the concluding section extend insights from neuroeconomics into applied and translational domains, demonstrating the field's relevance to real-world decision making. Topics include consumer behavior, marketing influence, prosocial giving, and public policy design. These chapters illustrate how neural and computational methods can forecast aggregate behaviors, reveal latent preferences, and guide interventions aimed at enhancing decision outcomes. Translational applications serve as practical tests for theoretical models, allowing researchers to evaluate the validity of neuroeconomic insights beyond controlled laboratory settings.

Throughout this volume, contributors were encouraged to integrate foundational knowledge with recent empirical developments, identify important open questions, and situate their research within broader theoretical and societal contexts. Although each chapter stands independently as a concise and accessible review, the overall structure emphasizes conceptual continuity across diverse research domains. Readers will encounter recurring themes—such as interactions between valuation and self-control systems, the role of uncertainty, and the pervasive influence of social cognition—that connect insights from individual choices to collective outcomes.

In summary, this volume aims to provide students, researchers, clinicians, and practitioners with a broad yet focused introduction to contemporary neuroeconomics. By highlighting current directions, synthesizing recent findings, and presenting a wide range of topics in an accessible form, we hope this book will serve as both a valuable reference and a catalyst for future interdisciplinary inquiry and collaboration, deepening engagement with the fundamental questions that define the field today.

Philadelphia, PA, USA David V. Smith
Birmingham, UK Patricia L. Lockwood
Garden City, NY, USA Dominic S. Fareri

Contents

About the Editors

David V. Smith is an Associate Professor of Psychology and Neuroscience at Temple University, where he directs the Neuroeconomics Laboratory and serves as Associate Director of Temple's Brain Research and Imaging Center. After earning his PhD in Cognitive Neuroscience from Duke University in 2012, he completed a postdoctoral fellowship at Rutgers University–Newark and launched his lab at Temple in 2017. His research examines the neural mechanisms that support social and economic decision-making across the lifespan, with particular interest in how people evaluate rewards, respond to incentives, and cooperate with others. Combining neuroimaging, computational modeling, and transcranial electrical stimulation, his work aims to advance basic understanding and inform translational applications in clinical and policy contexts. His research has been supported by the National Institute of Mental Health, the National Institute on Drug Abuse, and the National Institute on Aging. At Temple, he teaches courses on sensation and perception, decision neuroscience, behavioral economics, statistics, and functional neuroimaging. He has been named a Rising Star and a Fellow by the Association for Psychological Science. In addition, he has received Temple's Psychology Honors Excellence in Mentoring Award as well as the College of Liberal Arts' Presidential Faculty Award for excellence in teaching and mentoring. Finally, he also serves as an Associate Editor at *Collabra: Psychology* and is an active peer reviewer across leading journals in psychology, neuroscience, and decision science.

Patricia L. Lockwood is a Professor of Decision Neuroscience and Wellcome Trust/Royal Society Sir Henry Dale Fellow at the University of Birmingham, where she directs the Social Decision Neuroscience Lab (www.sdn-lab.org). She was previously a Junior Research Fellow at Christ Church, University of Oxford, and a Medical Research Council Fellow at the University of Birmingham, University of Oxford, and the University of Zurich. She holds a PhD in Psychology from University College London. Her lab investigates social learning and decision-making across the lifespan and in neurological and psychiatric disorders using a mixture of computational modeling, behavioral measures, self report, patient studies, and neuroimaging. She combines frameworks from psychology, economics, ecology, and decision neuroscience to capture how, when, and why people learn and make choices that have consequences for themselves and other people. She holds multiple nationalities (British, Portuguese, Brazilian), and she is a Fellow of the Association for Psychological Science, and a member of the Women of the Future Network. She has received the Janet Taylor Spence Award for Transformative Early Career Contributions from the Association for Psychological Science and the Society for Neuroeconomics Early Career Award.

Dominic S. Fareri is an Associate Professor of Psychology at Adelphi University where he directs the Fareri Lab for Social Learning and Decision-Making. He is also the Director of Undergraduate Psychology and Neuroscience at Adelphi University. He received his PhD in Psychology at Rutgers University-Newark in 2013, and completed his postdoctoral training at the University of California, Los Angeles and Columbia University in 2015 before joining the faculty of Adelphi University in 2015. His lab investigates how the social environment becomes integrated into value-based computations supporting decision-making. Implementing functional neuroimaging alongside behavioral and computational approaches, his research draws on frameworks from psychology, cognitive neuroscience, decision neuroscience to characterize how day-to-day social contexts change our decision-making and our subjective experiences of rewards. His work has been supported by the National Institute of Mental Health and the National Science Foundation. He serves on the editorial board for *Social Cognitive and Affective Neuroscience, Imaging Neuroscience and Scientific Reports.* He was named a Rising Star by the Association for Psychological Science, has received the Faculty Excellence Award for Scholarship from Adelphi University, and is serving as the President of the Social and Affective Neuroscience Society (SANS) for the 2025–2026 year.

Part I
Foundations of Neuroeconomics: History and Methods

David V. Smith, Patricia L. Lockwood, and Dominic S. Fareri

Neuroeconomics is an interdisciplinary field integrating neuroscience, psychology, and economics to understand how people make decisions. Departing from traditional economic models, which infer preferences solely from observed choices, neuroeconomics seeks to elucidate how the brain computes value, updates beliefs, manages uncertainty, and incorporates social information (Glimcher, 2010). This section introduces foundational studies that defined the early trajectory of the field, establishing core concepts, questions, and methodologies. It also presents key empirical and computational methods—such as neuroimaging, reinforcement learning models, lesion studies, and behavioral economic tasks—that have become standard approaches in neuroeconomics. Collectively, these contributions provide an essential starting point for understanding both the historical context and future directions of neuroeconomic research.

One of neuroeconomics' earliest and most influential findings is the discovery that midbrain dopamine neurons encode reward prediction errors—the discrepancy between expected and actual outcomes. Schultz (this volume) reviews foundational research demonstrating how these neural signals instantiate learning mechanisms described by temporal difference models, facilitating adaptive value updating over time (Schultz et al., 1997; Schultz, 2016). His chapter further details how the brain differentiates reward components, such as magnitude, probability, and delay, across distinct yet interacting valuation systems. This biologically grounded understanding of utility laid the groundwork for computational modeling in human decision-making research.

D. V. Smith (✉)
Temple University, Philadelphia, PA, USA
e-mail: david.v.smith@temple.edu

P. L. Lockwood
University of Birmingham, Birmingham, UK

D. S. Fareri
Adelphi University, Garden City, NY, USA

Building upon this neural understanding of value, Terry and Huettel (this volume) explore foundational experiments addressing uncertainty, specifically distinguishing between risk and ambiguity. They highlight findings that brain regions such as the anterior insula and dorsolateral prefrontal cortex exhibit differential sensitivity to uncertainty. Such sensitivity challenges economic perspectives that treat risk merely as a function of expected value, instead revealing that uncertainty engages distinct affective and cognitive neural systems (Loewenstein et al., 2008). This research advanced a richer conceptualization of decision making under uncertainty, recognizing it as a process shaped by emotional experiences and cognitive evaluations rather than purely statistical calculations.

A complementary methodological perspective emerges in Fellows (this volume), who leverages lesion studies to uncover the causal roles of prefrontal subregions in value-based decision making. Her chapter illustrates that orbitofrontal cortex damage specifically impairs option evaluation and comparison without broadly affecting executive functions, emphasizing the distinct contributions of prefrontal regions to economic behaviors. Lesion-based dissociations thus provide crucial insights, highlighting how economic choice involves distributed yet distinct components within the frontal lobe—insights not achievable through neuroimaging alone.

Cooper and McClure (this volume) further expand methodological approaches by synthesizing two decades of research on delay discounting. Their chapter traces how intertemporal choice paradigms have revealed distinct neural signatures differentiating immediate from delayed rewards. From a computational perspective, they argue discounting behaviors arise not from a singular self-control mechanism but from dynamic interactions among memory, valuation, and prospection systems. They identify new frontiers, including episodic memory's role and direct intracranial recording's potential to clarify dopamine's role in temporal decision-making processes.

Another foundational strand of neuroeconomic research examines how social contexts influence valuation and learning processes. Wu and O'Doherty (this volume) describe how individuals evaluate outcomes affecting others through neural circuits similar to those involved in self-relevant reward processing. Extending this theme, Ratala and Sanfey (this volume) analyze interactive decisions related to trust, fairness, and cooperation through structured tasks like the Trust and Ultimatum Games. They demonstrate that valuation systems dynamically interact with social cognition networks, notably the temporoparietal junction and medial prefrontal cortex, integrating economic and interpersonal motivations (Chang et al., 2011; Rilling & Sanfey, 2011). These insights establish social decision-making as a central and integral area within neuroeconomic inquiry, rather than a peripheral topic.

Finally, these core neuroeconomic mechanisms extend to practical, everyday decisions in consumer contexts. Reeck (this volume) applies neuroeconomic principles to consumer behavior, illustrating how neural responses to prices, branding, and product features encode subjective value. Her chapter highlights neural data's potential to predict preferences more accurately than traditional behavioral or self-report methods (Plassmann et al., 2007; Karmarkar & Plassmann, 2019), demonstrating neuroeconomics' relevance to real-world applications.

While this section emphasizes foundational contributions that shaped neuroeconomics, it also identifies emerging directions for future research. Key open questions include understanding how value representations evolve over developmental and contextual timescales, elucidating neural mechanisms underlying social learning in complex multi-agent settings, and leveraging neural models to inform economic policy and ecological decision-making contexts. These early studies laid a robust groundwork, continuing to inspire ongoing efforts to deepen our understanding of decision-making.

References

Chang, L. J., Smith, A., Dufwenberg, M., & Sanfey, A. G. (2011). Triangulating the neural, psychological, and economic bases of guilt aversion. *Neuron, 70*(3), 560–572. https://doi.org/10.1016/j.neuron.2011.02.056

Cooper, A., & McClure, S. M. (this volume). Delay discounting: Empirical foundations and future directions. In D. V. Smith, P. Lockwood, & D. S. Fareri (Eds.), *Neuroeconomics: Core topics and current directions*. Springer. Preprint: https://doi.org/10.31234/osf.io/twn3f_v1

Fellows, L. K. (this volume). What lesion studies have taught us about frontal lobe contributions to decision-making. In D. V. Smith, P. Lockwood, & D. S. Fareri (Eds.), *Neuroeconomics: Core topics and current directions*. Springer. Preprint: https://doi.org/10.31234/osf.io/geqnc_v1

Glimcher, P. W. (2010). *Foundations of neuroeconomic analysis*. Oxford University Press.

Karmarkar, U. R., & Plassmann, H. (2019). Consumer neuroscience: Past, present, and future. *Organizational Research Methods, 22*(1), 174–194. https://doi.org/10.1177/1094428117730598

Loewenstein, G., Rick, S., & Cohen, J. D. (2008). Neuroeconomics. *Annual Review of Psychology, 59*, 647–672. https://doi.org/10.1146/annurev.psych.59.103006.093710

Plassmann, H., O'Doherty, J., Shiv, B., & Rangel, A. (2007). Marketing actions can modulate neural representations of experienced pleasantness. *Proceedings of the National Academy of Sciences, 105*(3), 1050–1054. https://doi.org/10.1073/pnas.0706929105

Ratala, A., & Sanfey, A. G. (this volume). Foundations of social decision-making. In D. V. Smith, P. Lockwood, & D. S. Fareri (Eds.), *Neuroeconomics: Core topics and current directions*. Springer.

Reeck, C. (this volume). Foundations of consumer neuroscience. In D. V. Smith, P. Lockwood, & D. S. Fareri (Eds.), *Neuroeconomics: Core topics and current directions*. Springer. Preprint: https://doi.org/10.31234/osf.io/aeghp

Rilling, J. K., & Sanfey, A. G. (2011). The neuroscience of social decision-making. *Annual Review of Psychology, 62*, 23–48. https://doi.org/10.1146/annurev.psych.121208.131647

Schultz, W. (2016). Dopamine reward prediction-error signalling: A two-component response. *Nature Reviews Neuroscience, 17*(3), 183–195. https://doi.org/10.1038/nrn.2015.26

Schultz, W. (this volume). Reward learning and choice: From theory to neuronal signals. In D. V. Smith, P. Lockwood, & D. S. Fareri (Eds.), *Neuroeconomics: Core topics and current directions*. Springer. Preprint: https://ssrn.com/abstract=4890719

Schultz, W., Dayan, P., & Montague, P. R. (1997). A neural substrate of prediction and reward. *Science, 275*(5306), 1593–1599. https://doi.org/10.1126/science.275.5306.1593

Terry, J. M., & Huettel, S. A. (this volume). Resolving uncertainty: Foundational neuroeconomic studies of risk and ambiguity. In D. V. Smith, P. Lockwood, & D. S. Fareri (Eds.), *Neuroeconomics: Core topics and current directions*. Springer.

Wu, Q., & O'Doherty, J. P. (this volume). Neural mechanisms of reward, decision-making and learning in a social context: Insights from computational neuroimaging. In D. V. Smith, P. Lockwood, & D. S. Fareri (Eds.), *Neuroeconomics: Core topics and current directions*. Springer. Preprint: https://doi.org/10.31219/osf.io/ncz4j_v2

Reward Learning and Choice: From Theory to Neuronal Signals

Wolfram Schultz

Abstract Rewards satisfy the needs of biological agents for survival and gene propagation that result in evolutionary fitness. To do so, rewards elicit learning (positive reinforcement), approach behaviour, economic choice and emotions like pleasure, happiness and desire. Dopamine neurons signal reward prediction errors (RPE) that are suitable for learning according to reinforcement learning theory, including Pavlovian learning and temporal difference learning. As biological needs vary between individual agents, the value of reward is subjective, which can be assessed using principles of economic choice theory. Theory-driven tests ensure that animals make meaningful choices for inferring subjective reward value. Dopamine neurons code RPEs in the subjective value formalized as utility. Concepts of Revealed Preference Theory allow to extend these tests to rewards that typically contain multiple components. Neuronal responses in monkey orbitofrontal cortex follow preferences for multi-component rewards, even when preferred rewards have smaller individual components. These responses reflect the transformation of vectorial multi-component reward into scalar subjective value. In sum, concepts from animal learning theory and economic choice theory inform experimental designs of neurophysiological reward experiments that transcend ad-hoc situations and validate, extend and ultimately reduce theories of reinforcement learning and economic choice.

1 Introduction

The investigation of neuronal reward signals during the past 30 years has substantially benefitted from concepts of animal learning and economic choice theories. Reward is a theoretical concept that explains learning, approach and choice. However, there are no specific sensory receptors for rewards that provide a hardware

W. Schultz (✉)
Department of Physiology, Development and Neuroscience, University of Cambridge, Cambridge, UK
e-mail: ws42@pm.me

D. V. Smith et al. (eds.), *Neuroeconomics: Core Topics and Current Directions*,
https://doi.org/10.1007/978-3-032-02925-6_1

definition of their function. Rather, the function of rewards is defined by their value for the survival of humans and animals (collectively called agents henceforth) and the propagation of their genes. Thus, rewards serve evolutionary fitness by ensuring that agents survive and successfully propagate their genes into the next generation until that generation can propagate their genes again.

As the needs defining reward function differ between agents, reward value is subjective. Take the example of satiety; a pineapple loses value for an agent who has just eaten a pineapple, even though the pineapples may be physically exactly the same. And we have no way to compare subjective values between agents as there exists no inter-individual metric for subjective reward value. Thus, we need to address subjective reward value when we like to understand how the brain gets us the rewards we need.

While reinforcement learning serves reward acquisition by establishing predictions, economic choice theory provides estimation methods for subjective reward value. Already in 1738, Bernoulli postulated that the subjective gain from the same additional unit of money ('marginal utility') decreases with increasing wealth of the agent (Bernoulli, 1738). Since then, economic choice theory has become suitable for designing well-controlled behavioural animal studies for exploring neuronal signals for reward value and thus establish neuronal foundations for economic choice. This short review will present a few such examples from several laboratories, including the author's own laboratory, that have helped in our understanding of economic choice behaviour.

2 The Dopamine Reward Signal

Several mammalian brain structures contain neurons that process reward information distinct from sensory and movement processes. The structures include the dopamine neurons, whose large majority carries a rather specific signal reflecting reward prediction errors (RPE), and the orbitofrontal cortex, striatum and amygdala containing subsets of reward-sensitive neurons. Additional brain structures contain neurons that report specific aspects of rewards, including hypothalamic neurons signalling nutrients.

Reinforcement learning (RL) theory defines RPE as key variable for learning. Pavlovian learning of reward-predicting stimuli advances with RPEs: only rewards that are better or worse than predicted contribute to learning (Kamin, 1969; Rescorla & Wagner, 1972). A positive RPE, elicited by a reward that is better than predicted, increases the prediction of that reward, and a negative RPE, elicited by a reward that is worse than predicted, decreases the reward prediction (Fig. 1a, red vs. green). The process constitutes error-driven learning: adjusting the prediction via an error between the actually experienced reward and the prediction. However, RL is not an error-correcting mechanism, as it does not involve a fixed setpoint to which the system returns after an error has been corrected. Indeed, predictions in RL are basically unlimited: predictions can grow as long as rewards can increase enough to

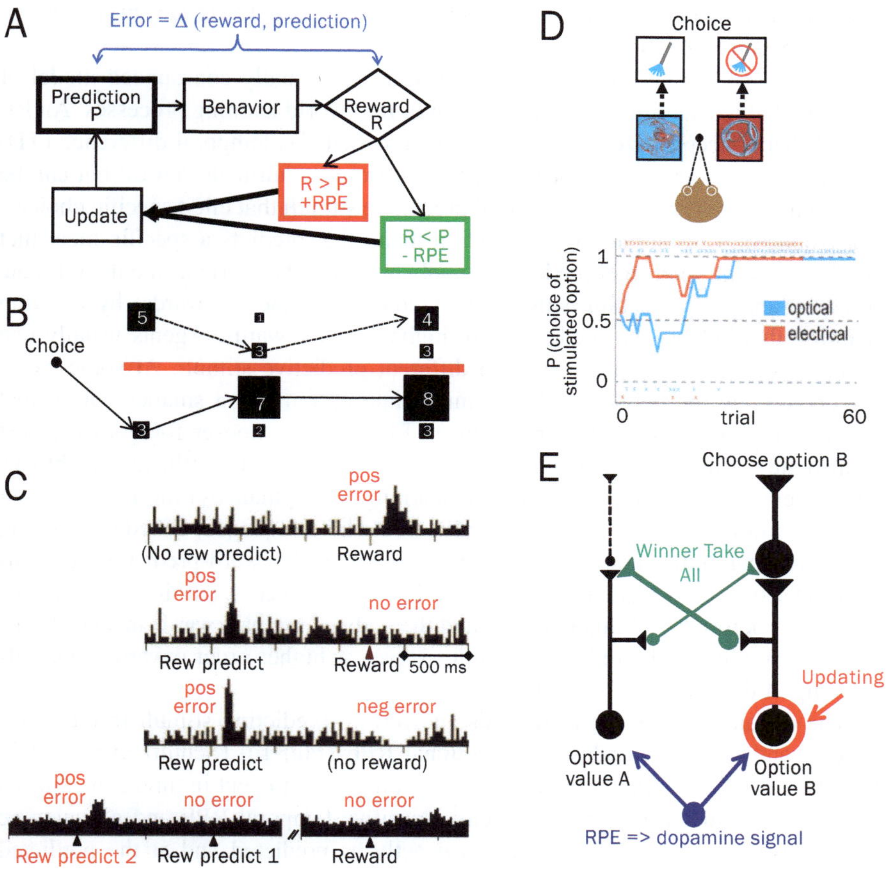

Fig. 1 Dopamine reward signal and reinforcement learning (RL) (**a**) RL block scheme. Positive reward prediction errors (+RPE: reward R exceeding prediction P) increase reward prediction, whereas negative RPEs decrease the prediction. Both predictions are unbounded, which distinguishes RL from error correction around a pre-determined setpoint (**b**) Maximization of state value rather than individual rewards using temporal difference (TD) learning of reward sequence. The agent learns the value of each primary or higher-order reward backwards after initial encounter of an ultimate primary reward. Overall state value is higher by foregoing an earlier larger reward for entering a sequence with later larger rewards (red line: sequence choice cannot be revisited). Numbers in black squares indicate values of primary and higher-order rewards (**c**) Dopamine RPE signal with reward and single or sequential conditioned stimuli (top three graphs, bottom graph). (Reproduced from Schultz et al., 1997 and Schultz, 1998) (**d**) Optogenetic stimulation of dopamine neurons in rhesus monkey. Repeated optical midbrain stimulation after choice of one option elicits increasing preference (P(choice)) of that option (blue fractal) to its alternative (red fractal). Midbrain electrical stimulation, directly opening voltage-gated sodium channels, has similar but stronger effects (bottom red). (Reproduced from Stauffer et al., 2016) (**e**) Winner-Take-All scheme of competitive choice between two options. See text for description

elicit positive RPEs; predictions can decrease as long as rewards can shrink enough to elicit negative RPEs.

Pavlovian learning provides the simplest and minimally confounded model of learning, but it is admittedly an abstraction of real-world learning processes. An RL model that conceptualizes real-world situations better is temporal difference (TD) learning. Agents are often several steps away from the ultimate reward but can be guided to that reward by a sequence of predictive stimuli that elicit specific choices. For example, a restaurant sign predicts a menu that predicts a specific meal that predicts specific foods that predict essential nutrients. The nutrients are the ultimate rewards, and the preceding stimuli are predictors of these rewards; by eliciting learning themselves, the predictors are higher-order rewards. Agents usually can choose at some or all steps between different predictive stimuli. At some steps, the choice may be between a better immediate option and a smaller option that predicts a later but much better option. In Fig. 1b, the chooser foregoes the best immediate option to obtain much larger rewards in the next steps (the red line blocks subsequent sequence change). Thus, TD learning conceptualizes how agents benefit from maximizing the overall reward value of a whole sequence, called state value, rather than maximizing only the immediate reward. Each reward-predicting stimulus in a TD sequence is learned using a basic Pavlovian process, starting with the last prediction before the ultimate reward and then advancing backwards in steps to the earliest stimulus; each predictive stimulus serves as higher-order reward that elicits learning of the preceding stimulus.

The dopamine response to rewards and reward-predicting stimuli in monkeys, rodents and fruit flies codes a bidirectional RPE (Fig. 1c) (Schultz et al., 1997; Cohen et al., 2012). An unpredicted reward at a given moment in time constitutes a better-than-predicted reward and elicits a positive temporal RPE and a dopamine excitation (top). A reward that occurs exactly as predicted and at the predicted time elicits no RPE and no dopamine response at all (middle high). A reward that fails to occur at its predicted time constitutes a worse-than-predicted reward and elicits a negative temporal RPE and a dopamine inhibition at the time at which the predicted reward would have occurred (middle low). Thus, dopamine neurons code a bidirectional temporal RPE. The bidirectional nature distinguishes the RPE from attention elicited by both positive and negative RPEs that would result in a common unidirectional neuronal response, usually an excitation. Further tests reveal that the dopamine response to reward-predicting stimuli itself constitutes also an RPE signal. Dopamine neurons are not only excited by a temporally unpredicted reward-predicting stimulus (left in both middle traces) but lose the response when that reward-predicting stimulus is predicted in time by an earlier predictive stimulus (bottom) (Schultz et al., 1993). Thus, the dopamine RPE response occurs with temporally unpredicted RPEs elicited by both ultimate rewards and reward-predicting stimuli. This feature resembles key characteristics of the TD RL model in which learning is mediated by RPEs at each rewarding event (Montague et al., 1996). Thus, the phasic dopamine response signals reward, and the form of the signal is not reward itself but the difference between reward and prediction across their respective time steps (Δr-p or dr/dt).

Besides the RPE signal, dopamine neurons also show a brief earlier unselective response to any event that reflects the novelty, sensory and motivational salience of the event (Schultz, 2016). Dopamine neurons are also activated with behavioural activation from movements of sufficient amplitude, speed or acceleration (Schultz, 2007; Cox & Witten, 2019). These multiple functions of phasic dopamine changes seem confusing when assuming that one brain system has only one function. However, pluripotent functions of individual neurons may be advantageous for an evolutionary ancient brain system.

The dopamine RPE signal registers rewards not only passively but also drives behaviour actively towards reward. Work in the laboratory of Roy Wise (Corbett & Wise, 1980) suggested that dopamine neurons are a major substrate for the intracranial electrical self-stimulation discovered earlier (Olds & Milner, 1954). Recent optogenetic experiments have confirmed that excitation of dopamine neurons drives monkeys and rodents towards behaviours that elicit dopamine excitation again (Tsai et al., 2009; Stauffer et al., 2016), for example, during choices (Fig. 1d). Correspondingly, inhibition of dopamine neurons drives animals away from behaviours eliciting dopamine inhibition (Tan et al., 2012). The artificially elicited dopamine excitations and inhibitions mimic positive and negative dopamine RPE signals. The observations suggest that dopamine RPE signals are causally sufficient for driving RL. It appears that animals seek better-than-predicted and avoid worse-than-predicted rewards in order to experience positive dopamine RPE signals and avoid negative RPE signals, although only work on humans may tell us about explicit intentional stance.

While a function in RL may seem abstract, the neuronal reinforcing effects of dopamine signals may consist of updating reward value in downstream neurons involved in economic choice. In a typical competitive Winner-Take-All model of binary choice, the stronger of two value inputs exerts stronger excitation on downstream neurons and stronger inhibition on neurons for the alternative option, as compared to the option with the weaker value input (Fig. 1e). A subsequent threshold mechanism assures that only the stronger of the two inputs reaches the output, and thus determines the choice, whereas the weaker input of the alternative option dissipates without consequence. Circuits implementing lateral inhibition are found in the striatum (caudate nucleus, putamen, nucleus accumbens) and orbitofrontal cortex (OFC). Dopamine release elicited by the RPE from the received reward is indiscriminately directed at basically all postsynaptic striatal neurons but only updates the option value at synapses whose previous activity has left an eligibility trace for modification (Fig. 1e red). Thus, according to this decision model, the widespread dopamine RPE signal updates the value in those neurons that contributed to the preceding rewarded choice, constituting a kind of neuronal RL mechanism.

3 Dopamine Neuroeconomics

Satiety makes a good case for the notion that rewards have value for satisfying the needs of individual agents; therefore, reward value is subjective. A candidate brain system for updating economic decision variables should code reward value on a subjective basis without requiring additional downstream mechanisms that translate objective into subjective value. Another case for subjective value is temporal discounting; the value of a later reward is lower than the value of an earlier but physically identical reward. While satiety and temporal distance are useful ad-hoc situations for identifying subjective value and its neuronal correlates, mathematically defined utility (Bernoulli, 1738) provides a metric for subjective value, defined as a typically nonlinear function of reward amount (Fig. 2a). A further basic determinant of subjective value is weighted reward probability (Kahneman & Tversky, 1979) that has recently been found to affect neurons in orbitofrontal cortex (Ferrari-Toniolo & Schultz, 2023) but will not be further elaborated here. While subjective reward value is determined by many other factors, utility and weighted probability are simple enough for testing in animals suitable for neurophysiological studies on reward.

An experimentally viable way to estimate utility functions employs choice under risk. Risk in economics concerns both gains and losses and is defined by the higher moments of reward distributions, namely variance, skewness and kurtosis. By contrast, the often-used definition of risk as probability of damage or loss fails to include gains that are usually the reason for engaging in risky choice. Intuition often seems to combine these two definitions.

Risk is conveniently investigated during choices involving two options. To avoid confounds and be well understood by animals and interpretable by experimenters, tasks should be simple and have straightforward parameters (notwithstanding the higher complexity of real-world choices):

- Options are distinct and well separated.
- Options are collectively exhaustive: all options are included in the presented option set.
- Options appear simultaneously: there are no intervening stimuli that may distract, require additional memory, or cause adaptation.
- Options are mutually exclusive: only one option can be chosen.
- Individual options may have two mutually exclusive reward amounts; only one amount is paid out with a specific probability (all reward probabilities within an option add to $P = 1.0$).
- The cost for obtaining an option remains constant throughout the test (but cost can differ between options).
- The reward in any option does not depend on other options.
- The reward of the chosen option is paid out immediately.

An important issue with animals is to assure they make meaningful choices. Three simple conditions can test this requirement. The first test for meaningful

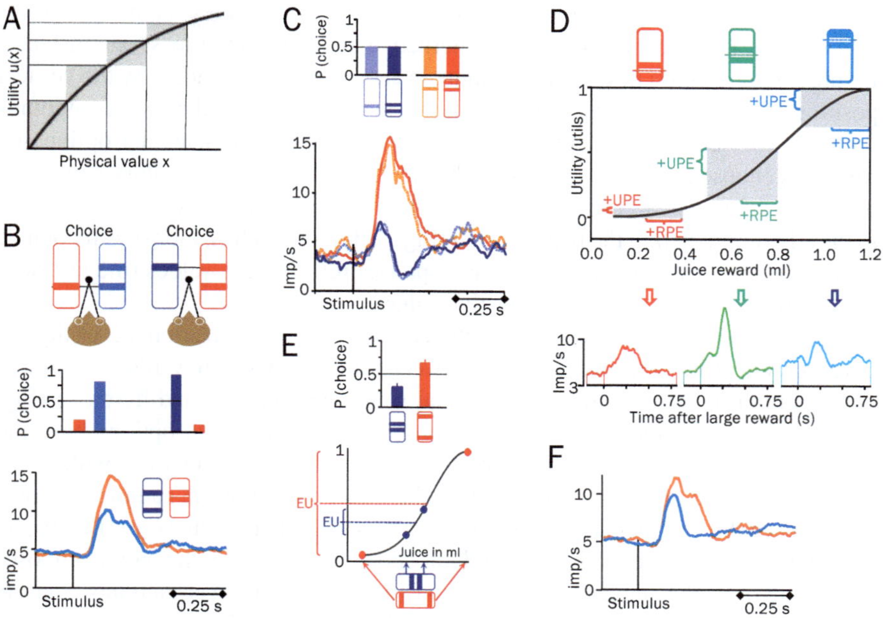

Fig. 2 Neuroeconomics of the dopamine reward signal (**a**) Bernoulli utility function. Decreasing height of grey squares indicates gradually declining marginal utility (du/dx) as a hallmark of subjective reward valuation (**b**) Assessment of meaningful choice with first-order stochastic dominance (FOSD). Top: choice between safe and risky rewards. Line height within each vertical rectangle indicates reward amount, higher is more; two lines inside the same rectangle indicate mutually exclusive reward amounts, each delivered with probability of $P = 0.5$. Blue options are better. Middle: monkey shows stochastic preference for better options, thus satisfying FOSD. Bottom: dopamine response to stimuli predicting the options is stronger for better reward (red) (**c**) 'Utility maximization' during choice between risky and safe rewards. At choice indifference (choice $P = 0.5$ of each of two options), subjective value is by definition the same in both options. Relative to the mean amount of a given gamble option, the safe reward is higher with small rewards (top left) but lower with large rewards (top right). Thus, the animal chooses the smaller amount in half the trials, indicating maximization of utility rather than amount. Bottom: the reward-sensitive dopamine neurons (red vs. blue) show the same response to each option of a given option set (blue and red), suggesting coding of subjective value rather than physical amount (**d**) Dopamine RPE signal codes utility. Top: three equiprobable gambles with same variance projected onto nonlinear utility function. The positive RPE of each gamble is identical (x-axis) but changes into varying positive utility prediction errors (UPE; y-axis) due to utility nonlinearity. Bottom: dopamine RPE signal varies with UPE, not physical RPE (**e**) Testing second-order stochastic dominance (SOSD) with mean-preserving spread. Despite same mean physical amounts, projection of the higher-variance gamble from the x-axis onto the y-axis results in higher utility (red) for the animal and thus predicts preference for that gamble, which is indeed the case (top) (**f**) Dopamine signal with SOSD. Despite same mean amounts, the higher-variance gamble elicits a stronger dopamine response compared to the lower-variance gamble shown in panel **e**, confirming neuronal coding of subjective value rather than physical amount. (Panels **b**, **d**, **e**, and **f** reproduced from Stauffer et al., 2014)

choice involves first-order stochastic dominance, which defines the better option: every option is at least as good as its alternative and better in at least one instance. Figure 2b (top) shows how reward amount in each option is predicted by the vertical position of a bar within a rectangle (higher is more; two bars indicate split amounts with $p = 0.5$ each); blue options are better and thus 'dominate' the option set. Monkeys choose the better options stochastically, with a few exploratory choices of the worse, 'dominated' option (Fig. 2b middle). Dopamine neurons respond strongly to the stimuli predicting the better option (Fig. 2b bottom). Thus, monkeys recognize what is better for them, and dopamine responses reflect these values (the test does not distinguish objective from subjective value).

The second test for meaningful choice involves maximization of utility. In psychophysically controlled choices between safe and risky rewards, the animal chooses both options equally frequently at some point (choice indifference) (Fig. 2c). As we infer subjective value from observable choice, both options have the same subjective value for the animal at choice indifference. However, with lower amounts, the safe reward amount exceeds the mean reward of the gamble at choice indifference (blue), whereas the converse is true with higher amounts (red). Thus, by choosing each option equally frequently at choice indifference, the animal chooses in half the trials the smaller reward amount (the gamble at lower amounts, and the safe reward at higher amounts; blue and red). It foregoes the larger amount on these trials as the smaller amount has the same subjective value. Thus, monkeys maximize subjective value (informal utility) rather than objective amount in a meaningful way. Dopamine responses show very similar responses to both options at choice indifference (while their increase with higher amounts attests to their sensitivity for value) (Fig. 2c bottom). Thus, dopamine neurons seem to code subjective value in this specific test.

An empirically estimated utility function would allow to test neuronal coding of subjective value on a more general basis. Utility functions can be estimated from risky choice using the so-called fractile procedure that employs a chain of indifference points with precisely defined utility, starting at 0.5 utils and working towards increasingly finer resolution. Utility functions are fitted from choices using splines or nonlinear functions. The obtained utility functions in monkeys are convex for small reward amounts and with increasing amounts become gradually more linear and finally concave (Fig. 2d, black curve) (Stauffer et al., 2014). Choices of monkeys between safe rewards analysed with the random utility model reveal similarly S-shaped utility functions with somewhat less nonlinearity (Bujold et al., 2022). For humans, the S-shaped pattern is referred to as gambling for peanuts (Weber & Chapman, 2005). When projecting three gambles with identical variance but different means onto different parts of this utility function, the positive RPEs (top amount minus mean; x-axis) are identical, whereas the positive utility prediction errors (UPE; y-axis) vary with the steepness of the utility function (Fig. 2d, grey rectangles). The varying dopamine responses to the larger reward follow the varying positive UPEs rather than the constant positive physical RPEs (Fig. 2d, bottom) (Stauffer et al., 2014), thus demonstrating RPE coding in the metric of subjective utility rather than objective physical reward amount.

The demonstration of dopamine utility coding allows to use second-order stochastic dominance as a third test for meaningful choice. The test uses the so-called mean-preserving spread that provides a minimal-confound definition of economic risk and tests risk sensitivity: two gambles with equal mean objective value differ only in the variance of their amounts (Rothschild & Stiglitz, 1970). A risk avoider would prefer the less risky gamble (smaller variance), whereas a risk taker would prefer the riskier gamble (larger variance). Monkeys prefer the riskier option (Fig. 2e top), thus confirming risk-taking with low amounts (Fig. 2c blue at top left). When projecting the two gambles onto the empirically estimated utility function for that monkey, the mean subjective value (expected utility, EU) of the riskier gamble exceeds the mean subjective value of the less risky gamble despite the same physical mean amount (Fig. 2e bottom), confirming dissociation between subjective and objective value. Thus, true to the predictive nature of mathematical functions, the utility function empirically estimated from measured choices predicts risk-taking choice (note that the test gambles differed from the gambles used for estimating the utility function and thus served also as out-of-sample test for the validity of the utility function). The dopamine responses to the two gambles followed the subjective value rather than the objective value, thus confirming nonlinear utility coding in this test (Fig. 2f). Although the dopamine response increases with risk, it is not a risk signal but a subjective value signal boosted by risk. Analogous utility prediction error signals are found in the human dopamine-receiving ventral striatum using functional magnetic resonance imaging (Konova et al., 2023). True risk signals, distinct from value (and not coding prediction errors), are found in monkey orbitofrontal cortex (O'Neill & Schultz, 2010).

Taken together, the meaningful choices under risk in rhesus monkeys demonstrate that subjective reward value, expressed as formal utility, is a key decision variable. It is coded by the RPE signal of dopamine neurons in the metric of formal utility, strictly spoken as UPE. The identification of subjective value coding would allow dopamine signals to participate in the updating of value signals for economic choice as proposed in Fig. 1e.

4 Neuronal Value Signals for Multi-component Rewards

Rewards are typically composed of several components. An apple has taste and size, a sausage contains calories and fat, and a coffee provides caffeine and water (and milk and sugar and maybe a few other ingredients). Each of these components has its own value that varies along a single dimension. A single-dimensional value is either low or high or somewhere in between and is therefore called a scalar. Rewards that have several components also have several values, and the value of each component is a distinct scalar. Thus, multi-component rewards are vectors composed of several scalars.

Even when rewards have multiple components, agents choose them only more or less along a single dimension, as if the value of even these rewards is a scalar.

Reward signals of individual neurons also vary along a single dimension; they can only be larger or smaller and thus are scalars just like reward value. The common scalar nature of reward value and neuronal reward signals encourages to ask the question how the single-dimensional scalar neuronal responses can reflect multi-component, vectorial rewards. Help comes from Revealed Preference Theory in economics (Samuelson, 1937) that derives scalar value from vectorial rewards. Once we understand that step, we can explore how such scalar value is encoded as scalar neuronal signals.

A convenient graphic demonstrates a solution for obtaining scalar subjective value from vectorial rewards. The scalar amounts of individual bundle components are plotted onto a two-dimensional x-y graph, and the positions of equally preferred bundles are connected as curves (Fig. 3a red, orange, green, purple). Due to their equal preference, all bundles on each curve have the same subjective value, and bundles on higher curves (farther away from the origin) are preferred to, and thus have higher subjective value than, bundles on lower curves. Note that the reward values are inferred from choice and therefore are subjective by definition, despite being set in the metric of physical amounts.

In the experiment, a monkey chooses between two visually similar two-component reward options ('bundles') (Fig. 3b). Each of the two bundles contains the same two reward types, like grape juice and blackcurrant juice, with individually set amounts (Fig. 3b) (Pastor-Bernier et al., 2017). In this stochastic design, individual bundle amounts are varied psychophysically until each bundle is chosen equally frequently (choice indifference). Systematic variation of initial bundle settings allows to estimate several indifference points that can be fitted by an indifference curve. Further variation of bundle settings results in different such indifference curves that can then be combined into an indifference map. The map demonstrates the vectorial-to-scalar transformation of value: same scalar value along curves, different scalar value across curves (Fig. 3c).

Now that bundles have integrated scalar value from both reward components, we can ask whether scalar neuronal reward responses can code such scalar values. This is indeed the case for a substantial population of neurons in the monkey orbitofrontal cortex (OFC). As Fig. 3d shows, the bundle with the highest value, on the higher indifference curve, elicits the strongest neuronal response (orange), whereas the two lower-valued bundles, on the lower indifference curve, elicit smaller responses (red, green). Note that the two lower-valued bundles, being positioned on the same curve, have similar value among them; accordingly, their neuronal responses fail to differ much from each other. Interestingly, when comparing across curves, the higher-valued bundle on the higher curve (Fig. 3c orange) elicits the stronger neuronal response, even though it contains less grape juice and less blackcurrant juice than the lower-valued bundles (red, green) on the lower indifference curve; it is the combined juice amounts that determine the integrated scalar value of the bundles and their neuronal responses (Pastor-Bernier et al., 2019). Thus, it seems that specific OFC reward neurons code the integrated scalar value of vectorial, multi-component rewards.

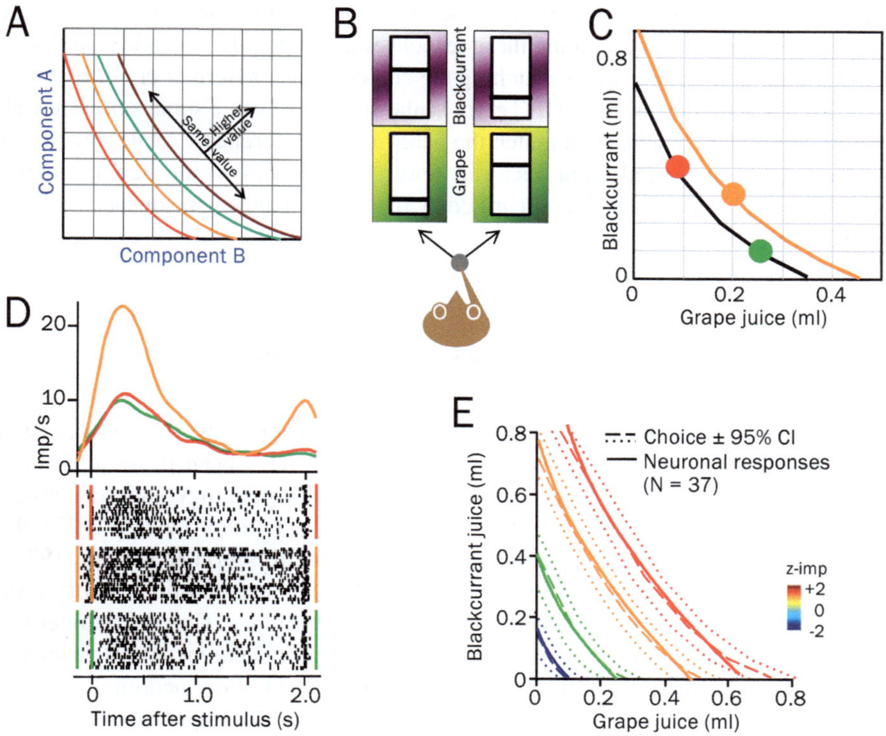

Fig. 3 Neuronal value signals for multi-component rewards in monkey orbitofrontal cortex (OFC) (**a**) Scheme of revealed preferences for two-component rewards. The amounts of the two reward components are plotted at their x-y coordinates; each point on the map represents a two-component reward. All points on the same curve, irrespective of their x-y positions, are equally preferred ('choice indifference') and thus have same subjective value; all points on higher curves are preferred to all points on lower curves and thus have higher subjective value (**b**) Experimental stimuli. The animal chooses between two options, each predicting the same two rewards, but at specific amounts (**c**) Empirically estimated choice indifference curves. Note that the orange two-component reward ('bundle') lies on a higher curve and thus is preferred to the red and green bundles, despite both orange bundle components having lower amounts (preference despite dominance of individual physical amounts) (**d**) Single neuron responses in OFC to bundles shown in panel **c**. Stronger response to higher valued bundle (orange) (**e**) Similar OFC population responses combine to neuronal indifference curves that match choice indifference curves within choice 95th confidence intervals. (Panels **b–d** reproduced from Pastor-Bernier et al., 2019)

Analysis at the OFC population level helps to better demonstrate the integrated neuronal coding of vectorial rewards. Similar (insignificantly differing) OFC responses can be combined into two-dimensional neuronal indifference curves by plotting and connecting the x-y positions of the bundles that elicit the responses. Indeed, the obtained neuronal indifference curves match the choice indifference curves within their 95th confidence intervals (Fig. 3e). Distinct from this neuronal code integrating several reward types, other OFC neurons are specific for individual reward types and show graded responses to only one specific juice used in the experiment; however, combining these responses into a population code results also in neuronal indifference maps matching the choices (Pastor-Bernier et al., 2019).

Taken together, neurons in OFC carry scalar subjective value signals of vectorial, multi-component bundles that match the behavioural preferences. Analogous scalar value signals integrated from vectorial rewards are found in the human ventral striatum, midbrain and OFC, as identified by functional magnetic resonance imaging (Seak et al., 2021).

5 Future Research and Unanswered Questions

The presented data demonstrate the value of scientific theories: the research, and the understanding of the obtained results, would not have been possible without precious theories elaborated by previous pioneers in psychology and economics. The theories concern, at the very least, Pavlovian conditioning (Pavlov, 1927), the role of prediction errors in conditioning (Kamin, 1969; Rescorla & Wagner, 1972), temporal difference learning (Sutton & Barto, 1981), expected utility theory (von Neumann & Morgenstern, 1944), and Revealed Preference Theory (Samuelson, 1937). Plenty of similar theories can provide guidance for planning new experiments that transcend individual ad-hoc designs and thus would allow neuronal research to validate these theories, enrich them, eliminate neuronally implausible concepts, and develop even simpler, reduced and more global integrated theories about evolutionary beneficial reward functions.

One of the successes of neuroeconomics derives from testing irregularities and irrationalities of choices with designs tested by behavioural economics. While this approach uncovered many aspects of human nature and impacts policymaking, it fails to go to the source of the behaviour that may go wrong. It is difficult to understand how a car engine works by only investigating its malfunctioning rather than understanding the physics of ignition and combustion. Going back to basic constructs of learning and choice theories, as proposed above, might help to address the issue. This approach requires to use animals in which high-resolution neurophysiology is possible, and humans whose brains present still another dimension of function beyond animals.

Recent advances in optogenetics have confirmed one of the most important findings in neuroscience in general and dopamine function in particular. Animals come back for more stimulation of their neurons (Olds & Milner, 1954), and the suggestion of dopamine neurons being the prime suspect (Corbett & Wise, 1980) has been confirmed using modern optogenetic methods (Witten et al. 2011). While simple reinforcement learning offers an observational interpretation, the underlying mechanism may well be related to the issue of Free Will that has been debated for centuries by philosophers. Are we slowly finding ground for this rather speculative and contentious topic that may tell us how deterministic life is? More work, including in humans as far as ethically acceptable, would be desirable and might put the theme on solid physical ground.

Acknowledgements Our work has been supported by the Wellcome Trust (095495, 106101), European Research Council (ERC, 293549) and NIH Conte Center at Caltech (P50MH094258).

References

Bernoulli, D. (1738). Specimen theoriae novae de mensura sortis. *Comentarii Academiae Scientiarum Imperialis Petropolitanae, 5*, 175–192. Translated as: Exposition of a new theory on the measurement of risk. *Econometrica* 22: 23–36, 1954.

Bujold, P. M., Ferrari-Toniolo, S., Seak, L. C. U., & Schultz, W. (2022). Comparing utility functions between risky and riskless choice in rhesus monkeys. *Animal Cognition, 25*, 385–399.

Cohen, J. Y., Haesler, S., Vong, L., Lowell, B. B., & Uchida, N. (2012). Neuron-type-specific signals for reward and punishment in the ventral tegmental area. *Nature, 482*, 85–88.

Corbett, D., & Wise, R. A. (1980). Intracranial self-stimulation in relation to the ascending dopaminergic systems of the midbrain: A moveable microelectrode study. *Brain Research, 185*, 1–15.

Cox, J., & Witten, I. B. (2019). Striatal circuits for reward learning and decision-making. *Nature Reviews. Neuroscience, 20*, 482–494.

Ferrari-Toniolo, S., & Schultz, W. (2023). Reliable population code for subjective economic value from heterogeneous neuronal signals in primate orbitofrontal cortex. *Neuron, 111*, 3683–3696.

Kahneman, D., & Tversky, A. (1979). Prospect theory: An analysis of decision under risk. *Econometrica, 47*, 263–291.

Kamin, L. J. (1969). Selective association and conditioning. In N. J. Mackintosh & W. K. Honig (Eds.), *Fundamental issues in instrumental learning* (pp. 42–64). Dalhousie University Press.

Konova, A. B., Ceceli, A. O., Horga, G., Moeller, S. J., Alia-Klein, N., & Goldstein, R. Z. (2023). Reduced neural encoding of utility prediction errors in cocaine addiction. *Neuron, 111*, 4058–4070.

Montague, P. R., Dayan, P., & Sejnowski, T. J. (1996). A framework for mesencephalic dopamine systems based on predictive Hebbian learning. *The Journal of Neuroscience, 16*, 1936–1947.

Olds, J., & Milner, P. (1954). Positive reinforcement produced by electrical stimulation of septal area and other regions of rat brain. *Journal of Comparative and Physiological Psychology, 47*, 419–427.

O'Neill, M., & Schultz, W. (2010). Coding of reward risk distinct from reward value by orbitofrontal neurons. *Neuron, 68*, 789–800.

Pastor-Bernier, A., Plott, C. R., & Schultz, W. (2017). Monkeys choose as if maximizing utility compatible with basic principles of revealed preference theory. *Proceedings of the National Academy of Sciences of the United States of America, 114*, E1766–E1775.

Pastor-Bernier, A., Stasiak, A., & Schultz, W. (2019). Orbitofrontal signals for two-component choice options comply with indifference curves of revealed preference theory. *Nature Communications, 10*, 4885.

Pavlov, P. I. (1927). *Conditioned reflexes*. Oxford University Press.

Rescorla, R. A., & Wagner, A. R. (1972). A theory of Pavlovian conditioning: Variations in the effectiveness of reinforcement and nonreinforcement. In A. H. Black & W. F. Prokasy (Eds.), *Classical conditioning II: Current research and theory* (pp. 64–99). Appleton Century Crofts.

Rothschild, M., & Stiglitz, J. E. (1970). Increasing risk: I. A definition. *Journal of Economic Theory, 2*, 225–243.

Samuelson, P. (1937). A note on measurement of utility. *The Review of Economic Studies, 4*, 155–161.

Schultz, W. (1998). Predictive reward signal of dopamine neurons. *Journal of Neurophysiology, 80*, 1–27.

Schultz, W. (2007). Multiple dopamine functions at different time courses. *Annual Review of Neuroscience (Palo Alto, CA), 30*, 259–288.

Schultz, W. (2016). Dopamine reward prediction error signalling: A two-component response. *Nature Reviews. Neuroscience, 17*, 183–195.

Schultz, W., Apicella, P., & Ljungberg, T. (1993). Responses of monkey dopamine neurons to reward and conditioned stimuli during successive steps of learning a delayed response task. *The Journal of Neuroscience, 13*, 900–913.

Schultz, W., Dayan, P., & Montague, R. R. (1997). A neural substrate of prediction and reward. *Science, 275*, 1593–1599.

Seak, L. C. U., Volkmann, K., Pastor-Bernier, A., Grabenhorst, F., & Schultz, W. (2021). Single-dimensional human brain signals for two-dimensional economic choice options. *The Journal of Neuroscience, 41*, 3000–3013.

Stauffer, W. R., Lak, A., & Schultz, W. (2014). Dopamine reward prediction error responses reflect marginal utility. *Current Biology, 24*, 2491–2500.

Stauffer, W. R., Lak, A., Yang, A., Borel, M., Paulsen, O., Boyden, E., & Schultz, W. (2016). Dopamine neuron-specific optogenetic stimulation in rhesus macaques. *Cell, 166*, 1564–1571.

Sutton, R. S., & Barto, A. G. (1981). Toward a modern theory of adaptive networks: Expectation and prediction. *Psychological Review, 88*, 135–170.

Tan, K. R., Yvon, C., Turiault, M., Mirzabekov, J. J., Doehner, J., Labouèbe, G., Deisseroth, K., Tye, K. M., & Lüscher, C. (2012). GABA neurons of the VTA drive conditioned place aversion. *Neuron, 73*, 1173–1183.

Tsai, H.-C., Zhang, F., Adamantidis, A., Stuber, G. D., Bonci, A., de Lecea, L., & Deisseroth, K. (2009). Phasic firing in dopaminergic neurons is sufficient for behavioral conditioning. *Science, 324*, 1080–1084.

von Neumann, J., & Morgenstern, O. (1944). *The theory of games and economic behavior*. Princeton University Press.

Weber, B. J., & Chapman, G. B. (2005). Playing for peanuts: Why is risk seeking more common for low-stakes gambles? *Organizational Behavior and Human Decision Processes, 97*, 31–46.

Witten, I. B., Steinberg, E. E., Lee, S. Y., Davidson, T. J., Zalocusky, K. A., Brodsky, M., Yizhar, O., Cho, S. L., Gong, S., Ramakrishnan, C., Stuber, G. D., Tye, K. M., Janak, P. H., & Deisseroth, K. (2011). Recombinase-driver rat lines: tools, techniques, and optogenetic application to dopamine-mediated reinforcement. *Neuron, 72*, 721–733, 2011.

Resolving Uncertainty: Foundational Neuroeconomic Studies of Risk and Ambiguity

Jade E. Terry and Scott A. Huettel

Abstract Understanding the neural mechanisms of decision-making under risk has been a fundamental topic within neuroeconomics. Early studies adopted paradigms from behavioral economics; many identified key brain regions that distinguished decisions about risky options from those involving certain outcomes, tracked the relative risk associated with different options, and that were modulated by risk preferences. Collectively, this work led to the general conception of a risk matrix that includes control-related regions in prefrontal, parietal, and insular cortices. Other key studies distinguished economic risk from related concepts like ambiguity, temporal discounting, and loss aversion—with concomitant distinctions in the supporting neural substrates. More recent work builds on these foundations by examining extensions to cognate areas of neuroscience, including developmental, social, and clinical domains. Future research on the neuroeconomics of risk should generalize beyond economic perspectives and connect other disciplines in the social sciences, to better represent the diverse forms of risky decisions encountered in everyday life.

1 Introduction

Few aspects of decision-making have been more central to research in neuroeconomics than the understanding of how decisions are influenced by *risk*. The continuing interest in risk—as well as in related forms of uncertainty—builds upon seminal work in probability and statistics (Kendall et al., 1961), economic theory (Haring & Smith, 1959), and behavioral economics (Jolls et al., 1998), along with considerable applied work examining real-world choices (Markowitz, 2010).

What is risk? Formal models consider risk to arise when a choice option has multiple potential outcomes that could occur with different probabilities. As an example, suppose that a gambler could win $100 by rolling a total of 7 across two

J. E. Terry · S. A. Huettel (✉)
Department of Psychology and Neuroscience, Duke University, Durham, NC, USA
e-mail: jade.terry@duke.edu; scott.huettel@duke.edu

D. V. Smith et al. (eds.), *Neuroeconomics: Core Topics and Current Directions*,
https://doi.org/10.1007/978-3-032-02925-6_2

dice (i.e., a 1-in-6 chance) at a potential cost of losing $25 otherwise (i.e., a 5-in-6 chance)—or they can play it safe by not rolling the dice at all. Similarly, an investor may direct their retirement savings into a stock index fund that tends to return about 8% annually but may gain or lose much more in any given year—or they could choose a money market fund with a lower fixed return of 4% annually. How people choose in such situations depends on their tendency toward risk-seeking (e.g., problem gamblers; Mishra et al., 2010), on their risk-aversion (e.g., conservative investing; Filbeck et al., 2005), and on the degree of risk associated with each outcome (e.g., variance, chance of a negative outcome, coefficient of variation, or another measure; Weber et al., 2004).

Yet, few decisions—outside of restricted contexts like some forms of investing and gambling—match well with formal definitions of risk. For example, we may evaluate the chance of rain when selecting a location for a vacation, but that evaluation will be based on a rough sense of likelihood, not a known probability. Such decisions with imperfect or hidden information about outcomes and their probabilities (Knight, 1921) are often considered to involve *ambiguity*. Moreover, neuroeconomic research on risk and ambiguity has long been intertwined with studies of related concepts (e.g., temporal discounting, loss aversion) as well as studies of how decision-making changes over contexts (e.g., from nonsocial to social settings) and across individuals (e.g., over development and in clinical disorders).

Accordingly, this review begins with foundational research on the neural mechanisms of risk and ambiguity and then connects to research on related phenomena in economic-value-based decision-making, prioritizing studies with human decision-makers but highlighting key examples where the same phenomena have been investigated in animal models (e.g., nonhuman primates). Finally, it concludes by considering extensions of that foundational work in neuroeconomics to new areas of scholarship in related fields.

2 Foundational Neuroeconomic Studies of Risk and Ambiguity

The earliest neuroeconomic studies of risky decision-making sought to compare conditions in which participants made decisions under risk to similar decisions involving certain outcomes. For example, Paulus et al. (2003) used functional magnetic resonance imaging (fMRI) to demonstrate increased activation in insular cortex when participants chose risky compared to safe options—an effect linked in subsequent work to aversive signals associated with loss prediction (Kuhnen & Knutson, 2005; Preuschoff et al., 2008).

While such results identified fundamental neural dissociations between processes related to risk and to certainty, different methods were needed for identifying effects specific to risk itself. A second approach involved identifying brain regions or networks whose activation tracked the amount of risk associated with different

Table 1 Key brain regions linked to risk, ambiguity, value, and social cognition

Neural activations	Example studies	MNI coordinates		
		x	y	z
Risk sensitive				
dlPFC	Huettel et al. (2005)	55	10	35
Insula	”	35	26	4
PPC	”	29	−70	48
Ambiguity > risk				
OFC	Hsu et al. (2005)	51	33	−6
Insula	Huettel et al. (2006)	41	19	9
dlPFC	Bach et al. (2009)	52	14	14
PPC	”	34	−50	50
Reward/value				
vmPFC	Bartra et al. (2013)	2	46	−8
PCC	”	−4	30	36
Insula	”	32	20	−6
Ventral striatum	”	12	10	−6
Social cognition				
dmPFC	McDonald et al. (2020)	14	30	52
Insula	Bellucci et al. (2017)	42	18	4
Amygdala	Santos et al. (2016)	21	−4	−17
PCC	”	−2	−46	20
TPJ	Smith et al. (2014)	−40	−60	32

Provided here are selected examples of the brain regions discussed within the chapter, with coordinates drawn from foundational studies. Regions are visualized in Fig. 2

decisions. Regions of dorsolateral prefrontal cortex (dlPFC) and posterior parietal cortex (PPC) were shown to track the probability of a risky outcome, with least activation for near-certain events (80% success) and most activation for most-uncertain (e.g., 50% success; Huettel et al., 2005; see Table 1). The observed parietal cortex activation echoed that in earlier studies in nonhuman primates, which found probability-related signals in homologous regions (Platt & Glimcher, 1999). A third approach identified neural markers of risk aversion; that is, brain regions whose activation to risky choices increases in those individuals who are most risk-averse; for example, Christopoulos et al. (2009) showed that activation in lateral PFC tracked inter-individual differences in risk aversion.

Note that all three of these approaches sought to identify neural correlates of risk itself, rather than of the overall *value* of a risky choice option. Subsequent work showed that subjective value of risky options is tracked by similar neural substrates as for other sorts of decision options (e.g., ventral striatum and ventromedial prefrontal cortex, vmPFC), supporting the conclusion that risk-specific computations from the regions described above feed into a domain-general system that represents the value of any potential reward (e.g., money, food, consumer goods) in a common

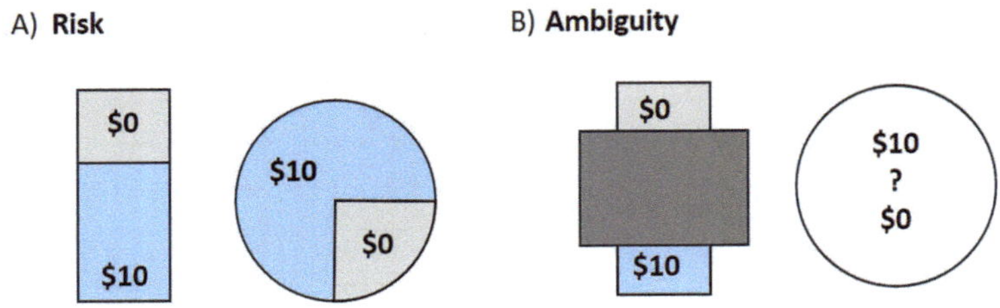

Fig. 1 Representing risk and ambiguity. (**a**) A choice option involves *risk* when, following the choice, multiple outcomes could occur with known probabilities. Shown here are two representations of a choice option that involves a 75% chance of winning $10 and a 25% chance of winning $0. (**b**) A choice option involves *ambiguity* if there is hidden information about potential outcomes and/or their probabilities. Shown are two representations of an ambiguous option in which either $10 or $0 could be received, although their respective probabilities are hidden from the decision-maker

currency (Bartra et al., 2013; Clithero & Rangel, 2014; Levy et al., 2009; see Table 1 and Fig. 2).

Concurrently, researchers also sought to dissociate economic risk from the related concept of economic ambiguity—with two early studies identifying different candidate neural mechanisms. By comparing tasks that involved ambiguity (e.g., choose from decks with unknown probabilities of good/bad outcomes) with similar tasks that involved risk (e.g., choose from decks with known equal probabilities of good/bad outcomes; see Fig. 1), Hsu et al. (2005) showed that activation in lateral orbitofrontal cortex (OFC) and amygdala was greater under conditions of ambiguity. Huettel et al. (2006) employed a similar comparison of economic gambles with ambiguity (i.e., unknown probabilities) or risk (i.e., known probabilities) and found that activation in dlPFC was greater for ambiguous gambles and increased with ambiguity preference, whereas PPC activation tracked risk preference. The differences in the two studies may reflect the different features of ambiguity they evoked. As examples, participants in the Hsu and colleagues' study were highly ambiguity averse, which led to activation in regions linked to affective components of uncertainty, whereas participants in the Huettel and colleagues' study were more ambiguity-neutral, leading to activation in regions associated with computations necessary for resolving ambiguity. Subsequent work by other groups revealed that ambiguity (i.e., information exists but is not currently known) could be distinguished from ignorance (i.e., complete absence of information), and that ambiguity but not unresolved ignorance leads to dlPFC and PPC activation (Bach et al., 2009; see Table 1 and Fig. 2).

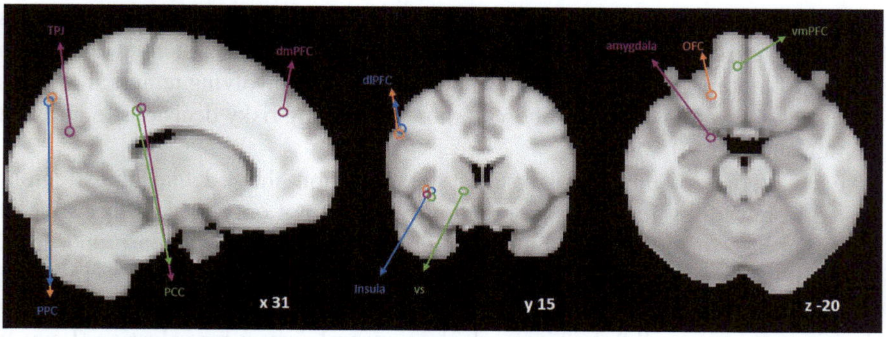

Fig. 2 Visualization of brain regions highlighted in this chapter. Shown are examples of the brain regions that have been linked to processing of risk (blue), ambiguity (orange), value (green), and social decision-making (purple). Coordinates are drawn from foundational studies identified in Table 1. We note that although both cerebral hemispheres have been linked to each of these processes, all coordinates are plotted in the right hemisphere for consistency of presentation

3 Separating Risk from Related Forms of Economic Decision-Making

In parallel to the research described above—which primarily identified neural mechanisms of decision-making under risk and ambiguity—other studies sought to distinguish those mechanisms from those that supported related aspects of decision-making. An important thrust of early neuroeconomic research examined temporal discounting, the phenomenon by which rewards typically diminish in value the farther they are in the future. Landmark papers had demonstrated that decisions involving temporal discounting rely on the same domain-general network for valuation (McClure et al., 2004; Kable & Glimcher, 2007), but whether risky decisions and temporal decisions reflected common or distinct inputs to that network remained an open question—with psychological theories supporting both perspectives (Chapman & Weber, 2006; Green & Myerson, 1996; Loewenstein & Prelec, 1991). Similarly, the phenomenon of loss aversion—such that the negative utility associated with potential losses is typically much larger than the positive utility associated with similar-magnitude gains (Kahneman & Tversky, 1979)—was thought to potentially share some underlying mechanisms with risk aversion.

Several foundational studies directly compared neural mechanisms of risky choices to temporal discounting. Across two early experiments, activation in the posterior cingulate cortex (PCC) increased for decisions involving delayed rewards (Weber & Huettel, 2008) and better tracked the valuation of delayed rewards (Peters & Buchel, 2009), when compared to similar-valued risky rewards. This PCC activation matched well to cognitive neuroscience research linking this region (as part of a larger network) to processes associated with prospection and mental time travel (Botzung et al., 2008; D'Argembeau et al., 2010); moreover, behavioral interventions that encourage future-oriented thinking both alter temporal discounting preferences (Weber et al., 2007) and modulate PCC activation (Peters & Buchel, 2010). Subsequent reviews have both supported this potential distinction while also noting that the neural substrates of decisions involving temporal rewards nevertheless overlap considerably with those supporting risky decisions (e.g., Frost & McNaughton, 2017).

Early studies of the neural basis of loss aversion showed that brain regions linked to reward evaluation (e.g., striatum, vmPFC) not only tracked the value of gains but also of losses, with individual differences in loss aversion proportional to the relative modulation of those regions (Tom et al., 2007). By themselves, these results support the theory that decisions involving potential losses (like risky decisions, more broadly) feed into a domain-general system for valuation (Levy & Glimcher, 2012). However, additional research has argued that appetitive/positive and aversive/negative elements of choice may be processed at least in part by distinct brain systems, with potential losses specifically associated with activation in regions outside of the traditionally conceived reward system (e.g., amygdala, insula; Kuhnen & Knutson, 2005; Sokol-Hessner & Rutledge, 2019).

Collectively, foundational work on risk, ambiguity, temporal discounting, loss aversion, and related phenomena has led to the general conception of a neural "risk matrix" that supports decisions that involve uncertain positive and negative outcomes (Knutson & Huettel, 2015). Hereafter, our review examines how this foundational work has been extended into new directions within developmental, social, and clinical domains.

4 Development of Neural Mechanisms for Risky Decision-Making

Though the perception and evaluation of risk changes over the lifespan (Josef et al., 2016), brain structures that influence risky decisions undergo the greatest changes throughout adolescence (Braams et al., 2015)—and many risky behaviors emerge in adolescence and early adulthood (Blakemore & Robbins, 2012). These developmental effects are often characterized in problematic terms that emphasize how adolescent risk-taking leads to dangerous, self-destructive outcomes. However, risk-taking is not always a problem; in many cases, being willing to take risks helps identify and generate new opportunities that lead to positive outcomes and personal growth (Duell & Steinberg, 2021). Given that risk-taking often involves such tradeoffs, researchers have sought to understand the trajectory of risk processing over development—with the eventual goal of identifying interventions that reduce the likelihood of negative life outcomes.

Adolescent brains are traditionally assumed to operate with a decreased capacity for self-regulation that leads to increased risk-taking (Yurgelun-Todd, 2007). However, these assumptions have been challenged by continuing research on neurobiological changes over adolescence (Braams et al., 2015), such as the rapid development of the reward system juxtaposed with the slower and more gradual development of the regions within the executive control system (e.g., PFC, PPC; Menon & D'Esposito, 2022). This work has led to dual-systems "imbalance models" (Casey et al., 2008; Steinberg, 2010) that shaped both developmental neuroscience and approaches to adolescent education and policy (Bailey et al., 2019; Chaplo & Fishbein, 2022; Meisel et al., 2019).

Additional research has noted that developmental changes in risk-taking behavior vary according to contexts. As an example, adolescents often perform much better than middle-aged and older adults on executive function tasks in the laboratory, yet perform more poorly in everyday settings (De Luca et al., 2003), particularly in emotionally or socially laden contexts (Defoe et al., 2015). Such observations have led to a flexible dual-systems model that accounts for the heterogeneity in observed risk-taking behavior across task demands by positing that adolescents selectively engage key behavioral control regions (e.g., prefrontal and cingulate cortices) according to contextual demands (Li, 2017; McCormick & Telzer, 2017). Because adolescents often have limited experiences with real-world risky decision-making, they may not have developed heuristics that provide computational shortcuts, identify patterns in their environments, and simplify complex choices. Limitations in adolescents' use of heuristics present an added layer of difficulty in deciding what choices offer the best outcomes (Blankenstein et al., 2021).

For adolescents, contexts that involve emotional and social factors are particularly salient (Li, 2017; Schonberg et al., 2011). While risky behavior is clearly maladaptive in some social contexts (e.g., the presence of peers leads to more aggressive/distracted driving), adolescents often receive social rewards for engaging in risky behaviors—potentiating those behaviors in social contexts (Chein et al., 2011; Smith et al., 2015; Trucco, 2020). Harnessing the power of social contexts to encourage positive behaviors can be particularly important for developing age-appropriate interventions (Ryzin et al., 2018; Schellenberg et al., 2022).

5 Interactions Between Social Contexts and Risky Decision-Making

Risk is endemic to social settings. Humans have always lived among other humans in families and communities, often facing the challenge of predicting other individuals' behavior in the face of limited information. Many common social interactions (e.g., bargaining, partner selection) rely on accurate evaluations of uncertain social situations, and thus social and adaptive fitness depends heavily on mechanisms for evaluating risks (Morelli et al., 2018; Tremblay et al., 2017; Trezza et al., 2011).

Previous literature has posited the evolutionary selection of a brain system for encoding and responding to social information (i.e., a "social brain"; Amodio & Frith, 2006; Tremblay et al., 2017). Key regions include the temporoparietal junction (TPJ), insula, posterior cingulate cortex (PCC), amygdala, and dorsomedial prefrontal cortex (dmPFC), among others (Adolphs, 2009; Santos et al., 2016; Smith et al., 2014; see Table 1 and Fig. 2). Research on the neuroscience of social decision-making has revealed both specific contributions of these regions as well as ways in which they interact with more general decision-making mechanisms, supporting trade-offs between benefits for oneself and for others (Pisauro et al., 2022; Ruff & Fehr, 2014).

Strategic decision-making, often modeled via game theory, involves two or more social agents playing out scenarios with potentially varying levels of gains or losses for each player (Lee, 2008). Successfully anticipating and responding to the actions of others is necessary to minimize individual risk and maximize rewards, particularly in cases where there is uncertainty about those actions (Hula et al., 2018). Brain regions associated with social cognition show selective engagement when decisions depend upon modeling others' behavior; for example, information carried by TPJ predicts subsequent bluffing decisions in a poker game, but only when playing against other humans and not computers (Carter et al., 2012). More dynamic games involve a continuous estimation of the opponent's intentions, often in the context of either cooperation or competition (Iqbal et al., 2019; Yoo et al., 2021). Neurons in dmPFC encode markers of those intentions as well as the optimal timing of actions (McDonald et al., 2020). Similar to results from simpler games, TPJ activation in advance of a decision is sensitive to the identity of a social partner (i.e., greater for human than computer opponents; McDonald et al., 2020; see Table 1 and Fig. 2).

In other instances, effective social interactions rely upon maximizing the rewards for an individual's group (Zhang et al., 2022) via actions like cooperation, trust, and reciprocity (Engemann et al., 2012; Van Dijk & De Dreu, 2021). Research using tasks from behavioral economics reveals that such decisions for others may elicit different risk preferences than decisions for oneself (Batteux et al., 2019; Fareri et al., 2022)—and those preferences depend on brain regions associated with social cognition, such as the TPJ and PCC (Bellucci et al., 2017; Tei et al., 2021). Similarly, preferences about outcomes for a group are tracked by different brain regions, notably the frontopolar cortex, than those that support preferences about one's own outcomes (vmPFC; Park et al., 2019). Moral principles and norms also permit social agents to optimize their behavior to support group outcomes, such as reducing the likelihood of a collective harm (Ellemers et al., 2019); this strategy corresponds to areas that subserve the social brain network (Qu et al., 2019; Yoder & Decety, 2018).

Decision-making in social contexts also relies on core computations—and brain regions—whose function is not specifically social (Behrens et al., 2008; Messimeris et al., 2023); such findings have fed into larger debates about the specificity of the social brain hypothesis (Ruff & Fehr, 2014). For instance, both the striatum and the vmPFC support learning about the value of rewards in nonsocial contexts (Balleine et al., 2007; Ciaramelli et al., 2021; Delgado et al., 2000), but they also interact with the regions discussed above when decisions involve social information (Izuma et al., 2008; Smith et al., 2014; see Table 1 and Fig. 2). One plausible interpretation of these interactions posits a two-step process, such that domain-specific information—whether social or nonsocial—converges into a domain-general network that supports valuation and subsequent choice (Harbaugh et al., 2007; Ruff & Fehr, 2014). Further exploration of the convergence and separability of domain-general and specifically social decision-making processes would be especially meaningful for understanding risk behaviors in real-world settings, many of which are disrupted in the clinical disorders considered in the next section.

6 Clinical Dysfunction and Potential Interventions

Researchers interested in how risky decision-making is altered in clinical conditions—notably addiction but also including a wide range of psychiatric and neurological disorders—have identified behavioral and neural markers of maladaptive risky choices. While such disorders are complex and often involve brain mechanisms well beyond those discussed earlier in this review, there has been a growing recognition that dysfunction in brain systems supporting the evaluation risks may be a common factor across multiple disorders (cf., the Research Domain Criteria, RDoC, framework of the National Institute of Mental Health).

Perhaps the most widely used paradigm for evaluating the effects of clinical conditions on risky decisions has been the Iowa Gambling Task (IGT), in which participants select among card decks that vary in their distributions of gains and losses (Bechara et al., 1994). Impaired performance in this task has been observed in a variety of conditions, including both focal damage to the prefrontal cortex (Fellows & Farah, 2005) and diffuse damage associated with mild traumatic brain injury (TBI; Cotrena et al., 2014). This task has also been used to probe function in healthy adults; for example, ACC activations were correlated with better decisions after suffering repeated losses when playing the IGT (Ma et al., 2015). Other tasks examine risky choices that involve keeping a sure reward or gambling for an even larger reward; for example, participants in the Balloon Analogue Risk Task (BART) earn rewards by pumping up a balloon but lose those rewards if it pops (Lejuez et al., 2002). BART performance is impaired when there is decreased sensitivity to loss, as often seen in individuals who engage in sensation-seeking behaviors (Xu et al., 2019). These behavioral effects are accompanied by parallel changes in brain function, such as decreased dlPFC activation in alcohol-dependent individuals when playing the BART (Burnette et al., 2021).

Collectively, work using these and similar risky decision-making tasks (Koffarnus & Kaplan, 2018) has revealed that repeated substance use alters neural systems that support sensitivity to rewards and integration of value-based information toward a decision; such findings have been shown in reviews (Bickel et al., 2014; Hüpen et al., 2023; Poudel et al., 2020) and empirical studies (Kohno et al., 2017; Kräplin et al., 2020; Liu et al., 2017). These systems—notably striatum and dlPFC—provide targets for potential clinical interventions. Recent work applying transcranial magnetic stimulation (TMS) and transcranial direct current stimulation (tDCS) to prefrontal cortex shows promise in targeting mechanisms of the decision-making process that lead to substance use (Chen et al., 2020; Maiti et al., 2017; Nardone et al., 2022; Pettorruso et al., 2020). Research on the neuroscience of risky decisions in addiction will likely have a bidirectional impact, as studies of the neural basis of risk processing and related decision-making provide deeper insights into dysfunction (Carland et al., 2019), and as studies on the neural basis of pathologies of decision-making reveal how those mechanisms can go awry (Ramey & Regier, 2019).

7 Conclusion: Moving Beyond *Economic* Risk

Neuroeconomic research has provided key insights into the mechanisms of risky decision-making, often through work combining the well-defined experimental paradigms of behavioral economics with computational modeling of brain data. Yet, the concept of risk extends well beyond decisions about economic gambles to include a much wider range of decisions: consumer choices, negotiations, personal safety, voting, ethical dilemmas, and legal settings, among many others. Such decisions rarely involve simple calculations with known probabilities about positive or negative outcomes—nor do they usually involve consequences only for oneself. Instead, the hallmarks of naturalistic decisions are ambiguity of outcomes; co-occurring changes in emotions, arousal, and/or body states; and social influences that alter how people perceive decisions and their consequences (Schonberg et al., 2011).

Understanding human behavior in such a diverse range of settings will require adoption of theories and research paradigms from a broad range of social sciences—and even extending to other fields like philosophy and cultural anthropology (Clithero et al., 2008; Levallois et al., 2012). Moreover, the richness of risky decision-making in natural settings requires a more extensive consideration of cultural factors that shape choice, potentially connecting neuroeconomics to the growing field of cultural neuroscience (Kim & Sasaki, 2014; Sasaki & Kim, 2017). Such work will necessarily require researchers to balance the competing demands of experimental control (e.g., well-defined measures in behavioral economic paradigms) and generalizability to real-world choices (e.g., complex naturalistic settings). In essence, future research may focus less on the topic of *risk* in isolation and more on the complex interactions through which uncertainty—in conjunction with many other factors—shapes the mechanisms of choice.

Conflicts of Interest We have no known conflicts of interest to disclose.

References

Adolphs, R. (2009). The social brain: Neural basis of social knowledge. *Annual Review of Psychology, 60*(1), 693–716. https://doi.org/10.1146/annurev.psych.60.110707.163514

Amodio, D. M., & Frith, C. D. (2006). Meeting of minds: The medial frontal cortex and social cognition. *Nature Reviews Neuroscience, 7*(4), 268–277. https://doi.org/10.1038/nrn1884

Bach, D. R., Seymour, B., & Dolan, R. J. (2009). Neural activity associated with the passive prediction of ambiguity and risk for aversive events. *The Journal of Neuroscience, 29*(6), 1648–1656. https://doi.org/10.1523/JNEUROSCI.4578-08.2009

Bailey, R., Meland, E. A., Brion-Meisels, G., & Jones, S. M. (2019). Getting developmental science Back into schools: Can what we know about self-regulation help change how we think about 'no excuses'? *Frontiers in Psychology, 10*, 1885. https://doi.org/10.3389/fpsyg.2019.01885

Balleine, B. W., Delgado, M. R., & Hikosaka, O. (2007). The role of the dorsal striatum in reward and decision-making. *Journal of Neuroscience, 27*(31), 8161–8165. https://doi.org/10.1523/JNEUROSCI.1554-07.2007

Bartra, O., McGuire, J. T., & Kable, J. W. (2013). The valuation system: A coordinate-based meta-analysis of BOLD fMRI experiments examining neural correlates of subjective value. *NeuroImage, 76*, 412–427. https://doi.org/10.1016/j.neuroimage.2013.02.063

Batteux, E., Ferguson, E., & Tunney, R. J. (2019). Do our risk preferences change when we make decisions for others? A meta-analysis of self-other differences in decisions involving risk. *PLoS One, 14*(5), e0216566. https://doi.org/10.1371/journal.pone.0216566

Bechara, A., Damasio, A. R., Damasio, H., & Anderson, S. W. (1994). Insensitivity to future consequences following damage to human prefrontal cortex. *Cognition, 50*(1–3), 7–15. https://doi.org/10.1016/0010-0277(94)90018-3

Behrens, T. E. J., Hunt, L. T., Woolrich, M. W., & Rushworth, M. F. S. (2008). Associative learning of social value. *Nature, 456*(7219), 245–249. https://doi.org/10.1038/nature07538

Bellucci, G., Chernyak, S. V., Goodyear, K., Eickhoff, S. B., & Krueger, F. (2017). Neural signatures of trust in reciprocity: A coordinate-based meta-analysis. *Human Brain Mapping, 38*(3), 1233–1248. https://doi.org/10.1002/hbm.23451

Bickel, W. K., Johnson, M. W., Koffarnus, M. N., MacKillop, J., & Murphy, J. G. (2014). The behavioral economics of substance use disorders: Reinforcement pathologies and their repair. *Annual Review of Clinical Psychology, 10*, 641–677. https://doi.org/10.1146/annurev-clinpsy-032813-153724

Blakemore, S.-J., & Robbins, T. W. (2012). Decision-making in the adolescent brain. *Nature Neuroscience, 15*(9), 1184–1191. https://doi.org/10.1038/nn.3177

Blankenstein, N. E., Huettel, S. A., & Li, R. (2021). Resolving ambiguity: Broadening the consideration of risky decision making over adolescent development. *Developmental Review, 62*, 100987. https://doi.org/10.1016/j.dr.2021.100987

Botzung, A., Denkova, E., & Manning, L. (2008). Experiencing past and future personal events: Functional neuroimaging evidence on the neural bases of mental time travel. *Brain and Cognition, 66*(2), 202–212. https://doi.org/10.1016/j.bandc.2007.07.011

Braams, B. R., Van Duijvenvoorde, A. C. K., Peper, J. S., & Crone, E. A. (2015). Longitudinal changes in adolescent risk-taking: A comprehensive study of neural responses to rewards, pubertal development, and risk-taking behavior. *The Journal of Neuroscience, 35*(18), 7226–7238. https://doi.org/10.1523/JNEUROSCI.4764-14.2015

Burnette, E. M., Grodin, E. N., Ghahremani, D. G., Galván, A., Kohno, M., Ray, L. A., & London, E. D. (2021). Diminished cortical response to risk and loss during risky decision making in alcohol use disorder. *Drug and Alcohol Dependence, 218*, 108391. https://doi.org/10.1016/j.drugalcdep.2020.108391

Carland, M. A., Thura, D., & Cisek, P. (2019). The urge to decide and act: Implications for brain function and dysfunction. *The Neuroscientist, 25*(5), 491–511. https://doi.org/10.1177/1073858419841553

Carter, R. M. K., Bowling, D. L., Reeck, C., & Huettel, S. A. (2012). A distinct role of the temporal-parietal junction in predicting socially guided decisions. *Science, 337*(6090), 109–111. https://doi.org/10.1126/science.1219681
Casey, B. J., Jones, R. M., & Hare, T. A. (2008). The adolescent brain. *Annals of the New York Academy of Sciences, 1124*, 111–126. https://doi.org/10.1196/annals.1440.010
Chaplo, S., & Fishbein, D. (2022). Capitalizing on neuroplasticity across development to redirect pathways from juvenile justice involvement. *Current Topics in Behavioral Neurosciences, 53*, 235–254. https://doi.org/10.1007/7854_2021_251
Chapman, G. B., & Weber, B. J. (2006). Decision biases in intertemporal choice and choice under uncertainty: Testing a common account. *Memory & Cognition, 34*(3), 589–602. https://doi.org/10.3758/BF03193582
Chein, J., Albert, D., O'Brien, L., Uckert, K., & Steinberg, L. (2011). Peers increase adolescent risk taking by enhancing activity in the brain's reward circuitry. *Developmental Science, 14*(2), F1–F10. https://doi.org/10.1111/j.1467-7687.2010.01035.x
Chen, J., Qin, J., He, Q., & Zou, Z. (2020). A meta-analysis of transcranial direct current stimulation on substance and food craving: What effect do modulators have? *Frontiers in Psychiatry, 11*, 598. https://doi.org/10.3389/fpsyt.2020.00598
Christopoulos, G. I., Tobler, P. N., Bossaerts, P., Dolan, R. J., & Schultz, W. (2009). Neural correlates of value, risk, and risk aversion contributing to decision making under risk. *The Journal of Neuroscience, 29*(40), 12574–12583. https://doi.org/10.1523/JNEUROSCI.2614-09.2009
Ciaramelli, E., De Luca, F., Kwan, D., Mok, J., Bianconi, F., Knyagnytska, V., Craver, C., Green, L., Myerson, J., & Shayna Rosenbaum, R. (2021). The role of ventromedial prefrontal cortex in reward valuation and future thinking during intertemporal choice. *eLife, 10*, e67387. https://doi.org/10.7554/eLife.67387
Clithero, J. A., & Rangel, A. (2014). Informatic parcellation of the network involved in the computation of subjective value. *Social Cognitive and Affective Neuroscience, 9*(9), 1289–1302. https://doi.org/10.1093/scan/nst106
Clithero, J. A., Tankersley, D., & Huettel, S. A. (2008). Foundations of neuroeconomics: from philosophy to practice. *PLoS Biology, 6*(11), e298.
Cotrena, C., Branco, L. D., Zimmermann, N., Cardoso, C. O., Grassi-Oliveira, R., & Fonseca, R. P. (2014). Impaired decision-making after traumatic brain injury: The Iowa gambling task. *Brain Injury, 28*(8), 1070–1075. https://doi.org/10.3109/02699052.2014.896943
D'Argembeau, A., Stawarczyk, D., Majerus, S., Collette, F., Van Der Linden, M., Feyers, D., Maquet, P., & Salmon, E. (2010). The neural basis of personal goal processing when envisioning future events. *Journal of Cognitive Neuroscience, 22*(8), 1701–1713. https://doi.org/10.1162/jocn.2009.21314
Defoe, I. N., Dubas, J. S., Figner, B., & van Aken, M. A. G. (2015). A meta-analysis on age differences in risky decision making: Adolescents versus children and adults. *Psychological Bulletin, 141*(1), 48–84. https://doi.org/10.1037/a0038088
Delgado, M. R., Nystrom, L. E., Fissell, C., Noll, D. C., & Fiez, J. A. (2000). Tracking the hemodynamic responses to reward and punishment in the striatum. *Journal of Neurophysiology, 84*(6), 3072–3077. https://doi.org/10.1152/jn.2000.84.6.3072
Duell, N., & Steinberg, L. (2021). Adolescents take positive risks, too. *Developmental Review, 62*, 100984. https://doi.org/10.1016/j.dr.2021.100984
Ellemers, N., Van Der Toorn, J., Paunov, Y., & Van Leeuwen, T. (2019). The psychology of morality: A review and analysis of empirical studies published from 1940 through 2017. *Personality and Social Psychology Review, 23*(4), 332–366. https://doi.org/10.1177/1088868318811759
Engemann, D. A., Bzdok, D., Eickhoff, S. B., Vogeley, K., & Schilbach, L. (2012). Games people play—Toward an enactive view of cooperation in social neuroscience. *Frontiers in Human Neuroscience, 6*. https://doi.org/10.3389/fnhum.2012.00148
Fareri, D. S., Stasiak, J. E., & Sokol-Hessner, P. (2022). Choosing for others changes dissociable computational mechanisms underpinning risky decision-making. *Scientific Reports, 12*(1), 14361. https://doi.org/10.1038/s41598-022-18437-9

Fellows, L. K., & Farah, M. J. (2005). Different underlying impairments in decision-making following ventromedial and dorsolateral frontal lobe damage in humans. *Cerebral Cortex, 15*(1), 58–63. https://doi.org/10.1093/cercor/bhh108

Filbeck, G., Hatfield, P., & Horvath, P. (2005). Risk aversion and personality type. *Journal of Behavioral Finance, 6*(4), 170–180. https://doi.org/10.1207/s15427579jpfm0604_1

Frost, R., & McNaughton, N. (2017). The neural basis of delay discounting: A review and preliminary model. *Neuroscience Biobehavioral Reviews, 79*, 48–65.

Green, L., & Myerson, J. (1996). Exponential versus hyperbolic discounting of delayed outcomes: Risk and waiting time. *American Zoologist, 36*(4), 496–505. https://doi.org/10.1093/icb/36.4.496

Harbaugh, W. T., Mayr, U., & Burghart, D. R. (2007). Neural responses to taxation and voluntary giving reveal motives for charitable donations. *Science, 316*(5831), 1622–1625. https://doi.org/10.1126/science.1140738

Haring, J., & Smith, G. (1959). Utility theory, decision theory, and profit maximization. *The American Economic Review, 49*(4), 566–583.

Hsu, M., Bhatt, M., Adolphs, R., Tranel, D., & Camerer, C. F. (2005). Neural systems responding to degrees of uncertainty in human decision-making. *Science, 310*(5754), 1680–1683. https://doi.org/10.1126/science.1115327

Huettel, S. A., Song, A. W., & McCarthy, G. (2005). Decisions under uncertainty: Probabilistic context influences activation of prefrontal and parietal cortices. *The Journal of Neuroscience, 25*(13), 3304–3311. https://doi.org/10.1523/JNEUROSCI.5070-04.2005

Huettel, S. A., Jill Stowe, C., Gordon, E. M., Warner, B. T., & Platt, M. L. (2006). Neural signatures of economic preferences for risk and ambiguity. *Neuron, 49*(5), 765–775. https://doi.org/10.1016/j.neuron.2006.01.024

Hula, A., Vilares, I., Lohrenz, T., Dayan, P., & Read Montague, P. (2018). A model of risk and mental state shifts during social interaction. *PLoS Computational Biology, 14*(2), e1005935. https://doi.org/10.1371/journal.pcbi.1005935

Hüpen, P., Habel, U., Votinov, M., Kable, J. W., & Wagels, L. (2023). A systematic review on common and distinct neural correlates of risk-taking in substance-related and non-substance related addictions. *Neuropsychology Review, 33*(2), 492–513. https://doi.org/10.1007/s11065-022-09552-5

Iqbal, S. N., Yin, L., Drucker, C. B., Kuang, Q., Gariépy, J.-F., Platt, M. L., & Pearson, J. M. (2019). Latent goal models for dynamic strategic interaction. *PLOS Computational Biology, 15*(3), e1006895. https://doi.org/10.1371/journal.pcbi.1006895. Edited by Jean Daunizeau.

Izuma, K., Saito, D. N., & Sadato, N. (2008). Processing of social and monetary rewards in the human striatum. *Neuron, 58*(2), 284–294. https://doi.org/10.1016/j.neuron.2008.03.020

Jolls, C., Sunstein, C. R., & Thaler, R. (1998). A behavioral approach to law and economics. *Stanford Law Review, 50*(5), 1471. https://doi.org/10.2307/1229304

Josef, A. K., Richter, D., Samanez-Larkin, G. R., Wagner, G. G., Hertwig, R., & Mata, R. (2016). Stability and change in risk-taking propensity across the adult life span. *Journal of Personality and Social Psychology, 111*(3), 430–450. https://doi.org/10.1037/pspp0000090

Kable, J. W., & Glimcher, P. W. (2007). The neural correlates of subjective value during intertemporal choice. *Nature Neuroscience, 10*(12), 1625–1633.

Kahneman, D., & Tversky, A. (1979). Prospect theory: An analysis of decision under risk. *Econometrica, 47*(2), 263–291. https://doi.org/10.2307/1914185

Kendall, M. G., Bernoulli, D., Allen, C. G., & Euler, L. (1961). Studies in the history of probability and statistics: XI. Daniel Bernoulli on maximum likelihood. *Biometrika, 48*(1/2), 1. https://doi.org/10.2307/2333125

Kim, H., & Sasaki, J. Y. (2014). Cultural neuroscience: Biology of the mind in cultural contexts. *Annual Review of Psychology, 65*, 487–514.

Knutson, B., & Huettel, S. A. (2015). The risk matrix. *Current Opinion in Behavioral Sciences, 5*, 141–146.

Knight, F. H. (1921). *Risk, uncertainty, and profit*. Houghton Mifflin.

Koffarnus, M. N., & Kaplan, B. A. (2018). Clinical models of decision making in addiction. *Pharmacology, Biochemistry, and Behavior, 164*, 71–83. https://doi.org/10.1016/j.pbb.2017.08.010

Kohno, M., Morales, A. M., Guttman, Z., & London, E. D. (2017). A neural network that links brain function, white-matter structure and risky behavior. *NeuroImage, 149*, 15–22. https://doi.org/10.1016/j.neuroimage.2017.01.058

Kräplin, A., Höfler, M., Pooseh, S., Wolff, M., Krönke, K.-M., Goschke, T., Bühringer, G., & Smolka, M. N. (2020). Impulsive decision-making predicts the course of substance-related and addictive disorders. *Psychopharmacology, 237*(9), 2709–2724. https://doi.org/10.1007/s00213-020-05567-z

Kuhnen, C. M., & Knutson, B. (2005). The neural basis of financial risk taking. *Neuron, 47*(5), 763–770. https://doi.org/10.1016/j.neuron.2005.08.008

Lee, D. (2008). Game theory and neural basis of social decision making. *Nature Neuroscience, 11*(4), 404–409. https://doi.org/10.1038/nn2065

Lejuez, C. W., Read, J. P., Kahler, C. W., Richards, J. B., Ramsey, S. E., Stuart, G. L., Strong, D. R., & Brown, R. A. (2002). Evaluation of a behavioral measure of risk taking: The balloon analogue risk task (BART). *Journal of Experimental Psychology. Applied, 8*(2), 75–84. https://doi.org/10.1037/1076-898x.8.2.75

Levy, D. J., & Glimcher, P. W. (2012). The root of all value: A neural common currency for choice. *Current Opinion in Neurobiology, 22*(6), 1027–1038. https://doi.org/10.1016/j.conb.2012.06.001

Levy, I., Snell, J., Nelson, A. J., Rustichini, A., & Glimcher, P. W. (2009). Neural representation of subjective value under risk and ambiguity. *Journal of Neurophysiology, 103*(2), 1036–1047. https://doi.org/10.1152/jn.00853.2009

Levallois, C., Clithero, J. A., Wouters, P., Smidts, A., & Huettel, S. A. (2012). Translating upwards: linking the neural and social sciences via neuroeconomics. *Nature Reviews Neuroscience, 13*(11), 789–797.

Li, R. (2017). Flexing dual-systems models: How variable cognitive control in children informs our understanding of risk-taking across development. *Developmental Cognitive Neuroscience, 27*, 91–98. https://doi.org/10.1016/j.dcn.2017.08.007

Liu, L., Xue, G., Potenza, M. N., Zhang, J.-T., Yao, Y.-W., Xia, C.-C., Lan, J., Ma, S.-S., & Fang, X.-Y. (2017). Dissociable neural processes during risky decision-making in individuals with internet-gaming disorder. *NeuroImage Clinical, 14*, 741–749. https://doi.org/10.1016/j.nicl.2017.03.010

Loewenstein, G., & Prelec, D. (1991). Negative time preference. *The American Economic Review, 81*(2), 347–352.

Luca, D., Cinzia, R., Wood, S. J., Anderson, V., Buchanan, J.-A., Proffitt, T. M., Mahony, K., & Pantelis, C. (2003). Normative data from the CANTAB. I: Development of executive function over the lifespan. *Journal of Clinical and Experimental Neuropsychology, 25*(2), 242–254. https://doi.org/10.1076/jcen.25.2.242.13639

Ma, S., Zang, Y., Cheung, V., & Chan, C. C. H. (2015). Importance of punishment frequency in the Iowa gambling task: An fMRI study. *Brain Imaging and Behavior, 9*(4), 899–909. https://doi.org/10.1007/s11682-015-9353-0

Maiti, R., Mishra, B. R., & Hota, D. (2017). Effect of high-frequency transcranial magnetic stimulation on craving in substance use disorder: A meta-analysis. *The Journal of Neuropsychiatry and Clinical Neurosciences, 29*(2), 160–171. https://doi.org/10.1176/appi.neuropsych.16040065

Markowitz, H. M. (2010). Portfolio theory: As I still see it. *Annual Review of Financial Economics, 2*(1), 1–23. https://doi.org/10.1146/annurev-financial-011110-134602

McClure, S. M., Laibson, D. I., Loewenstein, G., & Cohen, J. D. (2004). Separate neural systems value immediate and delayed monetary rewards. *Science, 306*(5695), 503–507.

McCormick, E. M., & Telzer, E. H. (2017). Adaptive adolescent flexibility: Neurodevelopment of decision-making and learning in a risky context. *Journal of Cognitive Neuroscience, 29*(3), 413–423. https://doi.org/10.1162/jocn_a_01061

McDonald, K. R., Pearson, J. M., & Huettel, S. A. (2020). Dorsolateral and dorsomedial prefrontal cortex track distinct properties of dynamic social behavior. *Social Cognitive and Affective Neuroscience, 15*(4), 383–393. https://doi.org/10.1093/scan/nsaa053

Meisel, S. N., Fosco, W. D., Hawk, L. W., & Colder, C. R. (2019). Mind the gap: A review and recommendations for statistically evaluating dual systems models of adolescent risk behavior. *Developmental Cognitive Neuroscience, 39*, 100681. https://doi.org/10.1016/j.dcn.2019.100681

Menon, V., & D'Esposito, M. (2022). The role of PFC networks in cognitive control and executive function. *Neuropsychopharmacology, 47*(1), 90–103. https://doi.org/10.1038/s41386-021-01152-w

Messimeris, D., Levy, R., & Le Bouc, R. (2023). Economic and social values in the brain: Evidence from lesions to the human ventromedial prefrontal cortex. *Frontiers in Neurology, 14*, 1198262. https://doi.org/10.3389/fneur.2023.1198262

Mishra, S., Lalumière, M. L., & Williams, R. J. (2010). Gambling as a form of risk-taking: Individual differences in personality, risk-accepting attitudes, and Behavioral preferences for risk. *Personality and Individual Differences, 49*(6), 616–621. https://doi.org/10.1016/j.paid.2010.05.032

Morelli, S. A., Knutson, B., & Zaki, J. (2018). Neural sensitivity to personal and vicarious reward differentially relates to prosociality and well-being. *Social Cognitive and Affective Neuroscience, 13*(8), 831–839. https://doi.org/10.1093/scan/nsy056

Nardone, R., Sebastianelli, L., Versace, V., Ferrazzoli, D., Brigo, F., Schwenker, K., Saltuari, L., & Trinka, E. (2022). TMS for the functional evaluation of cannabis effects and for treatment of cannabis addiction: A review. *Psychiatry Research, 310*, 114431. https://doi.org/10.1016/j.psychres.2022.114431

Park, S. A., Sestito, M., Boorman, E. D., & Dreher, J.-C. (2019). Neural computations underlying strategic social decision-making in groups. *Nature Communications, 10*(1), 5287. https://doi.org/10.1038/s41467-019-12937-5

Paulus, M. P., Rogalsky, C., Simmons, A., Feinstein, J. S., & Stein, M. B. (2003). Increased activation in the right insula during risk-taking decision making is related to harm avoidance and neuroticism. *NeuroImage, 19*(4), 1439–1448. https://doi.org/10.1016/S1053-8119(03)00251-9

Peters, J., & Buchel, C. (2009). Overlapping and distinct neural systems code for subjective value during intertemporal and risky decision making. *The Journal of Neuroscience, 29*(50), 15727–15734.

Peters, J., & Buchel, C. (2010). Episodic future thinking reduces reward delay discounting through an enhancement of prefrontal-mediotemporal interactions. *Neuron, 66*(1), 138–148.

Pettorruso, M., di Giannantonio, M., De Risio, L., Martinotti, G., & Koob, G. F. (2020). A light in the darkness: Repetitive transcranial magnetic stimulation (rTMS) to treat the hedonic dysregulation of addiction. *Journal of Addiction Medicine, 14*(4), 272–274. https://doi.org/10.1097/ADM.0000000000000575

Pisauro, M. A., Fouragnan, E. F., Arabadzhiyska, D. H., Apps, M. A. J., & Philiastides, M. G. (2022). Neural implementation of computational mechanisms underlying the continuous trade-off between cooperation and competition. *Nature Communications, 13*(1), 6873. https://doi.org/10.1038/s41467-022-34509-w

Platt, M. L., & Glimcher, P. W. (1999). Neural correlates of decision variables in parietal cortex. *Nature, 400*(6741), 233–238. https://doi.org/10.1038/22268

Poudel, R., Riedel, M. C., Salo, T., Flannery, J. S., Hill-Bowen, L. D., Eickhoff, S. B., Laird, A. R., & Sutherland, M. T. (2020). Common and distinct brain activity associated with risky and ambiguous decision-making. *Drug and Alcohol Dependence, 209*, 107884. https://doi.org/10.1016/j.drugalcdep.2020.107884

Preuschoff, K., Bossaerts, P., & Quartz, S. R. (2008). Human insula activation reflects risk prediction errors as well as risk. *The Journal of Neuroscience, 28*(11), 2745–2752.

Qu, C., Météreau, E., Butera, L., Villeval, M. C., & Dreher, J.-C. (2019). Neurocomputational mechanisms at play when weighing concerns for extrinsic rewards, moral values, and social image. *PLOS Biology, 17*(6), e3000283. https://doi.org/10.1371/journal.pbio.3000283. Edited by Matthew F. S. Rushworth.

Ramey, T., & Regier, P. S. (2019). Cognitive impairment in substance use disorders. *CNS Spectrums, 24*(1), 102–113. https://doi.org/10.1017/S1092852918001426

Ruff, C. C., & Fehr, E. (2014). The neurobiology of rewards and values in social decision making. *Nature Reviews Neuroscience, 15*(8), 549–562. https://doi.org/10.1038/nrn3776

Ryzin, V., Mark, J., & Roseth, C. J. (2018). Cooperative learning in middle school: A means to improve peer relations and reduce victimization, bullying, and related outcomes. *Journal of Educational Psychology, 110*(8), 1192–1201. https://doi.org/10.1037/edu0000265

Santos, S., Almeida, I., Oliveiros, B., & Castelo-Branco, M. (2016). The role of the Amygdala in facial trustworthiness processing: A systematic review and meta-analyses of fMRI studies. *PLOS One, 11*(11), e0167276. https://doi.org/10.1371/journal.pone.0167276. Edited by Hisao Nishijo.

Sasaki, J. Y., & Kim, H. (2017). Nature, nurture, and their interplay: A review of cultural neuroscience. *Journal of Cross-Cultural Psychology, 48*(1), 4–22.

Schellenberg, C., Steinebach, C., & Krauss, A. (2022). Empower peers 4 careers: Positive peer culture to prepare adolescents' career choices. *Frontiers in Psychology, 13*, 806103. https://doi.org/10.3389/fpsyg.2022.806103

Schonberg, T., Fox, C. R., & Poldrack, R. A. (2011). Mind the gap: Bridging economic and naturalistic risk-taking with cognitive neuroscience. *Trends in Cognitive Sciences, 15*(1), 11–19. https://doi.org/10.1016/j.tics.2010.10.002

Smith, D. V., Clithero, J. A., Boltuck, S. E., & Huettel, S. A. (2014). Functional connectivity with ventromedial prefrontal cortex reflects subjective value for social rewards. *Social Cognitive and Affective Neuroscience, 9*(12), 2017–2025. https://doi.org/10.1093/scan/nsu005

Sokol-Hessner, P., & Rutledge, R. B. (2019). The psychological and neural basis of loss aversion. *Current Directions Psychological Science, 28*(1), 20–27.

Smith, A. R., Steinberg, L., Strang, N., & Chein, J. (2015). Age differences in the impact of peers on adolescents' and adults' neural response to reward. *Developmental Cognitive Neuroscience, 11*, 75–82. https://doi.org/10.1016/j.dcn.2014.08.010

Steinberg, L. (2010). A dual systems model of adolescent risk-taking. *Developmental Psychobiology, 52*(3), 216–224. https://doi.org/10.1002/dev.20445

Tei, S., Fujino, J., Itahashi, T., Aoki, Y. Y., Ohta, H., Kubota, M., Sawajiri, S., et al. (2021). The right temporoparietal junction during a cooperation dilemma: An rTMS study. *Neuroimage Reports, 1*(3), 100033. https://doi.org/10.1016/j.ynirp.2021.100033

Tom, S. M., Fox, C. R., Trepel, C., & Poldrack, R. A. (2007). The neural basis of loss aversion in decision-making under risk. *Science, 315*(5811), 515–518.

Tremblay, S., Sharika, K. M., & Platt, M. L. (2017). Social decision-making and the brain: A comparative perspective. *Trends in Cognitive Sciences, 21*(4), 265–276. https://doi.org/10.1016/j.tics.2017.01.007

Trezza, V., Campolongo, P., & Vanderschuren, L. J. M. J. (2011). Evaluating the rewarding nature of social interactions in laboratory animals. *Developmental Cognitive Neuroscience, 1*(4), 444–458. https://doi.org/10.1016/j.dcn.2011.05.007

Trucco, E. M. (2020). A review of psychosocial factors linked to adolescent substance use. *Pharmacology, Biochemistry, and Behavior, 196*, 172969. https://doi.org/10.1016/j.pbb.2020.172969

Van Dijk, E., & De Dreu, C. K. W. (2021). Experimental games and social decision making. *Annual Review of Psychology, 72*(1), 415–438. https://doi.org/10.1146/annurev-psych-081420-110718

Weber, B. J., & Huettel, S. A. (2008). The neural substrates of probabilistic and intertemporal decision making. *Brain Research, 1234*, 104–115.

Weber, E. U., Shafir, S., & Blais, A.-R. (2004). Predicting risk sensitivity in humans and lower animals: Risk as variance or coefficient of variation. *Psychological Review, 111*(2), 430–445. https://doi.org/10.1037/0033-295X.111.2.430

Weber, E. U., Johnson, E. J., Milch, K. F., Chang, H., Brodscholl, J. C., & Goldstein, D. G. (2007). Asymmetric discounting in intertemporal choice: A query-theory account. *Psychological Science, 18*(6), 516–523. https://doi.org/10.1111/j.1467-9280.2007.01932.x

Xu, S., Luo, L., Xiao, Z., Zhao, K., Wang, H., Wang, C., & Rao, H. (2019). High sensation seeking is associated with Behavioral and neural insensitivity to increased negative outcomes during decision-making under uncertainty. *Cognitive, Affective, & Behavioral Neuroscience, 19*(6), 1352–1363. https://doi.org/10.3758/s13415-019-00751-x

Yoder, K. J., & Decety, J. (2018). The neuroscience of morality and social decision-making. *Psychology, Crime & Law, 24*(3), 279–295. https://doi.org/10.1080/1068316X.2017.1414817

Yoo, S. B. M., Hayden, B. Y., & Pearson, J. M. (2021). Continuous decisions. *Philosophical Transactions of the Royal Society B: Biological Sciences, 376*(1819), 20190664. https://doi.org/10.1098/rstb.2019.0664

Yurgelun-Todd, D. (2007). Emotional and cognitive changes during adolescence. *Current Opinion in Neurobiology, 17*(2), 251–257. https://doi.org/10.1016/j.conb.2007.03.009

Zhang, M., Zhang, X., Cunquan, Q., Wang, G., & Xin, L. (2022). The combination of social reward and punishment is conducive to the cooperation and heterogeneity of social relations. *Chaos: An Interdisciplinary Journal of Nonlinear Science, 32*(10), 103104. https://doi.org/10.1063/5.0102483

What Lesion Studies Have Taught Us About Frontal Lobe Contributions to Decision-Making

Lesley K. Fellows

Abstract As a relatively new field within cognitive neuroscience, neuroeconomics has tackled the general goal of defining the relevant component processes underlying evaluation, value-based decision-making, and related aspects of motivated ("economic") behavior, and determining the brain mechanisms that underlie those processes. Studies of humans with acquired focal brain injury contribute evidence about the causal contribution of the affected brain region to the behavior or cognitive process under study. Clinically, this method has guided localizationist understandings of brain function that are the basis of neurology and neuropsychology. Experimentally, it can be especially useful for validating brain-based models of the component processes of complex behavior, such as the behaviors that are the focus of neuroeconomics. Inferentially, evidence from lesion studies has particular strengths, supporting claims that a region is (or is not) necessary for a particular process, as well as demonstrating that putatively distinct component processes are, in fact, dissociable. This chapter provides an overview of findings from human lesion research relevant to neuroeconomics, with a focus on the frontal lobes. The strengths and weaknesses of the methodology are discussed, the more robust, replicated findings are highlighted, examples of how this work has helped to increasingly refine accounts of regional frontal lobe contributions to economic behavior are provided, and promising future directions are suggested.

1 The Effects of Brain Damage on Economic Behavior: From Anecdote to Experiment

The goal of neuroeconomics is to establish the brain basis of economic behavior in humans. There is a fast-expanding repertoire of human neuroscience tools and techniques available to study such questions. However, most of these provide corre-

L. K. Fellows (✉)
Department of Neurology & Neurosurgery, McGill University, Montréal, QC, Canada
e-mail: Lesley.fellows@mcgill.ca

D. V. Smith et al. (eds.), *Neuroeconomics: Core Topics and Current Directions*,
https://doi.org/10.1007/978-3-032-02925-6_3

lational evidence, showing, for example, that a particular pattern of neural activity is associated with a particular behavior. Even the most elegant experimental designs using such methods have the unavoidable inferential weakness that association does not establish causation. Studies of people with brain injuries or disorders can support causal inferences, addressing questions like "is this brain region *necessary* for a given behavior?" by assessing the effect of injury to a given brain region on a well-defined behavioral measure (Fellows et al., 2005; Verma & Kumar, 2022). This work, at the interface of clinical neuroscience and experimental psychology, is at the very foundation of cognitive neuroscience (Verma & Kumar, 2022).

Loss-of-function studies in humans are often called "natural experiments": researchers don't cause brain damage, of course, but rather examine the effects of naturally acquired brain lesions or another disease state with definable effects on the brain. These observations have limitations compared to the controlled lesion experiments possible in animal models (de Haan & Karnath, 2018; Mah et al., 2014), but nonetheless provide important causal evidence. That evidence can help in two (related) ways. First, it can be useful in validating a conceptual framework: helping us know that a working model of the relevant component processes of economic behavior is on the right track. Second, it can disentangle the distinct brain processes that underpin correlated behavioral processes. Such evidence is particularly helpful at the earlier stages of a new field, as in neuroeconomics. Economic behaviors are complex and may be described and measured in many ways. How do we know we have the "right" (most useful) explanatory framework?

Further, there are many correlated processes. For example, a risky choice might be associated with autonomic system activation (yielding sweaty palms and a racing heart) as well as with specific cognitive processes (i.e., value assessment, value comparison, response preparation). Studying the effects of damage to a particular brain region implicated by correlational evidence in risky choice, for example, the orbitofrontal cortex, and measuring these different processes can directly inform whether that damage affects risk-taking, the autonomic response, neither, or both. Such evidence simultaneously addresses the dissociability (or unity) of putative behavioral processes and supports causal inferences about their brain substrates.

There have been two major streams of lesion research relevant to neuroeconomics, to date: On the one hand, many decades of research in rodent and nonhuman primate models have yielded an extensive literature on the effects of focal brain injury on reinforcement learning (i.e., learning on a trial-by-trial basis from reward or punishment) (reviewed in (Fellows, 2018; Padoa-Schioppa & Schoenbaum, 2015)). On the other hand, there is a more recent literature addressing how economic "value" may be represented in the brain and how such information can come to influence decisions (Fellows, 2004; Kable & Glimcher, 2007). Both these streams of work have mainly focused on regions within the frontal lobes, particularly orbitofrontal and adjacent ventromedial prefrontal cortex (here referred to together in the coarser designation of the ventromedial frontal lobe (VMF)) and on related sub-cortical structures, including the striatum, hippocampus, and amygdala (Averbeck & O'Doherty, 2022; Biderman et al., 2020; Murray & Fellows, 2022).

Before considering the key findings from this work, it's useful to consider the various forms of human lesion evidence, each with strengths and limitations (Vaidya et al., 2019). Perhaps the most frequent use of lesion evidence is the clinical anecdote. Observing or questioning people living with a focal lesion to a part of the brain thought to be important in economic behavior would seem a very direct way of understanding brain-behavior relationships; indeed, it may trigger fresh insights straight from the experience of the person arguably best-placed to describe lesion effects: the patient (or their family) (Chatterjee, 2005). It can also help to close the translational research loop: ensuring that fundamental insights derived from research in clinical populations come back around to improve the diagnosis and care of such patients (Graham et al., 2006). Paradoxically, this is the weakest form of lesion evidence; the experience of a single individual can be difficult to interpret scientifically. Patients may lack insight into their difficulties (impaired insight is a frequent feature of frontal lobe damage, for example), and they or their families may not describe symptoms in ways that are easily mapped to formal models of economic behavior. To say, for example, that after frontal lobe injury a person is more "capricious" (as was said of perhaps the most famous frontal-lobe lesioned case of all time, Phineas Gage (Harlow, 1868)) is only modestly informative.

A more robust approach to studying lesion effects is to study groups of patients with damage to a particular brain region, using behavioral measures that are grounded in well-defined cognitive models. This requires comparison groups, typically demographically matched healthy people, and often a group of patients with damage due to similar causes but affecting other parts of the brain. This provides a tighter definition of the behavior, and stronger evidence that the observed behavioral pattern relies critically on the brain region in question, and is not an idiosyncratic individual difference or a nonspecific effect of serious illness. It also helps to validate the neurobiological plausibility of the conceptual framework that informed the choice of behavioral measure (Vaidya et al., 2019). The most common design of such group studies is to select participants on the basis of damage to a region of interest. Alternatives include a variety of lesion-symptom mapping approaches, which require much larger sample sizes, but have become more feasible with long-term lesion registries and multi-site collaborations (de Haan & Karnath, 2018; Fellows et al., 2008; Karnath et al., 2018; Mah et al., 2014; Vaidya et al., 2019). A still stronger design would involve assessing the person prior to the lesion, and then again after the brain damage, allowing them to serve as their own comparator, but, for obvious reasons, such opportunities rarely arise.

I will briefly review some examples of lesion results that have been important in advancing our understanding of what regions of the brain are necessary for particular aspects of economic behavior, focusing on the frontal lobes. In so doing, I will illustrate the challenges of identifying the relevant component processes of economic behavior and highlight some of the strengths and weaknesses of various experimental designs. Of course, any result needs to be interpreted in the context of other forms of evidence. Such converging evidence (also referred to as triangulation) is the best way to take advantage of the relative strengths and offset the different weaknesses of any given method.

2 Frontal-Executive Functions and Economic Behaviors

Although clinicians have long described the symptoms of people with frontal lobe damage in terms that seem relevant to economics, such as lack of foresight, impulsivity, or apathy, the notion that the frontal lobes contribute to value-based decision-making framed in economic terms is relatively recent. Instead, frontal lobe functions were more often considered in relation to other cognitive constructs such as problem-solving, planning, and working memory, often bundled under the generic heading "executive functions" (Alvarez & Emory, 2006; Barbey et al., 2012; Reber & Tranel, 2019). To put this another way, the study of the frontal lobes has emphasized its roles in the "how-to" of goal-directed behavior: i.e., the mechanisms involved in planning multiple steps, attending to task-relevant information, and staying on track over time and in the face of distractions to reach a distant goal.

In 1985, Eslinger and Damasio published a case study of a patient, EVR, who had behavioral changes after resection of a large brain tumor resulting in extensive damage to the bilateral VMF. EVR developed a range of behavioral symptoms, including a striking difficulty in making decisions, spending hours gathering information and deliberating over trivial consumer choices, for example, as well as displaying poor judgment in social contexts and a general lack of motivation, despite performing well on existing laboratory tests of ("how-to") executive function, such as the Wisconsin Card Sorting Task (Eslinger & Damasio, 1985). In subsequent work with EVR and other patients with similar patterns of frontal damage, Damasio and colleagues tried to define the behavioral deficits more specifically, developing a novel task in an effort to tap into the processes underlying the real-life decision-making impairments (Bechara et al., 1997).

That task, subsequently termed the Iowa Gambling Task (IGT), was a card game that required making sequential choices between decks of cards, with each card selected providing varying monetary wins or losses. Over the course of the task, healthy people were able to learn to steer clear of the "high-win" card decks, as these also occasionally led to even bigger losses, and so were disadvantageous overall. In contrast, those with frontal damage persisted in choosing from these riskier decks. This pattern has been replicated in other samples for people with ventromedial frontal lobe damage, although results have varied for those with dorsolateral damage, with some studies also finding deficits in those groups (Bechara et al., 1998; Clark & Manes, 2004; Fellows & Farah, 2005).

3 Risky Choice or Impaired Learning?

The IGT is a complex task, and there have been different interpretations of the observed impairment in frontal lobe-injured patients. The initial work proposed that VMF damage led to riskier choices due to a short-term focus on the prospect of high immediate reward (Bechara et al., 1997). Subsequent work reframed the IGT in terms of reinforcement learning rather than risk, pointing out that it required a form of flexible updating of stimulus-value associations known as reversal learning that research in nonhuman primates had previously shown was disrupted by ventral frontal lobe lesions (Butter, 1968; Dias et al., 1996). A typical reversal learning task involves choosing between two stimuli, one associated with a reward, the other, not. Once the initial association is learned, the contingencies are reversed, and the subject then needs to learn that the previously rewarded stimulus is now worthless, while the previously worthless stimulus is now valuable. We and others have shown deficits in reversal learning in humans with VMF damage (Fellows & Farah, 2003; Hornak et al., 2004; Tsuchida et al., 2010). We also showed that IGT deficits in patients with VMF damage were restored when the task was altered (i.e., the cards were re-ordered) so that the risk remained but the reversal learning requirement was removed (Fellows & Farah, 2005).

This work illustrates the challenges in moving from clinical descriptions of impaired behaviors to focused experimental tasks. It also demonstrates that drawing on research in animals, involving much simpler tasks, can be valuable in framing our thinking about the relevant component processes of behavior and their underlying brain mechanisms in human decision-making. Seeking reward and avoiding punishment in changing environments are fundamental aspects of any organism's behavior: it is reasonable to expect that there would be similarities in the brain circuitry involved across species, and that different aspects of value (e.g., food reward, social reward) might rely on shared circuitry (Machado & Bachevalier, 2007; Murray et al., 2007).

A great deal of research has subsequently built on these observations in a variety of ways, including drawing on computational modeling to formalize perturbations in reinforcement learning, more specifically defining the underlying anatomical circuits and cellular mechanisms, and seeking to understand how flexible value-driven learning may play out in real-world behaviors, such as so-called credit assignment (correctly ascribing a rewarding outcome to the predictive event, as a series of choices and outcomes plays out over time) (Murray & Rudebeck, 2018; Noonan et al., 2017), patch foraging (Rudebeck & Izquierdo, 2022), and social interactions (Delgado et al., 2016; Messimeris et al., 2023). The basic concept that circuitry within the ventral and medial frontal lobes is important in flexible, value-guided behavior has been amply supported. However, much more remains to be discovered.

4 Is Economic Value Represented in the Brain?

A largely separate stream of work has addressed whether behaviors defined in economic terms can be mapped to the brain. Subjective value, or utility, is a central economic concept: Does the brain represent value, in the same sense that it represents other behaviorally important variables such as perceptual inputs? This was a key early question for neuroeconomics, addressed mainly using functional

MRI in humans, and electrophysiological recordings in nonhuman primates (Bartra et al., 2013; Hunt et al., 2018; Kable & Glimcher, 2007; Padoa-Schioppa & Schoenbaum, 2015; Tom et al., 2007). Depending on how subjective value has been operationalized and on the neuroscience methods used to measure its brain correlates, evidence has been gathered arguing that subjective value is represented in specific brain regions: notably striatum, thalamus, ventromedial and/or orbitofrontal prefrontal cortex or that value is represented widely (or perhaps in multiple ways in the brain, rather than in a small set of anatomically segregated circuits (Hayden & Niv, 2021; Pool et al., 2019; Roesch & Olson, 2007; Stott & Redish, 2015).

As discussed above, showing that there are neural correlates of subjective value is informative, but far from definitive. Indeed, the work arguing that value is so general a construct that value-related signals can be found in many parts of the brain points to the importance of causal evidence to help define if, where, and how value-correlated signals are necessary for influencing specific value-related behaviors.

We and others have undertaken lesion studies to contribute loss-of-function evidence to this question. We found that patients with VMF damage, and (importantly) not patients with damage to other parts of the frontal lobes, had specific difficulties with value-based preference judgments. Value is, by definition, subjective, so how can we know experimentally that patients are making errors? We reasoned that if VMF value signals were necessary for value-based choice, then damage to that region would lead to increased variability or inconsistency in simple preference judgments. We asked participants to make repeated pair-wise choices between options, and inferred their order of preferences across the set of options from their choices. That is, if the participant chose an apple over a banana, and a banana over an orange, we would expect them to prefer apple over orange when offered that choice. If they instead chose orange over apple, this would be a violation of transitivity or a subjective value "error." If value signals in VMF were necessary for guiding choice, then we predicted that errors of preference transitivity would be more frequent in those with VMF damage. This is, in fact, what we observed, replicating this observation across several types of choices and in different samples of patients (Camille et al., 2011; Fellows & Farah, 2007; Henri-Bhargava et al., 2012). More recently, Yu and Kable also have replicated these findings, but more directly assessed whether the impairment reflects failures of transitivity of preferences, or simply more variability in preferences. They concluded that VMF damage leads to the latter, i.e., that intact VMF is "... necessary for having strong and reliable preferences" (Yu et al., 2022).

Selective satiety paradigms provide another window into flexible representations of subjective value. Widely studied in nonhuman primates, such paradigms first establish an animal's preference between two food rewards (e.g., banana vs peanut). The animals learn the association between arbitrary stimuli and each food, and then are offered choices between the stimuli. Rationally, they choose the stimulus associated with their preferred food most of the time, even when tested in extinction (i.e., the stimulus is not followed by the food reward). The animals are then provided with as much of their preferred food as they want to eat. Once they are sated on that specific food, they are returned to the test apparatus and offered the choice between the previously learned stimuli (in extinction). If subjective value representations are intact, the prediction is that the subjective value of the stimulus predicting the sated food is now reduced, so that they will prefer the other stimulus, i.e., the one they have learned as predictive of the non-sated, therefore now more valuable food reward. Animals with orbitofrontal lesions are not sensitive to this selective satiety manipulation: they continue to choose the initially preferred stimulus after satiety, suggesting that they cannot flexibly update their current subjective value representations of the stimuli to take account of a change in internal state (Rudebeck et al., 2017). An experiment using a variation of this design in humans with VMF damage drew similar conclusions (Reber et al., 2017).

One point worth highlighting from these various studies is that people with VMF damage can make choices. They do not generally suffer from "decision paralysis" (notwithstanding the description of the case of EVR, who did have difficulty arriving at decisions in real life). Indeed, in laboratory preference tasks, where decision time is measured to the millisecond, such patients make decisions as quickly as healthy controls (Henri-Bhargava et al., 2012). Rather, the bulk of the evidence suggests that OFC/VMF is necessary for using the expected subjective value of stimuli to guide choice. This is not a generic effect of brain damage: most of these studies included a control group of patients with damage to other parts of the brain and showed they performed similarly to healthy people. This work is a good example of how lesion evidence can be useful in sorting out whether signals observed with correlational methods are necessary for a given process. Signals correlated with value may be widely present in the brain, but damage to only some regions, and not others, leads to changes in specific value-sensitive behaviors.

Taken together, this work argues that VMF does contribute critically to the ability to flexibly track subjective value and use that information to make choices. However, the exact contribution of this region to this process remains a matter of debate. Over the last several years, we and others have been working on that question. We observed, quite by accident, that people with VMF damage might be systematically relying on different information while assessing subjective value (Xia et al., 2015). In trying to make sense of that observation, we considered that even simple pairwise preference judgments (Do you prefer an apple or a banana?) are potentially far from simple in terms of the underlying processes. Even in such mundane choices, each option has multiple attributes. What do we consider when estimating the value of a particular apple, or indeed, of apples in general? In principle, we could assess texture, flavor, sweetness, water content, nutritional value, and so on. Do we really weigh all that information? Do we assign a holistic value, capturing the emergent property of "appleness" that is perhaps different than the sum of the individual attributes? Do we focus on specific attributes? Do the attributes we consider vary depending on our state (hungry or sated?), our goals (weight loss?), or the other available options?

Behavioral economics and decision psychology researchers have long studied how multi-attribute decisions are solved but have typically approached these questions by explicitly defining the attributes (presented in tables, for example) and studying how people gather information to make a choice (Payne et al., 1992).

How attributes are considered, combined, or traded-off in more typical (and more ecologically valid) choices remains ill-defined in process terms, with work on the computational and neurobiological mechanisms ongoing (e.g., Fellows, 2006; Krajbich et al., 2012; Lee & Hare, 2023; Thomas et al., 2021).

We asked patients with focal frontal lobe lesions to make choices between pairs of political candidates in a simulated election. The candidates were shown as headshot photos, and otherwise were unknown to the participants. We subsequently asked them to rate each candidate on attributes known to be relevant to political choice in such paradigms, including attractiveness and (apparent) competence. Replicating prior work in young healthy people, we found that both attractiveness and competence were predictive of the pairwise voting choices. That is, people were more likely to have preferred one candidate over the other if they rated that candidate as either more attractive, more competent, or both. Patients with dorsolateral frontal damage performed normally on both the rating and voting task. While VMF patients could rate the attributes, their choices were predicted only by the relative attractiveness of the candidates; their votes were not predicted by relative competence (Xia et al., 2015). This raised the possibility that VMF damage did not prevent assigning value to complex stimuli, but instead changed what information was used to assess value. We found further evidence for this hypothesis in a study using very different, but also complex stimuli, in which participants were asked to rate the subjective (aesthetic) value of visual artworks. Those with VMF damage could assign value ratings and were relatively consistent in doing so over the course of the experiment, but they again seemed to prioritize different facets of the art compared to health people as well as to those with frontal damage sparing VMF (Vaidya & Fellows, 2015a; Vaidya et al., 2018).

Inspired by distinctions in how complex objects are represented in the temporal lobes, we pursued the possibility that the VMF was somehow important in interpreting the holistic value of multi-attribute options. That is, we tested the idea that this region was specifically needed to assign value when it emerged from the unique configuration of multiple attributes, and not when the overall value could be assessed simply by "adding up" the value of each attribute. We taught participants the value of novel pseudo-objects ("fribbles") made up of multiple attributes. In one condition, the value was linked to individual attributes; in the other, it related to the specific combination of two attributes. Once these values were learned, people were asked to choose between pairs of fribbles, aiming to select the ones worth the most money. We found that those with VMF damage were less likely to choose the more valuable fribble when its value emerged from the combination of attributes, and performed similarly to controls when value could be determined by summing up individual attributes, a pattern that supported our hypothesis (Pelletier & Fellows, 2019).

5 Value and Learning in the Real World

The body of work reviewed above supports the general idea that the VMF is necessary for value-related decisions but suggests that the critical contributions of this region are nuanced. Clearly, the VMF is not "the value center" in the brain. Rather the region becomes necessary for valuation only under specific circumstances: when subjective value is changing, as in reversal learning; when it needs to be interpreted in a particular internal context, as in selective satiety; or derived from multiple types of information interpreted with respect to each other, as in multi-attribute configurations. Similar descriptions of potential functions of OFC/VMF have been proposed in terms that are not inherently "economic." For example, Niv and colleagues have suggested the region provides a flexible, "state-space" representation of task-relevant information (Boorman et al., 2021; Hayden & Niv, 2021; Schuck et al., 2016), while a largely parallel literature, in part inspired by clinical observations of confabulation in some people after acute VMF damage, has implicated the same region in activating memory "schema" (Gilboa & Moscovitch, 2017). At least some of the work reviewed in this chapter could be interpreted as supporting such ideas (Yu et al., 2020).

However, if such theories are framed too generally, we end with a broader description of the relevant behaviors that would seem to circle back to more general conceptualizations of "executive function," i.e., the abilities to flexibly maintain and update task-relevant (contextualized) information across delays or in the face of distraction. This more general set of processes allows us to pursue goals that are distant in time and space, as well as to make the context-specific predictions that drive learning. Such capacities have long been linked to the frontal lobes more broadly. Returning to the observations of EVR that helped to launch the work reviewed here, we are reminded that he was intact on conventional tests of executive function, such as the Wisconsin card sorting task, which indeed require flexible behavioral adjustments—yet impaired on tasks that more specifically tapped flexible reward-related behaviors (Eslinger & Damasio, 1985). Many of the experiments reviewed in this chapter found different patterns of impairment in groups with frontal damage affecting VMF, compared to those with damage to other sectors of the frontal lobes, arguing for more specific accounts of brain-behavior relationships definable at the level of sub-regions within the frontal lobes (see also Reber & Tranel, 2019; Tsuchida & Fellows, 2013). While making a case for such component-process specificity, we also have the eventual challenge of putting these specific processes back together again to consider how they interact to support complex motivated behavior in the real world (Fig. 1).

6 Integrating Neuroeconomics and Cognitive Control

In considering value-based learning and decision-making in more real-world settings, it can be useful to broaden our view of executive functions. As mentioned above, such functions rely on processes such as working memory, selective attention, planning, monitoring, and sequencing multi-step behaviors. These are the capacities required to pursue a goal that is not directly perceivable or within physical reach. The behaviors required to achieve distant goals cannot be those triggered by the immediate environment; indeed, environmentally driven behaviors must be over-ridden to allow distant goals to be pursued. These abilities are believed to rely

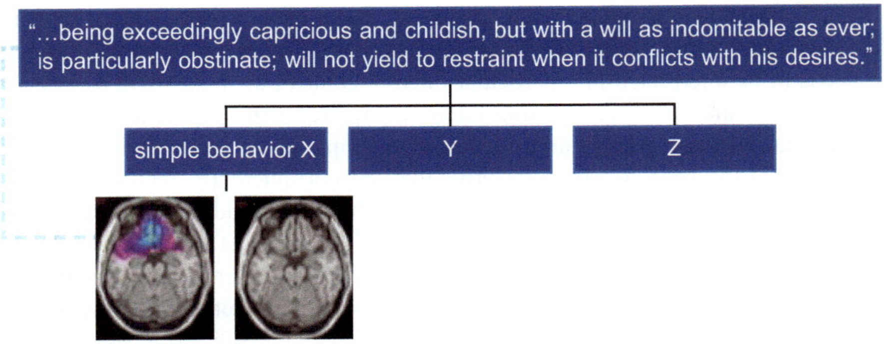

Fig. 1 A schematic showing how a clinical observation in a single case (in this example, the quote is an excerpt from Harlow's description of the behavioral changes observed in Phineas Gage after a penetrating head injury (Harlow, 1868)) can lead to models of potential component processes linked to the brain injury. These models can then be tested by developing experimental tasks that engage the process or processes of interest and comparing the performance of a group of patients with focal damage to the region under study, and another group without such damage, on these tasks. In this example, the extent of damage in a group of people with ventromedial frontal lesions is shown in an overlap image (left MRI image), while the lack of damage in this region in the comparison group is shown on the same slice in the image on the right. In this example, the comparison group instead had damage to more dorsal frontal regions (not shown). Ideally, the insights derived from the experimental studies can in turn help clinicians refine their assessment, understanding, and treatment of people presenting with behavioral symptoms

on the frontal lobes in general, in part because people with frontal lobe damage are far more likely to act in response to the immediate environment, showing so-called utilization behavior, for example, or having difficulty over-riding habitual behavioral repertoires (Fellows, 2019). The capacity to pursue an internally-set goal in the face of distraction is also referred to as cognitive control. A relative blind spot in the extensive work on the neural basis of cognitive control has been how the "goal" of goal-directed behavior is conceived of and selected. That's where the research reviewed in this chapter (and elsewhere in this book) comes in! In the lab, goals are typically assigned through task instructions, but in the real world, we are often establishing our goals based on an assessment of our anticipated needs: i.e., pursuing options with the greatest subjective value, or utility, given the context.

This value assessment must somehow come to influence behavior. This could be through biasing of cognitive control processes. For example, value could enhance selective attention, making us more likely to notice perceptual inputs that signal the availability of a motivationally important prospect (Anderson et al., 2011a, b; Grignolio et al., 2024). It could be that VMF plays a role in helping us perceive the world through a flexible, context-sensitive motivational filter, such that our behaviors are oriented towards noting and pursuing external opportunities that maximize subjective value (Vaidya & Fellows, 2015b, 2020), or even in imagining potential avenues to pursue to solve a given problem (Peters et al., 2017). There is growing work on the interplay between motivational/value processes and various aspects of cognitive control (Cools et al., 2019; Edelson & Hare, 2023; Krebs et al., 2011). This is an interesting and challenging future direction for neuroeconomics to help refine broader models of how the brain produces higher-order behavior.

More generally, this chapter has, I hope, helped to demonstrate how lesion evidence can contribute to our understanding of the brain basis of complex behavior, complementing the work in humans using correlational methods such as functional MRI. Given the importance of value-related behaviors, it is to be expected that the basic neural circuitry underlying many reward-related processes is likely to be similar across species, making animal models highly pertinent. Animal work may also help to inform the selection of candidate component processes and even tasks, further strengthening the potential for bringing findings from different research methods to address common questions. Such triangulation has been very informative in neuroeconomics to date, and has inferential strengths that are worth continuing to draw on.

7 Clinical Translation

I have shown here how people living with neurological disorders may help us better understand the healthy brain. In turn, this knowledge has the potential to shed light on how brain dysfunction leads to aberrant or maladaptive behavior, whether at the extreme ends of normal individual differences, or in mental illnesses or neurological disorders. Difficulties in highly context-sensitive, rapidly changing environments that require flexible motivated behavior, such as in social interactions, or navigating a world full of choices regarding how to spend our time and allocate our attention, are situations that are very challenging for many people with brain disorders. As we describe the neural systems, circuits, and cellular and molecular mechanisms underlying decision-making and other economic behaviors, we open the door to new ways of thinking about behavioral symptoms, new diagnostic approaches, and new treatments or rehabilitation strategies.

References

Alvarez, J. A., & Emory, E. (2006). Executive function and the frontal lobes: A meta-analytic review. *Neuropsychology Review, 16*(1), 17–42. https://doi.org/10.1007/s11065-006-9002-x

Anderson, B. A., Laurent, P. A., & Yantis, S. (2011a). Learned value magnifies salience-based attentional capture. *PLoS One, 6*(11), e27926. https://doi.org/10.1371/journal.pone.0027926

Anderson, B. A., Laurent, P. A., & Yantis, S. (2011b). Value-driven attentional capture. *Proceedings of the National Academy of Sciences of the United States of America, 108*(25), 10367–10371. https://doi.org/10.1073/pnas.1104047108

Averbeck, B., & O'Doherty, J. P. (2022). Reinforcement-learning in fronto-striatal circuits. *Neuropsychopharmacology, 47*(1), 147–162. https://doi.org/10.1038/s41386-021-01108-0

Barbey, A. K., Colom, R., Solomon, J., Krueger, F., Forbes, C., & Grafman, J. (2012). An integrative architecture for general intelligence and executive function revealed by lesion mapping. *Brain, 135*(Pt 4), 1154–1164. https://doi.org/10.1093/brain/aws021

Bartra, O., McGuire, J. T., & Kable, J. W. (2013). The valuation system: A coordinate-based meta-analysis of BOLD fMRI experiments examining neural correlates of subjective value. *NeuroImage, 76*, 412–427. https://doi.org/10.1016/j.neuroimage.2013.02.063

Bechara, A., Damasio, H., Tranel, D., & Damasio, A. R. (1997). Deciding advantageously before knowing the advantageous strategy. *Science, 275*(5304), 1293–1295. https://doi.org/10.1126/science.275.5304.1293

Bechara, A., Damasio, H., Tranel, D., & Anderson, S. W. (1998). Dissociation of working memory from decision making within the human prefrontal cortex. *The Journal of Neuroscience, 18*(1), 428–437. https://doi.org/10.1523/JNEUROSCI.18-01-00428.1998

Biderman, N., Bakkour, A., & Shohamy, D. (2020). What are memories for? The hippocampus bridges past experience with future decisions. *Trends in Cognitive Sciences, 24*(7), 542–556. https://doi.org/10.1016/j.tics.2020.04.004

Boorman, E. D., Witkowski, P. P., Zhang, Y., & Park, S. A. (2021). The orbital frontal cortex, task structure, and inference. *Behavioral Neuroscience, 135*(2), 291–300. https://doi.org/10.1037/bne0000465

Butter, C. (1968). Perseveration in extinction and in discrimination reversal tasks following selective frontal ablations in Macaca mulatta. *Physiology & Behavior, 4*, 163–171.

Camille, N., Griffiths, C. A., Vo, K., Fellows, L. K., & Kable, J. W. (2011). Ventromedial frontal lobe damage disrupts value maximization in humans. *The Journal of Neuroscience, 31*(20), 7527–7532. https://doi.org/10.1523/JNEUROSCI.6527-10.2011

Chatterjee, A. (2005). A madness to the methods in cognitive neuroscience? *Journal of Cognitive Neuroscience, 17*(6), 847–849. https://doi.org/10.1162/0898929054021085

Clark, L., & Manes, F. (2004). Social and emotional decision-making following frontal lobe injury. *Neurocase, 10*(5), 398–403. https://doi.org/10.1080/13554790490882799

Cools, R., Frobose, M., Aarts, E., & Hofmans, L. (2019). Dopamine and the motivation of cognitive control. *Handbook of Clinical Neurology, 163*, 123–143. https://doi.org/10.1016/B978-0-12-804281-6.00007-0

de Haan, B., & Karnath, H. O. (2018). A hitchhiker's guide to lesion-behaviour mapping. *Neuropsychologia, 115*, 5–16. https://doi.org/10.1016/j.neuropsychologia.2017.10.021

Delgado, M. R., Beer, J. S., Fellows, L. K., Huettel, S. A., Platt, M. L., Quirk, G. J., & Schiller, D. (2016). Viewpoints: Dialogues on the functional role of the ventromedial prefrontal cortex. *Nature Neuroscience, 19*(12), 1545–1552. https://doi.org/10.1038/nn.4438

Dias, R., Robbins, T. W., & Roberts, A. C. (1996). Dissociation in prefrontal cortex of affective and attentional shifts. *Nature, 380*(6569), 69–72. https://doi.org/10.1038/380069a0

Edelson, M. G., & Hare, T. A. (2023). Goal-dependent hippocampal representations facilitate self-control. *The Journal of Neuroscience, 43*(46), 7822–7830. https://doi.org/10.1523/JNEUROSCI.0951-22.2023

Eslinger, P. J., & Damasio, A. R. (1985). Severe disturbance of higher cognition after bilateral frontal lobe ablation: Patient EVR. *Neurology, 35*(12), 1731–1741. https://doi.org/10.1212/wnl.35.12.1731

Fellows, L. K. (2004). The cognitive neuroscience of human decision making: A review and conceptual framework. *Behavioral and Cognitive Neuroscience Reviews, 3*(3), 159–172. https://doi.org/10.1177/1534582304273251

Fellows, L. K. (2006). Deciding how to decide: Ventromedial frontal lobe damage affects information acquisition in multi-attribute decision making. *Brain, 129*(Pt 4), 944–952. https://doi.org/10.1093/brain/awl017

Fellows, L. K. (2018). The neuroscience of human decision-making through the lens of learning and memory. *Current Topics in Behavioral Neurosciences, 37*, 231–251. https://doi.org/10.1007/7854_2016_468

Fellows, L. K. (2019). The functions of the frontal lobes: Evidence from patients with focal brain damage. *Handbook of Clinical Neurology, 163*, 19–34. https://doi.org/10.1016/B978-0-12-804281-6.00002-1

Fellows, L. K., & Farah, M. J. (2003). Ventromedial frontal cortex mediates affective shifting in humans: Evidence from a reversal learning paradigm. *Brain, 126*(Pt 8), 1830–1837. https://doi.org/10.1093/brain/awg180

Fellows, L. K., & Farah, M. J. (2005). Different underlying impairments in decision-making following ventromedial and dorsolateral frontal lobe damage in humans. *Cerebral Cortex, 15*(1), 58–63. https://doi.org/10.1093/cercor/bhh108

Fellows, L. K., & Farah, M. J. (2007). The role of ventromedial prefrontal cortex in decision making: Judgment under uncertainty or judgment per se? *Cerebral Cortex, 17*(11), 2669–2674. https://doi.org/10.1093/cercor/bhl176

Fellows, L. K., Heberlein, A. S., Morales, D. A., Shivde, G., Waller, S., & Wu, D. H. (2005). Method matters: An empirical study of impact in cognitive neuroscience. *Journal of Cognitive Neuroscience, 17*(6), 850–858. https://doi.org/10.1162/0898929054021139

Fellows, L. K., Stark, M., Berg, A., & Chatterjee, A. (2008). Patient registries in cognitive neuroscience research: Advantages, challenges, and practical advice. *Journal of Cognitive Neuroscience, 20*(6), 1107–1113. https://doi.org/10.1162/jocn.2008.20065

Gilboa, A., & Moscovitch, M. (2017). Ventromedial prefrontal cortex generates pre-stimulus theta coherence desynchronization: A schema instantiation hypothesis. *Cortex, 87*, 16–30. https://doi.org/10.1016/j.cortex.2016.10.008

Graham, I. D., Logan, J., Harrison, M. B., Straus, S. E., Tetroe, J., Caswell, W., & Robinson, N. (2006). Lost in knowledge translation: Time for a map? *The Journal of Continuing Education in the Health Professions, 26*(1), 13–24. https://doi.org/10.1002/chp.47

Grignolio, D., Acunzo, D. J., & Hickey, C. (2024). Object-based attention is accentuated by object reward association. *Journal of Experimental Psychology. Human Perception and Performance, 50*(3), 280–294. https://doi.org/10.1037/xhp0001177

Harlow, J. (1868). Recovery from the passage of an iron bar through the head. *Massachusetts Medical Society, 2*, 327–347.

Hayden, B. Y., & Niv, Y. (2021). The case against economic values in the orbitofrontal cortex (or anywhere else in the brain). *Behavioral Neuroscience, 135*(2), 192–201. https://doi.org/10.1037/bne0000448

Henri-Bhargava, A., Simioni, A., & Fellows, L. K. (2012). Ventromedial frontal lobe damage disrupts the accuracy, but not the speed, of value-based preference judgments. *Neuropsychologia, 50*(7), 1536–1542. https://doi.org/10.1016/j.neuropsychologia.2012.03.006

Hornak, J., O'Doherty, J., Bramham, J., Rolls, E. T., Morris, R. G., Bullock, P. R., & Polkey, C. E. (2004). Reward-related reversal learning after surgical excisions in orbito-frontal or dorsolateral prefrontal cortex in humans. *Journal of Cognitive Neuroscience, 16*(3), 463–478. https://doi.org/10.1162/089892904322926791

Hunt, L. T., Malalasekera, W. M. N., de Berker, A. O., Miranda, B., Farmer, S. F., Behrens, T. E. J., & Kennerley, S. W. (2018). Triple dissociation of attention and decision computations across prefrontal cortex. *Nature Neuroscience, 21*(10), 1471–1481. https://doi.org/10.1038/s41593-018-0239-5

Kable, J. W., & Glimcher, P. W. (2007). The neural correlates of subjective value during intertemporal choice. *Nature Neuroscience, 10*(12), 1625–1633. https://doi.org/10.1038/nn2007

Karnath, H. O., Sperber, C., & Rorden, C. (2018). Mapping human brain lesions and their functional consequences. *NeuroImage, 165*, 180–189. https://doi.org/10.1016/j.neuroimage.2017.10.028

Krajbich, I., Lu, D., Camerer, C., & Rangel, A. (2012). The attentional drift-diffusion model extends to simple purchasing decisions. *Frontiers in Psychology, 3*, 193. https://doi.org/10.3389/fpsyg.2012.00193

Krebs, R. M., Boehler, C. N., Egner, T., & Woldorff, M. G. (2011). The neural underpinnings of how reward associations can both guide and misguide attention. *The Journal of Neuroscience, 31*(26), 9752–9759. https://doi.org/10.1523/JNEUROSCI.0732-11.2011

Lee, D. G., & Hare, T. A. (2023). Value certainty and choice confidence are multidimensional constructs that guide decision-making. *Cognitive, Affective, & Behavioral Neuroscience, 23*(3), 503–521. https://doi.org/10.3758/s13415-022-01054-4

Machado, C. J., & Bachevalier, J. (2007). The effects of selective amygdala, orbital frontal cortex or hippocampal formation lesions on reward assessment in nonhuman primates. *The European Journal of Neuroscience, 25*(9), 2885–2904. https://doi.org/10.1111/j.1460-9568.2007.05525.x
Mah, Y. H., Husain, M., Rees, G., & Nachev, P. (2014). Human brain lesion-deficit inference remapped. *Brain, 137*(Pt 9), 2522–2531. https://doi.org/10.1093/brain/awu164
Messimeris, D., Levy, R., & Le Bouc, R. (2023). Economic and social values in the brain: Evidence from lesions to the human ventromedial prefrontal cortex. *Frontiers in Neurology, 14*, 1198262. https://doi.org/10.3389/fneur.2023.1198262
Murray, E. A., & Fellows, L. K. (2022). Prefrontal cortex interactions with the amygdala in primates. *Neuropsychopharmacology, 47*(1), 163–179. https://doi.org/10.1038/s41386-021-01128-w
Murray, E. A., & Rudebeck, P. H. (2018). Specializations for reward-guided decision-making in the primate ventral prefrontal cortex. *Nature Reviews. Neuroscience, 19*(7), 404–417. https://doi.org/10.1038/s41583-018-0013-4
Murray, E. A., O'Doherty, J. P., & Schoenbaum, G. (2007). What we know and do not know about the functions of the orbitofrontal cortex after 20 years of cross-species studies. *The Journal of Neuroscience, 27*(31), 8166–8169. https://doi.org/10.1523/JNEUROSCI.1556-07.2007
Noonan, M. P., Chau, B. K. H., Rushworth, M. F. S., & Fellows, L. K. (2017). Contrasting effects of medial and lateral orbitofrontal cortex lesions on credit assignment and decision-making in humans. *The Journal of Neuroscience, 37*(29), 7023–7035. https://doi.org/10.1523/JNEUROSCI.0692-17.2017
Padoa-Schioppa, C., & Schoenbaum, G. (2015). Dialogue on economic choice, learning theory, and neuronal representations. *Current Opinion in Behavioral Sciences, 5*, 16–23. https://doi.org/10.1016/j.cobeha.2015.06.004
Payne, J. W., Bettman, J. R., Coupey, E., & Johnson, E. J. (1992). A constructive process view of decision making: Multiple strategies in judgment and choice. *Acta Psychologica, 80*(1), 107–141. https://doi.org/10.1016/0001-6918(92)90043-D
Pelletier, G., & Fellows, L. K. (2019). A critical role for human ventromedial frontal lobe in value comparison of complex objects based on attribute configuration. *The Journal of Neuroscience, 39*(21), 4124–4132. https://doi.org/10.1523/JNEUROSCI.2969-18.2019
Peters, S. L., Fellows, L. K., & Sheldon, S. (2017). The ventromedial frontal lobe contributes to forming effective solutions to real-world problems. *Journal of Cognitive Neuroscience, 29*(6), 991–1001. https://doi.org/10.1162/jocn_a_01088
Pool, E. R., Pauli, W. M., Kress, C. S., & O'Doherty, J. P. (2019). Behavioural evidence for parallel outcome-sensitive and outcome-insensitive Pavlovian learning systems in humans. *Nature Human Behaviour, 3*(3), 284–296. https://doi.org/10.1038/s41562-018-0527-9
Reber, J., & Tranel, D. (2019). Frontal lobe syndromes. *Handbook of Clinical Neurology, 163*, 147–164. https://doi.org/10.1016/B978-0-12-804281-6.00008-2
Reber, J., Feinstein, J. S., O'Doherty, J. P., Liljeholm, M., Adolphs, R., & Tranel, D. (2017). Selective impairment of goal-directed decision-making following lesions to the human ventromedial prefrontal cortex. *Brain, 140*(6), 1743–1756. https://doi.org/10.1093/brain/awx105
Roesch, M. R., & Olson, C. R. (2007). Neuronal activity related to anticipated reward in frontal cortex: Does it represent value or reflect motivation? *Annals of the New York Academy of Sciences, 1121*, 431–446. https://doi.org/10.1196/annals.1401.004
Rudebeck, P. H., & Izquierdo, A. (2022). Foraging with the frontal cortex: A cross-species evaluation of reward-guided behavior. *Neuropsychopharmacology, 47*(1), 134–146. https://doi.org/10.1038/s41386-021-01140-0
Rudebeck, P. H., Saunders, R. C., Lundgren, D. A., & Murray, E. A. (2017). Specialized representations of value in the orbital and ventrolateral prefrontal cortex: Desirability versus availability of outcomes. *Neuron, 95*(5), 1208–1220 e1205. https://doi.org/10.1016/j.neuron.2017.07.042
Schuck, N. W., Cai, M. B., Wilson, R. C., & Niv, Y. (2016). Human orbitofrontal cortex represents a cognitive map of state space. *Neuron, 91*(6), 1402–1412. https://doi.org/10.1016/j.neuron.2016.08.019

Stott, J. J., & Redish, A. D. (2015). Representations of value in the brain: An embarrassment of riches? *PLoS Biology, 13*(6), e1002174. https://doi.org/10.1371/journal.pbio.1002174

Thomas, A. W., Molter, F., & Krajbich, I. (2021). Uncovering the computational mechanisms underlying many-alternative choice. *eLife, 10*. https://doi.org/10.7554/eLife.57012

Tom, S. M., Fox, C. R., Trepel, C., & Poldrack, R. A. (2007). The neural basis of loss aversion in decision-making under risk. *Science, 315*(5811), 515–518. https://doi.org/10.1126/science.1134239

Tsuchida, A., & Fellows, L. K. (2013). Are core component processes of executive function dissociable within the frontal lobes? Evidence from humans with focal prefrontal damage. *Cortex, 49*(7), 1790–1800. https://doi.org/10.1016/j.cortex.2012.10.014

Tsuchida, A., Doll, B. B., & Fellows, L. K. (2010). Beyond reversal: A critical role for human orbitofrontal cortex in flexible learning from probabilistic feedback. *The Journal of Neuroscience, 30*(50), 16868–16875. https://doi.org/10.1523/JNEUROSCI.1958-10.2010

Vaidya, A. R., & Fellows, L. K. (2015a). Testing necessary regional frontal contributions to value assessment and fixation-based updating. *Nature Communications, 6*, 10120. https://doi.org/10.1038/ncomms10120

Vaidya, A. R., & Fellows, L. K. (2015b). Ventromedial frontal cortex is critical for guiding attention to reward-predictive visual features in humans. *The Journal of Neuroscience, 35*(37), 12813–12823. https://doi.org/10.1523/JNEUROSCI.1607-15.2015

Vaidya, A. R., & Fellows, L. K. (2020). Under construction: Ventral and lateral frontal lobe contributions to value-based decision-making and learning. *F1000Res, 9*, 10.12688/f1000research.21946.1.

Vaidya, A. R., Sefranek, M., & Fellows, L. K. (2018). Ventromedial frontal lobe damage alters how specific attributes are weighed in subjective valuation. *Cerebral Cortex, 28*(11), 3857–3867. https://doi.org/10.1093/cercor/bhx246

Vaidya, A. R., Pujara, M. S., Petrides, M., Murray, E. A., & Fellows, L. K. (2019). Lesion studies in contemporary neuroscience. *Trends in Cognitive Sciences, 23*(8), 653–671. https://doi.org/10.1016/j.tics.2019.05.009

Verma, K., & Kumar, S. (2022). How do we connect brain areas with cognitive functions? The past, the present and the future. *NeuroSci, 3*(3), 521–532. https://doi.org/10.3390/neurosci3030037

Xia, C., Stolle, D., Gidengil, E., & Fellows, L. K. (2015). Lateral orbitofrontal cortex links social impressions to political choices. *The Journal of Neuroscience, 35*(22), 8507–8514. https://doi.org/10.1523/JNEUROSCI.0526-15.2015

Yu, L. Q., Kan, I. P., & Kable, J. W. (2020). Beyond a rod through the skull: A systematic review of lesion studies of the human ventromedial frontal lobe. *Cognitive Neuropsychology, 37*(1–2), 97–141. https://doi.org/10.1080/02643294.2019.1690981

Yu, L. Q., Dana, J., & Kable, J. W. (2022). Individuals with ventromedial frontal damage display unstable but transitive preferences during decision making. *Nature Communications, 13*(1), 4758. https://doi.org/10.1038/s41467-022-32511-w

Delay Discounting: Empirical Foundations and Future Directions

Annie Cooper and Samuel M. McClure

Abstract Delay discounting refers to the set of cognitive and neural processes that underlie our ability to value rewards or punishments that will be received after a time delay. People and other animals prefer to receive rewards sooner and to delay punishments, all else being equal. Cognitive neuroscience has identified a set of brain areas that are involved in delay discounting and has documented how interactions between these brain areas support intertemporal decision-making. We describe these advances and argue that cognitive neuroscience has successfully identified the brain network involved in delay discounting, enabling current research focused on how discounting is altered in disorders such as addiction and attention deficit hyperactivity disorder (ADHD). Future advances in our understanding of the neural mechanisms of delay discounting will benefit from methodological advances, such as the use of novel techniques for measuring dopamine release in humans.

Discounting is a general term that refers to reducing subjective value based on some extrinsic factor. Probability discounting (Green & Myerson, 2010) is the reduction of value due to some likelihood (probability) of the outcome not occurring. It underlies risk preferences that are discussed elsewhere in this volume. Social discounting (Strombach et al., 2015) is the decreased value assigned to outcomes that occur to other people, as it depends on their social distance from you (i.e., close personal friend versus stranger). This chapter focuses on *delay discounting*, or how people discount subjective value due to the fact that outcomes will not be received until after some time delay. Decisions made between options that differ in their delay until receipt are known as *intertemporal choices*. Our goal is to provide the empirical background from which other chapters in this volume explain how delay discounting depends on specific cognitive processes (e.g., memory, chapter by Lempert; model-based decision-making, chapter by Wagner) and features in the choice options (chapter by Hare).

A. Cooper (✉) · S. M. McClure (✉)
Department of Psychology, Arizona State University, Tempe, AZ, USA
e-mail: annie.cooper@asu.edu; samuel.mcclure@asu.edu

D. V. Smith et al. (eds.), *Neuroeconomics: Core Topics and Current Directions*,
https://doi.org/10.1007/978-3-032-02925-6_4

In the laboratory, intertemporal decision tasks require people to make a series of choices between *smaller, sooner (SS)* and *larger, later (LL)* rewards or punishments. For example, you may be asked if you would prefer to receive \$10 right now or \$11 in 1 week. If you prefer the immediate \$10, then this implies that the discounted subjective value of the \$11 is less than \$10, or that the amount of discounting imposed for rewards delayed 1 week is greater than 9.1% (because (100%–9.1%) × \$11 ≈ \$10). If participants respond to a series of questions of this nature, then the precise amount of discounting that a person imposes at different time delays can be accurately estimated (Mazur, 1987).

Our use of US dollars to represent outcomes deserves some brief comment. First, of course, any currency can be used. However, discount rates tend to be smaller for larger valued rewards, so it is necessary to be careful with currency exchange rates and relative cost of living (Thaler, 1981). Second, money is a commonly used reward for delay discounting experiments, but it is not required. Money has the enviable characteristic of making sense at a near continuum of values. You can ask people to choose between \$10.28 and \$11.72, and this makes sense to them. Because of the level of precision afforded by money, the degree to which discounting can be estimated is superior for money than for most other types of outcomes. For example, we conducted a study in which thirsty subjects made choices between different volumes of fruit juice to be given to them after various time delays (McClure et al., 2007). However, we had people choose between different numbers of small squirts of juice (e.g., 1 or 3 squirts). It would not have made sense to ask people if they would prefer 1.25 squirts versus 2.63 squirts. Money is special in this sense and is the most commonly used outcome in studies of delay discounting as a consequence.

1 Exponential and Hyperbolic Discounting

There are a couple of initial questions to ask about delay discounting. First, do people discount subjective value based on anticipated future time delays? And second, if we do discount, then how do we do so?

The first question can be answered relatively trivially as "yes." For any person, we can ask if they would rather receive \$10 today or \$10 next week, and the answer will invariably be "today, please." In fact, questions of this sort are typically included in delay discounting tasks created for online data collection to serve as data quality checks (Mellis & Bickel, 2020). If people fail to choose the immediate reward, then they are excluded from subsequent analysis. The existence of delay discounting is therefore assumed.

The second question—how do we discount?—has a richer and more interesting history. One way to address it is normatively: how *should* we discount the future. The standard normative reasoning asserts that if we discount for delays, then we ought to do so in a manner that produces consistent preferences (Samuelson, 1937; Loewenstein & Thaler, 1989). An example helps. Consider the decision about whether to attend graduate school or not. This is an intertemporal choice that pits immediate earnings (lower for graduate school) against future job credentials (higher for graduate school). Whether you prefer to attend graduate school or not, rationality asserts that you should maintain that preference across time. Even if you later question your choice of enrolling in graduate school, your preference should never switch—it should never be the case that a current graduate student wishes to leave a PhD program in the absence of new information (e.g., bad advisor).

Mathematically, there is only one form of a discount function that dynamically maintains consistent preferences. It is the exponential function. If $V(r,t)$ is the subjective value of reward r at delay t, then we can write the generic version of delay discounting as

$$V(r, t) = U(r)D(t)$$

where U is the utility function that captures the (undiscounted) subjective value of the reward, and D is the amount of discounting applied for delay t. With an exponential function, $D(t)$ has the form

$$D(t) = \mathrm{e}^{-rt}$$

in which r is the discount rate.

Evidence against the exponential function as a model of human and animal behavior is now extensive. It dates back to studies performed with pigeons (Ainslie & Herrnstein, 1981). The birds were given the choice to peck on one of two keys. A choice of the first key opened the door to a food chamber for 2 s after a delay of d seconds. The second key gave twice the access to the food (4 s), but after a delay of $d + 4$ s. In one experimental condition, the choice began as half the food after 12 s or twice the food after 16 s. The pigeons preferred more food. The time delay was then reduced, first to 8 vs. 12 s, then to 6 vs. 10 s, and progressively to 2 vs. 6 s. The pigeons (by and large) preferred the larger reward in all of these cases. However, the final choice was between an immediate reward ($d = 0.1$ s) and twice the food 4 s later. For this choice, all of the pigeons selected the immediate (smaller, sooner) food. This pattern of behavior is known as a *preference reversal*, cannot be explained by exponential discounting, and is instead standardly used to argue that the discount function is hyperbolic, as expressed by

$$D(t) = 1/(1 + kt)$$

in which k is known as the *discount factor*.

People show evidence of preference reversals, and hence *hyperbolic discounting*, constantly. A quotidian example is our behavior around the alarm clock. We set an early alarm because we prefer to accomplish more work in the morning instead of sleeping longer. However, when the alarm sounds (think: $d = 0.1$ s), then we reverse and press the snooze bar. Also, we prefer to eat healthfully, only to succumb to the dessert when offered to us. We plan to catch up on work over the weekend, only to go out with friends when the moment arrives. Impulsive preference reversals are particularly germane to addiction. People may quit cigarettes only to reverse course when a smoking opportunity is immediately at hand. One trick that we use to avoid these preference reversals is to employ *precommitment devices*, or taking actions that make it difficult or impossible to change our minds. We place our alarm clocks on the other side of the room. Or, we pay money to join a weight loss program.

Examples aside, it is now clear that preference in intertemporal choices are dynamically inconsistent. A preference at one moment is susceptible to reversal when the smaller, sooner outcome becomes immediately available and tempting. The hyperbolic discounting function was developed to explain this phenomenon (Ainslie, 1975), but other forms of delay discounting function have since been developed (Loewenstein & Prelec, 1992; Laibson, 1997; van den Bos & McClure, 2013). The exact mathematical form of delay discounting does not matter too much to us here, except to the degree that the discount function is reflected in the brain structures that underlie intertemporal decision-making.

2 Cognitive Neuroscience of Delay Discounting

The psychological and neural basis of delay discounting had been anticipated for a long time. Plato wrote about *akrasia*, wondering how it is possible that we can knowingly act against our own better judgment. We can know that we should be working over the weekend, know that we will regret not doing so, and yet still choose leisure. A classic example from this comes from Homer's *Odyssey* in which Odysseus orders his crew to tie him to his ship's mast and to plug their ears with beeswax to resist the Sirens' temptation. Freud wrote about the *ego* and the *id*, according to which we can simultaneously maintain better judgment (ego) while being driven to violate it (id). In cognitive neuroscience terms, the implicit division in behavioral tendencies implies the existence of parallel decision systems in the brain that may produce conflicting actions. Of course, separate parallel systems are a common motif in neuroscience (Mishkin et al., 1983; McClelland et al., 1988; Hickok & Poeppel, 2004). For intertemporal choice, we date speculation about parallel decision systems to Richard Thaler who suggested that the drive for immediate temptation may arise from the limbic system, whereas better judgment derives from future planning and the prefrontal cortex (Shefrin & Thaler, 2004).

The first test of the existence of parallel decision processes for intertemporal choice had people perform the equivalent of the pigeon experiment from above while brain activity was measured with fMRI (McClure et al., 2004). Instead of time allotments for access to food, participants were offered different amounts of money. The time delays were also longer. The immediate payment could be received at the end of the experiments, and the longest delay was a couple of months after the end of the study. The key test was whether there were some brain areas that were only activated when decisions were made that involved immediate rewards, and whether there were separate areas that were equally involved in all choices as would be expected for future planning. Regions associated with the mesolimbic

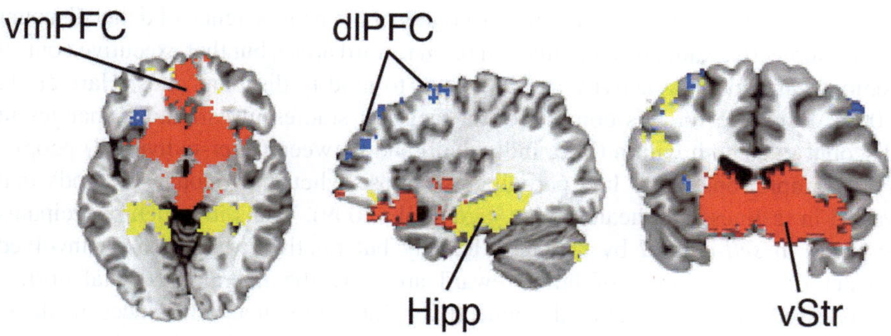

Fig. 1 Neuroanatomy of delay discounting. Subjective values are represented in brain reward areas (red), but may be modulated by executive control areas (blue) and regions associated with episodic future thinking (yellow). Key brain reward areas include the ventral striatum (vStr), ventromedial prefrontal cortex (vmPFC), and posterior cingulate cortex (not shown). Executive control areas include the dorsolateral prefrontal cortex (dlPFC) and posterior parietal cortex. The hippocampus (Hipp) interacts with the dlPFC to guide episodic future thinking. Brain regions were identified with neurosynth.org using the meta-analysis keywords "reward," "executive," and "episodic memory"

dopamine system—including the ventral striatum, ventromedial prefrontal cortex, and posterior cingulate cortex—were preferentially activated by the presence of immediate reward. Regions of the executive control system—in the lateral prefrontal cortex and posterior parietal cortex—were equally activated in all decisions (Fig. 1). It was concluded that our brain possesses (at least) two ways to make decisions about future rewards. One may be determined by direct access and learning about associations with reward through conditioning over short time delays (Schultz et al., 1997). The other depends on abstract future planning. This distinction is similar to theories about *model-based* and *model-free* modes of reinforcement learning (Daw et al., 2005; Smittenaar et al., 2013).

A challenge to the parallel systems view of delay discounting emerged from a study that directly compared hyperbolic discount functions to brain activity (Kable & Glimcher, 2007). The experiment required participants to choose between a fixed immediate reward of $20 and a larger reward delayed up to 6 months. Based on people's choices during the study, a best-fitting hyperbolic discount function was determined. This discount function was then used to predict the subjective value of the varying delayed reward for each choice (as $A \times D(t)$, where A is the amount of the delayed reward). It was found that a set of *brain reward areas*—ventral striatum, ventromedial prefrontal cortex, and posterior cingulate cortex—correlated with subjective value across choice. Activity from executive control regions were not needed to predict subjective value and hence behavior. In fact, these same brain reward areas predict subjective value and choices in a broad range of decision tasks (Bartra et al., 2013). Based on these findings, it was argued that brain reward areas calculate discounted value in total, as a single unitary system.

The resolution to these conflicting accounts of the neuroscience of delay discount is that subjective value is maintained in brain reward areas, but that executive control regions can modulate activity in these areas to modify discount rates (Hare et al., 2009). Evidence for this comes from a range of studies investigating changes in discount rates both within (intra-individual) and between (inter-individual) people. For example, one study had participants choose whether to consume foods that varied in tastiness and healthiness (Hare et al., 2009). Trials in which participants engaged in *self-control* by choosing healthy but relatively bland foods involved greater down-regulation of brain reward areas by the lateral prefrontal cortex, as assessed through functional connectivity. Interindividual differences in delay discounting are also related to connectivity between the lateral prefrontal cortex and regions of the dorsal striatum, as assessed using diffusion imaging and functional connectivity (van den Bos et al., 2014). Rates of delay discounting decrease across adolescence and are correlated with the development of functional connectivity between the lateral prefrontal cortex and striatum (van den Bos et al., 2015).

It was mentioned above that larger rewards are discounted with a smaller discount rate. This is known as the magnitude effect. We showed that the magnitude effect depends on differential recruitment of executive control regions (Ballard et al., 2017). Exerting control is costly, and a larger reward may signal that the decision is more important and increase executive control as a consequence (Shenhav et al., 2016). For the purposes of this chapter, it is important to note only that greater recruitment of executive control systems was associated with reduced rates of delay discounting. Applying low-frequency repetitive transcranial magnetic stimulation (rTMS) to the lateral prefrontal cortex also eliminated the magnitude effect in a separate study (Ballard et al., 2018).

Many other framing effects that alter delay discounting have been studied, and changes in discount rates have been alternately associated with differential recruitment of brain reward areas or the executive control system. We have reviewed these findings elsewhere (van den Bos & McClure, 2013). Overall, from the accumulation of these studies, we feel it is now safe to conclude that the neural basis of delay discounting is now largely agreed upon, and depends on the interaction of a well-defined set of brain areas (Fig. 1). We know that the subjective value of delayed rewards is represented in a set of brain reward areas associated with the mesolimbic dopamine system. We know that activity within brain reward areas can be modulated by the executive control system to change discount rates. It is also suspected that other brain systems may impinge on brain reward areas to shape delay discounting (Kurth-Nelson & Redish, 2012). In fact, it seems likely that a broad range of neural systems, and associated cognitive processes, may impact how we think about and hence evaluate the future. It may not be possible to intuit and design experiments to isolate all contributing cognitive processes. Ultimately, the best approach for fully identifying all of the neural contributions to delay discounting may be through the use of machine learning tools that are agnostic to component cognitive processes. Recent work has demonstrated the predictive power of this approach (Koban et al., 2023).

One system that has been a particular focus of research into the cognitive processes that shape delay discounting is the hippocampus and associated brain areas important for episodic memory and constructing plausible future events (Schacter & Addis, 2009; Schacter et al., 2012). When people are encouraged to think concretely about future time points, then hippocampal activity is increased and discount rates are reduced (Peters & Büchel, 2010). The intervention is rather straightforward. Consider a delay of 1 year. By default, we tend to think of events in one year in vague, general ways (Trope & Liberman, 2012). However, if we ask you to think specifically (and using your visual imagination as much as possible) about what you would like to be doing in 1 year, then you may imagine yourself on the beach in the midst of a summer vacation with your partner. During the delay discounting task, you are then presented with the word "beach" alongside the 1 year delay when making a choice. This manipulation is now known as *episodic future thinking* and is a highly studied tactic to combat impulsivity disorders.

3 ADHD and Addiction

A tremendous advantage of understanding the neurobiological basis of delay discounting is that it opens the door to studies of disorders associated with impulsivity. High delay discount rates characterize one type of impulsivity (Strickland & Johnson, 2020), and high discount rates are evident in people with ADHD (Swanson et al., 2007; Jackson & MacKillop, 2016), addiction (Bickel et al., 2007; MacKillop et al., 2011), and obesity (Epstein et al., 2010; Amlung et al., 2016). These findings have motivated targeted studies to determine if there are systematic differences in brain reward areas or in the executive control network that are associated with impulsivity in these disorders.

One approach that we have taken is to assess the structural integrity of connections within the brain reward system. The mesolimbic dopamine system originates in the midbrain substantia nigra and ventral tegmental area (SN/VTA). Dopamine axons project to the forebrain in an organized fashion. For example, in the striatum, there is a topographic organization by which more medial midbrain dopamine neurons project to the ventral striatum, and dopamine neurons located more laterally project to more dorsal regions of the striatum (Haber & Knutson, 2010). In turn, dorsal regions of the striatum are innervated by dorsolateral prefrontal cortex, and ventral regions by the ventromedial prefrontal cortex (van den Bos et al., 2014). Based on this, we used diffusion imaging to assess the structural integrity of midbrain projections to the ventral striatum (limbic) and dorsal striatum (executive) in groups of people with ADHD and typically developing (TD) controls. We found that impulsivity was directly related to the structural integrity of both of these axonal pathways (Elliott et al., 2022). The greater integrity of the midbrain-limbic pathway, the greater the assessed impulsivity, and vice versa for the midbrain-executive tract (Fig. 2).

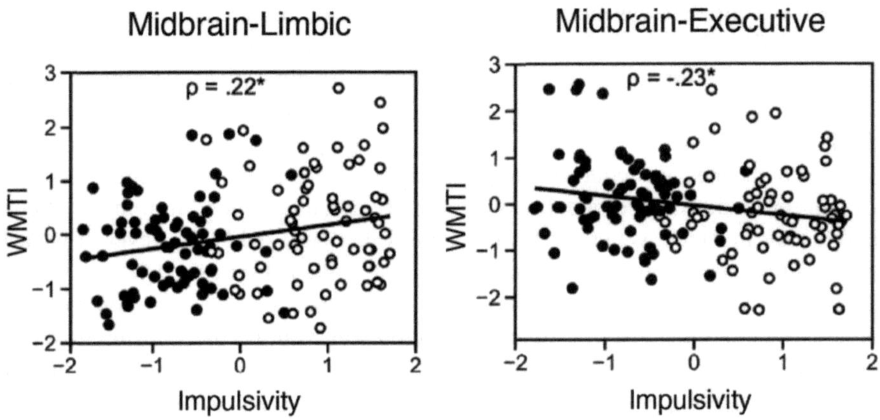

Fig. 2 Impulsivity and ADHD. Anatomical differences in midbrain-striatum pathways are related to impulsivity in people with ADHD (open circles) and controls (filled circles). Increased white matter tract integrity (WMTI) of axonal pathways between the SN/VTA and ventral striatum (limbic) correlated with greater impulsivity. The opposite pattern was evident for projections from the SN/VTA to the dorsal striatum (executive). (Adapted from Elliott et al., 2022)

Function MRI studies of delay discounting in addiction have also been guided by the parallel brain systems discussed above (McClure & Bickel, 2014). This has motivated a conceptual overview of addiction known as the competing neurobehavior decision systems framework (Bickel et al., 2016). Within this view, addiction may arise from either a hyperactive impulsive limbic system or hypoactive executive control. Experimental treatments based on this view began with cognitive training to increase executive functioning (Bickel et al., 2011). Unfortunately, these interventions are not robust, so more recent efforts have focused on episodic future thinking. For example, one recent study had people with alcohol use disorder complete a protracted intervention in which they were reminded of their future event cues twice daily for 2 weeks via text messages (Athamneh et al., 2022). Consistent with previous studies, episodic future thinking reduced the rates of delay discounting. Importantly, it also reduced the number of alcoholic drinks consumed daily from 6.72 per day to 4.54 per day—a 32% decrease. Continuing to develop and refine our understanding of the neuroscience of delay discounting may help motivate additional novel interventions to curb disorders of impulsivity.

4 A Call for Future Advances

Cognitive neuroscience is challenging. We aim to understand human brain function while limited to measurements and manipulations that have no long-term impact on the brain itself. This restriction severely limits the power of experiments that can be devised and conducted. Functional MRI is a powerful tool, but its resolution is limited to roughly a millimeter, and it measures blood flow, an indirect indicator of brain activity. In our view, the cognitive neuroscience of delay discounting has reached a state of maturity in which further advances from studies using MRI will be modest and incremental.

A specific roadblock may be worth exploring. The midbrain dopamine system is core to delay discounting and intertemporal choice. It forms the core of the brain reward areas and is also critical for the function of executive control areas (Ott & Nieder, 2019). Administration of the dopamine precursor L-Dopa to people increases activity in the ventral striatum and lateral prefrontal cortex, and increases rates of delay discounting (Pine et al., 2010). This leads naturally to questions about how dopamine contributes to delay discounting normally. Unfortunately, fMRI provides no direct information about dopamine release. However, recent work has begun using electrochemistry techniques on people during standard clinical care for epilepsy. Dopamine release in specific brain sites, like the anterior cingulate cortex, has been related to the arousal and valence of emotional stimuli (Batten et al., 2024). We expect it is only a matter of time before similar techniques are used to improve our understanding of how dopamine release across the forebrain is related to rates of delay discounting. A next generation of studies driven by intracranial recordings will guide revisions to our neurobiological understanding of delay discounting that has arisen from the use of MRI.

References

Ainslie, G. (1975). Specious reward: A behavioral theory of impulsiveness and impulse control. *Psychological Bulletin, 82*, 463–496.

Ainslie, G., & Herrnstein, R. J. (1981). Preference reversal and delayed reinforcement. *Animal Learning & Behavior, 9*, 476–482.

Amlung, M., Petker, T., Jackson, J., et al. (2016). Steep discounting of delayed monetary and food rewards in obesity: A meta-analysis. *Psychological Medicine, 46*, 2423–2434.

Athamneh, L. N., Brown, J., Stein, J. S., et al. (2022). Future thinking to decrease real-world drinking in alcohol use disorder: Repairing reinforcer pathology in a randomized proof-of-concept trial. *Experimental and Clinical Psychopharmacology, 30*, 326–337.

Ballard, I. C., Kim, B., Liatsis, A., et al. (2017). More is meaningful: The magnitude effect in intertemporal choice depends on self-control. *Psychological Science, 28*, 1443–1454.

Ballard, I. C., Aydogan, G., Kim, B., & McClure, S. M. (2018). Causal evidence for the dependence of the magnitude effect on dorsolateral prefrontal cortex. *Scientific Reports, 8*, 16545.

Bartra, O., McGuire, J. T., & Kable, J. W. (2013). The valuation system: A coordinate-based meta-analysis of BOLD fMRI experiments examining neural correlates of subjective value. *NeuroImage, 76*, 412–427.

Batten, S. R., Hartle, A., Barbosa, L. S., et al. (2024). Emotional words evoke region and valence-specific patterns of concurrent neuromodulator release in human thalamus and cortex. *Cell Reports, 44*, 115162.

Bickel, W. K., Miller, M. L., Yi, R., et al. (2007). Behavioral and neuroeconomics of drug addiction: Competing neural systems and temporal discounting processes. *Drug and Alcohol Dependence, 90*(Suppl 1), S85–S91.

Bickel, W. K., Yi, R., Landes, R. D., et al. (2011). Remember the future: Working memory training decreases delay discounting among stimulant addicts. *Biological Psychiatry, 69*, 260–265.

Bickel, W. K., Snider, S. E., Quisenberry, A. J., et al. (2016). Competing neurobehavioral decision systems theory of cocaine addiction: From mechanisms to therapeutic opportunities. *Progress in Brain Research, 223*, 269–293.

Daw, N., Niv, Y., & Dayan, P. (2005). Uncertainty-based competition between prefrontal and dorsolateral striatal systems for behavioral control. *Nature Neuroscience, 8*, 1704–1711.

Elliott, B. L., D'Ardenne, K., Mukherjee, P., et al. (2022). Limbic and executive Meso- and nigrostriatal tracts predict impulsivity differences in attention-deficit/hyperactivity disorder. *Biological Psychiatry: Cognitive Neuroscience and Neuroimaging, 7*, 415–423.

Epstein, L. H., Salvy, S. J., Carr, K. A., et al. (2010). Food reinforcement, delay discounting and obesity. *Physiology & Behavior, 100*, 438–445.

Green, L., & Myerson, J. (2010). Experimental and correlational analyses of delay and probability discounting. In G. J. Madden (Ed.), *Impulsivity: The behavioral and neurological science of discounting* (pp. 67–92). American Psychological Association.

Haber, S., & Knutson, B. (2010). The reward circuit: Linking primate anatomy and human imaging. *Neuropsychopharmacology, 35*, 4–26.

Hare, T. A., Camerer, C. F., & Rangel, A. (2009). Self-control in decision-making involves modulation of the vmPFC valuation system. *Science, 324*, 646–648.

Hickok, G., & Poeppel, D. (2004). Dorsal and ventral streams: A framework for understanding aspects of the functional anatomy of language. *Cognition, 92*, 67–99.

Jackson, J. N. S., & MacKillop, J. (2016). Attention-deficit/hyperactivity disorder and monetary delay discounting: A meta-analysis of case-control studies. *Biological Psychiatry: Cognitive Neuroscience and Neuroimaging, 1*, 316–325.

Kable, J., & Glimcher, P. (2007). The neural correlates of subjective value during intertemporal choice. *Nature Neuroscience, 10*, 1625–1633.

Koban, L., Lee, S., Schelski, D. S., et al. (2023). An fMRI-based brain marker of individual differences in delay discounting. *The Journal of Neuroscience, 43*, 1600–1613.

Kurth-Nelson, Z., & Redish, A. D. (2012). Don'T let me do that!—Models of precommitment. *Frontiers in Neuroscience, 6*, 138.

Laibson, D. (1997). Golden eggs and hyperbolic discounting. *Quarterly Journal of Economics, 112*, 443–478.

Loewenstein, G., & Prelec, D. (1992). Anomalies in intertemporal choice: Evidence and an interpretation. *Quarterly Journal of Economics, 107*, 573–597.

Loewenstein, G., & Thaler, R. H. (1989). Anomalies: Intertemporal choice. *The Journal of Economic Perspectives, 3*, 181–193.

MacKillop, J., Amlung, M. T., Few, L. R., et al. (2011). Delayed reward discounting and addictive behavior: A meta-analysis. *Psychopharmacology, 216*, 305–321.

Mazur, J. E. (1987). An adjusting procedure for studying delayed reinforcement. In M. L. Commons, J. E. Mazur, J. A. Nevin, & H. Rachlin (Eds.), *Quantitative analyses of behavior: Vol. 5. The effect of delay and of intervening events on reinforcement value* (pp. 55–73). Erlbaum.

McClelland, J. L., Rumelhart, D. E., & Hinton, G. E. (Eds.). (1988). *The appeal of parallel distributed processing*. Morgan Kaufmann.

McClure, S. M., & Bickel, W. K. (2014). A dual-systems perspective on addiction: Contributions from neuroimaging and cognitive training. *Annals of the New York Academy of Sciences, 1327*, 62–78.

McClure, S. M., Laibson, D. I., Loewenstein, G., & Cohen, J. D. (2004). Separate neural systems value immediate and delayed monetary rewards. *Science, 306*, 503–507.

McClure, S. M., Ericson, K. M., Laibson, D. I., et al. (2007). Time discounting for primary rewards. *The Journal of Neuroscience, 27*, 5796–5804.

Mellis, A. M., & Bickel, W. K. (2020). Mechanical Turk data collection in addiction research: Utility, concerns and best practices. *Addiction, 115*, 1960–1968.

Mishkin, M., Ungerleider, L. G., & Macko, K. A. (1983). Object vision and spatial vision: Two cortical pathways. *Trends in Neurosciences, 6*, 414–417.

Ott, T., & Nieder, A. (2019). Dopamine and cognitive control in prefrontal cortex. *Trends in Cognitive Sciences, 23*, 213–234.

Peters, J., & Büchel, C. (2010). Episodic future thinking reduces reward delay discounting through an enhancement of prefrontal-mediotemporal interactions. *Neuron, 66*, 138–148.

Pine, A., Shiner, T., Seymour, B., & Dolan, R. (2010). Dopamine, time, and impulsivity in humans. *The Journal of Neuroscience, 30*, 8888–8896.

Samuelson, P. A. (1937). A note on measurement of utility. *The Review of Economic Studies, 4*, 155.

Schacter, D. L., & Addis, D. R. (2009). On the nature of medial temporal lobe contributions to the constructive simulation of future events. *Philosophical Transactions of the Royal Society B, 364*, 1245–1253.

Schacter, D. L., Addis, D. R., Hassabis, D., et al. (2012). The future of memory: Remembering, imagining, and the brain. *Neuron, 76*, 677–694.

Schultz, W., Dayan, P., & Montague, P. R. (1997). A neural substrate of prediction and reward. *Science, 275*, 1593–1599.

Shefrin, H. M., & Thaler, R. H. (2004). Mental accounting, saving, and self-control. In C. F. Camerer, G. Loewenstein, & M. Rabin (Eds.), *Advances in Behavioral economics* (pp. 395–428). Princeton University Press.

Shenhav, A., Cohen, J. D., & Botvinick, M. M. (2016). Dorsal anterior cingulate cortex and the value of control. *Nature Neuroscience, 19*, 1286–1291.

Smittenaar, P., FitzGerald, T. H. B., Romei, V., et al. (2013). Disruption of dorsolateral prefrontal cortex decreases model-based in favor of model-free control in humans. *Neuron, 80*, 914–919.

Strickland, J., & Johnson, M. W. (2020). Rejecting impulsivity as a psychological construct: A theoretical, empirical, and sociocultural argument. *Psychological Review, 128*, 336.

Strombach, T., Weber, B., Hangebrauk, Z., et al. (2015). Social discounting involves modulation of neural value signals by temporoparietal junction. *Proceedings of the National Academy of Sciences of the United States of America, 112*, 1619–1624.

Swanson, J. M., Kinsbourne, M., Nigg, J., et al. (2007). Etiologic subtypes of attention-deficit/hyperactivity disorder: Brain imaging, molecular genetic and environmental factors and the dopamine hypothesis. *Neuropsychology Review, 17*, 39–59.

Thaler, R. (1981). Some empirical evidence on dynamic inconsistency. *Economics Letters, 8*, 201–207.

Trope, Y., & Liberman, N. (2012). Construal level theory. *Handbook of Theories of Social Psychology, 1*, 118–134.

van den Bos, W., & McClure, S. M. (2013). Towards a general model of temporal discounting. *Journal of the Experimental Analysis of Behavior, 99*, 58–73.

van den Bos, W., Rodriguez, C. A., Schweitzer, J. B., & McClure, S. M. (2014). Connectivity strength of dissociable striatal tracts predict individual differences in temporal discounting. *The Journal of Neuroscience, 34*, 10298–10310.

van den Bos, W., Rodriguez, C. A., Schweitzer, J. B., & McClure, S. M. (2015). Adolescent impatience decreases with increased frontostriatal connectivity. *Proceedings of the National Academy of Sciences of the United States of America, 112*, E3765–E3774.

Foundations of Social Decision-Making

Cătălina Elena Rățală and Alan Gerard Sanfey

Abstract This chapter provides an overview of some of the foundational work and methods on neuroeconomic approaches to studying social decision-making, that is, where individuals make choices in social contexts, balancing personal preferences with the values, beliefs, and expectations of others. Outlining three core aspects of social decision-making—trust, cooperation, and resource allocation—we discuss the experimental paradigms used to examine each, and highlight the insights gained into the psychological and neural mechanisms underlying social interactions. We conclude with a discussion of new approaches that have the potential to expand our current understanding of social decision-making.

1 Introduction and Overview

Imagine you are in a restaurant, choosing what to eat for lunch. Your choice will depend on your own personal preferences, constraints, values, and so on. But what if you were having lunch with a close friend, who asked you to order on their behalf? Or what if you were having a work lunch and were concerned about how you will be perceived by your dining partners? These latter examples illustrate situations when we must additionally incorporate social, other-regarding, preferences into our decision calculus. In fact, many of the important choices we face in life, from political choices to ethical judgments, happen within a social context or involve a social component. Here, we use the term *social decision-making* to refer specifically to instances where one needs to consider not only one's own values, beliefs, or expectations, but also those of others who may be impacted by the outcome of the decision (Rilling & Sanfey, 2011).

C. E. Rățală (✉)
Donders Institute (Radboud University), Nijmegen, Netherlands
e-mail: catalina.ratala@donders.ru.nl

A. G. Sanfey
Donders Institute and Behavioural Science Institute (Radboud University), Nijmegen, Netherlands

D. V. Smith et al. (eds.), *Neuroeconomics: Core Topics and Current Directions*,
https://doi.org/10.1007/978-3-032-02925-6_5

Social decision scenarios are often characterized by high complexity and uncertainty, sometimes providing us with incomplete or even contradictory pieces of information. Understanding how we decide in these contexts, and what factors play a role in these choices, is important not only on an individual level but also for society as a whole, as this can provide vital information for improving policy prescriptions across many domains (Stallen & Sanfey, 2015). By using methodology from different fields, primarily Psychology, Economics, and Neuroscience, the discipline of Neuroeconomics can offer useful insights into why, and how, we make our social decisions (Camerer, 2007; Sanfey et al., 2006).

The aim of the current chapter is to outline some of the foundational methodologies and findings across Neuroeconomics which capture the complexity of social choice. To do this, we will focus on three core concepts pertaining to choice in social contexts—trust, cooperation, and resource allocation—and describe both the experimental paradigms typically employed to address these issues, as well as some fundamental findings Neuroeconomics has uncovered to date.

2 Trust and Reciprocity

Trust refers to the expectation that another party will behave in a reliable manner in situations of social and economic exchange, especially situations when the other can abuse this trust for their own self-interest. For example, when you lend money to a colleague, you act in trust, expecting the colleague will repay their debt. Reciprocity extends trust to also involve the possibility of mutual exchange, where your initial act of trust is acknowledged and returned in a manner that demonstrates reliability. Both concepts are crucial in reinforcing initial trust and facilitating subsequent cooperation.

In order to study the dynamics of trust and reciprocity, researchers have typically used paradigms inspired by the mathematical framework of Game Theory (von Neumann & Morgenstern, 1944). One such experimental paradigm is the Trust Game (Berg et al., 1995) (see Fig. 1a for details). The amount of money the Investor offers quantifies the trust that has been placed in the Trustee: higher investments indicate higher trust. Similarly, the money returned by the Trustee represents their level of reciprocity. The decision to initially trust a partner is influenced by various cognitive-affective, social, and contextual factors, and is a function of the information available to the Investor as well as their ability to infer the trustworthiness (i.e., the future likelihood of reciprocation) of their interaction partner, taking into account the potential risks of trusting another (Balliet & Van Lange, 2013; Hancock et al., 2023). Standard experimental findings using the Trust Game illustrate a reasonably high level of trust in anonymous partners, shown by people's willingness to invest about half of their initial endowment on average (Berg et al., 1995). Trustees in turn typically return slightly less than half of the amount they receive, illustrating a general preference for at least some degree of reciprocity. These findings are quite robust (see Johnson & Mislin, 2011 for a meta-analysis) and

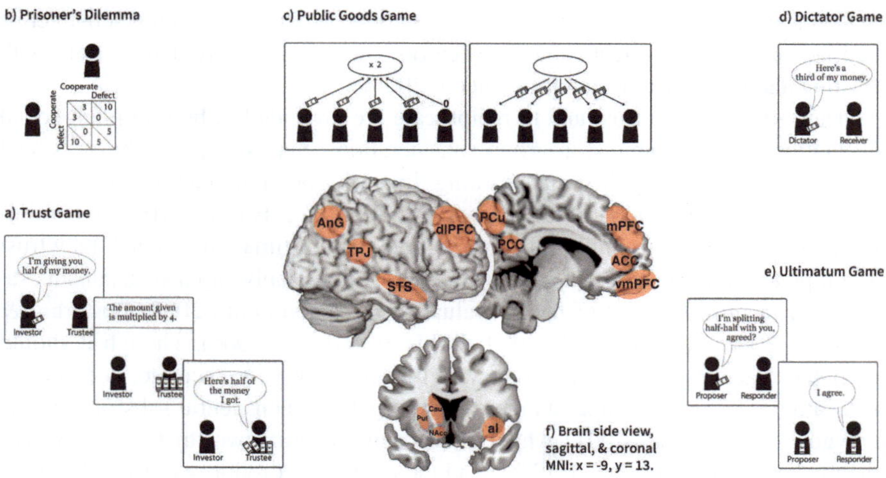

Fig. 1 Overview of classical experimental paradigms (**a**) **Trust Game**: This is a two-player game, where the Investor can send an amount of their money to the Trustee. The Trustee receives the quadrupled amount, and they can choose to reciprocate or not. The game isolates trust and reciprocity: the Investor must trust the Trustee to return a fair share, and the Trustee must decide whether or not to honor that trust. (**b**) **Prisoner's Dilemma.** This is a two-player game where each has two options: cooperate or defect. If both players cooperate, they both receive a moderate reward. If one player defects while the other cooperates, the defector gets a large reward and the cooperator gets no reward. The dilemma arises because the incentive to defect for personal gain can lead both to choose defection, resulting in a bad outcome for both. (**c**) **Public Goods Game.** It involves multiple players, who each receive an initial endowment. Players then decide how much to contribute to a public pot, keeping the rest for themselves. Contributions are multiplied by a specified factor, then shared *equally* among all. The dilemma is that optimal collective gains come from full contribution by all, but free-riders (who contribute nothing) can benefit from the public good while keeping their own initial amount. The game examines cooperative behavior and how individuals balance personal versus collective gain. (**d**) **Dictator Game.** This has two players. The Dictator is provided with an endowment and decides how much to share with the Recipient, retaining the remainder. The Recipient has no say in the matter. The game explores "pure" altruism, as the Dictator is under no obligation to share, but may choose to do so based on personal values or social norms. (**e**) **Ultimatum Game.** In this two-player game, the Proposer is given an endowment and proposes to the Responder how to divide this money. The Responder then must either accept or reject the offer. If the Responder accepts, the money is split as proposed. If the Responder rejects, neither player receives *anything*. In either event, the game is over. The game tests fairness and bargaining behavior, as the Proposer must make an offer that the Responder finds acceptable, balancing personal gain with the risk of rejection. (**f**) **Overview of brain areas associated with social decision-making**. In the side view: dorsal lateral prefrontal cortex (dmPFC), angular gyrus (AnG), temporal parietal junction (TPJ), and superior temporal sulcus (STS). In the sagittal view: medial prefrontal cortex (mPFC), anterior cingulate cortex (ACC), ventromedial prefrontal cortex (vmPFC), posterior cingulate cortex (PCC), and precuneus (PCu). In the coronal view: anterior insula (aI), putamen (Put), caudate (Cau), and nucleus accumbens (NAcc). The striatum comprises the putamen, caudate, and NACC. Insula and striatum are bilateral; display is unilateral for visual purposes

make clear that, contrary to the standard economic predictions of pure self-interest and hence zero investments and zero returns, players generally demonstrate both trusting and reciprocal behavior (Camerer, 2003).

These decisions to trust and to reciprocate are grounded in both implicit social evaluations (facial features, in-group considerations, biases, personality traits, and so on), as well as explicit social learning driven by previous interactions or overt social information (Mende-Siedlecki et al., 2013; see Fareri, 2019 for a mini-review). With regard to implicit factors, people often estimate how much they trust a complete stranger in just a glance, relying on extremely rapid mechanisms (at the level of milliseconds) of facial evaluation (Delgado et al., 2005; Oosterhof & Todorov, 2008; Todorov et al., 2009; Willis & Todorov, 2006). Though it should be noted that these rapid estimations are typically not very accurate, nonetheless neuroimaging evidence suggests these processes have their neural substrates in the amygdala (Engell et al., 2007), which is responsive to untrustworthy faces, as well as the medial prefrontal cortex (mPFC) and the precuneus (Todorov & Engell, 2008), which conversely show greater neural activity for (moderately) trustworthy faces. Crucially, when an Investor observes reciprocal behavior in their interaction partner, activation in the striatum, part of the brain's reward system, has been consistently recorded (Rilling et al., 2002), further reinforcing the initial trust decision.

Studies that focus on more explicit signals of trust and reciprocity often employ repeated-round Trust Games, where Investors have the opportunity to see exactly how their partners, the Trustees, respond to their trust over time, and vice versa. Factors such as prior trust experience or the level of trust bestowed by the Investor can substantially change the reciprocation response (Fareri, 2019). During the initial trust-building stages, the mPFC is involved in assessing trustworthiness and predicting future behavior while integrating information about social context and past experiences (Krueger et al., 2007; Krueger & Grafman, 2008). This process is in line with the idea that past occurrences serve as reliable indicators of future behavior (Axelrod & Hamilton, 1981; King-Casas et al., 2005). Furthermore, trust can be well captured via computational models utilizing prediction error-driven updates (Chang et al., 2010; Doll et al., 2009; Fareri et al., 2015). FMRI results have shown that the reputation of others can be encoded in the striatum, with reciprocity eliciting stronger responses than betrayal (Delgado, 2007; see Bellucci et al., 2017 for a meta-analysis). Across several interactions, the neural signals from the striatum shift earlier in time to help in anticipating the trust behavior of the interaction partner (King-Casas et al., 2005). Overall, reciprocity acts as both a socially rewarding experience and a learning signal to guide future trust decisions (Phan et al., 2010; Rilling et al., 2002), with areas such as the mPFC, striatum, and ACC being involved in positive reciprocation, while the anterior Insula seems to track negative reciprocation (see Fig. 1f). Since competing psychological accounts have been proposed for what may drive reciprocation behavior, for example, inequity aversion (Fehr & Schmidt, 1999; Tricomi et al., 2010) and guilt aversion (Battigalli & Dufwenberg, 2007; Dufwenberg & Gneezy, 2000), computational modelling offers a powerful framework for addressing these hypotheses. Chang and colleagues (2011) used a utility-based computational model (see Box 1) to illustrate how individual differences in beliefs and guilt sensitivity affect recruitment of the TPJ, insula, dACC, and dorsolateral PFC (see Fig. 1f). Further research using computational modelling and fMRI suggests the recruitment of several distinct strategies in accordance with those suggested above, as well as the ability to flexibly switch between these strategies in order to serve the twin goals of maintaining reputation while also maximizing one's own return (van Baar et al., 2019).

Box 1: Methodology

Computational Modeling. Computational modeling provides a quantifiable framework for studying individual choice in complex social environments. Mathematical models, often inspired from economics, are used to describe specific decision-related processes, such as learning, valuation, and social preferences. Different types of models can help capture links between psychological constructs, behavioral measures, and neural measurements. For instance, reinforcement learning models capture how individuals adjust their behavior over time based on rewards and punishments, shedding light on learning mechanisms in decision-making. Utility models quantify how individuals trade off social and monetary preferences in prosocial choices by assigning subjective values to different outcomes. Drift diffusion models describe the process of evidence accumulation, where information is integrated until a decision threshold is reached. Meanwhile, Bayesian models characterize how individuals update their beliefs about others' intentions by probabilistically incorporating new information as a function of prior knowledge or beliefs.

Physiological Measurements. Physiological responses provide valuable insights into the emotional and autonomic processes underlying decision-making. **Skin conductance response (SCR)** measures phasic changes in skin conductance driven by the sympathetic nervous system activity, reflecting emotional arousal. As a sensitive index of physiological arousal, SCR is widely used to study emotional engagement in decision-making and social interactions. **Heart rate variability (HRV)** indicates the balance between sympathetic and parasympathetic nervous system activity, which plays a role in stress regulation and decision-making under uncertainty. These physiological signals, when combined with computational models and/or neuroimaging, can offer a broader understanding of how affective and physiological states affect our choices. Techniques such as **eye-tracking** provide real-time insights into attention allocation, cognitive effort, and information processing. Eye-tracking measures fixation duration, saccades, and gaze patterns, revealing how individuals visually explore choice options and social cues.

Pharmacological Interventions. Pharmacological interventions involve the administration of drugs (e.g., methylphenidate, L-Dopa, atomoxetine) that

(continued)

modulate various neurotransmitters to examine their role in behavior and brain function. By comparing drug-induced changes across different populations, this approach provides insights into the neurochemical mechanisms underlying cognition and potential treatments for clinical conditions.

Electroencephalography (EEG): EEG is also a popular method in neuroeconomics, and it measures the electrical activity in the brain as a function of voltage changes in neurons. This method has excellent temporal resolution (on the millisecond scale), but poor spatial resolution. EEG can address questions related to tracking the time course of various neural processes, especially fast-changing ones like attention.

Magnetoencephalography (MEG): MEG measures the magnetic fields generated by neuronal electrical activity. Similar to EEG, MEG has excellent temporal resolution, but also better spatial resolution relative to EEG. Researchers will sometimes combine MEG and fMRI, to benefit from the complementary spatial and temporal information.

Functional Magnetic Resonance Imaging (fMRI). FMRI is one of the widely used methods in neuroeconomics. It measures metabolic changes, thus providing an indirect way to correlate mental processes with neural substrates. While it has good spatial resolution, its temporal resolution is limited. Various statistical approaches are used for analyzing fMRI data, focusing either on univariate average signal (univariate) or a distributed pattern (multivariate), making this method ideal for answering questions pertaining to where in the brain changes occur as a function of experimental task.

Positron Emission Tomography (PET). PET uses radioactive tracers to directly measure neurotransmitter function, receptor availability, and brain metabolism. It is essential in studying dopamine, serotonin, and opioid systems, which regulate reward processing.

Transcranial Magnetic Stimulation (TMS). TMS is a noninvasive brain stimulation technique that uses magnetic fields to induce electrical currents in targeted brain areas. This approach allows researchers to temporarily disrupt (or enhance) neural activity, making it a powerful tool for establishing causal relationships between brain regions and cognitive functions. TMS is frequently used in combination with neuroimaging to test specific hypotheses about brain function.

Transcranial Direct Current Stimulation (tDCS) and Transcranial Alternating Current Stimulation (tACS). Both tDCS and tACS are noninvasive neuromodulatory techniques that deliver low-intensity electrical currents to modulate neural activity. tDCS applies a constant direct current to alter cortical excitability, while tACS uses an oscillatory current at specific frequencies to influence brain rhythms. These techniques are often paired with behavioral tasks and neuroimaging (EEG, fMRI) to examine how targeted neural modulation affects decision-making, learning, and social cognition.

(continued)

Like TMS, they provide causal evidence for the role of specific brain regions in cognitive processes, making them valuable tools in experimental neuroeconomics.

Pharmacologically, a key neurotransmitter related to trust, trust learning, and reciprocity is dopamine. Primarily associated with reward receipt, reward anticipation, and prediction errors, several studies have shown that dopamine pathways become active during trust interactions, particularly in the striatum (Schultz, 2002, 2013; Izuma et al., 2008). Moreover, the evaluation of the trustworthiness of others has also been linked to the dopaminergic system (Krueger et al., 2007). In addition to dopamine, experimental work also connects other neurotransmitters, such as serotonin (Crockett et al., 2008, 2010), or hormones, such as testosterone and oxytocin, to trust behavior. However, findings to date are mixed, suggesting complex patterns of interaction are at play.

3 Cooperation, Coordination, and Competition

Cooperation refers to the joint investment of resources in order to achieve a mutually beneficial outcome. It is considered one of the cornerstones of human civilization (Tomasello & Vaish, 2013), and relies heavily on the ability to engage in mutual exchanges, which of course are potentially risky and not necessarily immediately or directly beneficial for oneself (Fehr & Fischbacher, 2003; see Stallen & Sanfey, 2013 for a review).

For instance, imagine you are working with a colleague on a project; each of you might have different strengths and weaknesses, and so collaboration can improve the resulting outcome relative to what could have been achieved by working alone. However, you might also want to guard against being exploited by your colleague, for example, if you do the lion's share of the work, but they take the credit. So, what are the factors that can facilitate or hinder cooperation? Why do people cooperate (or not), and how can Neuroeconomics help answer such questions?

To investigate cooperation experimentally, researchers have primarily used two classic paradigms or "coordination games": the Prisoner's Dilemma (Davis & Holt, 1993; Sally, 1995) and the Public Goods Game (Ledyard, 1995; see Fig. 1b, c for details). The Prisoner's Dilemma examines cooperative behavior as a function of joint outcomes. Findings here indicate that people prefer smaller mutual rewards gained via cooperation to higher personal rewards obtained by betraying the other (Davis & Holt, 1993), again contradicting theoretical predictions of pure self-interest (Sally, 1995). However, potential financial gains do matter, and both the increase of rewards or the introduction of punishment in this task has a moderate positive effect on cooperation (see Balliet et al., 2011 for a meta-analysis). Some of the primary factors that have been shown to impact cooperation in the Prisoner's Dilemma pertain to the payoff structure (Pruitt, 1967; Mengel, 2018; Gächter et al., 2021) and to how social values are weighted (see Bogaert et al., 2008 for a review). For example, if individuals find it more rewarding to behave prosocially (i.e., they value the benefit of others over their own financial gain), then they will cooperate more often. Additionally, when people interact together more than once, strategic considerations of reputation and reciprocity are observed. One of the most prominent and successful strategies in repeated interactions is the so-called "tit-for-tat," where each individual mirrors in the current round of the game the strategy their opponent chose in the previous round (Axelrod & Hamilton, 1981).

Experimental studies have explored the neural basis of cooperative behavior in the Prisoner's Dilemma (Rilling et al., 2008; Suzuki et al., 2011; Thompson et al., 2021; see Rilling & Sanfey, 2011 for a review). Accumulated evidence has identified the PFC, the ACC, the striatum, and the superior temporal sulcus (STS) as key brain regions involved in deciding whether to cooperate or defect. Cooperation in repeated games has been linked to activity in reward-related areas such as the striatum, ventromedial PFC, and orbitofrontal cortex (Haber & Knutson, 2010; Rilling et al., 2002). In addition, the posterior STS has been identified as crucial for mentalizing about the feelings, intentions, and strategies of others (Hampton et al., 2008; see Hein & Knight, 2008 for a review). Empathy has been linked with prosocial behavior (see Lockwood et al., 2016, for a review) and decisions to cooperate, involving areas such as insula, the ACC (Apps et al., 2016; Singer et al., 2004, 2006; see Lockwood & Wittmann, 2018 for a review), and the PFC (Masten et al., 2011).

Expanding on cooperative dynamics beyond two-player interactions, the Public Goods Game highlights social norms and collective action (Fehr & Gächter, 2000). While behavior here emphasizes the beneficial power of cooperation for the group (Stallen & Sanfey, 2015; Vives & FeldmanHall, 2018), decision-making is also susceptible to strategic considerations by individuals who prioritize personal gain over group welfare, often termed "free-riding" (Isaac & Walker, 1988). Experimental findings show collective investment behavior, with individuals on average contributing slightly less than half of their endowment to the common "pot" (Ledyard, 1995). This suggests motivations that go beyond purely financial incentives, and illustrates a willingness to contribute based on prosocial considerations (Fehr & Gächter, 2002; Rilling et al., 2002). A meta-analysis by Pletzer et al. (2018) has shown that this decision to cooperate is partly driven by the expectations about others' cooperative behavior. Simply making choices which improve the well-being of others recruits brain regions previously associated with socio-cognitive and empathic processes (see Bellucci et al., 2020 for a meta-analysis), such as the ventromedial PFC, the dorsolateral PFC, and the dorsal posterior cingulate cortex (see Fig. 1f). Additionally, accumulating evidence suggests that during various social interactions which involve mimicry or cooperation, there is also significant synchronization of brain activity in areas related to social cognition, for example, theory of mind or goal-oriented behavior (Dumas et al., 2010). Though evidence indicates that cooperative preferences are rather stable and generalized across various contexts (Peysakhovich et al., 2014), recent studies show that even when decision outcomes are similar, the brain mechanisms that underlie this behavior might differ (Rhoads et al., 2021; van Baar et al., 2019).

4 Resource Allocation

A final category of interactive scenarios that are important for understanding social choice involves decisions about the allocation and distribution of resources. Understanding how people trade off various competing motivations in these contexts offers crucial insights into the neural and psychological processes that underpin human social behavior (Rilling & Sanfey, 2011).

Imagine a scenario where an individual decides to donate a part of their salary to support the search for a disease treatment. This is considered altruistic behavior, as the donor themselves is unlikely to directly benefit from this. Among the several psychological factors that impact resource allocation, fairness and social preferences are some of the most commonly studied. The experimental paradigms used to investigate fairness employ tasks which attempt to tease apart how people balance social and monetary reward.

The Dictator Game (Kahneman et al., 1986) stands out as one of the most straightforward yet revealing paradigms in use (see Fig. 1d). Despite this simplicity, the Dictator Game has yielded substantial insights into human behavior. Contrary to the classical predictions of pure self-interest, most Dictators choose to give nonzero amounts to their game partners (Camerer, 1997, 2003; see Engel, 2011 for a meta-analysis), with a 50–50 split being the most frequent allocation (Camerer & Thaler, 1995; Fehr & Gächter, 2002). Although at first glance this finding seems to reveal an intrinsic preference for fairness that extends well beyond self-interest, literature suggests there are several other complex motivations at play (Andreoni & Bernheim, 2009; Murnighan et al., 2001).

One strand of research suggests that Dictators choose to share a part of their endowment due to a strong aversion to inequity or inequality (Fehr & Schmidt, 1999; Fehr et al., 2006). Krupka and Weber (2013) provide some evidence for these social preferences, showing that individuals exhibit a stable willingness to forfeit monetary gains to engender socially appropriate outcomes. Similarly, Schulz et al. (2014) illustrate that under high cognitive load (meant to decrease deliberation) participants choose fair allocations more often than under low cognitive load. This suggests fairness may well be the "default" option in the Dictator Game, with selfish behavior being the result of deliberative thinking. However, there appears to be substantial heterogeneity in allocation choices in this task. Meta-analytic findings indicate that fairness motivations are also moderated by socioeconomic factors and societal values: the more economically developed or individualistic a society, the less Dictators are willing to share (Engel, 2011). Furthermore, in the context of this paradigm, a significant number of Dictators tend to exploit any potential exculpatory mechanism available, such as ambiguity, plausible deniability, or willful ignorance, in order to appear fair without actually paying the monetary cost required by fairness

(Dana et al., 2007; Haisley & Weber, 2010). Thus, fairness seems not only subject to framing, contextual factors, and situational constraints (Andreoni & Rao, 2011; Camerer, 2003; Eckel & Grossman, 1996; Grossman, 2014), but also to individual tendencies for opportunism (Barron et al., 2022; van Baar et al., 2019).

The Ultimatum Game (Güth et al., 1982) is another experimental paradigm particularly useful in the investigation of fairness, focusing on the strategic element of distributive choice (see Fig. 1e). Relative to the Dictator Game, Ultimatum Game offers tend to be greater, likely due to the second-order preferences, beliefs, and strategic considerations of the allocator (Chen et al., 2017; Wells & Rand, 2013; Ruff et al., 2013; Sazhin et al., 2024; Spitzer et al., 2007). That is, concerns that a low offer will be immediately rejected by the Responder, thus costing the Proposer any gain. Similarly to the Dictator game, the Ultimatum Game has spawned numerous variations, each shedding light on different aspects of fairness considerations. For example, multi-round versions with changing roles (Nowak et al., 2000) have revealed how reputational concerns influence fairness behavior. The paradigm has been instrumental in showing also that people are willing to sacrifice personal gains in order to punish unfair behavior, a phenomenon that has been termed "altruistic punishment" (Fehr & Gächter, 2002). Importantly, varying the stakes (Cameron, 1999) illustrates that although people tend to increase their offers as stakes grow larger, the willingness to reject unfair offers (typically defined as lower than 20% of the endowed amount) persists even with substantial sums of money (see Larney et al., 2019 for a meta-analysis). In 2003, Sanfey and colleagues showed that the anterior insula is one of the key regions involved in the processing of unfair offers, with greater activation for unfair offers compared to fair offers predictive of the rejection of unfair offers. This finding suggests that the affective response to unfairness, potentially linked to feelings of disgust or anger, plays a crucial role in redistribution decisions (see Gabay et al., 2014 for a meta-analysis) and emphasizes the important role of emotions in social decision-making, challenging purely rational models of economic choice. Additionally, the dlPFC has been associated with cognitive control and the ability to override automatic responses in various contexts, including fairness decisions. More recent neuroimaging findings suggest that these strategic decisions activate the inferior frontal gyrus, and in particular, enhance its connectivity with the TPJ, whereas anterior insula-angular gyrus connectivity appears to modulate strategic behavior based on individual differences in reward (Sazhin et al., 2024) (see Fig. 1f). Several studies (van't Wout et al., 2005; Knoch et al., 2006) used transcranial magnetic stimulation (TMS) to disrupt dlPFC function and found that participants were more likely to accept unfair offers when this region was inhibited (see Box 1). Tricomi et al. (2010) illustrate how both advantageous and disadvantageous inequality can activate reward-related brain regions. Norm compliance likewise involves activity in cortical areas linked to reward processing. For instance, studies indicate that adherence to social norms under threat of punishment depends on the right lateral prefrontal cortex (Ruff et al., 2013). Experimentally increasing excitability in this region via anodal tDCS boosted sanction-induced norm compliance, whereas decreasing excitability dampens compliance. Complementary work demonstrated that rLPFC stimulation exerted stronger effects when social interactions are made explicit. Interestingly, studies have shown striatal activation not only when individuals receive rewards themselves, but also when they observe others receiving rewards, especially in fair contexts (Moll et al., 2006). This suggests a neural basis for the positive feelings associated with fair distributions and provides some support for the intrinsic value of equality.

5 Conclusion

Neuroeconomic approaches to social decision-making have provided valuable insights into the understanding of complex social behaviors, as well as the understanding of their neural underpinnings. By combining rigorous experimental paradigms with advanced neuroscientific techniques, researchers have started to unravel how the brain makes decisions about trust and reciprocity, cooperation, and resource allocation.

Future research in social decision-making will likely focus on integrating these findings into more comprehensive models of behavior. Additionally, the use of new approaches and technologies, such as wearable devices, large data sets, and virtual reality paradigms, to name but a few, all promise to add to the ecological validity of the current tasks and advance our current understanding of these complex phenomena.

Several directions of research are particularly promising, for instance the use of computational modelling where leveraging multimodal data and frameworks from machine learning and game theory can enhance prediction accuracy, theory development, and computational phenotyping across populations (Ahn & Busemeyer, 2016; Collins & Shenhav, 2022; Gillan & Whelan, 2017; Lee, 2013; Montague et al., 2012; see Brazil et al., 2018 for a review). Furthermore, the rapid developments in AI (see Harris, 2024 for a review) allow researchers to understand and predict social and economic behavior through the analysis of massive, highly complex data (see Aoujil et al., 2023 for a bibliographic analysis). Emerging research also highlights the gut-brain axis as a modulator of social decision-making, with gut microbiota influencing stress responses, emotional regulation, and reward processing.

Ultimately, the cumulative findings to date in social decision-making not only refine our understanding of the distinct motivations driving prosocial behavior (Schreuders et al., 2018) but also offer pathways for developing interventions for individuals with atypical social processing (Bellucci et al., 2020). Furthermore, this knowledge can be a powerful tool when applied to real-world settings, offering policymakers valuable insights in the design of evidence-based strategies aimed at promoting better, more sustainable decision-making and prosocial behavior in various societal contexts.

References

Ahn, W.-Y., & Busemeyer, J. R. (2016). Challenges and promises translating computational tools into clinical practice. *Current Opinion in Behavioral Sciences, 11*, 1–7. https://doi.org/10.1016/j.cobeha.2016.02.001

Andreoni, J., & Bernheim, B. D. (2009). Social image and the 50–50 norm: A theoretical and experimental analysis of audience effects. *Econometrica, 77*(5), 1607–1636. https://doi.org/10.3982/ECTA7384

Andreoni, J., & Rao, J. M. (2011). The power of asking: How communication affects selfishness, empathy, and altruism. *Journal of Public Economics, 95*(7), 513–520. https://doi.org/10.1016/j.jpubeco.2010.12.008

Aoujil, Z., Hanine, M., Flores, E. S., Samad, M. A., & Ashraf, I. (2023). Artificial intelligence and behavioral economics: A bibliographic analysis of research field. *IEEE Access, 11*, 139367–139394. https://doi.org/10.1109/ACCESS.2023.3339778

Apps, M. A., Rushworth, M. F., & Chang, S. W. (2016). The anterior cingulate gyrus and social cognition: Tracking the motivation of others. *Neuron, 90*(4), 692–707. https://doi.org/10.1016/j.neuron.2016.04.018

Axelrod, R., & Hamilton, W. D. (1981). The evolution of cooperation. *Science, 211*(4489), 1390–1396.

Balliet, D., & Van Lange, P. A. (2013). Trust, conflict, and cooperation: A meta-analysis. *Psychological Bulletin, 139*(5), 1090. https://doi.org/10.1037/a0030939

Balliet, D., Mulder, L. B., & Van Lange, P. A. M. (2011). Reward, punishment, and cooperation: A meta-analysis. *Psychological Bulletin, 137*(4), 594–615. https://doi.org/10.1037/a0023489

Barron, K., Stüber, R., & van Veldhuizen, R. (2022). *Motivated motive selection in the lying-dictator game* (CESifo Working Paper No. 9911). Available at SSRN. https://doi.org/10.2139/ssrn.4215030

Battigalli, P., & Dufwenberg, M. (2007). Guilt in games. *American Economic Review, 97*(2), 170–176. https://doi.org/10.1257/aer.97.2.170

Bellucci, G., Chernyak, S. V., Goodyear, K., Eickhoff, S. B., & Krueger, F. (2017). Neural signatures of trust in reciprocity: A coordinate-based meta-analysis. *Human Brain Mapping, 38*(3), 1233–1248. https://doi.org/10.1002/hbm.23451

Bellucci, G., Camilleri, J. A., Eickhoff, S. B., & Krueger, F. (2020). Neural signatures of prosocial behaviors. *Neuroscience and Biobehavioral Reviews, 118*, 186–195. https://doi.org/10.1016/j.neubiorev.2020.07.006

Berg, J., Dickhaut, J., & McCabe, K. (1995). Trust, reciprocity, and social history. *Games and Economic Behavior, 10*(1), 122–142. https://doi.org/10.1006/game.1995.1027

Bogaert, S., Boone, C., & Declerck, C. (2008). Social value orientation and cooperation in social dilemmas: A review and conceptual model. *British Journal of Social Psychology, 47*(3), 453–480. https://doi.org/10.1348/014466607X244970

Brazil, I. A., van Dongen, J. D., Maes, J. H., Mars, R. B., & Baskin-Sommers, A. R. (2018). Classification and treatment of antisocial individuals: From behavior to biocognition. *Neuroscience & Biobehavioral Reviews, 91*, 259–277. https://doi.org/10.1016/j.neubiorev.2016.10.010

Camerer, C. F. (1997). Progress in behavioral game theory. *Journal of Economic Perspectives, 11*(4), 167–188. https://doi.org/10.1257/jep.11.4.167

Camerer, C. F. (2003). *Behavioral game theory: Experiments in strategic interaction* (p. xv, 550).

Camerer, C. F. (2007). Neuroeconomics: Using neuroscience to make economic predictions. *The Economic Journal, 117*(519), C26–C42. https://doi.org/10.1111/j.1468-0297.2007.02033.x

Camerer, C., & Thaler, R. H. (1995). Anomalies: Ultimatums, dictators and manners. *Journal of Economic Perspectives, 9*(2), 209–219. https://doi.org/10.1257/jep.9.2.209

Cameron, L. A. (1999). Raising the stakes in the ultimatum game: Experimental evidence from Indonesia. *Economic Inquiry, 37*(1), 47–59. https://doi.org/10.1111/j.1465-7295.1999.tb01415.x

Chang, L. J., Doll, B. B., Van'T Wout, M., Frank, M. J., & Sanfey, A. G. (2010). Seeing is believing: Trustworthiness as a dynamic belief. *Cognitive Psychology, 61*(2), 87–105. https://doi.org/10.1016/j.cogpsych.2010.03.001

Chang, L. J., Smith, A., Dufwenberg, M., & Sanfey, A. G. (2011). Triangulating the neural, psychological, and economic bases of guilt aversion. *Neuron, 70*(3), 560–572. https://doi.org/10.1016/j.neuron.2011.02.056

Chen, Y.-H., Chen, Y.-C., Kuo, W.-J., Kan, K., Yang, C. C., & Yen, N.-S. (2017). Strategic motives drive proposers to offer fairly in ultimatum games: An fMRI study. *Scientific Reports, 7*(1), 527. https://doi.org/10.1038/s41598-017-00608-8

Collins, A. G. E., & Shenhav, A. (2022). Advances in modeling learning and decision-making in neuroscience. *Neuropsychopharmacology, 47*(1), 104–118. https://doi.org/10.1038/s41386-021-01126-y

Crockett, M. J., Clark, L., Tabibnia, G., Lieberman, M. D., & Robbins, T. W. (2008). Serotonin modulates behavioral reactions to unfairness. *Science, 320*(5884), 1739–1739. https://doi.org/10.1126/science.1155577

Crockett, M. J., Clark, L., Hauser, M. D., & Robbins, T. W. (2010). Serotonin selectively influences moral judgment and behavior through effects on harm aversion. *PNAS Proceedings of the National Academy of Sciences of the United States of America, 107*(40), 17433–17438. https://doi.org/10.1073/pnas.1009396107

Dana, J., Weber, R. A., & Kuang, J. X. (2007). Exploiting moral wiggle room: Experiments demonstrating an illusory preference for fairness. *Economic Theory, 33*(1), 67–80. https://doi.org/10.1007/s00199-006-0153-z

Davis, D., & Holt, C. (1993). *Experimental economics*. Princeton University Press. https://doi.org/10.1515/9780691233376

Delgado, M. R. (2007). Reward-related responses in the human striatum. *Annals of the New York Academy of Sciences, 1104*(1), 70–88. https://doi.org/10.1196/annals.1390.002

Delgado, M. R., Miller, M. M., Inati, S., & Phelps, E. A. (2005). An fMRI study of reward-related probability learning. *NeuroImage, 24*(3), 862–873. https://doi.org/10.1016/j.neuroimage.2004.10.002

Doll, B. B., Jacobs, W. J., Sanfey, A. G., & Frank, M. J. (2009). Instructional control of reinforcement learning: A behavioral and neurocomputational investigation. *Brain Research, 1299*, 74–94. https://doi.org/10.1016/j.brainres.2009.07.007

Dufwenberg, M., & Gneezy, U. (2000). Measuring beliefs in an experimental lost wallet game. *Games and Economic Behavior, 30*(2), 163–182. https://doi.org/10.1006/game.1999.0715

Dumas, G., Nadel, J., Soussignan, R., Martinerie, J., & Garnero, L. (2010). Inter-brain synchronization during social interaction. *PLoS One, 5*(8), e12166. https://doi.org/10.1371/journal.pone.0012166

Eckel, C. C., & Grossman, P. J. (1996). The relative price of fairness: Gender differences in a punishment game. *Journal of Economic Behavior & Organization, 30*(2), 143–158. https://doi.org/10.1016/S0167-2681(96)00854-2

Engel, C. (2011). Dictator games: A meta study. *Experimental Economics, 14*(4), 583–610. https://doi.org/10.1007/s10683-011-9283-7

Engell, A. D., Haxby, J. V., & Todorov, A. (2007). Implicit trustworthiness decisions: Automatic coding of face properties in the human amygdala. *Journal of Cognitive Neuroscience, 19*(9), 1508–1519. https://doi.org/10.1162/jocn.2007.19.9.1508

Fareri, D. S. (2019). Neurobehavioral mechanisms supporting trust and reciprocity. *Frontiers in Human Neuroscience, 13*. https://doi.org/10.3389/fnhum.2019.00271

Fareri, D. S., Chang, L. J., & Delgado, M. R. (2015). Computational substrates of social value in interpersonal collaboration. *The Journal of Neuroscience, 35*(21), 8170–8180. https://doi.org/10.1523/JNEUROSCI.4775-14.2015

Fehr, E., & Fischbacher, U. (2003). The nature of human altruism. *Nature, 425*(6960), 785–791. https://doi.org/10.1038/nature02043

Fehr, E., & Gächter, S. (2000). Cooperation and punishment in public goods experiments. *American Economic Review, 90*(4), 980–994. https://doi.org/10.1257/aer.90.4.980

Fehr, E., & Gächter, S. (2002). Altruistic punishment in humans. *Nature, 415*(6868), 137–140. https://doi.org/10.1038/415137a

Fehr, E., & Schmidt, K. M. (1999). A theory of fairness, competition, and cooperation. *The Quarterly Journal of Economics, 114*(3), 817–868.

Fehr, E., Naef, M., & Schmidt, K. M. (2006). Inequality aversion, efficiency, and maximin preferences in simple distribution experiments: Comment. *American Economic Review, 96*(5), 1912–1917. https://doi.org/10.1257/aer.96.5.1912

Gabay, A. S., Radua, J., Kempton, M. J., & Mehta, M. A. (2014). The ultimatum game and the brain: A meta-analysis of neuroimaging studies. *Neuroscience & Biobehavioral Reviews, 47*, 549–558. https://doi.org/10.1016/j.neubiorev.2014.10.014

Gächter, S., Lee, K., Sefton, M., & Weber, T. O. (2021). *Risk, temptation, and efficiency in the one-shot prisoner's dilemma* (CESifo working paper no. 9449). Available at SSRN. https://doi.org/10.2139/ssrn.3979354

Gillan, C. M., & Whelan, R. (2017). What big data can do for treatment in psychiatry. *Current Opinion in Behavioral Sciences, 18*, 34–42. https://doi.org/10.1016/j.cobeha.2017.07.003

Grossman, Z. (2014). Strategic ignorance and the robustness of social preferences. *Management Science, 60*(11), 2659–2665. https://doi.org/10.1287/mnsc.2014.1989

Güth, W., Schmittberger, R., & Schwarze, B. (1982). An experimental analysis of ultimatum bargaining. *Journal of Economic Behavior & Organization, 3*(4), 367–388. https://doi.org/10.1016/0167-2681(82)90011-7

Haber, S. N., & Knutson, B. (2010). The reward circuit: Linking primate anatomy and human imaging. *Neuropsychopharmacology, 35*(1), 4–26. https://doi.org/10.1038/npp.2009.129

Haisley, E. C., & Weber, R. A. (2010). Self-serving interpretations of ambiguity in other-regarding behavior. *Games and Economic Behavior, 68*(2), 614–625. https://doi.org/10.1016/j.geb.2009.08.002

Hampton, A. N., Bossaerts, P., & O'Doherty, J. P. (2008). Neural correlates of mentalizing-related computations during strategic interactions in humans. *Proceedings of the National Academy of Sciences, 105*(18), 6741–6746. https://doi.org/10.1073/pnas.0711099105

Hancock, P. A., Kessler, T. T., Kaplan, A. D., Stowers, K., Brill, J. C., Billings, D. R., Schaefer, K. E., & Szalma, J. L. (2023). How and why humans trust: A meta-analysis and elaborated model. *Frontiers in Psychology, 14*, 1081086. https://doi.org/10.3389/fpsyg.2023.1081086

Harris, L. T. (2024). The neuroscience of human and artificial intelligence presence. *Annual Review of Psychology, 75*(1), 433–466. https://doi.org/10.1146/annurev-psych-013123-123421

Hein, G., & Knight, R. T. (2008). Superior temporal sulcus—It's my area: Or is it? *Journal of Cognitive Neuroscience, 20*(12), 2125–2136. https://doi.org/10.1162/jocn.2008.20148

Isaac, R. M., & Walker, J. M. (1988). Group size effects in public goods provision: The voluntary contributions mechanism. *The Quarterly Journal of Economics, 103*(1), 179–199. https://doi.org/10.2307/1882648

Izuma, K., Saito, D. N., & Sadato, N. (2008). Processing of social and monetary rewards in the human striatum. *Neuron, 58*(2), 284–294. https://doi.org/10.1016/j.neuron.2008.03.020

Johnson, N. D., & Mislin, A. A. (2011). Trust games: A meta-analysis. *Journal of Economic Psychology, 32*(5), 865–889. https://doi.org/10.1016/j.joep.2011.05.007

Kahneman, D., Knetsch, J. L., & Thaler, R. (1986). Fairness as a constraint on profit seeking: Entitlements in the market. *The American Economic Review, 76*(4), 728–741.

King-Casas, B., Tomlin, D., Anen, C., Camerer, C. F., Quartz, S. R., & Montague, P. R. (2005). Getting to know you: Reputation and trust in a two-person economic exchange. *Science, 308*(5718), 78–83. https://doi.org/10.1126/science.1108062

Knoch, D., Pascual-Leone, A., Meyer, K., Treyer, V., & Fehr, E. (2006). Diminishing reciprocal fairness by disrupting the right prefrontal cortex. *Science, 314*(5800), 829–832. https://doi.org/10.1126/science.1129156

Krueger, F., & Grafman, J. (2008). The human prefrontal cortex stores structured event complexes. In T. Shipley & J. M. Zacks (Eds.), *Understanding events: From perception to action* (pp. 617–638). Oxford University Press.

Krueger, F., Moll, J., Zahn, R., Heinecke, A., & Grafman, J. (2007). Event frequency modulates the processing of daily life activities in human medial prefrontal cortex. *Cerebral Cortex, 17*(10), 2346–2353. https://doi.org/10.1093/cercor/bhl143

Krupka, E. L., & Weber, R. A. (2013). Identifying social norms using coordination games: Why does dictator game sharing vary? *Journal of the European Economic Association, 11*(3), 495–524. https://doi.org/10.1111/jeea.12006

Larney, A., Rotella, A., & Barclay, P. (2019). Stake size effects in ultimatum game and dictator game offers: A meta-analysis. *Organizational Behavior and Human Decision Processes, 151*, 61–72. https://doi.org/10.1016/j.obhdp.2019.01.002

Ledyard, J. O. (1995). *Public goods: A survey of experimental research*. https://authors.library.caltech.edu/records/6w2tn-cpt82.

Lee, D. (2013). Decision making: From neuroscience to psychiatry. *Neuron, 78*(2), 233–248. https://doi.org/10.1016/j.neuron.2013.04.008

Lockwood, P. L., & Wittmann, M. K. (2018). Ventral anterior cingulate cortex and social decision-making. *Neuroscience & Biobehavioral Reviews, 92*, 187–191. https://doi.org/10.1016/j.neubiorev.2018.05.030

Lockwood, P. L., Apps, M. A. J., Valton, V., & Roiser, J. P. (2016). Neurocomputational mechanisms of prosocial learning and links to empathy. *Proceedings of the National Academy of Sciences, 113*(35), 9763–9768. https://doi.org/10.1073/pnas.1603198113

Masten, C. L., Morelli, S. A., & Eisenberger, N. I. (2011). An fMRI investigation of empathy for "social pain" and subsequent prosocial behavior. *NeuroImage, 55*(1), 381–388. https://doi.org/10.1016/j.neuroimage.2010.11.060

Mende-Siedlecki, P., Baron, S. G., & Todorov, A. (2013). Diagnostic value underlies asymmetric updating of impressions in the morality and ability domains. *Journal of Neuroscience, 33*(50), 19406–19415. https://doi.org/10.1523/jneurosci.2334-13.2013

Mengel, F. (2018). Risk and temptation: A meta-study on prisoner's dilemma games. *The Economic Journal, 128*(616), 3182–3209. https://doi.org/10.1111/ecoj.12548

Moll, J., Krueger, F., Zahn, R., Pardini, M., de Oliveira-Souza, R., & Grafman, J. (2006). Human fronto–mesolimbic networks guide decisions about charitable donation. *Proceedings of the National Academy of Sciences, 103*(42), 15623–15628. https://doi.org/10.1073/pnas.0604475103

Montague, P. R., Dolan, R. J., Friston, K. J., & Dayan, P. (2012). Computational psychiatry. *Trends in Cognitive Sciences, 16*(1), 72–80. https://doi.org/10.1016/j.tics.2011.11.018

Murnighan, J. K., Oesch, J. M., & Pillutla, M. (2001). Player types and self-impression Management in Dictatorship Games: Two experiments. *Games and Economic Behavior, 37*(2), 388–414. https://doi.org/10.1006/game.2001.0847

Nowak, M. A., Page, K. M., & Sigmund, K. (2000). Fairness versus reason in the ultimatum game. *Science, 289*(5485), 1773–1775.

Oosterhof, N. N., & Todorov, A. (2008). The functional basis of face evaluation. *Proceedings of the National Academy of Sciences, 105*(32), 11087–11092. https://doi.org/10.1073/pnas.0805664105

Peysakhovich, A., Nowak, M. A., & Rand, D. G. (2014). Humans display a 'cooperative phenotype' that is domain general and temporally stable. *Nature Communications, 5*(1), 4939. https://doi.org/10.1038/ncomms5939

Phan, K. L., Sripada, C. S., Angstadt, M., & McCabe, K. (2010). Reputation for reciprocity engages the brain reward center. *Proceedings of the National Academy of Sciences, 107*(29), 13099–13104. https://doi.org/10.1073/pnas.1008137107

Pletzer, J. L., Balliet, D., Joireman, J., Kuhlman, D. M., Voelpel, S. C., & Van Lange, P. A. M. (2018). Social Value Orientation, Expectations, and Cooperation in Social Dilemmas: A Meta–Analysis. European Journal of Personality, 32(1), 62–83. https://doi.org/10.1002/per.2139.

Pruitt, D. G. (1967). Reward structure and cooperation: The decomposed prisoner's dilemma game. *Journal of Personality and Social Psychology, 7*(1, Pt.1), 21–27. https://doi.org/10.1037/h0024914

Rhoads, S. A., Cutler, J., & Marsh, A. A. (2021). A feature-based network analysis and fMRI meta-analysis reveal three distinct types of prosocial decisions. *Social Cognitive and Affective Neuroscience, 16*(12), 1214–1233. https://doi.org/10.1093/scan/nsab079

Rilling, J. K., & Sanfey, A. G. (2011). The neuroscience of social decision-making. *Annual Review of Psychology, 62*, 23–48. https://doi.org/10.1146/annurev.psych.121208.131647

Rilling, J. K., Gutman, D. A., Zeh, T. R., Pagnoni, G., Berns, G. S., & Kilts, C. D. (2002). A neural basis for social cooperation. *Neuron, 35*(2), 395–405. https://doi.org/10.1016/S0896-6273(02)00755-9

Rilling, J. K., Goldsmith, D. R., Glenn, A. L., Jairam, M. R., Elfenbein, H. A., Dagenais, J. E., Murdock, C. D., & Pagnoni, G. (2008). The neural correlates of the affective response to unreciprocated cooperation. *Neuropsychologia, 46*(5), 1256–1266. https://doi.org/10.1016/j.neuropsychologia.2007.11.033

Ruff, C. C., Ugazio, G., & Fehr, E. (2013). Changing social norm compliance with noninvasive brain stimulation. *Science, 342*(6157), 482–484. https://doi.org/10.1126/science.1241399

Sally, D. (1995). Conversation and cooperation in social dilemmas: A meta-analysis of experiments from 1958 to 1992. *Rationality and Society, 7*(1), 58–92. https://doi.org/10.1177/1043463195007001004

Sanfey, A. G., Rilling, J. K., Aronson, J. A., Nystrom, L. E., & Cohen, J. D. (2003). The neural basis of economic decision-making in the ultimatum game. *Science, 300*(5626), 1755–1758. https://doi.org/10.1126/science.1082976

Sanfey, A. G., Loewenstein, G., McClure, S. M., & Cohen, J. D. (2006). Neuroeconomics: Cross-currents in research on decision-making. *Trends in Cognitive Sciences, 10*(3), 108–116. https://doi.org/10.1016/j.tics.2006.01.009

Sazhin, D., Wyngaarden, J. B., Dennison, J. B., Zaff, O., Fareri, D., McCloskey, M. S., Alloy, L. B., Jarcho, J. M., & Smith, D. V. (2024). Trait reward sensitivity modulates connectivity with the temporoparietal junction and anterior insula during strategic decision making. *bioRxiv*. https://doi.org/10.1101/2023.10.19.563125

Schreuders, E., Klapwijk, E. T., Will, G. J., & Crone, E. A. (2018). Friend versus foe: Neural correlates of prosocial decisions for liked and disliked peers. *Cognitive, Affective, & Behavioral Neuroscience, 18*(1), 127–142. https://doi.org/10.3758/s13415-017-0557-1

Schultz, W. (2002). Getting formal with dopamine and reward. *Neuron, 36*(2), 241–263. https://doi.org/10.1016/s0896-6273(02)00967-4

Schultz, W. (2013). Updating dopamine reward signals. *Current Opinion in Neurobiology, 23*(2), 229–238. https://doi.org/10.1016/j.conb.2012.11.012

Schulz, J. F., Fischbacher, U., Thöni, C., & Utikal, V. (2014). Affect and fairness: Dictator games under cognitive load. *Journal of Economic Psychology, 41*, 77–87. https://doi.org/10.1016/j.joep.2012.08.007

Singer, T., Seymour, B., O'Doherty, J., Kaube, H., Dolan, R. J., & Frith, C. D. (2004). Empathy for pain involves the affective but not sensory components of pain. *Science, 303*(5661), 1157–1162. https://doi.org/10.1126/science.1093535

Singer, T., Seymour, B., O'Doherty, J. P., Stephan, K. E., Dolan, R. J., & Frith, C. D. (2006). Empathic neural responses are modulated by the perceived fairness of others. *Nature, 439*(7075), 466–469. https://doi.org/10.1038/nature04271

Spitzer, M., Fischbacher, U., Herrnberger, B., Grön, G., & Fehr, E. (2007). The neural signature of social norm compliance. *Neuron, 56*(1), 185–196. https://doi.org/10.1016/j.neuron.2007.09.011

Stallen, M., & Sanfey, A. G. (2013). The cooperative brain. *The Neuroscientist, 19*(3), 292–303. https://doi.org/10.1177/1073858412469728

Stallen, M., & Sanfey, A. G. (2015). The neuroscience of social conformity: Implications for fundamental and applied research. *Frontiers in Neuroscience, 9*, 337. https://doi.org/10.3389/fnins.2015.00337

Suzuki, S., Niki, K., Fujisaki, S., & Akiyama, E. (2011). Neural basis of conditional cooperation. *Social Cognitive and Affective Neuroscience, 6*(3), 338–347. https://doi.org/10.1093/scan/nsq042

Thompson, K., Nahmias, E., Fani, N., Kvaran, T., Turner, J., & Tone, E. (2021). The prisoner's dilemma paradigm provides a neurobiological framework for the social decision cascade. *PLoS One, 16*(3), e0248006. https://doi.org/10.1371/journal.pone.0248006

Todorov, A., & Engell, A. D. (2008). The role of the amygdala in implicit evaluation of emotionally neutral faces. *Social Cognitive and Affective Neuroscience, 3*(4), 303–312. https://doi.org/10.1093/scan/nsn033

Todorov, A., Pakrashi, M., & Oosterhof, N. N. (2009). Evaluating faces on trustworthiness after minimal time exposure. *Social Cognition, 27*(6), 813–833. https://doi.org/10.1521/soco.2009.27.6.813

Tomasello, M., & Vaish, A. (2013). Origins of human cooperation and morality. *Annual Review of Psychology, 64*, 231–255. https://doi.org/10.1146/annurev-psych-113011-143812

Tricomi, E., Rangel, A., Camerer, C. F., & O'Doherty, J. P. (2010). Neural evidence for inequality-averse social preferences. *Nature, 463*(7284), 1089–1091. https://doi.org/10.1038/nature08785

van Baar, J. M., Chang, L. J., & Sanfey, A. G. (2019). The computational and neural substrates of moral strategies in social decision-making. *Nature Communications, 10*(1), 1483. https://doi.org/10.1038/s41467-019-09161-6

van't Wout, M., Kahn, R. S., Sanfey, A. G., & Aleman, A. (2005). Repetitive transcranial magnetic stimulation over the right dorsolateral prefrontal cortex affects strategic decision-making. *Neuroreport, 16*(16), 1849–1852. https://doi.org/10.1097/01.wnr.0000183907.08149.14

Vives, M.-L., & FeldmanHall, O. (2018). Tolerance to ambiguous uncertainty predicts prosocial behavior. *Nature Communications, 9*(1), 2156. https://doi.org/10.1038/s41467-018-04631-9

von Neumann, J., & Morgenstern, O. (1944). *Theory of games and economic behavior* (pp. xviii–625). Princeton University Press.

Wells, J., & Rand, D. G. (2013). *Strategic self-interest can explain seemingly "fair" offers in the ultimatum game* (SSRN scholarly paper no. 2136707). https://doi.org/10.2139/ssrn.2136707

Willis, J., & Todorov, A. (2006). First impressions: Making up your mind after a 100-ms exposure to a face. *Psychological Science, 17*(7), 592–598. https://doi.org/10.1111/j.1467-9280.2006.01750.x

Neural Mechanisms of Reward, Decision-Making, and Learning in a Social Context: Insights from Computational Neuroimaging

Qianying Wu and John O'Doherty

Abstract In recent years, the study of the neural mechanisms underlying social behavior has gained significant attention within the field of neuroeconomics. During this time, core ideas and theoretical frameworks developed through the study of decision-making and learning in nonsocial contexts have been adapted and extended to social contexts. New paradigms have been introduced to measure social behaviors, new models and theories have been proposed to explain these behaviors, and experimental evidence has accumulated to validate these theories and elucidate the neural substrates underpinning social computations in humans.

This chapter highlights recent advances in the study of social processes using a neuroeconomics-based approach, focusing on important findings through computational modeling and neuroimaging methods. We begin by outlining the neural structures involved in processing various social rewards. Next, we discuss how values are computed and represented in the brain when making decisions that involve other people's outcomes. We then examine the motivations behind these decisions, in two specific cases: altruistic and fairness-driven decisions. Finally, we explore how individuals update their knowledge and beliefs about whom they interact with, and how they should interact—a process known as social learning. We review two broad types of social leaning: learning about others and learning from others, and end with a discussion about arbitrations between learning strategies.

1 Social Rewards

Similar to nonsocial rewards (Fareri & Delgado, 2014), some social stimuli can elicit a positive hedonic experience, motivate behaviors, and activate several regions that have been found to encode social rewards in a manner analogous to nonsocial rewards (Fig. 1), as discussed in the following section.

Q. Wu · J. O'Doherty (✉)
Division of the Humanities and Social Sciences, California Institute of Technology, Pasadena, CA, USA
e-mail: jdoherty@caltech.edu

D. V. Smith et al. (eds.), *Neuroeconomics: Core Topics and Current Directions*,
https://doi.org/10.1007/978-3-032-02925-6_6

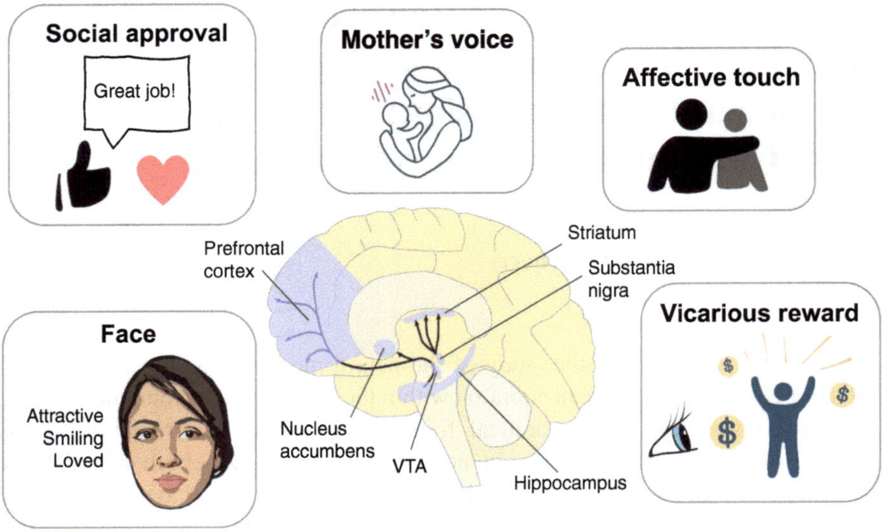

Fig. 1 Social reward. Social reward can activate several areas of the brain including the orbitofrontal cortex (part of the prefrontal cortex and the ventral striatum including the nucleus accumbens). Social reward has been studied in various forms, including viewing (attractive, smiling, or loved) faces, receiving social approval, hearing one's mother's voice, experiencing affective touch, and watching someone else getting reward. (Source of the brain image: "Dopamine pathways" by the National Institute of Health, retrieved from Wikimedia Commons: https://en.m.wikipedia.org/wiki/File:Dopamine_pathways.svg. Source of the face image: NIAID Visual & Medical Arts. (10/7/2024). Female Upper Profile Straight on View. NIAID NIH BIOART Source. bioart.niaid.nih.gov/bioart/548)

1.1 Rewarding Social Cues

A typical social reward is a face: simply seeing a face can be rewarding, and different faces have different rewarding values (O'Doherty et al., 2003; Strathearn et al., 2008; Mühlberger et al., 2011; Aharon et al., 2001). The rewarding nature of a face depends on several features, including attractiveness, facial expression, and likability (O'Doherty et al., 2003; Strathearn et al., 2008; Aharon et al., 2001; Aron et al., 2005; Hayden et al., 2007; Younger et al., 2010; Cooper et al., 2012). Humans are willing to exert greater physical effort to view attractive faces compared to less attractive ones (Aharon et al., 2001; Hayden et al., 2007), and the medial orbitofrontal cortex (OFC) has been found to track attractiveness ratings in a manner similar to monetary rewards—more attractive faces evoke greater brain responses (O'Doherty et al., 2003). In terms of facial expression, smiling faces are more rewarding than neutral or sad faces and recruit medial OFC (Strathearn et al., 2008). Interestingly, ventral striatum and medial OFC activity is also elicited when an angry face recedes from view, suggesting that the decay of an unpleasant, socially elicited emotion can be coded as a reward (Mühlberger et al., 2011). Additionally, faces of romantic partners are more rewarding than those of strangers or friends (Aron et al.,

2005; Younger et al., 2010; Cooper et al., 2012), which not only engage these same brain areas (Aron et al., 2005) but also reduce subjective feelings of physical pain (Younger et al., 2010).

Social approval is another frequently studied category of social reward. The need for acceptance, belonging, and approval lies on the third layer of Maslow's hierarchy of needs (Maslow, 1943). Anticipating and receiving social approval engages the dorsal and ventral striatum, ventromedial prefrontal cortex (vmPFC), and anterior cingulate cortex (ACC) (Cooper et al., 2012; Kohls et al., 2013; Sherman et al., 2016, 2018; Williams et al., 2020). Social approval can take many forms, including verbal praise (Williams et al., 2020), positive body gestures (Kohls et al., 2013; Williams et al., 2020), or receiving a "like" on social media (Sherman et al., 2016, 2018). Even sending a "like" activates the same brain regions as receiving one (Sherman et al., 2018). Social rejection—which can be elicited via the perception of certain body gestures from others or via exclusion by peers—produces a negative affective response that stands in stark contrast to that elicited by social rewards. On the other hand, the avoidance of social rejections is found to positively correlate with brain responses in the nucleus accumbens (NAcc) (Kohls et al., 2013; Masten et al., 2009).

Social reward manifests in various other forms. Sexual stimuli (Sescousse et al., 2010) can serve as potent rewards, engaging brain regions associated with reward across genders and sexual orientations (Stoléru et al., 2012; Sescousse et al., 2013). Similarly, physical touch, including mother-infant interaction and socially meaningful touch, is a powerful social reward (Panksepp, 1998; Rolls et al., 1999). Receiving pleasant affective touch—compared to neutral or painful touch—activates the OFC, NAcc, caudate, ACC, and anterior insula (Mielacher et al., 2024; Rolls et al., 2003). In addition, sounds sometimes convey important social cues. For instance, hearing a mother's voice triggers responses in the NAcc and vmPFC in a child's brain (Abrams et al., 2016, 2022).

Remarkably, experiencing a positive hedonic response does not always require direct personal experience of a reward—observing others receiving rewards can also be rewarding, a phenomenon known as "vicarious reward" (Ruff & Fehr, 2014; Lockwood, 2016). Vicarious rewards can take various forms similar to the experiential rewards, such as money, positive feedback, loving touch, success, and so on (Morelli et al., 2015).

1.2 Social and Nonsocial Rewards Share Functionality and Recruit Overlapping Neural Circuits

Social rewards serve similar functions as nonsocial rewards—they enhance motivation and incentivize behaviors. Numerous studies have demonstrated the reinforcing nature of social rewards, showing that social rewards as feedback reliably facilitate stimulus-value learning and improve performance (Lin et al., 2012; Spreckelmeyer et al., 2009; Wang et al., 2017). Whether social rewards are as effective as monetary rewards remains an open question: while some studies suggest that monetary rewards exert a stronger incentive effect than social rewards (Lin et al., 2012; Spreckelmeyer et al., 2009), others indicate that social rewards can be equally effective, particularly for children, adolescents, and women (Spreckelmeyer et al., 2009; Wang et al., 2017; Demurie et al., 2011).

Notably, the neural processing of social rewards is similar to that of nonsocial rewards. Several studies have revealed substantial overlap in the neural representation of different reward types, particularly in the vmPFC and the ventral striatum (Sescousse et al., 2013; Lin et al., 2012; Gu et al., 2019; Izuma et al., 2008). These shared neuroanatomical foundations support the notion of a common neural circuit for reward processing. Nevertheless, specialized reward-processing regions also exist for different rewards. For instance, a meta-analysis by Sescousse et al. identified erotic-specific reward areas including the extrastriate body area and hypothalamus (Sescousse et al., 2013). Additionally, a direct experimental comparison by Izuma et al. found stronger medial prefrontal cortex (mPFC) activation when participants received a positive social evaluation than when they received a monetary reward (Izuma et al., 2008). While a specific set of regions may enable the evaluation of different rewards using a "common currency," additional brain regions may contribute to the perception and processing of specific types of social rewards. This common currency mechanism allows individuals to compare different types of options (e.g., money vs. social approval) on a common scale, which is essential for guiding decisions across diverse contexts (Ruff & Fehr, 2014).

2 Social Decision-Making

When making decisions in nonsocial contexts, evidence suggests that individuals compute the subjective value of each available option to guide their behavioral preference (Von Neumann & Morgenstern, 1944; Savage, 1972). The same principle applies to social decisions—such as those that influence others' outcomes. Just as the vmPFC tracks the value of food and money (O'Doherty, 2004), it is also involved in representing the value of social options (Cooper et al., 2012; Hare et al., 2010; Lockwood et al., 2024). For example, in a speed dating experiment, participants viewed photos of potential dating partners, and vmPFC activity correlated with subjective ratings of physical attractiveness (Cooper et al., 2012). Similarly, when deciding how much to donate to charitable organizations, vmPFC activity encoded a subjective value signal that scaled proportionally with the perceived deservingness of the charity (Hare et al., 2010). What is perhaps unique about social decision-making compared to other forms of decision-making is that, while making social decisions, the vmPFC receives input from brain regions thought to be more specialized for social cognition (Box 1).

Box 1: Social Neuroscience: What We Learned About the Social Brain
The term "social neuroscience" was first introduced in the early 1990s (Cacioppo & Berntson, 1992). Back then, it was defined broadly as the study of the neural basis of social behaviors. Various techniques and analysis tools have been applied to characterize the social brain, that is, to determine which structures of the brain are specifically involved in social processes compared to nonsocial processes. Early on, human lesion studies identified multiple brain structures without which humans cannot perform certain social functions, such as the amygdala in relation to fear recognition, the insula in relation to empathy, and the temporal pole in relation to naming people (Kennedy & Adolphs, 2012). Later, with the emergence of fMRI, by comparing brain activity in response to social and nonsocial stimuli or conditions, specific brain regions have been implicated in social processes including the fusiform face area (FFA) for face processing, or the TPJ and mPFC in mentalizing. Now, with the emergence of network neuroscience, it is increasingly appreciated that social functions are not implemented by one brain region alone, but by distributed networks (Bassett & Sporns, 2017). Multiple brain regions coactivate under certain cognitive process, thus forming an interconnected network. With this notion in mind, several core social brain networks have been identified (Fig. 2) (Kennedy & Adolphs, 2012):

- Amygdala network mainly includes amygdala and vmPFC/OFC, computes social saliency, triggers emotional response, and produces social affiliative behaviors.
- Areas associated with mentalizing include the TPJ, STS, dmPFC, PCC, and the temporal pole. These areas can be activated when inferring the mental states of others.
- Areas associated with empathy include the insula and ACC, which are activated when representing the feelings of others (e.g., social pain).
- Areas known to contain mirror neurons include the premotor cortex, IFG, and IPL. These regions are recruited when observing and simulating others' motor actions.

Despite the identification of several sets of brain areas associated with social computations, it is still an ongoing debate as to whether there exists dedicated neural circuitry to deal with social activities or whether social processes share the computations and neural substrates with other "nonsocial" processes (Ruff & Fehr, 2014; Lockwood et al., 2020).

A parsimonious albeit cynical view on human behavior is the "self-interest hypothesis," which posits that people make decisions purely based on their own material self-interest (Smith, 1976; Goodwin et al., 2018). However, extensive

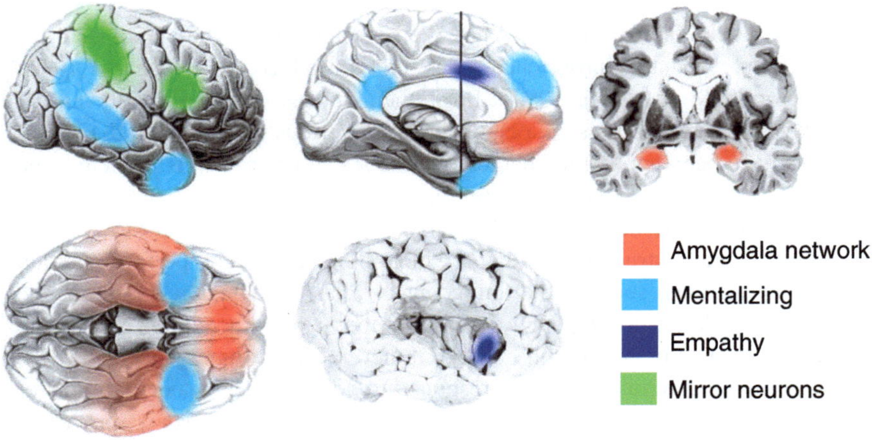

Fig. 2 The social brain networks. Brain areas implicated in four key social brain networks from the network perspective. (Reproduced from Kennedy & Adolphs, 2012)

evidence suggests that individuals are strongly motivated by other-regarding preferences (Fehr & Schmidt, 2003; Fehr & Fischbacher, 2003). In this section, we decompose these other-regarding, or "social preferences," through two common types of decisions: altruistic and fairness-driven decisions.

2.1 Altruistic Decision-Making

One fascinating aspect of human social life is that we care about others to the extent that we are willing to sacrifice our own benefits for their welfare—so-called altruistic behaviors (Batson & Powell, 2003).

Behavioral game theory has been widely used to illuminate and study social decision-making in the lab by simplifying complex social dilemmas into well-quantified and operationalized frameworks (Camerer & Hare, 2014) (Box 2). Out of the most extensively applied paradigms for studying altruistic decisions is the dictator game (DG), in which a "dictator" player unilaterally allocates a fixed amount of money between self and a "recipient" player (Guala & Mittone, 2010; Tusche & Bas, 2021). Even though the dictator player could easily maximize their own payoff by keeping all the money, empirical studies suggest that a considerable fraction of people chose to share—a meta study from Engel reported that, on average, dictators give away 28% of the total amount (Engel, 2011).

Box 2: Using Behavioral Game Theory to Study Social Interactions

Game theory in combination with simple economic games provides a framework for understanding real-life social dilemmas in a mathematically tractable manner (Camerer & Hare, 2014).

There are two broad categories of games: sequential and simultaneous games. In sequential games, players take turns to decide their actions. After the first player makes the decision, the second player knows that decision before their own decision is made. Many sequential games are applied to study strategic giving and to address how social preferences affect subjective utility. Table 1 shows three examples.

While many studies involve one-shot games (i.e., one participant interacts with different players in different rounds, thus their choices in the previous round won't affect the current opponent's behaviors), allowing repeated interactions adds the opportunity to study how reputations are built over time, and how people can learn about others' prosocial traits such as reciprocity and trustworthiness.

In simultaneous games, two (or more) players make decisions at the same time, and neither party has information about the other's moves before the decisions are revealed. Therefore, it is important to infer the choice of the opponent. Depending on the game outcomes (who gets how much when choosing what), sometimes players can be motivated to choose specific options. A useful class of models that infer the motivation is called "cognitive hierarchy" (CH) or level-k models. In the CH model, a player assumes other players to have at most (k-1)-th level of thinking, and that oneself is at the k-th level, thus iteratively optimizes own choice (e.g., a level-0 player may make random decisions). More details about the CH model can be found in (Camerer & Hare, 2014).

When simultaneous games are performed repeatedly, it is feasible to infer the other's decisions not only based on assumptions about their cognitive hierarchies but also by learning the personality, motives, and strategies through the decision histories (Hampton et al., 2008; Van Baar et al., 2021; Yoshida et al., 2010b). Several games are widely adapted to study repeated social interactions, especially to model cooperation and competition (Table 2).

Altruistic decisions have been characterized using formal models in which another's gain (Hutcherson et al., 2015; Tusche & Hutcherson, 2018) or loss (Hu et al., 2021) is integrated into one's own subjective value. In a modified DG experiment, Hutcherson et al. used functional MRI (fMRI) to examine participants' decisions to accept or reject proposed $100 splits between themselves and a stranger. Using a drift-diffusion model, the authors constructed decision values as a weighted sum of self- and other-payoff (Fig. 3a). fMRI data revealed distinct neural repre-

Table 1 Use sequential games to study strategic giving

Game	Dictator game	Ultimatum game	Trust game
Player 1 (P1)	Split x to keep y with self	Propose a split of x to keep y with self	Share z out of x with player 2
Player 2 (P2)	No available action	Accept or reject	Get $m \bullet z$, transfer y back to player 1
Payoff (P1, P2)	$(y, x - y)$	Accept: $(y, x - y)$ Reject: $(0, 0)$	$(x - z + y, m \bullet z - y)$
Used to study	Altruistic decision	Fairness-driven decision	Trust, trustworthiness learning
Real life example	Donation	Salary negotiation	Lend money to friend
Example studies	Hutcherson et al. (2015), Gao et al. (2018)	Holper et al. (2018), Qu et al. (2020)	King-Casas et al. (2005), Chang et al. (2010), Van Baar et al. (2019)

Comparison among three frequently used games: the dictator game, the ultimatum game, and the trust game

Table 2 Use simultaneous games to study cooperative and competitive behaviors

<table>
<tr><th>Game</th><th>Coordination game</th><th>Matching pennies</th><th>Stag hunt</th></tr>
<tr><td>Payoff matrix</td>
<td><table><tr><td></td><td>Left</td><td>Right</td></tr><tr><td>Left</td><td>1,1</td><td>0,0</td></tr><tr><td>Right</td><td>0,0</td><td>1,1</td></tr></table></td>
<td><table><tr><td></td><td>Head</td><td>Tail</td></tr><tr><td>Head</td><td>1,-1</td><td>-1,1</td></tr><tr><td>Tail</td><td>-1,1</td><td>1,-1</td></tr></table></td>
<td><table><tr><td></td><td>Stag</td><td>Hare</td></tr><tr><td>Stag</td><td>3,3</td><td>0,1</td></tr><tr><td>Hare</td><td>1,0</td><td>1,1</td></tr></table></td></tr>
<tr><td>Player motivations</td><td>Both players are motivated to match their choices</td><td>One player wants the choices to match, the other wants choices to mismatch</td><td>Stag is a better option, but hare is a safer option</td></tr>
<tr><td>Used to study</td><td>Cooperation</td><td>Competition</td><td>Cooperation, trust</td></tr>
<tr><td>Real-life example</td><td>Traffic coordination</td><td>Cyber security vs. hacker</td><td>Long-term collaboration vs. short-term benefit</td></tr>
<tr><td>Example studies</td><td>Van Baar et al. (2021)</td><td>Hampton et al. (2008), Van Baar et al. (2021), Seo et al. (2014)</td><td>Van Baar et al. (2021), Yoshida et al. (2010b)</td></tr>
</table>

Comparison among three frequently used games: the coordination game, the matching pennies game, and the stag hunt game

sentations of self-payoff (ventral striatum), other-payoff (temporoparietal junction, TPJ), and overall decision value (vmPFC).

Rather than increasing the other's payoff, Hu et al. examined altruistic decisions in a context where participants decided whether to forgo their own monetary gain to prevent a stranger from hearing a painful noise. Leveraging a value-based decision model, the authors found that participants integrated both self-cost and the other's suffering into their decisions; and that selfish and other-regarding motives were processed in the dACC and right inferior parietal lobule (IPL), respectively (Hu et al., 2021).

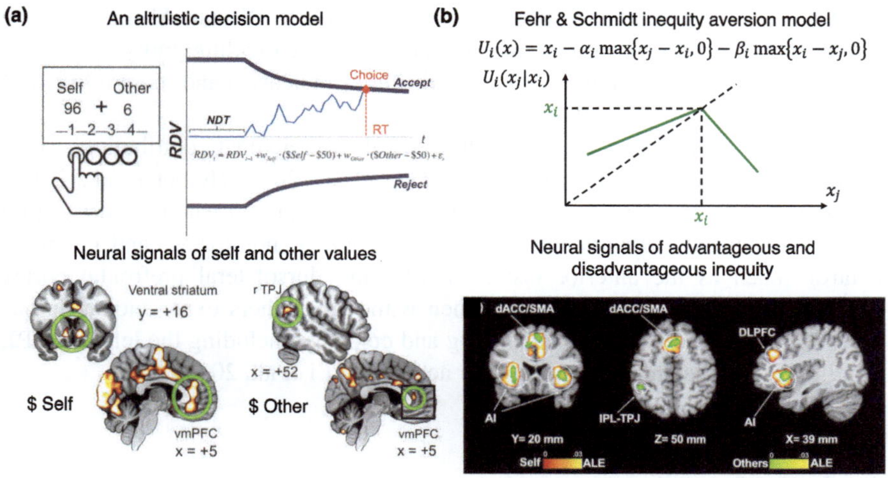

Fig. 3 Neurocomputations of value-based social decisions. (**a**) Altruistic decisions. Self-value and other-value are integrated in the decision signal and can be mapped to different brain regions. (Reproduced from Hutcherson et al., 2015). (**b**) Fairness-driven decisions. In the Fehr and Schmidt inequity aversion model, utility is a function of self-gain x_i and other-gain x_j. Conditioned on the self-gain, the utility reaches maximum when self-gain is equal to other-gain. Disadvantageous inequity and advantageous inequity trigger activations in partly overlapping regions. (Reproduced from Li et al., 2024). *vmPFC* ventromedial prefrontal cortex, *TPJ* temporoparietal junction, *AI* anterior insula, *dACC* dorsal anterior cingulate cortex, *SMA* supplementary motor area, *IPL* inferior parietal lobule, *DLPFC* dorsolateral prefrontal cortex, *ALE* activation likelihood estimation

2.2 Fairness-Driven Decisions

Imagine splitting $10 with a partner. The partner proposes taking $9, leaving you with $1. If you reject the proposal, both of you receive nothing. Will you accept?

In the above “Ultimatum Game” (Box 2), the optimal strategy is to accept any nonzero offer. However, in most cases, responders reject unfair offers (Güth et al., 1982), sacrificing their own benefit to punish the proposer’s greed. Fairness is a fundamental aspect of social life, and the brain responds accordingly: a study by Tricomi et al. found that both the ventral striatum and vmPFC exhibited stronger activation when a subject observes money transfer from a wealthy to a poor individual than the opposite direction, regardless of whether the individual was the sender or receiver (Tricomi et al., 2010).

To formally study how humans value fairness in decision-making, Fehr and Schmidt proposed the “inequity aversion” model to capture the willingness to “give up some material payoff to move in the direction of more equitable outcomes” (Fehr & Schmidt, 1999). Disadvantageous inequity (DI) occurs when an individual is worse off than another, whereas advantageous inequity (AI) occurs when the individual is better off. In the Fehr and Schmidt model, utility is calculated by subtracting both AI and DI from self-gain, with different weights (α for DI, β

for AI) (Fig. 3b). Importantly, people are more averse to DI than AI (i.e., $\alpha \geq \beta$). Interestingly, these weights are context-dependent: when feeling guilty toward the coplayer, one becomes more aversive to AI (i.e., α increases) and more tolerant of DI (i.e., β decreases) (Gao et al., 2018).

DI and AI have been successfully dissociated, with distinct contributions to the computation of an overall utility signal (Gao et al., 2018; Holper et al., 2018). According to a recent meta-analysis, when an individual is treated unfairly, brain regions associated with affective processing, conflict monitoring, and cognitive control—such as the anterior insula, dACC, and dorsolateral prefrontal cortex (dlPFC)—are recruited. In contrast, when witnessing others experiencing unfairness, brain areas involved in mentalizing and empathy, including the left IPL-TPJ, left dlPFC, and thalamus, show stronger activation (Li et al., 2024).

3 Social Learning

While interacting with the physical world, we associate stimuli and actions with outcomes and rewards. Through repeated interactions, we continuously update the associations based on discrepancies between expected and the actual outcome—a mechanism called "prediction error" (Joiner et al., 2017; Lockwood & Klein-Flügge, 2021).

Social learning is a specialized form of associative learning in which the target of learning is another individual, including their preferences, beliefs, and behaviors. A fundamental question concerns whether social learning relies on the same algorithm as nonsocial associative learning (i.e., a prediction error-based learning algorithm), and if so, where different types of social information are represented in the brain. We discuss this question by focusing on two broad types of social learning: learning about others and learning from others. The former helps us determine whom to interact with and how to interact with them, while the latter allows us to complete tasks more efficiently by observing others' actions.

3.1 Learning About Others

The ability to learn about others shapes how we cooperate, compete, and make decisions. Understanding others' personalities, preferences, and motives influences our strategies and responses across various contexts. The following subsections examine several critical aspects of social learning: how knowledge about others enhances our own outcomes through cooperation and competition, how it informs our expectations of others' reactions in novel social environments, and finally, how it shapes attitudes, preferences, and self-directed decisions.

3.1.1 Knowledge About Others Optimizes Cooperation and Competition

Learning about others helps us decide whom to cooperate with. Certain traits—trustworthiness, expertise, reputation, and generosity—are particularly valuable in cooperative (a.k.a. "prosocial") interactions and have thus been quantified using various decision-making tasks (King-Casas et al., 2005; Behrens et al., 2008; Chang et al., 2010; Boorman et al., 2013; Stanley, 2016).

Trustworthiness and reputation can be studied using the repeated trust game (Box 2), whereby an investor invests to a trustee to build reputation and the trustee shares the money back to gain trust. Chang et al. used this paradigm to model how an investor learns the trustworthiness of a trustee via a dynamic reinforcement learning approach (Chang et al., 2010). Meanwhile, King-Casas et al. examined how an investor's reputation is shaped based on their reciprocity in previous rounds, finding that trustees track unsigned prediction errors for the investor's reciprocity in the inferior frontal sulcus, superior frontal sulcus, thalamus, and inferior/superior colliculi (King-Casas et al., 2005). Another form of trustworthiness is expertise, operationalized as the accuracy of an individual's suggestions. In paradigms where a partner makes binary choices (or suggestions) between potentially profitable stimuli (e.g., an asset), participants have been found to track both the partner's predictions and their expertise through prediction errors, engaging brain regions such as the dorsomedial prefrontal cortex (dmPFC), middle temporal gyrus (MTG), superior temporal sulcus (STS), and TPJ (Behrens et al., 2008; Boorman et al., 2013). Finally, in an experiment where participants watched different individuals splitting money with others, a "generosity" prediction error—representing the difference between the expected and actual money split—was identified in the precuneus (Stanley, 2016).

Learning about others also helps us decide whom to compete with and how to compete effectively. Competition is fundamental to natural selection (Darwin & Kebler, 1859). When resources are limited, optimal competition strategies maximize gains (e.g., food, money, opportunities, social status) while minimizing investment.

To decide whom to compete with, individuals infer social dominance by observing previous competition interactions. Ligneul et al. had participants watch a player competing against multiple opponents. The player's social dominance status, defined by their rate of victory, was learned through a form of error signal related to the prediction of competitiveness. Both the competitive prediction error and the social dominance status were localized to the rostral medial prefrontal cortex (rmPFC) (Ligneul et al., 2016).

To decide how to compete, individuals mentalize their opponent's state of mind, learn their strategies, and finally react against them. For instance, Hampton et al. administered an "inspection game" (i.e., generalized "matching pennies game," Box 2, Fig. 4a) in which one player's gain leads to the other's loss (Hampton et al., 2008). Participants were found to use an "influence model," in which they inferred how their actions impacted the opponents' reactions and strategized accordingly. While mPFC tracked predicted future rewards for oneself, the posterior STS (pSTS)

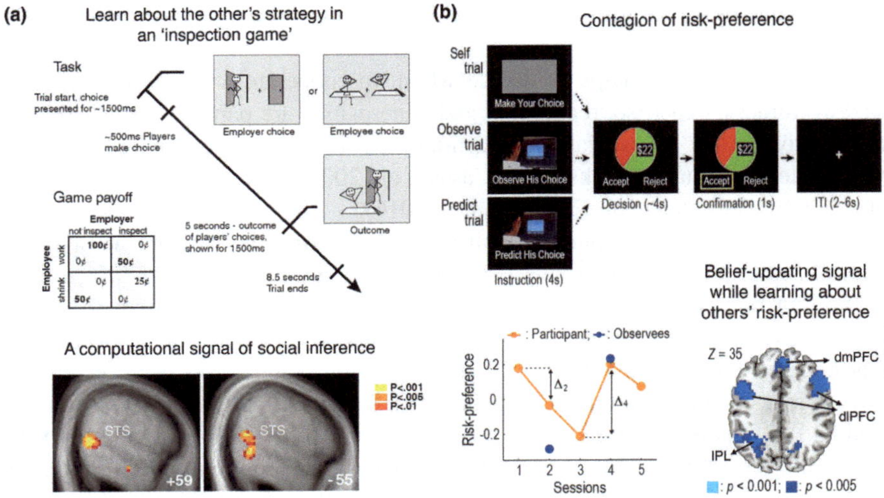

Fig. 4 Learn about others. (**a**) The "inspection game." Upper panel: Two interacting players were each individually given two action choices at the beginning of each trial. The "employer" could choose to inspect or not, and the "employee" could choose to work or shrink. After their choices, the actions of both players and the payoff (based on the payoff matrix) were shown. Lower panel: At the time of outcome, the influence update (i.e., how one's action influenced the opponent's reaction) of the inferred opponent's strategy shows significant correlations with activity in STS bilaterally. (Reproduced from Hampton et al., 2008. Copyright (2008) National Academy of Sciences). (**b**) Contagion of risk preference. Upper panel: On self-trials, participants decided whether to accept or reject a gamble (otherwise they took a guaranteed $10). On Observe trials, participants observed a choice of another person. On Predict trials, participants predicted the choice that the observee would make. Lower left panel: Change of an example participant's risk preference toward the observees. Lower right panel: Signals from the brain that significantly correlated with the belief updating signal while learning about the other's risk preference. *dmPFC* dorsomedial prefrontal cortex, *dlPFC* dorsolateral prefrontal cortex, *IPL* inferior parietal lobule. (Reproduced from Suzuki et al., 2016)

encoded the prediction error related to one's influence over the opponent (Hampton et al., 2008).

3.1.2 Knowledge About Others Can Generalize to New Contexts

Sometimes, our knowledge about others generalizes to broader contexts, allowing us to predict behavior in new settings. After learning a peer's preferences for various objects, one can successfully infer how much the peer would like a novel object. This new prediction relies on three components: the average preference within the population, the peer's preference for other objects, and the similarity between the new object and previously learned objects (Rosenblau et al., 2021).

Beyond preferences, it is also possible to predict others' future behaviors. One's past behaviors in a certain environment not only tell the actions per se but also reflect the causes—such as personality traits, desires, or attitudes that remain stable over time. A recent study by van Baar et al. modeled this process by examining how participants learned about opponents in a series of four economic games, each featuring different payoff structures and optimal strategies. Instead of focusing solely on the opponent's action history (e.g., cooperate, defect, etc.), participants generalized behavioral patterns to infer broader, stable traits including cooperativeness, greediness, risk-aversion, and utility maximization. While an opponent's actions varied depending on the specific game context, they consistently reflected these underlying traits (Van Baar et al., 2021). Consequently, knowing an opponent's desires and personalities, along with the structure of a new game, participants could effectively simulate the opponent's likely actions in a different context.

3.1.3 Knowledge About Others Can Influence Self-Decisions

Observing others' behaviors can influence our own actions. Emotion contagion is a well-known example—seeing someone laugh can make us laugh, even without knowing the cause. This contagion extends beyond emotions to other domains, such as risk-taking behavior (Albert et al., 2013; Izuma & Adolphs, 2013; Suzuki et al., 2016). Suzuki et al. found that observing others engage in risky gambles activates brain regions involved in learning the other's risk preferences, including the dlPFC, dmPFC, and IPL (Fig. 4b). Importantly, individuals adjusted their own risk preferences in the same direction as those they observed. Therefore, a risk-averse individual is more likely to invest in riskier gambles after watching a risk-seeking partner's choices (Suzuki et al., 2016). This phenomenon may explain why adolescents are more likely to engage in delinquent acts when in groups: observing the behavior of risk-seeking peers (as opposed to risk-averse adults) increases their willingness to take risks (Albert et al., 2013).

Another example concerns how another's valuation of an object can alter our own valuation. When individuals learn about another's preferences, the ventral striatum has been found to exhibit greater activity when their preferences align with others compared to when they diverge. Notably, individuals with stronger ventral striatum activation during preference agreement are more susceptible to social influence in shaping their opinions (Campbell-Meiklejohn et al., 2010). Interestingly, the direction of this influence depends on one's attitude toward the other person: people tend to adjust their preferences to align with those they like and to diverge from those they dislike (Izuma & Adolphs, 2013). Moreover, dmPFC activity increases when there is a mismatch between the object we like and the people we like—when the disliked others share the same preference with self, or when the liked others disagree with one's own preference (Izuma & Adolphs, 2013).

3.2 Learning from Others

In situations where direct interaction with the environment is costly or risky, acquiring information indirectly from others can be more efficient. Social learning provides an advantage by allowing individuals to optimize their decisions without firsthand experience. There are two general forms of social learning: observational learning and advice taking. The former relies on observing others' actions and consequences and does not require direct communication; the latter relies on cues or social communication and thus does not require observations.

3.2.1 Observational Learning

Observational learning is a broad category that varies in terms of what is known, observed, inferred, and learned. These variations lead to differences in the computational algorithms deployed and their neural correlates. Recent research has identified three key strategies of social learning (Charpentier & O'Doherty, 2021): vicarious reward learning, action imitation, and emulation. These three strategies differ in multiple aspects of the learning process (Table 3).

In vicarious reward learning, the learner observes another agent interact with the environment and tracks the agent's actions and outcomes. The underlying mechanism resembles direct experiential learning—the observer updates action or stimulus values by computing an observational reward prediction error (oRPE), which reflects the difference between the observee's actual outcome and the observer's expected outcome. Neuroimaging studies suggest that oRPE involves different neural circuits compared to experiential RPE (eRPE): while eRPE is frequently identified in the ventral striatum, oRPE is found in the vmPFC (Burke et al., 2010; Suzuki et al., 2012), rostrolateral prefrontal cortex (rlPFC) (Buritica, 2024), ACC (Hill et al., 2016; Lockwood et al., 2015, 2016), and dorsal striatum (Diaconescu et al., 2014), but not in the ventral striatum. This distinction suggests that individuals maintain different representations for self and others, even when using similar learning mechanisms. Indeed, observers encode separate representations of learning parameters (Buritica, 2024; Diaconescu et al., 2014), risk preferences (Suzuki et al., 2012), and stimulus values for themselves and others (Zhang & Gläscher, 2020). A recent study by Zhang and Gläscher explored this phenomenon by having participants play a gamble alongside four partners. Participants not only learned stimulus values from their choices but also tracked the reward histories and valuations of their partners. Similar to the finding that oRPE and eRPE are represented in different regions, here, vmPFC encoded self-value signals, whereas ACC encoded inferred value signals of others (Zhang & Gläscher, 2020).

Action imitation, or mimicry, refers to the copying of frequently observed actions without inferring the underlying goals or intentions. Imitation is perhaps the earliest form of observational learning expressed during child development: even before 1 year of age, infants begin to imitate adults' facial expressions, tongue movement,

Table 3 Three types of observational learning

	Vicarious reward learning	Action imitation	Emulation
Observe action?	Yes	Yes	Yes
Observe outcome?	Yes	Maybe	Maybe
What is learned	Action-outcome associations or stimulus-outcome association	Action of the observee	Knowledge about the environment + observee's mental states
Key computational variables	Observational reward prediction error (*oRPE*)	Action prediction error (*APE*)	Probability distributions of others' mental states (*m*) given action (*a*)
Example equation	$V_s(t) = V_s(t-1) + \alpha \bullet oRPE$ $oRPE = R_{obs} - V_s(t-1)$	$V_a(t) = V_a(t-1) + \alpha \bullet APE$ $APE = Action - V_a(t-1)$	$P(m\|a) = P(m) \bullet P(a\|m)$
Key neural correlates	vmPFC, ACC, dorsal striatum	PM, IPL, dmPFC, dlPFC	TPJ, STS, MPFC, ACC, etc.

Vicarious reward learning, action imitation, and emulation all involves observing the other's actions. However, they differ in many aspects, including whether the outcome is observed, what is learned, key computational variables, and key neural correlates. $V_s(t)$: estimation of the stimulus value at time t, $V_a(t)$: estimation of the action value at time t, α: learning rate, $P(s|a)$: posterior probability of the other's mental states given the observed action, $P(m)$: prior probability of the other's mental states, $P(a|m)$: likelihood of observing the action given a mental state

and interactions with objects (Abravanel et al., 1976; Field et al., 1982; Carpenter et al., 1998). This form of learning is suggested to be supported by mirror neurons (though see (Hickok, 2009) for an alternative perspective). Mirror neurons were first identified in the premotor cortex of macaque monkeys, which fire during an individual's own movements and when observing others' movements (di Pellegrino et al., 1992). In the human brain, areas proposed to contain mirror neurons include the ventral premotor cortex (PMv), inferior frontal gyrus (IFG), IPL, and intraparietal sulcus (IPS) (Yang & Hofmann, 2016). Computationally, during action imitation, individuals predict others' actions based on prior observations and update their knowledge by comparing expected actions with actual actions. As such, frequently chosen actions are positively reinforced. Note that action imitation does not require knowledge of action outcomes (Charpentier et al., 2020). A critical computation variable—action prediction error (APE)—has been detected in brain circuits partially overlapped with areas putatively containing mirror neurons, including premotor cortex, IPL, dmPFC, and dlPFC (Burke et al., 2010; Suzuki et al., 2012; Charpentier et al., 2020).

Emulation involves inferring an observee's goals, desires, and beliefs while observing them complete a task. Compared to imitation, emulation is thought to emerge later in development. At around age four to six, children begin to understand that others may hold beliefs different from their own. As they grow up, they develop the ability to manipulate others' beliefs through persuasion and deception (Frith & Frith, 2012). Emulation is more accurate and flexible than imitation but is also computationally more demanding. One commonly used framework for understanding emulation is Bayesian inference (Diaconescu et al., 2014; Charpentier et al., 2020; Baker et al., 2017; Collette et al., 2017; Charpentier et al., 2024). Under this framework, individuals estimate a posterior probability distribution of an observee's hidden mental states (e.g., goals, beliefs, desires) and/or environmental features. This posterior is updated trial-by-trial through Bayesian updating, where prior beliefs are weighted by new evidence (likelihood) to refine estimates. Finally, the inference can be translated to provide information relevant to the task at hand and guide one's decisions. For example, knowing a partner's reward preferences, one can learn about the environment (i.e., which option gives which reward) by observing the partner's choices and apply such knowledge to self-decisions (Charpentier et al., 2020; Collette et al., 2017; Wu et al., 2024). By tracking the accuracy of a partner's advice, one could infer the intention of the partner (i.e., truth or lie) and selectively make choices consistent or inconsistent with the advice (Diaconescu et al., 2014). Neuroimaging studies indicate that the mentalizing network, including the TPJ, STS, mPFC, and ACC, plays a crucial role in emulation. These brain regions appear to encode Bayesian learning update signals and are involved in simulating expected utilities from the observee's perspective (Charpentier et al., 2020; Collette et al., 2017).

3.2.2 Advice Taking

Sometimes, we receive direct advice from others about our choices (e.g., movie reviews, weather forecasts, stock recommendations); other times, we receive indirect advice through gossip. Compared to observational learning, advice taking transmits social information in a more reliable manner (Biele et al., 2011). A crucial decision in this process is determining when to take advice and when to trust our own judgment.

A classic paradigm for studying advice taking involves having participants first make an initial judgment (e.g., estimating the distance between two cities) and then providing them with a chance to revise their estimate based on external advice. The weight of advice (WOA) is calculated as the proportional shift in the judge's belief relative to the difference between the advice and initial judgment (Yaniv, 2004). People in general tend to weigh their own estimates more strongly than those of others, a phenomenon called "egocentric discounting" (Yaniv & Kleinberger, 2000). A recent meta-analysis pooling 129 independent datasets reported that the average WOA across studies is 39%, lower than the rational adjustment of 50% (Bailey et al., 2023; Larrick & Soll, 2006).

Choices are often biased regarding what advice to follow. Without objective proof, individuals prefer advice from certain sources. These biases include confirmation bias (discounting information that contradicts their own judgment) (Kappes et al., 2020), conformity bias (copying the majority decision) (Morgan et al., 2012; Leong & Zaki, 2018; Toyokawa & Gaissmaier, 2022), a preference for copying older individuals, and a preference for copying prestigious individuals (Laland, 2004; Brand et al., 2021). Perceived accuracy of advice, rather than actual accuracy, predicts whether people accept it (Bailey et al., 2023). Kappes et al. found that confirmation bias is associated with reduced neural sensitivity to the strength of others' opinions in the posterior medial prefrontal cortex (pMFC) only when others' opinions contradict one's own (Kappes et al., 2020); Meshi et al. found greater ventral striatum activity when participants expected advice from an expert rather than a novice (Meshi et al., 2012).

If individuals learn about advice quality through experience, this process can be formalized using reinforcement learning and Bayesian inference models (Behrens et al., 2008; Boorman et al., 2013; Diaconescu et al., 2014; Biele et al., 2011; Leong & Zaki, 2018). People update their estimation of an advisor's expertise by tracking prediction errors for advice accuracy (Boorman et al., 2013; Biele et al., 2011) or by incorporating the likelihood-based updates about an outcome (Behrens et al., 2008; Diaconescu et al., 2014; Leong & Zaki, 2018). A set of regions associated with mentalizing processes has been implicated in evaluating advice (e.g., dmPFC, IPL, medial temporal cortex, and IFG (Schilbach et al., 2013)) and in tracking relevant prediction errors (e.g., dmPFC, TPJ, STS, MTG (Behrens et al., 2008; Boorman et al., 2013)). Biele et al. found a specific neurocomputational mechanism during advice taking called an "outcome bonus"—when individuals follow recommended advice, reward prediction errors are updated with an additional bonus, regardless of whether the outcome is positive or negative. Consequently, good advice enhances learning (via a larger magnitude prediction error), while bad advice causes reduced learning (via a smaller magnitude of the prediction error). The septal area and left caudate have been identified as key neural regions encoding this outcome bonus signal (Biele et al., 2011).

3.3 Strategy Arbitrations During Social Learning

So far, we have reviewed the neuro-computational mechanisms of learning about others and learning from others. For experimental design purposes, these processes are often studied in isolation. Nevertheless, in real life, multiple strategies and learning processes can coexist within the same context. When different strategies are available, a higher-level controller may regulate which strategy to use. The process of evaluating which strategy is optimal in a given situation and then dynamically allocating control to a specific strategy or set of strategies is known as arbitration (O'Doherty et al., 2021).

In nonsocial experiential learning, research has shown that the brain tracks the reliability/uncertainty of different decision systems as environmental conditions change (O'Doherty et al., 2021; Dorfman & Gershman, 2019). According to the concept of reliability-based arbitration, a central arbitrator continuously monitors the reliability of each strategy and adaptively adjusts control of each system over behavior. The system with most reliable predictions exerts the greatest influence over decision-making (O'Doherty et al., 2021).

Over the past few years, the framework of arbitration has been applied to social learning as well. One example is arbitration between imitation and emulation during observational learning. Charpentier et al. found evidence to suggest that the brain uses the reliability of the emulation system to control the relative weighting of the two systems (Fig. 5a). When the emulation system generates more confident predictions, its reliability increases, and thus its prediction is favored in the final decision. This emulation reliability signal has been identified in the vlPFC, TPJ, and ACC (Charpentier et al., 2020).

Another example is arbitration between cooperation and competition. Whether to cooperate or compete is driven by social context (e.g., resource distribution, payoff structures) and the preferences or intentions of the interacting agents. Two studies have explored the computational basis of this arbitration: Pisauro et al. quantified cooperation and competition along a continuum and found that the TPJ encodes a social prediction error when predicting the other player's competitiveness, leading to changes in one's own tendency to compete (Pisauro et al., 2022). Philippe et al. modeled competition and cooperation as two expert systems and found that the difference in their reliabilities, tracked by the ventral striatum, mPFC, and posterior cingulate cortex (PCC), determines the direction of the arbitration (Philippe et al., 2024) (Fig. 5b).

Arbitration may also operate between social and experiential learning systems when different sources of information have varying levels of reliability. When

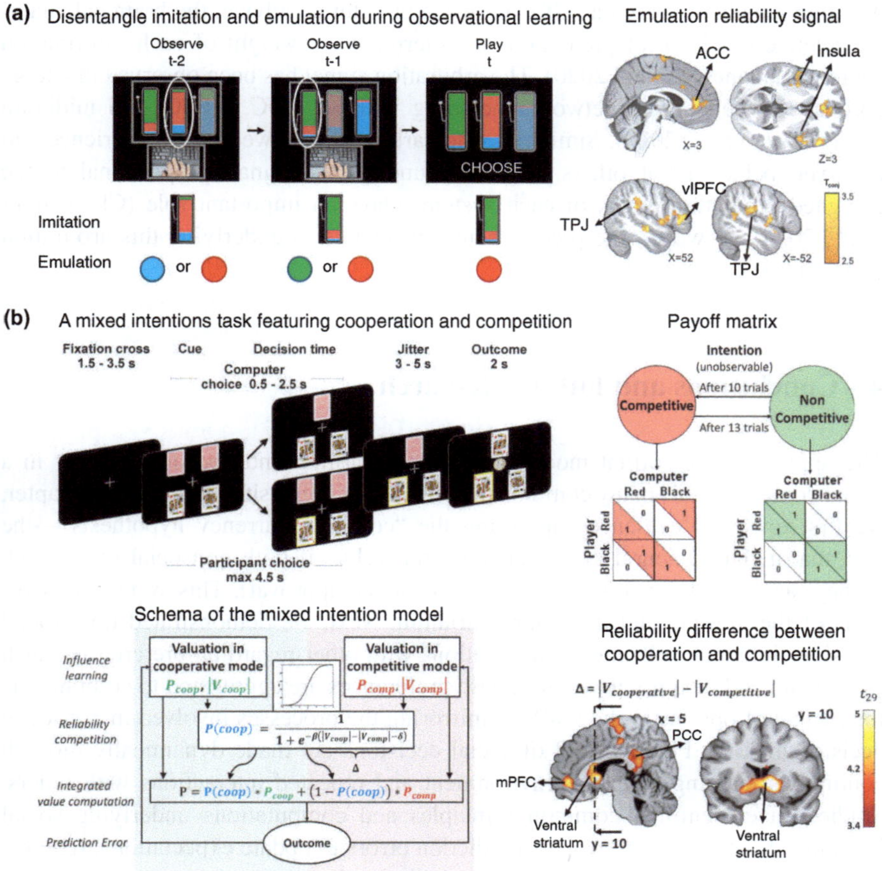

Fig. 5 Strategy arbitrations during social learning. (**a**) Arbitration between imitation and emulation during observational learning. Left panel: In an observational learning task, participants observed a partner make choices between slot machines that produce colored tokens (color proportions on the slot machine mapped to the probability of tokens). With imitation strategy, one would repeat the observed action (chosen slot machine), whereas with emulation strategy, one would make inferences about the partner's goals (which token is desired). (Reproduced from Wu et al., 2024). Right panel: Reliability of the emulation strategy correlated with the neural activities of several brain regions, including anterior cingulate cortex (ACC), insula, temporoparietal junction (TPJ), and ventrolateral prefrontal cortex (vlPFC). (Reproduced from Charpentier et al., 2020). (**b**) Arbitration between cooperation and competition in a mixed intention task. Upper left panel: In the task, the participant interacted with an opponent, and they both needed to choose one card (red or black). After their choices, the outcome was revealed. Upper right panel: The participant always wins if they choose the same card as the opponent, while the intention of the opponent may change between competitive (opponent wins if the cards do not match) and cooperative (opponent wins if the cards match). Lower left panel: Two influence models compute a value for choosing one specific target. A controller uses the difference between the absolute value of the value of each expert (i.e., reliability) to compute a probability that the other is cooperating. Then, the model weighs the values with the reliability and produces a final decision. Lower right: BOLD signal in ventral striatum, medial prefrontal cortex (mPFC), and posterior cingulate cortex (PCC) correlated with the difference in the reliability. (Reproduced from Philippe et al. 2024, licensed under CC BY 4.0 (https://creativecommons.org/licenses/by/4.0/). Changes were made to the original figures, including reorganization and annotation)

deciding whether to rely on self-experience or external advice, the brain calculates the relative precision of predictions to determine the weight of each information source (Diaconescu et al., 2020). The arbitration signal has been observed in a large cortical and subcortical network, including vmPFC, OFC, vlPFC, and midbrain (Diaconescu et al., 2020). Similarly, when arbitrating between self-experience and observed behaviors of others, a relative uncertainty signal, proportional to the unsigned prediction errors of each system, plays an important role (Charpentier et al., 2024). However, the precise neural mechanisms underlying this arbitration remain to be determined.

4 Conclusions and Future Research

The application of formal models to explain learning and decision making in a social context reveals many commonalities with nonsocial situations. In this chapter, we first reviewed evidence supporting the "common currency hypothesis"—the idea that a shared neural circuit encodes the value of both nonsocial (e.g., food, money) and social rewards (e.g., face, love, social approval). This system involves parts of the vmPFC and the ventral striatum. Next, we demonstrated how social decisions integrate self-interest motivations with other-regarding preferences, such as altruism and inequity aversion. These preferences are aggregated to compute the final value of options in the vmPFC, mirroring the processes involved in nonsocial decision-making. Finally, a lot of social decisions are made dynamically through continuous learning from the environment and repeated interactions with others. Studies have identified common principles and computations underlying social learning, showing that it involves prediction errors to update expectations, requires weighted attribute integration when assigning values to options, and allows for arbitration between multiple learning and decision systems to optimize behavior. Depending on the specific context and goal of learning, there is feature-selective involvement of various brain regions. For instance, brain regions implicated in mentalizing, including the mPFC, TPJ, STS, etc. track prediction error signals while learning about others' personality traits, infer others' goals and intentions during emulation learning, and evaluate the credibility of external advice. Regions of the human brain that have been found to contain mirror neurons in nonhuman primates (Rizzolatti & Craighero, 2004) facilitate imitation learning by tracking others' actions, whereas multiple reward-related areas encode observational reward prediction errors, enabling vicarious learning in a way similar to self-experiential reward learning.

Beyond these existing findings, increasing attention is being given to individual differences in social processes. While a lot of studies have developed experimental paradigms and computational models to quantify how social information is processed, valued, and integrated over time, an imminent future direction involves applying these models to characterize heterogeneity across populations. For example, Wu et al. utilized normative models to examine individuals' motives and preferences in social interactions—whether one prioritizes fairness, efficiency, or self-interest, as well as the context-dependency of the motives. Such research holds the promise for informing policy-making and promoting prosocial behavior at a societal scale (Wu et al., 2024b).

On the other hand, model-based analyses offer precision in distinguishing typical versus atypical social processes, extending into the psychiatric domain. While some individuals successfully perceive social information and engage in social behavior, others struggle to achieve desired social outcomes, such as those with autism spectrum disorder (ASD) or social anxiety disorder. By quantifying various aspects of social processing, including sensitivity to social reward/feedback, perspective-taking, and social information gathering, researchers can potentially identify different possible causes of an aberrant social behavior performance problem (Woodcock et al., 2020; Moran et al., 2011; Rosenthal et al., 2019; Yoshida et al., 2010a). Ultimately, this approach may facilitate personalized clinical assessment and intervention strategies.

References

Abrams, D. A., Chen, T., Odriozola, P., Cheng, K. M., Baker, A. E., Padmanabhan, A., Ryali, S., Kochalka, J., Feinstein, C., & Menon, V. (2016). Neural circuits underlying mother's voice perception predict social communication abilities in children. *Proceedings of the National Academy of Sciences of the United States of America, 113*, 6295–6300.

Abrams, D. A., Mistry, P. K., Baker, A. E., Padmanabhan, A., & Menon, V. (2022). A neurodevelopmental shift in reward circuitry from mother's to nonfamilial voices in adolescence. *The Journal of Neuroscience, 42*, 4164–4173.

Abravanel, E., Levan-Goldschmidt, E., & Stevenson, M. B. (1976). Action imitation: The early phase of infancy. *Child Development, 47*, 1032–1044.

Aharon, I., Etcoff, N., Ariely, D., Chabris, C. F., O'Connor, E., & Breiter, H. C. (2001). Beautiful faces have variable reward value: fMRI and Behavioral evidence. *Neuron, 32*, 537–551.

Albert, D., Chein, J., & Steinberg, L. (2013). Peer influences on adolescent decision making. *Current Directions in Psychological Science, 22*, 114–120.

Aron, A., Fisher, H., Mashek, D. J., Strong, G., Li, H., & Brown, L. L. (2005). Reward, motivation, and emotion systems associated with early-stage intense romantic love. *Journal of Neurophysiology, 94*, 327–337.

Bailey, P. E., Leon, T., Ebner, N. C., Moustafa, A. A., & Weidemann, G. (2023). A meta-analysis of the weight of advice in decision-making. *Current Psychology, 42*, 24516–24541.

Baker, C. L., Jara-Ettinger, J., Saxe, R., & Tenenbaum, J. B. (2017). Rational quantitative attribution of beliefs, desires and percepts in human mentalizing. *Nature Human Behaviour, 1*, 1–10.

Bassett, D. S., & Sporns, O. (2017). Network neuroscience. *Nature Neuroscience, 20*, 353–364.

Batson, C. D., & Powell, A. A. (2003). Altruism and prosocial behavior. In *Handbook of psychology: Personality and social psychology* (Vol. 5, pp. 463–484). Wiley.

Behrens, T. E. J., Hunt, L. T., Woolrich, M. W., & Rushworth, M. F. S. (2008). Associative learning of social value. *Nature, 456*, 245–249.

Biele, G., Rieskamp, J., Krugel, L. K., & Heekeren, H. R. (2011). The neural basis of following advice. *PLoS Biology, 9*, e1001089.

Boorman, E. D., O'Doherty, J. P., Adolphs, R., & Rangel, A. (2013). The Behavioral and neural mechanisms underlying the tracking of expertise. *Neuron, 80*, 1558–1571.

Brand, C. O., Mesoudi, A., & Morgan, T. J. H. (2021). Trusting the experts: The domain-specificity of prestige-biased social learning. *PLoS One, 16*, e0255346.

Buritica, J. M. R. (2024). *Observational reinforcement learning in children and young adults*. NPJ Science of Learning.

Burke, C. J., Tobler, P. N., Baddeley, M., & Schultz, W. (2010). Neural mechanisms of observational learning. *Proceedings of the National Academy of Sciences of the United States of America, 107*, 14431–14436.

Cacioppo, J. T., & Berntson, G. G. (1992). Social psychological contributions to the decade of the brain: Doctrine of multilevel analysis. *American Psychologist, 47*, 1019–1028.

Camerer, C. F., & Hare, T. A. (2014). The neural basis of strategic choice. In *Neuroeconomics* (pp. 479–492). Elsevier.

Campbell-Meiklejohn, D. K., Bach, D. R., Roepstorff, A., Dolan, R. J., & Frith, C. D. (2010). How the opinion of others affects our valuation of objects. *Current Biology, 20*, 1165–1170.

Carpenter, M., Akhtar, N., & Tomasello, M. (1998). Fourteen- through 18-month-old infants differentially imitate intentional and accidental actions. *Infant Behavior and Development, 21*, 315–330.

Chang, L. J., Doll, B. B., Van't Wout, M., Frank, M. J., & Sanfey, A. G. (2010). Seeing is believing: Trustworthiness as a dynamic belief. *Cognitive Psychology, 61*, 87–105.

Charpentier, C. J., & O'Doherty, J. P. (2021). Computational approaches to mentalizing during observational learning and strategic social interactions. In M. Gilead & K. N. Ochsner (Eds.), *The neural basis of Mentalizing* (pp. 489–501). Springer.

Charpentier, C. J., Iigaya, K., & O'Doherty, J. P. (2020). A neuro-computational account of arbitration between choice imitation and goal emulation during human observational learning. *Neuron, 106*, 687–699.e7.

Charpentier, C. J., Wu, Q., Min, S., Ding, W., Cockburn, J., & O'Doherty, J. P. (2024). Heterogeneity in strategy use during arbitration between experiential and observational learning. *Nature Communications, 15*, 4436.

Collette, S., Pauli, W. M., Bossaerts, P., & O'Doherty, J. (2017). Neural computations underlying inverse reinforcement learning in the human brain. *eLife, 6*, e29718.

Cooper, J. C., Dunne, S., Furey, T., & O'Doherty, J. P. (2012). Dorsomedial prefrontal cortex mediates rapid evaluations predicting the outcome of romantic interactions. *The Journal of Neuroscience, 32*, 15647–15656.

Darwin, C., & Kebler, L. (1859) On the origin of species by means of natural selection, or, The preservation of favoured races in the struggle for life. J. Murray.

Demurie, E., Roeyers, H., Baeyens, D., & Sonuga-Barke, E. (2011). Common alterations in sensitivity to type but not amount of reward in ADHD and autism spectrum disorders. *Journal of Child Psychology and Psychiatry, 52*, 1164–1173.

di Pellegrino, G., Fadiga, L., Fogassi, L., Gallese, V., & Rizzolatti, G. (1992). Understanding motor events: A neurophysiological study. *Experimental Brain Research, 91*, 176–180.

Diaconescu, A. O., Mathys, C., Weber, L. A. E., Daunizeau, J., Kasper, L., Lomakina, E. I., Fehr, E., & Stephan, K. E. (2014). Inferring on the intentions of others by hierarchical Bayesian learning. *PLoS Computational Biology, 10*, e1003810.

Diaconescu, A. O., Stecy, M., Kasper, L., Burke, C. J., Nagy, Z., Mathys, C., & Tobler, P. N. (2020). Neural arbitration between social and individual learning systems. *eLife, 9*, e54051.

Dorfman, H. M., & Gershman, S. J. (2019). Controllability governs the balance between Pavlovian and instrumental action selection. *Nature Communications, 10*, 5826.

Engel, C. (2011). Dictator games: A meta study. *Experimental Economics, 14*, 583–610.

Fareri, D. S., & Delgado, M. R. (2014). Social rewards and social networks in the human brain. *The Neuroscientist, 20*, 387–402.

Fehr, E., & Fischbacher, U. (2003). The nature of human altruism. *Nature, 425*, 785–791.

Fehr, E., & Schmidt, K. M. (1999). A theory of fairness, competition, and cooperation*. *The Quarterly Journal of Economics, 114*, 817–868.

Fehr, E., & Schmidt, K. M. (2003). Theories of fairness and reciprocity—Evidence and economic applications. In M. Dewatripont, L. P. Hansen, & S. J. Turnovsky (Eds.), *Advances in economics and econometrics, econometric society monographs, eighth world congress* (Vol. 1, pp. 208–257).

Field, T. M., Woodson, R., Greenberg, R., & Cohen, D. (1982). Discrimination and imitation of facial expression by neonates. *Science, 218*, 179–181.

Friston, K. J. (2009). Modalities, modes, and models in functional neuroimaging. *Science, 326*, 399–403.

Frith, C. D., & Frith, U. (2012). Mechanisms of social cognition. *Annual Review of Psychology, 63*, 287–313.

Gao, X., Yu, H., Sáez, I., Blue, P. R., Zhu, L., Hsu, M., & Zhou, X. (2018). Distinguishing neural correlates of context-dependent advantageous- and disadvantageous-inequity aversion. *Proceedings of the National Academy of Sciences of the United States of America*, 115.

Goodwin, N., Harris, J. M., Nelson, J. A., Rajkarnikar, P. J., Roach, B., & Torras, M. (2018). *Microeconomics in context*. Routledge.

Gu, R., Huang, W., Camilleri, J., Xu, P., Wei, P., Eickhoff, S. B., & Feng, C. (2019). Love is analogous to money in human brain: Coordinate-based and functional connectivity meta-analyses of social and monetary reward anticipation. *Neuroscience & Biobehavioral Reviews, 100*, 108–128.

Guala, F., & Mittone, L. (2010). Paradigmatic experiments: The dictator game. *Journal of Behavioral and Experimental Economics, 39*, 578–584.

Güth, W., Schmittberger, R., & Schwarze, B. (1982). An experimental analysis of ultimatum bargaining. *Journal of Economic Behavior & Organization, 3*, 367–388.

Hampton, A. N., Bossaerts, P., & O'Doherty, J. P. (2008). Neural correlates of mentalizing-related computations during strategic interactions in humans. *Proceedings of the National Academy of Sciences of the United States of America, 105*, 6741–6746.

Hare, T. A., Camerer, C. F., Knoepfle, D. T., O'Doherty, J. P., & Rangel, A. (2010). Value computations in ventral medial prefrontal cortex during charitable decision making incorporate input from regions involved in social cognition. *Journal of Neuroscience, 30*, 583–590.

Hayden, B. Y., Parikh, P. C., Deaner, R. O., & Platt, M. L. (2007). Economic principles motivating social attention in humans. *Proceedings of the Royal Society B, 274*, 1751–1756.

Hickok, G. (2009). Eight problems for the Mirror Neuron theory of action understanding in monkeys and humans. *Journal of Cognitive Neuroscience, 21*, 1229–1243.

Hill, M. R., Boorman, E. D., & Fried, I. (2016). Observational learning computations in neurons of the human anterior cingulate cortex. *Nature Communications, 7*, 12722.

Holper, L., Burke, C. J., Fausch, C., Seifritz, E., & Tobler, P. N. (2018). Inequality signals in dorsolateral prefrontal cortex inform social preference models. *Social Cognitive and Affective Neuroscience, 13*, 513–524.

Hu, J., Hu, Y., Li, Y., & Zhou, X. (2021). Computational and neurobiological substrates of cost-benefit integration in altruistic helping decision. *The Journal of Neuroscience, 41*, 3545–3561.

Hutcherson, C. A., Bushong, B., & Rangel, A. (2015). A Neurocomputational model of altruistic choice and its implications. *Neuron, 87*, 451–462.

Izuma, K., & Adolphs, R. (2013). Social manipulation of preference in the human brain. *Neuron, 78*, 563–573.

Izuma, K., Saito, D. N., & Sadato, N. (2008). Processing of social and monetary rewards in the human striatum. *Neuron, 58*, 284–294.

Joiner, J., Piva, M., Turrin, C., & Chang, S. W. C. (2017). Social learning through prediction error in the brain. *NPJ Science of Learning, 2*, 8.

Kappes, A., Harvey, A. H., Lohrenz, T., Montague, P. R., & Sharot, T. (2020). Confirmation bias in the utilization of others' opinion strength. *Nature Neuroscience, 23*, 130–137.

Kennedy, D. P., & Adolphs, R. (2012). The social brain in psychiatric and neurological disorders. *Trends in Cognitive Sciences, 16*, 559–572.

King-Casas, B., Tomlin, D., Anen, C., Camerer, C. F., Quartz, S. R., & Montague, P. R. (2005). Getting to know you: Reputation and trust in a two-person economic exchange. *Science, 308*, 78–83.

Kohls, G., Perino, M. T., Taylor, J. M., Madva, E. N., Cayless, S. J., Troiani, V., Price, E., Faja, S., Herrington, J. D., & Schultz, R. T. (2013). The nucleus accumbens is involved in both the pursuit of social reward and the avoidance of social punishment. *Neuropsychologia, 51*, 2062–2069.

Laland, K. N. (2004). Social learning strategies. *Learning & Behavior, 32*, 4–14.

Larrick, R. P., & Soll, J. B. (2006). Intuitions about combining opinions: Misappreciation of the avergaging principle. *Management Science, 52*, 111–127.

Leong, Y. C., & Zaki, J. (2018). Unrealistic optimism in advice taking: A computational account. *Journal of Experimental Psychology: General, 147*, 170–189.

Li, Q., Lai, X., Li, T., Madsen, K. H., Xiao, J., Hu, K., Feng, C., Fu, D., & Liu, X. (2024). Brain responses to self- and other- unfairness under resource distribution context: Meta-analysis of fMRI studies. *NeuroImage, 297*, 120707.

Ligneul, R., Obeso, I., Ruff, C. C., & Dreher, J.-C. (2016). Dynamical representation of dominance relationships in the human rostromedial prefrontal cortex. *Current Biology, 26*, 3107–3115.

Lin, A., Adolphs, R., & Rangel, A. (2012). Social and monetary reward learning engage overlapping neural substrates. *Social Cognitive and Affective Neuroscience, 7*, 274–281.

Lockwood, P. L. (2016). The anatomy of empathy: Vicarious experience and disorders of social cognition. *Behavioural Brain Research, 311*, 255–266.

Lockwood, P. L., & Klein-Flügge, M. C. (2021). Computational modelling of social cognition and behaviour—A reinforcement learning primer. *Social Cognitive and Affective Neuroscience, 16*, 761–771.

Lockwood, P. L., Apps, M. A. J., Roiser, J. P., & Viding, E. (2015). Encoding of vicarious reward prediction in anterior cingulate cortex and relationship with trait empathy. *The Journal of Neuroscience, 35*, 13720–13727.

Lockwood, P. L., Apps, M. A. J., Valton, V., Viding, E., & Roiser, J. P. (2016). Neurocomputational mechanisms of prosocial learning and links to empathy. *Proceedings of the National Academy of Sciences, 113*, 9763–9768.

Lockwood, P. L., Apps, M. A. J., & Chang, S. W. C. (2020). Is there a 'social' brain? Implementations and algorithms. *Trends in Cognitive Sciences, 24*, 802–813.

Lockwood, P. L., Cutler, J., Drew, D., Abdurahman, A., Jeyaretna, D. S., Apps, M. A. J., Husain, M., & Manohar, S. G. (2024). Human ventromedial prefrontal cortex is necessary for prosocial motivation. *Nature Human Behaviour, 8*, 1403–1416.

Maslow, A. H. (1943). A theory of human motivation. *Psychological Review, 50*, 370–396.

Masten, C. L., Eisenberger, N. I., Borofsky, L. A., Pfeifer, J. H., McNealy, K., Mazziotta, J. C., & Dapretto, M. (2009). Neural correlates of social exclusion during adolescence: Understanding the distress of peer rejection. *Social Cognitive and Affective Neuroscience, 4*, 143–157.

Meshi, D., Biele, G., Korn, C. W., & Heekeren, H. R. (2012). How expert advice influences decision making. *PLoS One*, 7.

Mielacher, C., Scheele, D., Kiebs, M., Schmitt, L., Dellert, T., Philipsen, A., Lamm, C., & Hurlemann, R. (2024). Altered reward network responses to social touch in major depression. *Psychological Medicine, 54*, 308–316.

Moran, J. M., Young, L. L., Saxe, R., Lee, S. M., O'Young, D., Mavros, P. L., & Gabrieli, J. D. (2011). Impaired theory of mind for moral judgment in high-functioning autism. *Proceedings of the National Academy of Sciences of the United States of America, 108*, 2688–2692.

Morelli, S. A., Sacchet, M. D., & Zaki, J. (2015). Common and distinct neural correlates of personal and vicarious reward: A quantitative meta-analysis. *NeuroImage, 112*, 244–253.

Morgan, T. J. H., Rendell, L. E., Ehn, M., Hoppitt, W., & Laland, K. N. (2012). The evolutionary basis of human social learning. *Proceedings of the Royal Society B, 279*, 653–662.

Mühlberger, A., Wieser, M. J., Gerdes, A. B. M., Frey, M. C. M., Weyers, P., & Pauli, P. (2011). Stop looking angry and smile, please: Start and stop of the very same facial expression differentially activate threat- and reward-related brain networks. *Social Cognitive and Affective Neuroscience, 6*, 321–329.

O'Doherty, J. P. (2004). Reward representations and reward-related learning in the human brain: Insights from neuroimaging. *Current Opinion in Neurobiology, 14*, 769–776.

O'Doherty, J., Winston, J., Critchley, H., Perrett, D., Burt, D. M., & Dolan, R. J. (2003). Beauty in a smile: The role of medial orbitofrontal cortex in facial attractiveness. *Neuropsychologia, 41*, 147–155.

O'Doherty, J. P., Lee, S. W., Tadayonnejad, R., Cockburn, J., Iigaya, K., & Charpentier, C. J. (2021). Why and how the brain weights contributions from a mixture of experts. *Neuroscience and Biobehavioral Reviews, 123*, 14–23.

Panksepp, J. (1998). *Affective neuroscience: The foundations of human and animal emotions*. Oxford University Press.

Philippe, R., Janet, R., Khalvati, K., Rao, R. P. N., Lee, D., & Dreher, J.-C. (2024). Neurocomputational mechanisms involved in adaptation to fluctuating intentions of others. *Nature Communications, 15*, 3189.

Pisauro, M. A., Fouragnan, E. F., Arabadzhiyska, D. H., Apps, M. A. J., & Philiastides, M. G. (2022). Neural implementation of computational mechanisms underlying the continuous trade-off between cooperation and competition. *Nature Communications, 13*, 6873.

Poldrack, R. A., Mumford, J. A., & Nichols, T. E. (2024). *Handbook of functional MRI data analysis*. Cambridge University Press.

Qu, C., Hu, Y., Tang, Z., Derrington, E., & Dreher, J.-C. (2020). Neurocomputational mechanisms underlying immoral decisions benefiting self or others. *Social Cognitive and Affective Neuroscience, 15*, 135–149.

Rizzolatti, G., & Craighero, L. (2004). The mirror-neuron system. *Annual Review of Neuroscience, 27*, 169–192.

Rolls, E. T., Critchley, H. D., Browning, A. S., Hernadi, I., & Lenard, L. (1999). Responses to the sensory properties of fat of neurons in the primate orbitofrontal cortex. *The Journal of Neuroscience, 19*, 1532–1540.

Rolls, E. T., O'Doherty, J., Kringelbach, M. L., Francis, S., Bowtell, R., & McGlone, F. (2003). Representations of pleasant and painful touch in the human orbitofrontal and cingulate cortices. *Cerebral Cortex, 13*, 308–317.

Rosenblau, G., Korn, C. W., Dutton, A., Lee, D., & Pelphrey, K. A. (2021). Neurocognitive mechanisms of social inferences in typical and autistic adolescents. *Biological Psychiatry: Cognitive Neuroscience and Neuroimaging, 6*, 782–791.

Rosenthal, I. A., Hutcherson, C. A., Adolphs, R., & Stanley, D. A. (2019). Deconstructing theory-of-mind impairment in high-functioning adults with autism. *Current Biology, 29*, 513–519.e6.

Ruff, C. C., & Fehr, E. (2014). The neurobiology of rewards and values in social decision making. *Nature Reviews. Neuroscience, 15*, 549–562.

Savage, L. J. (1972). *The foundations of statistics*. Courier Corporation.

Schilbach, L., Eickhoff, S. B., Schultze, T., Mojzisch, A., & Vogeley, K. (2013). *To you I am listening:* Perceived competence of advisors influences judgment and decision-making via recruitment of the amygdala. *Social Neuroscience, 8*, 189–202.

Seo, H., Cai, X., Donahue, C. H., & Lee, D. (2014). Neural correlates of strategic reasoning during competitive games. *Science, 346*, 340–343.

Sescousse, G., Redouté, J., & Dreher, J.-C. (2010). The architecture of reward value coding in the human orbitofrontal cortex. *Journal of Neuroscience, 30*, 13095–13104.

Sescousse, G., Caldú, X., Segura, B., & Dreher, J.-C. (2013). Processing of primary and secondary rewards: A quantitative meta-analysis and review of human functional neuroimaging studies. *Neuroscience & Biobehavioral Reviews, 37*, 681–696.

Sherman, L. E., Payton, A. A., Hernandez, L. M., Greenfield, P. M., & Dapretto, M. (2016). The power of the like in adolescence. *Psychological Science, 27*, 1027–1035.

Sherman, L. E., Hernandez, L. M., Greenfield, P. M., & Dapretto, M. (2018). What the brain 'likes': Neural correlates of providing feedback on social media. *Social Cognitive and Affective Neuroscience, 13*, 699–707.

Smith, A. (1976). In R. H. Campbell, A. S. Skinner, & W. B. Todd (Eds.), *An inquiry into the nature and causes of the wealth of nations*. Oxford University Press.

Spreckelmeyer, K. N., Krach, S., Kohls, G., Rademacher, L., Irmak, A., Konrad, K., Kircher, T., & Gründer, G. (2009). Anticipation of monetary and social reward differently activates mesolimbic brain structures in men and women. *Social Cognitive and Affective Neuroscience, 4*, 158–165.

Stanley, D. A. (2016). Getting to know you: General and specific neural computations for learning about people. *Social Cognitive and Affective Neuroscience, 11*, 525–536.

Stoléru, S., Fonteille, V., Cornélis, C., Joyal, C., & Moulier, V. (2012). Functional neuroimaging studies of sexual arousal and orgasm in healthy men and women: A review and meta-analysis. *Neuroscience & Biobehavioral Reviews, 36*, 1481–1509.

Strathearn, L., Li, J., Fonagy, P., & Montague, P. R. (2008). What's in a smile? Maternal brain responses to infant facial cues. *Pediatrics, 122*, 40–51.

Suzuki, S., Harasawa, N., Ueno, K., Gardner, J. L., Ichinohe, N., Haruno, M., Cheng, K., & Nakahara, H. (2012). Learning to simulate others' decisions. *Neuron, 74*, 1125–1137.

Suzuki, S., Jensen, E. L. S., Bossaerts, P., & O'Doherty, J. P. (2016). Behavioral contagion during learning about another agent's risk-preferences acts on the neural representation of decision-risk. *Proceedings of the National Academy of Sciences of the United States of America, 113*, 3755–3760.

Toyokawa, W., & Gaissmaier, W. (2022). Conformist social learning leads to self-organised prevention against adverse bias in risky decision making. *eLife, 11*, e75308.

Tricomi, E., Rangel, A., Camerer, C. F., & O'Doherty, J. P. (2010). Neural evidence for inequality-averse social preferences. *Nature, 463*, 1089–1091.

Tusche, A., & Bas, L. M. (2021). Neurocomputational models of altruistic decision-making and social motives: Advances, pitfalls, and future directions. *WIREs Cognitive Science*, 12.

Tusche, A., & Hutcherson, C. A. (2018). Cognitive regulation alters social and dietary choice by changing attribute representations in domain-general and domain-specific brain circuits. *eLife, 7*, e31185.

Van Baar, J. M., Chang, L. J., & Sanfey, A. G. (2019). The computational and neural substrates of moral strategies in social decision-making. *Nature Communications, 10*, 1483.

Van Baar, J. M., Nassar, M. R., Deng, W., & FeldmanHall, O. (2021). Latent motives guide structure learning during adaptive social choice. *Nature Human Behaviour, 6*, 404–414.

Von Neumann, J., & Morgenstern, O. (1944). *Theory of games and economic behavior*. Princeton University Press.

Wang, D., Liu, T., & Shi, J. (2017). Development of monetary and social reward processes. *Scientific Reports, 7*, 11128.

Williams, E. H., Bilbao-Broch, L., Downing, P. E., & Cross, E. S. (2020). Examining the value of body gestures in social reward contexts. *NeuroImage, 222*, 117276.

Woodcock, K. A., Cheung, C., González Marx, D., & Mandy, W. (2020). Social decision making in autistic adolescents: The role of theory of mind, executive functioning and emotion regulation. *Journal of Autism and Developmental Disorders, 50*, 2501–2512.

Wu, Q., Oh, S., Tadayonnejad, R., Feusner, J. D., Cockburn, J., O'Doherty, J. P., & Charpentier, C. J. (2024). Individual differences in autism-like traits are associated with reduced goal emulation in a computational model of observational learning. *Nature Mental Health, 2*(9), 1032–1044.

Wu, X., Ren, X., Liu, C., & Zhang, H. (2024b). The motive cocktail in altruistic behaviors. *Nature Computational Science, 4*, 659–676.

Yang, J., & Hofmann, J. (2016). Action observation and imitation in autism spectrum disorders: An ALE meta-analysis of fMRI studies. *Brain Imaging and Behavior, 10*, 960–969.

Yaniv, I. (2004). Receiving other people's advice: Influence and benefit. *Organizational Behavior and Human Decision Processes, 93*, 1–13.

Yaniv, I., & Kleinberger, E. (2000). Advice taking in decision making: Egocentric discounting and reputation formation. *Organizational Behavior and Human Decision Processes, 83*, 260–281.

Yoshida, W., Dziobek, I., Kliemann, D., Heekeren, H. R., Friston, K. J., & Dolan, R. J. (2010a). Cooperation and heterogeneity of the autistic mind. *The Journal of Neuroscience, 30*, 8815–8818.

Yoshida, W., Seymour, B., Friston, K. J., & Dolan, R. J. (2010b). Neural mechanisms of belief inference during cooperative games. *The Journal of Neuroscience, 30*, 10744–10751.

Younger, J., Aron, A., Parke, S., Chatterjee, N., & Mackey, S. (2010). Viewing pictures of a romantic partner reduces experimental pain: Involvement of neural reward systems. *PLoS One, 5*, e13309.

Zhang, L., & Gläscher, J. (2020). A brain network supporting social influences in human decision-making. *Science Advances, 6*(34), eabb4159.

Foundations of Consumer Neuroscience

Crystal Reeck

Abstract Consumer neuroscience is, to some extent, defined by its methodologies. Traditional consumer research relies upon traditional behavioral techniques, such as examining choice, asking people to provide self-report ratings, and measuring response time. Consumer neuroscience utilizes many of these tools but also employs neuroscientific methodologies, including eye-tracking, psychophysiology, neuroimaging, and circulating hormones. The present chapter examines how these tools have shed light on a range of domains. I focus here on brands, brand personalities, neuroforecasting, and purchase decisions. With respect to brands, I discuss findings from work comparing two brands (e.g., Coke and Pepsi) as well as multiple brands. For brand personalities, I discuss how brands are judged compared to people and novel neural techniques for capturing brand personality judgments. The neuroforecasting section discusses a range of types of predictions made using neural data. In investigating pricing and purchasing, I review a host of evidence from different paradigms. Across these topic areas, I review a range of types of evidence, including data from functional magnetic resonance imaging, circulating hormones, and lesion research, among others. In each case, evidence from neuroscience provides novel insight into a process that guides consumer behavior.

Consumer neuroscience is, to some extent, defined by its methodologies. Traditional consumer research relies upon traditional behavioral techniques, such as examining choice, asking people to provide self-report ratings, and measuring response time. Consumer neuroscience utilizes many of these tools but also employs neuroscientific methodologies, including eye-tracking, psychophysiology, neuroimaging, and circulating hormones (Fig. 1). These methods vary along numerous dimensions, including their time scale and the area of the body they examine (Clithero et al., 2024). One advantage of these techniques is that they may avoid some of the

C. Reeck (✉)
Department of Marketing, Temple University, Philadelphia, PA, USA
e-mail: crystal.reeck@temple.edu

D. V. Smith et al. (eds.), *Neuroeconomics: Core Topics and Current Directions*,
https://doi.org/10.1007/978-3-032-02925-6_7

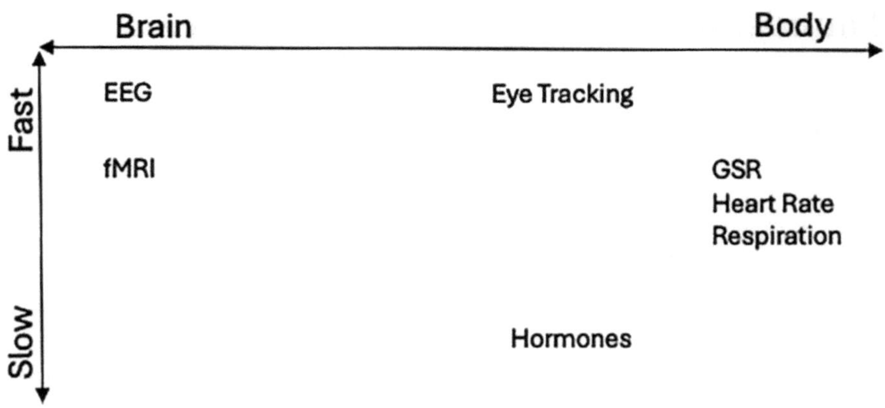

Fig. 1 A subset of the methods for studying consumer neuroscience

confounds or biases presented by traditional self-report (Karmarkar & Plassmann, 2019), such as a lack of insight into one's behavior or social desirability concerns. The neuroscientific methodologies are sometimes considered to provide a more unbiased signal that taps into the implicit or unconscious processes underlying consumer preferences. Prior research has implicated a network of brain regions in processes relevant to consumer behavior, including valuation, memory, and liking.

Consumer neuroscience has been used to study a host of topics in consumer behavior. I focus here on brands, brand personalities, neuroforecasting, and purchase decisions. In each case, evidence from neuroscience provides novel insight into a process that guides consumer behavior.

1 Brands and Product Preferences

People are susceptible to brand names, whether it is routinely choosing our preferred toothpaste or buying a new car. But how do brands influence decisions and how are they represented in the brain?

One of the most famous early consumer neuroscience experiments (Karmarkar & Plassmann, 2019) was built on the famous Pepsi Challenge to understand the neural representations underlying consumer brand preferences. McClure and colleagues asked participants to choose between two small cups of soda (McClure et al., 2004). In the anonymous taste test condition, the setup mimicked the traditional Pepsi challenge: the participants tasted from two unmarked cups that contained either Pepsi or Coca-Cola and chose which they preferred. In the semi-anonymous taste test, one cup was marked with the brand logo, and the other unmarked cup contained the same beverage (unbeknownst to participants). This allowed the experimenters to observe how brands influenced choices, and in this experiment, approximately half of the participants preferred Pepsi while half preferred Coca-Cola. Participants then completed the taste tests in the scanner while undergoing functional magnetic resonance imaging (fMRI). Overall, activation in the ventromedial prefrontal cortex tracked preferences in the anonymous taste test. This finding has since been interpreted as representing the subjective value of the beverages. Interestingly, when the Coca-Cola brand name was shown compared to when Pepsi was shown, there was greater activation in the hippocampus and dorsolateral prefrontal cortex, indicating these regions are involved in representing the brand.

McClure and colleagues showed there were differences in neural activation based on culturally significant labels (that is, brands). Erk and colleagues also showed neural differences in responses to products based on their cultural significance (Erk et al., 2002). Participants viewed different images of cars while undergoing fMRI. Behaviorally, participants rated sports cars as more attractive than limousines or small cars. Neurally, sports cars evoked greater activation in the ventral striatum and ventromedial prefrontal cortex than the other categories of cars. The authors interpret this as indicative of the cultural significance of sports cars, which are tied to higher status.

Deppe and colleagues have built on these findings related to brands, expanding them to other categories (Deppe et al., 2005). Instead of choosing between just two brands as McClure did with Pepsi and Coca-Cola, Deppe and colleagues asked people to make decisions between a host of different brands. Participants either made decisions about coffee or beer, but in each case they had a single preferred brand. While undergoing fMRI, participants made decisions between two beverages—either the target, preferred brand and a diverse brand, or two diverse brands. They found decreased activation in the dorsolateral prefrontal cortex when the favorite, target brand was presented, accompanied by increased ventromedial prefrontal cortical activation.

One challenge posed by neuroimaging experiments is they struggle to prove causality. The results are by their nature correlational and interpretation can run several risks, such as reverse inference. In the case of brand representation, though, there is evidence from lesion patients that points to a causal role for the ventromedial prefrontal cortex in guiding decisions around brands. Koenigs and Tranel recruited three groups of patients: those with brain damage to the ventromedial prefrontal cortex, those with brain damage elsewhere besides the ventromedial prefrontal cortex, and healthy controls. These participants then completed a blind taste test where they chose 20 times between two unmarked cups of soda that contained either Coca-Cola or Pepsi. They then also completed semi-blind taste tests where they chose between two cups, but one was indicated to be a specific brand. Overall, all three groups chose Pepsi more than 50% of the time in the blind taste tests. When the brand was indicated, though, a different pattern emerged. Healthy controls and those with brain damage outside the ventromedial prefrontal cortex were more likely to choose Coca-Cola than Pepsi, exhibiting what is known as the Pepsi Paradox. Stated a different way, the brand name guided their choices in significant ways. However, the ventromedial prefrontal lesion group did not exhibit this paradox and continued to show a preference for Pepsi. These findings indicate that, at least when two highly familiar brands are involved, the ventromedial prefrontal cortex plays an essential role in guiding decision making.

Brands may also need to consider how to present themselves to consumers. Advertisements, for example, can be functional (focusing on disseminating information), emotional (focusing on evoking feeling), or metaphorical (focusing on a metaphor relevant to the brand or product) (Beard et al., 2024). One experiment asked participants to review each of the three different types of advertisements, then try to recall them 1 week later while undergoing fMRI. Participants had better memory for metaphorical and emotional ads than for functional advertisements. This pattern mirrored hippocampal activation during recall, with the lowest hippocampal activation observed for functional advertisements. These behavioral and neuroimaging results complement one another and provide a more holistic understanding of the mnemonic disadvantage functional ads experience.

Circulating hormones may also alter reactions to brands, particularly when they relate to status as they did in the study by Erk and colleagues. One project by Nave and colleagues tested how administering testosterone to men altered their reactions to status goods (Nave et al., 2018). Compared to those given placebo, men administered testosterone showed higher preference for status brands, even compared to other high quality brands. Moreover, they exhibited more positive attitudes to goods that were described as status-enhancing compared to high quality. These findings highlight how circulating hormones can also influence consumer brand preferences.

2 Brand Personality

When people consider brands, a tradition in consumer behavior holds that they may imbue them with humanlike traits, therein granting them their own personalities (Aaker, 1997). The idea is intuitively appealing and spawns several natural lines of inquiry. For example, do people make judgments about brands and people using the same types of processes?

This question would be difficult to address purely with behavioral methods, but a consumer neuroscience approach allows for the examination of whether there is overlap in the neural regions involved in making these judgments. Yoon and colleagues deployed fMRI to determine whether judgments about people and judgments about brands relied on the same neurocircuitry (Yoon et al., 2006). Adapting a paradigm from social cognitive neuroscience, Yoon and colleagues instructed participants to view adjectives and determine if the applied to a given entity: either a person or a brand. Pretesting matched the items for familiarity and liking, and brands included both self-relevant and non-self-relevant exemplars. Analyses revealed greater activation in the medial prefrontal cortex when making judgments of people compared to brands, and greater activation in the left ventrolateral prefrontal cortex when making judgments about brands compared to people. This dissociation in activation, where one region is involved in one process and another is involved in a different process, was interpreted as evidence that different underlying neural processes support these two psychological processes. That is, judgments about people and brands do not appear to rely on the same brain systems and rather are dissociable.

It is worth noting, though, that participants in Yoon and colleagues' experiment were capable of assigning personalities to brands. Even if brand judgments and person judgments are not identical, one might still therefore ask where brand personality judgments come from and how they are represented in the brain. Chen and colleagues adopted a highly innovative approach to combine computational modeling with fMRI data in order to detect these representations (Chen et al., 2015). In their experiment, participants viewed 44 different brands while undergoing fMRI. Afterward, they rated the brands across a host of dimensions to assess the behavioral response to the brand and examined its relationship to neural responses. Five brand personality factors were identified in the behavioral data: Excitement, Competence, Sincerity, Ruggedness, and Sophistication. Machine learning was then applied to determine how each of the brands was represented in the brain, such that the maps that were developed could discriminate between each of the brands with high predictive validity (e.g., classifying Disney versus Gucci). One finding from this approach was that brands that were more similar psychologically (based on the behavioral ratings) were also more similar neurally (e.g., H&M looked similar to MTV in both metrics). Critically, these representations were widely distributed across the brain and not focused in any one region, and they suggest a map that exists for brand personalities in the brain.

3 Neuroforecasting

One of the key areas in which neuroscience data has helped improve market predictions is in the area of neuroforecasting (Knutson & Genevsky, 2018). In this line of research, data from a relatively small group of individuals is used to predict some kind of response across a larger, independent group (see Neuroforecasting chapter, this volume, for more details). Neuroforecasting has been used to generate these predictions of aggregated outcomes in a wide array of domains, from the success of music (Berns & Moore, 2012) to crowdfunding decisions (Genevsky & Knutson, 2015).

One of the first examples of neuroforecasting came from predicting the popularity of songs (Berns & Moore, 2012). Participants underwent fMRI while listening to different music clips. Two years later, the researchers examined how brain activity and ratings of liking predicted the number of downloads of each song. Activation in the nucleus accumbens (NAcc) and medial prefrontal cortex (MPFC) predicted song downloads while the ratings of liking did not. This study is an excellent example of the utility of neural methods in predicting aggregate choice.

One question that might arise is the relative utility of different methods in generating predictions for independent groups. In the study of songs, explicit ratings did not predict independent outcomes, but in other cases, they might have utility. One study examined whether one could predict advertising success in real markets from data collected in a smaller, independent group of participants (Venkatraman et al., 2015). Participants provided a range of data while viewing 30-s advertisements, including explicit ratings, implicit measures, eye-tracking, psychophysiology, EEG, and fMRI. These measures were then related to the market-level elasticities for the advertisements. fMRI—specifically activation in the NAcc—proved the best predictor in this neuroforecasting example, explaining the most variance. Indeed, NAcc activation predicted market-level elasticities over and above the other measures that were collected, including traditional explicit ratings.

As one final example of the value of fMRI data in neuroforecasting, researchers examined how neuroimaging could be used to predict the success of online microloans (Genevsky & Knutson, 2015). Participants reviewed microloan appeals while undergoing fMRI and providing their decisions of whether or not to fund a given loan. NAcc activation predicted not only individual lending choices but also the success of the loan appeal online. Importantly, the NAcc activation predicted the appeal's success better than the actual lending decisions from the smaller group did.

4 Pricing and Purchase Decisions

The key goal of most marketing actions is to promote consumer purchases. One of the main levers available to marketers to promote purchasing is price. As a result, consumer neuroscience has explored questions related to how people decide whether to make a purchase and how they process information about pricing.

One of the most famous consumer neuroscience experiments examined how people make decisions about whether or not to buy something. Knutson and colleagues developed the Save Holdings or Purchase (SHOP) Task, wherein on each trial participants viewed a particular product, then its price, then registered a decision whether to purchase at that price or not (Knutson et al., 2007). All of the items were heavily discounted to encourage purchasing, and participants underwent fMRI while making their decisions. Knutson and colleagues viewed purchase decisions as a function of consumer preference and price and sought to isolate neural activation related to both of these components. Activation in the ventral striatum tracked product preference, showing greater activation for purchases than non-purchases. Interestingly, two regions tracked information related to price: excessive prices activated the insula and deactivated the medial prefrontal cortex. Importantly, these neural signals predicted purchasing behavior over and above behavioral information, demonstrating the utility of neuroscience data in improving predictions.

Another famous early experiment examined responses to price more directly. Plassmann and colleagues asked participants to taste different wines and rate their liking while undergoing fMRI (Plassmann et al., 2008). In some cases, the wines were identical but presented with different price tags (e.g., \$10, \$90). Behaviorally, they replicated the common finding that people prefer consuming higher-priced goods. However, this behavioral finding could be confounded for several reasons: maybe participants felt compelled to rate higher-priced wines more highly to please the experimenter, or perhaps they said they liked it more because they felt they should even if they didn't actually like it any better. The neural findings helped address these potential confounds, as activation in the ventromedial prefrontal cortex was also higher for more highly priced wines. Thus, marketing actions—setting a particular price—could alter not only behavioral responses but also neural responses to the product.

Responses to prices are a form of marketing placebo effect (MPE), wherein marketing-based expectancies affect responses (Shiv et al., 2005). Subsequent work has examined how neural differences might explain individual variability in MPEs. Plassmann and Weber recruited participants who had structural brain scans and examined how the size of different brain regions correlated with MPEs (Plassmann & Weber, 2015). The size of the striatum and dorsomedial prefrontal cortex was positively correlated with people's MPE responsiveness scores, while the insula was negatively correlated with MPE responsiveness scores. These findings suggest that neural variability could be used to segment or classify consumers and shape decisions about how to market to those individuals.

One question that may arise is how and when people make decisions about products. Do people only represent information about a product's value when they make decisions about it, or do they do so more generally whenever they see a product? Tusche and colleagues were interested in this question and developed a creative technique using fMRI to examine it (Tusche et al., 2010). They developed a paradigm in which half of the participants viewed cars and rated their attractiveness, such that high attention was devoted to the attractiveness of the product. The other half were distracted while viewing the cars and therefore had low attention devoted to the attractiveness of the product. They then employed multivariate decoding to determine how the brain represented attractiveness across both groups. Interestingly, insula and medial prefrontal cortex tracked attractiveness in both cases, even when people were not attending to attractiveness. Stated a different way, attractiveness seems to be neurally represented in the brain even when it is not the focus of attention.

If value can be represented neurally even when only minimally attending to a product, the question then arises how attention and value representations interact in the moment. Milosavljevic and colleagues developed a tool to measure the visual salience of product displays to determine how visual attention guided choice (Milosavljevic et al., 2012). They then manipulated the conditions to observe how visual attention interacted with people's subjective preferences when making decisions. For fast decisions made under time pressure, visual salience guides choice more than preferences do. This effect is especially pronounced when people do not have strong preferences about the options. Thus, attention can guide decisions independently from consumer preferences.

People do not only make decisions about products in isolation but also make decisions about consuming several products at once. For example, a grocery shopper needs to decide whether to buy the same yogurt to eat every day that week or different yogurt flavors for each day. Generally, people seek variety when making multiple consumption decisions simultaneously (Simonson, 1990). One experiment sought to examine the brain regions that might be involved in building portfolios of products for consumption. In their task, participants chose from five items to create a basket of three items to potentially consume (Couwenberg et al., 2020). Participants were free to choose any combination of the items on each trial, including three separate flavors or all the same items. Analyses revealed that their search patterns and what they had already selected modulated activation in regions of the brain associated with valuation. Activation in the NAcc decreased when an item had previously been selected (which the authors interpret as a satiety signal) while activation in the medial PFC increased when other options had been selected (which the authors interpreted as a sign of novelty seeking). These findings suggest that the valuation system plays a central role in encouraging diversification when creating portfolios of products for consumption.

People do not always make the best decisions, and numerous efforts have been undertaken to shape consumer choice. One popular line of research explores choice architecture and how to alter the environment of the choice in order to promote more adaptive decisions (Thaler & Sunstein, 2008). One successful choice architecture intervention is precommitment, in which people precommit themselves to a course of action at a time when they are not tempted by inferior options (Ariely & Wertenbroch, 2002). Precommitment might include an alcoholic committing to remove alcohol from their home or someone who procrastinates adopting personal deadlines for projects. One fMRI investigation of precommitment found that activation in the lateral frontopolar cortex was associated with precommitment, rather than activation observed in the dorsolateral prefrontal cortex that was observed when people had to resist temptation on the fly and exert willpower. Investigations like these ones can help provide insight into the neural mechanisms supporting choice architecture interventions and can inform their use and scaling across the population.

5 Conclusions

Consumer neuroscience is a young and growing field. This chapter focused on brands and product preferences, brand personality, and pricing and purchasing behavior, but other domains of consumer behavior exist and are ripe for study. Importantly, several consumer neuroscience experiments have provided evidence that neural signals carry information that is distinct from traditional behavioral measures and can predict outcomes over and above the use of such traditional measures. The field shows signs of growing promise in the coming years.

References

Aaker, J. L. (1997). Dimensions of brand personality. *Journal of Marketing Research, 34*(3), 347–356.

Ariely, D., & Wertenbroch, K. (2002). Procrastination, deadlines, and performance: Self-control by precommitment. *Psychological Science, 13*(3), 219–224.

Beard, E., Henninger, N. M., & Venkatraman, V. (2024). Making ads stick: Role of metaphors in improving advertising memory. *Journal of Advertising, 53*(1), 86–103. https://doi.org/10.1080/00913367.2022.2089302

Berns, G. S., & Moore, S. E. (2012). A neural predictor of cultural popularity. *Journal of Consumer Psychology, 22*(1), 154–160.

Chen, Y.-P., Nelson, L. D., & Hsu, M. (2015). From "where" to "what": Distributed representations of brand associations in the human brain. *Journal of Marketing Research, 52*(4), 453–466.

Clithero, J. A., Karmarkar, U. R., Nave, G., & Plassmann, H. (2024). Reconsidering the path for neural and physiological methods in consumer psychology. *Journal of Consumer Psychology, 34*(1), 196–213.

Couwenberg, L. E., Boksem, M. A. S., Sanfey, A. G., & Smidts, A. (2020). Neural mechanisms of choice diversification. *Frontiers in Neuroscience, 14*, 502.

Deppe, M., Schwindt, W., Kugel, H., Plassmann, H., & Kenning, P. (2005). Nonlinear responses within the medial prefrontal cortex reveal when specific implicit information influences economic decision making. *Journal of Neuroimaging, 15*(2), 171–182.

Erk, S., Spitzer, M., Wunderlich, A. P., Galley, L., & Walter, H. (2002). Cultural objects modulate reward circuitry. *Neuroreport, 13*(18), 2499–2503.

Genevsky, A., & Knutson, B. (2015). Neural affective mechanisms predict market-level microlending. *Psychological Science, 26*(9), 1411–1422.

Karmarkar, U. R., & Plassmann, H. (2019). Consumer neuroscience: Past, present, and future. *Organizational Research Methods, 22*(1), 174–195.

Knutson, B., & Genevsky, A. (2018). Neuroforecasting aggregate choice. *Current Directions in Psychological Science, 27*(2), 110–115. https://doi.org/10.1177/0963721417737877

Knutson, B., Rick, S., Wimmer, G. E., Prelec, D., & Loewenstein, G. (2007). Neural predictors of purchases. *Neuron, 53*(1), 147–156.

McClure, S. M., Li, J., Tomlin, D., Cypert, K. S., Montague, L. M., & Montague, P. R. (2004). Neural correlates of behavioral preference for culturally familiar drinks. *Neuron, 44*(2), 379–387.

Milosavljevic, M., Navalpakkam, V., Koch, C., & Rangel, A. (2012). Relative visual saliency differences induce sizable bias in consumer choice. *Journal of Consumer Psychology, 22*(1), 67–74.

Nave, G., Nadler, A., Dubois, D., Zava, D., Camerer, C., & Plassmann, H. (2018). Single-dose testosterone administration increases men's preference for status goods. *Nature Communications, 9*(1), 2433.

Plassmann, H., & Weber, B. (2015). Individual differences in marketing placebo effects: Evidence from brain imaging and behavioral experiments. *Journal of Marketing Research, 52*(4), 493–510.

Plassmann, H., O'doherty, J., Shiv, B., & Rangel, A. (2008). Marketing actions can modulate neural representations of experienced pleasantness. *Proceedings of the National Academy of Sciences, 105*(3), 1050–1054.

Shiv, B., Carmon, Z., & Ariely, D. (2005). Placebo effects of marketing actions: Consumers may get what they pay for. *Journal of Marketing Research, 42*(4), 383–393.

Simonson, I. (1990). The effect of purchase quantity and timing on variety-seeking behavior. *Journal of Marketing Research, 27*(2), 150–162.

Thaler, R. H., & Sunstein, C. R. (2008). *Nudge: Improving decisions about health, weath, and happiness* (p. 190). Yale University Press.

Tusche, A., Bode, S., & Haynes, J.-D. (2010). Neural responses to unattended products predict later consumer choices. *Journal of Neuroscience, 30*(23), 8024–8031.

Venkatraman, V., Dimoka, A., Pavlou, P. A., Vo, K., Hampton, W., Bollinger, B., Hershfield, H. E., Ishihara, M., & Winer, R. S. (2015). Predicting advertising success beyond traditional measures: New insights from neurophysiological methods and market response modeling. *Journal of Marketing Research, 52*(4), 436–452.

Yoon, C., Gutchess, A. H., Feinberg, F., & Polk, T. A. (2006). A functional magnetic resonance imaging study of neural dissociations between brand and person judgments. *Journal of Consumer Research, 33*(1), 31–40.

Part II
Valuation, Learning, Risk, and Uncertainty

Patricia L. Lockwood, Dominic S. Fareri, and David V. Smith

Many of our decisions are guided by computations surrounding value, risk, and uncertainty. Neuroeconomists have therefore been very interested in measuring and quantifying these influences to describe behavioural preferences. This includes creating computational models of drift-diffusion, incorporating naturalistic methods, integrating animal models of economic choice, and examining how preferences can be changed in the absence of reinforcement. This section discusses the role of value, learning, risk, and uncertainty in choice. It brings together the most recent research to provide new insights and methodological approaches, setting the direction for future neuroeconomics research.

The section begins with a review of how abstract representations of economic choice are encoded in the primate orbitofrontal cortex (OFC) (Cai, this volume). The orbitofrontal cortex has long been associated with choice value in both human and animal models (Padoa-Schioppa & Assad, 2006). More recently, research has turned to understanding how OFC is involved in the processing of abstract representations that are common in real-world decision-making (Xie & Padoa-Schioppa, 2016). Cai (this volume) discusses how OFC neural circuitry may flexibly support common features in several choice tasks to optimize decision-making.

Next, the section outlines multiple perspectives on the role of information accumulation models in neuroeconomics. First, Krajbich et al. (this volume) introduce the concept of information accumulation and the foundational work and recent advances that led to drift-diffusion models of economic choice. They outline how DDMs have been updated to incorporate multiple attributes and responses along

P. L. Lockwood (✉)
University of Birmingham, Birmingham, UK
e-mail: p.l.lockwood@bham.ac.uk

D. S. Fareri
Adelphi University, Garden City, NY, USA

D. V. Smith
Temple University, Philadelphia, PA, USA

a continuum. They also consider how DDMs can be useful for new insights into social decision-making and in cognitive neuroscience to understand brain function. The overview suggests important applications and future avenues to parameterise and integrate information accumulation moving forward. Second, Leng et al. (this volume) put forward a perspective of information accumulation models as analogous to the concept of a tug of war. They consider these models as the dominant computational account of how decision-making unfolds. Instead, they argue that accounting for choice dynamics requires us to re-evaluate the tug of war approach and consider that there is greater independence for how evidence is accumulated across options. Overall, they suggest that including both independent and competitive aspects of evidence accumulation models is important.

Another key issue in neuroeconomics research is the generalizability of research findings from the lab to the wild. Various approaches have been suggested, including integrating theoretical frameworks from behavioural ecology (Kolling et al., 2012; Gabay & Apps, 2021; Mobbs et al., 2018). Choice stimuli can also be used that have real-world relevance, or lab-based measures can be applied in real-world settings. Ceceli and Tricomi (this volume) consider the latter of these two approaches in the context of habit research. They suggest that integrating naturalistic methods can bridge the gap between experimental and real-world applications of motivational control research.

Finally, the section reviews approaches to behavioural choice outside of the dominant understanding and research focus of appetitive external rewards. Lopez-Guzman and Atlas (this volume) highlight how, in addition to being strongly guided by rewards, behavioural choice also must consider possible aversive outcomes. This is the case when considering pain processing in the field of neuroeconomics. This chapter reviews the crucial influence of pain on decision-making and learning and suggests possible applications for important challenges such as the opioid pandemic. Cantor et al. (this volume) introduce the reader to the concept of non-reward reinforced preference change. Specifically, they consider non-reinforced preferences as those that occur without external reinforcement. They argue that it is crucial to understand the role of internal feedback and how integrating this perspective will be essential for future neuroeconomics research.

Overall, this section emphasizes several themes throughout the book, including different perspectives on evidence accumulation, the important role of non-human animal research in understanding economic choice, and the usefulness of considering approaches beyond appetitive reward, including aversive stimuli and internal reinforcement. By bridging these different concepts, we can provide a fuller illumination of what drives behavioural choice in the lab and in the wild.

References

Cai, X. (this volume). Abstract representations of economic choice in the primate orbitofrontal cortex. In D. V. Smith, P. L. Lockwood, & D. S. Fareri (Eds.), *Neuroeconomics: Core topics and current directions*. Springer. Preprint: https://papers.ssrn.com/sol3/papers.cfm?abstract_id=4984901

Cantor, A., Bar Or, M., Itzkovitch, A., & Schonberg, T. (this volume). Non-reinforced preference change. In D. V. Smith, P. L. Lockwood, & D. S. Fareri (Eds.), *Neuroeconomics: Core topics and current directions*. Springer. Preprint: https://osf.io/preprints/psyarxiv/b6nkv_v1

Ceceli, A. O., & Tricomi, E. (this volume). Enhancing ecological validity in habit research via naturalistic methods. In D. V. Smith, P. L. Lockwood, & D. S. Fareri (Eds.), *Neuroeconomics: Core topics and current directions*. Springer.

Gabay, A. S., & Apps, M. A. (2021). Foraging optimally in social neuroscience: Computations and methodological considerations. *Social Cognitive and Affective Neuroscience, 16*(8), 782–794.

Kolling, N., Behrens, T. E., Mars, R. B., & Rushworth, M. F. (2012). Neural mechanisms of foraging. *Science, 336*(6077), 95–98. https://doi.org/10.1126/science.1216930

Krajbich I, Fernandez K, Yang X. (this volume). Decomposing economic choices with drift-diffusion models. In D. V. Smith, P. L. Lockwood, & D. S. Fareri (Eds.), *Neuroeconomics: Core topics and current directions*. Springer.

Leng, X., Su, Y., Frömer, R., & Shenhav, A. (this volume). Breaking the tug-of-war: What neuroeconomics can gain by moving past competition-only models. In D. V. Smith, P. L. Lockwood, & D. S. Fareri (Eds.), *Neuroeconomics: Core topics and current directions*. Springer. Preprint: https://doi.org/10.31234/osf.io/g2va5_v1

Lopez-Guzman, S., & Atlas, L. Y. (this volume). Pain and value-based decision-making: An introduction and implications for neuroeconomics. In D. V. Smith, P. L. Lockwood, & D. S. Fareri (Eds.), *Neuroeconomics: Core topics and current directions*. Springer. Preprint: https://doi.org/10.31234/osf.io/nwu5j

Mobbs, D., Trimmer, P. C., Blumstein, D. T., & Dayan, P. (2018). Foraging for foundations in decision neuroscience: Insights from ethology. *Nature Reviews Neuroscience, 19*(7), 419–427.

Padoa-Schioppa, C., & Assad, J. A. (2006). Neurons in the orbitofrontal cortex encode economic value. *Nature, 441*(7090), 223–226.

Xie, J., & Padoa-Schioppa, C. (2016). Neuronal remapping and circuit persistence in economic decisions. *Nature Neuroscience, 19*(6), 855–861.

Abstract Representations of Economic Choice in the Primate Orbitofrontal Cortex

Xinying Cai

Abstract The orbitofrontal cortex (OFC) of primates is crucial for economic choice. Neurons in the OFC, comprising various functional types, integrate diverse attributes, ensure consistent value representation, and adapt to different decision contexts to guide choices. These neurons support subjective value computation and comparison in an abstract space, independent of sensorimotor contingencies. The choice outcome is then transformed into actions, potentially involving the lateral prefrontal cortex (LPFC), the anterior cingulate cortex (ACC), and the striatum. Furthermore, the OFC exhibits the property of task abstraction, maintaining abstract representations that can be flexibly applied across various situations. Understanding the OFC's role in both abstract decision formation and task abstraction enhances our knowledge of the neural basis of economic decision-making, providing deeper insights into the neural mechanisms supporting higher cognitive functions.

1 Good-Based Model of Economic Choice

1.1 Introduction

An economic choice occurs when individuals make decisions based on subjective preferences. Economic and psychological research shows that individuals evaluate various attributes of goods and assign subjective values, which they then use to compare and make decisions (Niehans, 1990; Padoa-Schioppa, 2011; Rangel et al., 2008; Wallis, 2007). For example, consumers may consider factors such as taste, price, and nutritional content when choosing breakfast cereals, ultimately selecting the one that best matches their personal preferences. Over the past two decades, a good-based model has been proposed to describe the mental operations involved in economic choice (Padoa-Schioppa, 2011; Padoa-Schioppa & Cai, 2011; Padoa-

X. Cai (✉)
NYU Shanghai, Shanghai, China
e-mail: xinying.cai@nyu.edu

D. V. Smith et al. (eds.), *Neuroeconomics: Core Topics and Current Directions*,
https://doi.org/10.1007/978-3-032-02925-6_8

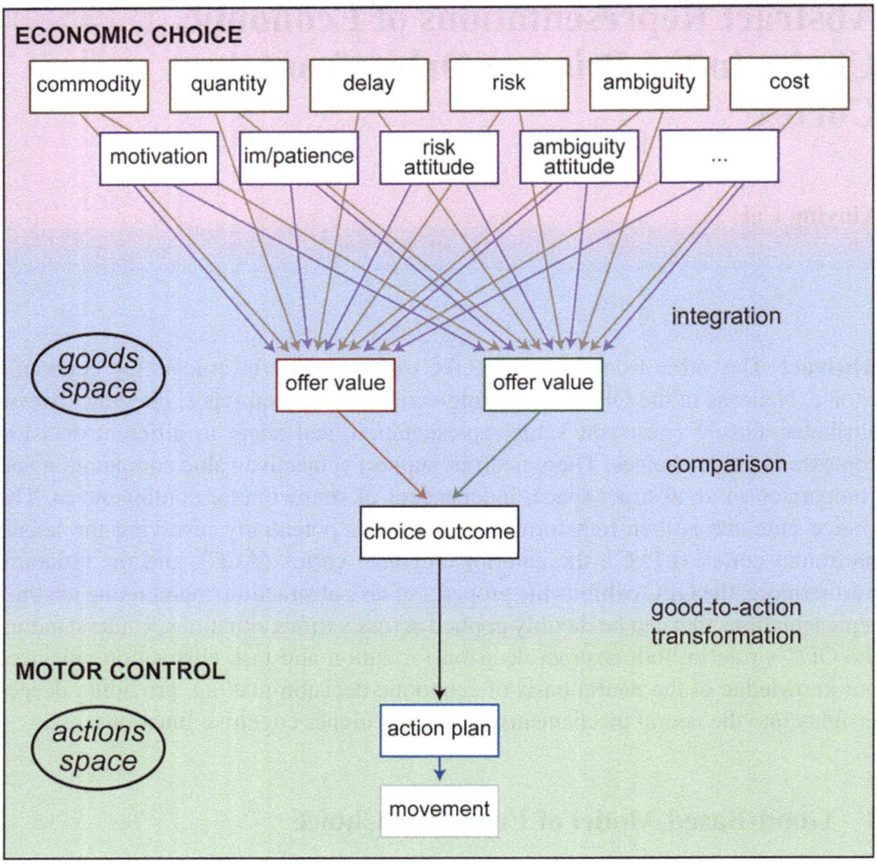

Fig. 1 Good-based model. The value of each good is computed by integrating multiple determinants, of which some are external (e.g., commodity, quantity, etc.), and others are internal (motivation, risk attitude, etc.). The values of different goods are computed independently of one another and then compared to make a decision. This comparison takes place within the space of goods. The choice outcome (chosen good and chosen value) then guides an action plan through the good-to-action transformation. (Figure credit: Camillo Padoa-Schioppa)

Schioppa & Conen, 2017); For discussions on alternative models, see Cisek (2012) and Chen and Stuphorn (2015). According to this model (Fig. 1), a "good" is defined by multiple determinants, which can be both external (e.g., quantity of commodities) and internal (e.g., risk attitude). The value of each good is computed at the time of choice by integrating these determinants. Crucially, the good-based model posits that the brain maintains an abstract representation of goods (i.e., independent of the sensorimotor contingencies of choice), and the choice process—the computation and comparison of values—occurs within this abstract space. At the neural level, the good-based model suggested that both the computation and comparison of values occur within prefrontal regions and possibly relevant subcortical areas, particularly the OFC, ventromedial prefrontal cortex (vmPFC), and the amygdala. The outcome of the choice (i.e., the identity and value of the chosen good) then guides the selection of an appropriate action plan through a process known as good-to-action transformation (Fig. 1).

1.2 Value Computation in the Abstract Reference Frame

A fundamental pursuit in identifying the neural mechanisms underlying good-based economic choice is mapping neuronal activity onto different components of the good-based model. Many studies have significantly advanced our understanding of neural computation at various stages of this model, primarily through single-unit recording studies in nonhuman primates. These studies have provided detailed insights into how neurons in specific brain regions integrate various determinants to compute the value of goods and facilitate the comparison and selection processes central to economic decision-making.

The proposal that value computation and comparison may occur in a goods space was initially motivated by the findings of Padoa-Schioppa and Assad (2006). In their experiment, monkeys chose between juices that varied in two dimensions: juice taste and quantity (Fig. 2a). Monkeys displayed a preference for one juice when offered in equal amounts but chose the less preferred juice when it was offered in a sufficiently larger quantity (Fig. 2b). This behavior suggests that monkeys made a trade-off between taste and quantity, and the relative value of the juices can be inferred from the indifference point—the quantity ratio at which monkeys chose either juice with equal frequency (Fig. 2b).

Using this paradigm, neuronal recordings were performed in the central OFC (cOFC, Fig. 2c), which refers to Brodmann areas 13 m and 11 l (Ongur & Price, 2000). Three groups of neurons in the OFC were identified: offer value cells, chosen value cells, and chosen juice cells. Offer value cells encode the value of individual juices (Fig. 2d, e), chosen value cells encode the value of the selected offer (Fig. 2f), and chosen juice cells encode the binary decision outcome (Fig. 2g). Importantly, value representation in the OFC does not depend on either the spatial configuration of the offer or the specific action associated with the choice (Padoa-Schioppa & Assad, 2006). The activity pattern of offer value neurons shows how goods are mapped onto an attribute space defined by two dimensions: juice taste and quantity (Fig. 2d, e). Note that to distinguish between the two goods for choice, each good must be identified by a "label." In this task, taste was taken by the brain as the label for the two pools of neurons whose firing rate is correlated with the quantity of each juice (Fig. 2d, e). Thus, in this decision context, each offer is neurally instantiated on the 2D attribute space through neuronal pool assignment based on taste (binary coding) and the level of neuronal activity (monotonic coding) of each pool.

To make a decision, these two distinct attributes were converted into a common currency of value for comparison. This is evidenced by the observation that the firing pattern of chosen value neurons aligns with the behavioral trade-off between

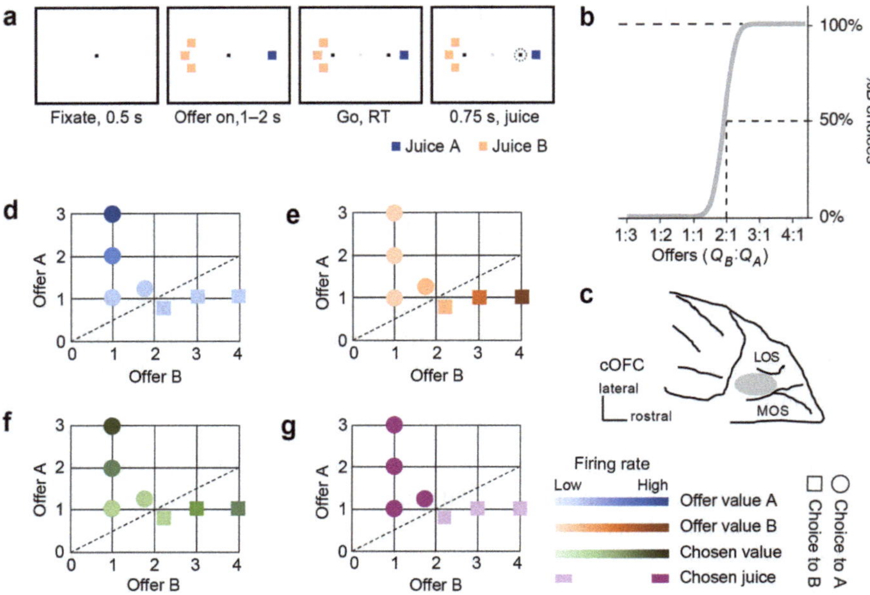

Fig. 2 Neural representations for economic choice. (**a**) Economic choice task. In each session, the monkey chose between two juices, labeled A and B, with A preferred. At the beginning of each trial, the animal was fixed at a central position on the monitor. After 0.5 s, two sets of colored squares appeared on the two sides of the center fixation. The two sets of squares represent the two offers, with the color associated with a particular juice type and the number of squares indicating the juice quantity. The animal maintained center fixation for a randomly variable delay (1–2 s), at which point the center fixation point was extinguished (go signal). The animal revealed its choice by making a saccade toward one of two targets located by the offers, and peripheral fixation was maintained for an additional 0.75 s before the chosen juice was delivered. (**b**) Illustration of a typical choice pattern with an indifference point of 1A = 2B for one session. The percentage of B choices is plotted against the ratio Q_B:Q_A, where Q_A and Q_B are the quantities of juice A and B, respectively. (**c**) Recording area, the central OFC. MOS, medial orbital sulcus; LOS, lateral orbital sulcus. (**d**) Schematic of the activity pattern of an example neuron encoding the value of offer A with a positive encoding slope with the animal's choice pattern consistent with that in (**b**). The animal chooses offer A for offer types above the dotted indifference line and chooses offer B for offer types below the indifference line. At offer type 1A:2B, the animal chose either offer with the same frequency. The encoding of the offer value is independent of the animal's choice. A darker shade indicates a higher firing rate. (**e**) Schematic of the activity pattern of an example neuron encoding the value of offer B with a positive slope. Same convention as in (**c**). (**f**) Schematic of the activity pattern of neurons encoding the value of the chosen offer (chosen value) with positive encoding slopes. At offer type 1A:2B, the animal chose either offer with the same frequency (therefore, there are two data points for offer type 1A:2B), and the firing rate for choosing either offer was the same. (**g**) Schematic of the activity pattern of a neuron encoding the animal's choice (chosen juice). The neuronal responses differ between choices to offer A and B with the same firing rate for choices to offer A and a different level of the same firing rate for choices to offer B (binary coding)

juice quantity and taste (Fig. 2f). For example, with a behavioral indifference point of 1A = 2B (Fig. 2b), chosen value neurons exhibit the same firing rate when 1A or 2B is chosen or when 2A or 4B is chosen. This neural and behavioral match suggests that OFC activity reflects the decision-maker's subjective preference. Ferrari-Toniolo and Schultz (2023) also recently reported that although reward neuronal signals in the monkey OFC are heterogeneous, they combine to form a reliable population code that closely matches the animal's choices. In a computational modeling study, Rustichini and Padoa-Schioppa (2015) proposed that the representation of chosen value involves a competitive selection process. Consistent with this notion, Rich and Wallis (2016) demonstrated, using population decoding analysis, that the OFC shows state alternation between the chosen and unchosen values during a single trial, reflecting an internal deliberation process.

A broad range of studies have also identified several key features of value coding in the OFC that are important for economic decision-making. The first key feature is domain-general, meaning that the integration of attributes is consistent across various types of attributes. Many studies have demonstrated that the integration of attributes extends beyond initially reported juice taste and quantity. Adding reward probability as an additional determinant of value Raghuraman and Padoa-Schioppa (2014) reported that central OFC neurons compute value by integrating all three determinants. Roesch and Olson (2005) and Setogawa et al. (2019) showed that neurons in the primate OFC are modulated by both the quantity of juice and the delay leading to juice delivery. Consistent with the representation of value, the two modulations have opposite signs. Similarly, Morrison and Salzman (2009) reported that rewarding and aversive outcomes often converge on single neurons and antagonistically modulate their activity. In addition, Wallis and colleagues found that along with other frontal areas, the activity of neurons in the OFC depends not only on the quantity of juice delivered to a monkey but also on the probability and the physical effort exerted by the animal to obtain the juice (Kennerley et al., 2009; Kennerley & Wallis, 2009). Using a choice task, they also found that neurons in this area encode the subjective value defined by the trade-off between juice quantity and delay as well as juice quantity and physical effort (Hosokawa et al., 2013). Interestingly, the integration of value determinants also includes social information and context (Azzi et al., 2012). Moreover, OFC activity also reflects the internal state of the choosers, consistent with the notion of subjective economic reward value coding in OFC neurons (Pastor-Bernier et al., 2021).

The second key feature of value coding in the OFC is range adaptation. Specifically, in any behavioral context, the firing rate of OFC neurons is approximately a linear function of the offered values, and their tuning slope is inversely proportional to the range of values available in that context (Kobayashi et al., 2010; Padoa-Schioppa, 2009). A later study combining experimental and theoretical approaches demonstrated that quasilinear but range-adapting tuning functions are sufficient to ensure close-to-optimal choice behavior (Rustichini et al., 2017).

The third key feature is menu invariance, meaning that neuronal responses in the OFC encoding the value of a particular juice do not depend on which juice was offered as an alternative. Menu invariance and choice transitivity are closely related because preferences can violate transitivity only if values depend on the menu (Grace, 1993; Tversky & Simonson, 1993). Thus, menu invariance observed in the OFC may explain why human and animal choices generally satisfy preference transitivity, a fundamental principle of rational choice theory (Choi et al., 2007; Harbaugh et al., 2001; Sumpter et al., 1999).

Importantly, recent work employing electrical stimulation has demonstrated that neuronal activity in the OFC causally contributes to value computation and comparison. In the studies by Ballesta et al. (2020, 2022), varying magnitudes of electrical currents were applied to OFC neurons. This manipulation revealed three distinct behavioral effects: range-dependent bias, changes in order bias, and increased choice variability. These effects vary depending on the level of electrical current. Overall, the use of electrical stimulation has provided causal evidence for the involvement of OFC neurons in both the computation and comparison of economic values.

1.3 Abstract Decision Formation

Intuitively, the three groups of neurons described above—offer value cells, chosen value cells, and chosen juice cells—seem sufficient to characterize and potentially generate a decision in an abstract reference frame. Offer value cells capture the input to the decision process, while chosen value and chosen juice cells capture the value and identity of the chosen good, thus representing the decision outcome. However, encoding abstract post-decision variables (chosen value and chosen juice) does not directly prove that the decision is resolved in the space of goods. An alternative hypothesis is that neurons in other brain areas, particularly motor regions, participate in or even determine the decision, subsequently informing the activity of chosen value or chosen juice neurons. To disentangle these possibilities, Cai and Padoa-Schioppa (2014) designed an economic choice task that encouraged but did not enforce good-based decisions. Specifically, they allowed monkeys to choose between two juices offered in variable amounts while dissociating the presentation of the offers and the indication of the action associated with each offer in both time and space. Choices were ultimately revealed through eye movement.

Neuronal responses in the OFC, which encode the choice outcome long before the presentation of the saccade targets, suggested that decisions were indeed made in goods space. At the population level, the fraction of neurons encoding chosen value and chosen juice during the decision period was strikingly similar to that in tasks where offers and actions were spatially associated and temporally coupled (Fig. 3d in Cai & Padoa-Schioppa, 2014). This finding reinforces the notion that the OFC plays a central role in economic choice by operating within an abstract space of goods, independent of the specific motor actions required to execute the choice.

In a follow-up study, Cai and Padoa-Schioppa (2019) further tested the abstract nature of the choice process in the OFC by imposing asymmetrical action costs on the two goods offered to the animals, reasoning that effort cost necessarily reflects aspects of the action. In their experiments, monkeys made economic choices between goods defined by three attributes: juice taste, quantity, and action cost (Fig. 3a). Importantly, all three attributes were communicated to the animals early in the trial during the offer presentation, well before the location of the eventual saccade targets was indicated. This design allowed the animals to make their decisions independently of action planning. The action cost induced a noticeable effect on the animal's choice pattern.

The neuronal analysis focused on cells encoding the chosen juice (Fig. 3c) and chosen cost (Activity profiles not shown). The choice-correlated separation of activity was quantified using receiver operating characteristic (ROC) analysis, which returned the area under the curve (AUC). Importantly, the ROC analysis was performed only on "cost-overt" trial types, where the animal chose the low-cost offer more frequently (more than 10%) than the high-cost offer, provided the animal chose each option at least twice. This ensures behavioral verification that the animal indeed considered action costs when making decisions. Across the population of chosen juice cells, the distribution of the AUC was significantly above the chance level in each of the three non-overlapping time windows (250 ms) following the offer presentation but before target presentation, confirming that cost-integrated decisions were made before action planning (Fig. 3d). A similar analysis of the population of neurons encoding the chosen cost showed the same results (Fig. 3e).

In summary, both chosen juice and chosen cost neurons revealed the binary choice outcome well before target presentation, indicating that decisions were made in goods space. These findings provide compelling evidence that the OFC is capable of abstractly encoding economic choices, even when effort costs need to be integrated. Note that OFC does not seem to be the only brain region where abstract choices are represented. Other studies indicate that abstract coding of economic choices exists in other reward structures, in particular, the amygdala (Grabenhorst et al., 2012; Grabenhorst et al., 2023; Hernadi et al., 2015; Jezzini & Padoa-Schioppa, 2020) and potentially the ventral medial prefrontal cortex (Strait et al., 2014; Yamada et al., 2018) but see discussion in Ballesta and Padoa-Schioppa (2019).

1.4 Good-to-Action Transformation

While choices may be resolved within the space of goods, they ultimately need to be translated into suitable actions. This transformation from goods space to action space involves neural circuits that bridge the gap between decision outcomes and action planning.

The central OFC lacks direct anatomical connections with motor structures (Carmichael & Price, 1995). However, it projects extensively to the ventral portion of the lateral prefrontal cortex (LPFCv) (Petrides & Pandya, 2006; Saleem et al., 2013), which in turn is connected to the dorsal portion of the lateral prefrontal cortex (LPFCd), known for its dense connections with motor systems (Lu et al.,

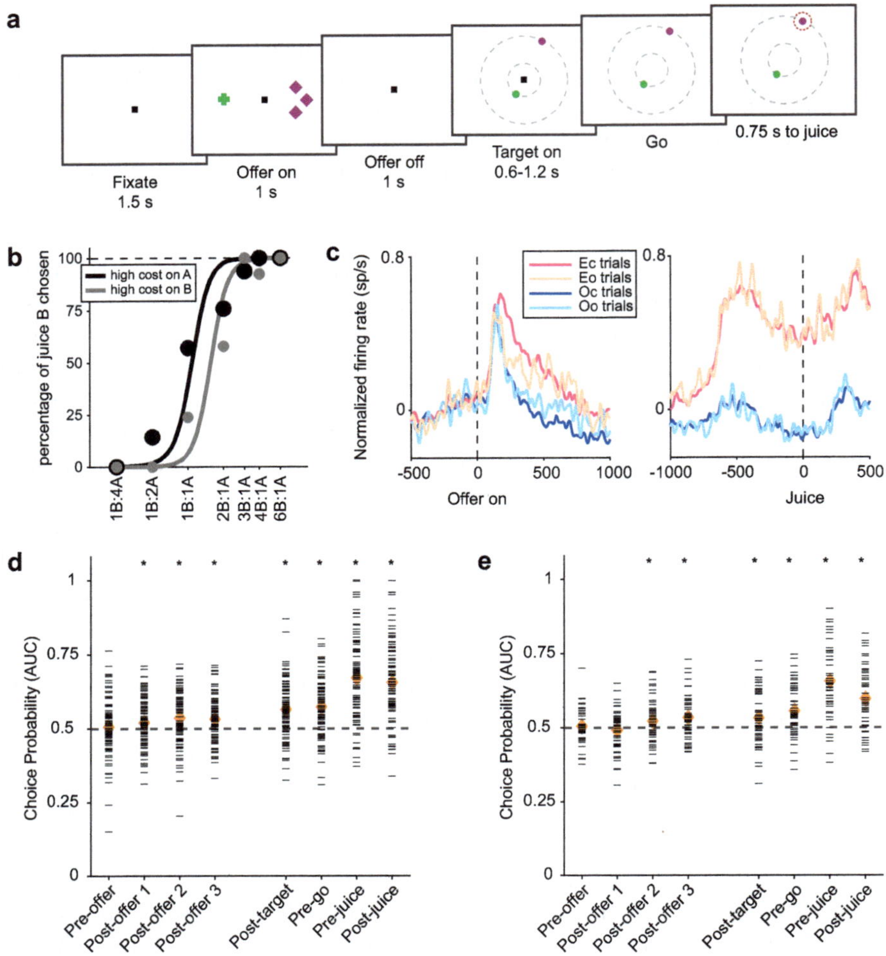

Fig. 3 Good-based economic decisions under variable action costs. (**a**) Choice task structure. At the beginning of the trial, the monkey fixated a center point on the monitor. After 1.5 s, two offers appeared to the left and right of the fixation point. The offers are represented by sets of color symbols, with the color indicating the juice type, the number of squares indicating the juice amount, and the shape of the symbols indicating the action cost associated with the offer (squares, low cost; diamonds, high cost). The offers remained on the monitor for 1 s and then disappeared. The monkey continued fixating on the center point for another 1 s. At the end of this delay, two saccade targets (two color dots) appeared. The two saccade targets were located on two (invisible) concentric rings centered on the fixation point. In each trial, one of the saccade targets was placed on the low-cost (small radius) ring, whereas the other saccade target was placed on high-cost (large radius) ring. The two targets were always placed on opposite sides of the center fixation. The angle defining their position was selected on every trial among four possible values, corresponding to 22.5°, 157.5°, 202.5°, and 337.5° from azimuth. Thus, for each juice, there were eight possible saccade target positions (two distances x four angles). The animal maintained fixation for a randomly variable delay (0.6–1.2 s) before the center fixation point was extinguished (go signal), at which point the monkey indicated its choice with a saccade. Importantly, the association between the symbols (square,

1994; Takada et al., 2004; Takahara et al., 2012). This anatomical connectivity pattern suggests that LPFCv/d may participate in the transformation of decisions from goods space to action space.

To investigate this hypothesis, Cai and Padoa-Schioppa (2014) recorded neural activity from the LPFCv/d while monkeys performed an economic choice task. They observed that before target presentation, neurons in the LPFCv/d encoded the choice outcome in goods space. However, following target presentation, these neurons gradually begin to encode the spatial location of the saccade targets and subsequently the planned actions. This pattern of neural activity suggested the involvement of the LPFCv/d in transforming decision outcomes into action plans. Additionally, consistent with the anatomical connections, they found that the LPFCv leads the LPFCd in the representation of spatial and action-related information.

To elucidate the network mechanisms for such transformation, Yim et al. (2019) constructed a neural circuit model consisting of modules representing choice, integration of choice with target locations, and the final action plan. Their model with heterogeneous synaptic connectivity predicts the coexistence of three types of neurons with distinct functions. These types of neurons were confirmed by analyzing the neural activity in the LPFC recorded in the study of Cai and Padoa-Schioppa (2014). The analysis revealed a much more distinct classification of functional neuron types in the ventral than the dorsal region of LPFC. These results suggest that the action plan may be initially generated in the ventral LPFC.

Another brain region likely involved in this transformation is the dorsal anterior cingulate cortex (ACCd) (Hadland et al., 2003). To investigate the role of ACC in economic choice, Cai and Padoa-Schioppa (2012) recorded activity from both the dorsal and ventral banks of ACC. They found that both regions encode chosen value but not offer value, with the dorsal bank also modulated by the animal's

---

Fig. 3 (continued) diamond) and the saccade amplitudes (short, long) was known for the monkey. (**b**) Typical choice patterns, one session. The percentage of B choices is plotted against the ratio #B:#A, where #A and #B are quantities of juice A and B, respectively. Trials were separated into two groups depending on the action cost for juice A. The choice pattern obtained when juice A had a high cost (black) was displaced to the left (lower indifference point) compared with the choice pattern obtained when juice A had a low cost (gray). Offer types were defined as "cost-overt" if the animal chose the low-cost offer more frequently (more than 10%) than the high-cost offer, provided the animal chose each option at least twice (e.g., 1B:1A). Conversely, "cost-covert" offer types were those where the animal consistently chose the same option independently of the action cost (e.g., 1B:4A). (**c**) Chosen juice cells ($N = 97$). Trials were divided depending on whether the animal chose the juice encoded by the cell (E) or the other juice (O) and on whether the different action costs produced observable behavior effects (covert (c) or overt (o)). (**d**) ROC analysis for 97 chosen juice cells in cost-overt trials. Each horizontal gray line represents one neuronal response. Each orange diamond represents the mean AUC, which was significantly >0.5 in each of the three nonoverlapping time windows before target presentation. (**e**) ROC analysis for 51 chosen cost cells in cost-overt trials. Similar to **c**, across the population, the mean (AUC) was significantly greater than 0.5 in two of the three time windows before target. In c and d, an asterisk (*) indicates $p < 0.02$, t-test. AUC, area under the curve; ROC, receiver operating characteristic. (The figure was reproduced from Cai & Padoa-Schioppa, 2019)

saccade directions. Using the same task designs as in their OFC study (Cai & Padoa-Schioppa, 2019), they later expanded the ACC study by including effort cost as an additional attribute of the offers. They discovered that neurons in the ACCd do not encode pre-decision variables such as offer value even when effort cost is integrated when computing value. Instead, they represent variables related to the choice outcome in multiple reference frames, including chosen value, chosen offer location, chosen target location, chosen cost, and chosen juice (Cai & Padoa-Schioppa, 2021). This multiplexing of post-decision information in ACCd suggests its role as a hub for broadcasting information related to the choice outcome across different reference frames, including spatial and action-based ones.

A third candidate brain region is the striatum, which receives direct projections from the OFC for reward processing (Haber et al., 2006). Neurons in the dorsal striatum encode economic value and choice outcomes in the action-based reference frame (Cai et al., 2011; Samejima et al., 2005; Shin et al., 2021). Additionally, in a rodent version of the primate economic choice task, the choice signal in the striatum follows that in the OFC, suggesting that information about the choice outcome flows from OFC projection neurons to the striatum, which is involved in action planning and execution (Gore et al., 2023).

Taken together, the LPFCv/d, ACCd, and the striatum serve as key nodes in the neural network responsible for transforming abstract decision outcomes into action plans.

2 Task Abstraction

The contexts in which choices are made can vary significantly in real-world decision-making scenarios. The abstract property of the orbitofrontal neuronal coding may equip the OFC with the ability to navigate these diverse contexts effectively. In other words, the OFC should maintain representations of mental operations that can be transferred to new situations, demonstrating the ability of abstraction. Research has shown that neurons in the OFC play an active role in maintaining such representations. Specifically, studies have shown that OFC neurons persistently encode the value of goods even when the sensory attributes of those goods are replaced. For instance, experiments involving monkeys making economic choices between different amounts of two juices, A and B, revealed that OFC neurons encode task-related variables such as offer value, chosen value, and the chosen juice associated with these specific juices (Xie & Padoa-Schioppa, 2016). Strikingly, when the two original juices are replaced with two different juices, C and D, the same neurons continue to encode these task-related variables for the new set of goods. This result indicates that the representation of value in the OFC is independent of the specific sensory features of the goods involved. In other words, OFC neurons generalize their value coding across different sensory contexts.

This finding was further supported by subsequent discoveries in Xie and Padoa-Schioppa (2016). In their follow-up experiment, initially, juices A and B were presented in session 1, with juice A being preferred. In session 2, however, the juices were switched to C and A, with C now the preferred juice. Remarkably, most neurons encoding the offer value of juice A in session 1 remapped to encode the value of juice C in session 2. Conversely, most neurons encoding the offer value of juice B in session 1 remapped to encode the quantity of juice A in session 2. This remapping phenomenon suggests that the attribute of taste is likely encoded along an abstract dimension reflecting taste preference ranking.

The abstract nature of the remapping described above hints at the possibility that the corresponding neural circuitry may not be dedicated to any specific task but rather be designed to flexibly support a variety of tasks with common features (Cai, 2021). This implies a potential functional specialization of OFC neuronal ensembles for economic decision-making. To investigate this idea, Xu et al. (2024) conducted a study in which monkeys alternated between two choice tasks: a multi-attribute choice (MC) task and a single-attribute choice (SC) task. In the MC task, monkeys chose between two juices that varied in both taste and quantity, with their choices reflecting a subjective tradeoff between these attributes. In contrast, in the SC task, monkeys chose between different quantities of the same juice, consistently favoring the larger quantity. These distinct choice patterns suggest that different cognitive processes underlie the two tasks. Neurophysiologically, in the MC task, OFC neurons primarily encoded task-related variables within the goods space. However, when transitioning to the SC task, many active neurons during the MC task disengaged. Instead, different neurons emerged to encode SC-related variables in a spatial reference frame.

In essence, the findings from Chen et al. suggest that choice tasks involving different cognitive processes are represented by dissociable neuronal modules in the OFC. By maintaining dissociable modules for two different choice tasks, the brain may optimize decision-making processes by efficiently leveraging specialized circuits tailored to different task demands.

3 Summary and Future Directions

In this chapter, I reviewed the abstract representation of economic choice in the OFC. I introduced the good-based model for economic choice and explained how neural activities in the OFC correspond to various components of this model. Essentially, neurons in the OFC are categorized into functional types that support subjective value computation and comparison in an abstract space. The choice outcome is then transformed into actions likely involving the LPFC, ACC, and striatum. I also reviewed studies providing causal evidence linking the OFC to economic decision-making processes. Finally, I discussed emerging evidence demonstrating task abstraction and the likely existence of functionally specialized neuronal modules in the OFC.

A key objective for future research is to dissect the structure and functioning of the choice network in the brain. At the local level, critical unresolved questions include elucidating the connections and interactions among various functional cell types within the OFC and understanding the mechanisms that facilitate information transfer and neuron recruitment for diverse tasks.

At the network level, we know quite a bit about how the output of abstract decisions might be transformed into an action plan. However, little is known about how information on multiple attributes potentially converges in the OFC for utility computation. One of the candidate regions that could relay good attributes is the gustatory cortex (GC). However, when the animals were performing the same choice task as described in Padoa-Schioppa and Assad (2006), few neurons in GC encoded pre-decision variables (*offer value A* or *offer value B*), while most task-related neurons encoded post-decision variables (*chosen value*, *chosen number*, and *chosen juice*) (Jezzini & Padoa-Schioppa, 2024). These signals may help shape the offer value signals in the OFC, while the animals establish the association between the visual stimulus and the attributes they represent. Conversely, GC may be informed by the OFC of subjective value signals for consummatory appraisals. Further studies, especially those with causal manipulations, will better address these questions.

We also need more investigations to understand the interdependence or lack thereof of brain regions implicated in abstract economic choice. For example, regarding the interaction between the OFC and the amygdala, it has been shown that excitotoxic lesions of the amygdala did not impair the animal's choice of different magnitudes of reward, and it only led to a mild reduction of reward signals in the OFC (Rudebeck et al., 2013). The caveat is that Rudebeck's study only examined the effect of amygdala lesions to reward magnitude discrimination but not subjective value computation involving multiple attributes. Further studies are needed to explore the importance of the interaction between the OFC and amygdala in abstract economic decision-making.

References

Azzi, J. C., Sirigu, A., & Duhamel, J. R. (2012). Modulation of value representation by social context in the primate orbitofrontal cortex. *Proceedings of the National Academy of Sciences of the United States of America, 109*, 2126–2131.

Ballesta, S., & Padoa-Schioppa, C. (2019). Economic decisions through circuit inhibition. *Current Biology, 29*, 3814–3824 e3815.

Ballesta, S., Shi, W., Conen, K. E., & Padoa-Schioppa, C. (2020). Values encoded in orbitofrontal cortex are causally related to economic choices. *Nature, 588*, 450–453.

Ballesta, S., Shi, W., & Padoa-Schioppa, C. (2022). Orbitofrontal cortex contributes to the comparison of values underlying economic choices. *Nature Communications, 13*, 4405.

Cai, X. (2021). The neural instantiation of an abstract cognitive map for economic choice. *Neuroscience, 477*, 106–114.

Cai, X., & Padoa-Schioppa, C. (2012). Neuronal encoding of subjective value in dorsal and ventral anterior cingulate cortex. *The Journal of Neuroscience, 32*, 3791–3808.

Cai, X., & Padoa-Schioppa, C. (2014). Contributions of orbitofrontal and lateral prefrontal cortices to economic choice and the good-to-action transformation. *Neuron, 81*, 1140–1151.

Cai, X., & Padoa-Schioppa, C. (2019). Neuronal evidence for good-based economic decisions under variable action costs. *Nature Communications, 10*, 393.

Cai, X., & Padoa-Schioppa, C. (2021). Neuronal activity in dorsal anterior cingulate cortex during economic choices under variable action costs. *eLife, 10*, e71695.

Cai, X., Kim, S., & Lee, D. (2011). Heterogeneous coding of temporally discounted values in the dorsal and ventral striatum during intertemporal choice. *Neuron, 69*, 170–182.

Carmichael, S. T., & Price, J. L. (1995). Sensory and premotor connections of the orbital and medial prefrontal cortex of macaque monkeys. *The Journal of Comparative Neurology, 363*, 642–664.

Chen, X., & Stuphorn, V. (2015). Sequential selection of economic good and action in medial frontal cortex of macaques during value-based decisions. *eLife, 4*, e09418.

Choi, S., Fisman, R., Gale, D., & Kariv, S. (2007). Consistency and heterogeneity of individual behavior under uncertainty. *The American Economic Review, 97*, 1858–1876.

Cisek, P. (2012). Making decisions through a distributed consensus. *Current Opinion in Neurobiology, 22*, 927–936.

Ferrari-Toniolo, S., & Schultz, W. (2023). Reliable population code for subjective economic value from heterogeneous neuronal signals in primate orbitofrontal cortex. *Neuron, 111*, 3683–3696 e3687.

Gore, F., Hernandez, M., Ramakrishnan, C., Crow, A. K., Malenka, R. C., & Deisseroth, K. (2023). Orbitofrontal cortex control of striatum leads economic decision-making. *Nature Neuroscience, 26*, 1566–1574.

Grabenhorst, F., Hernadi, I., & Schultz, W. (2012). Prediction of economic choice by primate amygdala neurons. *Proceedings of the National Academy of Sciences of the United States of America, 109*, 18950–18955.

Grabenhorst, F., Ponce-Alvarez, A., Battaglia-Mayer, A., Deco, G., & Schultz, W. (2023). A view-based decision mechanism for rewards in the primate amygdala. *Neuron, 111*(3871–3884), e3814.

Grace, R. C. (1993). Violations of transitivity: Implications for a theory of contextual choice. *Journal of the Experimental Analysis of Behavior, 60*, 185–201.

Haber, S. N., Kim, K. S., Mailly, P., & Calzavara, R. (2006). Reward-related cortical inputs define a large striatal region in primates that interface with associative cortical connections, providing a substrate for incentive-based learning. *The Journal of Neuroscience, 26*, 8368–8376.

Hadland, K. A., Rushworth, M. F., Gaffan, D., & Passingham, R. E. (2003). The anterior cingulate and reward-guided selection of actions. *Journal of Neurophysiology, 89*, 1161–1164.

Harbaugh, W. T., Krause, K., & Berry, T. R. (2001). GARP for kids: On the development of rational choice behavior. *The American Economic Review, 91*, 1539–1545.

Hernadi, I., Grabenhorst, F., & Schultz, W. (2015). Planning activity for internally generated reward goals in monkey amygdala neurons. *Nature Neuroscience, 18*, 461–469.

Hosokawa, T., Kennerley, S. W., Sloan, J., & Wallis, J. D. (2013). Single-neuron mechanisms underlying cost-benefit analysis in frontal cortex. *The Journal of Neuroscience, 33*, 17385–17397.

Jezzini, A., & Padoa-Schioppa, C. (2020). Neuronal activity in the primate amygdala during economic choice. *The Journal of Neuroscience, 40*, 1286–1301.

Jezzini, A., & Padoa-Schioppa, C. (2024). Neuronal activity in the gustatory cortex during economic choice. *The Journal of Neuroscience, 44*, e2150232024.

Kennerley, S. W., & Wallis, J. D. (2009). Evaluating choices by single neurons in the frontal lobe: Outcome value encoded across multiple decision variables. *The European Journal of Neuroscience, 29*, 2061–2073.

Kennerley, S. W., Dahmubed, A. F., Lara, A. H., & Wallis, J. D. (2009). Neurons in the frontal lobe encode the value of multiple decision variables. *Journal of Cognitive Neuroscience, 21*, 1162–1178.

Kobayashi, S., Pinto de Carvalho, O., & Schultz, W. (2010). Adaptation of reward sensitivity in orbitofrontal neurons. *The Journal of Neuroscience, 30*, 534–544.

Lu, M. T., Preston, J. B., & Strick, P. L. (1994). Interconnections between the prefrontal cortex and the premotor areas in the frontal lobe. *The Journal of Comparative Neurology, 341*, 375–392.

Morrison, S. E., & Salzman, C. D. (2009). The convergence of information about rewarding and aversive stimuli in single neurons. *The Journal of Neuroscience, 29*, 11471–11483.

Niehans, J. (1990). *A history of economic theory: Classic contributions, 1720–1980*. Johns Hopkins University Press.

Ongur, D., & Price, J. L. (2000). The organization of networks within the orbital and medial prefrontal cortex of rats, monkeys and humans. *Cerebral Cortex, 10*, 206–219.

Padoa-Schioppa, C. (2009). Range-adapting representation of economic value in the orbitofrontal cortex. *The Journal of Neuroscience, 29*, 14004–14014.

Padoa-Schioppa, C. (2011). Neurobiology of economic choice: A good-based model. *Annual Review of Neuroscience, 34*, 333–359.

Padoa-Schioppa, C., & Assad, J. A. (2006). Neurons in orbitofrontal cortex encode economic value. *Nature, 441*, 223–226.

Padoa-Schioppa, C., & Cai, X. (2011). The orbitofrontal cortex and the computation of subjective value: Consolidated concepts and new perspectives. *Annals of the New York Academy of Sciences, 1239*, 130–137.

Padoa-Schioppa, C., & Conen, K. E. (2017). Orbitofrontal cortex: A neural circuit for economic decisions. *Neuron, 96*, 736–754.

Pastor-Bernier, A., Stasiak, A., & Schultz, W. (2021). Reward-specific satiety affects subjective value signals in orbitofrontal cortex during multicomponent economic choice. *Proceedings of the National Academy of Sciences of the United States of America, 118*, e2022650118.

Petrides, M., & Pandya, D. N. (2006). Efferent association pathways originating in the caudal prefrontal cortex in the macaque monkey. *The Journal of Comparative Neurology, 498*, 227–251.

Raghuraman, A. P., & Padoa-Schioppa, C. (2014). Integration of multiple determinants in the neuronal computation of economic values. *The Journal of Neuroscience, 34*, 11583–11603.

Rangel, A., Camerer, C., & Montague, P. R. (2008). A framework for studying the neurobiology of value-based decision making. *Nature Reviews. Neuroscience, 9*, 545–556.

Rich, E. L., & Wallis, J. D. (2016). Decoding subjective decisions from orbitofrontal cortex. *Nature Neuroscience, 19*, 973–980.

Roesch, M. R., & Olson, C. R. (2005). Neuronal activity in primate orbitofrontal cortex reflects the value of time. *Journal of Neurophysiology, 94*, 2457–2471.

Rudebeck, P. H., Mitz, A. R., Chacko, R. V., & Murray, E. A. (2013). Effects of amygdala lesions on reward-value coding in orbital and medial prefrontal cortex. *Neuron, 80*, 1519–1531.

Rustichini, A., & Padoa-Schioppa, C. (2015). A neuro-computational model of economic decisions. *Journal of Neurophysiology, 114*, 1382–1398.

Rustichini, A., Conen, K. E., Cai, X., & Padoa-Schioppa, C. (2017). Optimal coding and neuronal adaptation in economic decisions. *Nature Communications, 8*, 1208.

Saleem, K. S., Miller, B., & Price, J. L. (2013). Subdivisions and connectional networks of the lateral prefrontal cortex in the macaque monkey. *The Journal of Comparative Neurology, 522*, 1641–1690.

Samejima, K., Ueda, Y., Doya, K., & Kimura, M. (2005). Representation of action-specific reward values in the striatum. *Science, 310*, 1337–1340.

Setogawa, T., Mizuhiki, T., Matsumoto, N., Akizawa, F., Kuboki, R., Richmond, B. J., & Shidara, M. (2019). Neurons in the monkey orbitofrontal cortex mediate reward value computation and decision-making. *Communications Biology, 2*, 126.

Shin, E. J., Jang, Y., Kim, S., Kim, H., Cai, X., Lee, H., Sul, J. H., Lee, S. H., Chung, Y., Lee, D., et al. (2021). Robust and distributed neural representation of action values. *eLife, 10*, e53045.

Strait, C. E., Blanchard, T. C., & Hayden, B. Y. (2014). Reward value comparison via mutual inhibition in ventromedial prefrontal cortex. *Neuron, 82*, 1357–1366.

Sumpter, C., Temple, W., & Foster, T. (1999). The transitivity of choices between different response requirements. *Journal of the Experimental Analysis of Behavior, 72*, 235–249.

Takada, M., Nambu, A., Hatanaka, N., Tachibana, Y., Miyachi, S., Taira, M., & Inase, M. (2004). Organization of prefrontal outflow toward frontal motor-related areas in macaque monkeys. *The European Journal of Neuroscience, 19*, 3328–3342.

Takahara, D., Inoue, K. I., Hirata, Y., Miyachi, S., Nambu, A., Takada, M., & Hoshi, E. (2012). Multisynaptic projections from the ventrolateral prefrontal cortex to the dorsal premotor cortex in macaques—anatomical substrate for conditional visuomotor behavior. *The European Journal of Neuroscience, 36*, 3365–3375.

Tversky, A., & Simonson, I. (1993). Context-dependent preferences. *Management Sciences, 39*, 117–185.

Wallis, J. D. (2007). Orbitofrontal cortex and its contribution to decision-making. *Annual Review of Neuroscience, 30*, 31–56.

Xie, J., & Padoa-Schioppa, C. (2016). Neuronal remapping and circuit persistence in economic decisions. *Nature Neuroscience, 19*, 855–861.

Xu, C., Ni, K., & Cai, X. (2024). Dissociable neuronal modules in the orbitofrontal cortex representing single and multi-attribute choice tasks. *bioRxiv*, 2022–2007.

Yamada, H., Louie, K., Tymula, A., & Glimcher, P. W. (2018). Free choice shapes normalized value signals in medial orbitofrontal cortex. *Nature Communications, 9*, 162.

Yim, M. Y., Cai, X., & Wang, X. J. (2019). Transforming the choice outcome to an action plan in monkey lateral prefrontal cortex: A neural circuit model. *Neuron, 103*, 520–532 e525.

Decomposing Economic Choices with Drift-Diffusion Models

Ian Krajbich, Kianté Fernandez, and Xiaozhi Yang

Abstract Many decisions arise from a dynamic process of information accumulation and comparison. Thus, to fully understand decision-making, we must decompose the choice process into its parts. Here, we review foundational work and recent advances in the use of dynamic—e.g., drift-diffusion (DDMs)—models to study economic/value-based decision-making. We first introduce the DDM, which uses choice and response times (RT) to distinguish between the effects of choice options versus choice context. We then review modifications to the DDM for multi-attribute settings, where attributes are considered at different points in time, showing how these models can be related to traditional dual-process theories. Beyond choices and RT, DDMs also account for the fact that overt visual attention amplifies option- and attribute-level information, resulting in links between gaze and choice. While most DDMs are applied to binary choice, a growing literature extends these models to scenarios with many options or a continuum of responses (e.g., Likert ratings). We also consider work in adjacent domains, such as social applications, where people may infer information from observing each other's decision processes, and in cognitive neuroscience, which has helped us understand how this dynamic decision process plays out in the brain. We conclude by highlighting emerging trends and future directions, including reinforcement learning, natural language processing, and applications in field settings.

I. Krajbich (✉) · K. Fernandez
Department of Psychology, University of California Los Angeles, Los Angeles, CA, USA
e-mail: krajbich@ucla.edu

X. Yang
MindCORE, University of Pennsylvania, Philadelphia, PA, USA

D. V. Smith et al. (eds.), *Neuroeconomics: Core Topics and Current Directions*,
https://doi.org/10.1007/978-3-032-02925-6_9

1 Introduction

Most decisions do not happen instantly. Instead, they arise from a noisy, dynamic process. As a result, choices are not due solely to underlying utilities/values—they can also reflect context and manipulations. Thus, to understand preferences, we must study the choice process and decompose it into its constituent mechanisms.

Many decisions are well captured by a class of models known as sequential-sampling (Usher & McClelland, 2001; Ratcliff & Smith, 2004) or evidence-accumulation models (Heathcote & Matzke, 2022), the most well-known of which is the drift-diffusion (or diffusion-decision) model (DDM; Ratcliff & McKoon, 2008). These models assume that the decision-maker evaluates the options in front of them, gathering evidence to support one over the other(s). When the net evidence favoring one option reaches a pre-specified boundary, a choice is made (Fig. 1a, b). These models capture links between underlying values, choice probabilities, and response times (RT) (and other measures, see below). For instance, choices between two options with very different values are fast and consistent, while choices between two options with similar values are slow and inconsistent. Drift-diffusion models thus allow us to estimate preferences (i.e., values) from both choice and RT data (Konovalov & Krajbich, 2019; Clithero, 2018a, b; Berlinghieri et al., 2023).

The DDM has several appealing features. First, under basic assumptions, it is the optimal decision-making rule in the sense that it minimizes RT for a desired accuracy rate (Wald, 1945; Degroot, 1970; Bogacz et al., 2006; Bhui, 2019a, b; but see Fudenberg et al., 2018, Tajima et al., 2016). Second, empirically it provides a tight match to choice and RT distributions across many domains of cognition including memory, perception, and preference (Ratcliff et al., 2016; Evans & Wagenmakers, 2020; Busemeyer et al., 2019). Third, it has substantial support from neuroscience, where decision-related neurons appear to ramp up their activity until they hit a threshold just prior to choice (Gold & Shadlen, 2007; Forstmann et al., 2016, but see Latimer et al., 2015, Zoltowski et al., 2019). The DDM has a lot going for it.

In recent years, the DDM (and other related models) have been used to uncover the dynamics of economic/value-based choice (Krajbich & Smith, 2015; Clithero, 2018a, b). This research is exciting, in part because it has attracted researchers beyond psychology (Dai & Busemeyer, 2014), from economics (Krajbich et al. 2014; Fudenberg et al., 2018; Webb, 2019; Baldassi et al., 2020; Alós-Ferrer et al., 2021; Hébert & Woodford, 2023; Berlinghieri et al., 2023), marketing (Xiang Chiong et al., 2024), finance (Frydman & Nave, 2017), and neuroscience (Shadlen & Shohamy, 2016). In this chapter, we will discuss this work and highlight the different ways in which the DDM has been used to gain novel insight into the economic choice process.

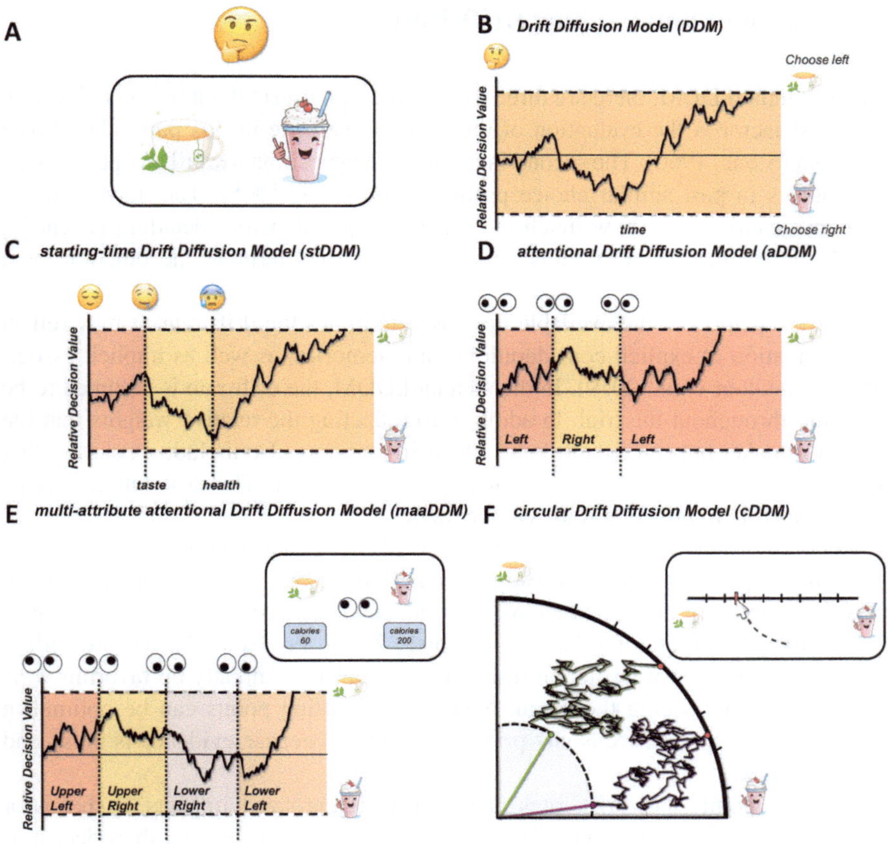

Fig. 1 A range of DDMs have been applied to economic choice. (**a**) Example of a typical economic choice between two goods. (**b**) The decision-maker samples evidence in favor of both options until they hit a decision boundary. (**c**) In the stDDM, different attributes enter the drift rate at different times. In this example, initially there are no attributes, then taste enters in, and later health enters in. (**d**) In the aDDM, shifts in gaze between options alter the drift rate, favoring the fixated option. (**e**) In the maaDDM, shifts in gaze between options and attributes alter the drift rate, favoring the fixated element. (**f**) The circular DDM captures continuous rather than discrete choices; in this case, the strength of preference between the two options. In all these models, the decision-maker may start with information that biases them toward preferring one option over another. For example, when in a coffee shop, people may be predisposed to favor tea (green line), but when in an ice cream parlor, they may be predisposed to favor a milkshake (purple line), regardless of the actual options on the menu (which determine the drift rate). In this example, the gray area represents evidence that is there before the decision, i.e., the starting point

2 Starting Points Versus Drift Rates

In the standard DDM, there are three factors that can affect which option is favored. The first factor is the evaluation of the options available in this particular choice problem (Stone, 1960). The second factor is a predisposition toward an option based on choices in past similar choice problems (Edwards, 1965). The third factor is response caution, i.e., how much evidence is required before deciding (Pachella, 1973). These three factors, plus noise, combine to determine the outcome of a particular choice.

The evaluation of the available options determines the drift rate. It may reflect a combination of explicit considerations and memories, as well as implicit associations (Bakkour et al., 2019). In the standard DDM, the drift rate is assumed to be constant throughout the trial. In addition to reflecting the relative weights that the decision-maker puts on the attributes, the drift rate may also include a constant that reflects a preference toward one category over the other. The drift rate serves an analogous role to utility in static choice models.

A decision-maker's predisposition is reflected in the starting point of the decision process, i.e., evidence that is already there before the options have been presented. If there is any prior or contextual information, the decision-maker may already have reason to favor an option (Desai & Krajbich, 2022). For example, when deciding what drink to order at an Italian restaurant, a diner may already be favoring wine over beer before seeing the menu. Non-neutral starting points can be optimal in environments where there is any prior information, because evidence is noisy and time is costly.

Starting points and drift rates can have different-sized impacts on behavior depending on time constraints. When time is tight or very costly, then decision-makers will exert less response caution, resulting in narrower decision boundaries. In such cases, decision-makers will spend less time evaluating their options and thus be more influenced by their predispositions.

Whether a behavioral effect is due to drift rates or starting points is an important distinction. Drift rates reflect how a person perceives the options, while starting points reflect considerations before seeing the options. In one example, Izakson et al. (2024) investigated decoy effects, finding that nearby consumer products influenced subjects' values for target products. However, upon further analysis, the authors found that this was due to starting points, not drift rates—the effect disappeared for slower decisions. Thus, in that setting, decoys did not systematically change how subjects perceived the target items. Instead, the decoys provided contextual information that subjects used to get a head start on their decisions. Rather than requiring new theories about contextual value distortions, this alternative interpretation of the choice data relied on a simple, adaptive mechanism.

Researchers have started to use the DDM to study the role of starting points, drift rates, and other DDM parameters on economic choice (Enax et al., 2016; Zhao et al., 2020; Yu et al., 2021; Desai & Krajbich, 2022). In one notable example, Zhao et al. (2022a) decomposed a battery of 19 common behavioral nudges and classified them based on the mechanism through which they operate. For instance, price information and cognitive load operate primarily through starting point; default options, accept/reject framing, and accountability operate primarily through drift rate; social norms and recommendations use both. Overall, this line of research is demonstrating that economic phenomena can be attributed to a mix of drift rates and starting points.

Next, we will consider innovations in drift-diffusion models that have emerged as a result of their application to value-based choice.

3 Latency Effects

One development in value-based DDMs is the consideration of multi-attribute choice. Unlike standard perceptual and memory tasks, most value-based decisions involve trading off different attributes of the options. For example, when choosing what to eat, a decision-maker has to decide how much to consider the tastiness versus the healthiness of the options. These attribute weights add another layer of complexity to the choice process.

In recent years, the field has developed DDMs that allow for the drift rate to change by adding in attributes over time (Fig. 1c). Attributes that enter the drift rate faster have earlier and potentially larger effects on decisions.

Attribute latencies have been estimated in a couple different ways. One way is through formal modeling of choice and RT distributions using a time-varying DDM (Sullivan & Huettel, 2021; Maier et al., 2020; Barakchian et al., 2021). The other way is with mouse-tracking experiments, where a decision-maker's hand movements while selecting an option reveal their attribute latencies (Sullivan et al., 2015; Chen et al., 2024a, b). Chen et al. (2024b) demonstrated that the mouse trajectory-derived latency is highly correlated with the estimated latency from the time-varying DDM, indicating a strong correspondence between mouse movements and decision mechanisms.

These latency models have provided a richer understanding of the choice process in a variety of domains. Sullivan and Huettel (2021) and Maier et al. (2020) applied latency models to dietary choice and found that when health considerations are delayed relative to taste considerations, differences in weights on taste and health matter less for choice (Fig. 2a). In intertemporal choice, Amasino et al. (2019) and Chen et al. (2024a) showed that when decision-makers consider reward information before time information, they tend to be more patient. Chen et al. (2024b) also applied a latency model to social decisions. They found that selfish individuals take longer to process others' payoffs, whereas the opposite was true for prosocial individuals. In the domain of risky choice, Chen et al. (2022) found age differences in the order of attribute processing, with older adults considering probability information earlier.

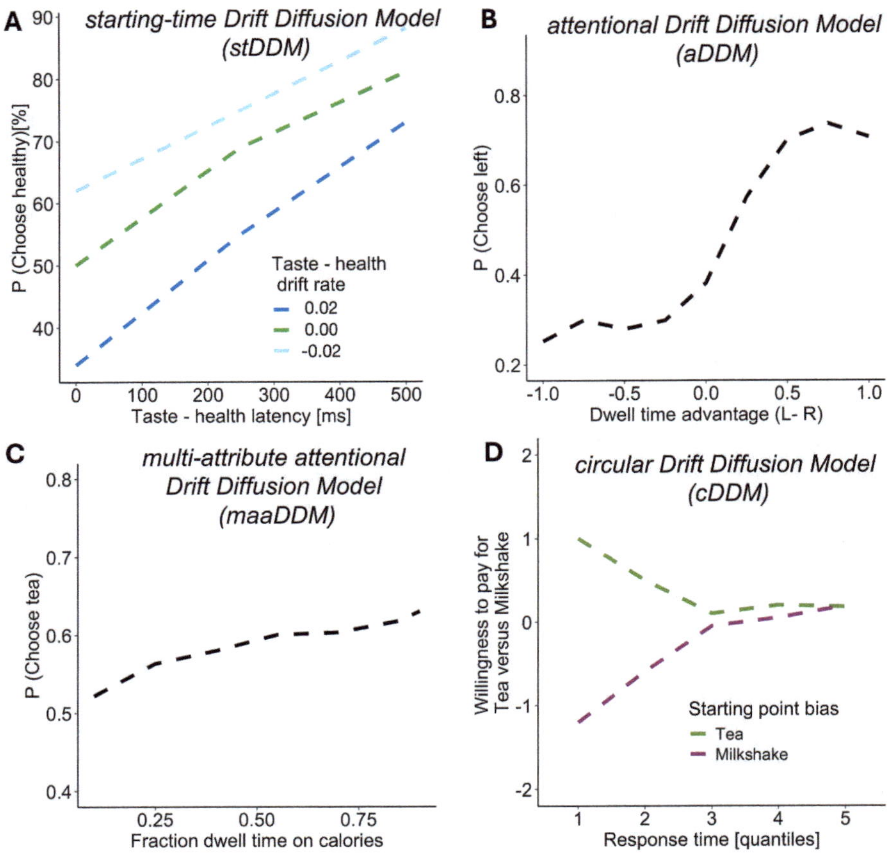

Fig. 2 Results derived from the models in Fig. 1. (**a**) The DDM with attribute latencies correctly predicts that when the latencies are more staggered, the attribute information has less impact on the drift rate. In this example, based on data from Sullivan and Huettel (2021), choices are less sensitive to the predicted drift rates (based on taste-health information), as the latencies diverge. This can be seen by noting the shrinking distances between the three curves from left to right. (**b**) In the aDDM, a relative dwell time advantage toward one option (e.g., left) correlates with increased choice of that option. (**c**) In the maaDDM, a relative dwell time advantage toward one attribute (e.g., calories in the Fig. 1 example) correlates with increased choice of the option that is better on that attribute (in this case, tea instead of milkshake). (**d**) In the circular DDM, here based on data from Izakson et al., 2024, a starting point bias toward an option leads to a higher value for that option, but only for the fastest decisions

Manipulating the order in which attributes are considered can alter the decision process, demonstrating that attribute latency causally influences choice. For example, Chen et al. (2024a) had decision-makers process the rewards/times first, which forced rewards/times to enter the decision process earlier, resulting in more/less patient choices. Maier et al. (2020) found that cueing attention to healthiness can both shorten the health latency and increase the weight on health, while Sullivan and Huettel (2021) only found effects on the attribute weights.[1] Additionally, Maier et al. (2020) found that electrical stimulation over the left dorsolateral prefrontal cortex (DLPFC) caused a change in the relative weights on taste and healthiness, but not in their latencies.

This work on latency effects explains how the timing of information processing can shape decisions (for related work on Query Theory, see Weber et al., 2007). It also suggests a way to understand how "intuitive" and "deliberative" processes might contribute to decisions, as explored in the next section.

4 Relation to Dual-Process Models

Drift-diffusion models provide a framework to think about dual processes. Dual-process models assume that an intuitive system (System 1) and a deliberative system (System 2) interact to reach a decision (Kahneman, 2011). Within the DDM framework, dual processes can be conceptualized in at least two ways. In one approach, System 1 influences the starting point of the decision process, while System 2 influences the drift rate. In another approach, System 1 and System 2 both contribute to the drift rate, perhaps with different latencies (e.g., System 1 has a shorter latency and thus influences the decision earlier).

Some proponents of dual-process theories have argued that time pressure promotes intuitive responding (e.g., Rand et al., 2012). Using the DDM, Chen and Krajbich (2018) found that time pressure indeed puts more emphasis on the starting point. They showed that decision-makers vary in their starting point toward selfish or altruistic choices, and that time pressure accentuates the effects of these starting points, making prosocial people more prosocial and selfish people more selfish. However, subsequent work has argued that these effects may also/instead be due to attentional priorities (Teoh et al., 2020; 2022) or attribute latencies (Chen et al., 2024b). These competing mechanisms highlight the complexity belied by the standard dual-process account.

Alternatively, one can also think of System 1 and System 2 as different drift rates, either operating in parallel (Alós-Ferrer et al., 2021) or in sequence (Diederich & Trueblood, 2018). In the Diederich and Trueblood model, System 1 drives evidence accumulation at the beginning of a decision, while System 2 takes over later in the decision. Such a model can also explain why intuitive decisions tend to happen earlier and more often under time pressure.

While the DDM framework paves the way for more detailed modeling of dual-process cognition, the DDM goes well beyond it. A dual-process model is merely a special case of a more general multi-stage decision model, where the number of

---

[1] It is unclear why cueing affected latencies in one paradigm and not the other, but perhaps it is due to the fact that the cues were manipulated within-subject in the Maier task but between-subject in the Sullivan and Huettel task.

stages aligns with the number of attributes, options, or ways that attention can be allocated, as we describe next.

5 The Attentional DDM

Another development in DDMs allows for the drift rate to shift back and forth over the course of a decision as the decision-maker shifts their attention between options (or attributes, see below) (Fig. 1d). This idea builds on an older literature on Decision Field Theory (Busemeyer & Townsend, 1993; Diederich, 1997), except that now attention is linked to eye movements. These gaze-informed DDMs, like the attentional DDM (aDDM; Krajbich et al., 2010) and the gaze-weighted linear accumulator model (GLAM; Thomas et al. 2019; Molter et al., 2019), assume that the fixated information gets more weight in the drift rate than non-fixated information.

There is both correlational and causal support for these models. Studies have repeatedly shown correlations between relative dwell time, final fixation location, and choice (Fig. 2b; see Krajbich 2019 for a review). Decisions with high overall value display shorter RT and larger dwell advantage effects, which is consistent with the aDDM's assumption of a multiplicative (Smith & Krajbich, 2019; Krajbich et al., 2021), rather than additive (Cavanagh et al., 2014; Stewart et al., 2016), effect of gaze (see Westbrook et al., 2020 for a hybrid model). More recent models have argued that gaze amplifies goal-consistent information, rather than value per se (Kovach et al., 2014; Sepulveda et al., 2020; see also Frömer et al., 2019).

There is also now a wealth of evidence that gaze manipulations can bias choice (Bhatnagar & Orquin, 2022), contrary to earlier claims (Mormann & Russo, 2021). Some manipulations are more effective than others, and, because of covert attention, shifting a person's gaze does not guarantee that you have shifted their attention. Manipulations that bias relative dwell time are more effective than those that bias first fixations, consistent with the aDDM (Bhatnagar & Orquin, 2022).

Gaze-informed DDMs have been used to explain a variety of behavioral phenomena/models including risk preference (Sheng et al., 2020; Zilker & Pachur, 2022, 2023; Glickman et al., 2019; He & Bhatia, 2023), the endowment effect (Ashby et al., 2012; Sheng et al., 2023), preference reversals (Kim et al., 2012), opportunity-cost neglect (Smith et al., 2024), prosociality (Krajbich et al., 2015), morality (Pärnamets et al. 2015), brand preference (Yang et al., 2024), etc.

There is still an open question of what influences gaze. Several articles have proposed models where gaze is drawn to options that the decision-maker is least certain about (Song et al., 2019; Callaway et al., 2021; Jang et al., 2021; Li & Ma, 2021) or, in multi-alternative choice, to options that the decision-maker is favoring (Callaway et al., 2021). There is also empirical evidence that, in some cases, gaze is drawn to the preferred option over time (Sepulveda et al., 2020; Westbrook et al., 2020; Gluth et al., 2020). However, our own unpublished findings have found it to occur primarily in experiments with small sets of repeated/overlearned options, suggesting that gaze may be drawn to specific stimuli rather than value per se (Cavanagh et al., 2019). This issue requires further investigation (for reviews, see Wedel et al., 2023, Ting & Gluth, 2024).

6 Multi-attribute aDDM

When options have multiple attributes, gaze can shift between those attributes, which can also affect the drift rate. The multi-attribute aDDM (maaDDM) assumes that the value of an *attribute* is also amplified by gaze (Yang & Krajbich, 2023; Fisher, 2021), and so attributes that are fixated longer receive more weight in the decision process (Fig. 2c). Thus, the drift rate is determined by both the option and the attribute currently being fixated on (Fig. 1e). Yang & Krajbich, (2023) found that in most of their cases, unattended attributes of unattended options are discounted twice, once for each factor. This aligns with the idea that attention is a gradient, with items farther from the focal point yielding less information (LaBerge & Brown, 1989).

The work on the maaDDM addresses some questions raised by the aDDM but also poses many new ones. It helps to address the debate about additive versus multiplicative gaze effects (Cavanagh et al., 2014; Smith & Krajbich, 2019; Westbrook et al., 2020). The additive models interpret the correlation between gaze and choice as resulting from the decision-maker shifting their attention to the likely choice. But this model cannot explain why the allocation of gaze between attributes would matter. That would require a new, separate mechanism. Gaze amplifying fixated information is a more parsimonious explanation for both option-wise and attribute-wise gaze effects on choice.

On the other hand, the issue of causality is murkier with multi-attribute gaze. The results from Yang & Krajbich (2023) indicate that the allocation of gaze between attributes helps to explain choice, above and beyond a decision-maker's average weight on the attributes. However, it is also true that decision-makers gaze longer at attributes that they on average put more weight on. Future work will need to investigate causal manipulations of attribute-wise attention to determine how much it affects choice.

7 Multiple Options and Continuous Responses

While much of the DDM literature has focused on binary choice, many decisions in life involve more than two options (Towal et al. 2013). Other decisions require us to assign a value to an item, e.g., making an offer on a car or house (Johnson & Busemeyer, 2005). These more complex decisions are underexplored but are gaining momentum.

A core issue when extending the DDM to multiple options is the nature of the comparison process. Do decision-makers continue to compare options, or do they evaluate the options independently? If they compare, do they compare all pairs of options, the best options, or one pair at a time (Cerreia-Vioglio et al., 2023)?

There are several good arguments for a model that primarily compares the best options. The "max versus next" model is asymptotically optimal in the sense that it minimizes RT for near-perfect accuracy. Moreover, Krajbich & Rangel, 2011 found that the best versus next model outperformed the "max versus average" model in predicting behavior out of sample. A variant of this model also performed well when explaining choices in sets of 9, 16, 25, and 36 options (Thomas et al., 2021). Callaway et al., 2021 and Sepulveda et al., 2020 argue that a model that focuses attention on the best and second-best options is optimal, and Callaway et al. (2021) showed that such a model can account for several interesting patterns in trinary-choice data, such as longer fixations as the decision progresses and more fixation time on higher value items. Thus, there is a lot of support for models that increasingly focus on the best options over time (see also Gluth et al., 2020).

On the other hand, there is some evidence that the choice process may shift away from a relative comparison model as the number of options grows. Choice overload is a well-known phenomenon where people find it more difficult to choose from larger sets (Chernev et al., 2015; Iyengar & Lepper, 2000; but see Scheibehenne et al., 2010). An issue that surely lies at the heart of choice overload is limited working memory. As sets grow, it becomes increasingly difficult to keep track of multiple options. Thomas et al. (2021) found that for the larger set sizes, the best-fitting model shifted from a relative comparison model to a probabilistic satisficing model (Reutskaja et al., 2011). Consistent with limited working memory, they also found that subjects looked at a smaller fraction of items, and thus a roughly constant number of items, as set size grew.

In multi-option, multi-attribute settings, decoy effects can occur when the addition of a third option distorts a decision-maker's preference between two existing options (Roe et al., 2001; Trueblood 2022). Models that incorporate gaze processes can help to explain decoy effects (Gluth et al., 2018; Molter et al., 2022; Noguchi & Stewart, 2018; Marini et al., 2020). For example, Noguchi and Stewart (2014) found that eye movements and choices were most consistent with a process where people compare pairs of alternatives on single dimensions, which led them to propose a model where evidence is accumulated based on pairwise ordinal comparisons of attribute values (Noguchi & Stewart, 2018). They also found that more similar options receive greater attention, making them more likely to be selected for comparison. Gluth et al. (2018) argued that high-value decoys attract attention away from the best options, leading to slower and less accurate choices. Molter et al. (2022) further confirmed the importance of gaze-dependent evidence accumulation but also found support for similarity-dependent inhibition mechanisms.

Another exciting area of research is modeling decisions on continuous scales (for a review, see Rasanan et al. (2024)). Consider someone deciding how much they are willing to pay (WTP) for a car. Like with the DDM, we can think of this decision as a dynamic process that starts from a prior expectation and evolves as the decision-maker evaluates the car. We can model the decision variable as starting at the center of a circle (or circular sector) and accumulating evidence over time until it reaches the edge of the circle (Fig. 1f). The angle of that intersection point determines the decision-maker's reported value (Kvam & Busemeyer, 2020; Kvam & Turner, 2021). In these models, the drift angle reflects the decision-maker's evaluation of the good.

A circular diffusion model was recently applied in an economic setting by Kvam and Busemeyer (2020), who asked participants for their WTP to play a gamble and their willingness to accept (WTA) to give up that chance. Their model captured the reported price distributions, including the right-skew for low-probability gambles and left-skew for high-probability gambles. The model also provided an explanation for the difference in skew between WTP and WTA conditions, attributing it to variations in the starting points for each price type; buyers typically start low, while sellers generally start high. This offers a unified, process-level account that explains WTP and WTA disparities using a simple model.

Another notable example is Izakson et al. (2024), discussed above, which provides a glimpse at how these continuous diffusion models could be used to study decoy effects in a WTP task. Consistent with what the authors found in binary choice, decoy values had an attractive force on the WTP for the target items, but only for the fastest decisions (Fig. 2d). The authors found that this pattern could only be explained by a model where decoys affect the starting point of the process and not the drift rates of the goods.

The use of sequential sampling models to understand multi-option and continuous economic decisions is still a nascent, exciting topic. There is still much to learn about how these models accommodate larger choice sets and explain distributions of continuous responses, such as the apparent discretization of WTP (Krajbich et al. 2012).

8 Social Use of DDM

Researchers can learn a lot from a decision-maker's RT. A related question is whether decision-makers learn from each other's RT. For example, one might be able to infer another's strength-of-preference or belief from their RT (Gates et al., 2021; Bavard et al., 2024). This RT inference could occur in either cooperative or competitive settings. In the former, people could learn more from each other. In the latter, people could access private information to gain a competitive advantage.

There are now a few examples in cooperative settings. Frydman & Krajbich (2016, 2022) and Tump et al. (2024) both examined information cascades games where people are trying to guess the state of the world. Participants received private information but also saw other participants' guesses. Frydman & Krajbich showed that players are slower to guess when they have conflicting information, and so subsequent players are less likely to follow someone who guessed slowly.

Similarly, Tump et al. showed that players with better information guess earlier in the sequence, providing guidance to those with worse information.

In competitive settings, Konovalov & Krajbich (2017, 2023), and Cotet et al. (2025) have studied RT inference in bargaining. When bargaining, a quick rejection likely signals that your offer was way off, while a slow rejection likely signals that you were close to acceptance. Konovalov & Krajbich (2023) showed that laboratory bargainers indeed adjust their second offers more after being rejected quickly versus slowly, and that they prefer to reject others more quickly. Cotet et al. (2025) showed that the same information is present in RT data from eBay bargaining, but that bargainers there instead seem to be discouraged by slow rejections.

Together, these findings highlight how decision-makers use RT as a cue for interpreting others' intentions or information. We now switch gears to discuss the underlying neural mechanisms behind the decision processes discussed throughout the chapter.

9 Neural Basis of Economic DDM

Research in neuroeconomics has largely focused on utility/value/reward representations in the brain (e.g., Schultz et al., 1997; Plassmann et al., 2007; Bartra et al., 2013; Castegnetti et al., 2021). However, without a model of the choice process, it is challenging to further decompose these brain regions into distinct components or to know how these utility representations determine choice. The DDM offers a framework for addressing these issues.

Much of the research connecting the DDM to decision neuroscience originated from studies of perceptual decision-making (for reviews, see Shadlen & Kiani, 2013, and O'Connell et al., 2012; 2018). Using a dot-motion task, researchers showed that motion-sensitive neurons display steady activity proportional to motion strength, while eye-movement neurons ramp up their activity over time with a slope proportional to motion strength (the macaque decision-makers choose with their eyes). Moreover, these latter neurons reach a threshold level of activity before triggering a choice. These patterns align well with the DDM (for a review, see Gold & Shadlen, 2007). This evidence is supported by studies not only in monkeys (Roitman & Shadlen, 2002; De Lafuente et al., 2015; Jun et al., 2021) but also in mice (Hanks et al., 2015), rats (Brunton et al., 2013), and humans (Kelly & O'Connell, 2013; O'Connell et al., 2012; Gherman et al., 2024; Gluth et al., 2013; Twomey et al., 2015; Lui et al., 2021). Much of the work in humans supports the existence of a localized centroparietal source that tracks evidence accumulation (O'Connell et al., 2018). Consequently, a growing body of human research continues to map key parameters of the DDM onto MEG/EEG signatures (for reviews, see Nunez et al., 2024; Harris & Hutcherson, 2021). However, more work is needed to determine if scalp potentials encode evidence accumulation, correlated processes, or mixtures of signals (Philiastides et al., 2014; Frömer et al., 2024).

Compared to perceptual choice, where decision-makers sample external evidence from the stimuli, in economic choice decision-makers often sample internal representations of utility/value/reward (Hare et al. 2011a, b; Shadlen & Shohamy, 2016). Thus, reward regions like the ventral striatum and ventromedial prefrontal cortex (VMPFC) serve an analogous role to the motion-sensitive neurons in the dot-motion task. These regions provide input to evidence accumulation regions elsewhere in the cortex.

While less suited for studying neural dynamics, fMRI has been used to study economic evidence accumulation in the brain. Several studies have identified the DLPFC, insula, pre-supplementary motor area (preSMA), and intraparietal sulci (IPS) as potential accumulation regions (Basten et al., 2010; Hare et al. 2011a, b; Gluth et al., 2012; Rodriguez et al., 2014; Magrabi et al., 2022). Notably, Hare et al. (2011b) used effective connectivity analyses to identify a chain of decision-related activity that starts with VMPFC sending input to preSMA and IPS, followed by preSMA, IPS, and DLPFC sending excitatory/inhibitory signals to the left/right motor cortices for right/left choices, respectively. These results paint a rich picture of how economic decisions unfold in the brain.

For a more granular picture of how economic decisions unfold, it is useful to combine the temporal resolution of EEG/MEG with the spatial resolution of fMRI. Research with EEG/MEG aligns well with the fMRI results described above, as it has identified both parietal and frontal regions that track economic evidence accumulation (Hunt et al., 2012; Polanía et al., 2014). Some of the best research on economic evidence accumulation has used concurrent fMRI and EEG (Pisauro et al., 2017), revealing centroparietal activity that closely matched the ramping signature predicted by DDM, which could be traced to activity in the SMA, which in turn showed functional connectivity with the VMPFC and striatum. Similar results have been found in monkeys, where orbitofrontal neurons reflect relative stimulus values while anterior cingulate neurons integrate the evidence (Hunt et al., 2018). Causal evidence from human lesion patients further supports the importance of dorsomedial PFC in the decision process (Vaidya & Fellows, 2015).

If people are sampling internal evidence to make their decisions, then memory should play a role (Shadlen & Shohamy, 2016). In an fMRI study, Bakkour et al. (2019) found that longer decision times engage the hippocampus more, leading to the conclusion that in economic decisions, memory-related regions are more active during decisions with longer RT. This finding was further supported by evidence showing that patients with hippocampal damage are selectively impaired in economic decisions but not in perceptual decisions. Related research with the aDDM/GLAM has found that people accumulate evidence for options even when they are invisible and need to be retrieved from memory. In such cases, attentional discount factors are even more extreme (Weilbächer et al., 2021; Eum et al., 2023).

There has also been some investigation into the neural effects of attention in DDM processes. Lim et al. (2011) demonstrated that reward regions (i.e., striatum and VMPFC) are modulated by attention, showing opposite correlations with value difference (left-right) depending on fixation location, as predicted by the aDDM. Shevlin et al., (2025) extended these results, showing evidence for gaze modulation of accumulated evidence in pre-SMA. Similarly, in monkeys, value-related activity in the orbitofrontal cortex is amplified when subjects fixate near the stimulus (McGinty et al. 2016; Hunt et al., 2018). Krajbich et al. (2021) used transcranial magnetic stimulation (TMS) to disrupt brain activity crucial for guiding visual attention. As a result, subjects' high-value choices were slower and less correlated with relative gaze time, consistent with less attentional discounting in the aDDM.

Another advantage of taking a DDM perspective is that it can provide new interpretations of findings in the literature. In one persuasive example, Hutcherson & Tusche (2022) argued that DLPFC activity during self-control tasks, which has often been associated with impulse inhibition, is better explained by there being more evidence accumulation in conflicting decisions. They showed that the typical correlation between DLPFC activity and self-control is reversed when self-control is the typical response.

Admittedly, while significant progress has been made in understanding the neural mechanisms underlying preference formation during decision-making, incorporating relevant neuroscience and mathematical frameworks from psychology is still in its early stages. More work is needed to integrate neurophysiological data into economic decision models.

10 Emerging Trends

10.1 Reinforcement Learning with DDM

A major topic in decision neuroscience is reinforcement learning (RL). But like with utility/value models, work in RL tends to ignore how values lead to choices. Recent research has sought to enhance RL research by integrating the learning models into DDM-like choice models (Ratcliff & Frank, 2012; Pedersen et al., 2017; Fontanesi et al., 2019; Miletić et al., 2020, 2021). By doing so, these RLDDMs allow us to explain learning effects, such as the increased probability of choosing the option with the higher utility/value and the corresponding decrease in RT across trials, as well as perseveration. Other research has exploited the dynamics of the choice process, using eye-tracking and mouse-tracking to better understand RL (Cavanagh et al., 2014) and the distinction between model-free and model-based learning (Konovalov & Krajbich, 2016, 2020). With the practical advantages of better parameter recovery relative to standard RL models (Shahar et al., 2019) and freely available software packages designed to fit RLDDMs (Pedersen & Frank, 2020; Fengler et al., 2022), we believe that more great work is yet to come.

10.2 Natural Language Processing as Input to DDM

The advancement of powerful machine learning techniques has enabled the extraction of detailed quantitative representations of words and sentences. Recently, researchers have begun integrating these representations from large-scale digital data into cognitive-process models (Bhatia & Aka, 2022). For instance, Zhao et al. (2022b) utilized a thought listing paradigm to examine memory-based decision-making. They obtained vector representations of decision-related thoughts and grouped these thoughts into distinct clusters. These thought clusters were then used as inputs for sequential sampling models for decisions made from memory. Instead of using eye-tracking to observe information sampling, this approach demonstrates the value of using semantic vectors to model thought content during naturalistic decision-making.[2]

10.3 Field Settings

One critique of research on the DDM is that it has, until very recently, been limited to fast, repeated decisions in the lab. Thus, a critical step forward is determining whether the framework extends to slow, infrequent decisions in the field. In principle, there is nothing limiting the DDM to fast decisions—the model is agnostic to the time scale of evidence accumulation. From a practical perspective, it may seem implausible that people are accumulating evidence over minutes, hours, or days. Nevertheless, there are now a few examples of DDM-like behavior in the field.

One example was described above (Social Use of DDM), where Cotet et al. (2025) found that sellers bargaining on eBay take hours longer to reject higher offers and to accept lower offers on their products. Buyers exhibit analogous behavior in response to sellers' counteroffers. Clearly, there is information in RT in real markets, contrary to many economists' intuition that people should know ahead of time what offers they will accept. In another field setting, Card et al. (2024) examined editors' RT at top Economics journals. They found that papers with slower revise-and-submit decisions are less likely to be ultimately accepted at the journal and receive fewer future citations. In contrast, papers with slower rejection decisions receive more future citations. Finally, Chiong et al. (2024) examined users' RT to mobile ads and found that the RTs predicted out-of-sample ad-engagement better than choices.

We anticipate that this is just the tip of the iceberg in terms of the ways in which the DDM can enrich our understanding of behavior outside the lab.

[2] Limitations of this approach are the interpretability of the high-dimensional vector representations and the question of whether representations derived solely from language models are sufficient for characterizing mental representation.

11 Conclusion

All told, there are countless exciting developments occurring in the realm of sequential sampling models. This chapter merely scratches the surface of the theoretical developments and applications of these models to decision-making problems. We look forward to the next advancements in the field and hope that this chapter has given the reader some new ideas for their own research.

Acknowledgments We thank Sebastian Gluth for comments and the National Science Foundation (Award #2333979) for financial support.

References

Alós-Ferrer, C., Fehr, E., & Netzer, N. (2021). Time will tell: Recovering preferences when choices are noisy. *Journal of Political Economy, 129*(6), 1828–1877. https://doi.org/10.1086/713732

Amasino, D. R., Sullivan, N. J., Kranton, R. E., & Huettel, S. A. (2019). Amount and time exert independent influences on intertemporal choice. *Nature Human Behaviour, 3*(4), 383–392. https://doi.org/10.1038/s41562-019-0537-2

Ashby, N. J. S., Dickert, S., & Glöckner, A. (2012). Focusing on what you own: Biased information uptake due to ownership. *Judgment and Decision Making, 7*(3), 254–267. https://doi.org/10.1017/S1930297500002230

Bakkour, A., Palombo, D. J., Zylberberg, A., Kang, Y. H., Reid, A., Verfaellie, M., Shadlen, M. N., & Shohamy, D. (2019). The hippocampus supports deliberation during value-based decisions. *eLife, 8*, e46080. https://doi.org/10.7554/eLife.46080

Baldassi, C., Cerreia-Vioglio, S., Maccheroni, F., Marinacci, M., & Pirazzini, M. (2020). A Behavioral characterization of the drift diffusion model and its multialternative extension for choice under time pressure. *Management Science, 66*(11), 5075–5093. https://doi.org/10.1287/mnsc.2019.3475

Barakchian, Z., Beharelle, A. R., & Hare, T. A. (2021). Healthy decisions in the cued-attribute food choice paradigm have high test-retest reliability. *Scientific Reports, 11*(1), Article 1. https://doi.org/10.1038/s41598-021-91933-6

Bartra, O., McGuire, J. T., & Kable, J. W. (2013). The valuation system: A coordinate-based meta-analysis of BOLD fMRI experiments examining neural correlates of subjective value. *NeuroImage, 76*, 412–427. https://doi.org/10.1016/j.neuroimage.2013.02.063

Basten, U., Biele, G., Heekeren, H. R., & Fiebach, C. J. (2010). How the brain integrates costs and benefits during decision making. *Proceedings of the National Academy of Sciences, 107*(50), 21767–21772. https://doi.org/10.1073/pnas.0908104107

Bavard, S., Stuchlý, E., Konovalov, A., & Gluth, S. (2024). Humans can infer social preferences from decision speed alone. *PLoS Biology, 22*(6), e3002686. https://doi.org/10.1371/journal.pbio.3002686

Berlinghieri, R., Krajbich, I., Maccheroni, F., Marinacci, M., & Pirazzini, M. (2023). Measuring utility with diffusion models. *Science Advances, 9*(34), eadf1665. https://doi.org/10.1126/sciadv.adf1665

Bhatia, S., & Aka, A. (2022). Cognitive Modeling with representations from large-scale digital data. *Current Directions in Psychological Science, 31*(3), 207–214. https://doi.org/10.1177/09637214211068113

Bhatnagar, R., & Orquin, J. L. (2022). A meta-analysis on the effect of visual attention on choice. *Journal of Experimental Psychology: General, 151*(10), 2265–2283. https://doi.org/10.1037/xge0001204

Bhui, R. (2019a). A statistical test for the optimality of deliberative time allocation. *Psychonomic Bulletin & Review, 26*(3), 855–867. https://doi.org/10.3758/s13423-018-1555-1

Bhui, R. (2019b). Testing optimal timing in value-linked decision making. *Computational Brain & Behavior, 2*(2), 85–94. https://doi.org/10.1007/s42113-019-0025-9

Bogacz, R., Brown, E., Moehlis, J., Holmes, P., & Cohen, J. (2006). The physics of optimal decision making: A formal analysis of models of performance in two-alternative forced-choice tasks. *Psychological Review, 113*(4), 700–765. https://doi.org/10.1037/0033-295X.113.4.700

Brunton, B. W., Botvinick, M. M., & Brody, C. D. (2013). Rats and humans can optimally accumulate evidence for decision-making (, 340(6128), 95–98. https://doi.org/10.1126/science.1233912

Busemeyer, J., & Townsend, J. (1993). Decision field theory: A dynamic-cognitive approach to decision making in an uncertain environment. *Psychological Review, 100*(3), 432–459. https://doi.org/10.1037/0033-295X.100.3.432

Busemeyer, J. R., Gluth, S., Rieskamp, J., & Turner, B. M. (2019). Cognitive and neural bases of multi-attribute, multi-alternative, value-based decisions. *Trends in Cognitive Sciences, 23*(3), 251–263. https://doi.org/10.1016/j.tics.2018.12.003

Callaway, F., Rangel, A., & Griffiths, T. L. (2021). Fixation patterns in simple choice reflect optimal information sampling. *PLoS Computational Biology, 17*(3), e1008863. https://doi.org/10.1371/journal.pcbi.1008863

Card, D., DellaVigna, S., Jiang, C., & Taubinsky, D. (2024). *Understanding expert choices using decision time* (Working paper 32515). National Bureau of Economic Research. https://doi.org/10.3386/w32515

Castegnetti, G., Zurita, M., & De Martino, B. (2021). How usefulness shapes neural representations during goal-directed behavior. *Science Advances, 7*(15), eabd5363. https://doi.org/10.1126/sciadv.abd5363

Cavanagh, J. F., Wiecki, T. V., Kochar, A., & Frank, M. J. (2014). Eye tracking and pupillometry are indicators of dissociable latent decision processes. *Journal of Experimental Psychology: General, 143*, 1476–1488. https://doi.org/10.1037/a0035813

Cavanagh, S. E., Malalasekera, W. M. N., Miranda, B., Hunt, L. T., & Kennerley, S. W. (2019). Visual fixation patterns during economic choice reflect covert valuation processes that emerge with learning. *Proceedings of the National Academy of Sciences of the United States of America, 116*(45), 22795–22801. https://doi.org/10.1073/pnas.1906662116

Cerreia-Vioglio, S., Maccheroni, F., Marinacci, M., & Rustichini, A. (2023). Multinomial logit processes and preference discovery: Inside and outside the black box. *The Review of Economic Studies, 90*(3), 1155–1194.

Chen, F., & Krajbich, I. (2018). Biased sequential sampling underlies the effects of time pressure and delay in social decision making. *Nature Communications, 9*(1), 3557. https://doi.org/10.1038/s41467-018-05994-9

Chen, H.-Y., Lombardi, G., Li, S.-C., & Hare, T. A. (2022). Older adults process the probability of winning sooner but weigh it less during lottery decisions. *Scientific Reports, 12*(1), 11381.

Chen, F., Zheng, J., Wang, L., & Krajbich, I. (2024a). Attribute latencies causally shape intertemporal decisions. *Nature Communications, 15*(1), 2948.

Chen, F., Zhu, Z., Shen, Q., Krajbich, I., & Hare, T. A. (2024b). Intrachoice dynamics shape social decisions. *Management Science, 70*(2), 1137–1153. https://doi.org/10.1287/mnsc.2023.4732

Chernev, A., Böckenholt, U., & Goodman, J. (2015). Choice overload: A conceptual review and meta-analysis. *Journal of Consumer Psychology, 25*(2), 333–358. https://doi.org/10.1016/j.jcps.2014.08.002

Clithero, J. A. (2018a). Improving out-of-sample predictions using response times and a model of the decision process. *Journal of Economic Behavior & Organization, 148*, 344–375. https://doi.org/10.1016/j.jebo.2018.02.007

Clithero, J. A. (2018b). Response times in economics: Looking through the lens of sequential sampling models. *Journal of Economic Psychology, 69*, 61–86. https://doi.org/10.1016/J.JOEP.2018.09.008

Cotet, M., Zhao, W. J., & Krajbich, I. (2025). Deliberation during online bargaining reveals strategic information. *Proceedings of the National Academy of Sciences, 122*(7), e2410956122.

Dai, J., & Busemeyer, J. R. (2014). A probabilistic, dynamic, and attribute-wise model of intertemporal choice. *Journal of Experimental Psychology: General, 143*, 1489–1514. https://doi.org/10.1037/a0035976

de Lafuente, V., Jazayeri, M., & Shadlen, M. N. (2015). Representation of accumulating evidence for a decision in two parietal areas. *The Journal of Neuroscience: The Official Journal of the Society for Neuroscience, 35*(10), 4306–4318. https://doi.org/10.1523/JNEUROSCI.2451-14.2015

Degroot, M. H. (1970). *Optimal statistical decisions* (1st ed.). Mcgraw-Hill Book Company.

Desai, N., & Krajbich, I. (2022). Decomposing preferences into predispositions and evaluations. *Journal of Experimental Psychology: General, 151*(8), 1883–1903. https://doi.org/10.1037/xge0001162

Diederich, A. (1997). Dynamic stochastic models for decision making under time constraints. *Journal of Mathematical Psychology, 41*(3), 260–274.

Diederich, A., & Trueblood, J. S. (2018). A dynamic dual process model of risky decision making. *Psychological Review, 125*(2), 270–292. https://doi.org/10.1037/rev0000087

Edwards, W. (1965). Optimal strategies for seeking information: Models for statistics, choice reaction times, and human information processing. *Journal of Mathematical Psychology, 2*(2), 312–329. https://doi.org/10.1016/0022-2496(65)90007-6

Enax, L., Krajbich, I., & Weber, B. (2016). Salient nutrition labels increase the integration of health attributes in food decision-making. *Judgment and Decision Making, 11*(5), 460–471. https://doi.org/10.1017/S1930297500004563

Eum, B., Dolbier, S., & Rangel, A. (2023). Peripheral visual information halves attentional choice biases. *Psychological Science, 34*(9), 984–998. https://doi.org/10.1177/09567976231184878

Evans, N. J., & Wagenmakers, E.-J. (2020). Evidence accumulation models: Current limitations and future directions. *The Quantitative Methods for Psychology, 16*(2), 73–90. https://doi.org/10.20982/tqmp.16.2.p073

Fengler, A., Bera, K., Pedersen, M. L., & Frank, M. J. (2022). Beyond drift diffusion models: Fitting a broad class of decision and reinforcement learning models with HDDM. *Journal of Cognitive Neuroscience, 34*(10), 1780–1805. https://doi.org/10.1162/jocn_a_01902

Fisher, G. (2021). A multiattribute attentional drift diffusion model. *Organizational Behavior and Human Decision Processes, 165*, 167–182. https://doi.org/10.1016/j.obhdp.2021.04.004

Fontanesi, L., Gluth, S., Spektor, M. S., & Rieskamp, J. (2019). A reinforcement learning diffusion decision model for value-based decisions. *Psychonomic Bulletin & Review, 26*(4), 1099–1121. https://doi.org/10.3758/s13423-018-1554-2

Forstmann, B. U., Ratcliff, R., & Wagenmakers, E.-J. (2016). Sequential sampling models in cognitive neuroscience: Advantages, applications, and extensions. *Annual Review of Psychology, 67*(1), 641–666. https://doi.org/10.1146/annurev-psych-122414-033645

Frömer, R., Dean Wolf, C. K., & Shenhav, A. (2019). Goal congruency dominates reward value in accounting for behavioral and neural correlates of value-based decision-making. *Nature Communications, 10*(1), Article 1. https://doi.org/10.1038/s41467-019-12931-x

Frömer, R., Nassar, M. R., Ehinger, B. V., et al. (2024). Common neural choice signals can emerge artefactually amid multiple distinct value signals. *Nature Human Behaviour*. https://doi.org/10.1038/s41562-024-01971-z

Frydman, C., & Krajbich, I. (2016). *Using response times to infer others' beliefs: An application to information cascades* (Working paper 2817026). SSRN. https://scholar.google.com/scholar?cluster=8231454400084874942&hl=en&oi=scholarr

Frydman, C., & Krajbich, I. (2022). Using response times to infer others' private information: An application to information cascades. *Management Science, 68*(4), 2970–2986. https://doi.org/10.1287/mnsc.2021.3994

Frydman, C., & Nave, G. (2017). Extrapolative beliefs in perceptual and economic decisions: Evidence of a common mechanism. *Management Science, 63*(7), 2340–2352. https://doi.org/10.1287/mnsc.2016.2453

Fudenberg, D., Strack, P., & Strzalecki, T. (2018). Speed, accuracy, and the optimal timing of choices. *American Economic Review, 108*(12), 3651–3684. https://doi.org/10.1257/aer.20150742

Gates, V., Callaway, F., Ho, M. K., & Griffiths, T. L. (2021). A rational model of people's inferences about others' preferences based on response times. *Cognition, 217*, 104885. https://doi.org/10.1016/j.cognition.2021.104885

Gherman, S., Markowitz, N., Tostaeva, G., Espinal, E., Mehta, A. D., O'Connell, R. G., Kelly, S. P., & Bickel, S. (2024). Intracranial electroencephalography reveals effector-independent evidence accumulation dynamics in multiple human brain regions. *Nature Human Behaviour, 8*(4), 758–770. https://doi.org/10.1038/s41562-024-01824-9

Glickman, M., Sharoni, O., Levy, D. J., Niebur, E., Stuphorn, V., & Usher, M. (2019). The formation of preference in risky choice. *PLoS Computational Biology, 15*(8), e1007201. https://doi.org/10.1371/journal.pcbi.1007201

Gluth, S., Rieskamp, J., & Büchel, C. (2012). Deciding when to decide: Time-variant sequential sampling models explain the emergence of value-based decisions in the human brain. *Journal of Neuroscience, 32*(31), 10686–10698. https://doi.org/10.1523/JNEUROSCI.0727-12.2012

Gluth, S., Rieskamp, J., & Büchel, C. (2013). Classic EEG motor potentials track the emergence of value-based decisions. *NeuroImage, 79*, 394–403.

Gluth, S., Spektor, M. S., & Rieskamp, J. (2018). Value-based attentional capture affects multi-alternative decision making. *eLife, 7*, e39659. https://doi.org/10.7554/eLife.39659

Gluth, S., Kern, N., Kortmann, M., & Vitali, C. L. (2020). Value-based attention but not divisive normalization influences decisions with multiple alternatives. *Nature Human Behaviour, 4*(6), 634–645. https://doi.org/10.1038/s41562-020-0822-0

Gold, J. I., & Shadlen, M. N. (2007). The neural basis of decision making. *Annual Review of Neuroscience, 30*, 535–574. https://doi.org/10.1146/annurev.neuro.29.051605.113038

Hanks, T. D., Kopec, C. D., Brunton, B. W., Duan, C. A., Erlich, J. C., & Brody, C. D. (2015). Distinct relationships of parietal and prefrontal cortices to evidence accumulation. *Nature, 520*(7546), 220–223. https://doi.org/10.1038/nature14066

Hare, T. A., Malmaud, J., & Rangel, A. (2011a). Focusing attention on the health aspects of foods changes value signals in vmPFC and improves dietary choice. *Journal of Neuroscience, 31*(30), 11077–11087. https://doi.org/10.1523/JNEUROSCI.6383-10.2011

Hare, T. A., Schultz, W., Camerer, C. F., O'Doherty, J. P., & Rangel, A. (2011b). Transformation of stimulus value signals into motor commands during simple choice. *Proceedings of the National Academy of Sciences, 108*(44), 18120–18125. https://doi.org/10.1073/pnas.1109322108

Harris, A., & Hutcherson, C. A. (2021). Temporal dynamics of decision making: A synthesis of computational and neurophysiological approaches. *WIREs. Cognitive Science, 13*(3), e1586. https://doi.org/10.1002/wcs.1586

He, L., & Bhatia, S. (2023). Complex economic decisions from simple neurocognitive processes: The role of interactive attention. *Proceedings of the Royal Society B: Biological Sciences, 290*(1992), 20221593. https://doi.org/10.1098/rspb.2022.1593

Heathcote, A., & Matzke, D. (2022). Winner takes all! What are race models, and why and how should psychologists use them? *Current Directions in Psychological Science, 31*(5), 383–394.

Hébert, B., & Woodford, M. (2023). Rational inattention when decisions take time. *Journal of Economic Theory, 208*(C) https://ideas.repec.org//a/eee/jetheo/v208y2023ics002205312300008x.html

Hunt, L. T., Kolling, N., Soltani, A., Woolrich, M. W., Rushworth, M. F. S., & Behrens, T. E. J. (2012). Mechanisms underlying cortical activity during value-guided choice. *Nature Neuroscience, 15*(3), 470–476. https://doi.org/10.1038/nn.3017

Hunt, L. T., Malalasekera, W. M. N., de Berker, A. O., Miranda, B., Farmer, S. F., Behrens, T. E. J., & Kennerley, S. W. (2018). Triple dissociation of attention and decision computations across prefrontal cortex. *Nature Neuroscience, 21*(10), 1471–1481. https://doi.org/10.1038/s41593-018-0239-5

Hutcherson, C. A., & Tusche, A. (2022). Evidence accumulation, not "self-control", explains dorsolateral prefrontal activation during normative choice. *eLife, 11*, e65661. https://doi.org/10.7554/eLife.65661

Iyengar, S. S., & Lepper, M. R. (2000). When choice is demotivating: Can one desire too much of a good thing? *Journal of Personality and Social Psychology, 79*(6), 995–1006. https://doi.org/10.1037/0022-3514.79.6.995

Izakson, L., Yoo, M., Hakim, A., Krajbich, I., Webb, R., & Levy, D. J. (2024). Valuations of target items are drawn towards unavailable decoy items due to prior expectations. *PNAS Nexus, 3*(7), pgae232. https://doi.org/10.1093/pnasnexus/pgae232

Jang, A. I., Sharma, R., & Drugowitsch, J. (2021). Optimal policy for attention-modulated decisions explains human fixation behavior. *eLife, 10*, e63436. https://doi.org/10.7554/eLife.63436

Johnson, J. G., & Busemeyer, J. R. (2005). A dynamic, stochastic, computational model of preference reversal phenomena. *Psychological Review, 112*(4), 841–861. https://doi.org/10.1037/0033-295X.112.4.841

Jun, E. J., Bautista, A. R., Nunez, M. D., Allen, D. C., Tak, J. H., Alvarez, E., & Basso, M. A. (2021). Causal role for the primate superior colliculus in the computation of evidence for perceptual decisions. *Nature neuroscience, 24*(8), 1121–1131

Kahneman, D. (2011). *Thinking, fast and slow* (p. 499). Farrar, Straus and Giroux.

Kelly, S. P., & O'Connell, R. G. (2013). Internal and external influences on the rate of sensory evidence accumulation in the human brain. *Journal of Neuroscience, 33*(50), 19434–19441. https://doi.org/10.1523/JNEUROSCI.3355-13.2013

Kim, B. E., Seligman, D., & Kable, J. W. (2012). Preference reversals in decision making under risk are accompanied by changes in attention to different attributes. *Frontiers in Neuroscience, 6*, 109. https://doi.org/10.3389/fnins.2012.00109

Konovalov, A., & Krajbich, I. (2016). Gaze data reveal distinct choice processes underlying model-based and model-free reinforcement learning. *Nature Communications, 7*(1), 12438.

Konovalov, A., & Krajbich, I. (2017). On the strategic use of response times. *SSRN Electronic Journal.*

Konovalov, A., & Krajbich, I. (2019). Revealed strength of preference: Inference from response times. *Judgment and Decision Making, 14*(4), 381–394. https://doi.org/10.1017/S1930297500006082

Konovalov, A., & Krajbich, I. (2020). Mouse tracking reveals structure knowledge in the absence of model-based choice. *Nature Communications, 11*(1), 1893.

Konovalov, A., & Krajbich, I. (2023). Decision times reveal private information in strategic settings: Evidence from bargaining experiments. *The Economic Journal, 133*(656), 3007–3033.

Kovach, C. K., Sutterer, M. J., Rushia, S. N., Teriakidis, A., & Jenison, R. L. (2014). Two systems drive attention to rewards. *Frontiers in Psychology, 5*. https://doi.org/10.3389/fpsyg.2014.00046

Krajbich, I. (2019). Accounting for attention in sequential sampling models of decision making. *Current Opinion in Psychology, 29*, 6–11.

Krajbich, I., & Rangel, A. (2011). Multialternative drift-diffusion model predicts the relationship between visual fixations and choice in value-based decisions. *Proceedings of the National Academy of Sciences, 108*(33), 13852–13857. https://doi.org/10.1073/pnas.1101328108

Krajbich, I., Lu, D., Camerer, C., & Rangel, A. (2012). The attentional drift-diffusion model extends to simple purchasing decisions. *Frontiers in Psychology, 3*, 193.

Krajbich, I., Oud, B., & Fehr, E. (2014). Benefits of neuroeconomic modeling: New policy interventions and predictors of preference. *American Economic Review, 104*(5), 501–506

Krajbich, I., & Smith, S. M. (2015). Modeling eye movements and response times in consumer choice. *Journal of Agricultural & Food Industrial Organization, 13*, 55–72. https://doi.org/10.1515/jafio-2015-0016

Krajbich, I., Armel, C., & Rangel, A. (2010). Visual fixations and the computation and comparison of value in simple choice. *Nature Neuroscience, 13*(10), Article 10. https://doi.org/10.1038/nn.2635

Krajbich, I., Hare, T., Bartling, B., Morishima, Y., & Fehr, E. (2015). A common mechanism underlying food choice and social decisions. *PLoS Computational Biology, 11*(10), e1004371. https://doi.org/10.1371/journal.pcbi.1004371

Krajbich, I., Mitsumasu, A., Polania, R., Ruff, C. C., & Fehr, E. (2021). A causal role for the right frontal eye fields in value comparison. *eLife, 10*, e67477. https://doi.org/10.7554/eLife.67477

Kvam, P. D., & Busemeyer, J. R. (2020). A distributional and dynamic theory of pricing and preference. *Psychological Review, 127*(6), 1053–1078. https://doi.org/10.1037/rev0000215

Kvam, P. D., & Turner, B. M. (2021). Reconciling similarity across models of continuous selections. *Psychological Review, 128*(4), 766–786. https://doi.org/10.1037/rev0000296

LaBerge, D., & Brown, V. (1989). Theory of attentional operations in shape identification. *Psychological Review, 96*(1), 101–124. https://doi.org/10.1037/0033-295X.96.1.101

Latimer, K. W., Yates, J. L., Meister, M. L. R., Huk, A. C., & Pillow, J. W. (2015). Single-trial spike trains in parietal cortex reveal discrete steps during decision-making. *Science, 349*(6244), 184–187. https://doi.org/10.1126/science.aaa4056

Li, Z.-W., & Ma, W. J. (2021). An uncertainty-based model of the effects of fixation on choice. *PLoS Computational Biology, 17*(8), e1009190.

Lim, S.-L., O'Doherty, J. P., & Rangel, A. (2011). The decision value computations in the vmPFC and striatum use a relative value code that is guided by visual attention. *The Journal of Neuroscience, 31*(37), 13214–13223. https://doi.org/10.1523/JNEUROSCI.1246-11.2011

Lui, K. K., Nunez, M. D., Cassidy, J. M., Vandekerckhove, J., Cramer, S. C., & Srinivasan, R. (2021). Timing of readiness potentials reflect a decision-making process in the human brain. *Computational Brain & Behavior, 4*(3), 264–283. https://doi.org/10.1007/s42113-020-00097-5

Magrabi, A., Ludwig, V. U., Stoppel, C. M., Paschke, L. M., Wisniewski, D., Heekeren, H. R., & Walter, H. (2022). Dynamic computation of value signals via a common neural network in multi-attribute decision-making. *Social Cognitive and Affective Neuroscience, 17*(7), 683–693. https://doi.org/10.1093/scan/nsab125

Maier, S. U., Raja Beharelle, A., Polanía, R., Ruff, C. C., & Hare, T. A. (2020). Dissociable mechanisms govern when and how strongly reward attributes affect decisions. *Nature Human Behaviour, 4*(9), 949–963. https://doi.org/10.1038/s41562-020-0893-y

Marini, M., Ansani, A., & Paglieri, F. (2020). Attraction comes from many sources: Attentional and comparative processes in decoy effects. *Judgment and Decision Making, 15*(5), 704–726. https://doi.org/10.1017/S1930297500007889

McGinty, V. B., Rangel, A., & Newsome, W. T. (2016). Orbitofrontal cortex value signals depend on fixation location during free viewing. *Neuron, 90*(6), 1299–1311.

Miletić, S., Boag, R. J., & Forstmann, B. U. (2020). Mutual benefits: Combining reinforcement learning with sequential sampling models. *Neuropsychologia, 136*, 107261. https://doi.org/10.1016/j.neuropsychologia.2019.107261

Miletić, S., Boag, R. J., Trutti, A. C., Stevenson, N., Forstmann, B. U., & Heathcote, A. (2021). A new model of decision processing in instrumental learning tasks. *eLife, 10*, e63055.

Molter, F., Thomas, A. W., Heekeren, H. R., & Mohr, P. N. C. (2019). GLAMbox: A python toolbox for investigating the association between gaze allocation and decision behaviour. *PLoS One, 14*(12), e0226428. https://doi.org/10.1371/journal.pone.0226428

Molter, F., Thomas, A. W., Huettel, S. A., Heekeren, H. R., & Mohr, P. N. C. (2022). Gaze-dependent evidence accumulation predicts multi-alternative risky choice behaviour. *PLoS Computational Biology, 18*(7), e1010283. https://doi.org/10.1371/journal.pcbi.1010283

Mormann, M., & Russo, J. E. (2021). Does attention increase the value of choice alternatives? *Trends in Cognitive Sciences, 25*(4), 305–315. https://doi.org/10.1016/j.tics.2021.01.004

Noguchi, T., & Stewart, N. (2014). In the attraction, compromise, and similarity effects, alternatives are repeatedly compared in pairs on single dimensions. *Cognition, 132*(1), 44–56. https://doi.org/10.1016/j.cognition.2014.03.006

Noguchi, T., & Stewart, N. (2018). Multialternative decision by sampling: A model of decision making constrained by process data. *Psychological Review, 125*(4), 512–544. https://doi.org/10.1037/rev0000102

Nunez, M. D., Fernandez, K., Srinivasan, R., & Vandekerckhove, J. (2024). A tutorial on fitting joint models of M/EEG and behavior to understand cognition. *Behavior Research Methods.* https://doi.org/10.3758/s13428-023-02331-x

O'Connell, R. G., Dockree, P. M., & Kelly, S. P. (2012). A supramodal accumulation-to-bound signal that determines perceptual decisions in humans. *Nature Neuroscience, 15*(12), 1729–1735. https://doi.org/10.1038/nn.3248

O'Connell, R. G., Shadlen, M. N., Wong-Lin, K., & Kelly, S. P. (2018). Bridging neural and computational viewpoints on perceptual decision-making. *Trends in Neurosciences, 41*(11), 838–852. https://doi.org/10.1016/j.tins.2018.06.005

Pachella, R. G. (1973). The interpretation of reaction time in information-processing research 1. In *Human information processing* (pp. 41–82). Routledge.

Pärnamets, P., Johansson, P., Hall, L., Balkenius, C., Spivey, M. J., & Richardson, D. C. (2015). Biasing moral decisions by exploiting the dynamics of eye gaze. *Proceedings of the National Academy of Sciences, 112*(13), 4170–4175. https://doi.org/10.1073/pnas.1415250112

Pedersen, M. L., & Frank, M. J. (2020). Simultaneous hierarchical Bayesian parameter estimation for reinforcement learning and drift diffusion models: A tutorial and links to neural data. *Computational Brain & Behavior, 3*(4), 458–471. https://doi.org/10.1007/s42113-020-00084-w

Pedersen, M. L., Frank, M. J., & Biele, G. (2017). The drift diffusion model as the choice rule in reinforcement learning. *Psychonomic Bulletin & Review, 24*(4), 1234–1251. https://doi.org/10.3758/s13423-016-1199-y

Philiastides, M. G., Heekeren, H. R., & Sajda, P. (2014). Human scalp potentials reflect a mixture of decision-related signals during perceptual choices. *The Journal of Neuroscience: The Official Journal of the Society for Neuroscience, 34*(50), 16877–16889. https://doi.org/10.1523/JNEUROSCI.3012-14.2014

Pisauro, M. A., Fouragnan, E., Retzler, C., & Philiastides, M. G. (2017). Neural correlates of evidence accumulation during value-based decisions revealed via simultaneous EEG-fMRI. *Nature Communications, 8*(1), Article 1. https://doi.org/10.1038/ncomms15808

Plassmann, H., O'Doherty, J., & Rangel, A. (2007). Orbitofrontal cortex encodes willingness to pay in everyday economic transactions. *Journal of Neuroscience, 27*(37), 9984–9988. https://doi.org/10.1523/JNEUROSCI.2131-07.2007

Polanía, R., Krajbich, I., Grueschow, M., & Ruff, C. C. (2014). Neural oscillations and synchronization differentially support evidence accumulation in perceptual and value-based decision making. *Neuron, 82*(3), 709–720. https://doi.org/10.1016/j.neuron.2014.03.014

Rand, D. G., Greene, J. D., & Nowak, M. A. (2012). Spontaneous giving and calculated greed. *Nature, 489*(7416), 427–430. https://doi.org/10.1038/nature11467

Roe, R. M., Busemeyer, J. R., & Townsend, J. T. (2001). Multialternative decision field theory: A dynamic connectionst model of decision making. *Psychological Review, 108*(2), 370.

Rasanan, A. H. H., Evans, N. J., Fontanesi, L., Manning, C., Huang-Pollock, C., Matzke, D., Heathcote, A., Rieskamp, J., Speekenbrink, M., Frank, M. J., Palminteri, S., Lucas, C. G., Busemeyer, J. R., Ratcliff, R., & Rad, J. A. (2024). Beyond discrete-choice options. *Trends in Cognitive Sciences, 28*(9), 857–870. https://doi.org/10.1016/j.tics.2024.07.004

Ratcliff, R., & Frank, M. J. (2012). Reinforcement-based decision making in corticostriatal circuits: Mutual constraints by neurocomputational and diffusion models. *Neural Computation, 24*(5), 1186–1229. https://doi.org/10.1162/NECO_a_00270

Ratcliff, R., & McKoon, G. (2008). The diffusion decision model: Theory and data for two-choice decision tasks. *Neural Computation, 20*(4), 873–922.

Ratcliff, R., & Smith, P. L. (2004). A comparison of sequential sampling models for two-choice reaction time. *Psychological Review, 111*(2), 333–367. https://doi.org/10.1037/0033-295X.111.2.333

Ratcliff, R., Smith, P. L., Brown, S. D., & McKoon, G. (2016). Diffusion decision model: Current issues and history. *Trends in Cognitive Sciences, 20*(4), 260–281. https://doi.org/10.1016/j.tics.2016.01.007

Reutskaja, E., Nagel, R., Camerer, C. F., & Rangel, A. (2011). Search dynamics in consumer choice under time pressure: An eye-tracking study. *American Economic Review, 101*(2), 900–926. https://doi.org/10.1257/aer.101.2.900

Rodriguez, C. A., Turner, B. M., & McClure, S. M. (2014). Intertemporal choice as discounted value accumulation. *PLoS One, 9*(2), e90138. https://doi.org/10.1371/journal.pone.0090138

Roitman, J. D., & Shadlen, M. N. (2002). Response of neurons in the lateral intraparietal area during a combined visual discrimination reaction time task. *The Journal of Neuroscience: The Official Journal of the Society for Neuroscience, 22*(21), 9475–9489. https://doi.org/10.1523/JNEUROSCI.22-21-09475.2002

Scheibehenne, B., Greifeneder, R., & Todd, P. M. (2010). Can there ever be too many options? A meta-analytic review of choice overload. *Journal of Consumer Research, 37*(3), 409–425. https://doi.org/10.1086/651235

Schultz, W., Dayan, P., & Montague, P. R. (1997). A neural substrate of prediction and reward. *Science, 275*(5306), 1593–1599. https://doi.org/10.1126/science.275.5306.1593

Sepulveda, P., Usher, M., Davies, N., Benson, A. A., Ortoleva, P., & De Martino, B. (2020). Visual attention modulates the integration of goal-relevant evidence and not value. *eLife, 9*, e60705. https://doi.org/10.7554/eLife.60705

Shadlen, M. N., & Kiani, R. (2013). Decision making as a window on cognition. *Neuron, 80*(3), 791–806. https://doi.org/10.1016/j.neuron.2013.10.047

Shadlen, M. N., & Shohamy, D. (2016). Decision making and sequential sampling from memory. *Neuron, 90*(5), 927–939. https://doi.org/10.1016/j.neuron.2016.04.036

Shahar, N., Hauser, T. U., Moutoussis, M., Moran, R., Keramati, M., Consortium, N., & Dolan, R. J. (2019). Improving the reliability of model-based decision-making estimates in the two-stage decision task with reaction-times and drift-diffusion modeling. *PLoS Computational Biology, 15*(2), e1006803. https://doi.org/10.1371/journal.pcbi.1006803

Sheng, F., Ramakrishnan, A., Seok, D., Zhao, W. J., Thelaus, S., Cen, P., & Platt, M. L. (2020). Decomposing loss aversion from gaze allocation and pupil dilation. *Proceedings of the National Academy of Sciences, 117*(21), 11356–11363. https://doi.org/10.1073/pnas.1919670117

Sheng, F., Wang, R., Liang, Z., Wang, X., & Platt, M. L. (2023). The art of the deal: Deciphering the endowment effect from traders' eyes. *Science Advances, 9*(34), eadf2115. https://doi.org/10.1126/sciadv.adf2115

Shevlin, B. R. K., Gwinn, R. E., Makwana, A., & Krajbich, I. (2025). Overt visual attention modulates decision-related signals in ventral and dorsal medial prefrontal cortex. *eLife, 14*, RP103846.

Smith, S. M., & Krajbich, I. (2019). Gaze amplifies value in decision making. *Psychological Science, 30*, 116–128. https://doi.org/10.1177/0956797618810521

Smith, S. M., Krajbich, I., & Spiller, S. A. (2024). The role of visual attention in opportunity cost neglect and consideration. *Cognition, 261*, 106145.

Song, M., Wang, X., Zhang, H., & Li, J. (2019). Proactive information sampling in value-based decision-making: Deciding when and where to saccade. *Frontiers in Human Neuroscience, 13*. https://doi.org/10.3389/fnhum.2019.00035

Stewart, N., Hermens, F., & Matthews, W. J. (2016). Eye movements in risky choice. *Journal of Behavioral Decision Making, 29*(2–3), 116–136. https://doi.org/10.1002/bdm.1854

Stone, M. (1960). Models for choice-reaction time. *Psychometrika, 25*, 251–260. https://doi.org/10.1007/BF02289729

Sullivan, N. J., & Huettel, S. A. (2021). Healthful choices depend on the latency and rate of information accumulation. *Nature Human Behaviour, 5*(12), 1698–1706. https://doi.org/10.1038/s41562-021-01154-0

Sullivan, N., Hutcherson, C., Harris, A., & Rangel, A. (2015). Dietary self-control is related to the speed with which attributes of healthfulness and tastiness are processed. *Psychological Science, 26*(2), 122–134. https://doi.org/10.1177/0956797614559543

Tajima, S., Drugowitsch, J., & Pouget, A. (2016). Optimal policy for value-based decision-making. *Nature Communications, 7*(1), 12400. https://doi.org/10.1038/ncomms12400

Teoh, Y. Y., & Hutcherson, C. A. (2022). The games we play: Prosocial choices under time pressure reflect context-sensitive information priorities. *Psychological Science, 33*(9), 1541–1556. https://doi.org/10.1177/09567976221094782

Teoh, Y. Y., Yao, Z., Cunningham, W. A., & Hutcherson, C. A. (2020). Attentional priorities drive effects of time pressure on altruistic choice. *Nature Communications, 11*(1), 3534. https://doi.org/10.1038/s41467-020-17326-x

Thomas, A. W., Molter, F., Krajbich, I., Heekeren, H. R., & Mohr, P. N. (2019). Gaze bias differences capture individual choice behaviour. *Nature Human Behaviour, 3*(6), 625–635.

Thomas, A. W., Molter, F., & Krajbich, I. (2021). Uncovering the computational mechanisms underlying many-alternative choice. *eLife, 10*. https://doi.org/10.7554/elife.57012

Ting, C. C., & Gluth, S. (2024). Unraveling information processes of decision-making with eye-tracking data. *Frontiers in Behavioral Economics, 3*, 1384713.

Trueblood, J. S. (2022). Theories of context effects in multialternative, multiattribute choice. *Current Directions in Psychological Science, 31*(5), 428–435. https://doi.org/10.1177/09637214221109587

Towal, R. B., Mormann, M., & Koch, C. (2013). Simultaneous modeling of visual saliency and value computation improves predictions of economic choice. *PNAS Proceedings of the National Academy of Sciences of the United States of America, 110*(40), E3858–E3867. https://doi.org/10.1073/pnas.1304429110

Tump, A. N., Deffner, D., Pleskac, T. J., Romanczuk, P., & Kurvers, R. H. J. M. (2024). A cognitive computational approach to social and collective decision-making. *Perspectives on Psychological Science, 19*(2), 538–551. https://doi.org/10.1177/17456916231186964

Twomey, D. M., Murphy, P. R., Kelly, S. P., & O'Connell, R. G. (2015). The classic P300 encodes a build-to-threshold decision variable. *The European Journal of Neuroscience, 42*(1), 1636–1643. https://doi.org/10.1111/ejn.12936

Usher, M., & McClelland, J. L. (2001). The time course of perceptual choice: The leaky, competing accumulator model. *Psychological Review, 108*(3), 550–592. https://doi.org/10.1037/0033-295X.108.3.550

Vaidya, A. R., & Fellows, L. K. (2015). Testing necessary regional frontal contributions to value assessment and fixation-based updating. *Nature Communications, 6*(1), 10120.

Wald, A. (1945). Sequential tests of statistical hypotheses. *The Annals of Mathematical Statistics, 16*(2), 117–186.

Webb, R. (2019). The (neural) dynamics of stochastic choice. *Management Science, 65*(1), 230–255. https://doi.org/10.1287/mnsc.2017.2931

Weber, E. U., Johnson, E. J., Milch, K. F., Chang, H., Brodscholl, J. C., & Goldstein, D. G. (2007). Asymmetric discounting in intertemporal choice: A query-theory account. *Psychological Science, 18*(6), 516–523.

Wedel, M., Pieters, R., & van der Lans, R. (2023). Modeling eye movements during decision making: A review. *Psychometrika, 88*(2), 697–729.

Weilbächer, R. A., Krajbich, I., Rieskamp, J., & Gluth, S. (2021). The influence of visual attention on memory-based preferential choice. *Cognition, 215*, 104804. https://doi.org/10.1016/j.cognition.2021.104804

Westbrook, A., van den Bosch, R., Määttä, J. I., Hofmans, L., Papadopetraki, D., Cools, R., & Frank, M. J. (2020). Dopamine promotes cognitive effort by biasing the benefits versus costs of cognitive work. *Science, 367*(6484), 1362–1366. https://doi.org/10.1126/science.aaz5891

Xiang Chiong, K., Shum, M., Webb, R., & Chen, R. (2024). Combining choice and response time data: A drift-diffusion model of mobile advertisements. *Management Science, 70*(2), 1238–1257. https://doi.org/10.1287/mnsc.2023.4738

Yang, X., & Krajbich, I. (2023). A dynamic computational model of gaze and choice in multi-attribute decisions. *Psychological Review, 130*(1), 52–70. https://doi.org/10.1037/rev0000350

Yang, X., Retzler, C., Krajbich, I., Ratcliff, R., & Philiastides, M. G. (2024). Attention to brand labels affects, and is affected by, evaluations of product attractiveness. *Frontiers in Behavioral Economics, 2*, 1274815.

Yu, H., Siegel, J. Z., Clithero, J. A., & Crockett, M. J. (2021). How peer influence shapes value computation in moral decision-making. *Cognition, 211*, 104641.

Zhao, W. J., Walasek, L., & Bhatia, S. (2020). Psychological mechanisms of loss aversion: A drift-diffusion decomposition. *Cognitive Psychology, 123*, 101331. https://doi.org/10.1016/j.cogpsych.2020.101331

Zhao, W. J., Coady, A., & Bhatia, S. (2022a). Computational mechanisms for context-based behavioral interventions: A large-scale analysis. *Proceedings of the National Academy of Sciences, 119*(15), e2114914119. https://doi.org/10.1073/pnas.2114914119

Zhao, W. J., Richie, R., & Bhatia, S. (2022b). Process and content in decisions from memory. *Psychological Review, 129*(1), 73–106. https://doi.org/10.1037/rev0000318

Zilker, V., & Pachur, T. (2022). Nonlinear probability weighting can reflect attentional biases in sequential sampling. *Psychological Review, 129*(5), 949.

Zilker, V., & Pachur, T. (2023). Attribute attention and option attention in risky choice. *Cognition, 236*, 105441.

Zoltowski, D. M., Latimer, K. W., Yates, J. L., Huk, A. C., & Pillow, J. W. (2019). Discrete stepping and nonlinear ramping dynamics underlie spiking responses of LIP neurons during decision-making. *Neuron, 102*(6), 1249–1258.e10. https://doi.org/10.1016/j.neuron.2019.04.031

Breaking the Tug-of-War: What Neuroeconomics Can Gain by Moving Past Competition-Only Models

Xiamin Leng, Yi-Hsin Su, Romy Frömer, and Amitai Shenhav

Abstract Popular models of decision-making characterize choice dynamics as a tug-of-war process, where evidence for competing options accumulates until a threshold is reached. While these models capture several key aspects of decision-making, they have difficulty accounting for others, including the influence of individual option values on choice speed and the subjective cost of choosing, as well as the fact that choice competition is itself malleable. This chapter argues for a broader framework that moves beyond a strict tug-of-war approach, by incorporating both independent and competitive evidence accumulation. We show that adopting this alternate framework not only better accounts for the evaluative processes underlying decisions about which option is best but also generalizes more effectively to other types of decisions—including those involving different evaluation criteria (e.g., choosing the worst option), integrating over options (e.g., appraising the overall set), deciding whether and when to choose (e.g., deferral or opt-out), and determining how many options to choose.

Shared authorship: Romy Frömer and Amitai Shenhav

X. Leng · Y.-H. Su
Department of Psychology, University of California, Berkeley, CA, USA

R. Frömer
School of Psychology, Centre for Human Brain Health, University of Birmingham, Birmingham, UK

A. Shenhav (✉)
Department of Psychology, University of California, Berkeley, CA, USA

Helen Wills Neuroscience Institute, University of California, Berkeley, CA, USA
e-mail: amitai@berkeley.edu

D. V. Smith et al. (eds.), *Neuroeconomics: Core Topics and Current Directions*,
https://doi.org/10.1007/978-3-032-02925-6_10

1 Introduction

If pressed to produce a single metaphor to describe the process of making a decision, researchers might converge on something approximating a tug-of-war. A tug-of-war summarizes three core intuitions about the dynamics underlying decision-making: when faced with a set of options, there is a force pulling in support of each option (evidence being accumulated); moving toward choosing one option means moving away from the other, with the rate of movement proportional to the relative strength between options (relative evidence); and the winner is determined by whichever option nets enough movement to surpass a predetermined criterion (threshold). In spite of its simplicity, the tug-of-war has been remarkably successful as a framework for understanding decision-making, not only at the metaphorical level but also as a dominant computational account of the process by which decision-making unfolds.

The tug-of-war framework is perhaps best embodied in the most popular model of decision dynamics: the drift-diffusion model (DDM, Ratcliff, 1978; Ratcliff et al., 2016). According to the DDM, decisions are made by repeatedly sampling information about the available options (e.g., related to their perceptual features, or value-related information in memory; Gold & Shadlen, 2001; Krajbich et al., 2010; Shadlen & Shohamy, 2016); comparing the evidence at each time point; and integrating this relative evidence over time until a predefined level of evidence (threshold) is reached, at which point a decision for the winning option is made. Within research on value-based decision-making, the DDM has proven successful at capturing changes in behavior, neural activity, and metacognition as a function of the difference in values between one's options (i.e., the value-based equivalent of net force; Calder-Travis et al., 2024; Frank et al., 2015; Hare et al., 2011; Lee & Usher, 2023; Pisauro et al., 2017; Polanía et al., 2014; see Krajbich et al., 2025). For instance, this model captures the finding that choices become faster and more accurate as this value difference increases (Collins & Shenhav, 2022; Rangel, 2009).

The DDM is a member of a broader class of *sequential sampling models*, all of which share the properties of evidence accumulating over time and a threshold that marks the end point of a decision (Busemeyer & Townsend, 1993; Forstmann et al., 2016; Ratcliff & Smith, 2004). As suggested above, a core feature of the DDM, which is not shared by all sequential sampling models, is that the evidence that is accumulated is purely *relative*. This means that, to the model, a decision between two bad options and two good options is the same as long as the difference between the option values is the same—just like in a tug-of-war, the winner is determined by the relative strength between the two sides, not their absolute strength.

Not all sequential sampling models share the features of the DDM that make it a direct analog of the tug-of-war. In fact, the DDM can be viewed as one extreme along an axis of competition (Fig. 1a), whereby evidence across options competes entirely and no information is retained about the values of individual options beyond that point (e.g., at the response level). At the other extreme is the race model, which consists of entirely *independent* evidence accumulators (Busemeyer et al., 2019; Fig. 1c). In a race model, evidence accumulates across options separately, and the

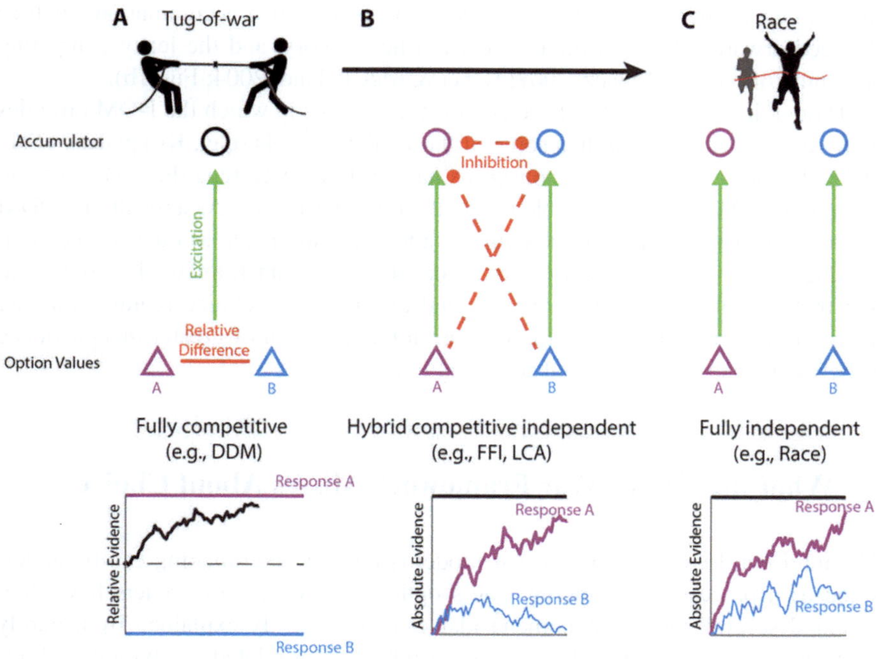

Fig. 1 A spectrum of sequential sampling models. All sequential sampling models involve a transformation from inputs (e.g., values of left and right options) to responses (e.g., responding left or right). One key dimension along which these models vary is the extent to which these inputs and responses compete with one another. (**a**) At one end of this spectrum are fully competitive models like the drift-diffusion model (DDM), which are most similar to a tug-of-war. The DDM accumulates support for the responses strictly in terms of the difference in evidence for the two options, which collectively determine the relative evidence in favor of choosing, e.g., left vs. right. (**b**) At intermediate points along this spectrum are models that incorporate elements of competitive accumulation along with elements of independent accumulation, for instance, by allowing evidence to flow directly from individual options to individual responses. In these hybrid models, competition can be implemented in the form of inhibition from one input to an opposing response (as in the feedforward inhibition models, FFI) and/or mutual inhibition between the responses (as in the leaky competing accumulation models, LCA). (**c**) At the other end of this spectrum are fully independent models like the race model. As the name suggests, this process operates strictly as a race to the finish line across potential responses, each accumulating evidence directly from a corresponding input, with no competition between the accumulators. Lower panels represent examples of accumulation dynamics for an individual decision. (Adapted from Bogacz, 2007)

decision ends when sufficient evidence is accumulated for *any one of them* to surpass the decision threshold ("finish line"). Whereas the DDM only carries information about the difference in value across options and not the individual option values, the race model does the opposite: carrying information about individual options but lacking any competition to generate information about relative value (Teodorescu & Usher, 2013). Intermediate to these two extremes are varieties of models that incorporate elements of both independent and competitive accumulation, such as the feedforward inhibition model, decision field theory, and the leaky competing accumulator model (Bogacz, 2007; Usher & McClelland, 2004; Fig. 1b).

Though individual studies have demonstrated cases in which the DDM provides a poorer fit to choice data than some of these other models (e.g., Kirkpatrick et al., 2021; Tsetsos et al., 2012), a consensus has yet to emerge regarding whether and when the DDM is a more suitable model than its alternatives. As a result, the DDM and the wider tug-of-war framework remain dominant as an account of decision-making. In this chapter, we will argue that this dominant framework is deficient, and that accounting for the dynamics and experience of choice requires moving away from the extreme of the tug-of-war, in the direction of greater independence in how evidence is accumulated across options.

2 What the Tug-of-War Framework Misses About Choice

The intuition driving the tug-of-war model is that decision-making can be understood as a process of comparing one's options. However, researchers have documented several choice phenomena that cannot be easily explained by a purely comparison-based account. Below, we introduce three related behavioral findings that cannot be straightforwardly accounted for by a tug-of-war.

2.1 The Influence of Overall Value on Choice Behavior

Past work has consistently shown that choice behavior varies with the difference in value between one's options (relative value), in particular that decision-makers are faster and more likely to choose a given option the greater the difference between that option and the lesser-valued alternatives. A tug-of-war model naturally accounts for these relative value findings but assumes that this is the only way in which value will influence behavior.

This prediction has been roundly disconfirmed by evidence that choice behavior (particularly choice speed) also varies with the overall (average) value of one's options, holding relative value constant (Frömer et al., 2019; Hunt et al., 2012; Sepulveda et al., 2020; Shenhav & Buckner, 2014; Smith & Krajbich, 2019; Teodorescu et al., 2016; see Frömer & Shenhav, 2022 for a review). For instance, a person tasked with choosing which option they like better will choose faster when faced with two high-value options than when choosing between two low-value options, even if they prefer the better option to the same extent in both cases (e.g., 6 vs. 8, compared to 4 vs. 2; Fig. 2a). Such overall value (or magnitude) effects can be seen in both value-based and perceptual choices (e.g., Pirrone et al., 2018), including in the level of confidence a person feels after making their choice (e.g., Peters et al., 2017; Sepulveda et al., 2020). Accounting for these observations requires that at least some information about the absolute value of each of the options makes it into

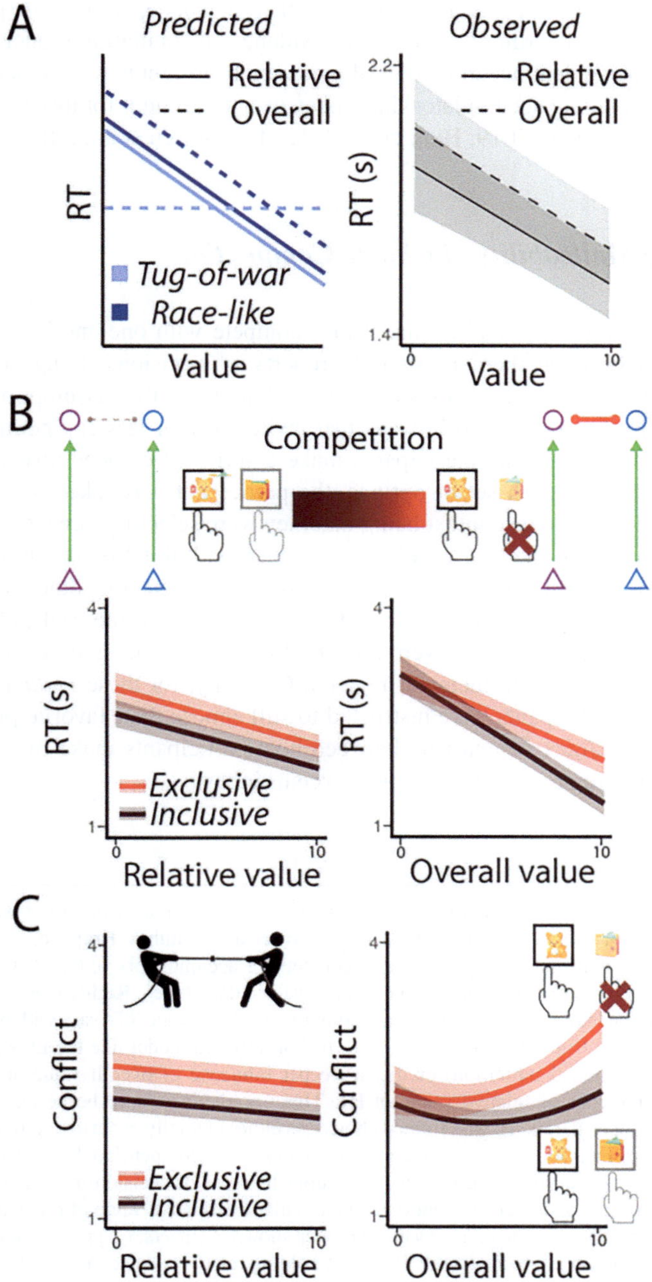

Fig. 2 The tug-of-war account fails to account for key elements of choice. (**a**) **Left**: Competition-only models (light blue) like the traditional DDM predict that response times (RTs) to make a

the stream of evidence being accumulated, rather than being entirely marginalized as in a tug-of-war. Accordingly, variants of evidence accumulation models that allow for some independent accumulation of evidence from inputs to responses (such as the leaky competing accumulator) have been used to account for these overall value effects (Frömer et al., 2019; Hunt et al., 2012; Teodorescu et al., 2016).

2.2 *The Malleability of Choice Competition*

A tug-of-war not only implies that values compete with one another, but that the competition between them is a fixed property of decisions. Leng et al. (2024) recently performed a series of studies that challenged this assumption, showing that decisions can be framed in ways that invite more or less competition into the choice process. They had participants make a series of choices involving sets of four products. For each set of options, the participant was asked to choose their favorite option, with the context being that they were selecting items to add to a gift registry. For half of the trials, their choice ended with this first selection, meaning that selecting a given option excluded the possibility of selecting others. Contrasting with these *mutually exclusive* choices, for the other half of the trials, after making their first choice, participants were allowed to choose as many other options from the product set to add to their gift registry. Critically, for these *mutually inclusive* choice trials, participants were instructed to still choose their favorite product first, allowing for a direct comparison between how participants make this first choice when options are presented as exclusive versus inclusive.

Fig. 2 (continued) decision should vary only with the relative value of one's options (solid line) and not with their overall value (dashed line; though see also Smith & Krajbich, 2019, discussed below). Models with both independent and competitive accumulators (e.g., LCA, FFI) predict that RTs will vary both with relative and overall value (dark blue). **Right:** Consistent with the predictions of the latter accounts, empirical findings show that choice RTs vary with both measures of value. (Adapted from Frömer et al., 2024). (**b**) For a hybrid model like the LCA, lower levels of competition (mutual inhibition) predict a steeper influence of overall value on RT, without altering the influence of relative value on RT. These predictions were borne out by empirical findings by Leng et al. (2024), who framed choices as either mutually exclusive (promoting greater competition; red lines) or as mutually inclusive (promoting less competition; black lines). (**c**) A tug-of-war framework focuses questions about the subjective cost of decision-making around factors related to the competition between options, in particular the relative value of one's options. **Left:** Consistent with this intuition, relative value has been shown to (inversely) predict subjective choice costs (e.g., feelings of conflict). **Right:** However, holding relative value constant, choice costs also vary in a U-shaped fashion with overall value: highest for choices involving the worst options (lose-lose) or the best options (win-win). The costs associated with making win-win choices decrease when making inclusive relative to exclusive choices. (Panels **b** and **c** are adapted from Leng et al., 2024)

Despite the choice being otherwise identical, the study found that an inclusive choice frame made people choose this first option more efficiently relative to an exclusive frame—that is, participants were still choosing their favorite option at a similar rate, but they were faster to do so. Moreover, choice inclusivity altered the influence of overall value on behavior, such that higher valued options sped up choice to an even greater degree for inclusive relative to exclusive choices (Fig. 2b, bottom). These behavioral patterns were best accounted for by not only assuming that evidence can accumulate about options independently (to account for the overall value effects, as discussed above), but that the level of competition between one's options (parameterized as the degree to which candidate responses inhibit one another) can change depending on the level of inclusivity (Fig. 2b, top). Behavior during exclusive choices reflected higher levels of competition (more tug-of-war-like), whereas behavior during inclusive choices reflected lower levels of competition (more race-like). Collectively, this work demonstrated that choice competition is not only malleable, but that lower levels of competition are in fact normatively preferable, in terms of maximizing the efficiency of choice.

Notably, this study also revealed that relative value effects on choice (the sole predictions of a tug-of-war model) were poor indices of the degree to which options were competing with one another. The influence of relative value on choices and response times remained the same across levels of choice exclusivity. Focusing too narrowly on this traditional choice metric would therefore also lead one to neglect factors that contribute to variability in choice competition.

2.3 The Subjective Costs of Choosing

The tug-of-war perspective also shapes our intuitions for which choices should feel most challenging, namely that these subjective costs should scale with factors that make it hard for a single option to edge out against others. Consistent with this intuition, choices are harder when the options are more similar, when there are more options to compare, and when the comparison is more complex (Schwartz, 2004). However, holding these factors constant, it has been shown that other elements of a decision can make a choice more costly, which are not readily accounted for by a tug-of-war.

For instance, when varying both overall value and relative value, Shenhav and Buckner (2014) found that people experienced choices between high-value options ("win-win" choices) as more anxiety-provoking than choices between mediocre (mid-valued) options (see also Chernev & Hamilton, 2009; Iyengar & Lepper, 2000). Subsequent work showed that the same was true on the opposite end of the overall value spectrum—choosing between options with very low value ("lose-lose" choices) also produces greater anxiety than choosing between mediocre options (e.g., Shenhav et al., 2018; see also Botti et al., 2009). Thus, experiences of choice costs follow a U-shaped pattern with respect to overall value (Fig. 2c), highest when choosing between options that a person wants most and when choosing between options that they want least.

By varying mutual exclusivity as described in the previous section, Leng et al. (2024) showed that subjective choice costs also vary with the level of competition between one's options. They found that inclusive choices (those for which participants could later return and select more items) were experienced as less costly than exclusive choices (those for which participants could only select one option). This benefit of inclusivity on choice costs also interacted with the overall value of the choice set, such that participants benefited most from inclusivity when making high-value (win-win) choices (i.e., when they would be most prone to choose additional options if offered the opportunity; Fig. 2c).

Collectively, these findings suggest that choice costs not only reflect how difficult it is to compare between options (as would be suggested by a tug-of-war model), but that they further reflect the overall stakes of the choice, and how these stakes are altered by the competitive structure of the choice. Choice costs increase when the choice is a win-win, potentially because of the size of the sacrifices incurred by choosing; eliminating the need for sacrifice by making choices inclusive eliminates those costs. Choice costs also increase when the choice is a lose-lose, potentially because of the negative value obtained as a result; allowing participants to select more than one of these options does not improve their situation because they would rather not have any of them.

3 New Horizons Beyond Traditional Choice Comparison

Once we move past the tug-of-war model, we are able to look at decision-making from a more holistic perspective and identify important avenues for inquiry that fell outside the remit of this old paradigm. These new avenues include examining how people evaluate options as a *whole* (rather than only as competing parts); how people decide *whether* and *when* (rather than only how) to decide; and how people decide *how many* options to choose (rather than only which one).

3.1 Integrating Over Options

It's natural to think about competitive selection when considering what people do when faced with a set of options, but anyone who has browsed a catalog or gone window shopping knows that people are equally capable of evaluating options in a way that doesn't necessarily involve selection. Cases like these, where a person might *appraise* a set of options to determine how much they like them on the whole, require integration rather than competition across the option values. While such computations are impossible with a purely competitive architecture (i.e., tug-of-war), they can be accommodated by process models that include a combination of independent and competitive accumulators.

For instance, Frömer and colleagues (2022) showed that a model incorporating both of these elements (implemented as feedforward excitation and mutual inhibition) could be configured to either integrate information about values in an option set (appraisal) or select a single option from that set (choice) (Fig. 3a). Within this model, independent accumulation is critical both when selecting an option (as in the models described earlier) and for allowing information about all of the options to be considered when rating the option set as a whole (e.g., how good or bad the options are on a 5-point scale). Inhibition allows for competition across choice options (with evidence for some options counting as evidence against others) as well as for a graded form of competition across appraisal ratings (with evidence for the ratings on one end of the rating scale counting as strong evidence against ratings on the other end, and weak evidence against ratings in the middle of the scale). The latter dynamic predicts faster appraisals when the options are especially high or low in value (i.e., at the extremes of the scale) compared to when options are of middling value (i.e., closer to the middle). This prediction has been borne out in empirical findings, with appraisal ratings demonstrating an inverse U-shaped relationship with response time (Clairis & Pessiglione, 2022; Lebreton et al., 2015; Shenhav & Karmarkar, 2019).

3.2 *Choosing Whether (and/or When) to Choose*

The same integrative mechanism that allows a person to evaluate how good their options are can, in principle, address a further question: are these options worth choosing among? Needless to say, this question is beyond the scope of a tug-of-war model, which is premised on a person already being committed to making a choice. However, before making such a commitment, a person needs to decide whether and when to decide.

Recent work suggests that these decisions are heavily determined by the overall value of one's options. When people are unhappy with their options, they are more likely to delay or avoid making a choice (Berens & Funke, 2020; Bhatia & Mullett, 2016; Tversky & Shafir, 1992; see Anderson, 2003 for a review).

Su and Shenhav (2024) recently showed that this may account for different forms of opt-out behavior that are prevalent among would-be voters—when faced with undesirable candidates, voters are prone to opt out of voting for one of them (by opting not to vote or indicating that they are undecided), even when they have a preference between the candidates (Fig. 3b). Collectively, these patterns of decision avoidance may reflect an appraisal-like mechanism as described above (whereby participants are categorizing an option set as good or bad), which then feeds into a determination of whether to engage in choosing between the options. It's possible that such a process can even account for dynamics of recurring decision avoidance, such as those that characterize procrastination (e.g., Le Bouc & Pessiglione, 2022; Zhang & Ma, 2020).

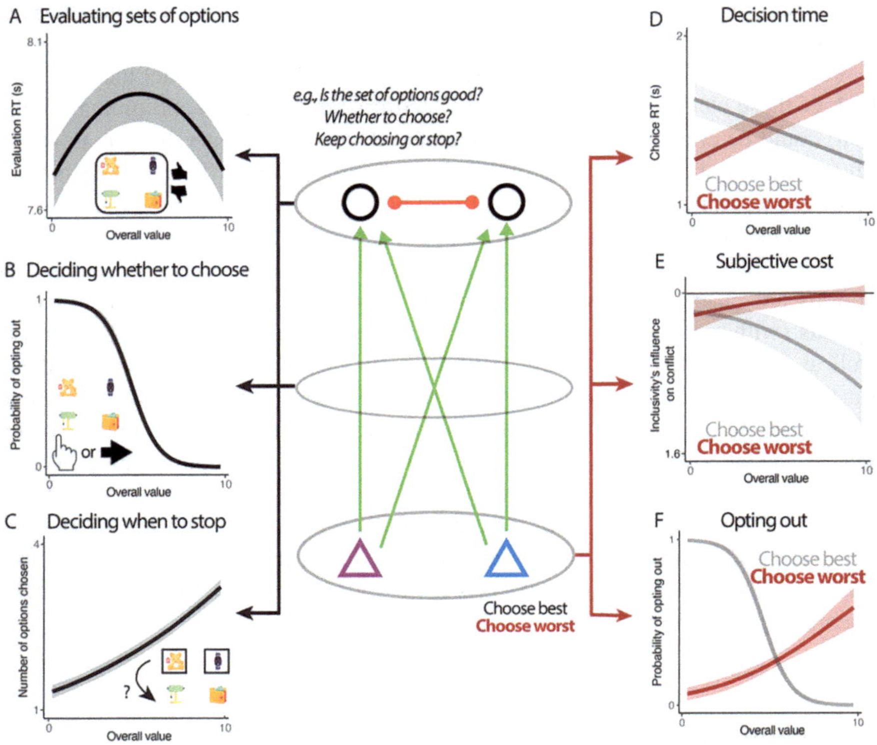

Fig. 3 Decisions beyond traditional choice comparisons. A given set of option values can be processed in ways that give rise to multiple different kinds of decisions, beyond traditional choice comparison. (**a**) Rather than comparing the options to select the best one, a person can integrate across the option values to appraise the option set as a whole. This process can be parameterized as a combination of independent accumulation of values to relevant responses (e.g., appraisal ratings) and graded competition across responses as a function of their distance from one another (e.g., along the rating scale; Frömer et al., 2022). This hybrid model predicts that appraisals will be fastest when they fall closer to the end of the scale rather than toward the middle, as shown here empirically (adapted from Frömer et al., 2024). (**b**) A person can also decide to opt out of choosing any options. One way to make this opt-out decision would be to appraise the option values to determine whether they are of sufficient value to choose among. Consistent with predictions of such a model, decisions to opt out have been shown to be driven heavily by the overall value of one's options, as shown here for a recent study of voting choice (in which participants were most likely to opt out of choosing between bad candidates). Adapted from Su & Shenhav, 2024. (**c**) People often have the freedom to choose how many options to select, resulting in iterative decisions about whether to continue selecting items or not. These choices have also been shown to scale with the overall value of one's options. Adapted from Leng et al., 2024. (**d–f**) Inverting one's choice goal, from choosing the best option to choosing the worst option (e.g., as the option to remove or reject from the set), flips the influence of overall value on choice speed (**d**), subjective experience (**e**), and the probability of opting out (**f**). (Adapted from Frömer et al., 2019; Leng et al., 2024; and Su & Shenhav, 2024)

3.3 Choosing When to Stop Choosing

As noted in the previous section, beyond the options they can select from a given choice set, people often have the option of not choosing (at least not yet). By the same token, people often have the option to choose more than one option from a set, a situation also not captured by a traditional tug-of-war. This was, in fact, the situation participants found themselves in for mutually inclusive choices in the study by Leng and colleagues (2024; cf. Figure 2b, c); for these choices, participants could choose as many options from the four-item choice set as they wanted. This provided the experimenters with an opportunity to additionally examine how these participants decided *how many* options to select (Fig. 3c).

The findings showed that participants seemed to follow a systematic evaluation rule to determine whether to continue selecting options from the set: are any of the remaining options at least as valuable as the average items seen throughout the experiment (i.e., the global average of item values)? Thus, participants were more likely to choose three or four items from the choice set when the overall value of that set was high (i.e., more above-average items) and more likely to only choose one or two items when the overall value was low (i.e., more below-average items). This selection rule is reminiscent of stopping rules prescribed by the marginal value theorem within foraging theory (Charnov, 1976; Pirrone et al., 2018; Wolfe, 2013), suggesting a potential generalization of this choice rule across decision contexts. As with decision avoidance, it is possible that these foraging-like selection decisions emerge from a combination of integrative and competitive decision mechanisms (cf. Kane et al., 2022).

3.4 Choosing How to Accumulate Evidence

Until now, we have assumed that a person's goal in making a decision is to evaluate how good the options are, and thereby determine which option is best; how good the options are as a whole; whether to choose; and if so how many to choose. But sometimes a person may instead seek to evaluate their options along an inverted axis: how bad they are rather than how good they are. This is the case, for instance, when choosing which option is the worst one in a set. As trivial of an adjustment as this may seem, it turns out that adjusting a person's decision process to this value inversion has significant consequences for each of the domains outlined above.

To accommodate a change in one's choice goal—toward choosing the worst option rather than the best one—the decision process must be reconfigured such that the relationship between option values and response evidence is inverted (i.e., low-value inputs now generate the strongest evidence, and high-value inputs now generate the weakest evidence). Frömer et al. (2019) showed that a particularly striking impact of this change is to invert the influence of overall set value on choice speed, such that low-value choices become the fastest and high-value choices the slowest (Fig. 3d, red); this was the mirror-opposite of the pattern that had typically observed in value-based choice studies up to this point (compare Fig. 3d, gray, Fig. 2b), which predominantly instructed their participants to choose the best option from a set (De Martino & Cortese, 2023; Frömer & Shenhav, 2022).

Adjusting one's choice goal also affects the cost of choosing. As noted earlier, participants experience the greatest choice costs for lose-lose and win-win choices. When allowed to select one or more options to acquire from a set (starting with the best one), inclusivity reduces these costs, but only for win-win choices (Leng et al., 2024; Fig. 2c). When participants in a separate experiment instead started each trial by owning all of the options in a choice set and were then allowed them to remove one or more options (starting with the worst), inclusivity had the opposite effect: lose-lose choices were now *least* costly rather than most costly, presumably because participants could remove as many of the undesirable options as they desired (Fig. 3e).

For a similar reason, whether a person decides to avoid making a choice is heavily dependent on whether they are evaluating their options positively (as in the selection case above) or negatively (as in the removal case). Whereas people are prone to avoid lose-lose choices in cases where they are asked to choose the best option, this effect reverses when the choice goal is inverted. For instance, Su and Shenhav's (2024) voting choice study observed that participants were most likely to avoid choosing when faced with the political equivalent of a lose-lose choice (a pair of undesirable candidates; Fig. 3b). When asked instead to choose which candidate they liked *less* (i.e., which candidate they wanted to reject), opt-out rates for low-value options went down drastically, and these choice sets were if anything less likely to be avoided than high-value sets (Fig. 3f), as expected given the reversals observed in choice speed under the different choice goals.

4 Concluding Remarks

The idea that decisions are made by an internal process akin to a tug-of-war is so deeply intuitive that it pervades folk understanding and symbology surrounding decision-making. These deep intuitions have also been reflected in the overwhelming focus of research in this area, with decisions being predominantly modeled as a direct competition between evidence accumulated in favor of each of one's options. Models like the DDM that center this assumption have been tremendously successful at accounting for the influence of value difference on choice. However, they have also left critical gaps in our understanding of how choices are made, how choices are experienced, and how evaluations of options inform decisions beyond traditional choice comparisons.

We reviewed work demonstrating such deficits when trying to explain the fact that option values affect choice speed independent of relative value; that competition is not a fixed property of evidence accumulation but can be modified by context; and that, holding relative value constant, the costs of choosing scale (non-monotonically) with the overall value of one's options. While the DDM can be modified to account for some of these behaviors (e.g., the attentionally weighted DDM can account for overall value effects on RT; Shevlin et al., 2022; Smith & Krajbich, 2019), it cannot straightforwardly account for most others (e.g., appraisal). Collectively, these findings support decision-making frameworks (and concomitant models) that are not strictly competitive (as in a tug-of-war; Fig. 1a) but rather incorporate elements of both independent and competitive accumulation of evidence (e.g., Fig. 1b, c). In addition to accounting for these deficits in accounts of choice comparison (e.g., when required to select one out of a set of options), we show that adopting such alternate models opens the door to account for how people make decisions in more typical situations where they have the flexibility to choose not to decide (e.g., defer making a decision or otherwise opt-out); to choose to select more than one option; to appraise a set of options as a whole; and/or to focus their evaluation on how bad their options are rather how good they are.

Acknowledgments The work was funded by a grant from the National Science Foundation (Collaborative Research in Computational Neuroscience Award #2309022) to A.S. and R.F.

References

Anderson, C. J. (2003). The psychology of doing nothing: Forms of decision avoidance result from reason and emotion. *Psychological Bulletin, 129*(1), 139–167.

Berens, S., & Funke, J. (2020). A vignette study of option refusal and decision deferral as two forms of decision avoidance: Situational and personal predictors. *PLoS One, 15*(10), e0241182.

Bhatia, S., & Mullett, T. L. (2016). The dynamics of deferred decision. *Cognitive Psychology, 86*, 112–151.

Bogacz, R. (2007). Optimal decision-making theories: Linking neurobiology with behaviour. *Trends in Cognitive Sciences, 11*(3), 118–125.

Botti, S., Orfali, K., & Iyengar, S. S. (2009). Tragic choices: Autonomy and emotional responses to medical decisions. *Journal of Consumer Research, 36*(3), 337–352.

Busemeyer, J. R., & Townsend, J. T. (1993). Decision field theory: A dynamic-cognitive approach to decision making in an uncertain environment. *Psychological Review, 100*(3), 432–459.

Busemeyer, J. R., Gluth, S., Rieskamp, J., & Turner, B. M. (2019). Cognitive and neural bases of multi-attribute, multi-alternative, value-based decisions. *Trends in Cognitive Sciences, 23*(3), 251–263.

Calder-Travis, J., Charles, L., Bogacz, R., & Yeung, N. (2024). Bayesian confidence in optimal decisions. *Psychological Review, 131*(5), 1114–1160.

Charnov, E. L. (1976). Optimal foraging, the marginal value theorem. *Theoretical Population Biology, 9*(2), 129–136.

Chernev, A., & Hamilton, R. (2009). Assortment size and option attractiveness in consumer choice among retailers. *Journal of Marketing Research, 46*(3), 410–420.

Clairis, N., & Pessiglione, M. (2022). Value, confidence, deliberation: A functional partition of the medial prefrontal cortex demonstrated across rating and choice tasks. *The Journal of Neuroscience, 42*(28), 5580–5592.

Collins, A. G. E., & Shenhav, A. (2022). Advances in modeling learning and decision-making in neuroscience. *Neuropsychopharmacology, 47*(1), 104–118.

De Martino, B., & Cortese, A. (2023). Goals, usefulness and abstraction in value-based choice. *Trends in Cognitive Sciences, 27*(1), 65–80.

Forstmann, B. U., Ratcliff, R., & Wagenmakers, E.-J. (2016). Sequential sampling models in cognitive neuroscience: Advantages, applications, and extensions. *Annual Review of Psychology, 67*, 641–666.

Frank, M. J., Gagne, C., Nyhus, E., Masters, S., Wiecki, T. V., Cavanagh, J. F., & Badre, D. (2015). fMRI and EEG predictors of dynamic decision parameters during human reinforcement learning. *Journal of Neuroscience, 35*(2), 485–494.

Frömer, R., & Shenhav, A. (2022). Filling the gaps: Cognitive control as a critical lens for understanding mechanisms of value-based decision-making. *Neuroscience & Biobehavioral Reviews, 134*, 104483.

Frömer, R., Dean Wolf, C. K., & Shenhav, A. (2019). Goal congruency dominates reward value in accounting for behavioral and neural correlates of value-based decision-making. *Nature Communications, 10*(1), 4926.

Frömer, R., Nassar, M. R., Ehinger, B. V., & Shenhav, A. (2024). Common neural choice signals can emerge artefactually amid multiple distinct value signals. *Nature Human Behaviour, 8*(11), 2194–2208.

Frömer, R., Gluth, S., & Shenhav, A. (2022). Hidden knobs: Representations for flexible goal-directed decision-making. In Proceedings of the 5th Multidisciplinary Conference on Reinforcement Learning and Decision Making. Providence, RI.

Gold, J. I., & Shadlen, M. N. (2001). Neural computations that underlie decisions about sensory stimuli. *Trends in Cognitive Sciences, 5*(1), 10–16.

Hare, T. A., Schultz, W., Camerer, C. F., O'Doherty, J. P., & Rangel, A. (2011). Transformation of stimulus value signals into motor commands during simple choice. *Proceedings of the National Academy of Sciences, 108*(44), 18120–18125.

Hunt, L. T., Kolling, N., Soltani, A., Woolrich, M. W., Rushworth, M. F. S., & Behrens, T. E. J. (2012). Mechanisms underlying cortical activity during value-guided choice. *Nature Neuroscience, 15*(3), 470–476.

Iyengar, S. S., & Lepper, M. R. (2000). When choice is demotivating: Can one desire too much of a good thing? *Journal of Personality and Social Psychology, 79*(6), 995–1006.

Kane, G. A., James, M. H., Shenhav, A., Daw, N. D., Cohen, J. D., & Aston-Jones, G. (2022). Rat anterior cingulate cortex continuously signals decision variables in a patch foraging task. *The Journal of Neuroscience, 42*(29), 5730–5744.

Kirkpatrick, R. P., Turner, B. M., & Sederberg, P. B. (2021). Equal evidence perceptual tasks suggest a key role for interactive competition in decision-making. *Psychological Review, 128*(6), 1051–1087.

Krajbich, I., Armel, C., & Rangel, A. (2010). Visual fixations and the computation and comparison of value in simple choice. *Nature Neuroscience, 13*(10), 1292–1298.

Krajbich, I., Fernandez, K., & Yang, X. (2025). Decomposing economic choices with drift-diffusion models. In D. V. Smith, P. L. Lockwood, & D. S. Fareri (Eds.), *Neuroeconomics: Core topics and current directions*. Springer. https://doi.org/10.31234/osf.io/4wsrp_v1

Le Bouc, R., & Pessiglione, M. (2022). A neuro-computational account of procrastination behavior. *Nature Communications, 13*(1), 5639.

Lebreton, M., Abitbol, R., Daunizeau, J., & Pessiglione, M. (2015). Automatic integration of confidence in the brain valuation signal. *Nature Neuroscience, 18*(8), 1159–1167.

Lee, D. G., & Usher, M. (2023). Value certainty in drift-diffusion models of preferential choice. *Psychological Review, 130*(3), 790–806.

Leng, X., Frömer, R., Summe, T., & Shenhav, A. (2024). Mutual inclusivity improves decision-making by smoothing out choice's competitive edge. *Nature Human Behaviour*, 1–13.

Peters, M. A. K., Thesen, T., Ko, Y. D., Maniscalco, B., Carlson, C., Davidson, M., Doyle, W., Kuzniecky, R., Devinsky, O., Halgren, E., & Lau, H. (2017). Perceptual confidence neglects decision-incongruent evidence in the brain. *Nature Human Behaviour, 1*(7), 1–8.

Pirrone, A., Azab, H., Hayden, B. Y., Stafford, T., & Marshall, J. A. R. (2018). Evidence for the speed–value trade-off: Human and monkey decision making is magnitude sensitive. *Decision, 5*(2), 129–142.

Pisauro, M. A., Fouragnan, E., Retzler, C., & Philiastides, M. G. (2017). Neural correlates of evidence accumulation during value-based decisions revealed via simultaneous EEG-fMRI. *Nature Communications, 8*(1), 15808.

Polanía, R., Krajbich, I., Grueschow, M., & Ruff, C. C. (2014). Neural oscillations and synchronization differentially support evidence accumulation in perceptual and value-based decision making. *Neuron, 82*(3), 709–720.

Rangel, A. (2009). Chapter 28: The computation and comparison of value in goal-directed choice. In P. W. Glimcher, C. F. Camerer, E. Fehr, & R. A. Poldrack (Eds.), *Neuroeconomics* (pp. 425–440). Academic.

Ratcliff, R. (1978). A theory of memory retrieval. *Psychological Review, 85*(2), 59–108.

Ratcliff, R., & Smith, P. L. (2004). A comparison of sequential sampling models for two-choice reaction time. *Psychological Review, 111*(2), 333–367.

Ratcliff, R., Smith, P. L., Brown, S. D., & McKoon, G. (2016). Diffusion decision model: Current issues and history. *Trends in Cognitive Sciences, 20*(4), 260–281.

Schwartz, B. (2004). *The paradox of choice: Why more is less* (pp. xi–265). HarperCollins Publishers.

Sepulveda, P., Usher, M., Davies, N., Benson, A. A., Ortoleva, P., & De Martino, B. (2020). Visual attention modulates the integration of goal-relevant evidence and not value. *eLife, 9*, e60705.

Shadlen, M. N., & Shohamy, D. (2016). Decision making and sequential sampling from memory. *Neuron, 90*(5), 927–939.

Shenhav, A., & Buckner, R. L. (2014). Neural correlates of dueling affective reactions to win–win choices. *Proceedings of the National Academy of Sciences, 111*(30), 10978–10983.

Shenhav, A., & Karmarkar, U. R. (2019). Dissociable components of the reward circuit are involved in appraisal versus choice. *Scientific Reports, 9*(1), 1958.

Shenhav, A., Dean Wolf, C. K., & Karmarkar, U. R. (2018). The evil of banality: When choosing between the mundane feels like choosing between the worst. *Journal of Experimental Psychology. General, 147*(12), 1892–1904.

Shevlin, B. R., Smith, S. M., Hausfeld, J., & Krajbich, I. (2022). High-value decisions are fast and accurate, inconsistent with diminishing value sensitivity. Proceedings of the National Academy of Sciences, *119*(6), e2101508119.

Smith, S. M., & Krajbich, I. (2019). Gaze amplifies value in decision making. *Psychological Science, 30*(1), 116–128.

Su, Y.-H., & Shenhav, A. (2024). Rejection-based choices discourage voters from opting out. *Social Science Research Network*. https://doi.org/10.2139/ssrn.4866114

Teodorescu, A. R., & Usher, M. (2013). Disentangling decision models: From independence to competition. *Psychological Review, 120*(1), 1–38.

Teodorescu, A. R., Moran, R., & Usher, M. (2016). Absolutely relative or relatively absolute: Violations of value invariance in human decision making. *Psychonomic Bulletin & Review, 23*(1), 22–38.

Tsetsos, K., Gao, J., McClelland, J. L., & Usher, M. (2012). Using time-varying evidence to test models of decision dynamics: Bounded diffusion vs. the leaky competing accumulator model. *Frontiers in Neuroscience, 6*.

Tversky, A., & Shafir, E. (1992). Choice under conflict: The dynamics of deferred decision. *Psychological Science, 3*(6), 358–361.

Usher, M., & McClelland, J. L. (2004). Loss aversion and inhibition in dynamical models of multialternative choice. *Psychological Review, 111*(3), 757–769.

Wolfe, J. M. (2013). When is it time to move to the next raspberry bush? Foraging rules in human visual search. *Journal of Vision, 13*(3), 10.

Zhang, P., & Ma, W. J. (2020). A process model of procrastination. *Proceedings of the Annual Meeting of the Cognitive Science Society, 42*. https://escholarship.org/uc/item/03z6r5bk

Enhancing Ecological Validity in Habit Research via Naturalistic Methods

Ahmet O. Ceceli and Elizabeth Tricomi

Abstract The motivational control of action refers to the idea that actions are motivated by desired outcomes (goals) or triggered by antecedent cues (habit), governing how we adaptively execute value calculations to navigate our environment. Most research on habits has relied on well-controlled laboratory settings that parse the sensitivity of an instrumental response based on changes in reinforcer value. While significantly contributing to our understanding of motivational control, restricting habit estimates to the lab may be limiting the ecological relevance and generalizability of the findings. This chapter reviews recent advancements in neurobehavioral research that extend the study of habits beyond the lab. We highlight innovative implementations of stimuli with real-world relevance in controlled experiments, as well as efforts to apply lab-based habit measures to real-world settings. We further discuss designs that leverage remote data collection, which show promise in improving the validity of motivational control measures. While capturing human habits and translating this knowledge into clinical practice has been challenging, emerging evidence suggests that increasing ecological validity by harnessing naturalistic methods may bridge the gap between experimental and real-world applications of motivational control research. As these methodologies advance, the potential for developing more robust models of habit formation and clinical interventions grows, offering new opportunities for harnessing habit science in translational applications.

Habitual behaviors are cue-driven and performed in absence of their desired outcomes (Dickinson, 1985). For example, imagine taking the usual route to your office on a day off, flicking a light switch when entering a dark room despite a power outage, or entering your email app at the sight of the icon moments after

---

A. O. Ceceli (✉)
Department of Psychiatry, Icahn School of Medicine at Mount Sinai, New York, NY, USA
e-mail: ahmet.ceceli@mssm.edu

E. Tricomi
Department of Psychology, Rutgers University, Newark, Newark, NJ, USA

D. V. Smith et al. (eds.), *Neuroeconomics: Core Topics and Current Directions*,
https://doi.org/10.1007/978-3-032-02925-6_11

exiting it. On the other hand, goal-directed actions are performed deliberately to attain desirable outcomes, in congruence with their availability and value. This dual process account of how we interact with our environment emphasizes the balance between goals and habits. Referred to as the motivational control of action, this account characterizes our actions as either motivated by desirable outcomes (goal-directed) or antecedent cues (habitual) (Dickinson & Balleine, 1994). The field of neuroeconomics is tightly linked to motivational control. The goal-directed system, which learns quickly and adjusts to changes in the environment or internal states, assigns a high value to the outcome to guide actions in pursuit of maximal gain. Habits, on the other hand, are learned slowly and, therefore, cannot adapt to environmental or internal changes (Rangel et al., 2008). Thus, complex neuroeconomical and motivational processes underpin the everyday interactions we have with our environment, as motivational control arbitrates whether to engage goal-directed behaviors that require more cognitive resources to compute value updates, or habits that can be deployed readily at the expense of potentially inappropriate responses that do not consider changes to outcome value.

Habits have also been studied from the perspective of attentional and memory systems. In attentional accounts, goal-directed and habitual behaviors are posited to align with controlled and automatic processing (Ashby et al., 2010): resource-consuming and reflective (controlled) behaviors contrast with less effortful (automatic) actions that can be performed simultaneously with other tasks. The memory account invokes explicit processes such as declarative episodes and rules, contrasting with implicit processes such as skills and conditioning, that are thought to align with goal-directed and habitual behaviors (Mishkin et al., 1984), respectively. While there exists substantial overlap between the motivational, attentional, and memory-based perspectives on how we control our interactions with the environment, the motivational account is the focus of this chapter, as it operates on value-based origins of habits and goals.

A great deal of what we know about motivational control stems from studies in rodents that utilized operant conditioning paradigms. Typically, in rodent models of motivational control, an animal is presented with a stimulus (e.g., neutral tone) that, upon a lever press (i.e., response), signals the availability of a reward (i.e., outcome). Initially, the response-outcome association guides behavior, with the outcome motivating the instrumental lever press. With training, the stimulus-response association is strengthened, as the response is carried out even in extinction (e.g., when the outcome is devalued, such as following selective consumption to satiety) (Dickinson et al., 1995). This stimulus-response habit is thus impervious to its consequences and is attributed to the enhanced salience of the associated cue. These studies have identified the dorsomedial striatum to mediate goal-directed actions, as reported in Yin et al. (2005a, b), where DMS lesion/deactivation resulted in lower outcome sensitivity (here, fewer presses for the still-valued reward following selective satiety). In the same vein, the deactivation of the dorsolateral striatum has been shown to decrease the emergence of post-devaluation outcome insensitivity (i.e., slowing or preventing the transition from goal-directed to habitual actions) (Yin et al., 2004).

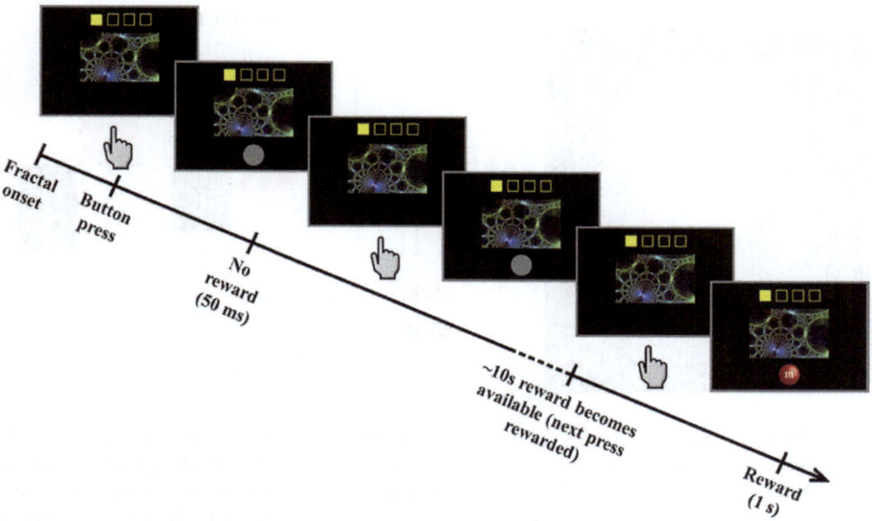

Fig. 1 A free-operant outcome devaluation task with food rewards, as featured in Ceceli et al. (2020). Two distinct fractal images serve as stimuli for stimulus-response-outcome training. Contingent on a button press response (indicated by a yellow filled box), the stimuli yield a food reward outcome (here, either M&M or Goldfish) on average every 10 s. Following training, participants consume their earned rewards, and one of the two food outcomes is made available for consumption to achieve selective satiety (i.e., outcome devaluation). An extinction test follows, whereby the fractals are presented without reward to test for outcome sensitivity. Goal-directed control is reflected by decreases in response rate for the devalued outcome but not the still-valued outcome. Habitual control is indicated by sustained response rate to the devalued outcome. (The figure is used with permission Ceceli et al., 2020)

Analogous operant conditioning tasks have been used to investigate motivational control in humans. These tasks have included training stimulus-response-outcome associations using primary (e.g., food (Ceceli et al., 2020; Tricomi et al., 2009), as depicted in Fig. 1) or secondary (e.g., money/points (Ersche et al., 2016)) rewards, followed by devaluation procedures and extinction tests to probe outcome sensitivity. For detailed descriptions of tasks that are based on this and similar premises, we refer readers to our previous review (Ceceli & Tricomi, 2018). Studies that have employed such tasks have offered valuable insights into the neural, behavioral, computational, and clinical substrates of habits and goals. Notably, they have been used to map the cortico-striatal substrates of motivational control: goal-directed behaviors rely on the caudate (analogous to the dorsomedial striatum in rodents) and ventromedial prefrontal cortex (vmPFC), and habits have been associated with the putamen (analogous to the dorsolateral striatum) (Balleine & O'Doherty, 2010). The neurobiology of motivational control has also been tested under endocrinological perturbations (de Wit, Standing, et al., 2012a) and in various clinical populations, including people with obsessive-compulsive disorder (Gillan et al., 2015), alcohol use disorder (Sjoerds et al., 2013), and schizophrenia (Morris et al., 2015). For example, we investigated the neural substrates of motivational

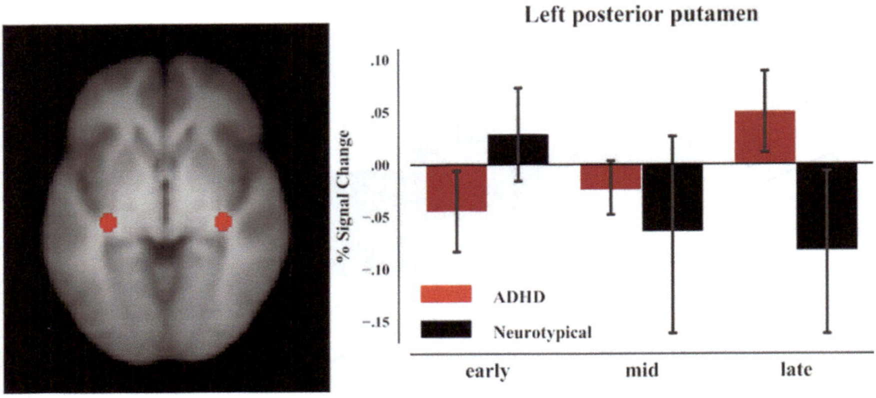

Fig. 2 Posterior putamen activity as a function of stimulus-response-outcome training length in people with attention-deficit/hyperactivity disorder as compared to neurotypical controls, as featured in Ceceli et al. (2020). The posterior putamen (dorsolateral striatum homologue that is associated with habitual control) showed higher engagement in response to the stimulus in people with attention-deficit/hyperactivity disorder over the course of moderate training, consistent with this region's engagement following over-training in the general population. The moderate training length did not significantly recruit the posterior putamen in neurotypical participants. (The figure is adapted from Ceceli et al. (2020) with permission)

control in attention-deficit/hyperactivity disorder by moderately training stimulus-response-outcome associations (see Fig. 1) during fMRI over a single session (as opposed to prolonged training over multiple sessions as reported in Tricomi et al. (2009)). Our results supported the putative cortico-striatal bases of motivational control: posterior putamen activity in individuals with attention-deficit/hyperactivity disorder was enhanced as a function of training as compared to controls (Fig. 2). We also revealed impairments in functional connectivity between the putamen and the prefrontal cortex in this group, suggesting cortico-striatal abnormalities that are relevant for regulating goal-directed control (Ceceli et al., 2020).

Recent reviews and meta-analyses have discussed the underlying neurobiology of motivational control (Ameden & Tricomi, 2024; Ceceli & Tricomi, 2018), the insufficient and narrow operationalization of habits in the field and potential remedies (Watson et al., 2022), mechanisms of habit formation and disruption as informed by lab-based and limited real-world efforts (Buabang et al., 2025), and the striatal bases of natural behaviors that are thought to be habit-driven (e.g., walking, driving, reading, etc.) (Guida et al., 2022). An emerging theme in habit science is the difficulty of accurately and reliably estimating habits in humans (Ceceli & Tricomi, 2018; Watson et al., 2022; Watson & de Wit, 2018), potentially due to limitations imposed by lab-based, cross-sectional, or narrow longitudinal designs that attempt to capture habits via over-training. In this chapter, we will focus on the characterization of goals and habits using approaches that are aimed at enhancing the ecological validity of motivational control estimates. Specifically, we will emphasize neurobehavioral human habit research with novel task paradigms or study designs that facilitate a naturalistic scaling of objective motivational control estimates, including gamified tasks and remote data collection.

1 Capturing Real-World Habits in the Laboratory

While training novel stimulus-response-outcome associations has been widely used in rodent and human studies to study habits, this approach has shown limited success in the latter. First, the novel associations that are developed in a laboratory context may not be representative of the habits that we aim to understand in daily life, suggesting low ecological validity. Next, the human-analog of the tasks traditionally used in rodents may not be taking into account the potential individual differences in humans. For instance, some people may experience social pressure when asked to eat a food to satiety in an outcome devaluation task, potentially interfering with the outcome sensitivity test. Individual differences in stress or anxiety have also been demonstrated to influence outcome sensitivity tests. Thus, high-stress individuals express habitual control earlier than those with low stress in a free-operant task with food outcomes (Pool et al., 2022), and individuals who undergo a stress induction manipulation show attenuated sensitivity to reward contingency changes in a two-step decision task (Cremer et al., 2021).

To circumvent these disadvantages in studying motivational control, a potential remedy is using associations with real-world relevance. For instance, the classic slips-of-action task with fractal stimuli (de Wit, Watson, et al., 2012b) has been modified to include as stimuli or outcome images of snacks. In a study that directly compared these task versions, it was revealed that actions when snack images served as outcomes were more resistant to change using a behavioral strategy adopted to disrupt habits (i.e., implementation intentions (Gollwitzer, 1999)) as compared to abstract stimuli as outcomes (Verhoeven et al., 2018). Extending these efforts further, we focused on stimulus-response associations that have been well-established outside of the laboratory to better model real-world habits. Specifically, we capitalized on familiar color-response associations (i.e., red-stop, green-go) as compared to novel associations (i.e., purple-stop, blue-go) contextualized in traffic light stimuli. We found that errors of commission were significantly lower to the familiar associations of red-stop/green-go vs. green-stop/red-go, suggesting outcome-sensitive, habitual control, with no differential commission error changes to the novel color-response mappings (see Fig. 3). This pattern is in line with an adoption of red and green traffic light stimuli for the purpose of estimating inhibitory control (a related cognitive process) as a function of familiarity in a stop-signal task. Here, the red traffic light stimulus was associated with shorter stop-signal response time (the hallmark estimate of inhibitory control, with shorter response time indicating better stopping performance) compared to the green traffic light stimulus (as well as a control black stimulus). These results suggest that existing associations (with minimal in-lab training) can be used to estimate motivational control.

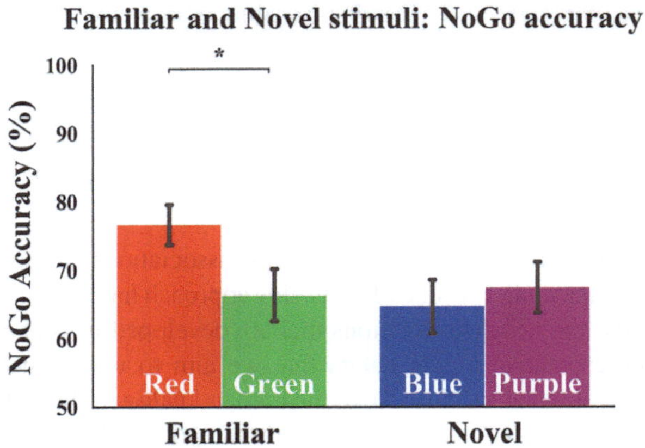

Fig. 3 Errors of commission in response to familiar and novel color-response mappings presented in a traffic light context. Participants made fewer errors of commission (i.e., responding when instructed to withhold) when the no-go signal was red compared to green. A control condition with novel color-response mappings (no-go as purple or blue) did not yield significant differences in errors of commission, suggesting that existing, familiar associations (e.g., red-stop) may elicit outcome-insensitive actions. Color of bars depicts no-go stimulus mappings. (The figure is adapted from Ceceli et al. (2019) with permission)

2 Sampling Real-World Habits Outside of the Laboratory

In addition to bringing real-world habits into the lab, researchers have also extended their habit measures to the real world. In a naturalistic study of habitual control, participants were provided new key covers for their home, which, after a short (3-week) or long (6-week) training phase, were switched with a dummy key. Participants recorded their performance (e.g., effort, time, mistakes, attentional measures) after their key selection attempts to open the door over the course of the study. Thus, the participants formed and trained a stimulus (key cover)—response (use key)—outcome (open door) association in their daily lives. Importantly, outcome sensitivity was tested by switching out the key cover, necessitating participants to override a newly developed habit to reach their goal of opening the door. As participants learned the stimulus-response-outcome association, they recruited less attention, but after the key cover switch, their mistakes, time, and effort expended to select the correct key increased, in line with what one would expect from an outcome-insensitive habit (Linnebank et al., 2018).

In another study that intersected lab-based measures with real-world behaviors to inspect the neurobehavioral substrates of motivational control, participants completed an outcome devaluation (slips-of-action) task in the lab, underwent structural neuroimaging, and were instructed to begin taking a vitamin pill (a placebo, unbeknownst to the participants) in one of the two time windows (every morning or evening) over the course of 3 weeks (i.e., the routine formation phase). Next, participants were instructed to switch to the other time window (i.e., the switch phase), which served as the test for outcome sensitivity, with switch phase pills taken in the time window that was assigned for the routine formation phase indicating habit formation. While participants took significantly fewer pills in their designated time window during the switch phase as compared to the final week of the routine formation phase on average, only a minority of participants took the pill in the previous time window, suggesting that the evidence for action slips in real-world behaviors following 3 weeks of training was weak. The measure of habitual action slips in the lab-based task did not correlate with pill intake but showed a correlation with a subjective report of automatization, such that those making more habitual responses in the task also reported performing the pill-taking behavior as more automatic. The expected correlation between habitual pill intake or self-reported automatization and posterior putamen–premotor cortex connectivity was not evident. Instead, pill intake in the determined time window and self-reported automatization significantly correlated with white matter tracts between the caudate and the prefrontal cortex (including the vmPFC) (van de Vijver et al., 2023). It is difficult to infer, however, whether successful pill intake reflects goal-directed or habitual processes in this sample. For instance, given the lack of variability in phase duration and the low rates of action slips in pill taking during the switch phase, whether prolonged training in a real-world behavior can induce habit formation remains unknown. Taken together, these results suggest that strong goal-directed control may hinder habit formation, though more research is needed to better understand the role of training in developing habits in the real world.

3 Intersecting Lab-Based and Naturalistic Methods Using Online Data Collection

The studies we have featured so far have relied on lab-based paradigms that capitalize on stimuli with real-world meanings or estimated real-world behaviors outside of the lab under a motivational control framework. Researchers have also utilized smartphones to sample behaviors using well-controlled experimental paradigms outside of the lab, intersecting traditional and naturalistic methods to estimate motivational control measures. In a recent study, participants installed a novel, gamified free-operant task into their smartphones, which they freely entered to play a space treasure game (Gera et al., 2024). As depicted in Fig. 4, participants were instructed enter the app whenever desired to search for gold that could later be converted into real money by making a series of pressing actions on the screen. Thus, the app icon served as the stimulus, yielding a screen in which the participants executed the response, resulting in a reward indicator outcome (or no-reward). This training phase was followed by an outcome devaluation manipulation, whereby the participant was informed that the warehouse that stored the accumulated gold became full until the next day. A control manipulation informed the participant that the warehouse was at half capacity (i.e., accepting gold and thus still-valued).

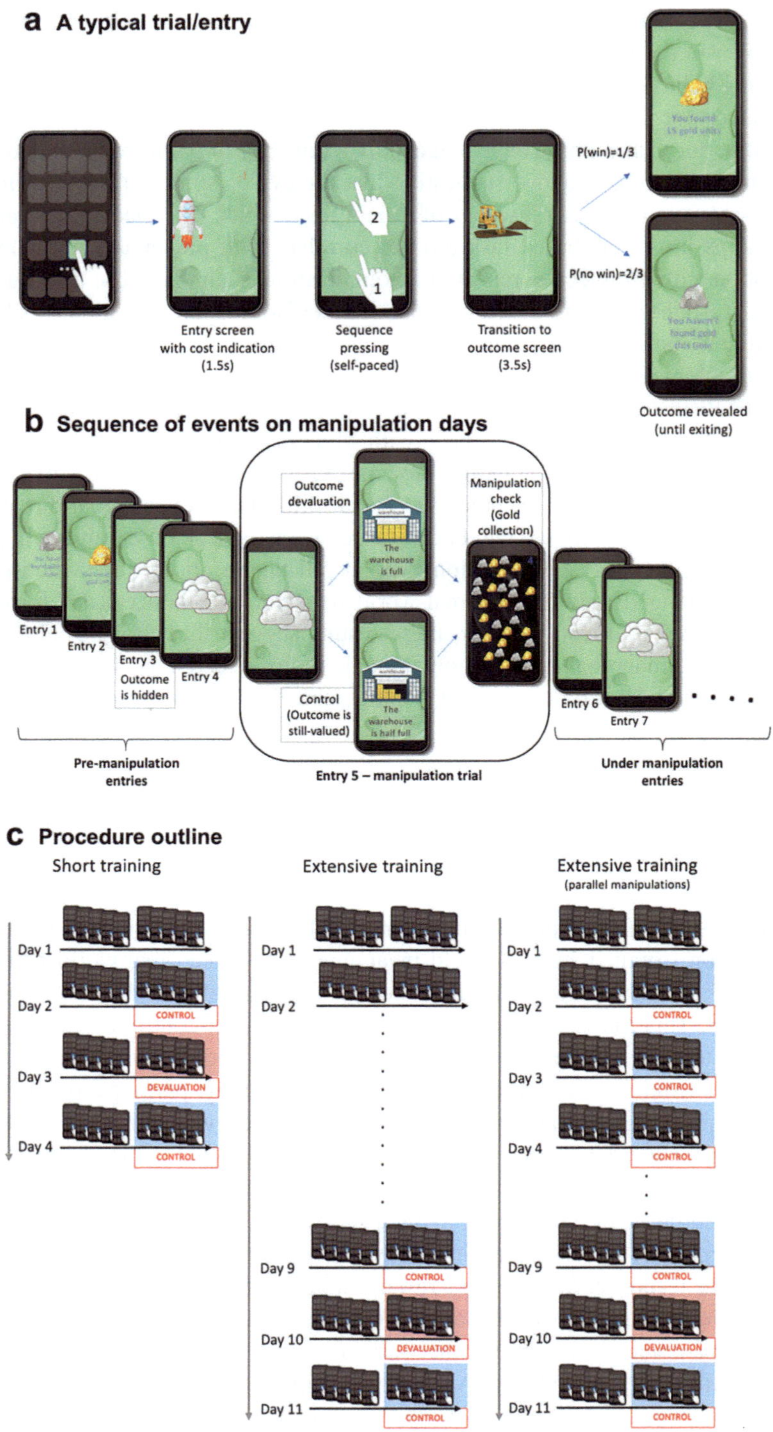

Fig. 4 A smartphone-gamified habit task developed by Gera and colleagues. (**a**) Participants enter the app to search for gold by making a sequence of pressing responses, yielding a reward

Therefore, outcome-insensitive habits could be tracked via the number of app entries after the devaluation procedure, as compared to the app entries following the control manipulation. By varying the amount of training (i.e., 3 or 10 days), it was revealed that individuals were more likely to make habitual responses if they were in the 10-day training group, suggesting that the novel task was sensitive to over-training effects that are otherwise difficult to produce in other human habit tasks (de Wit et al., 2018; Pool et al., 2022).

While the above results underscore the utility of smartphones in examining motivational control, an imperative point to address is whether the well-controlled lab-based habit estimates can be replicated when administered remotely. To this end, Hartley and colleagues deployed a two-step decision task using an online platform to track motivational control over the course of development in a cross-sectional study (Nussenbaum et al., 2020). Here, they replicated their original (in-lab) finding (Decker et al., 2016) that the recruitment of a goal-directed (model-based) choice strategy strengthens over the course of development. While children, adolescents, and adults all used a hybrid choice strategy (i.e., recruiting habits and goals, as seen in the general population (Daw et al., 2011)), the model-based choice estimate in particular increased as a function of age. It should be noted that while model-based choice in the two-step task has overlapped with goal-directed control in a cross-validation study, its counterpart, model-free choice, has not similarly aligned with habitual control, suggesting that the two-step task may be more sensitive to goal-directed control (and its impairments) (Sjoerds et al., 2016). Importantly, their remotely collected data indicated minimal differences with in-lab data, and a majority of the participants provided quality data per measures of low non-task engagement, correct responses to task comprehension questions, low missed trials, and few exceedingly high response time trials. These results align with previous remote validations of neuropsychological tasks (Chaytor et al., 2021), including related efforts in reinforcement learning (Weydmann et al., 2023), that suggest comparable findings as compared to in-lab administrations. Overall, these data support the utility of expanding beyond the constraints of lab-based motivational control testing to maximize ecological validity and generalizability of our inferences.

Fig. 4 (continued) indicated by a gold image, or no reward as indicated by a rock image. (**b**) On the fifth entry of each manipulation day, participants were notified that the warehouse is full and cannot store any more gold until the next day (i.e., outcome devaluation) or that it is at half capacity (control condition, still valued). The outcome was concealed from the participants on the rest of the day's trials (similar to extinction), and responses in this phase served as the test for outcome sensitivity (i.e., app entries following devaluation vs. control manipulations). (**c**) Participants underwent short (3-day) or extensive (10-day) training. The "extensive training with parallel manipulations" group controlled for the earlier encounters with devaluation and control manipulations as experienced in the short training. This design permitted the naturalistic study of habit formation as a function of training length. (The figure is adapted from Gera et al. (2024) with permission)

4 Future Directions and Conclusions

Despite the exciting advances reviewed here, such as the incorporation of lab-based measures in naturalistic settings (and the use of naturalistic stimuli in lab settings), several gaps remain in our understanding of habits. First, while rodent studies have commonly induced habitual responses by over-training stimulus-response associations (Adams, 1982; Dickinson et al., 1983, 1995; Or & Klavir, 2021), suggesting that prolonged experience is conducive to habit formation, habit formation has proven more difficult to address in human studies. To this end, naturalistic and longitudinal designs may offer insights into real-world factors that facilitate (or hinder) the emergence of an outcome-insensitive habit with everyday life relevance. Second, perhaps because we do not have strong examples of experimentally induced habits, we have yet to elucidate how unwanted habits can be disrupted or substituted by more adaptive alternatives. Capitalizing on the innovative designs discussed in this chapter (e.g., measuring real-world habits in the lab) may yield strong habits that can be tested for their rigidity in well-controlled experimental settings. Next, cross-sectional and narrow longitudinal studies limit our ability to track dynamic patterns in goals and habits over the course of time, which can have translational implications. For instance, in an experiment at the intersection of computational psychiatry and neuroeconomics, Konova et al. longitudinally sampled a risky decision making task to observe behavioral fluctuations over time via computational modeling, demonstrating that changes in risky decision-making can predict prospective relapse in opioid use disorder (2019). Although this study did not estimate habits, a comparable study design that is sensitive to potential fluctuations in motivational control over time could help track severity and vulnerability in neuropsychiatric disorders that accompany aberrant motivational control, such as addiction. While influential models of drug addiction posit that this devastating brain disease is underpinned by a transition from goal-directed to habitual and eventually compulsive drug seeking (Everitt & Robbins, 2005), whether the primarily preclinical evidence translates to people with substance use disorders is heavily debated (Everitt & Robbins, 2016; Hogarth, 2020). Thus, enhancing the validity and generalizability of the tools in our disposal is imperative for addressing basic and translational gaps in the role of habits in our lives, in disease and health.

A large body of work encompasses the naturalistic sampling of self-estimated habitual tendencies. These studies include intensive sampling of self-selected behaviors (e.g., exercise, drinking more water, and social contact) to examine how habits are formed over time and repetition (Lally et al., 2010; van der Weiden et al., 2020), and the inspection of specific behaviors, such as a nutrition habit (i.e., increasing servings of vegetables consumed) (Kilb & Labudek, 2022), or reducing sedentary behavior with momentary reminders (Maher & Dunton, 2020). While these studies use well-validated subjective measures of habit strength (e.g., the Self-Report Habit Index (Verplanken & Orbell, 2003)), we did not focus on them in our discussion of enhancing the validity of habit estimates because they do not capture an objective marker of the balance between goal-directed and habitual behaviors. Importantly, subjective measures of habit strength have been shown to be uncorrelated with various objective, lab-based estimates of habitual control (Nebe et al., 2024), suggesting that lab-based tasks are capturing aspects of habitual control that are not readily accessible by self-report, or that lab-based estimates translate poorly to real-world scenarios of habit. Nonetheless, the adoption of naturalistic and longitudinal designs in investigating self-estimated habits supports the notion that objective habit measures can be extended to the real world for enhanced translational impact.

While we frequently rely on habits to navigate our environment, most of what we know about habits is restricted to lab-based models that are limited in their ecological relevance. In this chapter, we reviewed recent advances in neurobehavioral research that underscore the value of extending the estimation of habits beyond the lab. Notably, we presented evidence for successful implementations of stimuli with real-world relevance to well-controlled lab-based efforts. In the reverse direction, we also featured studies with lab tasks that were sampled in everyday settings. Importantly, we illustrated that innovative designs that capitalize on remote data collection show promise in enhancing the validity of our measures of motivational control. While capturing the elusive habit in humans and translating it for clinical benefit has been a challenge, the evidence as reviewed here suggests that enhancing ecological validity in our efforts may bridge the gap between experimental findings and real-world applications. Incorporating more naturalistic stimuli and leveraging remote data collection technologies may yield more robust models of how habits form and operate in daily life. As these methods are refined, the potential for clinical applications (e.g., behavioral interventions for relevant psychiatric conditions) may grow, offering a pathway for more effective and individualized treatment strategies.

References

Adams, C. D. (1982). Variations in the sensitivity of instrumental responding to reinforcer devaluation. *The Quarterly Journal of Experimental Psychology Section B, 34*(2), 77–98. https://doi.org/10.1080/14640748208400878

Ameden, W. C., & Tricomi, E. (2024). Historical and modern perspectives on the neuroscience of habits. In Y. Vandaele (Ed.), *Habits: Their definition, neurobiology, and role in addiction* (pp. 227–247). Springer. https://doi.org/10.1007/978-3-031-55889-4_10

Ashby, F. G., Turner, B. O., & Horvitz, J. C. (2010). Cortical and basal ganglia contributions to habit learning and automaticity. *Trends in Cognitive Sciences, 14*(5), 208–215. https://doi.org/10.1016/j.tics.2010.02.001

Balleine, B. W., & O'Doherty, J. P. (2010). Human and rodent homologies in action control: Corticostriatal determinants of goal-directed and habitual action. *Neuropsychopharmacology, 35*(1), 48–69. https://doi.org/10.1038/npp.2009.131

Buabang, E. K., Donegan, K. R., Rafei, P., & Gillan, C. M. (2025). Leveraging cognitive neuroscience for making and breaking real-world habits. *Trends in Cognitive Sciences, 29*(1), 41–59. https://doi.org/10.1016/j.tics.2024.10.006

Ceceli, A. O., & Tricomi, E. (2018). Habits and goals: A motivational perspective on action control. *Current Opinion in Behavioral Sciences, 20*, 110–116. https://doi.org/10.1016/j.cobeha.2017.12.005

Ceceli, A. O., Myers, C. E., & Tricomi, E. (2019). Demonstrating and disrupting well-learned habits. *bioRxiv, 745356*. https://doi.org/10.1101/745356

Ceceli, A. O., Natsheh, J. Y., Cruz, D., & Tricomi, E. (2020). The neurobehavioral mechanisms of motivational control in attention-deficit/hyperactivity disorder. *Cortex, 127*, 191–207. https://doi.org/10.1016/j.cortex.2020.02.009

Chaytor, N. S., Barbosa-Leiker, C., Germine, L. T., Fonseca, L. M., McPherson, S. M., & Tuttle, K. R. (2021). Construct validity, ecological validity and acceptance of self-administered online neuropsychological assessment in adults. *The Clinical Neuropsychologist, 35*(1), 148–164. https://doi.org/10.1080/13854046.2020.1811893

Cremer, A., Kalbe, F., Gläscher, J., & Schwabe, L. (2021). Stress reduces both model-based and model-free neural computations during flexible learning. *NeuroImage, 229*, 117747. https://doi.org/10.1016/j.neuroimage.2021.117747

Daw, N. D., Gershman, S. J., Seymour, B., Dayan, P., & Dolan, R. J. (2011). Model-based influences on humans' choices and striatal prediction errors. *Neuron, 69*(6), 1204–1215. https://doi.org/10.1016/j.neuron.2011.02.027

de Wit, S., Standing, H. R., DeVito, E. E., Robinson, O. J., Ridderinkhof, K. R., Robbins, T. W., & Sahakian, B. J. (2012a). Reliance on habits at the expense of goal-directed control following dopamine precursor depletion. *Psychopharmacology, 219*(2), 621. https://doi.org/10.1007/s00213-011-2563-2

de Wit, S., Watson, P., Harsay, H. A., Cohen, M. X., van de Vijver, I., & Ridderinkhof, K. R. (2012b). Corticostriatal connectivity underlies individual differences in the balance between habitual and goal-directed action control. *The Journal of Neuroscience: The Official Journal of the Society for Neuroscience, 32*(35), 12066–12075. https://doi.org/10.1523/JNEUROSCI.1088-12.2012

de Wit, S., Kindt, M., Knot, S. L., Verhoeven, A. A. C., Robbins, T. W., Gasull-Camos, J., Evans, M., Mirza, H., & Gillan, C. M. (2018). Shifting the balance between goals and habits: Five failures in experimental habit induction. *Journal of Experimental Psychology. General, 147*(7), 1043–1065. https://doi.org/10.1037/xge0000402

Decker, J. H., Otto, A. R., Daw, N. D., & Hartley, C. A. (2016). From creatures of habit to goal-directed learners: Tracking the developmental emergence of model-based reinforcement learning. *Psychological Science, 27*(6), 848–858. https://doi.org/10.1177/0956797616639301

Dickinson, A. (1985). Actions and habits: The development of behavioural autonomy. *Royal Society of London Philosophical Transactions Series B, 308*, 67–78. https://doi.org/10.1098/rstb.1985.0010

Dickinson, A., & Balleine, B. (1994). Motivational control of goal-directed action. *Animal Learning & Behavior, 22*(1), 1–18. https://doi.org/10.3758/BF03199951

Dickinson, A., Nicholas, D. J., & Adams, C. D. (1983). The effect of the instrumental training contingency on susceptibility to reinforcer devaluation. *The Quarterly Journal of Experimental Psychology Section B, 35*(1), 35–51. https://doi.org/10.1080/14640748308400912

Dickinson, A., Balleine, B., Watt, A., Gonzalez, F., & Boakes, R. A. (1995). Motivational control after extended instrumental training. *Animal Learning & Behavior, 23*(2), 197–206. https://doi.org/10.3758/BF03199935

Ersche, K. D., Gillan, C. M., Jones, P. S., Williams, G. B., Ward, L. H. E., Luijten, M., de Wit, S., Sahakian, B. J., Bullmore, E. T., & Robbins, T. W. (2016). Carrots and sticks fail to change behavior in cocaine addiction , 352(6292), 1468–1471. https://doi.org/10.1126/science.aaf3700.

Everitt, B. J., & Robbins, T. W. (2005). Neural systems of reinforcement for drug addiction: From actions to habits to compulsion. *Nature Neuroscience, 8*(11), 1481–1489. https://doi.org/10.1038/nn1579

Everitt, B. J., & Robbins, T. W. (2016). Drug addiction: Updating actions to habits to compulsions ten years on. *Annual Review of Psychology, 67*(1), 23–50. https://doi.org/10.1146/annurev-psych-122414-033457

Gera, R., Barak, S., & Schonberg, T. (2024). A novel free-operant framework enables experimental habit induction in humans. *Behavior Research Methods, 56*(4), 3937–3958. https://doi.org/10.3758/s13428-023-02263-6
Gillan, C. M., Apergis-Schoute, A. M., Morein-Zamir, S., Urcelay, G. P., Sule, A., Fineberg, N. A., Sahakian, B. J., & Robbins, T. W. (2015). Functional neuroimaging of avoidance habits in obsessive-compulsive disorder. *The American Journal of Psychiatry, 172*(3), 284–293. https://doi.org/10.1176/appi.ajp.2014.14040525
Gollwitzer, P. M. (1999). Implementation intentions: Strong effects of simple plans. *The American Psychologist, 54*(7), 493–503. https://doi.org/10.1037/0003-066X.54.7.493
Guida, P., Michiels, M., Redgrave, P., Luque, D., & Obeso, I. (2022). An fMRI meta-analysis of the role of the striatum in everyday-life vs laboratory-developed habits. *Neuroscience and Biobehavioral Reviews, 141*, 104826. https://doi.org/10.1016/j.neubiorev.2022.104826
Hogarth, L. (2020). Addiction is driven by excessive goal-directed drug choice under negative affect: Translational critique of habit and compulsion theory. *Neuropsychopharmacology, 45*(5), 720–735. https://doi.org/10.1038/s41386-020-0600-8
Kilb, M., & Labudek, S. (2022). Effects of behavioral performance, intrinsic reward value, and context stability on the formation of a higher-order nutrition habit: An intensive longitudinal diary study. *The International Journal of Behavioral Nutrition and Physical Activity, 19*(1), 105. https://doi.org/10.1186/s12966-022-01343-8
Konova, A. B., Lopez-Guzman, S., Urmanche, A., Ross, S., Louie, K., Rotrosen, J., & Glimcher, P. W. (2019). Computational markers of risky decision-making for identification of temporal windows of vulnerability to opioid use in a real-world clinical setting. *JAMA Psychiatry*. https://doi.org/10.1001/jamapsychiatry.2019.4013
Lally, P., van Jaarsveld, C. H. M., Potts, H. W. W., & Wardle, J. (2010). How are habits formed: Modelling habit formation in the real world. *European Journal of Social Psychology, 40*(6), 998–1009. https://doi.org/10.1002/ejsp.674
Linnebank, F. E., Kindt, M., & de Wit, S. (2018). Investigating the balance between goal-directed and habitual control in experimental and real-life settings. *Learning & Behavior, 46*(3), 306–319. https://doi.org/10.3758/s13420-018-0313-6
Maher, J. P., & Dunton, G. F. (2020). Editor's choice: Dual-process model of older adults' sedentary behavior: An ecological momentary assessment study. *Psychology & Health, 35*(5), 519–537. https://doi.org/10.1080/08870446.2019.1666984
Mishkin, M., Malamut, B., & Bachevalier, J. (1984). Memories and habits: Two neural systems. In G. Lynch, J. L. McGaugh, & N. M. Weinberger (Eds.), *Neurobiology of learning and memory* (pp. 65–77). Guilford.
Morris, R. W., Quail, S., Griffiths, K. R., Green, M. J., & Balleine, B. W. (2015). Corticostriatal control of goal-directed action is impaired in schizophrenia. *Biological Psychiatry, 77*(2), 187–195. https://doi.org/10.1016/j.biopsych.2014.06.005
Nebe, S., Kretzschmar, A., Brandt, M. C., & Tobler, P. N. (2024). Characterizing human habits in the lab. *Collabra. Psychology, 10*(1), 92949. https://doi.org/10.1525/collabra.92949
Nussenbaum, K., Scheuplein, M., Phaneuf, C. V., Evans, M. D., & Hartley, C. A. (2020). Moving developmental research online: Comparing in-lab and web-based studies of model-based reinforcement learning. *Collabra. Psychology, 6*(1), 17213. https://doi.org/10.1525/collabra.17213
Or, M. B., & Klavir, O. (2021). The differential effects of the amount of training on sensitivity of distinct actions to reward devaluation. *Brain Sciences, 11*(6), 732. https://doi.org/10.3390/brainsci11060732
Pool, E. R., Gera, R., Fransen, A., Perez, O. D., Cremer, A., Aleksic, M., Tanwisuth, S., Quail, S., Ceceli, A. O., Manfredi, D. A., Nave, G., Tricomi, E., Balleine, B., Schonberg, T., Schwabe, L., & O'Doherty, J. P. (2022). Determining the effects of training duration on the behavioral expression of habitual control in humans: A multilaboratory investigation. *Learning & Memory (Cold Spring Harbor, N.Y.), 29*(1), 16–28. https://doi.org/10.1101/lm.053413.121
Rangel, A., Camerer, C., & Montague, P. R. (2008). Neuroeconomics: The neurobiology of value-based decision-making. *Nature Reviews. Neuroscience, 9*(7), 545–556. https://doi.org/10.1038/nrn2357

Sjoerds, Z., de Wit, S., van den Brink, W., Robbins, T. W., Beekman, A. T. F., Penninx, B. W. J. H., & Veltman, D. J. (2013). Behavioral and neuroimaging evidence for overreliance on habit learning in alcohol-dependent patients. *Translational Psychiatry, 3*(12), e337. https://doi.org/10.1038/tp.2013.107

Sjoerds, Z., Dietrich, A., Deserno, L., de Wit, S., Villringer, A., Heinze, H.-J., Schlagenhauf, F., & Horstmann, A. (2016). Slips of action and sequential decisions: A cross-validation study of tasks assessing habitual and goal-directed action control. *Frontiers in Behavioral Neuroscience, 10*, 234. https://doi.org/10.3389/fnbeh.2016.00234

Tricomi, E., Balleine, B. W., & O'Doherty, J. P. (2009). A specific role for posterior dorsolateral striatum in human habit learning. *European Journal of Neuroscience, 29*(11), 2225–2232. https://doi.org/10.1111/j.1460-9568.2009.06796.x

van de Vijver, I., Verhoeven, A. A. C., & de Wit, S. (2023). Individual differences in Corticostriatal white-matter tracts predict successful daily-life routine formation. *Journal of Cognitive Neuroscience, 35*(4), 571–587. https://doi.org/10.1162/jocn_a_01967

van der Weiden, A., Benjamins, J., Gillebaart, M., Ybema, J. F., & de Ridder, D. (2020). How to form good habits? A longitudinal field study on the role of self-control in habit formation. *Frontiers in Psychology, 11*, 560. https://doi.org/10.3389/fpsyg.2020.00560

Verhoeven, A. A. C., Kindt, M., Zomer, C. L., & de Wit, S. (2018). An experimental investigation of breaking learnt habits with verbal implementation intentions. *Acta Psychologica, 184*, 124–136. https://doi.org/10.1016/j.actpsy.2017.05.008

Verplanken, B., & Orbell, S. (2003). Reflections on past behavior: A self-report index of habit Strength1. *Journal of Applied Social Psychology, 33*(6), 1313–1330. https://doi.org/10.1111/j.1559-1816.2003.tb01951.x

Watson, P., & de Wit, S. (2018). Current limits of experimental research into habits and future directions. *Current Opinion in Behavioral Sciences, 20*, 33–39. https://doi.org/10.1016/j.cobeha.2017.09.012

Watson, P., O'Callaghan, C., Perkes, I., Bradfield, L., & Turner, K. (2022). Making habits measurable beyond what they are not: A focus on associative dual-process models. *Neuroscience and Biobehavioral Reviews, 142*, 104869. https://doi.org/10.1016/j.neubiorev.2022.104869

Weydmann, G., Palmieri, I., Simões, R. A. G., Centurion Cabral, J. C., Eckhardt, J., Tavares, P., Moro, C., Alves, P., Buchmann, S., Schmidt, E., Friedman, R., & Bizarro, L. (2023). Switching to online: Testing the validity of supervised remote testing for online reinforcement learning experiments. *Behavior Research Methods, 55*(7), 3645–3657. https://doi.org/10.3758/s13428-022-01982-6

Yin, H. H., Knowlton, B. J., & Balleine, B. W. (2004). Lesions of dorsolateral striatum preserve outcome expectancy but disrupt habit formation in instrumental learning. *The European Journal of Neuroscience, 19*(1), 181–189.

Yin, H. H., Knowlton, B. J., & Balleine, B. W. (2005a). Blockade of NMDA receptors in the dorsomedial striatum prevents action-outcome learning in instrumental conditioning. *The European Journal of Neuroscience, 22*(2), 505–512. https://doi.org/10.1111/j.1460-9568.2005.04219.x

Yin, H. H., Ostlund, S. B., Knowlton, B. J., & Balleine, B. W. (2005b). The role of the dorsomedial striatum in instrumental conditioning. *The European Journal of Neuroscience, 22*(2), 513–523. https://doi.org/10.1111/j.1460-9568.2005.04218.x

Pain and Value-Based Decision-Making: An Introduction and Implications for Neuroeconomics

Silvia Lopez-Guzman and Lauren Y. Atlas

Abstract Pain is central to numerous clinical conditions and is the main reason that individuals seek medical treatment. Although pain involves subjective decisions and the psychophysics and neural basis of pain have been studied for decades, few have applied neuroeconomic approaches to the study of pain and its modulation. In this chapter, we provide an introduction to pain for those in the field of neuroeconomics. We focus on acute pain, which can be experimentally manipulated and studied using neuroscience tools. We first review aspects of pain-related decisions and the circuits that support different aspects of pain. We then discuss the common neural pathways that support pain and reward, as well as evidence of their interactions. Next, we introduce how pain might be viewed from a neuroeconomic perspective, focusing on the value of pain. Finally, we address hypotheses and outstanding questions for those at the intersection of pain and neuroeconomics. By providing this introduction, our goal is to inspire a new generation of studies that bridge neuroeconomics and pain to tackle critical challenges such as those brought about by the opioid epidemic.

1 Introduction

Neuroeconomics has provided powerful insights on the brain mechanisms of decision-making across a wide variety of domains, including primary and secondary

S. Lopez-Guzman (✉)
National Institute of Mental Health, National Institutes of Health, Bethesda, MD, USA

National Institute on Drug Abuse, National Institutes of Health, Baltimore, MD, USA
e-mail: silvia.lopezguzman@nih.gov

L. Y. Atlas (✉)
National Institute of Mental Health, National Institutes of Health, Bethesda, MD, USA

National Institute on Drug Abuse, National Institutes of Health, Baltimore, MD, USA

National Center for Complementary and Integrative Health, National Institutes of Health, Bethesda, MD, USA
e-mail: lauren.atlas@nih.gov

D. V. Smith et al. (eds.), *Neuroeconomics: Core Topics and Current Directions*,
https://doi.org/10.1007/978-3-032-02925-6_12

reinforcers (Delgado et al., 2006; Sescousse et al., 2013), social learning and cognition (Fehr & Camerer, 2007; Hackel & Amodio, 2018), and even political preferences (Krastev et al., 2016). Although it is widely accepted that a fundamental drive for all organisms is to seek pleasure and avoid pain and many have reviewed general relationships between pain and reward (Navratilova & Porreca, 2014; Leknes & Tracey, 2008), researchers have begun linking the vast literature on neuroeconomics with the diverse neuroscience of pain only recently. We believe that this is an essential path forward and that neuroeconomic perspectives can provide important insights on pain and pain-related decision-making, similar to the recent emphasis in the field of computational psychiatry. Reward, pain, and decision-making can interact in ways that can be either devastating (e.g., the opioid crisis) or beneficial (e.g., placebo effects). In this chapter, our goal is to address ways in which economic theory can provide insights on the clinical problem of pain and motivate those in the field of neuroeconomics to consider how their findings may be relevant for pain. We first provide an introduction to pain for those in the field of neuroeconomics, with a focus on pain-related decisions. We next address the relationship between the neural systems that support pain and reward and their interaction, laying a foundation for using economic theory to understand pain decision-making. From here, we explore how the fundamentals of neuroeconomics are relevant to the study of pain from a behavioral perspective and highlight key studies that have evaluated the value function of pain. Finally, we discuss outstanding questions and implications of this work for future studies and the amelioration of pain and related clinical conditions.

2 Pain-Related Decision-Making and Relationship with Neural Circuits

While most of us have been asked by a clinician to rate our pain on a simple 0–10 scale ranging from "no pain" to "most pain imaginable," the psychophysics of pain and its measurement has been the subject of extensive study and debate for over 50 years. In fact, even the definition of pain remains hotly debated and regularly revised (Loeser & Treede, 2008; Raja et al., 2020). For the purpose of this chapter, we focus on pain as defined most recently by the International Association for the Study of Pain (IASP) (Raja et al., 2020): "An unpleasant sensory and emotional experience associated with, or resembling that associated with, actual or potential tissue damage." Pain is different from nociception, which is defined as "the neural process of encoding and processing noxious stimuli." (Loeser & Treede, 2008). Thus, from an economic standpoint, nociception is most analogous to objective value, while the conscious pain experience is most related to subjective value. Like valuation and perception, the mapping between objective noxious input (e.g., temperature in thermal pain; Fig. 1a) and subjective pain appears to be a nonlinear and described by a power function (Stevens, 1957), although the exact form of

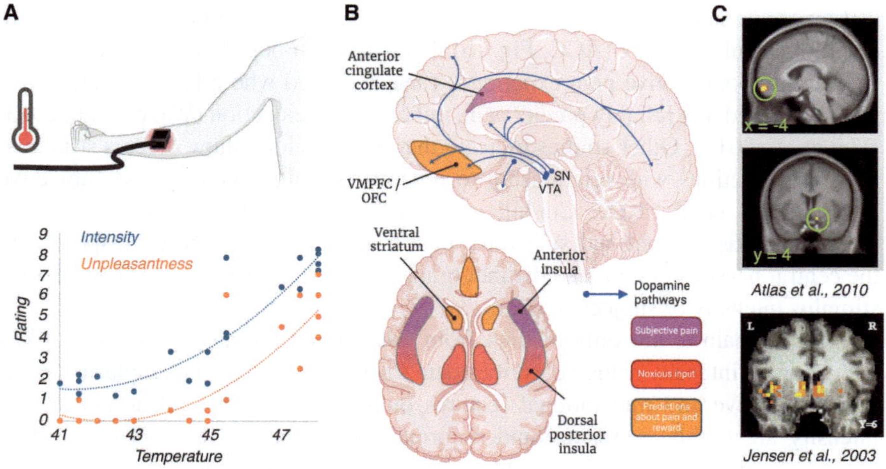

Fig. 1 Pain psychophysics and associated brain pathways. (**a**) *Top:* Studies of acute pain use specialized devices to deliver noxious stimuli. Here, a thermode delivers heat at different temperatures to a participant's forearm. *Bottom:* Participants are asked to rate pain in response to changes in noxious stimulus intensity. This figure depicts example data from a healthy participant who separately rated the intensity and unpleasantness of an 8-second thermal pain stimulus applied to the forearm at different temperatures. Both aspects of pain are fit with a power function. (**b**) Cortical regions such as the anterior cingulate cortex and the insula differentiate between subjective pain (purple) and noxious input (red). Regions involved in subjective pain overlap with those involved in pain affect and affective processing more generally. In addition, targets of dopamine pathways such as the ventral striatum and the ventromedial prefrontal cortex (VMPFC) and orbitofrontal cortex (OFC) are involved in updating predictions about pain, similar to their role in error-driven reward learning. (**c**) Traditional reward networks exhibit activation during anticipation of pain. *Top:* The ventral striatum and VMPFC show elevated activation in response to cues that predict high pain, relative to low pain (reprinted with permission from Atlas et al., 2010). *Bottom:* The ventral striatum also shows elevated activation in response to conditioned cues that predict unpleasant electric shocks relative to neutral cues (reprinted with permission from Jensen et al., 2003). (Figure created in part using biorender.com)

the power function can differ depending on many factors, including unnecessary aspects of the noxious stimulus itself (e.g., modality, duration) and how pain is measured (Nielsen et al., 2005; Price et al., 1992). Unlike economic value, which is often evaluated with behavioral measures (e.g., choices, willingness-to-pay, etc.) and sometimes subjective ratings (e.g., desirability ratings), pain is based entirely on patients' subjective ratings. Even the revised definition of pain (Raja et al., 2020) states "A person's report of an experience as pain should be respected" while noting "Verbal description is only one of several behaviors to express pain; inability to communicate does not negate the possibility that a human or a nonhuman animal experiences pain." We return to the measurement of pain and relationship with economic principles later in Sect. 4.

The distinction between pain and nociception is accompanied by segregation in the central nervous system. Several studies (Craig et al., 2000; Atlas et al.,

2014a; Apkarian et al., 1999) have revealed gradients within the insula and anterior cingulate cortex (ACC; Fig. 1b) that map from objective nociceptive input (posterior) to subjective pain experience (anterior), and whole brain patterns have been identified that reliably predict pain, rather than nociception (Wager et al., 2013; Woo et al., 2017). Other studies indicate that brain regions show diverse stimulus-response functions with relationship to noxious stimulus intensity and subjective pain (Büchel et al., 2002; Coghill et al., 1999). Thus, the distinction between objective noxious input and subjective pain is meaningful, based on responses in the central nervous system and distinct brain circuits that respond to either noxious stimulus intensity or subjective pain.

Notably, pain is not only distinct from nociception but also contains two key components: intensity, which describes pain's sensory features, and unpleasantness, or the affective and motivational features of pain. Although unpleasantness and intensity are highly correlated under normal circumstances (e.g., Fig. 1a), these two can be dissociated behaviorally and anatomically. For example, intensity and unpleasantness have distinct stimulus-response functions (Price, 2000; Rainville et al., 1992; Price et al., 1987), and opioid analgesics (Atlas et al., 2014b; Price et al., 1985), hypnosis (Rainville et al., 1999), unpleasant odors (Villemure & Bushnell, 2009), and some forms of meditation (Perlman et al., 2010) and emotion regulation (Salomons et al., 2014) can all influence pain unpleasantness while leaving pain intensity intact. Conversely, attention can alter pain intensity without affecting unpleasantness (Villemure & Bushnell, 2009). These dissociations are also present in the central nervous system. Studies generally suggest that a medial pain system (e.g., ACC, insula) is involved primarily in affective components of pain, while a lateral pain system (primary and secondary somatosensory cortex) is required for sensory components of pain. Variations in unpleasantness, controlling for intensity, are correlated with activation in the ACC (Tolle et al., 1999; Rainville et al., 1997), while variations in intensity, controlling for unpleasantness, are associated with activation in primary somatosensory cortex (S1) (Hofbauer et al., 2001). ACC and insula lesions are associated with altered affective responses to pain (Foltz & White, 1962; Berthier et al., 1988), while patients with somatosensory lesions find noxious stimuli unpleasant without being able to localize the stimulus (Ploner et al., 1999). The mechanisms that support pain unpleasantness are likely to be shared across areas of negative affect, consistent with recent fMRI work that used machine learning to identify a brain signature to predict negative affect across modalities, including thermal pain, pressure pain, aversive sound, and aversive images (Čeko et al., 2022). Responses in midline, forebrain, insular, and somatosensory regions predicted unpleasantness regardless of modality, while stimulus-specific negative affect was predicted by responses in primary and secondary sensory cortices. In the final section of this chapter, we address how future work should use neuroeconomic theory to gain insights on the functional significance of the dissociation between pain intensity and pain unpleasantness, similar to work on wanting versus liking (Berridge & Robinson, 2016; Dayan, 2022).

Together, these studies indicate that there are important distinctions between (a) pain and nociception and (b) sensory and affective components of pain and that pain-related decision-making involves both specialized brain circuits (e.g., those that contain nociceptive neurons) and systems that are likely to be conserved across domains of negative affect. From this foundation, we now consider how neuroeconomic approaches may be useful for the study of pain.

3 Pain and Reward: A Common Currency?

Many brain regions involved in reward and appetitive learning and decision-making play similar roles in pain and aversive learning (for reviews, see Navratilova & Porreca, 2014; Leknes & Tracey, 2008). In particular, dopaminergic mesolimbic/corticostriatal circuits involved in value processing, including the ventral striatum/nucleus accumbens (VS/NAc), ventromedial prefrontal cortex (VMPFC), and orbitofrontal cortex (OFC; Fig. 1b), are implicated in both acute and chronic pain. Predictions about painful stimuli are associated with anticipatory responses in the VS (Atlas et al., 2010; Jensen et al., 2003, 2007; Delgado et al., 2008; Seymour et al., 2004, 2005) and VMPFC/OFC (Atlas et al., 2010; Jensen et al., 2007) and value signals and prediction errors in these regions update during aversive learning in ways that are consistent with quantitative models of reinforcement learning (Jensen et al., 2007; Delgado et al., 2008; Seymour et al., 2004, 2005), similar to reward learning. Studies that compare pain with monetary reward indicate that the VS responds similarly to the anticipation of pain and reward, while the OFC and anterior insula seem to separately code valence (Jensen et al., 2007). Of course, value circuits are highly heterogeneous, and thus it is quite possible that pain and reward processing remain segregated at the level of individual neurons, even if they influence similar structures.

However, even if the neuronal underpinnings of appetitive and aversive value processing remain separate, there are functional interactions between pain and reward processing at the level of both brain and behavior. This bidirectional, interactive relationship is evident in both basic laboratory studies of acute and chronic pain and when we consider the high rates of anhedonia in chronic pain (Borsook et al., 2016) and comorbidity between chronic pain and depression (Lerman et al., 2015). Furthermore, although we can independently manipulate pain and reward, the two are actually intrinsically linked, as pain relief is rewarding (i.e., negative reinforcement; for review, see Volkow et al., 2018). Relief can be operationalized experimentally as transient relief from tonic pain (Seymour et al., 2005), responses at pain offset (Baliki et al., 2010; Becerra & Borsook, 2008), or relative relief based on the context surrounding pain, e.g., receiving a moderately painful stimulus when expecting high pain (Leknes et al., 2013, 2011). In healthy volunteers, pain relief is associated with activation in the VS and OFC/VMPFC (Seymour et al., 2005; Baliki et al., 2010; Becerra & Borsook, 2008; Leknes et al., 2013, 2011). Relative to healthy volunteers, people with chronic pain show reduced responses to pain offset in the NAc (Baliki et al., 2010) and the dopaminergic ventral tegmental area (VTA) (Loggia et al., 2014). Animal models confirm that pain offset engages VTA dopamine neurons (Brischoux et al., 2009) and dopamine release in the NAc shell (Budygin et al., 2012) and that inhibiting VTA dopamine prevents relief-related conditioned place preference (Navratilova et al., 2012). Notably, reward is also associated with relief via analgesic treatments: Healthy individuals who report higher trait reward responsivity experience stronger opioid analgesia (Wanigasekera et al., 2012), and individuals who show stronger NAc responses to monetary reward experience stronger placebo analgesia and have higher placebo-induced dopamine binding in the VS (Scott et al., 2007).

Together this work highlights the many ways in which pain and reward are intrinsically linked. From this foundation, we now turn to direct consideration of how the basic tenets of neuroeconomics have the potential to lead to new insights in our understanding of pain.

4 Neuroeconomic Approaches and Their Utility to the Study of Pain

Two main aspects of neuroeconomic theory are particularly helpful in conceptualizing decision-making in the context of pain and reward. First, value serves as a unifying signal for decision-making because it straddles both ends of the valence dimension (see Box 1). In other words, value is not only reward: the value of a stimulus or an action can be negative when the outcome is a loss and positive when the outcome is a gain. While most outcomes in behavioral economics studies are monetary losses and gains, the results from these studies could be generalized to aversive outcomes like pain when conceptualized as a loss or diminishment in well-being. Moreover, appetitive and aversive outcome values may be integrated through a cost-benefit computation into a net decision value (Pessiglione & Delgado, 2015). A unified monotonic value signal has been consistently found to be encoded in the VMPFC and anterior part of VS (Bartra et al., 2013; Peters & Büchel, 2010). Critically, the brain's valuation signal in VMPFC appears to be invariant to the specific type of good or outcome that is evaluated (FitzGerald et al., 2009; Chib et al., 2009) (from monetary to consummatory rewards or consumer goods), while OFC seems to preserve information about the specific identity of the outcome (Rudebeck & Murray, 2014; Howard et al., 2015; Howard & Kahnt, 2021). This evaluation happens automatically even in the absence of a behavioral output or an overt choice, such as when passively viewing stimuli (Lebreton et al., 2009; Levy et al., 2011).

The second aspect is that neuroeconomics rests on the premise that the brain "transforms" the objective value of the stimuli or options available (e.g., monetary rewards or consummable goods) into a *subjective value* representation, which is used for decision-making. This is relevant to pain because it is also characterized by a *subjective* experience that is distinct from the objective parameters of the painful stimulus itself, such as its intensity or nature. The mapping between objective and subjective value is referred to in neuroeconomics as the *utility function* and is central to understanding both individual differences in decision-making and making predictions about choice behavior under uncertainty (Box 1). Drawing parallels between pain and reward-related decision-making is possible, and application of approaches from neuroeconomics, as we will discuss below, have started to expand our understanding of the similarities and differences between the two.

5 The Value of Pain

Using a neuroeconomics approach to the study of pain implies inferring the subjective valuation process by presenting the decision-maker with options of varying objective value. Park et al. (2011) directly derived a participant-specific stimulus-response function by fitting a power model to pain ratings obtained as participants received shocks of increasing intensity. They found the shape of the objective stimulus to subjective experience function to be nonlinear, such that increases in the intensity of the pain stimulus lead to exponentially increasing perceived pain (Box 1, panel A, left). Using this function to personalize the pairing of monetary reward and pain, they tested different subjective value models against behavior and neural activity in VMPFC, OFC, and subgenual ACC. In all cases, the best model was one that performed a nonlinear integration of reward and pain. Evidence of pain and reward value integration as a cost-benefit computation has also been reported in the ACC and VS, such that these areas' responses to reward were attenuated in the presence of pain (Talmi et al., 2009). More recently, Slimani et al. (2022) used the same approach as Park et al. (2011) to personalize painful stimuli according to the participant-derived stimulus-response function and created trials that tiled the space between pain detection and tolerance thresholds. They then evaluated the participants' willingness to accept monetary rewards paired with different levels of painful electrical stimulation and estimated the utility function of pain from the points of indifference (50% probability of accepting a specific offer of combined monetary reward and pain). The best pain utility model was a quadratic function of experienced pain level (Box 1, panel A, right).

The shape of this utility function for pain has implications for decision-making under risk (see Box 1). A concave utility function means that the expected utility (probability * utility) of a lottery is significantly lower than the utility of its expected value (probability * objective value), leading to the preference of a smaller more certain offer over a higher riskier one. Conversely, convexity is equivalent to the presence of risk-seeking preferences. As follows, a convex pain value function would predict that the individual is willing to take risk in order to avoid pain. Interestingly, while the outcomes in the Slimani et al. study were not probabilistic, the curvature of the pain value function they derived suggests a tendency for risk aversion for pain (willingness to accept some pain with certainty but avoid options that could lead to more pain with some probability <1). While nonlinearities have also been described for the utility function for monetary losses (Box 1, panel B), risk tolerance rather than risk aversion has been typically observed for monetary reinforcers. Few empirical studies have directly investigated the shape of the utility function for pain, but there is evidence that risk preferences may vary depending on the attribute of pain (intensity versus duration in a cold-pressor task) (Schosser et al., 2016), with individuals exhibiting risk aversion (concave function) for pain intensity but risk seeking (convex function) based on pain duration. This divergence has been considered in prior work that posits that the additional affective value of a reinforcer (such as pain) could be a separate input into the decision, thereby biasing it from its purely cognitive value (Loewenstein et al., 2001; Lerner et al., 2015). There is also evidence that risk preferences for pain may vary depending on the object of pain, with individuals being more risk averse when pain is to be experienced by a stranger compared to themselves or their loved ones (Loued-Khenissi et al., 2022), contrary to what has been reported for monetary outcomes. More research is needed to resolve to what extent there are general or distinct mechanisms of valuation for aversive outcomes of different modalities such as primary (e.g., pain) and secondary (e.g., monetary losses) reinforcers and how affective and social factors modulate these valuation processes.

Consistent with other predictions from the neuroeconomic framework, the pain value function seems to be affected by monetary range (reward context) (Slimani et al., 2022), such that availability of larger rewards decreased the willingness to accept pain offers that were previously tolerated and accepted. This feature of pain valuation recapitulates predictions from the efficient coding hypothesis, how optimal neuronal encoding according to reward availability in the environment leads to contextual changes in valuation, a phenomenon known as *range adaptation* (Rustichini et al., 2017; Padoa-Schioppa, 2009). That reward context impacts the pain value function in this way further supports the notion that pain and reward interact and are integrated into a cost-benefit computation that guides choice behavior (Talmi et al., 2009). Range adaptation and another contextual mechanism known as reference-dependence (Box 1) have been proposed to explain irrational decision-making behavior (Bavard et al., 2018) and have been identified in previous studies of appetitive and aversive value learning (Palminteri et al., 2015).

It is also possible to conceptualize the presence of pain as affecting the sensitivity to reward. This is consistent with the phenomenon of reference dependence as described in Prospect Theory (Box 1, Panel B). Pain could thus shift the frame of reference to a worse state of the world. A leftward shift in the utility function could lead to a relative change in the utility perceived to be derived from the same reward (Schwartz et al., 2008; Reck & Seibold, 2023), potentially increasing reward seeking. Several studies have explored the effect of acute pain on economic choice and identified a tendency for acute pain to lead to increase in risk tolerance and delay discounting (Koppel et al., 2017; Chen et al., 2021; Pertl et al., 2024). Moreover, individuals with chronic pain exhibit difficulties in distinguishing high versus low risks (Apkarian et al., 2004) and have an increased preference for immediate rewards (Herman & Stanton, 2022; Craft et al., 2022; Mistretta et al., 2023). One study has reported that a prospective procedure aimed at reducing delay discounting—episodic future thinking—may lower pain severity in chronic pain patients through a decrease in impulsive decision-making (Craft et al., 2020). The neural mechanisms of these pain-induced shifts in reward related behavior are not known. One possibility is that pain is impacting the valuation itself through a sort of gain modulation, as has been posited for other internal states such as craving (Biernacki et al., 2022). Future work could address this question by combining careful neuroeconomic modeling combined with brain measures of valuation.

Box 1 : Pain and Prospect Theory

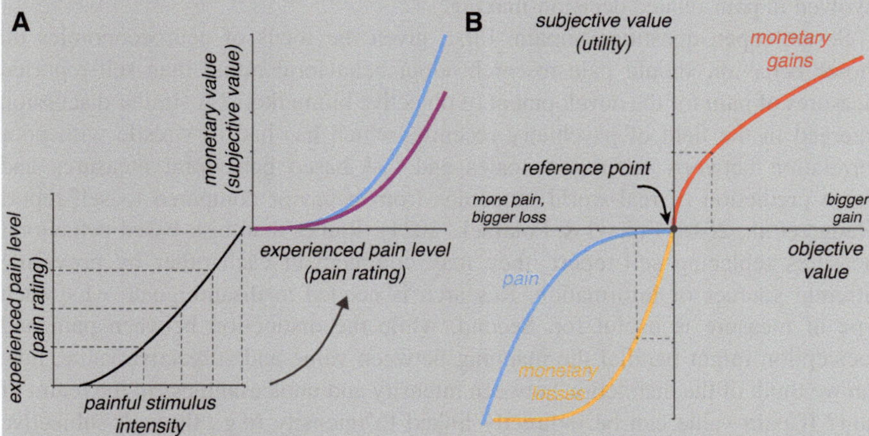

(**A**) *Left:* Pain stimulus-response function mapping the objective stimulus intensity (nociception) to the subjective pain experience. *Right:* Resulting pain ratings can be used to derive a theoretical utility function that maps the subjective experience of pain to monetary value (adapted from Park et al., 2011; Slimani et al., 2022). The pain value function can vary across individuals (blue vs. purple). (**B**). Utility functions for monetary gains (red) and losses (yellow) as described by Prospect Theory (Kahneman & Tversky, 1979). Utility function for pain (blue) as derived from (**a**) Slimani et al. (2022)

While Prospect Theory's prediction of a "reflection effect" for monetary outcomes (concavity for gains and convexity for losses) seems to generalize to feelings (Charpentier et al., 2016), the pain utility function described in panel A (right) does not show the same pattern. The shape of the pain function indicates that willingness to accept decreases with each increase in pain level (more and more money is needed to accept). Individuals thus seem to exhibit risk aversion for experienced pain level (Schosser et al., 2016; Attema et al., 2016).

6 Conclusions and Future Directions

The central thesis that avoiding pain and procuring reward are two essential motivators of animal behavior is present in philosophy, economics, and neuroscience. In the utilitarian theory advanced by Jeremy Bentham, the pains and pleasures we experience should not only be quantifiable but, most importantly, contribute jointly to constructing the organizing force of behavior—utility (Bentham, 1789). Recent behavioral and neural studies have started to test the limits of this old idea and begun to paint a more complex picture of interrelated affective and cognitive processes involved in pain-related decision-making.

Several open questions remain. First, given the focus of neuroeconomics on choice behavior, should pain research adopt behavioral rather than self-reported measures of pain for the development of objective biomarkers? A similar discussion emerged in the field of psychiatry recently, which has had to wrestle with poor correlation between self-report scales and task-based behavioral measures and worse prediction of real-world outcomes from behavior compared to self-report (Hedge et al., 2018; Enkavi & Poldrack, 2021). Rather than task-based behavioral measures replacing self-report, they may complement each other by providing different sources of information. Research is needed to disambiguate what each type of measure is useful for. Second, while the distinction between pain and nociception might parallel the mapping between value and subjective value, how can we think of the distinction between intensity and unpleasantness in the realm of pain? If pain value can be indirectly linked to intensity (e.g., through subjective pain ratings), is the derived value function also capturing unpleasantness or the hedonic aspect of pain? Neuroeconomics has encountered a similar dichotomy in thinking about the difference between "wanting" and "liking" a reward (Berridge, 2009; Berridge & Robinson, 2016). Just as pain intensity and unpleasantness can be uncorrelated, a reinforcer can be intensely "wanted" but not necessarily "liked" or vice versa. Clinical examples of this difference are often seen in individuals with chronic substance use disorders whose drug consumption is sought (wanted) despite no longer leading to a positive hedonic experience (Morales & Berridge, 2020). Neuroeconomists have proposed that this divergence can be resolved through multiple utility functions (Kahneman et al., 1997) or posited "liking" as an early stage of value learning (Dayan, 2022), but this remains an issue of active inquiry and interest in the field. Finally, given the interrelated nature of pain relief and reward, it is fruitful to measure how interventions that specifically target pain unpleasantness, such as opioid analgesics, impact economic decisions. Mu-opioid receptor activation has been shown to increase choices for high-value/high-probability targets (Eikemo et al., 2017), and individuals receiving treatment for opioid use disorder with medications such as buprenorphine or methadone become more tolerant of uncertainty prior to a reuse event (Konova et al., 2020). More research is needed to bridge the neuroeconomics of pain relief with that of opioid addiction.

The devastating consequences of the opioid use disorder epidemic in the United States and the lack of tangible solutions for the safe treatment of chronic pain conditions have made more urgent the call for progress in this research. Similar to recent efforts in computational psychiatry that apply neuroeconomics and computational neuroscience frameworks to clinical problems in mental health, a new generation of studies should use these methods to advance our understanding of pain and pain-related behaviors.

Acknowledgments This research was supported by the Intramural Research Program of the National Institutes of Health (NIH), specifically the NIMH Intramural Research Program (SLG: ZIA-MH002981) and the NCCIH Intramural Research Program (LYA: ZIA-AT000030). The contributions of the NIH authors were made as part of their official duties as NIH federal employees, are in compliance with agency policy requirements, and are considered Works of the United States Government. However, the findings and conclusions presented in this paper are those of the author(s) and do not necessarily reflect the views of the NIH or the U.S. Department of Health and Human Services. Figure 1 includes data from NIH IRB protocol 15-AT-0132 (clinicaltrials.gov identifier NCT02446262) and was created in part using Biorender.com.

References

Apkarian, A. V., Darbar, A., Krauss, B. R., Gelnar, P. A., & Szeverenyi, N. M. (1999). Differentiating cortical areas related to pain perception from stimulus identification: Temporal analysis of fMRI activity. *Journal of Neurophysiology, 81*, 2956–2963.

Apkarian, A. V., et al. (2004). Chronic pain patients are impaired on an emotional decision-making task. *Pain, 108*, 129–136.

Atlas, L. Y., Bolger, N., Lindquist, M. A., & Wager, T. D. (2010). Brain mediators of predictive cue effects on perceived pain. *Journal of Neuroscience, 30*, 12964–12977.

Atlas, L., Lindquist, M., Bolger, N., & Wager, T. (2014a). Brain mediators of the effects of noxious heat on pain. *Pain, 155*, 1632–1648.

Atlas, L. Y., Wielgosz, J., Whittington, R. A., & Wager, T. D. (2014b). Specifying the nonspecific factors underlying opioid analgesia: expectancy, attention, and affect. *Psychopharmacology, 231*, 813–823.

Attema, A. E., Brouwer, W. B. F., l'Haridon, O., & Pinto, J. L. (2016). An elicitation of utility for quality of life under prospect theory. *Journal of Health Economics, 48*, 121–134.

Baliki, M. N., Geha, P. Y., Fields, H. L., & Apkarian, A. V. (2010). Predicting value of pain and analgesia: Nucleus accumbens response to noxious stimuli changes in the presence of chronic pain. *Neuron, 66*, 149–160.

Bartra, O., McGuire, J. T., & Kable, J. W. (2013). The valuation system: A coordinate-based meta-analysis of BOLD fMRI experiments examining neural correlates of subjective value. *NeuroImage, 76*, 412–427.

Bavard, S., Lebreton, M., Khamassi, M., Coricelli, G., & Palminteri, S. (2018). Reference-point centering and range-adaptation enhance human reinforcement learning at the cost of irrational preferences. *Nature Communications, 9*, 4503.

Becerra, L., & Borsook, D. (2008). Signal valence in the nucleus accumbens to pain onset and offset. *European Journal of Pain (London, England), 12*, 866–869.

Bentham, J. (1789). *An introduction to the principles of morals and legislation.* Clarendon.

Berridge, K. C. (2009). Wanting and liking: Observations from the neuroscience and psychology laboratory. *Inquiry, 52*, 378–398.

Berridge, K. C., & Robinson, T. E. (2016). Liking, wanting and the incentive-sensitization theory of addiction. *The American Psychologist, 71*, 670–679.

Berthier, M., Starkstein, S., & Leiguarda, R. (1988). Asymbolia for pain: A sensory-limbic disconnection syndrome. *Annals of Neurology, 24*, 41–49.

Biernacki, K., et al. (2022). A neuroeconomic signature of opioid craving: How fluctuations in craving bias drug-related and nondrug-related value. *Neuropsychopharmacology, 47*, 1440–1448.

Borsook, D., et al. (2016). Reward deficiency and anti-reward in pain chronification. *Neuroscience & Biobehavioral Reviews, 68*, 282–297.

Brischoux, F., Chakraborty, S., Brierley, D. I., & Ungless, M. A. (2009). Phasic excitation of dopamine neurons in ventral VTA by noxious stimuli. *Proceedings of the National Academy of Sciences of the United States of America, 106*, 4894–4899.

Büchel, C., et al. (2002). Dissociable neural responses related to pain intensity, stimulus intensity, and stimulus awareness within the anterior cingulate cortex: A parametric single-trial laser functional magnetic resonance imaging study. *The Journal of neuroscience : the official journal of the Society for Neuroscience, 22*, 970–976.

Budygin, E. A., et al. (2012). Aversive stimulus differentially triggers subsecond dopamine release in reward regions. *Neuroscience, 201*, 331–337.

Čeko, M., Kragel, P. A., Woo, C.-W., López-Solà, M. & Wager, T. D. (2022). Common and stimulus-type-specific brain representations of negative affect. *Nature Neuroscience, 25*, 760–770.

Charpentier, C. J., De Neve, J.-E., Li, X., Roiser, J. P., & Sharot, T. (2016). Models of affective decision making. *Psychological Science, 27*, 763–775.

Chen, J., et al. (2021). Does pain make people short-sighted? The impacts of physical and psychological pains on intertemporal choice. *Judgment and Decision Making, 16,* 1097–1112.

Chib, V. S., Rangel, A., Shimojo, S., & O'Doherty, J. P. (2009). Evidence for a common representation of decision values for dissimilar goods in human ventromedial prefrontal cortex. *Journal of Neuroscience, 29*, 12315–12320.

Coghill, R. C., Sang, C. N., Maisog, J. M., & Iadarola, M. J. (1999). Pain intensity processing within the human brain: a bilateral, distributed mechanism. *Journal of Neurophysiology, 82*, 1934–1943.

Craft, W. H., Tegge, A. N., & Bickel, W. K. (2020). Episodic future thinking reduces chronic pain severity: A proof of concept study. *Drug and Alcohol Dependence, 215*, 108250.

Craft, W. H., Tegge, A. N., & Bickel, W. K. (2022). Chronic pain and COVID-19: The association of delay discounting with perceived stress and pain severity. *Experimental and Clinical Psychopharmacology, 30*, 692–700.

Craig, A. D., Chen, K., Bandy, D., & Reiman, E. M. (2000). Thermosensory activation of insular cortex. *Nature Neuroscience, 3*, 184–190.

Dayan, P. (2022). "Liking" as an early and editable draft of long-run affective value. *PLOS Biology, 20*, e3001476.

Delgado, M. R., Labouliere, C. D., & Phelps, E. A. (2006). Fear of losing money? Aversive conditioning with secondary reinforcers. *Social Cognitive and Affective Neuroscience, 1*, 250–259.

Delgado, M. R., Li, J., Schiller, D., & Phelps, E. A. (2008). The role of the striatum in aversive learning and aversive prediction errors. *Philosophical Transactions of the Royal Society B: Biological Sciences, 363*, 3787–3800.

Eikemo, M., Biele, G., Willoch, F., Thomsen, L., & Leknes, S. (2017). Opioid modulation of value-based decision-making in healthy humans. *Neuropsychopharmacology, 42*, 1833–1840.

Enkavi, A. Z., & Poldrack, R. A. (2021). Implications of the lacking relationship between cognitive task and self-report measures for psychiatry. *Biological Psychiatry: Cognitive Neuroscience and Neuroimaging, 6*, 670–672.

Fehr, E., & Camerer, C. F. (2007). Social neuroeconomics: The neural circuitry of social preferences. *Trends in Cognitive Sciences, 11*, 419–427.

FitzGerald, T. H. B., Seymour, B., & Dolan, R. J. (2009). The role of human orbitofrontal cortex in value comparison for incommensurable objects. *The Journal of Neuroscience: The Official Journal of the Society for Neuroscience, 29*, 8388–8395.

Foltz, E. L., & White, L. E. (1962). Pain "Relief" by frontal cingulumotomy. *Journal of Neurosurgery, 19*, 89–100.

Hackel, L. M., & Amodio, D. M. (2018). Computational neuroscience approaches to social cognition. *Current Opinion in Psychology. Social Neuroscience, 24*, 92–97.

Hedge, C., Powell, G., & Sumner, P. (2018). The reliability paradox: Why robust cognitive tasks do not produce reliable individual differences. *Behavior Research Methods, 50*, 1166–1186.

Herman, A. M., & Stanton, T. R. (2022). Delay and effort-based discounting, and the role of bodily awareness, in people experiencing long-term pain: A cross-sectional study. *The Journal of Pain, 23*, 487–500.

Hofbauer, R. K., Rainville, P., Duncan, G. H., & Bushnell, M. C. (2001). Cortical representation of the sensory dimension of pain. *Journal of Neurophysiology, 86*, 402–411.

Howard, J. D., Gottfried, J. A., Tobler, P. N., & Kahnt, T. (2015). Identity-specific coding of future rewards in the human orbitofrontal cortex. *Proceedings of the National Academy of Sciences, 112*, 5195–5200.

Howard, J. D., & Kahnt, T. (2021). To be specific: The role of orbitofrontal cortex in signaling reward identity. *Behavioral Neuroscience, 135*, 210–217.

Jensen, J., et al. (2003). Direct activation of the ventral striatum in anticipation of aversive stimuli. *Neuron, 40*, 1251–1257.

Jensen, J., et al. (2007). Separate brain regions code for salience vs. valence during reward prediction in humans. *Human Brain Mapping, 28*, 294–302.

Kahneman, D., & Tversky, A. (1979). Prospect theory: An analysis of decision under risk. *Econometrica, 47*, 263–291.

Kahneman, D., Wakker, P. P., & Sarin, R. (1997). Back to Bentham? Explorations of experienced utility. *The Quarterly Journal of Economics, 112*, 375–406.

Konova, A. B., et al. (2020). Computational markers of risky decision-making for identification of temporal windows of vulnerability to opioid use in a real-world clinical setting. *JAMA Psychiatry, 77*, 368–377.

Koppel, L., et al. (2017). The effect of acute pain on risky and intertemporal choice. *Experimental Economics, 20*, 878–893.

Krastev, S., et al. (2016). Do political and economic choices rely on common neural substrates? A systematic review of the emerging neuropolitics literature. *Frontiers in Psychology, 7*, 264.

Lebreton, M., Jorge, S., Michel, V., Thirion, B., & Pessiglione, M. (2009). An automatic valuation system in the human brain: Evidence from functional neuroimaging. *Neuron, 64*, 431–439.

Leknes, S., Lee, M., Berna, C., Andersson, J., & Tracey, I. (2011). Relief as a reward: Hedonic and neural responses to safety from pain. *PLoS ONE, 6* , e17870 (ed Izquierdo, I.).

Leknes, S., & Tracey, I. (2008). A common neurobiology for pain and pleasure. *Nature Reviews Neuroscience, 9*, 314–320.

Leknes, S., et al. (2013). The importance of context: When relative relief renders pain pleasant. *Pain, 154*, 402–410.

Lerman, S. F., Rudich, Z., Brill, S., Shalev, H., & Shahar, G. (2015). Longitudinal associations between depression, anxiety, pain, and pain-related disability in chronic pain patients. *Psychosomatic Medicine, 77*, 333–341.

Lerner, J. S., Li, Y., Valdesolo, P., & Kassam, K. S. (2015) Emotion and decision making. *Annual Review of Psychology, 66*, 799–823.

Levy, I., Lazzaro, S. C., Rutledge, R. B., & Glimcher, P. W. (2011). Choice from non-choice: Predicting consumer preferences from blood oxygenation level-dependent signals obtained during passive viewing. *The Journal of Neuroscience, 31*, 118–125.

Loeser, J. D., & Treede, R.-D. (2008) The Kyoto protocol of IASP basic pain terminology. *PAIN, 137*, 473–477.

Loewenstein, G. F., Weber, E. U., Hsee, C. K., & Welch, N. (2001). Risk as feelings. *Psychological Bulletin, 127*, 267–286.

Loggia, M. L., et al. (2014). Disrupted brain circuitry for pain-related reward/punishment in fibromyalgia. *Arthritis & rheumatology (Hoboken, N.J.), 66*, 203–212.

Loued-Khenissi, L., Martin-Brevet, S., Schumacher, L., & Corradi-Dell'Acqua, C. (2022). The effect of uncertainty on pain decisions for self and others. *European Journal of Pain, 26*, 1163–1175.
Mistretta, E. G., Craft, W. H., Davis, M. C., McClure, S. M., & Bickel, W. K. (2023). Delay discounting in adults with and without chronic pain: Differentiation across commodity and sign. *Experimental and Clinical Psychopharmacology, 31*, 745–755.
Morales, I., & Berridge, K. C. (2020). 'Liking' and 'wanting' in eating and food reward: Brain mechanisms and clinical implications. *Physiology & Behavior, 227*, 113152.
Navratilova, E., & Porreca, F. (2014) Reward and motivation in pain and pain relief. *Nature Neuroscience, 17*, 1304–1312.
Navratilova, E., et al. (2012). Pain relief produces negative reinforcement through activation of mesolimbic reward–valuation circuitry. *Proceedings of the National Academy of Sciences of the United States of America, 109*, 20709–20713.
Nielsen, C. S., Price, D. D., Vassend, O., Stubhaug, A., & Harris, J. R. (2005). Characterizing individual differences in heat-pain sensitivity. *Pain, 119*, 65–74.
Padoa-Schioppa, C. (2009). Range-adapting representation of economic value in the orbitofrontal cortex. *The Journal of Neuroscience, 29*, 14004–14014.
Palminteri, S., Khamassi, M., Joffily, M., & Coricelli, G. (2015). Contextual modulation of value signals in reward and punishment learning. *Nature Communications, 6*, 8096.
Park, S. Q., Kahnt, T., Rieskamp, J., & Heekeren, H. R. (2011). Neurobiology of value integration: When value impacts valuation. *Journal of Neuroscience, 31*, 9307–9314.
Perlman, D. M., Salomons, T. V., Davidson, R. J., & Lutz, A. (2010). Differential effects on pain intensity and unpleasantness of two meditation practices. *Emotion, 10*, 65–71.
Pertl, S. M., Srirangarajan, T., & Urminsky, O. (2024). A multinational analysis of how emotions relate to economic decisions regarding time or risk. *Nature Human Behaviour, 8*, 1–17.
Pessiglione, M., & Delgado, M. R. (2015). The good, the bad and the brain: Neural correlates of appetitive and aversive values underlying decision making. *Current Opinion in Behavioral Sciences, 5*, 78–84.
Peters, J., & Büchel, C. (2010). Neural representations of subjective reward value. *Behavioural Brain Research, 213*, 135–141.
Ploner, M., Freund, H.-J., & Schnitzler, A. (1999). Pain affect without pain sensation in a patient with a postcentral lesion. *Pain, 81*, 211–214.
Price, D. D. (2000). Psychological and neural mechanisms of the affective dimension of pain. *Science, 288*, 1769–1772.
Price, D. D., Harkins, S. W., & Baker, C. (1987). Sensory-affective relationships among different types of clinical and experimental pain. *Pain, 28*, 297–307.
Price, D. D., McHaffie, J. G., & Stein, B. E. (1992). The psychophysical attributes of heat-induced pain and their relationships to neural mechanisms. *Journal of Cognitive Neuroscience, 4*, 1–14.
Price, D. D., Von Der Gruen, A., Miller, J., Rafii, A., & Price, C. (1985). A psychophysical analysis of morphine analgesia. *Pain, 22*, 261–269.
Rainville, P., Duncan, G. H., Price, D. D., Carrier, B., & Bushnell, M. C. (1997). Pain affect encoded in human anterior cingulate but not somatosensory cortex. *Science, 277*, 968–971.
Rainville, P., Carrier, B., Hofbauer, R. K., Bushnell, C. M., & Duncan, G. H. (1999). Dissociation of sensory and affective dimensions of pain using hypnotic modulation. *Pain, 82*, 159–171.
Rainville, P., Feine, J. S., Bushnell, M. C., & Duncan, G. H. (1992). A psychophysical comparison of sensory and affective responses to four modalities of experimental pain. *Somatosensory & Motor Research, 9*, 265–277.
Raja, S. N., et al. (2020). The revised international association for the study of pain definition of pain: Concepts, challenges, and compromises. *Pain, 161*, 1976–1982. ISBN: 0304-3959.
Reck, D., & Seibold, A. (2023). *The Welfare Economics of Reference Dependence*. Working Paper.
Rudebeck, P. H., & Murray, E. A. (2014). The orbitofrontal oracle: cortical mechanisms for the prediction and evaluation of specific behavioral outcomes. *Neuron, 84*, 1143–1156.
Rustichini, A., Conen, K. E., Cai, X., & Padoa-Schioppa, C. (2017). Optimal coding and neuronal adaptation in economic decisions. *Nature Communications, 8*, 1208.

Salomons, T. V., Moayedi, M., Erpelding, N., & Davis, K. D. (2014). A brief cognitive-behavioural intervention for pain reduces secondary hyperalgesia. *Pain, 155,* 1446–1452.

Schosser, S., Trarbach, J. N., & Vogt, B. (2016). How does the perception of pain determine the selection between different treatments?: Experimental evidence for convex utility functions over pain duration and concave utility functions over pain intensity. *Journal of Economic Behavior & Organization. Experimental and Behavioral Economics of Healthcare, 131*, 174–182.

Schwartz, A., Goldberg, J., & Hazen, G. (2008). Prospect theory, reference points, and health decisions. *Judgment and Decision Making, 3*, 174–180.

Scott, D. J., et al. (2007). Individual differences in reward responding explain placebo-induced expectations and effects. *Neuron, 55*, 325–336.

Sescousse, G., Caldú, X., Segura, B., & Dreher, J.-C. (2013). Processing of primary and secondary rewards: A quantitative meta-analysis and review of human functional neuroimaging studies. *Neuroscience & Biobehavioral Reviews, 37*, 681–696.

Seymour, B., et al. (2004). Temporal difference models describe higher-order learning in humans. *Nature, 429*, 664–667.

Seymour, B., et al. (2005). Opponent appetitive-aversive neural processes underlie predictive learning of pain relief. *Nature Neuroscience, 8*, 1234–1240.

Slimani, H., Rainville, P., & Roy, M. (2022). The aversive value of pain in human decision-making. *European Journal of Pain, 26*, 668–679.

Stevens, S. S. (1957). On the psychophysical law. *Psychological Review, 64*, 153–181.

Talmi, D., Dayan, P., Kiebel, S. J., Frith, C. D., & Dolan, R. J. (2009). How humans integrate the prospects of pain and reward during choice. *The Journal of Neuroscience: The Official Journal of the Society for Neuroscience, 29*, 14617–14626.

Tolle, T. R., et al. (1999). Region-specific encoding of sensory and affective components of pain in the human brain: A positron emission tomography correlation analysis. *Annals of Neurology, 45*, 40–47.

Villemure, C., & Bushnell, M. C. (2009). Mood influences supraspinal pain processing separately from attention. *The Journal of Neuroscience, 29*, 705–715.

Volkow, N., Benveniste, H., & McLellan, A. T. (2018). Use and misuse of opioids in chronic pain. *Annual Review of Medicine, 69*, 451–465.

Wager, T. D., et al. (2013). An fMRI-based neurologic signature of physical pain. *New England Journal of Medicine, 368*, 1388–1397.

Wanigasekera, V., et al. (2012). Baseline reward circuitry activity and trait reward responsiveness predict expression of opioid analgesia in healthy subjects. *Proceedings of the National Academy of Sciences, 109*, 17705–17710.

Woo, C.-W., et al. (2017). Quantifying cerebral contributions to pain beyond nociception. *Nature Communications, 8*, 14211.

Non-reinforced Preference Change

Adi Cantor, Maya Bar Or, Alon Itzkovitch, and Tom Schonberg

Abstract This chapter explores non-reinforced preference change, examining how preferences shift in the absence of external reinforcement. We differentiate between various types of reinforcements, including external, non-external, and internal reinforcement, to understand their roles in shaping preferences. The chapter highlights the cognitive and neural mechanisms underlying preference changes, emphasizing how internal reinforcement influences decision-making processes. This synthesis of literature aims to advance our understanding of how preferences can evolve in the absence of overt rewards, shedding light on the complexities of human decision-making and the underlying cognitive mechanisms.

1 Preference Changes

1.1 External Reinforcement

Behavioral change refers to shifts in actions, thoughts, and emotions observed in individuals over time (Bandura & Adams, 1977; Duckworth & Gross, 2020; Lench et al., 2011). These changes can be manifested in various ways, from subtle alterations in daily habits to significant transformations in personality or

A. Cantor · M. Bar Or
Sagol School of Neuroscience, Tel Aviv University, Tel Aviv, Israel
e-mail: adicantor@mail.tau.ac.il

A. Itzkovitch
School of Neurobiology, Biochemistry and Biophysics, Faculty of Life Science, Tel Aviv University, Tel Aviv, Israel

T. Schonberg (✉)
School of Neurobiology, Biochemistry and Biophysics, Faculty of Life Science, Tel Aviv University, Tel Aviv, Israel

Sagol School of Neuroscience, Tel Aviv University, Tel Aviv, Israel
e-mail: schonberg@tauex.tau.ac.il

D. V. Smith et al. (eds.), *Neuroeconomics: Core Topics and Current Directions*,
https://doi.org/10.1007/978-3-032-02925-6_13

lifestyle (Cash et al., 2023; Duckworth & Gross, 2020). Reinforcement has been a fundamental concept in behavioral psychology, shaping behavior through external rewards or the removal of aversive stimuli (Jones et al., 1977). The development of external reinforcement theory was further advanced by researchers like Edward Thorndike, who formulated the Law of Effect (Thorndike, 1927). This concept, which later became a foundation for B.F. Skinner's operant conditioning theory, has evolved significantly since its inception (Skinner, 1963). This principle states that behaviors followed by satisfying consequences are more likely to be repeated, while those followed by unpleasant consequences are less likely to recur. This formulation has laid the foundation for understanding how external reinforcement shapes behavior and was used in most behavioral change interventions to change preferences (Cameron & Pierce, 1994).

Building on this foundation, recent research continued to explore and apply reinforcement principles. For example, external reinforcement principles were used to develop deposit contract methods to encourage smoking abstinence, demonstrating the practical applications of these concepts in health interventions (Naslund et al., 2017). Another example is contingency management (CM), an effective behavioral intervention based on operant conditioning principles of reinforcing positive behaviors through tangible rewards (Petry, 2011).

While CM shows relatively effective results in treating substance use disorders (Bolívar et al., 2021), external reinforcement interventions have limitations that can impact the success of the intervention. CM relies on rewards to reinforce positive behaviors, such as abstinence, but once incentives are removed, individuals may struggle to sustain behavior change. This approach can also lead to dependency on external rewards rather than intrinsic motivation, reducing long-term effectiveness. A study that tested the effect of CM combined with cognitive behavioral treatment (CBT) on participants with substance use disorder (SUD) found that adding CM to CBT improved smoking abstinence rates among individuals with SUD at the end of treatment. However, the positive effects declined over time, showing no sustained impact beyond treatment (Secades-Villa et al., 2022). Moreover, the evidence regarding the effectiveness of CM in assisting other domains is less obvious. One meta-analysis that tested the long-term effects of CM on different health-related behaviors (weight change, physical activity, medication/device adherence, and viral load) found mixed results, which suggests that the effectiveness of CM tends to fade once the incentives are no longer provided (Ellis et al., 2021). Furthermore, researchers have found that providing external rewards or incentives for an activity that is already intrinsically rewarding can diminish its intrinsic appeal. It was found that although extrinsic rewards can boost motivation for straightforward or repetitive tasks, they might weaken intrinsic motivation in more complex or creative endeavors. The results indicate that offering external incentives can result in a decline in intrinsic interest and engagement over time (Boedecker et al., 2013). Thus, it seems that other alternatives for preference change should be put in place to achieve long-term preference change without external reinforcements.

1.2 Preference Changes Without External Reinforcement

The earliest example of such preference change dates back to the mere-exposure effect (Zajonc, 1968). In this work, the author demonstrated how looking at Chinese-like characters and faces led to greater preference for items that had more exposure. Interestingly, Zajonc himself explained this phenomenon using reinforcement theory, arguing that because nothing negative had occurred while looking at them, the connections were reinforced.

This work laid the foundation for further experiments in the field. One study investigated the relationship between attention and choice preference and found that manipulating visual attention can bias simple choices (Armel et al., 2008). They found that it was possible to increase the likelihood of an item being chosen by altering the relative amount of time subjects visually fixate on that item during the decision-making process. In another study, a computational model was created to explain complex relationships between fixation patterns and choices (Krajbich et al., 2010). Using the model, the authors found that visual attention plays a significant role in the decision-making process. Items that had longer fixation durations were more likely to be chosen, unless they were aversive.

1.3 Preference Changes Without External Reinforcement: Errors Leading to Internal Feedback

Different behavioral tasks, mainly from the research field of impulsivity, have been used to induce preference change. These tasks could be considered as non-reinforced preference change tasks as they do not include external reinforcements per se. However, the structure of these tasks could lead to the creation of internal feedback that leads to preference change (Table 1):

A. The Go/NoGo (GNG) task examines the capacity to suppress prepotent responses (Lappin & Eriksen, 1966). This task is a fundamental paradigm in cognitive psychology and was first introduced by Donders (1969). This method assesses response inhibition by requiring participants to respond quickly to certain stimuli (Go) while withholding responses to others (NoGo) (Gomez et al., 2007). The Go/NoGo task has made significant contributions to the behavioral field, enabling researchers to quantify inhibitory control across various populations and conditions (Young et al., 2018). It has been particularly valuable in studying developmental disorders, addiction, and neurological conditions characterized by impaired impulse control (Ding et al., 2014; Dong et al., 2010; Ladouceur et al., 2006; Trommer et al., 1988; Uzefovsky et al., 2016; Waters & Valvoi, 2009; Yerys et al., 2013). In the context of preference change, it has been demonstrated that GNG training could effectively modify food preferences, with participants showing a preference for Go food items over

Table 1 Key methods to enhance change preference without external reinforcement

Method	Description	Advantages	Real-life applications
Mere Exposure Effect	Individuals develop a preference for stimuli simply due to repeated exposure, leading to increased liking and familiarity without conscious awareness (Zajonc, 1968)	MEE increases liking and preference for stimuli that are repeatedly encountered (Bornstein & Craver-Lemley, 2022)	MEE can be used for marketing and advertising (Zajonc, 1968), as well as in other promotional efforts, such as political campaigns (Grush et al., 1978)
Choice Leads to Choice	Making a choice between similarly valued options creates psychological discomfort that is resolved by reevaluating the chosen option more positively and the rejected option more negatively, leading to a change in preferences that aligns with the decision made (Festinger., 1957)	This effect helps individuals maintain cognitive consistency between their choices and preferences (Sharot et al., 2012)	"Choice leads to choice" can help in various fields, such as with sales (Sharot et al., 2012) and advertising (Nakamura & Kawabata, 2013)
Go/NoGo Task (GNG)	Assesses response inhibition by requiring quick responses to some stimuli (Go) and inhibition to others (NoGo) (Gomez et al., 2007; Veling et al., 2021)	Measures response inhibition, effective in studying impulse control, can influence behavior (e.g., food preference)	Go/NoGo training targeting high-calorie foods has shown promise as a tool to facilitate short-term weight loss in some studies. The training may help individuals reduce consumption of unhealthy foods as part of weight management efforts (Veling et al., 2017b)

Approach-Avoidance Task (AAT)	Measures automatic behavioral tendencies by requiring participants to approach or avoid stimuli using a joystick (Strack & Deutsch, 2004; Wiers et al., 2007)	Captures automatic behavioral tendencies. Useful in studying implicit attitudes, applied in addiction and anxiety research	AAT has shown promise in treating various addictions. For alcohol addiction, AAT training has been used to reduce approach bias toward alcohol-related stimuli, potentially supporting treatment outcomes (Sharbanee et al., 2014). Similar applications are being explored for other substance use disorders and behavioral addictions like problem gambling (Hoppe et al., 2022)
Stop-Signal Task (SST)	Measures response inhibition, a critical component of executive function. In this task, participants must quickly respond to a primary stimulus but inhibit their response when presented with a stop signal, allowing researchers to quantify the speed and efficiency of the inhibitory process (Logan et al., 1984; Verbruggen & Logan, 2009)	Offers several advantages, including its adaptability to various experimental designs, its ability to provide quantitative measures of inhibitory control, and its widespread use in both cognitive psychology and neuroscience research (Poulton et al., 2022; Wang et al., 2018)	Potential applications include assessing impulse control in clinical populations, evaluating the effectiveness of interventions targeting executive function, and investigating the neural mechanisms underlying decision-making and behavioral regulation in everyday situation (Aksiotis et al., 2023; Verbruggen & Logan, 2008)
Cue-Approach Training (CAT)	Induces preference changes toward associated stimuli through repeated pairings of a neutral cue with a speeded response (Salomon et al., 2018, 2022; Schonberg et al., 2014; Xia et al., 2024)	Modifies preferences without external reinforcements. Can be applied to various domains (e.g., leadership, positive traits)	Several real-life applications, such as food intake reduction and cognitive assessment. Has the potential to help elicit food-related preference change (Itzkovitch et al., 2022). CAT is increasingly used as a cognitive assessment tool to measure implicit behavioral tendencies that may not be captured by self-report measures (Itzkovitch et al., 2022)

NoGo food items in subsequent choice tasks (Veling et al., 2021). This task can be seen as inducing internal feedback, since participants recognize when they make an error or respond correctly, such as making a "Go" response when a "NoGo" was required and vice versa.

B. The Approach-Avoidance Task (AAT) (Neumann & Strack, 2000) has been used to study automatic approach-avoidance behaviors. AAT was developed by researchers studying implicit attitudes and automatic behavioral tendencies and has gained relative prominence in the 2000s (Strack & Deutsch, 2004). The AAT typically involves participants responding to stimuli by either pulling a joystick toward themselves (approach) or pushing it away (avoidance) (van Dantzig et al., 2008). The task measures the automatic tendencies to approach positive stimuli and avoid negative ones, providing insights into underlying cognitive and emotional processes. A significant application of the AAT was used by Rinck and Becker in 2007, who used it to study automatic avoidance tendencies in spider-fearful individuals (Rinck & Becker, 2007). Their work demonstrated that the AAT could measure implicit behavioral tendencies that were not captured by explicit self-reports. In the following years, other researchers used the AAT in addiction research, particularly in studying and modifying approach biases toward alcohol-related stimuli (Wiers et al., 2007). Their work has shown that training individuals to avoid alcohol-related cues could reduce drinking behavior. Current research continues to explore the applications of AAT in various domains, including anxiety disorders, addiction, and eating behaviors (Kakoschke et al., 2019; Kim & Lee, 2015). Importantly, Veling et al. found that both GNG and AAT influenced food choice in similar ways, with participants preferring Go over NoGo items and Approach over Avoidance items (Veling et al., 2021). We can regard this task as inducing internal feedback as participants are aware when they have made a mistake, i.e., they make an approach/avoidance response when the other response was required.
C. The Stop-Signal Task (SST) measures the ability to suppress ongoing motor responses (Logan et al., 1984). The SST was developed as a method to study response inhibition, a crucial component of cognitive control. The task requires participants to perform a primary task (e.g., pressing a button in response to a stimulus) but to inhibit their response when a stop signal is presented shortly after the primary stimulus. Logan's work provided a quantitative framework for response inhibition, demonstrating that it is an active process that can be measured and quantified. Since its introduction, the SST has become a widely used paradigm in cognitive psychology and neuroscience (Bartholdy et al., 2016; Lipszyc & Schachar, 2010; Li et al., 2006). It has been used for behavioral change as it supports one's ability to adjust to a changing environment (Verbruggen & Logan, 2008). We can regard this task as inducing internal feedback since participants are aware when they have made a mistake, i.e., they do not stop on time for the stop signal. Support for this assertion comes from a study that showed that reward prospects influence strategic adjustments in the stop-signal task, with subjects adapting their behavior according to varying reward conditions (Giuffrida et al., 2023).

1.3.1 Preference Change Without External Feedback and No Errors

Another type of task that leads to preference changes does not include performance errors and, thus, does not lead to internal feedback due to the task structure.

A. The "choice leads to choice" phenomenon describes how initial choices increase the chance of choosing them again, creating a self-reinforcing cycle that progressively narrows the range of selected choices (Sharot et al., 2010). This concept relies on Festinger's theory of cognitive dissonance (Festinger, 1957). The cognitive dissonance theory offers significant insights into the relationship between choice and subsequent changes in preferences and attitudes. Festinger proposed that when we make a difficult decision between two similarly desirable options, we experience cognitive dissonance. To reduce this dissonance, we tend to either enhance the positive aspects of the chosen option or downplay the positive aspects of the rejected option. This phenomenon is known as "spreading of alternatives" and has been demonstrated in various studies (Harmon-Jones & Mills, 2019).
B. The cue-approach training (CAT) task is a relatively novel behavioral training paradigm that induces preference changes without external reinforcements (Schonberg et al., 2014). This is a multiphase task (see Fig. 1). In the initial phase, participants express their baseline preferences for items. In the training phase, individual pictures of items are presented on a screen and participants are requested to press a button whenever a cue is presented (either auditory or visual). In the original CAT task, the mere association of snack food pictures with a neutral auditory cue and a speeded motor response led to enhanced preferences toward the items that were associated with the stimuli (Schonberg et al., 2014). Preference change was measured in a subsequent binary choice phase, in which couples of items with similar initial value, one associated with the cue and the other not, were presented together and participants were requested to choose one of the presented items. Results indicate that participants preferred the items previously associated with the Go cue. Multiple studies have shown consistently that CAT induced behavioral changes, whereby Go items were chosen significantly above chance (Bakkour et al., 2016; Botvinik-Nezer et al., 2020, 2021; Salomon et al., 2018, 2020, 2022; Schonberg et al., 2014). The CAT effect had since been demonstrated with various cues (visual or auditory), as well as various stimuli types (snacks, unfamiliar faces, fractal art images, and positive International Affective Picture System (IAPS) images) (Salomon et al., 2018). It has also been demonstrated using fruits and vegetables (Veling et al., 2017a). The effect of behavioral changes has been shown to be preserved up to 6 months after initial training (Botvinik-Nezer et al., 2020), indicating that a memory element is putatively incorporated in this effect.

   Most of CAT studies included an initial choice phase, and so to address a "choice leads to choice" effect, one study showed that preference modifications were still found, even if training occurred on 1 day and participants performed the choice session only 3 days later (Botvinik-Nezer et al., 2021). However, the

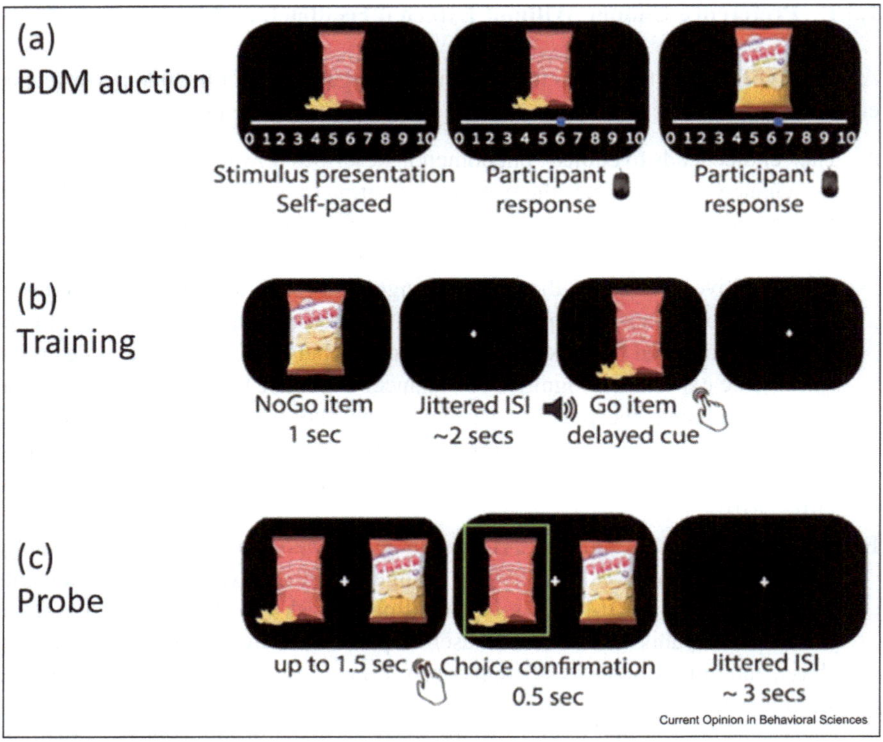

Fig. 1 Overview of the three main procedural components of the cue-approach training (CAT) paradigm. (**a**) Initial preference evaluation: baseline preferences for all stimuli were assessed using either a Becker-DeGroot-Marschak (BDM) auction for consumable items or a forced-choice binary ranking task for nonconsumable items (Becker et al., 1964). (**b**) Training phase: stimuli were displayed for 1 s, and participants were instructed to respond as quickly as possible when a cue appeared (around 750 ms into the trial). Go items were paired with both the cue and a motor response, while NoGo items were not. (**c**) Probe phase: preference changes were assessed by presenting participants with a binary forced choice between pairs of stimuli of similar initial value, comparing one Go stimulus to one NoGo stimulus (Itzkovitch et al., 2022). (Adapted from Botvinik-Nezer et al., 2020)

CAT effect seems to be limited to neutral and positive items as when Go items were negative IAPS images no choice effect was found (Salomon et al., 2018). Additionally, response time is another factor, as the behavioral effect was not observed when participants had longer choice times (Veling et al., 2017a).

Over the past few years, new applications of CAT have been discovered across various research laboratories. One notable example is the work of Yao et al., who integrated CAT with sleep-based targeted memory reactivation (TMR) to enhance positive self-referential processing (Yao et al., 2024). Their study demonstrated that this combined approach not only facilitated quicker endorsement of positive traits immediately after training but also resulted in participants identifying more positive traits as self-descriptive a week later. Another context in which CAT has been applied is leadership development (Cavarretta, 2024). The researchers used CAT to modify implicit attitudes toward leadership tasks, particularly focusing on enhancing positive associations with managerial activities. Their findings demonstrated that CAT could effectively help aspiring leaders develop a more positive "taste" for the often-challenging aspects of leadership, such as dealing with interpersonal issues and managing team dynamics. Participants who underwent CAT showed increased positive associations with managerial activities, suggesting potential improvements in their motivation and effectiveness in leadership roles.

A recent study applied a computational Bayesian learning model to data from 28 CAT samples and identified that a shift in reaction time during the training phase predicted the subsequent behavioral-change effect (Salomon et al., 2022). Specifically, participants who pressed the button earlier in the trial as training progressed subsequently showed greater preference changes in the probe phase.

During the training phase, participants learn to associate a stimulus with a cue. While proceeding in training, they begin to press the button earlier, sometimes even before the cue appears. When participants start responding to the stimuli instead of the cue, and the cue eventually appears, it acts as internal feedback, similarly to missed trials in Go/NoGo or SST. This internal feedback further enhances the learning process via regular feedback mechanisms. Additionally, this learning could also rely on a stronger memory trace and enhanced attentional processes, which are difficult to discriminate from the overall learning process.

1.4 Short-Term Versus Long-Term Change

The timescale of preference change varies significantly across different experimental paradigms. Preference changes induced by mere exposure, choice leads to choice, or cue-approach training (CAT) operate on different timescales. MEE usually lasts for a short term in a span of minutes or hours (Bornstein & Craver-Lemley, 2022; de Zilva et al., 2013), CAT's behavioral effects have been observed to persist for months (Salomon et al., 2018), and choice-induced preference changes can last for years (Sharot et al., 2012). This disparity in timescales suggests the possibility of distinct underlying mechanisms: long-term preference changes might rely more heavily on memory-related processes for value assignment (Chammat et al., 2017; Voigt et al., 2019), while short-term effects could involve other mechanisms. However, differences in timescale might be a consequence of experimental design rather than indicative of fundamentally different mechanisms, and it remains unclear whether the differences in the timescale of effects across paradigms truly reflect distinct underlying mechanisms or are simply artifacts of varying experimental designs.

1.5 Neural Basis of Non-reinforced Preference Change

The neural basis of preference change has been studied primarily in the context of external reinforcement and through the lens of reinforcement learning models (Elliott & Deakin, 2005; Folia & Petersson, 2014; Kustubayeva et al., 2022; Stanger & Garavan, 2015). This process has been associated with the dopaminergic nuclei in the midbrain and their connections to the ventral and dorsal striatum (Schultz et al., 1997). The representation of subjective value assigned to stimuli or responses is closely linked to the activity in the ventromedial prefrontal cortex (vmPFC) and the orbitofrontal cortex (OFC), which are believed to encode value in a unified and common currency (Bartra et al., 2013; Levy & Glimcher, 2012). The vmPFC and OFC are thought to be interpreted by downstream areas that guide value-based decisions and actions, such as the anterior cingulate cortex and related regions within the frontoparietal network (Gluth et al., 2012; Hare et al., 2011; Kennerley et al., 2006; Pisauro et al., 2017; Rich & Wallis, 2016; Rodriguez et al., 2015).

When comparing brain regions activated by various internal reinforcement techniques, both similarities and differences emerge. A notable similarity is the engagement of the prefrontal cortex (PFC) across all tasks, though it manifests in distinct ways for each technique (Eagle et al., 2008; Schonberg & Katz, 2020; Simmonds et al., 2008). Go/NoGo and SST focus more on response inhibition and selection, engaging prefrontal-parietal networks such as right inferior frontal gyrus and dorsolateral prefrontal cortex (Goghari & MacDonald, 2009; Hughes et al., 2014). Furthermore, both tasks share activation in the pre-supplementary motor area (pre-SMA) (Simmonds et al., 2008; Tabu et al., 2011). AAT has been shown to involve emotional and motivational circuits, including the amygdala and striatum (McDermott et al., 2022).

Another study explored the neural mechanisms underlying choice-induced preference formation, emphasizing the relationship between resting-state brain connectivity and preference changes (Voigt et al., 2020). Their findings indicated that diminished resting-state excitation between the left dorsolateral prefrontal cortex (dlPFC) and the vmPFC was linked to changes in preferences, while increased resting-state excitation from the anterior cingulate cortex (ACC) to the left hippocampus and vmPFC correlated with preference stability (Voigt et al., 2020). These dynamic network configurations suggested that resting-state brain activity significantly influenced both the formation and consistency of preferences during decision-making processes. Additionally, it was proposed that the act of making choices could increase the perceived value of noninstrumental information. By fostering a sense of control, individuals became more engaged in learning, leading to an increase in their appreciation of noninstrumental information. This highlights how decision-making could actively shape individuals' evaluations and preferences (Jiwa et al., 2021). Collectively, these studies contribute to our understanding of the interplay between cognitive processes and neural activity in the context of preference formation without external reinforcements.

Recent findings suggest that sensitivity to intrinsic rewards is domain-general, applying across various types of rewards such as visual, cognitive, and social stimuli (Blain & Sharot, 2021; Blain et al., 2023). In their recent research on intrinsic reward, Blain et al. found that approximately 40% of response variance across different stimuli in the experiment was explained by a general sensitivity to all rewards, indicating a common underlying mechanism for reward processing. A significant finding was the association between affective aspects of mental health and sensitivity to intrinsic rewards, but not monetary rewards (Blain et al., 2023). A related study used a neurocomputational model to understand and measure the inherent value of self-rewarding experiences, such as playing music without errors (Chew et al., 2021). It highlighted the vmPFC as a critical brain region involved in processing these intrinsic rewards. Through neuroimaging and computational modeling, the researchers demonstrated those intrinsic rewards significantly influence subjective affective states, even in the absence of direct instrumental value. Additionally, the study compared intrinsic rewards with extrinsic ones, revealing that some individuals are more motivated by the intrinsic rewards than by external incentives, like money.

Functional MRI (fMRI) studies with CAT revealed enhanced activations in the vmPFC, during choices of Go items in the binary-choice phase (Bakkour et al., 2016; Salomon et al., 2018; Schonberg et al., 2014). A few CAT versions added an additional measurement of the change in neural activations for presented items, called passive viewing task (Botvinik-Nezer et al., 2020). During this task, individual stimulus was presented on the screen before and after training, without any response from the participants. For Go items, BOLD activations during this task were found to increase after training in the lateral occipital cortex (Botvinik-Nezer et al., 2020). An fMRI study focused on the training phase found that BOLD activity in the striatum during the last training run was correlated with individual subsequent preference modification during probe (Salomon et al., 2020). This suggests the involvement of the reward system, even in the absence of external reinforcements, during the training phase. The results indicate that CAT task induces a robust effect that influences preferences for nonnegative stimuli and suggests that the effect is driven by the involvement of attention, memory, motor, and reward systems.

Relatedly, for participants who made more Go over NoGo choices in the binary-choice probe phase, a greater functional connectivity was found between higher-level visual regions, prefrontal cortices, and the striatum (Salomon et al., 2020). These results were replicated for the GNG task (Wu et al., 2023; Xia et al., 2024), suggesting that attention processes took place in non-reinforcement learning procedures. These findings also align with Lim's work, which demonstrated that visual attention guides the relative value coding in the vmPFC and striatum during decision-making. This study tested how these brain regions compute relative decision values by comparing the value of the attended item to the value of the unattended item (Lim et al., 2011). The results suggest that attention biases the valuation process toward the attended item, making it more likely to be chosen. The modulation of value by attention indirectly impacts choice outcomes by favoring the item in focus. These results are consistent with earlier behavioral studies (Armel et al., 2008) and suggest that attention plays a crucial role in value-based decision-making, providing a potential mechanism for how CAT and GNG tasks influence preferences through attentional processes.

To test overlapping regions active during CAT, we performed a meta-analysis of studies that used CAT task scanned in fMRI. We used uncorrected maps uploaded to NeuroVault and computed an overlap of activation maps to analyze their aggregated activation patterns. We tested group-level statistical maps of the activation during the training and probe phases. We used the meta-analysis toolbox taken from Nilearn package in Python, with a smoothing of FWHM = 6 mm and a threshold of $|z| \geq 2.3$ at the group level. For the training phase, we found results from two studies displaying the fMRI response to "Go" items and not responses for "NoGo" items on the last training run (Salomon et al., 2020; Schonberg et al., 2014). Results showed repeated activation in the prefrontal cortex, cingulate cortex, and striatum. The probe sessions' data included the results of "Go" minus "NoGo" items from three studies (Botvinik-Nezer et al., 2020; Salomon et al., 2020; Schonberg et al., 2014). The analysis showed repeated activation in the prefrontal cortex as well as temporal and parietal areas (see Fig. 2). Unthresholded statistical images of the meta-analysis results are available on NeuroVault (Gorgolewski et al., 2015): https://identifiers.org/neurovault.collection:19619.

2 The Dorsal Value Pathway or How Can Neuroscience Data Inform Behavioral Models?

The above-mentioned neural mechanisms engage reward-related and habit-forming regions like vmPFC, striatum, and insula, all related to dopaminergic circuits that could be named ventral value pathway (VVP) (Schonberg & Katz, 2020). The VVP relies on a feedback loop of reward-based learning, a mechanism that involves updating the expected value of actions based on recent outcomes (Rangel et al., 2008). The VVP shapes preferences through direct reinforcement from external outcomes (O'Doherty et al., 2017).

The same authors proposed that another neural pathway underlies the types of non-external reinforcement preference changes reviewed in this chapter. They termed it the dorsal value pathway (DVP) (Schonberg & Katz, 2020) (see Fig. 3). The DVP involves several key brain regions, including the striatum and the vmPFC, as well as higher perceptual and attentional areas such as the frontal eye field (FEF) and the posterior parietal cortex (PPC) (Schonberg & Katz, 2020). Building on the involvement of attention in the value construction process, the pathway flows by integrating information from these regions, with the striatum that plays a crucial role in valuation and reward systems, the vmPFC associated with decision-making, and the higher perceptual and attentional regions contribute to selective attention and item evaluation (Schonberg & Katz, 2020). It has been offered that the DVP is either integrated with or segregated from the well-established VVP, which is

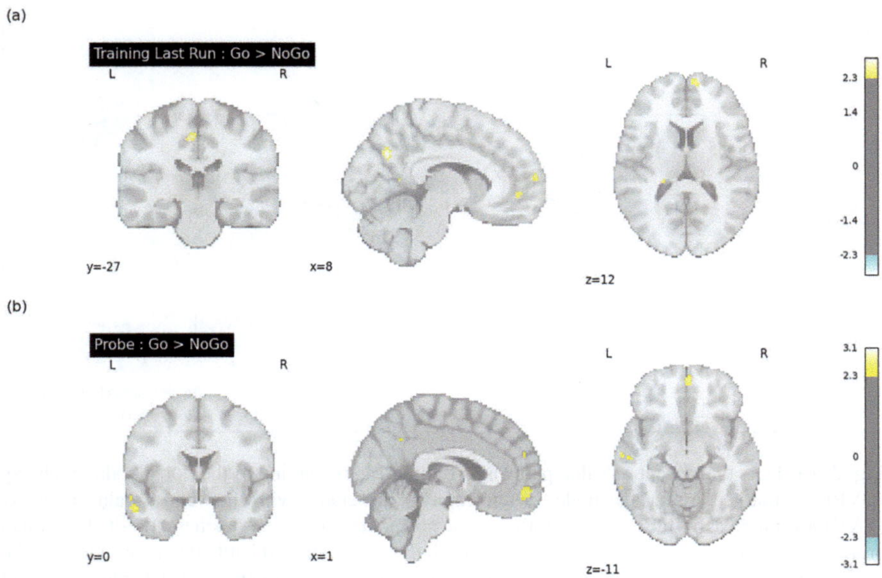

Fig. 2 Overlapping activation maps from two fMRI studies using CAT task. Uncorrected statistical maps were obtained from NeuroVault and analyzed for common activation patterns using the Nilearn package in Python. (**a**) Training phase results based on two studies (Salomon et al., 2020; Schonberg et al., 2014) show consistent activation in the prefrontal cortex, cingulate cortex, and striatum during responses to "Go" items and not responses for "NoGo" items. (**b**) Probe phase results based on three studies (Botvinik-Nezer et al., 2020; Salomon et al., 2020; Schonberg et al., 2014) reveal overlapping activation in the prefrontal cortex and temporal areas for "Go" minus "NoGo" trials. Warmer colors indicate regions of greater overlap across the above-cited studies

primarily related to external reinforcement learning, through the association with dopaminergic regions and projections.

The DVP plays a critical role in attention, salience, and reward processing, which is essential for understanding changes in preferences that are not externally reinforced (Schonberg & Katz, 2020). The interaction of different structures within the DVP, such as the ventral striatum's role in reward signaling, the dorsal striatum's role in habitual behavior (Yin & Knowlton, 2006), and the PFC's role in decision-making (Miller & Cohen, 2001), may support the stabilization of these preferences. Furthermore, the PFC integrates information from various neural circuits, allowing for complex decision-making and adaptive behavior, which is vital for the stabilization of newly formed preferences.

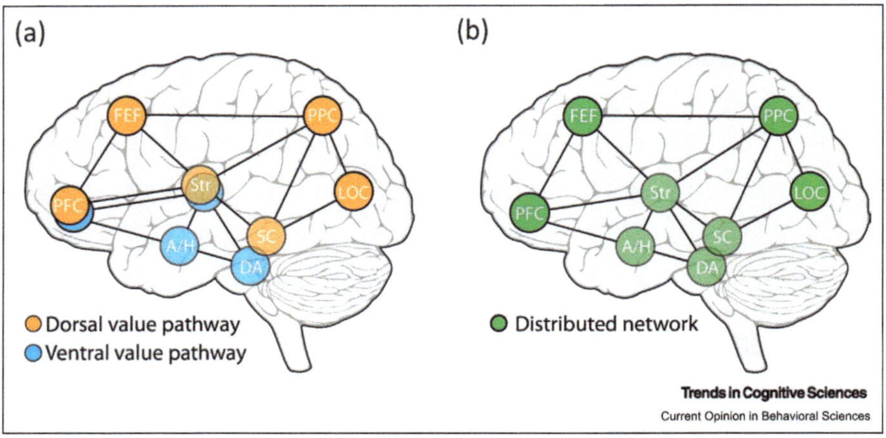

Fig. 3 (**a**) Dorsal and ventral value pathways: brain areas belonging to the dorsal value pathway (DVP) are marked by orange nodes, while regions associated with the ventral value pathway (VVP) are marked by blue nodes. Both sets of nodes are shown on a schematic of the human brain, with transparency indicating regions beneath the neocortex. (**b**) Unified network view: in the unified model, these areas form a single, distributed network. Abbreviations: *A/H* amygdala and hippocampus, *DA* midbrain dopamine neurons, *FEF* frontal eye field, *LOC* lateral occipital cortex, *PFC* prefrontal cortex, *PPC* posterior parietal cortex, *SC* superior colliculus, *Str* striatum. (CC- by Schonberg & Katz, 2020)

3 Conclusions

In this chapter, we explored the diverse tasks through which preference changes can occur without external reinforcement. From the effects of mere exposure and cognitive dissonance to more structured interventions like AAT and cue-approach training (CAT), we see that putatively internal feedback plays a significant role in shaping preferences. We highlighted the neural mechanisms underlying non-reinforced preference change that involve a network of regions named the dorsal value pathway (DVP).

The findings presented here point to promising avenues for future research. Investigating the long-term effects of non-reinforced preference changes, especially in applied contexts like health behavior modification and therapeutic interventions, may provide valuable insights. Furthermore, advancements in neuroimaging and computational modeling will allow for deeper exploration of the DVP and its interactions with the VVP, which has been extensively studied as its underlying reinforced change.

In summary, the study of non-reinforced preference change broadens our understanding of decision-making processes and highlights the potential for neuroscience to inform models that capture the intrinsic dynamics of preference formation. As research in this field progresses, we may uncover innovative ways to influence preferences sustainably and ethically, leveraging our understanding of internal motivations and cognitive control.

References

Aksiotis, V., Myachykov, A., & Tumyalis, A. (2023). Stop-signal delay reflects response selection duration in stop-signal task. *Attention, Perception & Psychophysics, 85*(6), 1976–1989. https://doi.org/10.3758/s13414-023-02752-y

Armel, K. C., Beaumel, A., & Rangel, A. (2008). Biasing simple choices by manipulating relative visual attention. *Judgment and Decision making, 3*(5), 396–403. https://doi.org/10.1017/S1930297500000413

Bakkour, A., Leuker, C., Hover, A. M., Giles, N., Poldrack, R. A., & Schonberg, T. (2016). Mechanisms of choice behavior shift using Cue-approach training. *Frontiers in Psychology, 7*, 421. https://doi.org/10.3389/fpsyg.2016.00421

Bandura, A., & Adams, N. E. (1977). Analysis of self-efficacy theory of behavioral change. *Cognitive Therapy and Research, 1*(4), 287–310. https://doi.org/10.1007/BF01663995

Bartholdy, S., Dalton, B., O'Daly, O. G., Campbell, I. C., & Schmidt, U. (2016). A systematic review of the relationship between eating, weight and inhibitory control using the stop signal task. *Neuroscience and Biobehavioral Reviews, 64*, 35–62. https://doi.org/10.1016/j.neubiorev.2016.02.010

Bartra, O., McGuire, J. T., & Kable, J. W. (2013). The valuation system: A coordinate-based meta-analysis of BOLD fMRI experiments examining neural correlates of subjective value. *NeuroImage, 76*, 412–427. https://doi.org/10.1016/j.neuroimage.2013.02.063

Becker, G. M., DeGroot, M. H., & Marschak, J. (1964). Measuring utility by a single-response sequential method. *Behavioral Science, 9*(3), 226–232. https://doi.org/10.1002/bs.3830090304

Blain, B., & Sharot, T. (2021). Intrinsic reward: Potential cognitive and neural mechanisms. *Current Opinion in Behavioral Sciences, 39*, 113–118. https://doi.org/10.1016/j.cobeha.2021.03.008

Blain, B., Pinhorn, I., & Sharot, T. (2023). Sensitivity to intrinsic rewards is domain general and related to mental health. *Nature Mental Health, 1*(9), 679–691. https://doi.org/10.1038/s44220-023-00116-x

Boedecker, J., Lampe, T., & Riedmiller, M. (2013). Modeling effects of intrinsic and extrinsic rewards on the competition between striatal learning systems. *Frontiers in Psychology, 4*, 739. https://doi.org/10.3389/fpsyg.2013.00739

Bolívar, H. A., Klemperer, E. M., Coleman, S. R. M., DeSarno, M., Skelly, J. M., & Higgins, S. T. (2021). Contingency Management for Patients Receiving Medication for opioid use disorder: A systematic review and meta-analysis. *JAMA Psychiatry, 78*(10), 1092–1102. https://doi.org/10.1001/jamapsychiatry.2021.1969

Bornstein, R. F., & Craver-Lemley, C. (2022). Mere exposure effect. In *Cognitive illusions: Intriguing phenomena in thinking, judgment, and memory* (pp. 241–258). Routledge. https://doi.org/10.4324/9781003154730-18

Botvinik-Nezer, R., Salomon, T., & Schonberg, T. (2020). Enhanced bottom-up and reduced top-down fMRI activity is related to long-lasting nonreinforced Behavioral change. *Cerebral Cortex, 30*(3), 858–874. https://doi.org/10.1093/cercor/bhz132

Botvinik-Nezer, R., Bakkour, A., Salomon, T., Shohamy, D., & Schonberg, T. (2021). Memory for individual items is related to nonreinforced preference change. *Learning & Memory, 28*(10), 348–360. https://doi.org/10.1101/lm.053411.121

Cameron, J., & Pierce, W. D. (1994). Reinforcement, reward, and intrinsic motivation: A meta-analysis. *Review of Educational Research, 64*(3), 363–423. https://doi.org/10.3102/00346543064003363

Cash, P., Wrobel, A., Maier, A., & Hansen, J. P. (2023). Understanding long-term behaviour change techniques: A mixed methods study. *Journal of Engineering Design, 34*(5–6), 383–410. https://doi.org/10.1080/09544828.2023.2227933

Cavarretta, F. (2024). Weary of the harsh realities of people management? Leadership development as cultivating a taste for muddy situations. *Organizational Dynamics, 101028.* https://doi.org/10.1016/j.orgdyn.2024.101028

Chammat, M., Karoui, I. E., Allali, S., Hagège, J., Lehongre, K., Hasboun, D., Baulac, M., Epelbaum, S., Michon, A., Dubois, B., Navarro, V., Salti, M., & Naccache, L. (2017). Cognitive dissonance resolution depends on episodic memory. *Scientific Reports, 7*, 41320. https://doi.org/10.1038/srep41320

Chew, B., Blain, B., Dolan, R. J., & Rutledge, R. B. (2021). A neurocomputational model for intrinsic reward. *The Journal of Neuroscience, 41*(43), 8963–8971. https://doi.org/10.1523/JNEUROSCI.0858-20.2021

de Zilva, D., Vu, L., Newell, B. R., & Pearson, J. (2013). Exposure is not enough: Suppressing stimuli from awareness can abolish the mere exposure effect. *PLoS One, 8*(10), e77726. https://doi.org/10.1371/journal.pone.0077726

Ding, W., Sun, J., Sun, Y.-W., Chen, X., Zhou, Y., Zhuang, Z., Li, L., Zhang, Y., Xu, J., & Du, Y. (2014). Trait impulsivity and impaired prefrontal impulse inhibition function in adolescents with internet gaming addiction revealed by a Go/NoGo fMRI study. *Behavioral and Brain Functions, 10*, 20. https://doi.org/10.1186/1744-9081-10-20

Donders, F. C. (1969). On the speed of mental processes. *Acta Psychologica, 30*, 412–431. https://doi.org/10.1016/0001-6918(69)90065-1

Dong, G., Zhou, H., & Zhao, X. (2010). Impulse inhibition in people with internet addiction disorder: Electrophysiological evidence from a Go/NoGo study. *Neuroscience Letters, 485*(2), 138–142. https://doi.org/10.1016/j.neulet.2010.09.002

Duckworth, A. L., & Gross, J. J. (2020). Behavior Change. *Organizational Behavior and Human Decision Processes, 161*(Suppl), 39–49. https://doi.org/10.1016/j.obhdp.2020.09.002

Eagle, D. M., Baunez, C., Hutcheson, D. M., Lehmann, O., Shah, A. P., & Robbins, T. W. (2008). Stop-signal reaction-time task performance: Role of prefrontal cortex and subthalamic nucleus. *Cerebral Cortex, 18*(1), 178–188. https://doi.org/10.1093/cercor/bhm044

Elliott, R., & Deakin, B. (2005). *Role of the orbitofrontal cortex in reinforcement processing and inhibitory control: Evidence from functional magnetic resonance imaging studies in healthy human subjects* (Vol. 65, pp. 89–116). Elsevier. https://doi.org/10.1016/S0074-7742(04)65004-5

Ellis, J. D., Struble, C. A., Fodor, M. C., Cairncross, M., Lundahl, L. H., & Ledgerwood, D. M. (2021). Contingency management for individuals with chronic health conditions: A systematic review and meta-analysis of randomized controlled trials. *Behaviour Research and Therapy, 136*, 103781. https://doi.org/10.1016/j.brat.2020.103781

Festinger, L. (1957). *A theory of cognitive dissonance*. Stanford University Press. https://www.sup.org/books/sociology/theory-cognitive-dissonance

Folia, V., & Petersson, K. M. (2014). Implicit structured sequence learning: An fMRI study of the structural mere-exposure effect. *Frontiers in Psychology, 5*, 41. https://doi.org/10.3389/fpsyg.2014.00041

Giuffrida, V., Marc, I. B., Ramawat, S., Fontana, R., Fiori, L., Bardella, G., Fagioli, S., Ferraina, S., Brunamonti, E., & Pani, P. (2023). Reward prospect affects strategic adjustments in stop signal task. *Frontiers in Psychology, 14*, 1125066. https://doi.org/10.3389/fpsyg.2023.1125066

Gluth, S., Rieskamp, J., & Büchel, C. (2012). Deciding when to decide: Time-variant sequential sampling models explain the emergence of value-based decisions in the human brain. *The Journal of Neuroscience, 32*(31), 10686–10698. https://doi.org/10.1523/JNEUROSCI.0727-12.2012

Goghari, V. M., & MacDonald, A. W. (2009). The neural basis of cognitive control: Response selection and inhibition. *Brain and Cognition, 71*(2), 72–83. https://doi.org/10.1016/j.bandc.2009.04.004

Gomez, P., Ratcliff, R., & Perea, M. (2007). A model of the Go/NoGo task. *Journal of Experimental Psychology: General, 136*(3), 389–413. https://doi.org/10.1037/0096-3445.136.3.389

Gorgolewski, K. J., Varoquaux, G., Rivera, G., Schwarz, Y., Ghosh, S. S., Maumet, C., Sochat, V. V., Nichols, T. E., Poldrack, R. A., Poline, J.-B., Yarkoni, T., & Margulies, D. S. (2015). NeuroVault.org: a web-based repository for collecting and sharing unthresholded statistical maps of the human brain. *Frontiers in Neuroinformatics, 9*, 8. https://doi.org/10.3389/fninf.2015.00008

Grush, J. E., McKeough, K. L., & Ahlering, R. F. (1978). Extrapolating laboratory exposure research to actual political elections. *Journal of Personality and Social Psychology, 36*(3), 257–270. https://doi.org/10.1037/0022-3514.36.3.257

Hare, T. A., Schultz, W., Camerer, C. F., O'Doherty, J. P., & Rangel, A. (2011). Transformation of stimulus value signals into motor commands during simple choice. *Proceedings of the National Academy of Sciences of the United States of America, 108*(44), 18120–18125. https://doi.org/10.1073/pnas.1109322108

Harmon-Jones, E., & Mills, J. (2019). An introduction to cognitive dissonance theory and an overview of current perspectives on the theory. In E. Harmon-Jones (Ed.), *Cognitive dissonance: Reexamining a pivotal theory in psychology* (2nd ed., pp. 3–24). American Psychological Association. https://doi.org/10.1037/0000135-001

Hoppe, J. M., et al. (2022). Internet-delivered approach-avoidance conflict task shows temporal stability and relation to trait anxiety. *Learning and Motivation, 80*, 101848.

Hughes, M. E., Budd, T. W., Fulham, W. R., Lancaster, S., Woods, W., Rossell, S. L., & Michie, P. T. (2014). Sustained brain activation supporting stop-signal task performance. *The European Journal of Neuroscience, 39*(8), 1363–1369. https://doi.org/10.1111/ejn.12497

Itzkovitch, A., Bar Or, M., & Schonberg, T. (2022). Cue-approach training for food behavior. *Current Opinion in Behavioral Sciences, 47*, 101202. https://doi.org/10.1016/j.cobeha.2022.101202

Jiwa, M., Cooper, P. S., Chong, T. T.-J., & Bode, S. (2021). Choosing increases the value of non-instrumental information. *Scientific Reports, 11*(1), 8780. https://doi.org/10.1038/s41598-021-88031-y

Jones, R. T., Nelson, R. E., & Kazdin, A. E. (1977). The role of external variables in self-reinforcement. *Behavior Modification, 1*(2), 147–178. https://doi.org/10.1177/014544557712001

Kakoschke, N., Albertella, L., Lee, R. S. C., & Wiers, R. W. (2019). Assessment of automatically activated approach–avoidance biases across appetitive substances. *Current Addiction Reports, 1–10*. https://doi.org/10.1007/s40429-019-00254-2

Kennerley, S. W., Walton, M. E., Behrens, T. E. J., Buckley, M. J., & Rushworth, M. F. S. (2006). Optimal decision making and the anterior cingulate cortex. *Nature Neuroscience, 9*, 940–947. https://doi.org/10.1038/nn1724

Kim, D.-Y., & Lee, J.-H. (2015). Development of a virtual approach-avoidance task to assess alcohol cravings. *Cyberpsychology, Behavior and Social Networking, 18*(12), 763–766. https://doi.org/10.1089/cyber.2014.0490

Krajbich, I., Armel, C., & Rangel, A. (2010). Visual fixations and the computation and comparison of value in simple choice. *Nature Neuroscience, 13*(10), 1292–1298. https://doi.org/10.1038/nn.2635

Kustubayeva, A. M., Nelson, E. B., Smith, M. L., Allendorfer, J. B., & Eliassen, J. C. (2022). Functional MRI study of feedback-based reinforcement learning in depression. *Frontiers in Neuroinformatics, 16*, 1028121. https://doi.org/10.3389/fninf.2022.1028121

Ladouceur, C. D., Dahl, R. E., Williamson, D. E., Birmaher, B., Axelson, D. A., Ryan, N. D., & Casey, B. J. (2006). Processing emotional facial expressions influences performance on a Go/NoGo task in pediatric anxiety and depression. *Journal of Child Psychology and Psychiatry, and Allied Disciplines, 47*(11), 1107–1115. https://doi.org/10.1111/j.1469-7610.2006.01640.x

Lappin, J. S., & Eriksen, C. W. (1966). Use of a delayed signal to stop a visual reaction-time response. *Journal of Experimental Psychology, 72*(6), 805–811. https://doi.org/10.1037/h0021266

Lench, H. C., Flores, S. A., & Bench, S. W. (2011). Discrete emotions predict changes in cognition, judgment, experience, behavior, and physiology: A meta-analysis of experimental emotion elicitations. *Psychological Bulletin, 137*(5), 834–855. https://doi.org/10.1037/a0024244

Levy, D. J., & Glimcher, P. W. (2012). The root of all value: A neural common currency for choice. *Current Opinion in Neurobiology, 22*(6), 1027–1038. https://doi.org/10.1016/j.conb.2012.06.001

Li, C. R., Huang, C., Constable, R. T., & Sinha, R. (2006). Imaging response inhibition in a stop-signal task: Neural correlates independent of signal monitoring and post-response processing. *The Journal of Neuroscience, 26*(1), 186–192. https://doi.org/10.1523/JNEUROSCI.3741-05.2006

Lim, S.-L., O'Doherty, J. P., & Rangel, A. (2011). The decision value computations in the vmPFC and striatum use a relative value code that is guided by visual attention. *The Journal of Neuroscience, 31*(37), 13214–13223. https://doi.org/10.1523/JNEUROSCI.1246-11.2011

Lipszyc, J., & Schachar, R. (2010). Inhibitory control and psychopathology: A meta-analysis of studies using the stop signal task. *Journal of the International Neuropsychological Society, 16*(6), 1064–1076. https://doi.org/10.1017/S1355617710000895

Logan, G. D., Cowan, W. B., & Davis, K. A. (1984). On the ability to inhibit simple and choice reaction time responses: A model and a method. *Journal of Experimental Psychology: Human Perception and Performance, 10*(2), 276–291. https://doi.org/10.1037/0096-1523.10.2.276

McDermott, T. J., Berg, H., Touthang, J., Akeman, E., Cannon, M. J., Santiago, J., Cosgrove, K. T., Clausen, A. N., Kirlic, N., Smith, R., Craske, M. G., Abelson, J. L., Paulus, M. P., & Aupperle, R. L. (2022). Striatal reactivity during emotion and reward relates to approach-avoidance conflict behaviour and is altered in adults with anxiety or depression. *Journal of Psychiatry & Neuroscience, 47*(5), E311–E322. https://doi.org/10.1503/jpn.220083

Miller, E. K., & Cohen, J. D. (2001). An integrative theory of prefrontal cortex function. *Annual Review of Neuroscience, 24*, 167–202. https://doi.org/10.1146/annurev.neuro.24.1.167

Nakamura, K., & Kawabata, H. (2013). I choose, therefore I like: Preference for faces induced by arbitrary choice. *PLoS One, 8*(8), e72071. https://doi.org/10.1371/journal.pone.0072071

Naslund, J. A., Kim, S. J., Aschbrenner, K. A., McCulloch, L. J., Brunette, M. F., Dallery, J., Bartels, S. J., & Marsch, L. A. (2017). Systematic review of social media interventions for smoking cessation. *Addictive Behaviors, 73*, 81–93. https://doi.org/10.1016/j.addbeh.2017.05.002

Neumann, R., & Strack, F. (2000). Approach and avoidance: The influence of proprioceptive and exteroceptive cues on encoding of affective information. *Journal of Personality and Social Psychology, 79*(1), 39–48. https://doi.org/10.1037/0022-3514.79.1.39

O'Doherty, J. P., Cockburn, J., & Pauli, W. M. (2017). Learning, reward, and decision making. *Annual Review of Psychology, 68*, 73–100. https://doi.org/10.1146/annurev-psych-010416-044216

Petry, N. M. (2011). Contingency management: What it is and why psychiatrists should want to use it. *The Psychiatrist, 35*(5), 161–163. https://doi.org/10.1192/pb.bp.110.031831

Pisauro, M. A., Fouragnan, E., Retzler, C., & Philiastides, M. G. (2017). Neural correlates of evidence accumulation during value-based decisions revealed via simultaneous EEG-fMRI. *Nature Communications, 8*, 15808. https://doi.org/10.1038/ncomms15808

Poulton, A., Chen, L. P. E., Dali, G., Fox, M., & Hester, R. (2022). Web-based independent versus laboratory-based stop-signal task performance: Within-subjects counterbalanced comparison study. *Journal of Medical Internet Research, 24*(5), e32922. https://doi.org/10.2196/32922

Rangel, A., Camerer, C., & Montague, P. R. (2008). A framework for studying the neurobiology of value-based decision making. *Nature Reviews. Neuroscience, 9*(7), 545–556. https://doi.org/10.1038/nrn2357

Rich, E. L., & Wallis, J. D. (2016). Decoding subjective decisions from orbitofrontal cortex. *Nature Neuroscience, 19*(7), 973–980. https://doi.org/10.1038/nn.4320

Rinck, M., & Becker, E. S. (2007). Approach and avoidance in fear of spiders. *Journal of Behavior Therapy and Experimental Psychiatry, 38*(2), 105–120. https://doi.org/10.1016/j.jbtep.2006.10.001

Rodriguez, C. A., Turner, B. M., Van Zandt, T., & McClure, S. M. (2015). The neural basis of value accumulation in intertemporal choice. *The European Journal of Neuroscience, 42*(5), 2179–2189. https://doi.org/10.1111/ejn.12997

Salomon, T., Botvinik-Nezer, R., Gutentag, T., Gera, R., Iwanir, R., Tamir, M., & Schonberg, T. (2018). The Cue-approach task as a general mechanism for long-term non-reinforced Behavioral change. *Scientific Reports, 8*(1), 3614. https://doi.org/10.1038/s41598-018-21774-3

Salomon, T., Botvinik-Nezer, R., Oren, S., & Schonberg, T. (2020). Enhanced striatal and prefrontal activity is associated with individual differences in nonreinforced preference change for faces. *Human Brain Mapping, 41*(4), 1043–1060. https://doi.org/10.1002/hbm.24859

Salomon, T., Itzkovitch, A., Daw, N. D., & Schonberg, T. (2022). A passive computational marker for individual differences in non-reinforced learning. *BioRxiv.*https://doi.org/10.1101/2022.03.20.484477

Schonberg, T., & Katz, L. N. (2020). A neural pathway for nonreinforced preference change. *Trends in Cognitive Sciences, 24*(7), 504–514. https://doi.org/10.1016/j.tics.2020.04.002

Schonberg, T., Bakkour, A., Hover, A. M., Mumford, J. A., Nagar, L., Perez, J., & Poldrack, R. A. (2014). Changing value through cued approach: An automatic mechanism of behavior change. *Nature Neuroscience, 17*(4), 625–630. https://doi.org/10.1038/nn.3673

Schultz, W., Dayan, P., & Montague, P. R. (1997). A neural substrate of prediction and reward. *Science, 275*(5306), 1593–1599. https://doi.org/10.1126/science.275.5306.1593

Secades-Villa, R., Aonso-Diego, G., & González-Roz, A. (2022). A randomized controlled trial of contingency management for smoking cessation in substance use treatment patients. *International Journal of Clinical and Health Psychology, 22*(3), 100314. https://doi.org/10.1016/j.ijchp.2022.100314

Sharbanee, J. M., Hu, L., Stritzke, W. G. K., Wiers, R. W., Rinck, M., & MacLeod, C. (2014). The effect of approach/avoidance training on alcohol consumption is mediated by change in alcohol action tendency. *PLoS One, 9*(1), e85855. https://doi.org/10.1371/journal.pone.0085855

Sharot, T., Velasquez, C. M., & Dolan, R. J. (2010). Do decisions shape preference? Evidence from blind choice. *Psychological Science, 21*(9), 1231–1235. https://doi.org/10.1177/0956797610379235

Sharot, T., Fleming, S. M., Yu, X., Koster, R., & Dolan, R. J. (2012). Is choice-induced preference change long lasting? *Psychological Science, 23*(10), 1123–1129. https://doi.org/10.1177/0956797612438733

Simmonds, D. J., Pekar, J. J., & Mostofsky, S. H. (2008). Meta-analysis of Go/No go tasks demonstrating that fMRI activation associated with response inhibition is task-dependent. *Neuropsychologia, 46*(1), 224–232. https://doi.org/10.1016/j.neuropsychologia.2007.07.015

Skinner, B. F. (1963). Operant behavior. *American Psychologist, 18*(8), 503–515. https://doi.org/10.1037/h0045185

Stanger, C., & Garavan, H. (2015). Neural influences in contingency management for adolescent substance use. In S. W. F. Ewing, K. Witkiewitz, & F. M. Filbey (Eds.), *Neuroimaging and psychosocial addiction treatment* (pp. 244–256). Palgrave Macmillan UK. https://doi.org/10.1057/9781137362650_16

Strack, F., & Deutsch, R. (2004). Reflective and impulsive determinants of social behavior. *Personality and Social Psychology Review, 8*(3), 220–247. https://doi.org/10.1207/s15327957pspr0803_1

Tabu, H., Mima, T., Aso, T., Takahashi, R., & Fukuyama, H. (2011). Functional relevance of pre-supplementary motor areas for the choice to stop during stop signal task. *Neuroscience Research, 70*(3), 277–284. https://doi.org/10.1016/j.neures.2011.03.007

Thorndike, E. L. (1927). The law of effect. *The American Journal of Psychology, 39*(1/4), 212. https://doi.org/10.2307/1415413

Trommer, B. L., Hoeppner, J. A., Lorber, R., & Armstrong, K. J. (1988). The go-no go paradigm in attention deficit disorder. *Annals of Neurology, 24*(5), 610–614. https://doi.org/10.1002/ana.410240504

Uzefovsky, F., Allison, C., Smith, P., & Baron-Cohen, S. (2016). Brief report: The Go/NoGo task online: Inhibitory control deficits in autism in a large sample. *Journal of Autism and Developmental Disorders, 46*(8), 2774–2779. https://doi.org/10.1007/s10803-016-2788-3
van Dantzig, S., Pecher, D., & Zwaan, R. A. (2008). Approach and avoidance as action effects. *Quarterly Journal of Experimental Psychology, 61*(9), 1298–1306. https://doi.org/10.1080/17470210802027987
Veling, H., Chen, Z., Tombrock, M. C., Verpaalen, I. A. M., Schmitz, L. I., Dijksterhuis, A., & Holland, R. W. (2017a). Training impulsive choices for healthy and sustainable food. *Journal of Experimental Psychology. Applied, 23*(2), 204–215. https://doi.org/10.1037/xap0000112
Veling, H., Lawrence, N. S., Chen, Z., van Koningsbruggen, G. M., & Holland, R. W. (2017b). What is trained during food Go/NoGo training? A review focusing on mechanisms and a research agenda. *Current Addiction Reports, 4*(1), 35–41. https://doi.org/10.1007/s40429-017-0131-5
Veling, H., Verpaalen, I. A. M., Liu, H., Mosannenzadeh, F., Becker, D., & Holland, R. W. (2021). How can food choice best be trained? Approach-avoidance versus Go/NoGo training. *Appetite, 163*, 105226. https://doi.org/10.1016/j.appet.2021.105226
Verbruggen, F., & Logan, G. D. (2008). Response inhibition in the stop-signal paradigm. *Trends in Cognitive Sciences, 12*(11), 418–424. https://doi.org/10.1016/j.tics.2008.07.005
Verbruggen, F., & Logan, G. D. (2009). Models of response inhibition in the stop-signal and stop-change paradigms. *Neuroscience and Biobehavioral Reviews, 33*(5), 647–661. https://doi.org/10.1016/j.neubiorev.2008.08.014
Voigt, K., Murawski, C., Speer, S., & Bode, S. (2019). Hard decisions shape the neural coding of preferences. *The Journal of Neuroscience, 39*(4), 718–726. https://doi.org/10.1523/JNEUROSCI.1681-18.2018
Voigt, K., Murawski, C., Speer, S., & Bode, S. (2020). Effective brain connectivity at rest is associated with choice-induced preference formation. *Human Brain Mapping, 41*(11), 3077–3088. https://doi.org/10.1002/hbm.24999
Wang, W., Hu, S., Ide, J. S., Zhornitsky, S., Zhang, S., Yu, A. J., & Li, C.-S. R. (2018). Motor preparation disrupts proactive control in the stop signal task. *Frontiers in Human Neuroscience, 12*, 151. https://doi.org/10.3389/fnhum.2018.00151
Waters, A. M., & Valvoi, J. S. (2009). Attentional bias for emotional faces in paediatric anxiety disorders: An investigation using the emotional Go/No Go task. *Journal of Behavior Therapy and Experimental Psychiatry, 40*(2), 306–316. https://doi.org/10.1016/j.jbtep.2008.12.008
Wiers, R. W., Bartholow, B. D., van den Wildenberg, E., Thush, C., Engels, R. C. M. E., Sher, K. J., Grenard, J., Ames, S. L., & Stacy, A. W. (2007). Automatic and controlled processes and the development of addictive behaviors in adolescents: A review and a model. *Pharmacology, Biochemistry, and Behavior, 86*(2), 263–283. https://doi.org/10.1016/j.pbb.2006.09.021
Wu, Q., Xia, H., Shields, G. S., Nie, H., Li, J., Chen, H., & Yang, Y. (2023). Neural correlates underlying preference changes induced by food Go/NoGo training. *Appetite, 186*, 106578. https://doi.org/10.1016/j.appet.2023.106578
Xia, H., Wu, Q., Shields, G. S., Nie, H., Hu, X., Liu, S., Zhou, Z., Chen, H., & Yang, Y. (2024). Neural activity and connectivity are related to food preference changes induced by food Go/NoGo training. *Neuropsychologia, 201*, 108919. https://doi.org/10.1016/j.neuropsychologia.2024.108919
Yao, Z., Xia, T., Wei, J., Zhang, Z., Lin, X., Zhang, D., Qin, P., Ma, Y., & Hu, X. (2024). Reactivating cue approached positive personality traits during sleep promotes positive self-referential processing. *IScience, 27*(7), 110341. https://doi.org/10.1016/j.isci.2024.110341
Yerys, B. E., Kenworthy, L., Jankowski, K. F., Strang, J., & Wallace, G. L. (2013). Separate components of emotional Go/NoGo performance relate to autism versus attention symptoms in children with autism. *Neuropsychology, 27*(5), 537–545. https://doi.org/10.1037/a0033615
Yin, H. H., & Knowlton, B. J. (2006). The role of the basal ganglia in habit formation. *Nature Reviews. Neuroscience, 7*(6), 464–476. https://doi.org/10.1038/nrn1919

Young, M. E., Sutherland, S. C., & McCoy, A. W. (2018). Optimal Go/NoGo ratios to maximize false alarms. *Behavior Research Methods, 50*(3), 1020–1029. https://doi.org/10.3758/s13428-017-0923-5

Zajonc, R. B. (1968). Attitudinal effects of mere exposure. *Journal of Personality and Social Psychology, 9*(2, Pt.2), 1–27. https://doi.org/10.1037/h0025848

Part III
Temporal Discounting, Effort-Based Choice, and Self-Control

Dominic S. Fareri, Patricia L. Lockwood, and David V. Smith

How do we weigh the tradeoff between rewards that are available now relative to in the future? What are the neurocomputational mechanisms that allow us to arrive at the most valued option? And how does individual variability in memory, motivation, and self-regulation shape these processes? This section bridges recent advances in the neuroeconomics of temporal discounting, effort-based choice, and self-control, offering a multifaceted view of how people navigate tradeoffs between immediacy, effort, and value in their decision calculus. While self-control and executive function have long been considered central mechanisms governing discounting behavior, more recent work supports a more nuanced understanding of these kinds of choices in one in which memory and motivational processes may be similarly important.

There is increasing recognition that discounting of rewards is not governed by a single mechanism, but rather invokes multiple cognitive and neural processes. Wagner et al. (this volume) survey foundational models of and methods for investigating intertemporal choice. They highlight the roles of dopaminergic systems, emotion, and prospective thought, while also stressing the utility of computational approaches like drift diffusion modeling for capturing individual differences in component cognitive processes supporting discounting behavior. Building on this, Hare (this volume) challenges classical theories of self-control and response inhibition in temporal discounting, arguing that many intertemporal decisions may instead reflect considerations of the uncertainty within the choice environment and the importance of the specific choice attributes (i.e., amount, time). These perspectives instead

D. S. Fareri (✉)
Adelphi University, Garden City, NY, USA
e-mail: dfareri@adelphi.edu

P. L. Lockwood
University of Birmingham, Birmingham, UK

D. V. Smith
Temple University, Philadelphia, PA, USA

emphasize the role that attention and valuation play in discounting rewards over time as opposed to this being a failure to regulate impulsive choice.

Recent evidence also implicates memory capabilities as playing a key role in temporal discounting. Lempert (this volume) argues that while prefrontally mediated executive processes like self-control may be involved in discounting of rewards over time, they are not *necessary*. Instead, this chapter argues for the important contributions of both episodic and semantic memory systems. Episodic memory is related to our ability to prospectively think into the future and is often associated with reduced discounting. Semantic fluency (Akilov & Lempert, 2025) and thickness of the entorhinal cortex (Garzón et al., 2023) appear to also predict individual variability in discounting behavior. This perspective further supports the need for novel frameworks surrounding temporal discounting, suggesting that memory systems may be key components of our ability to imagine the value of future rewards.

Discounting of rewards does not solely occur with respect to their time of delivery. The amount of effort and motivation required to obtain a reward is another key factor determining choice. Scholey et al. (this volume) review a rich literature regarding how effort costs shape the subjective value of pursuing a reward. These processes are conceptualized within neurocomputational frameworks implicating components of frontostriatal circuits (e.g., dorsal anterior cingulate cortex, anterior insula, ventral striatum) playing important roles in both domain-general and context-specific effort-cost computations. This chapter highlights links between these processes with impairments in motivation in clinical and subclinical samples characterized by motivational deficits (e.g., Parkinson's disease, depression), concluding with a push to incorporate alternative methods and frameworks (i.e., behavioral ecology) to more comprehensively delineate the ways in which motivation and effort shape valuation.

While much of the work covered in this section focuses on the idea that discounting behavior encompasses much more than self-control, that is not to say that self-control processes do not play significant roles in value-based choice. Raio (this volume) explores this idea through the lens of dietary decision-making, examining whether the role of the prefrontal cortex in such decisions is to compute the value of the effort needed to exert control or to execute control-related decisions computed elsewhere (e.g., dorsal anterior cingulate). Cognitive regulation strategies mediated by prefrontal circuits have been shown to enhance or diminish self-control during decision-making, depending upon one's goals, and computational costs of exerting self-control may be lessened by engaging in strategies like precommitment (Kurth-Nelson & Redish, 2012; Crockett et al., 2013) to limit potential effort costs (Raio & Glimcher, 2021). Raio thus provides a practical and dynamic framework within which to consider how self-control shapes value-based choice.

Together, chapters in this section offer comprehensive frameworks for considering how discounting of time and effort, as well as how/when to exert control, impacts reward-based decision-making. Decisions to opt for immediately available rewards or low effort courses of action are not necessarily always maladaptive of failures of self-control. Rather, they may emerge as the rational result of value-based computations in conjunction with other cognitive processes, highlighting the continued need for interdisciplinary approaches to understand discounting behavior.

References

Akilov, D., & Lempert, K. M. (2025). Semantic fluency is associated with reduced temporal discounting. *Journal of Memory and Language, 142*, 104616. https://doi.org/10.1016/j.jml.2025.104616

Crockett, M. J., Braams, B., & Clark, L. (2013). Restricting temptations: Neural mechanisms of pre-commitment. *Neuron, 79*, 391–401. https://doi.org/10.1016/j.neuron.2013.05.028

Garzón, B., Kurth-Nelson, Z., Bäckman, L., Nyberg, L., & Guitart-Masip, M. (2023). Investigating associations of delay discounting with brain structure, working memory, and episodic memory. *Cerebral Cortex., 33*(5), 1669–1678. https://doi.org/10.1093/cercor/bhac164

Hare, T. A. (this volume). Intertemporal choices are influenced by many things beyond time itself. In D. V. Smith, P. Lockwood, & D. S. Fareri (Eds.), *Neuroeconomics: Core topics and current directions*. Springer.

Kurth-Nelson, Z., & Redish, A. D. (2012). Don't let me do that!-models of pre-commitment. *Frontiers in Neuroscience, 6*, 138. https://doi.org/10.3389/fnbeh.2010.00184

Lempert, K. M. (this volume). The role of memory in temporal discounting. In D. V. Smith, P. Lockwood, & D. S. Fareri (Eds.), *Neuroeconomics: Core topics and current directions*. Springer. Preprint: https://osf.io/preprints/psyarxiv/3t5c4_v1

Raio, C. M. (this volume). Self-control and decision-making. In D. V. Smith, P. Lockwood, & D. S. Fareri (Eds.), *Neuroeconomics: Core topics and current directions*. Springer.

Raio, C. M., & Glimcher, P. G. (2021). Quantifying the subjective costs of self-control. *National Academy of Sciences of the United States of America, 118*(35), e2018726118.

Scholey, E., Lugymeijer, S., & Apps, M. A. (this volume). The neuroeconomics of work: Computational and neural mechanisms of the dynamics of effort-based decisions. In D. V. Smith, P. Lockwood, & D. S. Fareri (Eds.), *Neuroeconomics: Core topics and current directions*. Springer. Preprint: https://osf.io/preprints/psyarxiv/3t5c4_v1

Wagner, B. J., Knauth, K., & Peters, J. (this volume). Intertemporal choice. In D. V. Smith, P. Lockwood, & D. S. Fareri (Eds.), *Neuroeconomics: Core topics and current directions*. Springer. Preprint: https://osf.io/preprints/psyarxiv/n9evr_v1

Intertemporal Choice

Ben J. Wagner, Kilian Knauth, and Jan Peters

Abstract Intertemporal choice involves decisions between rewards available at an earlier point in time and delayed rewards (delay discounting). Such decisions are common in everyday life, influencing outcomes from personal health decisions to broader societal and political developments. Humans typically prefer smaller-but-sooner over larger-but-later rewards, although substantial inter-individual variability exists. Extreme preferences for immediate or delayed rewards are associated with various mental disorders, ranging from addictions to anorexia nervosa. Preferences are further affected by pharmacological manipulations affecting, e.g., the dopamine system, as well as factors such as episodic future thinking, emotional cueing, or other state-dependent factors. We review individual differences with clinical significance, state-dependent modulators of intertemporal choice and highlight potential intervention targets. We then introduce methods to quantify behavior on intertemporal choice tasks. First, we highlight simple, model-agnostic methods (choice proportions, area under the empirical discounting curve, AUC). We then introduce computational approaches, such as exponential and hyperbolic discounting models, and highlight recent methodological advances that integrate discounting models with evidence accumulation frameworks such as the drift diffusion model (DDM), jointly accounting for choices and response times. These approaches allow

Authors Ben J. Wagner and Kilian Knauth have equally contributed to this chapter.

B. J. Wagner (✉)
Chair of Cognitive Computational Neuroscience, TU Dresden, Dresden, Germany

Developmental Computational Psychiatry Lab, University Hospital Tübingen, Tübingen, Germany
e-mail: ben.wagner@tuebingen.mpg.de

K. Knauth
Department of Psychology, Biological Psychology, University of Cologne, Cologne, Germany

Department of Psychiatry and Psychotherapy, University of Bonn, Bonn, Germany

J. Peters
Department of Psychology, Biological Psychology, University of Cologne, Cologne, Germany

D. V. Smith et al. (eds.), *Neuroeconomics: Core Topics and Current Directions*,
https://doi.org/10.1007/978-3-032-02925-6_14

for deeper insights into underlying cognitive mechanisms. We conclude with an outlook that highlights potential future directions and open questions in the field.

1 Introduction

1.1 Everyday Examples and Broader Implications

We face a myriad of choices 24/7. The first might arise early in the morning: Should I hit the snooze button for a few extra minutes of sleep, or get up on time to set out for a productive day working on a book chapter? Later, I might have to decide whether to spend my money on that new fancy gadget or put it into my savings plan, and in the evening, I might have to consider whether small immediate "pleasures" like smoking or feasting outweigh the greater long-term benefits of good health. Decisions that involve outcomes at different points in time are referred to as *inter-temporal choices*. Some of these decisions may turn out to be irrelevant, while others may significantly shape our future. Even when the outcomes of our choices are uncertain, our inter-temporal decision-making tendencies, whether long or short term, can significantly affect our lives over time (Kable, 2014; Keidel et al., 2021).

In such situations, humans and many other animals exhibit a tendency toward myopia. That is, when given the choice, we often lean toward the selection of smaller sooner (SS) over larger later (LL) rewards (Kalenscher & Pennartz, 2008). The reasons for this preference may be manifold: the perceived cost of waiting, the more immediate appeal of instant gratification over a larger but delayed reward, and/or the uncertainty associated with future outcomes. Either way, future rewards are discounted in value, referred to as *delay* or *temporal discounting* (TD) (Ainslie, 1975; Frederick et al., 2002; Mazur & Coe, 1987).

TD is relevant on the individual level but may also affect societies at large (Frederick et al., 2002; Golsteyn et al., 2014). The question of how much we invest for the future, how much we consider and anticipate future states and circumstances, appears crucial to many policy areas, including education, health-care, retirement planning, energy, and climate change (Ericson & Laibson, 2018). Given the importance and prevalence of intertemporal decisions, it is unsurprising that they have been extensively researched across various academic disciplines, including psychology, cognitive neuroscience, behavioral economics, and public policy (Frederick et al., 2002; Urminsky & Zauberman, 2015).

1.2 Testing Future Reward Devaluation in the Lab

In experimental studies, intertemporal choice is often evaluated using pen-and-paper questionnaires (Kirby & Maraković, 1996) or computerized tasks, e.g., (Peters & Büchel, 2009), in which individuals make a series of choices between a smaller reward (e.g., money or food), which can be received sooner and a larger reward associated with a delay (e.g., "Would you prefer 30 $ today or 45 $ in 40 days?" (Myerson & Green, 1995; Weinsztok et al., 2021). By varying temporal delays and reward amounts, researchers can measure how much an individual discounts reward value over time. Various models and measures exist to quantify this behavior, the most common being a *discount rate* that describes the steepness in value reduction over time (Scholten et al., 2019) (see methods section).

1.3 Clinical Significance of (Stable) Intertemporal Preference

Inter-temporal preferences are fairly stable within an individual, such that one's discount rate today will likely be relatively stable over time periods ranging from months to years (Bruder et al., 2021; Kirby, 2009; Odum, 2011; Wagner et al., 2023), with a potential genetic component (Anokhin et al., 2015; MacKillop, 2013). TD is associated with numerous health-related behaviors, such as a reduced probability of monitoring blood pressure and cholesterol, texting while driving, seatbelt nonadherence, working out, receiving vaccinations, and adhering to medical treatment (Bickel et al., 2019; Bickel, 2015). TD is also implicated in various mental disorders (Bickel, 2015), in particular disorders along the impulsive-compulsive spectrum such as addictions, making TD a candidate transdiagnostic process (Lempert et al., 2019). Due to its involvement in multiple disorders, this shared behavioral pattern may offer new insights into their common underlying features (Amlung et al., 2019).

1.4 State Effects on Temporal Discounting

Although TD exhibits considerable trait-like stability on individual levels (Gelino et al., 2024), it is also susceptible to within-subject change (Lempert & Phelps, 2016; Lempert et al., 2019), such that several experimental manipulations can lead to short-term changes in TD. These are often seen in experimental settings where trial-specific cues or other manipulations are used to modulate choices within a session, e.g., various kinds of pharmacological agents (e.g., dopaminergic drugs), environmental stimuli, and a variety of visual cues (Dixon et al., 2006; Foerde et al., 2016; Lempert & Phelps, 2016; Miedl et al., 2014; Mathar et al., 2022b, 2023a; Wagner et al., 2020, 2023). We illustrate this literature via three examples: (1) the effects of dopaminergic drugs, (2) the effects of episodic future thinking, and (3) the effects of emotional cues.

Dopamine (DA) is implicated in reward processing, cognitive control, and action selection. Multiple studies have investigated TD in the context of dopaminergic modulation. While one highly cited earlier study reported an increase in discounting under L-DOPA (Pine et al., 2010), other studies reported consistent decreases under pharmacologically elevated DA levels (Wagner et al., 2020; Erfanian Abdoust et al., 2024; Smith et al., 2024; de Wit, 2002) or an absence of effects (Acheson & de Wit,

2008). Thus, the overall evidence suggests that increasing DA neurotransmission, if anything, likely reduces discounting of delayed rewards.

Episodic future thinking (Peters & Büchel, 2010b; Rösch et al., 2022), also referred to as prospection, describes the mental process of imagining or projecting oneself into the future to pre-experience events or situations. TD has been consistently shown to be reduced under conditions of vivid future thinking (Rösch et al., 2022), potentially by increasing the concreteness of an abstract distant future (Lempert et al., 2019). On the neuronal level, this is accompanied by increased activity in pathways associated with the processing of uncertainty, memory, and future imagination (Peters & Büchel, 2010b; Rösch et al., 2022; Cona et al., 2023). However, cues about occurring events could also carry over incidental emotions associated with these events which in turn might affect TD (Liu et al., 2013).

Effects of emotional visual cues on TD have been extensively studied (van den Bergh et al., 2007; Kim & Zauberman, 2013; Knauth & Peters, 2022; Mathar et al., 2023a; Wilson & Daly, 2014). Here, experimental designs usually involve passive or active visual cue processing before or during TD tasks. Early studies often observed increased TD following presentation of highly arousing erotic cues, in particular in men (van den Bergh et al., 2007; Kim & Zauberman, 2013; Wilson & Daly, 2014). However, more recent work has begun to question the universality and magnitude of these effects (Mathar et al., 2023a; Simmank et al., 2015). Studies have raised also doubt regarding previously proposed mechanisms, such as upregulated neuronal reward- (Knauth et al., 2023) or autonomic arousal systems (Knauth & Peters, 2022), which were thought to promote myopic behaviors, such that additional research is needed to capture underlying drivers of emotional or affective stimuli in the above-mentioned decision-making contexts.

Taken together, a better understanding of state effects is crucial to identify and understand situational conditions that might exacerbate short-sighted (impulsive) behaviors. This is especially relevant for psychiatric disorders like substance-related or behavioral addictions. For example, field studies have shown that TD is substantially elevated when individuals suffering from disordered gambling are exposed to gambling venues (Dixon et al., 2006; Wagner et al., 2022). Such cue exposure is often associated with both subjective (Wagner et al., 2022) and psychophysiological cue-reactivity responses, craving (urge for consumption), and an increased relapse probability during abstinence (Robinson & Berridge, 1993; Vafaie & Kober, 2022).

1.5 Training and Intervention Effects

While state effects are important (see previous section), several attempts have been made to achieve longer-lasting changes in TD behavior. For example, EFT can also be structured as a training program, where participants engage in regular sessions of future-oriented thinking (Stein et al., 2016; Sze et al., 2017; Hollis-Hansen et al., 2019). This training aims to fundamentally restructure cognitive processes related to future planning and delay gratification. Working memory training may also improve TD. For example, Bickel et al. (2011) showed that, following a working memory training, discount rates were significantly reduced in individuals undergoing treatment for stimulant use. Similar results were observed by Felton et al. (2019), who examined working memory training in at-risk individuals recruited from a community center serving low-income and homeless individuals. Whether such interventions can lead to enduring changes in the baseline level of reward devaluation, thus influencing trait-like aspects of TD, is an active area of research.

2 Methods

For any psychological construct to be useful, adequate methods and/or models are essential for construct quantification. Intertemporal choice can be quantified using a variety of approaches, including model-agnostic approaches and model-based approaches that specify underlying computational mechanisms. We will start with more straightforward methods and then move on to more recent developments in computational modeling. Readers already familiar with the subject may therefore wish to skip the following more introductory paragraphs.

2.1 Model-Agnostic Measures

One simple but straightforward method is to examine choice proportions of smaller sooner (SS) or larger later (LL) rewards:

$$P(LL) = \frac{n(LL)}{n(SS) + n(LL)} \tag{1}$$

Here, *P(LL)* is the proportion of LL choice, *n(LL)* is the number of times the LL option was selected, and *n(SS)* is the number of times the SS option was chosen.

This metric provides a normalized value that allows for comparisons across different participants, groups, or conditions. One participant may select the LL reward in 80% of trials, whereas another only does so in 20% of trials (Wagner et al., 2022). However, this metric does not offer any insights into how the perceived value of rewards decreases over time, such as whether this decrease follows a linear trend or any other form. It also does not formulate any data generating process that might provide further insights into human cognition.

The area under the curve (AUC) provides a measure of an individual's overall level of TD across the range of delays investigated in a study (Myerson et al., 2001). AUC computation involves two steps. The first step is the computation of so-called *indifference points*, separately for each delay. An indifference point is the point of subjective equivalence in value between SS and LL rewards, which can be

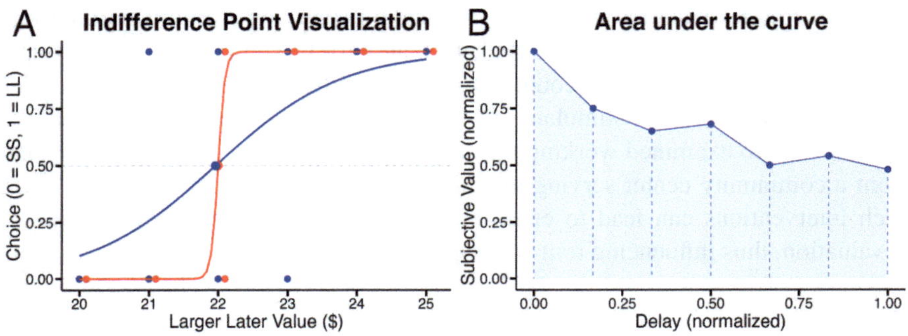

Fig. 1 Indifference points and the area under the curve. (**a**) illustrating the fitting of a logistic regression, a specific type of sigmoid function (Eqs. 2 and 3), to SS and LL choices within one delay. The red sigmoid function, characterized by a nearly stepwise and steep slope, represents a participant with low decision noise, where a clear preference shift from SS to LL is observed. The blue function, with its shallower slope, indicates noisier decision-making, i.e., this participant's choices are inconsistent, and preferences seem to switch non-systematically at the $21, $22, and $23 LL values. However, both participants' ID points are nearly identical. (**b**) Multiple connected indifference points each for a different delay. Note, to get a normalized AUC, the x- (delay) and y-axes (subjective value) are both normalized. The blue-shaded trapezoids are integrated across the whole distance. This area can range between 0 and 1, while areas approximating 0 correspond to steep discounting and areas approximating 1 correspond to shallow discounting

obtained via logistic regression. Consider a scenario with a delay of 12 days, where the immediate SS reward is $20 and the LL rewards range from $21 to $25 (e.g., Fig. 1a). The logistic function models the probability of selecting the LL reward as

$$p(LL) = \frac{1}{1 + b^{-cx}} \tag{2}$$

where x is the LL amount and c is the slope of the sigmoid. Choices are coded as 0 when the SS is chosen and 1 when the LL is chosen (see the red and blue dots as examples for two different participants in Fig. 1a). Here, the indifference point corresponds to the amount x where $P(LL) = 0.5$ (the participant is equally likely to choose SS or LL). The indifference point can thus be obtained by setting $P(LL)$ to 0.5 and solving for x:

$$x = \frac{\log\left(\frac{\left(\frac{1}{0.5} - 1\right)}{b}\right)}{c} \tag{3}$$

This yields the point of subjective equivalence. Having identified the indifference point for each delay, these points can be connected to form an empirical discounting curve (Fig. 1b), reflecting how a participant's valuation of the LL reward changes over time. To yield a normalized measure that ranges between 0 and 1, the indifference points and delays are normalized relative to their maximum values. The AUC then constitutes the area under the normalized empirical discounting curve, which is computed by computed by summing the shaded trapezoids in Fig. 1b (Myerson et al., 2001).

A larger AUC indicates a lower degree of discounting (less impulsive behavior). Conversely, a smaller AUC reflects a steeper discounting (more impulsive behavior). The AUC can be used to compare discounting behavior across different groups of participants or across different experimental conditions. However, it makes no assumptions about the functional form of the relationship between value and time and thus provides a theoretically agnostic way of how value is discounted over time. This model-agnostic nature of the AUC is one of its strengths and a reason why it is still widely used today, despite the availability of more sophisticated methods. However, it is also one of its weaknesses because it does not formulate a process model that allows for prediction and can be tested against empirical data. We will discuss this in more detail in the section on the Model-Based Area Under The Curve.

2.2 *Computational Model-Based Approaches*

Computational models not only fit observed data but can also specify a data-generating process that can be used to predict unobserved data in untested scenarios, allowing the generalizability of specific formulations of the cognitive mechanisms underlying decision-making. Computational models incorporate theoretical frameworks with testable assumptions and can be adapted or extended based on specific research questions. Recent advances in computational modeling allow for an investigation of the underlying processes with increasing sophistication, thereby providing further explanatory power (Farrell & Lewandowsky, 2018).

2.2.1 Exponential Discounting

Originally conceptualized by Samuelson (1937) within the framework of classical utility theory, modeling of TD has come a long way since the first half of the twentieth century. From this early perspective, the utility (value) of a good decreases exponentially over time, such that the subjective value of the LL reward, the SV(LL), is modeled as an exponential function (Fig. 2a):

$$SV\left(LL_t\right) = A_t * e^{-k*D_t} \tag{4}$$

Here, k is a discount factor modeling the steepness of the function, D is the delay, and A is the objective magnitude of the LL reward in trial t (Eq. 4). The exponential model assumes that preferences are time consistent, such that adding a constant delay to both the SS and LL options should not change preferences. However, human and animal behavior often deviates from this assumption. Humans

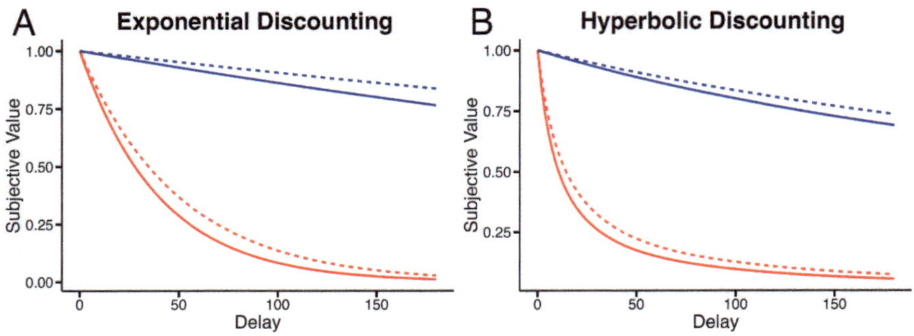

Fig. 2 Exponential and hyperbolic discounting. This figure illustrates (**a**) exponential, and (**b**) hyperbolic discounting curves with blue and red lines representing individuals with shallow and steep discounting rates, respectively. The solid and dashed lines within each color category highlight individual differences in discounting behavior across different contexts, such as varying environmental or psychological conditions (e.g., Wagner et al., 2022). Note how the red exponential function decays at a constant rate, whereas the hyperbolic function places greater weight on shorter delays compared to longer delays

often show a decreasing rate of discounting with increasing delay, contradicting the assumptions of the exponential model, leading to *preference reversals* (Green & Myerson, 1996; Mazur & Biondi, 2009; McKerchar et al., 2009; Green et al., 1994; Peters & Büchel, 2010a).

For example, an individual might prefer an LL reward with a delay of 100 days over an SS reward with a delay of 90 days but switch her preference when the choice is between the same LL reward with a delay of 12 days and an *SS* reward with a delay of 2 days. This phenomenon dates back to observations in pigeons almost 50 years ago (Ainslie, 1974, 1975).

2.2.2 Hyperbolic Discounting

The hyperbolic discounting model (Fig. 2b), in contrast, suggests the rate of discounting decreases as the delay to a reward increases. That is, the hyperbolic model can account for preference reversals and is therefore often the standard model in intertemporal choice (Mazur & Coe, 1987; Peters & Büchel, 2010a). Formally, a simple hyperbolic function describes how value changes as a function of delay (Eq. 5).

$$SV\left(LL_t\right) = \frac{A_t}{1 + k * D_t} \tag{5}$$

Here, A is the objective reward amount of the LL option in trial t. D is the corresponding delay in days and k the hyperbolic discounting rate.

To link Eqs. 4 or 5 to the observed choices, a standard logistic action-selection function (softmax) can be applied to model the probability for choosing the *LL* reward function of the difference between the subjective *SS* and *LL* values (Sutton & Barto, 2018):

$$P\left(LL_t\right) = \frac{e^{(\beta * SV(LLt))}}{e^{(\beta * SV(SSt))} + e^{(\beta * SV(LLt))}} \tag{6}$$

Here, β models choice stochasticity (i.e., decision noise relative to the discounting model). As β increases, choices become more dependent on the subjective values. As β approaches zero, choices become more random relative to the model.

2.2.3 Model-Based Area Under the Curve

While traditional AUC measures do not depend explicitly on a particular underlying discounting model (such as the hyperbolic or exponential model), they nonetheless rest on specific assumptions. For instance, consider an intertemporal choice task with delays set at 1, 2, 3, 10, 20, and 50 days. Computing a standard AUC, one must integrate the area between the normalized indifference points for each delay (see Fig. 1b). Therefore, one usually assumes linear interpolation between these points (as illustrated by the straight lines connecting indifference points in Fig. 1b). In consequence, we implicitly assume the same linear relationship between days 1 and 2 as between days 20 and 50 or any arbitrary consecutive delays. However, such an assumption might misrepresent the processes underlying discounting behavior. Specifically, this linear interpolation may fail to capture established insights about the functional shapes of discounting curves (i.e., hyperbolic) or may introduce bias due to the selection of delays, especially when relatively few data points are available. Another limitation arises from other task design decisions related to delay, such as the overall temporal horizon. For example, a standard AUC calculated using delays ranging from 1 to 30 days cannot be directly compared to an AUC computed with delays spanning from 1 to 3000 days. This is because longer delay intervals inherently produce smaller areas, reflecting greater reward devaluation, further complicating cross-study comparisons of the AUC.

A recent methodological proposal designed to address these limitations and enhance cross-study comparability is the model-based area under the discounting curve (MB-AUC; Gilroy & Hantula, 2018, 2023). This approach combines the intuitive interpretability of an area-based metric with the theoretical foundations of established computational models and introduces a method for standardizing the area over a defined range of delays. To compute the MB-AUC, one defines the maximum undiscounted reward value (A) and both the initial ($T_{i=1}$) and final delays ($T_{i=2}$) of interest. Using, for example, the hyperbolic discounting model, the MB-AUC is analytically derived via the following equations:

$$AUC_{T_i} = \frac{(A * LN\,(1 + k * T_i))}{k} \tag{7}$$

The normalized MB-AUC value across a specific delay range is then determined as:

$$MB - AUC_T = \frac{\left(AUC_{T_1} - AUC_{T_2}\right)}{A * (T_2 - T_1)} \tag{8}$$

This method does provide some kind of standardized scale that can be applied across different discounting models or used with the best-fitting model identified through model comparison (Gilroy & Hantula, 2018). Researchers can also select specific delays of choice, i.e., by setting T_2 to a delay that is available in both or all studies of interest. In doing so, this method allows for robust cross-study comparisons. Moreover, by explicitly incorporating theoretical assumptions about the shape of discounting functions (using the hyperbolic or exponential model), the MB-AUC offers improved interpretability and avoids several pitfalls discussed above, such as issues with linear interpolation and sensitivity to delay selection (Gilroy & Hantula, 2023).

2.2.4 Temporal Discounting Drift Diffusion Models

One problem with the softmax model (see Eq. 6) is that the interpretation of β is not always straightforward, as it can be interpreted as decision noise (value insensitivity), or the trade-off between exploration (switching between options) and exploitation (sticking to one option) (Sutton & Barto, 2018). More generally, β is directly related to model fit, such that lower values of β generally reflect variance in choices not accounted for by the specific discounting model under consideration. The softmax model also does not account for response times (RTs) or processes that likely modulate RTs (Forstmann et al., 2016). A key feature of sequential sampling models, such as the drift diffusion model (DDM) (Ratcliff & Smith, 2004), is that they account for both choices and RT distributions. Such models offer a more nuanced approach by decomposing measures into different underlying computational processes, including motor preparation, response bias, information quality, and the speed-accuracy trade-off (Forstmann et al., 2016). The DDM assumes decisions to arise from the accumulation of noisy evidence that proceeds until a decision threshold is reached. The foundational work for this process traces back to modeling Brownian motion as random fluctuations of physical particles at specific temperatures (Brown, 1828) and was significantly advanced by Norbert

Wiener's mathematical exploration of continuous and noisy stochastic processes (Nelson, 1967). In 1978, Roger Ratcliff built on these findings to propose the first mathematical diffusion model in psychology capable of predicting RT distributions in the context of memory retrieval (Ratcliff, 1978). Interestingly, neural correlates of gradual evidence accumulation in monkeys (Selen et al., 2012; Bichot et al., 2001) and humans (Selen et al., 2012; Shadlen & Kiani, 2013) support some of the DDM's assumptions of how decisions relate to neural activity (Shadlen & Kiani, 2013).

In its simplest form, the standard DDM has four parameters:

1. The drift rate (ν), which models the rate of evidence accumulation.
2. The starting point (ξ), which models a bias toward either response boundary.
3. The boundary separation (α), which captures the distance between the response boundaries and adjusts the speed-accuracy trade-off, i.e., the amount of evidence needed before a boundary is reached.
4. The nondecision time (τ), which captures aspects of the RT related to stimulus encoding (e.g., sensory processing) and motor preparation. That is, τ reflects all aspects of the RT that are unrelated to the evidence accumulation process (see Fig. 3, i.e., on the x-axis, where the blue and red lines begin to diffuse toward the response boundaries).

One recent advancement was the combination of the DDM with standard models, e.g., with reinforcement learning (Fontanesi et al., 2019; Chakroun et al., 2023; Pedersen et al., 2017) or TD models (Peters & D'Esposito, 2020; Wagner et al., 2020; Rodriguez et al., 2015). Here, the boundaries correspond to the selection of specific options (e.g., SS vs. LL options, see Fig. 3 for an illustration). A choice is made as soon as the accumulated evidence crosses one of the boundaries (Peters and D'Esposito 2020).

However, in its standard form, the evidence accumulation process in the DDM is constant across trials (or varies as a function of experimental condition, e.g., due to differences in stimulus quality in the context of perceptual decision-making; see Ratcliff & McKoon, 2008).

The key insight is that decisions should be faster and more accurate as the value difference between options increases. This can be implemented by making the trials-wise drift rate proportional to the value difference between options (as per e.g. Eqs. 9 or 10), combining the strengths of the well-replicated hyperbolic model with the advantages of the DDM. In this combined model, the evidence accumulation process is a function of subjective value differences as per the discounting model. Model comparison has revealed that the relationship between RTs and decision conflict is better captured by a nonlinear (sigmoid) function (Fontanesi et al., 2019) than a linear function of value differences (Peters & D'Esposito, 2020; Wagner et al., 2020, 2022; Mathar et al., 2022a; for an example, see Fig. 4). That is, in the linear mapping scheme (Pedersen et al., 2017), the trial-wise drift rate v_t corresponds to:

$$v_t = v_{\text{coeff}} * (SV(LL_t) - SV(SS_t)) \tag{9}$$

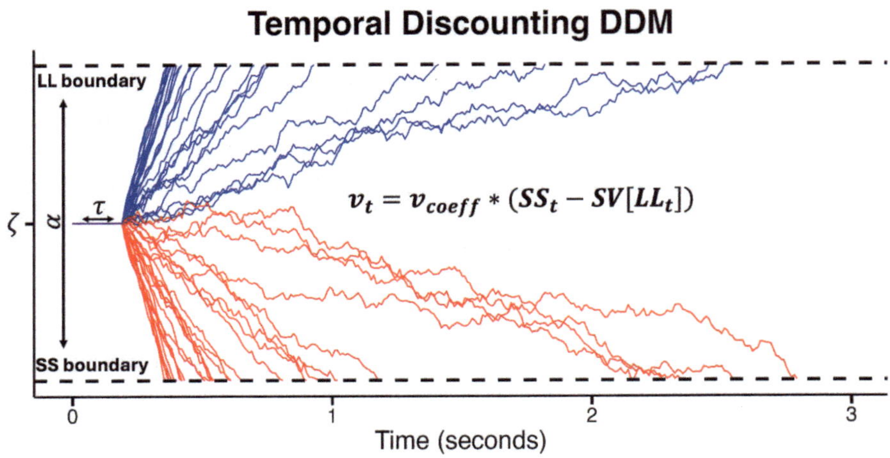

Fig. 3 Temporal discounting drift diffusion model. The x-axis represents time in seconds, where under standard conditions the participant is confronted with a choice at t = 0. The y-axis stretches from the decision boundary for the SS option to the LL option and represents the evidence space captured by the boundary separation parameter α. The starting point parameter ξ captures the starting point and can be interpreted as a bias to either response boundary, notably before any evidence is evaluated. For example, a starting point situated at $\xi = 0.5$ is equidistant between both response boundaries (1/2 α) for SS and LL rewards and can be interpreted as no bias for either option. The time between t = 0 and the initiation of the evidence accumulation process (on the x-axis where both the blue and red walks start to diffuse to the response boundaries) is captured by the nondecision time parameter τ. This parameter captures the time for stimulus encoding and motor preparation. The blue and red random walks illustrate example paths of the evidence accumulation process for SS versus LL options, driven by the rate of evidence accumulation (drift rate, v). In this case, the drift rate v_t is influenced by the linear difference (or sigmoid function) between the trial-wise values of SS and LL options and determines the rate of evidence accumulation as a function of value differences. Models that use a sigmoid mapping typically outperform models using a linear mapping, as demonstrated in studies by Fontanesi et al. (2019) for reinforcement learning and Peters and D'Esposito (2020) and Wagner et al. (2020); Wagner et al. (2022) for intertemporal choice

In the nonlinear mapping scheme (Fontanesi et al., 2019), this scaled difference is additionally passed through a sigmoid:

$$v_t = \frac{2 * (v_{max})}{1 + e^{(-(v_{coeff} * (SV(LL_t) - SV(SS_t))))}} - (v_{max}) \tag{10}$$

Here, v_{max} models the asymptote of the drift rate mapping function. The RT on each trial t is then modeled as the Wiener first passage time[*wfpt*] with parameters boundary separation (α), nondecision time (τ), bias (z), and trial-wise drift rate (v_t):

$$RT_t \sim wfpt\,(\alpha, \tau, z, v_t) \tag{11}$$

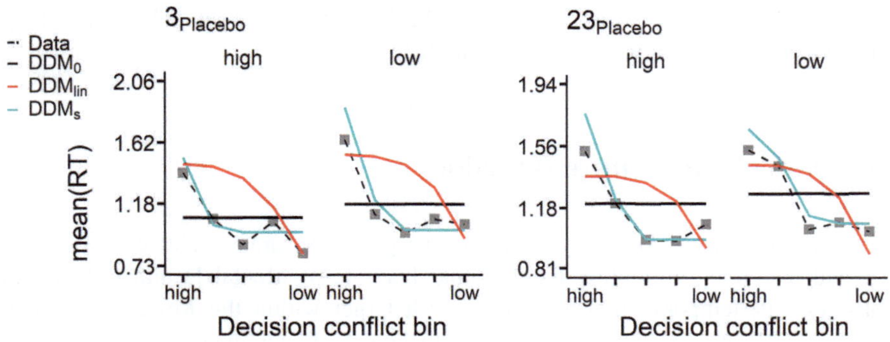

Fig. 4 Posterior predictive checks for two example participants (3 and 23) from the placebo group in Wagner et al. (2020). Shown are mean observed RTs (data; dotted lines) and model-predicted RTs from 10,000 simulations from the posterior distribution of three temporal discounting DDMs: a model without value information (DDM_0, black), a model where the drift rate is modeled as a linear function of scaled value differences (DDM_{lin}, red; see Eq. 9), and a model where value differences map on the drift rate through a sigmoid (DDM_s, light blue; see Eq. 10). Left and right facets show the high and low magnitude task conditions, respectively (for details, see Wagner et al., 2020). Trials were binned into five equal-sized bins based on decision conflict, defined as the absolute difference between subjective LL and SS values. Lower value differences (i.e., higher conflict) appear on the left side of each facet. Note that while the DDM_0 predicts mean RTs independent of decision difficulty, both the DDM_{lin} and DDM_s capture the pattern that RTs increase when the subjective values of LL and SS are more similar, and this pattern is much better accounted for by the DDM_s. (Figure adapted from Wagner et al. 2020)

Note that one can easily adapt the other DDM parameters to change as a function of various variables of interest. For instance, we can implement a model where the boundary separation changes as a function of task progression, assuming that participants lower their threshold as they become more experienced with the task (Fontanesi et al., 2019). This combined TD DDM can offer new insights into underlying processes.

For example, one study found that the DA D2 antagonist amisulpride may act differently on different subcomponents of the decision process (Soutschek & Tobler, 2023). Another study found that an increase in DA neurotransmission was associated with both a decrease in nondecision time (τ) and a decrease in the discount rate, supporting the idea that DA may impact upon both motor- and non-motor aspects of behavior in TD tasks (Wagner et al., 2020). A study using cue reactivity and erotic stimuli showed that participants' bias (ξ) toward SS rewards was increased following cue exposure (Mathar et al., 2023b). Notably, additional modeling work revealed that this bias effect was most pronounced directly after cue exposure, and subsequently declined. Furthermore, the DA precursor tyrosine reduced the boundary separation (α) (Mathar, Abdoust, et al., 2022a), consistent with some DA effects in reinforcement learning (Chakroun et al., 2023). In summary, the integration of TD models and DDMs not only provides a good account of both choices and RTs but furthermore allows for a more fine-grained analysis of underlying processes (Miletić et al., 2020). This may be a valuable tool to help understand pharmacological and clinical effects in TD.

3 Outlook and Future Directions

Extreme degrees of TD (e.g., very steep or very shallow) are associated with several mental disorders, prompting the National Institute of Mental Health (NIMH) to propose intertemporal choice as a research target within the Research Domain Criteria (RDOC), as part of its long-term goal to better understand the overall processes that go awry in mental illness (Lempert et al., 2019; Lempert & Phelps, 2016; Bickel et al., 2019). Steep discounting is consistently associated with linked to behavioral addictions like disordered gambling (Reynolds, 2006; Wiehler et al., 2015), substance use disorders (Bickel et al., 2014; Ciccarelli et al., 2019) likely increases with further comorbidities (Moody et al., 2016), and is often detected in major depressive disorder (Pulcu et al., 2013; Alloy et al., 2016), ADHD (Jackson & MacKillop, 2016), and other mental disorders (Steinglass et al., 2017; Sohn et al., 2014; Lempert et al., 2019).

Here, Bayesian perspectives on intertemporal choice may also provide novel insights. Gershman and Bhui (2020) recently revisited the idea that discounting of future rewards may be linked to the uncertainty associated with future outcomes. In a Bayesian view, the precision with which future rewards can be represented may be inversely related to their delay. Preferences for SS rewards may therefore arise from increased simulation noise associated with LL rewards (Gershman & Bhui, 2020), a finding that fits well with the modulating effects of EFT discussed above (Peters & Büchel, 2010b; Brands et al., 2024; Rösch et al., 2022). EFT may counteract this simulation noise by increasing the boundary separation in a TD-DDM (Brands et al., 2024), again highlighting the utility of adopting an evidence accumulation perspective.

More generally, several mental disorders are associated with a loss of precision in mental simulation (Friston, 2023), i.e., the inability to integrate uncertainty in one's predictions about the future. For instance, depression and anxiety involve uncertainty and negative expectations about the future (Clark et al., 2018). Here, steep discounting might reflect a rational response to anticipated instability in the environment. Similarly, in substance use disorders, high levels of uncertainty about future rewards may drive individuals toward immediate gratification, akin to a coping mechanism (Dingle et al., 2018; Kober, 2014). TD may be related to both internal and external sources of uncertainty (Bulley & Schacter, 2020). Unpredictable events in the outside world may favor impulsive choice, as shown in a recent simulation study (Burk & Averbeck, 2023). Factoring in both internal uncertainty (e.g., noise in value representations) and external uncertainty (e.g., unpredictable environmental conditions; socioeconomic status) might therefore provide novel insights into links between TD, future simulation (Gershman & Bhui, 2020; Bulley and Schacter), and mental disorders.

References

Acheson, A., & de Wit, H. (2008). Bupropion improves attention but does not affect impulsive behavior in healthy young adults. *Experimental and Clinical Psychopharmacology, 16*(2), 113–123. https://doi.org/10.1037/1064-1297.16.2.113

Ainslie, G. W. (1974). Impulse control in pigeons. *Journal of the Experimental Analysis of Behavior, 21*(3), 485–489. https://doi.org/10.1901/jeab.1974.21-485

Ainslie, G. (1975). Specious reward: A behavioral theory of impulsiveness and impulse control. *Psychological Bulletin, 82*(4), 463–496. https://doi.org/10.1037/h0076860

Alloy, L. B., Olino, T., Freed, R. D., & Nusslock, R. (2016). Role of reward sensitivity and processing in major depressive and bipolar Spectrum disorders. *Behavior Therapy, 47*(5), 600–621. https://doi.org/10.1016/j.beth.2016.02.014

Amlung, M., Marsden, E., Holshausen, K., Morris, V., Patel, H., Vedelago, L., et al. (2019). Delay discounting as a transdiagnostic process in psychiatric disorders: A meta-analysis. *JAMA Psychiatry, 76*(11), 1176–1186. https://doi.org/10.1001/jamapsychiatry.2019.2102

Anokhin, A. P., Grant, J. D., Mulligan, R. C., & Heath, A. C. (2015). The genetics of impulsivity: Evidence for the heritability of delay discounting. *Biological Psychiatry, 77*(10), 887–894. https://doi.org/10.1016/j.biopsych.2014.10.022

Bichot, N. P., Rao, S. C., & Schall, J. D. (2001). Continuous processing in macaque frontal cortex during visual search. *Neuropsychologia, 39*(9), 972–982. https://doi.org/10.1016/S0028-3932(01)00022-7

Bickel, W. K. (2015). Discounting of delayed rewards as an endophenotype. *Biological Psychiatry, 77*(10), 846–847. https://doi.org/10.1016/j.biopsych.2015.03.003

Bickel, W. K., Yi, R., Landes, R. D., Hill, P. F., & Baxter, C. (2011). Remember the future: Working memory training decreases delay discounting among stimulant addicts. *Biological Psychiatry, 69*(3), 260–265. https://doi.org/10.1016/j.biopsych.2010.08.017

Bickel, W. K., Koffarnus, M. N., Moody, L., & Wilson, A. G. (2014). The behavioral- and neuro-economic process of temporal discounting: A candidate behavioral marker of addiction. *Neuropharmacology, 76*(Pt B), 518–527. https://doi.org/10.1016/j.neuropharm.2013.06.013

Bickel, W. K., Athamneh, L. N., Basso, J. C., Mellis, A. M., DeHart, W. B., Craft, W. H., & Pope, D. (2019). Excessive discounting of delayed reinforcers as a trans-disease process: Update on the state of the science. *Current Opinion in Psychology, 30*, 59–64. https://doi.org/10.1016/j.copsyc.2019.01.005

Brands, A., Shebotinova, E., Herbertz, A., Knauth, K., Marteau, A., & Peters, J. (2024). *Future and past episodic cues reduce temporal discounting and increase control allocation during inter-temporal choice*. Center for Open Science.

Brown, R. (1828). *XXVII. A brief account of microscopical observations made in the months of June, July and August 1827, on the particles contained in the pollen of plants; and* ... Available online at https://www.tandfonline.com/doi/pdf/10.1080/14786442808674769?casa_token=agrmy1gnilaaaaaa:zq5uh8s2tqcnhg3izncevng5iw8gsf0w2icxyux6on57o00ljw0wia2eejv_ecxptpkvsfowvftpng

Bruder, L. R., Scharer, L., & Peters, J. (2021). Reliability assessment of temporal discounting measures in virtual reality environments. *Science Reports, 11*(1), 1–16. https://doi.org/10.1038/s41598-021-86388-8

Bulley, A., & Schacter, D. L. (2020). Deliberating trade-offs with the future. *Nature Human Behaviour, 4*(3), 238–247. https://doi.org/10.1038/s41562-020-0834-9

Burk, D. C., & Averbeck, B. B. (2023). Environmental uncertainty and the advantage of impulsive choice strategies. *PLoS Computational Biology, 19*(1), e1010873. https://doi.org/10.1371/journal.pcbi.1010873

Chakroun, K., Wiehler, A., Wagner, B. J., Mathar, D., Ganzer, F., et al. (2023). Dopamine regulates decision thresholds in human reinforcement learning in males. *Nature Communications, 14*(1), 1–14. https://doi.org/10.1038/s41467-023-41130-y

Ciccarelli, M., Cosenza, M., D'Olimpio, F., Griffiths, M. D., & Nigro, G. (2019). An experimental investigation of the role of delay discounting and craving in gambling chasing behavior. *Addictive Behaviors, 93*, 250–256. https://doi.org/10.1016/j.addbeh.2019.02.002

Clark, J. E., Watson, S., & Friston, K. J. (2018). What is mood? A computational perspective. *Psychological Medicine, 48*(14), 2277–2284. https://doi.org/10.1017/s0033291718000430

Cona, G., Santacesaria, P., & Scarpazza, C. (2023). Envisioning the future: An ALE meta-analysis on neural correlates of future thinking, prospective memory and delay discounting. *Neuroscience & Biobehavioral Reviews, 0149–7634*(153), 105355. https://doi.org/10.1016/j.neubiorev.2023.105355

de Wit, H. (2002). Acute administration of d-amphetamine decreases impulsivity in healthy volunteers. *Neuropsychopharmacology, 27*(5), 813–825. https://doi.org/10.1016/S0893-133X(02)00343-3

Dingle, G. A., Da Neves, D. C., Alhadad, S. S. J., & Hides, L. (2018). Individual and interpersonal emotion regulation among adults with substance use disorders and matched controls. *The British Journal of Clinical Psychology, 57*(2), 186–202. https://doi.org/10.1111/bjc.12168

Dixon, M. R., Jacobs, E. A., & Sanders, S. (2006). Contextual control of delay discounting by pathological gamblers. *Journal of Applied Behavior Analysis, 39*(4), 413–422. https://doi.org/10.1901/jaba.2006.173-05

Erfanian Abdoust, M., Froböse, M. I., Schnitzler, A., Schreivogel, E., & Jocham, G. (2024). Dopamine and acetylcholine have distinct roles in delay- and effort-based decision-making in humans. *PLoS Biology, 22*(7), e3002714. https://doi.org/10.1371/journal.pbio.3002714

Ericson, K. M., & Laibson, D. (2018). *Intertemporal choice*. National Bureau of Economic Research.

Farrell, S., & Lewandowsky, S. (2018). *Computational modeling of cognition and behavior*. Cambridge University Press.

Felton, J. W., Collado, A., Ingram, K. M., Doran, K., & Yi, R. (2019). Improvement of working memory is a mechanism for reductions in delay discounting among mid-age individuals in an urban medically underserved area. *Annals of Behavioral Medicine, 53*(11), 988–998. https://doi.org/10.1093/abm/kaz010

Foerde, K., Figner, B., Doll, B. B., Woyke, I. C., Braun, E. K., Weber, E. U., & Shohamy, D. (2016). Dopamine modulation of intertemporal decision-making: Evidence from Parkinson disease. *Journal of Cognitive Neuroscience, 28*(5), 657–667. https://doi.org/10.1162/jocn_a_00929

Fontanesi, L., Gluth, S., Spektor, M. S., & Rieskamp, J. (2019). A reinforcement learning diffusion decision model for value-based decisions. *Psychonomic Bulletin & Review, 26*(4), 1099–1121. https://doi.org/10.3758/s13423-018-1554-2

Forstmann, B. U., Ratcliff, R., & Wagenmakers, E.-J. (2016). Sequential sampling models in cognitive neuroscience: Advantages, applications, and extensions. *Annual Review of Psychology, 67*, 641–666. https://doi.org/10.1146/annurev-psych-122414-033645

Frederick, S., Loewenstein, G., & O'Donoghue, T. (2002). Time discounting and time preference: A critical review. *Journal of Economic Literature, 40*(2), 351–401. https://doi.org/10.1257/002205102320161311

Friston, K. (2023). Computational psychiatry: From synapses to sentience. *Molecular Psychiatry, 28*(1), 256–268. https://doi.org/10.1038/s41380-022-01743-z

Gelino, B. W., Schlitzer, R. D., Reed, D. D., & Strickland, J. C. (2024). A systematic review and meta-analysis of test–retest reliability and stability of delay and probability discounting. *Journal of the Experimental Analysis of Behavior, 121*(3), 358–372. https://doi.org/10.1002/jeab.910

Gershman, S. J., & Bhui, R. (2020). Rationally inattentive intertemporal choice. *Nature Communications, 11*(1), 3365. https://doi.org/10.1038/s41467-020-16852-y

Gilroy, S. P., & Hantula, D. A. (2018). Discounting model selection with area-based measures: A case for numerical integration. *Journal of the Experimental Analysis of Behavior, 109*(2), 433–449. https://doi.org/10.1002/jeab.318

Gilroy, S. P., & Hantula, D. A. (2023). Exact solutions for area-based delay discounting analyses. *Experimental and Clinical Psychopharmacology, 31*(4), 849–860. https://doi.org/10.1037/pha0000638

Golsteyn, B. H. H., Grönqvist, H., & Lindahl, L. (2014). Adolescent time preferences predict lifetime outcomes. *The Economic Journal, 124*(580), F739–F761. https://doi.org/10.1111/ecoj.12095

Green, L., & Myerson, J. (1996). Exponential versus hyperbolic discounting of delayed outcomes: Risk and waiting time. *Integrative and Comparative Biology, 36*(4), 496–505. https://doi.org/10.1093/icb/36.4.496

Green, L., Fristoe, N., & Myerson, J. (1994). Temporal discounting and preference reversals in choice between delayed outcomes. *Psychonomic Bulletin & Review, 1*(3), 383–389. https://doi.org/10.3758/BF03213979

Hollis-Hansen, K., O'Donnell, S. E., Seidman, J. S., Brande, S. J., & Epstein, L. H. (2019). Improvements in episodic future thinking methodology: Establishing a standardized episodic thinking control. *PLoS One, 14*(3), e0214397. https://doi.org/10.1371/journal.pone.0214397

Jackson, J. N. S., & MacKillop, J. (2016). Attention-deficit/hyperactivity disorder and monetary delay discounting: A meta-analysis of case-control studies. *Biological Psychiatry: Cognitive Neuroscience and Neuroimaging, 1*(4), 316–325. https://doi.org/10.1016/J.BPSC.2016.01.007

Kable, J. W. (2014). Valuation, intertemporal choice, and self-control. *Neuroeconomics*, 173–192. https://doi.org/10.1016/B978-0-12-416008-8.00010-3

Kalenscher, T., & Pennartz, C. (2008). Is a bird in the hand worth two in the future? The neuroeconomics of intertemporal decision-making. *Progress in Neurobiology, 84*(3), 284–315. https://doi.org/10.1016/j.pneurobio.2007.11.004

Keidel, K., Rramani, Q., Weber, B., Murawski, C., & Ettinger, U. (2021). Individual differences in intertemporal choice. *Frontiers in Psychology, 12*. https://doi.org/10.3389/fpsyg.2021.643670

Kim, B. K., & Zauberman, G. (2013). Can Victoria's secret change the future? A subjective time perception account of sexual-cue effects on impatience. *Journal of Experimental Psychology: General, 142*(2), 328–335. https://doi.org/10.1037/a0028954

Kirby, K. N. (2009). One-year temporal stability of delay-discount rates. *Psychonomic Bulletin & Review, 16*(3), 457–462. https://doi.org/10.3758/PBR.16.3.457

Kirby, K. N., & Maraković, N. N. (1996). Delay-discounting probabilistic rewards: Rates decrease as amounts increase. *Psychonomic Bulletin & Review, 3*(1), 100–104. https://doi.org/10.3758/BF03210748

Knauth, K., & Peters, J. (2022). Trial-wise exposure to visual emotional cues increases physiological arousal but not temporal discounting. *Psychophysiology, 59*(4), e13996. https://doi.org/10.1111/psyp.13996

Knauth, K., Mathar, D., Kuzmanovic, B., Tittgemeyer, M., & Peters, J. (2023). Erotic cue exposure increases neural reward responses without modulating temporal discounting. *Imaging Neuroscience, 1*, 1–25. https://doi.org/10.1162/imag_a_00008

Kober, H. (2014). *Emotion regulation in substance use disorders*. The Guilford Press. Available online at http://www.gruberpeplab.com/teaching/psych3303_spring2019/documents/6.2_koberboling2014.pdf

Lempert, K. M., & Phelps, E. A. (2016). The malleability of intertemporal choice. *Trends in Cognitive Sciences, 20*(1), 64–74. https://doi.org/10.1016/j.tics.2015.09.005

Lempert, K. M., Steinglass, J. E., Pinto, A., Kable, J. W., & Simpson, H. B. (2019). Can delay discounting deliver on the promise of RDoC? *Psychological Medicine, 49*(2), 190–199. https://doi.org/10.1017/S0033291718001770

Liu, L., Feng, T., Chen, J., & Li, H. (2013). The value of emotion: How does episodic prospection modulate delay discounting? *PLoS One, 8*(11), e81717. https://doi.org/10.1371/journal.pone.0081717

MacKillop, J. (2013). Integrating behavioral economics and behavioral genetics: Delayed reward discounting as an endophenotype for addictive disorders. *Journal of the Experimental Analysis of Behavior, 99*(1), 14–31. https://doi.org/10.1002/jeab.4

Mathar, D., Abdoust, M. E., Marrenbach, T., Tuzsus, D., & Peters, J. (2022a). The catecholamine precursor tyrosine reduces autonomic arousal and decreases decision thresholds in reinforcement learning and temporal discounting. *PLoS Computational Biology, 18*(12), e1010785. https://doi.org/10.1371/journal.pcbi.1010785

Mathar, D., Erfanian Abdoust, M., Marrenbach, T., Tuzsus, D., & Peters, J. (2022b). The catecholamine precursor tyrosine reduces autonomic arousal and decreases decision thresholds in reinforcement learning and temporal discounting. *PLoS Computational Biology, 18*(12), e1010785. https://doi.org/10.1371/journal.pcbi.1010785

Mathar, D., Wiebe, A., Tuzsus, D., Knauth, K., & Peters, J. (2023a). Erotic cue exposure increases physiological arousal, biases choices toward immediate rewards, and attenuates model-based reinforcement learning. *Psychophysiology, 60*(12), e14381. https://doi.org/10.1111/psyp.14381

Mathar, D., Wiebe, A., Tuzsus, D., Knauth, K., & Peters, J. (2023b). Erotic cue exposure increases physiological arousal, biases choices toward immediate rewards, and attenuates model-based reinforcement learning. *Psychophysiology, 60*(12). https://doi.org/10.1111/psyp.14381

Mazur, J. E., & Biondi, D. R. (2009). Delay-amount tradeoffs in choices by pigeons and rats: Hyperbolic versus exponential discounting. *Journal of the Experimental Analysis of Behavior, 91*(2), 197–211. https://doi.org/10.1901/jeab.2009.91-197

Mazur, J. E., & Coe, D. (1987). Tests of transitivity in choices between fixed and variable reinforcer delays. *Journal of the Experimental Analysis of Behavior, 47*(3), 287–297. https://doi.org/10.1901/jeab.1987.47-287

McKerchar, T. L., Green, L., Myerson, J., Pickford, T. S., Hill, J. C., & Stout, S. C. (2009). A comparison of four models of delay discounting in humans. *Behavioural Processes, 81*(2), 256–259. https://doi.org/10.1016/j.beproc.2008.12.017

Miedl, S. F., Büchel, C., & Peters, J. (2014). Cue-induced craving increases impulsivity via changes in striatal value signals in problem gamblers. *The Journal of Neuroscience, 34*(13), 4750–4755. https://doi.org/10.1523/JNEUROSCI.5020-13.2014

Miletić, S., Boag, R. J., & Forstmann, B. U. (2020). Mutual benefits: Combining reinforcement learning with sequential sampling models. *Neuropsychologia, 136*, 107261. https://doi.org/10.1016/j.neuropsychologia.2019.107261

Moody, L., Franck, C., & Bickel, W. K. (2016). Comorbid depression, antisocial personality, and substance dependence: Relationship with delay discounting. *Drug and Alcohol Dependence, 160*, 190–196. https://doi.org/10.1016/j.drugalcdep.2016.01.009

Myerson, J., & Green, L. (1995). Discounting of delayed rewards: Models of individual choice. *Journal of the Experimental Analysis of Behavior, 64*(3), 263–276. https://doi.org/10.1901/jeab.1995.64-263

Myerson, J., Green, L., & Warusawitharana, M. (2001). Area under the curve as a measure of discounting. *Journal of the Experimental Analysis of Behavior, 76*(2), 235–243. https://doi.org/10.1901/jeab.2001.76-235

Nelson, E. (1967). *Dynamical theories of Brownian motion* (Mathematical notes, 106). Princeton University Press.

Odum, A. L. (2011). Delay discounting: I'm a k, you're a k. *Journal of the Experimental Analysis of Behavior, 96*(3), 427–439. https://doi.org/10.1901/jeab.2011.96-423

Pedersen, M. L., Frank, M. J., & Biele, G. (2017). The drift diffusion model as the choice rule in reinforcement learning. *Psychonomic Bulletin & Review, 24*(4), 1234–1251. https://doi.org/10.3758/s13423-016-1199-y

Peters, J., & Büchel, C. (2009). Overlapping and distinct neural systems code for subjective value during intertemporal and risky decision making. *International Journal of Neuroscience, 29*(50), 15727–15734. https://doi.org/10.1523/JNEUROSCI.3489-09.2009

Peters, J., & Büchel, C. (2010a). Neural representations of subjective reward value. *Behavioural Brain Research, 213*(2), 135–141. https://doi.org/10.1016/j.bbr.2010.04.031

Peters, J., & Büchel, C. (2010b). Episodic future thinking reduces reward delay discounting through an enhancement of prefrontal-mediotemporal interactions. *Neuron, 66*(1), 138–148. https://doi.org/10.1016/j.neuron.2010.03.026

Peters, J., & D'Esposito, M. (2020). The drift diffusion model as the choice rule in inter-temporal and risky choice: A case study in medial orbitofrontal cortex lesion patients and controls. *PLoS Computational Biology, 16*(4), e1007615. https://doi.org/10.1371/journal.pcbi.1007615
Pine, A., Shiner, T., Seymour, B., & Dolan, R. J. (2010). Dopamine, time, and impulsivity in humans. *The Journal of Neuroscience, 30*(26), 8888–8896. https://doi.org/10.1523/JNEUROSCI.6028-09.2010
Pulcu, E., Trotter, P. D., Thomas, E. J., McFarquhar, M., Juhasz, G., Sahakian, B. J., et al. (2013). Temporal discounting in major depressive disorder. *Psychological Medicine, 44*(9), 1825–1834. https://doi.org/10.1017/s0033291713002584
Ratcliff, R. (1978). A theory of memory retrieval. *Psychological Review, 85*(2), 59–108. https://doi.org/10.1037/0033-295x.85.2.59
Ratcliff, R., & Smith, P. L. (2004). A comparison of sequential sampling models for two-choice reaction time. *Psychological Review, 111*(2), 333–367. https://doi.org/10.1037/0033-295X.111.2.333
Ratcliff, R., & McKoon, G. (2008). The diffusion decision model: theory and data for two-choice decision tasks. *Neural Computing, 20*(4), 873–922. https://doi.org/10.1162/neco.2008.12-06-420
Reynolds, B. (2006). A review of delay-discounting research with humans: Relations to drug use and gambling. *Behavioural Pharmacology, 17*(8), 651–667. https://doi.org/10.1097/FBP.0b013e3280115f99
Robinson, T., & Berridge, K. C. (1993). The neural basis of drug craving: An incentive-sensitization theory of addiction. *Brain Research Reviews, 18*(3), 247–291. https://doi.org/10.1016/0165-0173(93)90013-P
Rodriguez, C. A., Turner, B. M., van Zandt, T., & McClure, S. M. (2015). The neural basis of value accumulation in intertemporal choice. *The European Journal of Neuroscience, 42*(5), 2179–2189. https://doi.org/10.1111/ejn.12997
Rösch, S. A., Stramaccia, D. F., & Benoit, R. G. (2022). Promoting farsighted decisions via episodic future thinking: A meta-analysis. *Journal of Experimental Psychology. General, 151*(7). https://doi.org/10.1037/xge0001148
Samuelson, P. A. (1937). A note on measurement of utility. *The Review of Economic Studies, 4*(2), 155. https://doi.org/10.2307/2967612
Scholten, H., Scheres, A., de Water, E., Graf, U., Granic, I., & Luijten, M. (2019). Behavioral trainings and manipulations to reduce delay discounting: A systematic review. *Psychonomic Bulletin & Review, 26*(6), 1803–1849. https://doi.org/10.3758/s13423-019-01629-2
Selen, L. P. J., Shadlen, M. N., & Wolpert, D. M. (2012). Deliberation in the motor system: Reflex gains track evolving evidence leading to a decision. *International Journal of Neuroscience, 32*(7), 2276–2286. https://doi.org/10.1523/JNEUROSCI.5273-11.2012
Shadlen, M. N., & Kiani, R. (2013). Decision making as a window on cognition. *Neuron, 80*(3), 791–806. https://doi.org/10.1016/j.neuron.2013.10.047
Simmank, J., Murawski, C., Bode, S., & Horstmann, A. (2015). Incidental rewarding cues influence economic decisions in people with obesity. *Frontiers in Behavioral Neuroscience, 9*. https://doi.org/10.3389/fnbeh.2015.00278
Smith, E., Theis, H., van Eimeren, T., Knauth, K., Tuzsus, D., Zhang, L., et al. (2024). Dopamine and temporal discounting: Revisiting pharmacology and individual differences. *bioRxiv*, 2024–2008.
Sohn, S. Y., Kang, J. I., Namkoong, K., & Kim, S. J. (2014). Multidimensional measures of impulsivity in obsessive-compulsive disorder: Cannot wait and stop. *PLoS One, 9*(11), e111739. https://doi.org/10.1371/journal.pone.0111739
Soutschek, A., & Tobler, P. N. (2023). A process model account of the role of dopamine in intertemporal choice. *eLife, 12*. https://doi.org/10.7554/eLife.83734
Stein, J. S., Wilson, A. G., Koffarnus, M. N., Daniel, T. O., Epstein, L. H., & Bickel, W. K. (2016). Unstuck in time: Episodic future thinking reduces delay discounting and cigarette smoking. *Psychopharmacology, 233*(21–22), 3771–3778. https://doi.org/10.1007/s00213-016-4410-y

Steinglass, J. E., Lempert, K. M., Choo, T.-H., Kimeldorf, M. B., Wall, M., Walsh, B. T., et al. (2017). Temporal discounting across three psychiatric disorders: Anorexia nervosa, obsessive compulsive disorder, and social anxiety disorder. *Depression and Anxiety, 34*(5), 463–470. https://doi.org/10.1002/da.22586

Sutton, R. S., & Barto, A. (2018). *Reinforcement learning. An introduction* (Adaptive computation and machine learning) (2nd ed.). The MIT Press.

Sze, Y. Y., Stein, J. S., Bickel, W. K., Paluch, R. A., & Epstein, L. H. (2017). Bleak present, bright future: Online episodic future thinking, scarcity, delay discounting, and food demand. *Clinical Psychological Science, 5*(4), 683–697. https://doi.org/10.1177/2167702617696511

Urminsky, O., & Zauberman, G. (2015). The psychology of intertemporal preferences. In *The Wiley Blackwell handbook of judgment and decision making* (pp. 141–181). Wiley.

Vafaie, N., & Kober, H. (2022). Association of drug cues and craving with drug use and relapse: A systematic review and meta-analysis. *JAMA Psychiatry, 79*(7), 641–650. https://doi.org/10.1001/jamapsychiatry.2022.1240

van den Bergh, B., Dewitte, S., & Warlop, L. (2007). Bikinis instigate generalized impatience in intertemporal choice. *SSRN Journal*. https://doi.org/10.2139/ssrn.1094711

Wagner, B. J., Clos, M., Sommer, T., & Peters, J. (2020). Dopaminergic modulation of human intertemporal choice: A diffusion model analysis using the D2-receptor antagonist haloperidol. *The Journal of Neuroscience, 40*(41), 7936–7948. https://doi.org/10.1523/JNEUROSCI.0592-20.2020

Wagner, B. J., Mathar, D., & Peters, J. (2022). Gambling environment exposure increases temporal discounting but improves model-based control in regular slot-machine gamblers. *Computational Psychiatry, 6*(1). https://doi.org/10.5334/cpsy.84

Wagner, B. J., Schüller, C. B., Schüller, T., Baldermann, J. C., Kohl, S., Visser-Vandewalle, V., et al. (2023). Chronic deep brain stimulation of the human nucleus Accumbens region disrupts the stability of intertemporal preferences. *The Journal of Neuroscience, 43*(43), 7175–7185. https://doi.org/10.1523/JNEUROSCI.0138-23.2023

Weinsztok, S., Brassard, S., Balodis, I., Martin, L. E., & Amlung, M. (2021). Delay discounting in established and proposed Behavioral addictions: A systematic review and meta-analysis. *Frontiers in Behavioral Neuroscience, 15*. https://doi.org/10.3389/fnbeh.2021.786358

Wiehler, A., Bromberg, U., & Peters, J. (2015). The role of prospection in steep temporal reward discounting in gambling addiction. *Frontiers in Psychiatry, 6*, 112. https://doi.org/10.3389/fpsyt.2015.00112

Wilson, M., & Daly, M. (2014). Do pretty women inspire men to discount the future? *Proceedings of the Royal Society of London. Series B: Biological Sciences*. https://doi.org/10.1098/rsbl.2003.0134

Intertemporal Choices Are Influenced by Many Things Beyond Time Itself

Todd A. Hare

Abstract Intertemporal choices balance immediate rewards against future outcomes across various life domains. This chapter explores temporal discounting, where future rewards are devalued compared to immediate ones. This chapter examines the usefulness and shortcomings of common temporal discounting functions. It also investigates how attribute-based models, which consider specific features of rewards complement process-tracing methods and other neuroeconomic tools. I discuss how the clarity and certainty of future rewards, along with individual differences in time perception, influence intertemporal choices. I also discuss the role of self-control in intertemporal choices, questioning whether selecting a smaller, sooner reward necessarily indicates a lack of self-control. The chapter highlights recent work on the psychology and neurobiology of intertemporal choice, highlighting its complexity and the need for nuanced interpretations.

Intertemporal tradeoffs are a pervasive aspect of decision-making in daily life, influencing choices that span across domains including finance, health, climate change, and personal relationships. These tradeoffs require decision makers to find their preferred balance between short-term gratification and long-term outcomes—be they benefits or negative consequences. Given their ubiquity and direct connections to many types of well-being, intertemporal choices have been investigated and considered in the sciences and philosophy for millennia (Cohen et al., 2020; Kable, 2014; Lempert & Phelps, 2016). This chapter focuses on recent advances in neuroeconomics and related disciplines, highlighting unresolved questions.

T. A. Hare (✉)
Department of Economics, University of Zurich, Zurich, Switzerland
e-mail: todd.hare@econ.uzh.ch

D. V. Smith et al. (eds.), *Neuroeconomics: Core Topics and Current Directions*,
https://doi.org/10.1007/978-3-032-02925-6_15

1 Temporal Discounting Functions

Humans and nonhuman animals prioritize immediate rewards over delayed ones. Future rewards are discounted compared to immediate ones (see also other chapters by Wagner et al., this volume; Lempert, this volume, and Su & Lockwood, this volume). Temporal discounting makes sense as the future is uncertain, and normative decision theories indicate that decision makers should discount future rewards. Samuelson (1937) proposed an exponentially decreasing function (Eq. 1) as one possible functional form for temporal discounting, and later work (Fishburn & Rubinstein, 1982; Strotz, 1955) showed this form is consistent with core assumptions in economics and other decision sciences. Specifically, it is consistent with the assumption that decision makers maximize a consistent utiliity function, and thus choices reveal preferences. However, there is now a long history of empirical studies showing that humans and other animals don't choose according to exponential discounting when faced with intertemporal choices (Ainslie & Herrnstein, 1981; Cohen et al., 2020; Kable, 2014; Mazur, 1987).

Equations for common temporal discounting functions	
Eq. 1: $DU(x,t) = \delta^t U(x)$ with $0 \leq \delta \leq 1$	Exponential discounting
Eq. 2: $DU\,(x,t) = \frac{U(x)}{1+kt}$ with $k \geq 0$	Hyperbolic discounting
Eq. 3: $DU(x,t) = \beta\delta^t U(x)$ with $0 \leq \beta \leq 1$ for all $t > 0$, and $0 \leq \delta \leq 1$	Quasi-hyperbolic discounting

Intertemporal choices are better explained by hyperbolic or quasi-hyperbolic discounting than by exponential models. We will later address whether any of these functions fully describe behavior. First, we review an instructive neuroeconomic debate: do neural activity patterns measured with fMRI support hyperbolic or quasi-hyperbolic discounting (Eqs. 2 & 3)? While these functions yield similar choices (Fig. 1), they can be interpreted as implying different neural mechanisms.

The quasi-hyperbolic model, if conceived of as a dual process model in the brain, suggests that one system responds to immediate rewards, while another uses a constant exponential discount function across delays. The beta term in the quasi-hyperbolic function (Eq. 3) represents a present bias that boosts the utility of immediate rewards. Two fMRI studies (McClure et al., 2004, 2007) tested for evidence of separate "beta systems" and "delta systems" in the brain using primary and secondary rewards. Both studies varied the time to the earliest reward (today vs 2 weeks for gift cards, and 0, 10, or 20 min for juice rewards). Critically, the same reward magnitudes were used in both time conditions. Comparing trials with immediate rewards (today or 0 min) to trials with delayed rewards, the authors found greater activity in brain regions like the posterior cingulate cortex (PCC), ventromedial prefrontal cortex (vmPFC), and ventral striatum (VS). These regions were referred to as "beta system" regions because they were more active

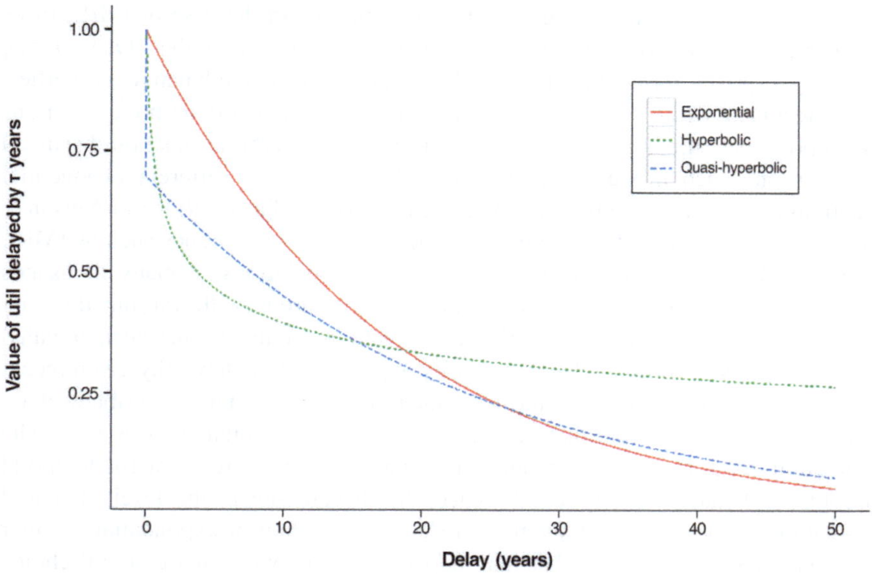

Figure 2. The Shapeof Different Discount Functions

Notes: Plots of the exponential δ^{τ} with $\delta = 0.944$, hyperbolic $(1 + \alpha\tau)^{-\gamma/\alpha}$ with $\alpha = 4$, $\gamma = 1$, and quasi-hyperbolic$\beta^{1(\tau>0)}\,\delta^{\tau}$ with $\beta = 0.7, \delta = 0.957$.

Fig. 1 From Cohen et al. (2020). This figure illustrates the shape of the different discount functions listed in Eqs. 1–3

for immediate rewards. However, this pattern of activity isn't unique to regions implementing the beta term. It's consistent with temporal discounting in general because rewards available sooner are valued more.

The hyperbolic model differs from the quasi-hyperbolic model in that it can be implemented in a common valuation system across all time points. A hyperbolic discounting function fitted to each participant's behavior correlates with BOLD activity in the PCC, vmPFC, and VS at all delays (Kable & Glimcher, 2007, 2010). This suggests a single valuation system that evaluates all rewards, regardless of delay, contradicting the idea that these regions are a "beta system" that selectively responds to immediate rewards. Supporting the single process hypothesis, these brain regions reflect the subjective value of various goods in different contexts and temporal scales (Bartra et al., 2013; Clithero & Rangel, 2013), and several studies have replicated the correlation between BOLD activity and values predicted by a hyperbolic discount function across time delays (Hare et al., 2014; Kable et al., 2017; Peters & Büchel, 2009; Sripada et al., 2011). Together, this neuroimaging evidence indicates that intertemporal tradeoffs are evaluated by the same neural decision circuitry as other choices without explicit temporal components.

The precise nature of the discounting function(s) within the brain is still an open question. It is important to remember that these neuroimaging data are based on aggregated signals from large numbers of neurons. In the case of fMRI, these aggregate signals are further convoluted due to their reliance on the BOLD signal, which is a proxy for neural activity that depends on hemodynamic properties. These technical details are relevant because mixtures of non-hyperbolic functions can approximate the hyperbolic function. The quasi-hyperbolic function discussed above is one such mixture, but mixtures of two or more different exponential functions can also approximate a hyperbolic function (Azfar, 1999; Kurth-Nelson & Redish, 2009; Sozou, 1998). Noninvasive neuroimaging techniques such as fMRI, fNIRS, and EEG measure a mixed signal from populations of many thousands of neurons. This means that a local mixture of neurons with varying discount functions—none of which are hyperbolic in isolation—could produce an aggregated response that is hyperbolic. This possibility is particularly noteworthy given recent findings that dopamine neurons have a range of reward expectations (Dabney et al., 2020), making it plausible that neurons have varying discount rates as well. The population code, hyperbolic or otherwise, may be what is relevant for temporal discounting behavior. However, if hyperbolic discounting at the level of neural populations or behavior is implemented by a combination of exponential or other functions in single neurons, that may explain, in part, why intertemporal choice behavior doesn't fully conform to key predictions of the hyperbolic function.

Intertemporal choice patterns often violate predictions derived from either the exponential or hyperbolic discounting functions. Researchers across fields have generally acknowledged that humans and nonhuman animals are not exponential discounters at the level of observed behavior, i.e., they violate the axioms presented in Fishburn and Rubinstein (1982). A common assumption is that they are instead (quasi-)hyperbolic discounters. However, a less-widely acknowledged fact is that *behavior deviates quantitatively and qualitatively from the predictions of the hyperbolic discounting function too* (Cohen et al., 2020; Frederick et al., 2002; Imas et al., 2022; Kable & Glimcher, 2010; Luhmann, 2013; Read, 2001; Rubinstein, 2003; Stevens, 2016; van den Bos & McClure, 2013). Nevertheless, summarizing a person's intertemporal choice behavior by fitting it with a hyperbolic function has proven to be useful even if it is not the true functional form or process underlying such decisions.

Fitted hyperbolic discount function parameters explain individual variability in cognitive abilities, personality traits, psychopathologies, and socioeconomic factors. Correlations between hyperbolic discount rates and these factors are small but reliable (Cohen et al., 2020; Keidel et al., 2021). The directionality of these associations is unknown, but they've been replicated in various contexts. Hyperbolic discounting parameters can be seen as an approximate summary of intertemporal choice patterns, like BMI is a rough guide to a person's morphology. Both measures compress complex information into a single number, though imperfectly, and are useful summaries. The hyperbolic model of discounting is useful as an approximate functional form, but it's important to continue searching for reasons why and how intertemporal choices deviate from its predictions. Discovering which assumptions about intertemporal choice mechanisms to maintain and revise is a challenge and an opportunity for neuroeconomics.

2 Attribute-Based Examinations of Intertemporal Choices

Neuroeconomic approaches and tools are well-suited to examining intertemporal choices from an attribute-based perspective. The exponential, hyperbolic, and quasi-hyperbolic discounting functions all inherently assume that the reward's magnitude and its delay are combined into an integrated discounted value. In contrast, attribute-based models provide a different perspective on the reasons for and mechanisms of temporal discounting. Several different attribute-based models of temporal discounting have been proposed (Amasino et al., 2019; Dai & Busemeyer, 2014; He et al., 2023; Marzilli Ericson et al., 2015; Read et al., 2013; Scholten & Read, 2010; Wulff & van den Bos, 2018). Studies that directly compare attribute-based and discounting models show that both classes of models do reasonably well in explaining choice behavior, which may explain why hyperbolic discounting has proven to be useful in explaining individual differences in many domains. However, attribute-based models appear to have a slight advantage over discounting models in explaining behavioral patterns overall, although there is considerable heterogeneity across individuals and contexts (Amasino et al., 2019; Dai & Busemeyer, 2014; He et al., 2023; Wulff & van den Bos, 2018). An advantage of attribute-based models of intertemporal choice is that they pair well with process-tracking methods.

Recent work has shown that the sequence in which information is acquired and attributes are considered influences intertemporal choices. Process tracking methods (e.g., eye-tracking or computer cursor/mouse tracking) have become a core tool in neuroeconomics and related disciplines. Reeck et al. (2017) had human participants make intertemporal choices for which they had to click into squares in four different regions of a computer screen to reveal the amounts and delays for the two available options. The researchers recorded the cursor movements participants made while searching for information and found that two different information search patterns were associated with either more choices for larger later or more choices for smaller sooner rewards (Fig. 2). A comparative search pattern that involved moving horizontally between the amount and delay information across options was associated with making more larger-later choices, while an integrative pattern of moving vertically between the amount and delay within an option was associated with more smaller-sooner choices. Critically, Reeck et al. ran a separate experiment in which they used subtle manipulations in the time it took to reveal the information hidden in a square after a mouse click to induce participants to use comparative or integrative search patterns. These results show that comparative search relative to integrative patterns *causes* more larger-later choices in their experimental context. This is important because it establishes that search patterns causally affect intertemporal choices. Subsequent work replicated the association between search patterns and intertemporal choices while also showing that the effects of search on choice depend on the sequence of information received rather than comparative or integrative strategies per se (Khaw et al., 2018).

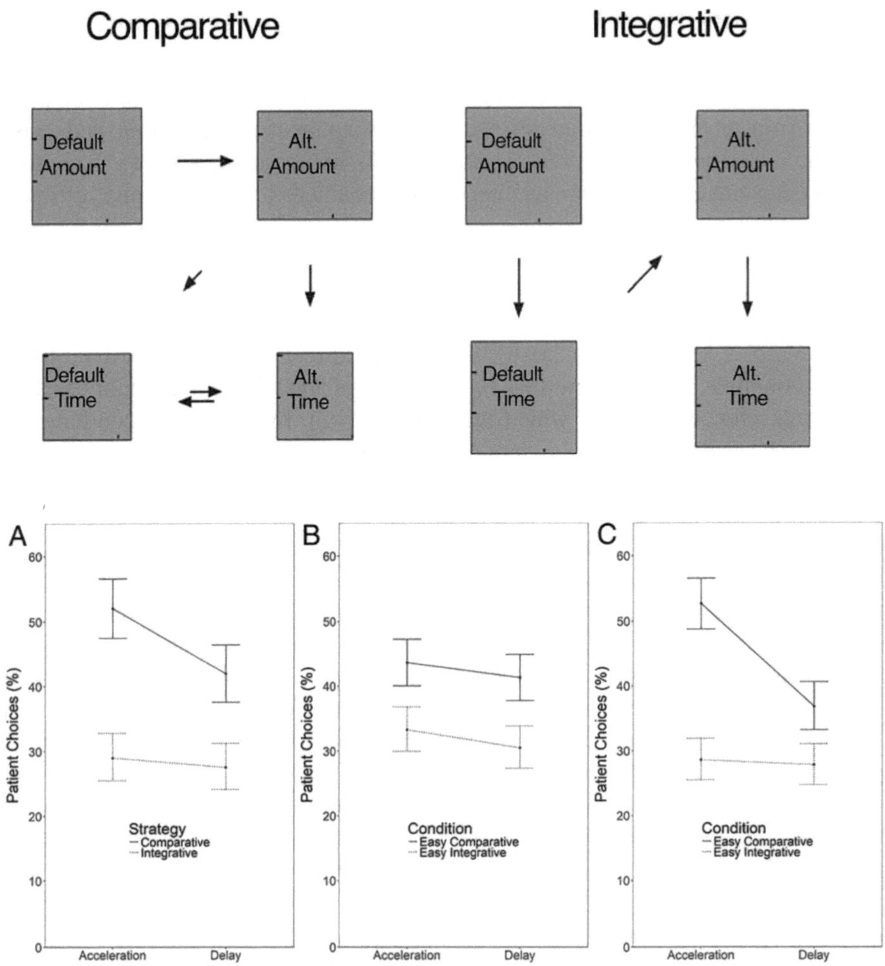

Fig. 2 From Reeck et al., 2017. Top panel: Icon plots displaying search behavior for the two groups identified by the k-means cluster analysis. Each box represents one of the four pieces of information in each trial. Box height indicates how many times participants acquired that piece of information on average (each tick marks 0.5 transitions). Box width indicates the duration of the time participants spent viewing that piece of information on average (each tick marks 500 ms). Arrows indicate the average transitions on a given trial, with arrow length representing how commonly the transition occurred. Transitions occurring less than 2.5% of the time are omitted to improve display legibility. "Alt." refers to the alternative, nondefault option. (Left) Comparative searchers were more likely to engage in attribute-based searching. (Right) Integrative searchers were more likely to engage in alternative-based searching Bottom panels: Percentage of larger/later options chosen by participants. (**a**) Experiment 1: Comparative searchers were more likely overall to select the larger/later option than integrative searchers and were more likely to exhibit an effect of acceleration framing. (**b**) Experiment 2: All trials. Those in the easy comparative condition were more patient overall than those in the easy integrative condition. (**c**) Experiment 2: Restricted to trials in which search behavior matched the strategy encouraged by manipulation. Those in the easy comparative condition were both more patient and exhibited a significant effect of acceleration versus delay framing. Error bars represent ±1SEM

3 Noise and Uncertainty in Representations of Reward and Time

In addition to the information acquisition sequence, the nature and clarity of the information used to make intertemporal choices will affect their outcomes as well. The future is less concrete than the present, a potential reason why delayed rewards are discounted. Linking a delayed reward to concrete episodes in the future (e.g., a vacation) reduces temporal discounting (Peters et al., 2020; Peters & Büchel, 2010). This process, episodic future thinking, is linked to hippocampal function (Lebreton et al., 2013). Episodic future thinking reveals a link between reward mental representations and preferences. During most decisions, benefits must be forecasted, which relies on intact hippocampal function even for simple choices over two immediately available food rewards (Bakkour et al., 2019). While temporal availability may differ between options, benefits are generally consumed after the decision (with a delay). The more distant the future, the harder it is to imagine precisely (Arnold et al., 2011; D'Argembeau & Van der Linden, 2004), suggesting a role for imagination or mental simulation in intertemporal choices. Indeed, greater similarity in vmPFC activity during imagined versus actual consumption of juice rewards (i.e., more accurate reward simulation) correlates with less temporal discounting (Hakimi & Hare, 2015). Moreover, a randomized-controlled trial showed that training in imagining future events leads to a decrease in temporal discounting rates in both internet addicts and healthy controls (Li, 2021). Therefore, the ability to imagine and represent future rewards seems to play an important role in intertemporal choices.

There are at least two theories explaining how less precise delayed reward representations may lead to a preference for sooner rewards, independent of any subjective decrease in future outcome value. Gershman and Bhui (2020) extended Gabaix and Laibson (2017) to demonstrate how noisier future reward representations, combined with prior beliefs sensitive to information costs, can lead to apparent discounting of delayed rewards (Fig. 3). The idea is that a person has a prior expectation about reward magnitudes. Perceived rewards are noisy, and the decision signal is the combination of prior and noisy percept, regularized toward the prior mean. The strength of regularization increases with percept noise. Noisier larger-later rewards are regularized more than less noisy smaller-sooner rewards. Assuming the prior mean is smaller than the average larger-later reward, many larger-later rewards are perceived as less valuable, favoring smaller-sooner rewards. Gershman and Bhui also explain the magnitude effect in intertemporal choices via rational inattention. Decision makers allocate more attention to high-stakes trials to avoid large-magnitude mistakes, leading to less noisy reward representations and less discounting of larger-later rewards. Episodic tags for future payoffs may also decrease noisiness and temporal discounting. Noisier future reward representations are a key reason why delayed rewards are discounted relative to immediate rewards.

In addition to being represented less precisely, the future is also inherently less certain than the present. For example, something unfortunate may happen to the

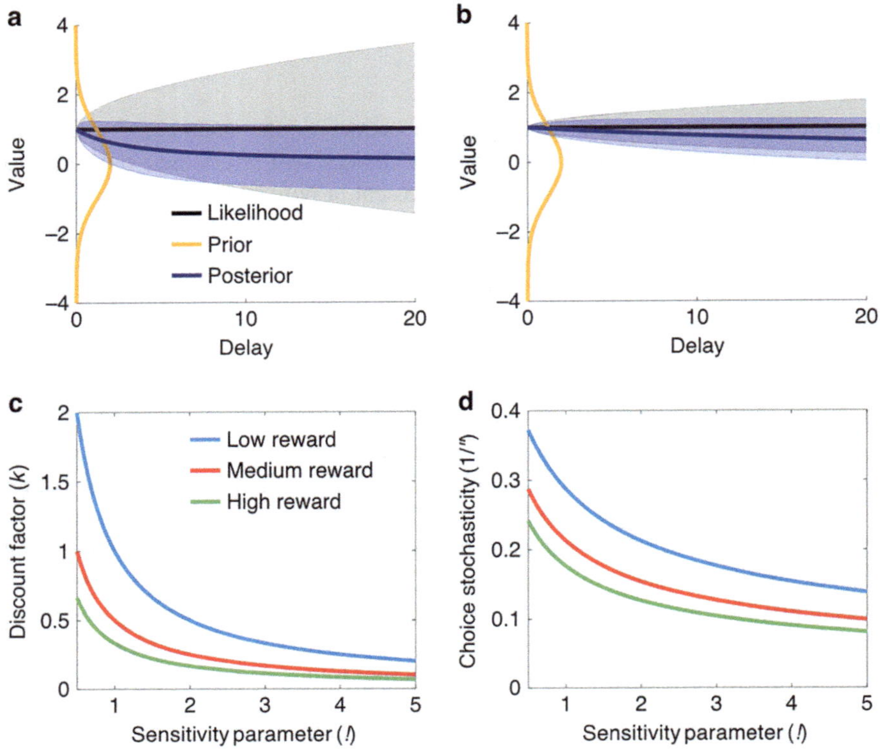

Fig. 3 From Gershman & Bhui, 2020. Illustration of rational discounting model. (**a**) Illustration of discounting as Bayesian inference. Mental simulation generates a noisy value signal (st = 1), which is combined with a prior distribution to form a posterior distribution over value. The black line denotes the true underlying value *u*, and the shaded region shows the standard deviation of the noise corrupting the value at each possible delay *t* (in arbitrary units). Similarly, the blue line is the posterior mean, and the shaded region is the posterior standard deviation. Because the standard deviation of st (simulation noise) increases with delay, the posterior mean is pulled more strongly toward the prior mean for longer delays. (**b**) Same simulation as in (a) but with a smaller signal variance, demonstrating a reduction in discounting. (**c**) Discount factor under the rational inattention model as a function of the sensitivity parameter (β) and reward magnitude. (**d**) Choice stochasticity under the rational inattention model as a function of the sensitivity parameter (β) and reward magnitude

source of the delayed reward or the person themselves that prevents the delivery or use of the reward in the future. Thus, risk preferences are likely to contribute to intertemporal choices. Lopez-Guzman et al. (2018) showed time preferences are distorted if risk preferences are not taken into account. Epper and Fehr-Duda (Epper & Fehr-Duda, 2024) provide further evidence that risk and time preferences are intertwined and propose a unifying model accounting for both that explains several commonly observed but perplexing behavioral results including delay-dependent risk tolerance, preferences for the timing of the resolution of uncertainty, the differential discounting of risky and certain outcomes, hyperbolic discounting, subadditive discounting, and others. This interdependence between time and uncertainty/risk is an important consideration for any model of intertemporal choices because time is a central model input in all cases.

Time perception is crucial in decisions between temporally distributed alternatives. Animals, including humans, don't always perceive time linearly, and interval durations adapt to the scale (Tsao et al., 2022). Measures of intertemporal preferences are dependent on time as an input, and distortions in the subjective perception or prospection of time will distort inferences about those preferences if not accounted for (Rachlin, 2006; Takahashi, 2005, 2009; Vieider, 2021). Ray and Bossaerts (2011) demonstrated that, in theory, apparent present biases in intertemporal choices can be explained by subjective distortions in time perception and, when time distortions are accounted for, discounting behavior is exponential, i.e., it matches normative predictions. Other theoretical work shows that seemingly hyperbolic discounting of delayed rewards may even arise from the scalar property of time perception without any additional temporal discounting of value itself (Cui, 2011). Consistent with a causal effect of time perception on intertemporal choices, experimental manipulations of time sensitivity differentially affect valuations in the near versus far future (Ebert & Prelec, 2007). Similarly, Chen and Zhao (2024) have also shown causal effects of time perception on intertemporal choice by demonstrating that increases in agents' subjective time durations (i.e., perceived time is longer than objective time) lead to more choices in favor of smaller-sooner rewards.

Interestingly, the neural systems linked to time and interval perception overlap with those implicated in valuation and choice (Merchant et al., 2013; Tsao et al., 2022). Studies show that BOLD activity in the vmPFC and VS, reward-sensitive regions, correlate with self-reports of a temporal delay (Cooper et al., 2013). This activity—measured in the absence of any choice—correlates with intertemporal choices made later. These results suggest that reward and time attributes may be integrated within the brain's valuation system as has been shown for various factors (e.g., risk & magnitude, self & other payoffs, pleasure & health) in choices involving nontemporal tradeoffs (Bartra et al., 2013; Clithero & Rangel, 2013). This is another example of how intertemporal choices share psychological and neurobiological mechanisms that support value-based and perceptual decision-making more generally.

In contrast, one aspect that is often thought to distinguish intertemporal choices from other types of decisions is a requirement for self-control. It's assumed that the decision maker must override a natural impulse for the proximal reward to choose the objectively larger delayed reward. The last portion of this chapter will discuss the validity of this assumption.

4 Do Intertemporal Choices Measure Self-Control?

Choices between smaller-sooner and larger-later rewards don't necessarily measure self-control. McGuire and Kable's work on choices with unknown larger-later rewards challenges the notion that temporal discounting is inherently linked with self-control (McGuire & Kable, 2013, 2015). This work focuses on the delay of gratification paradigm, popularized by Walter Mischel and colleagues (Mischel et al., 1972; Mischel & Ebbesen, 1970). In these tests, participants can take a small reward now or wait for an unknown larger reward later. Unlike binary and irreversible intertemporal choices, participants can decide to stop waiting and take the smaller reward anytime during the delay. The length of time children wait correlates with skill and success later in life (Benjamin et al., 2020; Eigsti et al., 2006). Failing to wait the whole period is considered a self-control failure. McGuire and Kable show that a person's belief about waiting time distribution determines whether to wait and when to switch to the smaller option. Heavy-tailed waiting time distributions make it optimal to wait initially and switch to the smaller option if the larger reward hasn't arrived. These switches aren't self-control failures but rational actions based on initial beliefs and acquired evidence. Many people have heavy-tailed beliefs about waiting times in experiments and real life, such as achieving weight loss or mastering a skill (McGuire & Kable, 2013, 2015). Lempert et al. (2023) used pupillometric data to show that people are more surprised by rewards after longer delays, consistent with heavy-tailed waiting-time expectations. Furthermore, effort signals didn't increase as quitting times got longer, contrary to the self-control model of delay of gratification. Instead, effort signals increased when quit decisions were made at unusual times, like much earlier or later than usual. Overall, the data indicate that people are not trying to wait as long as possible before "giving up" in delay of gratification tasks.

Beliefs significantly impact self-control assessment during delayed gratification. To determine if individuals lacked the willpower to wait as expected, we need to understand their beliefs about delay length and how they change during waiting. Delay of gratification task choices correlate with other behaviors and skills, but these associations may not be solely due to shared reliance on self-control or willpower. Further research with precise measures of beliefs and different self-control assays is needed to identify the underlying reasons for these associations.

Choices favoring smaller, sooner rewards over later, larger ones may not indicate self-control even with known delays. This is because subjective time perception differs from objective duration, time and risk attitudes are intertwined, delayed rewards may be perceived less precisely than sooner ones, and attention and search alter intertemporal choices. These factors, and likely others, should be considered before we make any attempt to rank people's self-control abilities based on intertemporal choices. Despite these caveats, one may argue that if one assumes that self-control is at least partially responsible for monetary intertemporal choices, then behavior on such tasks should carry at least a noisy signal of this ability. There is reason to question this self-control assumption though.

Do people use self-control during explicit-delay monetary intertemporal choice tasks? A common definition of self-control is the ability to override temptations or counterproductive impulses to achieve one's goal. Data from response times and confidence ratings call the assumption that people have a goal to wait for the larger-later reward into question. Bulley et al. (2022) found that confidence ratings in having made the "correct" choice correspond to decision difficulty rather than choosing the larger-later reward. People reported higher confidence on trials with greater discriminability[1] and faster decision times. Confidence wasn't higher when choosing larger-later rewards. This confidence result contradicts the assumption that participants have a goal to choose the larger-later reward, which would imply that selecting larger-later is the correct choice. *If choosing the larger-later reward is not the goal, then intertemporal choices don't require self-control.*

Response time data from other experiments support the findings from explicit reports of confidence. Data on confidence in intertemporal choices are still rare, but the relationship between confidence, choice discriminability, and decision times is well established (De Martino et al., 2013; Fleming & Daw, 2017; Kiani et al., 2014; Ratcliff & Starns, 2009). Thus, for now, we might cautiously use decision times as a rough approximation to confidence, with faster decision times indicating higher confidence. Intertemporal choices show the same response time regularities present in other choices (Amasino et al., 2019; Chabris et al., 2008; Dai et al., 2018; Dai & Busemeyer, 2014; Konovalov & Krajbich, 2019; Krajbich et al., 2015). Greater option discriminability is closely associated with faster decision times. Importantly, decision times don't differ between smaller-sooner and larger-later choices at the same level of discriminability (Krajbich et al., 2015). Thus, decision time data from many independent intertemporal choice studies suggest that confidence relates to discriminability during intertemporal choices rather than being highest on larger-later decisions, consistent with direct reports of confidence (Bulley et al., 2022). However, more data on confidence reports after intertemporal choices are needed to test if people may have goals to choose larger-later options and rely on self-control in specific contexts. The current data suggest reconsidering the assumption that monetary intertemporal choice tasks measure self-control.

5 Summary

Despite long-standing interest in intertemporal decision-making, much remains to be learned. Recent findings show that intertemporal choices are influenced by various skills and preferences beyond any underlying change in the subjective value of rewards across time. The multifaceted nature of temporal discounting behavior may be why it is a transdiagnostic marker of psychiatric disorders and correlates

[1] For intertemporal choices, discriminability is the absolute value of the difference between the utilities of the smaller-sooner and larger-later options.

with cognitive abilities, personality traits, and socioeconomic factors. Future work in neuroeconomics may aim to identify how the various underlying drivers of intertemporal choices are associated with specific disorders, abilities, or traits linked to temporal discounting.

References

Ainslie, G., & Herrnstein, R. J. (1981). Preference reversal and delayed reinforcement. *Animal Learning & Behavior, 9*(4), 476–482. https://doi.org/10.3758/BF03209777

Amasino, D. R., Sullivan, N. J., Kranton, R. E., & Huettel, S. A. (2019). Amount and time exert independent influences on intertemporal choice. *Nature Human Behaviour, 3*(4), Article 4. https://doi.org/10.1038/s41562-019-0537-2

Arnold, K. M., McDermott, K. B., & Szpunar, K. K. (2011). Imagining the near and far future: The role of location familiarity. *Memory & Cognition, 39*(6), 954–967. https://doi.org/10.3758/s13421-011-0076-1

Azfar, O. (1999). Rationalizing hyperbolic discounting. *Journal of Economic Behavior & Organization, 38*(2), 245–252. https://doi.org/10.1016/S0167-2681(99)00009-8

Bakkour, A., Palombo, D. J., Zylberberg, A., Kang, Y. H., Reid, A., Verfaellie, M., Shadlen, M. N., & Shohamy, D. (2019). The hippocampus supports deliberation during value-based decisions. *eLife, 8*, e46080. https://doi.org/10.7554/eLife.46080

Bartra, O., McGuire, J. T., & Kable, J. W. (2013). The valuation system: A coordinate-based meta-analysis of BOLD fMRI experiments examining neural correlates of subjective value. *NeuroImage, 76*, 412–427. https://doi.org/10.1016/j.neuroimage.2013.02.063

Benjamin, D. J., Laibson, D., Mischel, W., Peake, P. K., Shoda, Y., Wellsjo, A. S., & Wilson, N. L. (2020). Predicting mid-life capital formation with pre-school delay of gratification and life-course measures of self-regulation. *Journal of Economic Behavior & Organization, 179*, 743–756. https://doi.org/10.1016/j.jebo.2019.08.016

Bulley, A., Lempert, K. M., Conwell, C., Irish, M., & Schacter, D. L. (2022). Intertemporal choice reflects value comparison rather than self-control: Insights from confidence judgements. *Philosophical Transactions of the Royal Society B: Biological Sciences, 377*(1866), 20210338. https://doi.org/10.1098/rstb.2021.0338

Chabris, C. F., Laibson, D., Morris, C. L., Schuldt, J. P., & Taubinsky, D. (2008). *Measuring intertemporal preferences using response times* (Working paper no. 14353). National Bureau of Economic Research. https://doi.org/10.3386/w14353

Chen, X., & Zhao, X. (2024). How time flies: Time perception and intertemporal choice. *Journal of Behavioral and Experimental Economics, 109*, 102160. https://doi.org/10.1016/j.socec.2023.102160

Clithero, J. A., & Rangel, A. (2013). Informatic parcellation of the network involved in the computation of subjective value. *Social Cognitive and Affective Neuroscience, 9*(9), 1289–1302.

Cohen, J., Ericson, K. M., Laibson, D., & White, J. M. (2020). Measuring time preferences. *Journal of Economic Literature, 58*(2), 299–347. https://doi.org/10.1257/jel.20191074

Cooper, N., Kable, J. W., Kim, B. K., & Zauberman, G. (2013). Brain activity in valuation regions while thinking about the future predicts individual discount rates. *Journal of Neuroscience, 33*(32), 13150–13156. https://doi.org/10.1523/JNEUROSCI.0400-13.2013

Cui, X. (2011). Hyperbolic discounting emerges from the scalar property of interval timing. *Frontiers in Integrative Neuroscience, 5*. https://doi.org/10.3389/fnint.2011.00024

D'Argembeau, A., & Van der Linden, M. (2004). Phenomenal characteristics associated with projecting oneself back into the past and forward into the future: Influ-

ence of valence and temporal distance. *Consciousness and Cognition, 13*(4), 844–858. https://doi.org/10.1016/j.concog.2004.07.007
Dabney, W., Kurth-Nelson, Z., Uchida, N., Starkweather, C. K., Hassabis, D., Munos, R., & Botvinick, M. (2020). A distributional code for value in dopamine-based reinforcement learning. *Nature, 577*(7792), Article 7792. https://doi.org/10.1038/s41586-019-1924-6
Dai, J., & Busemeyer, J. R. (2014). A probabilistic, dynamic, and attribute-wise model of intertemporal choice. *Journal of Experimental Psychology: General, 143*(4), 1489–1514. https://doi.org/10.1037/a0035976
Dai, J., Pleskac, T. J., & Pachur, T. (2018). Dynamic cognitive models of intertemporal choice. *Cognitive Psychology, 104*, 29–56. https://doi.org/10.1016/j.cogpsych.2018.03.001
De Martino, B., Fleming, S. M., Garrett, N., & Dolan, R. J. (2013). Confidence in value-based choice. *Nature Neuroscience, 16*(1), Article 1. https://doi.org/10.1038/nn.3279
Ebert, J. E. J., & Prelec, D. (2007). The fragility of time: Time-insensitivity and valuation of the near and far future. *Management Science, 53*(9), 1423–1438. https://doi.org/10.1287/mnsc.1060.0671
Eigsti, I.-M., Zayas, V., Mischel, W., Shoda, Y., Ayduk, O., Dadlani, M. B., Davidson, M. C., Aber, J. L., & Casey, B. J. (2006). Predicting cognitive control from preschool to late adolescence and young adulthood. *Psychological Science, 17*(6), 478–484. https://doi.org/10.1111/j.1467-9280.2006.01732.x
Epper, T. F., & Fehr-Duda, H. (2024). Risk in time: The intertwined nature of risk taking and time discounting. *Journal of the European Economic Association, 22*(1), 310–354. https://doi.org/10.1093/jeea/jvad041
Fishburn, P. C., & Rubinstein, A. (1982). Time preference. *International Economic Review, 23*(3), 677–694. https://doi.org/10.2307/2526382
Fleming, S. M., & Daw, N. D. (2017). Self-evaluation of decision-making: A general Bayesian framework for metacognitive computation. *Psychological Review, 124*(1), 91–114. https://doi.org/10.1037/rev0000045
Frederick, S., Loewenstein, G., & O'Donoghue, T. (2002). Time discounting and time preference: A critical review. *Journal of Economic Literature, 40*(2), 351–401. https://doi.org/10.1257/002205102320161311
Gabaix, X., & Laibson, D. (2017). *Myopia and discounting* (working paper no. 23254). National Bureau of economic. *Research*. https://doi.org/10.3386/w23254
Gershman, S. J., & Bhui, R. (2020). Rationally inattentive intertemporal choice. *Nature. Communications, 11*(1), 3365. https://doi.org/10.1038/s41467-020-16852-y
Hakimi, S., & Hare, T. A. (2015). Enhanced neural responses to imagined primary rewards predict reduced monetary temporal discounting. *Journal of Neuroscience, 35*(38), 13103–13109. https://doi.org/10.1523/JNEUROSCI.1863-15.2015
Hare, T., Hakimi, S., & Rangel, A. (2014). Activity in dlPFC and its effective connectivity to vmPFC are associated with temporal discounting. *Frontiers in Neuroscience, 8*, 50. https://doi.org/10.3389/fnins.2014.00050
He, L., Wall, D., Reeck, C., & Bhatia, S. (2023). Information acquisition and decision strategies in intertemporal choice. *Cognitive Psychology, 142*, 101562. https://doi.org/10.1016/j.cogpsych.2023.101562
Imas, A., Kuhn, M. A., & Mironova, V. (2022). Waiting to choose: The role of deliberation in intertemporal choice. *American Economic Journal: Microeconomics, 14*(3), 414–440. https://doi.org/10.1257/mic.20180233
Kable, J. W. (2014). Chapter 10—Valuation, intertemporal choice, and self-control. In P. W. Glimcher & E. Fehr (Eds.), *Neuroeconomics* (2nd ed., pp. 173–192). Academic. https://doi.org/10.1016/B978-0-12-416008-8.00010-3
Kable, J. W., & Glimcher, P. W. (2007). The neural correlates of subjective value during intertemporal choice. *Nature Neuroscience, 10*(12), 1625–1633. https://doi.org/10.1038/nn2007
Kable, J. W., & Glimcher, P. W. (2010). An "as soon as possible" effect in human intertemporal decision making: Behavioral evidence and neural mechanisms. *Journal of Neurophysiology, 103*(5), 2513–2531. https://doi.org/10.1152/jn.00177.2009

Kable, J. W., Caulfield, M. K., Falcone, M., McConnell, M., Bernardo, L., Parthasarathi, T., Cooper, N., Ashare, R., Audrain-McGovern, J., Hornik, R., Diefenbach, P., Lee, F. J., & Lerman, C. (2017). No effect of commercial cognitive training on brain activity, choice behavior, or cognitive performance. *Journal of Neuroscience, 37*(31), 7390–7402. https://doi.org/10.1523/JNEUROSCI.2832-16.2017

Keidel, K., Rramani, Q., Weber, B., Murawski, C., & Ettinger, U. (2021). Individual differences in intertemporal choice. *Frontiers in Psychology, 12*. https://doi.org/10.3389/fpsyg.2021.643670

Khaw, M. W., Li, Z., & Woodford, M. (2018). Temporal discounting and search habits: Evidence for a task-dependent relationship. *Frontiers in Psychology, 9*. https://doi.org/10.3389/fpsyg.2018.02102

Kiani, R., Corthell, L., & Shadlen, M. N. (2014). Choice certainty is informed by both evidence and decision time. *Neuron, 84*(6), 1329–1342. https://doi.org/10.1016/j.neuron.2014.12.015

Konovalov, A., & Krajbich, I. (2019). Revealed strength of preference: Inference from response times. *Judgment & Decision Making, 14*(4).

Krajbich, I., Bartling, B., Hare, T., & Fehr, E. (2015). Rethinking fast and slow based on a critique of reaction-time reverse inference. *Nature Communications, 6*, 7455. https://doi.org/10.1038/ncomms8455

Kurth-Nelson, Z., & Redish, A. D. (2009). Temporal-difference reinforcement learning with distributed representations. *PLoS One, 4*(10), e7362. https://doi.org/10.1371/journal.pone.0007362

Lebreton, M., Bertoux, M., Boutet, C., Lehericy, S., Dubois, B., Fossati, P., & Pessiglione, M. (2013). A critical role for the hippocampus in the valuation of imagined outcomes. *PLoS Biology, 11*(10), e1001684. https://doi.org/10.1371/journal.pbio.1001684

Lempert, K. M. (this volume). The role of memory in temporal discounting. In D. V. Smith, P. Lockwood, & D. S. Fareri (Eds.), *Neuroeconomics: Core topics and current directions*. Springer.

Lempert, K. M., & Phelps, E. A. (2016). The malleability of intertemporal choice. *Trends in Cognitive Sciences, 20*(1), 64–74. https://doi.org/10.1016/j.tics.2015.09.005

Lempert, K. M., Carballeira, C., Sehgal, S., & Kable, J. W. (2023). Pupillometric evidence for a temporal expectations-based account of persistence under temporal uncertainty. *Cognitive, Affective, & Behavioral Neuroscience, 23*(3), 705–717. https://doi.org/10.3758/s13415-023-01100-9

Li, H. (2021). Imagining the future: Future imagination training decreases delay discounting among internet addicts and non-problematic users. *Frontiers in Psychology, 12*. https://doi.org/10.3389/fpsyg.2021.731708

Lopez-Guzman, S., Konova, A. B., Louie, K., & Glimcher, P. W. (2018). Risk preferences impose a hidden distortion on measures of choice impulsivity. *PLoS One, 13*(1), e0191357. https://doi.org/10.1371/journal.pone.0191357

Luhmann, C. C. (2013). Discounting of delayed rewards is not hyperbolic. *Journal of Experimental Psychology: Learning, Memory, and Cognition, 39*(4), 1274–1279. https://doi.org/10.1037/a0031170

Marzilli Ericson, K. M., White, J. M., Laibson, D., & Cohen, J. D. (2015). Money earlier or later? Simple heuristics explain intertemporal choices better than delay discounting does. *Psychological Science, 26*(6), 826–833. https://doi.org/10.1177/0956797615572232

Mazur, J. E. (1987). An adjusting procedure for studying delayed reinforcement. In *The effect of delay and of intervening events on reinforcement value* (pp. 55–73). Lawrence Erlbaum Associates, Inc.

McClure, S. M., Laibson, D. I., Loewenstein, G., & Cohen, J. D. (2004). Separate neural systems value immediate and delayed monetary rewards. *Science, 306*(5695), 503–507. https://doi.org/10.1126/science.1100907

McClure, S. M., Ericson, K. M., Laibson, D. I., Loewenstein, G., & Cohen, J. D. (2007). Time discounting for primary rewards. *Journal of Neuroscience, 27*(21), 5796–5804. https://doi.org/10.1523/JNEUROSCI.4246-06.2007

McGuire, J. T., & Kable, J. W. (2013). Rational temporal predictions can underlie apparent failures to delay gratification. *Psychological Review, 120*(2), 395–410. https://doi.org/10.1037/a0031910

McGuire, J. T., & Kable, J. W. (2015). Medial prefrontal cortical activity reflects dynamic re-evaluation during voluntary persistence. *Nature Neuroscience, 18*(5), 760–766. https://doi.org/10.1038/nn.3994

Merchant, H., Harrington, D. L., & Meck, W. H. (2013). Neural basis of the perception and estimation of time. *Annual Review of Neuroscience, 36*, 313–336. https://doi.org/10.1146/annurev-neuro-062012-170349

Mischel, W., & Ebbesen, E. B. (1970). Attention in delay of gratification. *Journal of Personality and Social Psychology, 16*(2), 329–337. https://doi.org/10.1037/h0029815

Mischel, W., Ebbesen, E. B., & Raskoff Zeiss, A. (1972). Cognitive and attentional mechanisms in delay of gratification. *Journal of Personality and Social Psychology, 21*(2), 204–218. https://doi.org/10.1037/h0032198

Peters, J., & Büchel, C. (2009). Overlapping and distinct neural systems code for subjective value during intertemporal and risky decision making. *The Journal of Neuroscience, 29*(50), 15727–15734. https://doi.org/10.1523/JNEUROSCI.3489-09.2009

Peters, J., & Büchel, C. (2010). Episodic future thinking reduces reward delay discounting through an enhancement of prefrontal-Mediotemporal interactions. *Neuron, 66*(1), 138–148. https://doi.org/10.1016/j.neuron.2010.03.026

Peters, J., Brassen, S., Bromberg, U., Büchel, C., Sasse, L., & Wiehler, A. (2020). Effects of episodic future thinking on temporal discounting: A re-analysis of six data sets using hierarchical Bayesian parameter estimation and compilation of effect sizes. *bioRxiv*, 2020.03.24.005892. https://doi.org/10.1101/2020.03.24.005892

Rachlin, H. (2006). Notes on discounting. *Journal of the Experimental Analysis of Behavior, 85*(3), 425–435. https://doi.org/10.1901/jeab.2006.85-05

Ratcliff, R., & Starns, J. J. (2009). Modeling confidence and response time in recognition memory. *Psychological Review, 116*(1), 59–83. https://doi.org/10.1037/a0014086

Ray, D., & Bossaerts, P. (2011). Positive temporal dependence of the biological clock implies hyperbolic discounting. *Frontiers in Neuroscience, 5*. https://doi.org/10.3389/fnins.2011.00002

Read, D. (2001). Is time-discounting hyperbolic or subadditive? *Journal of Risk and Uncertainty, 23*(1), 5–32. https://doi.org/10.1023/A:1011198414683

Read, D., Frederick, S., & Scholten, M. (2013). DRIFT: An analysis of outcome framing in intertemporal choice. *Journal of Experimental Psychology: Learning, Memory, and Cognition, 39*(2), 573–588. https://doi.org/10.1037/a0029177

Reeck, C., Wall, D., & Johnson, E. J. (2017). Search predicts and changes patience in intertemporal choice. *Proceedings of the National Academy of Sciences, 114*(45), 11890–11895. https://doi.org/10.1073/pnas.1707040114

Rubinstein, A. (2003). "Economics and psychology"? The case of hyperbolic discounting. *International Economic Review, 44*(4), 1207–1216. https://doi.org/10.1111/1468-2354.t01-1-00106

Samuelson, P. A. (1937). A note on measurement of utility. *The Review of Economic Studies, 4*(2), 155–161. https://doi.org/10.2307/2967612

Scholten, M., & Read, D. (2010). The psychology of intertemporal tradeoffs. *Psychological Review, 117*(3), 925–944. https://doi.org/10.1037/a0019619

Sozou, P. D. (1998). On hyperbolic discounting and uncertain hazard rates. *Proceedings of the Royal Society B: Biological Sciences, 265*(1409), 2015–2020. https://doi.org/10.1098/rspb.1998.0534

Sripada, C. S., Gonzalez, R., Luan Phan, K., & Liberzon, I. (2011). The neural correlates of intertemporal decision-making: Contributions of subjective value, stimulus type, and trait impulsivity. *Human Brain Mapping, 32*(10), 1637–1648. https://doi.org/10.1002/hbm.21136

Stevens, J. R. (2016). Intertemporal similarity: Discounting as a last resort. *Journal of Behavioral Decision Making, 29*(1), 12–24. https://doi.org/10.1002/bdm.1870

Strotz, R. H. (1955). Myopia and inconsistency in dynamic utility maximization. *The Review of Economic Studies, 23*(3), 165–180. https://doi.org/10.2307/2295722

Su, Z., & Lockwood, P. L. (this volume). The neuroeconomics of social influence and contagion. In D. V. Smith, P. Lockwood, & D. S. Fareri (Eds.), *Neuroeconomics: Core topics and current directions*. Springer.
Takahashi, T. (2005). Loss of self-control in intertemporal choice may be attributable to logarithmic time-perception. *Medical Hypotheses, 65*(4), 691–693. https://doi.org/10.1016/j.mehy.2005.04.040
Takahashi, T. (2009). Theoretical frameworks for neuroeconomics of intertemporal choice. *Journal of Neuroscience, Psychology, and Economics, 2*(2), 75–90. https://doi.org/10.1037/a0015463
Tsao, A., Yousefzadeh, S. A., Meck, W. H., Moser, M.-B., & Moser, E. I. (2022). The neural bases for timing of durations. *Nature Reviews Neuroscience, 23*(11), 646–665. https://doi.org/10.1038/s41583-022-00623-3
van den Bos, W., & McClure, S. M. (2013). Towards a general model of temporal discounting. *Journal of the Experimental Analysis of Behavior, 99*(1), 58–73. https://doi.org/10.1002/jeab.6
Vieider, F. (2021). *Noisy coding of time and reward discounting* (Working papers of Faculty of Economics and Business Administration, Ghent University, Belgium). Ghent University, Faculty of Economics and Business Administration. https://econpapers.repec.org/paper/rugrugwps/21_2f1036.htm
Wagner, B. J., Knauth, K., & Peters, J. (this volume). Intertemporal choice. In D. V. Smith, P. Lockwood, & D. S. Fareri (Eds.), *Neuroeconomics: Core topics and current directions*. Springer.
Wulff, D. U., & van den Bos, W. (2018). Modeling choices in delay discounting. *Psychological Science, 29*(11), 1890–1894. https://doi.org/10.1177/0956797616664342

The Role of Memory in Temporal Discounting

Karolina M. Lempert

Abstract A widely observed phenomenon in intertemporal choice is temporal discounting; people prefer to have rewards sooner rather than later, even if the delayed rewards are larger. Despite the universality of temporal discounting, people vary substantially in their temporal discounting rates, with some people being much more willing to wait for delayed rewards than others. What are the neurocognitive mechanisms underlying these individual differences? This chapter presents the field's current understanding of the roles of major cognitive capacities (specifically: inhibitory control, episodic memory, and semantic memory) in determining temporal discounting rates. The chapter also argues that whereas there are circumstances under which inhibitory control and episodic memory retrieval are deployed during intertemporal choice, semantic memory retrieval may actually be the key cognitive contributor to variability in intertemporal preferences. Inhibitory control is likely to be engaged during intertemporal choice when choices are especially difficult or high-stakes. Episodic retrieval is likely to be involved when people are cued to imagine the future before they make these choices. Neither of these processes appears to be necessary for future-oriented choice, however, and there is growing evidence that the structural integrity of the anterior temporal lobe—a hub for semantic memory—is associated with temporal discounting. More research is needed to test this theory and to understand the precise mechanism by which semantic memory supports future-oriented choice. Further research is also needed to test the generalizability of these conclusions to other kinds of intertemporal choices, such as those involving primary rewards, punishments, or temporal uncertainty.

K. M. Lempert (✉)
Gordon F. Derner School of Psychology, Adelphi University, Garden City, NY, USA
e-mail: kmlempert@adelphi.edu

D. V. Smith et al. (eds.), *Neuroeconomics: Core Topics and Current Directions*,
https://doi.org/10.1007/978-3-032-02925-6_16

1 Introduction: Understanding Variability in Temporal Discounting

Temporal discounting[1] refers to the tendency for individuals to devalue delayed outcomes, as the delay to receiving those outcomes increases (Samuelson, 1937; Strotz, 1956). For example, people often spend money on luxury goods now, rather than invest that money to enjoy more of it in the future. Although temporal discounting is observed almost universally in the context of intertemporal choice, people vary substantially in their temporal discounting *rate,* or the extent to which they discount delayed outcomes (Keidel et al., 2021; Peters & Büchel, 2011). The higher someone's temporal discounting rate is, the more likely they are to prefer proximal rewards at their long-term expense.

Temporal discounting rates derived from laboratory intertemporal choice tasks have high test-retest reliability (Anokhin et al., 2015; Gelino et al., 2024; Kirby, 2009) and good ecological validity: people with relatively high discount rates are more likely to experience depression (Amlung et al., 2019; Pulcu et al., 2014), substance use disorder (Amlung et al., 2017), and pathological gambling (Reynolds, 2006). They also tend to have worse financial and health outcomes (Bartels et al., 2023; Brown et al., 2024; Chabris et al., 2008). On the other hand, people with relatively low discount rates—those who are willing to give up small rewards now to obtain larger rewards in the future—tend to show better academic performance (Kirby et al., 2005; Wang et al., 2017) and take care of their health more (Bartels et al., 2023; Tate, 2015).

In light of this evidence that variation in temporal discounting is associated with real-world outcomes, many neuroeconomists seek to understand the cognitive processes and neural mechanisms that lead people to value immediate rewards so much more than future rewards. Historically, individual differences in temporal discounting have been attributed to individual differences in executive function, in general, and to inhibitory control, in particular. The mechanism behind that proposed association is that people who are better able to inhibit the impulse to choose tempting immediate rewards are more likely to select larger, later rewards. The evidence that executive function underlies individual differences in temporal discounting is mixed, however. More recent research suggests that the declarative memory system, which is comprised of both episodic and semantic memory, may be more important for establishing and expressing intertemporal preferences. This chapter will summarize that literature. Possible mechanisms by which memory promotes more future-oriented choice will also be discussed.

[1] The term "temporal discounting" will be used throughout this chapter, but this concept is synonymous with "delay discounting" (which is more commonly used in the clinical psychology literature) and "time preference" (which is more commonly used in the economics literature).

1.1 The Scope of the Chapter

Intertemporal choices can take many different forms, and temporal discounting can be assessed and quantified in several different ways, but the studies presented in this chapter have almost exclusively used binary intertemporal choice tasks. These tasks require people to make a series of decisions between smaller monetary rewards that they could receive immediately (e.g., $10 today) and larger monetary rewards that they could receive after some delay (e.g., $30 in 60 days). These delays are usually in the range of days to years (e.g., Kirby et al., 1999; Senecal et al., 2012). From these choice tasks, a temporal discounting rate can be derived. Sometimes these discount rates are derived through a model-fitting procedure. For example, one can assume that the value of a future reward drops *hyperbolically* as a function of delay (Mazur, 1987),[2] and then estimate the slope of that hyperbola. If the slope is high, then the discount rate is high (value drops off rapidly as delay increases), and if the slope is low, then the discount rate is low (value does not change much as delay increases). For the purposes of this chapter, however, the precise functional form of the discounting function does not matter so much. In fact, if an intertemporal choice task is well-designed, even a coarse metric, like the proportion of smaller, sooner reward choices that people make out of all choices, can be a good proxy of discount rate. Given that all of the research presented in this chapter uses these monetary intertemporal choice tasks, the conclusions may apply only to situations where people are (1) making decisions between smaller, sooner, and larger, later rewards (not punishments), (2) making decisions where the delays are known, and are on the timescale of days (rather than seconds), and (3) making decisions when both rewards are of the same type (i.e., monetary rewards). It is important for the reader to keep in mind that the mechanisms underlying intertemporal choices that involve punishments, temporal uncertainty, smaller timescales, and/or primary rewards might be different.

Another caveat is that this chapter is not meant to provide a comprehensive overview of the neural mechanisms of intertemporal choice. For example, individual differences in the process of valuation will not be considered, even though the valuation network (anchored by the ventral striatum and ventromedial prefrontal cortex; Fig. 1a) is reliably involved in intertemporal decision-making (Kable & Glimcher, 2007, 2010), just as it is in other forms of value-based decision-making (Bartra et al., 2013; Clithero & Rangel, 2014). These regions are necessary to evaluate delayed rewards properly; damage to the ventromedial or orbitofrontal

[2] The hyperbolic function is the most commonly used function in psychology and neuroscience, since it has been shown to fit data well, and it yields a single discount rate parameter (usually called k) that is easy to interpret: higher k values indicate less tolerance of delay. Therefore, it is a good descriptive model of temporal discounting, although some other, more complex forms have been proposed, too. The hyperbolic model is not a *normative* model, however, since it implies that people treat short delays differently from long ones, which may lead them to reverse their preferences in a way that is incompatible with the axioms of expected utility theory. Thus, an exponential model is more commonly used in the economics literature.

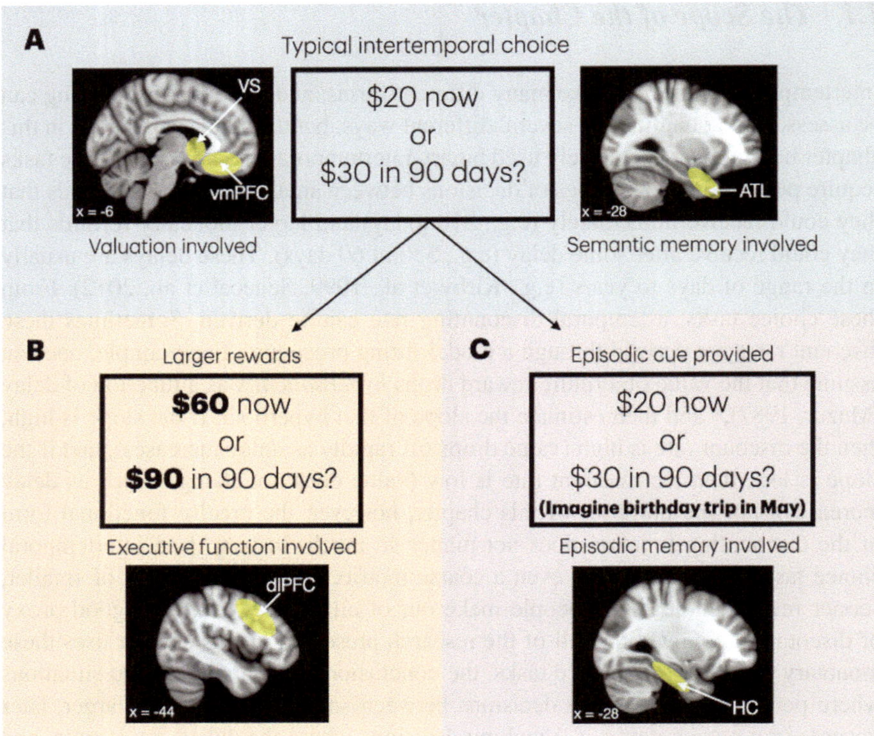

Fig. 1 Summary fig. (**a**) Under typical circumstances, the valuation network, including the ventral striatum (VS) and ventromedial prefrontal cortex (vmPFC), and the semantic memory system, including the anterior temporal lobe (ATL), are most likely necessary and sufficient for choosing larger, delayed rewards. (**b**) When the magnitudes of the rewards on offer are higher, people tend to choose delayed rewards more, an effect that is mediated by activation of the dorsolateral prefrontal cortex (dlPFC) and the executive function system. (**c**) When episodic cues are provided to encourage people to engage in prospection, people tend to choose delayed rewards more, an effect that is mediated by activation of the hippocampus (HC) and episodic memory system

cortex increases temporal discounting (Ciaramelli et al., 2021; Peters & D'Esposito, 2016), and stimulation of the orbitofrontal cortex reduces discounting (Moro et al., 2023). Thus, individual differences in the valuation network will almost certainly contribute to individual differences in temporal discounting. This chapter will focus instead on cognitive processes that have more controversial roles in intertemporal choice: executive function, episodic memory, and semantic memory.

2 Executive Function and Temporal Discounting

In their search for the neurocognitive mechanisms underlying intertemporal choice, it is no surprise that neuroeconomists zeroed in on executive function first. Executive function refers to the collection of processes that people use to deliberately control their behavior and achieve their goals (Jurado & Rosselli, 2007). These processes include selective attention, cognitive flexibility, working memory, and inhibitory control.[3] Brain regions putatively involved in implementing executive function are mostly in the frontal lobe (Alvarez & Emory, 2006; Stuss & Alexander, 2000), with the dorsolateral prefrontal cortex (dlPFC) being most important. Executive function was thought to be essential for future-oriented decision-making because it was assumed that people needed to exert inhibitory control in order to override the impulse to select immediate rewards. This notion is implicit in many dual-systems models, which propose that immediate rewards elicit more emotion, so a "cool," deliberative system must be engaged in order for people to pursue long-term goals (Figner et al., 2009; Metcalfe & Mischel, 1999).

It is beyond the scope of this chapter to review all of the neuroeconomics literature that has examined the role of executive function in temporal discounting, but much of the evidence comes from functional magnetic resonance imaging (fMRI) studies that have shown that the dlPFC is more active when people choose larger, later rewards over smaller, sooner rewards (McClure et al., 2004; Weber & Huettel, 2008; Wesley & Bickel, 2014). These fMRI studies are merely correlational, though; studies that attempt to manipulate the involvement of inhibitory control and/or dlPFC activity would be more conclusive. There are a handful of transcranial magnetic stimulation (TMS) and transcranial direct current stimulation (TDCS) studies that have shown that facilitating the dlPFC leads to more delayed reward choices (Figner et al., 2010; He et al., 2016; Hecht et al., 2013). However, the results of these experiments are mixed. For example, Figner et al. (2010) found that disrupting the left, but not the right, lateral prefrontal cortex increased discounting. On the other hand, Hecht et al. (2013) found that *facilitating* the left dlPFC increased discounting. Yet another group found that disrupting the right dlPFC reduced discounting (Cho et al., 2010). Recent meta-analyses have concluded that noninvasive stimulation of the dlPFC does significantly influence executive functions, but the effects on discounting are small and not significant (de Boer et al., 2021; Lin & Feng, 2024). Moreover, patients with lesions to the dlPFC have similar discounting rates to healthy controls (Fellows & Farah, 2005).

In addition to noninvasive stimulation of dlPFC, other manipulations of executive function have been attempted in the context of intertemporal choice. One manipulation that impairs executive function is acute stress (Arnsten, 2009), but a recent meta-analysis concluded that acute stress does not influence temporal discounting

[3] The terms "executive function" and inhibitory control" may be used interchangeably in this chapter, but most scientists would consider inhibitory control to be just one process under the umbrella of executive function.

(Forbes et al., 2024). Cognitive load should also disrupt the ability to use inhibitory control, but cognitive load does not always increase discounting, and may instead just make choices noisier (Franco-Watkins et al., 2010; Olschewski et al., 2018). Finally, a large study using both behavioral and neuroimaging methods found that there was no effect of executive function training on temporal discounting (Kable et al., 2017), although this kind of cognitive training has been shown to reduce discounting in some groups, such as in individuals with low socioeconomic status (Felton et al., 2019; Zhao et al., 2022) or with stimulant addiction (Bickel et al., 2011).

Taken together, this literature suggests that dlPFC-mediated executive functions do not play a causal role in temporal discounting, but there is ample evidence that executive function is *involved* in intertemporal choice. One possibility is that executive function is more engaged during intertemporal choice when decisions are especially difficult (Jimura et al., 2018). That is, it could be that inhibitory control is only brought to bear when a person is torn between two options—in those cases, inhibitory control might help the person to go against their usual tendency. For most people, that tendency is to choose the immediate reward, but that does not necessarily have to be the case. Another fascinating line of research has found that the dlPFC may underlie the *magnitude effect* (Ballard et al., 2017, 2018; Fig. 1b). The magnitude effect is the finding that when the rewards on offer are larger overall, people tend to be more willing to wait for delayed rewards (Green et al., 1997). The fact that the dlPFC mediates the magnitude effect suggests that its function might be to flexibly change intertemporal preferences when the stakes are high. The dlPFC may even facilitate the memory search process that often accompanies value-based decisions (Bakkour et al., 2019). Thus, the dlPFC might help support future-oriented choice, but only under limited circumstances. The basic phenomenon of temporal discounting cannot be explained by variance in dlPFC function or inhibitory control alone.

3 Episodic Memory and Temporal Discounting

A compelling alternative mechanism underlying temporal discounting is that it depends on the extent to which people can imagine future outcomes. This process is called *episodic future thinking* or *prospection.* Instead of steep temporal discounting being caused by an inability to control impulses, it may be caused by an inability to visualize future outcomes vividly enough to evaluate them properly. This theory emerged in the late 2000s, after an influential set of studies found that people with amnesia due to hippocampal damage also had difficulty imagining the future (Hassabis et al., 2007), and the brain regions involved in episodic memory retrieval and episodic future thinking were highly overlapping (Schacter et al., 2007). Boyer (2008) argued that people may have evolved the ability to engage in "mental time travel" so that they could make choices that are better in the long term. Since the episodic memory system is recruited in service of prospection, this work set the stage for research into the relationship between episodic memory and intertemporal choice.

The idea that prospection could nudge people toward making more patient intertemporal choices was first tested in a pair of very similar neuroimaging studies (Benoit et al., 2011; Peters & Büchel, 2010). In both studies, participants were cued to engage in episodic future thinking, either by being shown phrases describing their own future plans (e.g., "vacation in Paris") or by being invited to imagine plausible future scenarios (e.g., going to a museum). Participants made more choices of larger, later rewards after prospection, and this effect was mediated by functional connectivity between frontal regions and medial temporal lobe regions (hippocampus and amygdala). Since those seminal studies were published, many follow-up experiments replicated the behavioral effect that episodic future thinking reduces temporal discounting (Rösch et al., 2022). This manipulation even works when the studies are done online (Sze et al., 2017), and the effects do not appear to be driven by experimenter demand (Rung & Madden, 2019). Even when prospection is negatively valenced, it can encourage more patience (Ballance et al., 2022; Bulley et al., 2019; but see Liu et al., 2013). The hippocampus is critical for this effect, as damage to the hippocampus eliminates the effect of prospection on discounting (Palombo et al., 2015; Fig. 1c). Together, this experimental evidence confirms that episodic future thinking is helpful for promoting more patient choices.

The observation that prospection can be deployed to flexibly change intertemporal choices does not imply, however, that individual differences in temporal discounting stem from individual differences in episodic future thinking (or in episodic memory function in general). The answer to that question is still unknown. Whereas a handful of studies have shown that the vividness of prospection was associated with lower temporal discounting across people (Bromberg et al., 2015; Guo et al., 2022; McCue et al., 2019), one study showed that people with better visualization abilities actually make more *impulsive* choices (Parthasarathi et al., 2017). Furthermore, despite showing deficits in imagining positive future events (Keeler et al., 2022), people with anorexia nervosa tend to have *lower* discount rates than healthy controls (Steinglass et al., 2012, 2017). One way to explain these discrepancies is that the vividness of future thinking may not be as important as the *frequency* of future-directed thoughts; indeed, people with lower discount rates do tend to have more thoughts of the future when prompted (Eivazi et al., 2022).

Unlike prospection ability, episodic memory ability can be measured using more objective methods (e.g., neuropsychological tests of verbal memory). The relationship between episodic memory and temporal discounting has been studied mostly in older adults, due to the substantial variability in the structural integrity of their episodic memory systems. Older adults with mild cognitive impairment—a prodromal phase of Alzheimer's disease, defined by episodic memory deficits—show steeper temporal discounting than cognitively normal older adults do (Coelho et al., 2016; Geng et al., 2020; Lempert et al., 2020b; Lindbergh et al., 2014). A few studies of patients with Alzheimer's disease (AD) found that they have higher discount rates than healthy controls do as well (El Haj et al., 2020a, b; Lebreton et al., 2013). These studies do not provide definitive evidence, however, since mild cognitive impairment and AD impact executive function as well; in fact, one study found that people with AD show less of a magnitude effect, perhaps due to changes in executive function (Mariano et al., 2020).

When examining only cognitively normal older adults, so far no reliable association between episodic memory ability and temporal discounting rate has been found (Boyle et al., 2012; Lempert et al., 2020b; Seinstra et al., 2015). In a large sample of adults of various ages, the association between verbal memory performance and temporal discounting was also null (Garzón et al., 2023). However, these studies were limited through their use of very basic measures of episodic memory. It is possible that more fine-grained aspects of episodic memory, such as those that are carried out by the entorhinal cortex, are associated with temporal discounting. Indeed, there is a correlation between entorhinal cortical structural integrity and temporal discounting, even in healthy adults (Lempert et al., 2020a, b; Owens et al., 2017), such that more entorhinal cortical thickness is associated with lower discounting. The entorhinal cortex is involved in memory for time details (Kitamura et al., 2015; Montchal et al., 2019; Tsao et al., 2018), and the extent to which people remember time details in their autobiographical memories is associated with temporal discounting (Lempert et al., 2020a), so perhaps people with better memory for time details in particular are more far-sighted decision makers. This possibility has yet to be formally tested.

In sum, whether and how episodic memory function is related to individual differences in temporal discounting is still unclear. Future studies using more sophisticated episodic memory paradigms—such as paradigms that specifically target entorhinal cortex functions—may shed more light on this issue. For now, it is apparent that the hippocampus is needed for people to flexibly use prospection in order to lower their temporal discounting.[4] The next section will argue that some of the inconsistencies in the literature on memory and temporal discounting might be resolved if we consider that *semantic* memory might be sufficient for making decisions about the future.

4 Semantic Memory and Temporal Discounting

There are a few findings that challenge the theory that episodic memory function drives future-oriented decision-making. First of all, whereas episodic future thinking reduces temporal discounting, episodic memory retrieval prior to intertemporal choice does not (Duff et al., 2024; Lempert et al., 2024). Therefore, activating episodic memory retrieval circuitry is not adequate to attenuate discounting. Even more damning evidence, however, comes from hippocampal lesion studies. At

[4] The hippocampus probably also plays a bigger role in intertemporal choices when the delays are much shorter (e.g., seconds) and actually experienced (Patt et al., 2023), but these kinds of choices are not the focus of the current chapter.

baseline, people with amnesia due to hippocampal damage have discount rates that are no different from those of healthy controls (Craver et al., 2014; Kwan et al., 2012). There is no significant relationship between hippocampal volume and temporal discounting rate, either (Garzón et al., 2023; Lempert et al., 2020b; Owens et al., 2017). And even the finding that people with Alzheimer's disease have higher discount rates than healthy controls has often not been replicated (Beagle et al., 2020; Bertoux et al., 2015; Chiong et al., 2016; Mariano et al., 2020). It is hard to make the case that episodic memory is necessary for choosing delayed rewards when people who cannot form new episodic memories are still capable of choosing delayed rewards.

One explanation for these results is that semantic memory, which remains unimpaired in people with damage restricted to the hippocampus, is sufficient for considering the future in decision making. Semantic memory is not entirely distinct from episodic memory; episodic memories become "semanticized" over time (Renoult et al., 2019; Sekeres et al., 2018). Semantic memories can be distinguished from episodic memories, though, because they are not confined to a single time and place (Tulving, 1987). Rather, semantic memory comprises categories, concepts, and schemas that are learned over time. These schemas can include general facts about the self—a category of knowledge known as "personal semantics" (Renoult et al., 2012). Schemas stored in semantic memory provide the scaffolding for prospection (Irish et al., 2012), since imagined future events, no matter how vividly they can be conjured, lack specific time and place details by definition. Therefore, if someone is able to draw on semantic knowledge—that is, knowledge of how things *usually are*—then they will be able to make decisions in their long-term best interest, even if they cannot envision future events in detail.

The strongest evidence for this idea is neuropsychological. The brain region that is most crucial for semantic memory is the anterior temporal lobe (Lambon Ralph et al., 2012). The anterior temporal lobe is also the region where gray matter volume is most robustly and uniquely associated with temporal discounting (Garzón et al., 2023): people with more intact anterior temporal lobes have lower discount rates (Fig. 2a). Furthermore, people who are known to have extensive damage to the anterior temporal lobe, such as those with semantic variant primary progressive aphasia (svPPA), do show steep discounting, both compared to healthy controls and to people with other neurodegenerative diseases (Beagle et al., 2020; Chiong et al., 2016). The defining feature of svPPA is a loss of semantic knowledge, but patients with this disorder also have a more severe deficit in episodic future thinking than in episodic memory (Irish et al., 2012; Irish & Piolino, 2016). At the time of writing this chapter, only one study attempted to relate performance on a measure of semantic memory (specifically, semantic fluency) to temporal discounting in a general population. There was a significant association, such that people who can more easily access their semantic store are also more willing to wait for larger, later rewards (Akilov & Lempert, 2025; Fig. 2b). Together, there is some promising preliminary evidence that semantic memory is integral to the expression of intertemporal preferences.

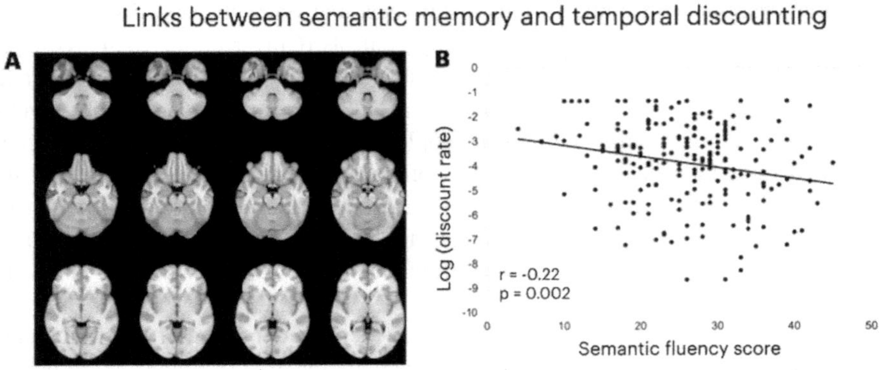

Fig. 2 Links between semantic memory and temporal discounting. Panel (**a**) is adapted from Supplementary Fig. 2 in Garzón et al. (2023), *Cereb Cortex*. It shows a voxel-wise analysis of gray matter volume, using data from the Human Connectome Project ($n = 1196$). In blue are brain regions where gray matter volume was significantly negatively associated with temporal discounting after controlling for linear and quadratic terms for age, sex, education, and income (FWER corrected, $p < 0.05$). Regions in red are from the same analysis, additionally controlling for general intelligence. These temporal lobe regions are also critical for the retrieval of semantic memory. (**b**) Convergent behavioral evidence from Akilov and Lempert (2025; $n = 203$). Semantic fluency involves naming as many category exemplars as possible in a limited amount of time (e.g., animals and vegetables). Good performance on this task, which requires the anterior temporal lobe, is negatively associated with temporal discounting

5 Conclusions

Intertemporal choices are complex value-based decisions. When making them, people recruit several cognitive processes and brain networks, including valuation, prospection, and executive function networks (Godefroy et al., 2023; Lempert et al., 2019; Peters & Büchel, 2011). In the last 15 years or so, there has been an explosion of research into precisely which cognitive functions are more efficient in people who have low discount rates. From these studies, we can conclude the following (see Fig. 1 for a summary). Executive function abilities can serve to reduce discounting but only under circumstances where inhibitory control is needed, such as when the potential rewards are very large, or the choices are very difficult. Episodic future thinking, which depends on the hippocampus, can certainly lessen discounting, but people are capable of making future-oriented choices even without engaging in prospection (or having a functional hippocampus). The latest research suggests that semantic memory, including personal semantic knowledge, may be sufficient for choosing larger, later rewards, and it might be deployed more in people with lower discount rates. More work is needed in this area, however, to delineate the precise roles of episodic and semantic memory in temporal discounting (see Future Research Questions box). Temporal discounting has been shown to predict many important real-world outcomes, such as health and financial security. Therefore, the extraordinary recent progress in understanding the neural mechanisms of temporal discounting may soon help us to develop interventions to promote positive behavior change.

6 Future Research Questions

- Under what conditions is executive function necessary for future-oriented choice?
- Working memory is considered an executive function that is separate from episodic memory. However, working memory is involved in the formation of episodic memories, and working memory training does diminish temporal discounting in some groups. What role does working memory play in intertemporal choice?
- Interventions that encourage episodic future thinking have been shown to reduce temporal discounting, but not all populations respond well to these manipulations (e.g., older adults: Mok et al., 2020; Sasse et al., 2017). What are some other priming manipulations that might be as effective but more generalizable?
- What are the precise functions of the hippocampus, entorhinal cortex, and anterior temporal lobe in the formation and maintenance of intertemporal preferences?
- How do episodic and semantic memory contribute to intertemporal choice when those choices involve punishments, primary rewards, temporal uncertainty, or very short delays?

References

Akilov, D., & Lempert, K. M. (2025). Semantic fluency is associated with reduced temporal discounting. *Journal of Memory and Language, 142* 104646. https://doi.org/10.1016/j.jml.2025.10461

Alvarez, J. A., & Emory, E. (2006). Executive function and the frontal lobes: A meta-analytic review. *Neuropsychology Review, 16*(1), 17–42. https://doi.org/10.1007/s11065-006-9002-x

Amlung, M., Vedelago, L., Acker, J., Balodis, I., & MacKillop, J. (2017). Steep delay discounting and addictive behavior: A meta-analysis of continuous associations. *Addiction, 112*(1), 51–62. https://doi.org/10.1111/add.13535

Amlung, M., Marsden, E., Holshausen, K., Morris, V., Patel, H., Vedelago, L., Naish, K. R., Reed, D. D., & McCabe, R. E. (2019). Delay discounting as a Transdiagnostic process in psychiatric disorders: A meta-analysis. *JAMA Psychiatry, 76*(11), 1176–1186. https://doi.org/10.1001/jamapsychiatry.2019.2102

Anokhin, A. P., Golosheykin, S., & Mulligan, R. C. (2015). Long-term test-retest reliability of delayed reward discounting in adolescents. *Behavioural Processes, 111*, 55–59. https://doi.org/10.1016/j.beproc.2014.11.008

Arnsten, A. F. T. (2009). Stress signalling pathways that impair prefrontal cortex structure and function. *Nature Reviews. Neuroscience, 10*(6), 410–422. https://doi.org/10.1038/nrn2648

Bakkour, A., Palombo, D. J., Zylberberg, A., Kang, Y. H., Reid, A., Verfaellie, M., Shadlen, M. N., & Shohamy, D. (2019). The hippocampus supports deliberation during value-based decisions. *eLife, 8*, e46080. https://doi.org/10.7554/eLife.46080

Ballance, B. C., Tuen, Y. J., Petrucci, A. S., Orwig, W., Safi, O. K., Madan, C. R., & Palombo, D. J. (2022). Imagining emotional events benefits future-oriented decisions. *Quarterly Journal of Experimental Psychology, 75*(12), 2332–2348. https://doi.org/10.1177/17470218221086637

Ballard, I. C., Kim, B., Liatsis, A., Aydogan, G., Cohen, J. D., & McClure, S. M. (2017). More is meaningful: The magnitude effect in intertemporal choice depends on self-control. *Psychological Science, 28*(10), 1443–1454. https://doi.org/10.1177/0956797617711455

Ballard, I. C., Aydogan, G., Kim, B., & McClure, S. M. (2018). Causal evidence for the dependence of the magnitude effect on dorsolateral prefrontal cortex. *Scientific Reports, 8*(1), 16545. https://doi.org/10.1038/s41598-018-34900-y

Bartels, D. M., Li, Y., & Bharti, S. (2023). How well do laboratory-derived estimates of time preference predict real-world behaviors? Comparisons to four benchmarks. *Journal of Experimental Psychology: General, 152*(9), 2651–2665. https://doi.org/10.1037/xge0001380

Bartra, O., McGuire, J. T., & Kable, J. W. (2013). The valuation system: A coordinate-based meta-analysis of BOLD fMRI experiments examining neural correlates of subjective value. *NeuroImage, 76*, 412–427. https://doi.org/10.1016/j.neuroimage.2013.02.063

Beagle, A. J., Zahir, A., Borzello, M., Kayser, A. S., Hsu, M., Miller, B. L., Kramer, J. H., & Chiong, W. (2020). Amount and delay insensitivity during intertemporal choice in three neurodegenerative diseases reflects dorsomedial prefrontal atrophy. *Cortex, 124*, 54–65. https://doi.org/10.1016/j.cortex.2019.10.009

Benoit, R. G., Gilbert, S. J., & Burgess, P. W. (2011). A neural mechanism mediating the impact of episodic prospection on farsighted decisions. *The Journal of Neuroscience, 31*(18), 6771–6779. https://doi.org/10.1523/JNEUROSCI.6559-10.2011

Bertoux, M., de Souza, L. C., Zamith, P., Dubois, B., & Bourgeois-Gironde, S. (2015). Discounting of future rewards in behavioural variant frontotemporal Dementia & Alzheimer's disease. *Neuropsychology*. https://doi.org/10.1037/neu0000197

Bickel, W. K., Yi, R., Landes, R. D., Hill, P. F., & Baxter, C. (2011). Remember the future: Working memory training decreases delay discounting among stimulant addicts. *Biological Psychiatry, 69*(3), 260–265. https://doi.org/10.1016/j.biopsych.2010.08.017

Boyer, P. (2008). Evolutionary economics of mental time travel? *Trends in Cognitive Sciences, 12*(6), 219–224. https://doi.org/10.1016/j.tics.2008.03.003

Boyle, P. A., Yu, L., Segawa, E., Wilson, R. S., Buchman, A. S., Laibson, D. I., & Bennett, D. A. (2012). Association of cognition with temporal discounting in community based older persons. *BMC Geriatrics, 12*(1), 48. https://doi.org/10.1186/1471-2318-12-48

Bromberg, U., Wiehler, A., & Peters, J. (2015). Episodic future thinking is related to impulsive decision making in healthy adolescents. *Child Development*. https://doi.org/10.1111/cdev.12390

Brown, J. M., Sofis, M., Zimmer, S., & Kaplan, B. A. (2024). Delay discounting is associated with addiction and mental health measures while controlling for health behaviors and health barriers in a large US sample. *Addictive Behaviors Reports, 19*, 100545. https://doi.org/10.1016/j.abrep.2024.100545

Bulley, A., Miloyan, B., Pepper, G. V., Gullo, M. J., Henry, J. D., & Suddendorf, T. (2019). Cuing both positive and negative episodic foresight reduces delay discounting but does not affect risk-taking. *Quarterly Journal of Experimental Psychology, 72*(8), 1998–2017. https://doi.org/10.1177/1747021818819777

Chabris, C. F., Laibson, D., Morris, C. L., Schuldt, J. P., & Taubinsky, D. (2008). Individual laboratory-measured discount rates predict field behavior. *Journal of Risk and Uncertainty, 37*(2–3), 237–269. https://doi.org/10.1007/s11166-008-9053-x

Chiong, W., Wood, K. A., Beagle, A. J., Hsu, M., Kayser, A. S., Miller, B. L., & Kramer, J. H. (2016). Neuroeconomic dissociation of semantic dementia and behavioural variant frontotemporal dementia. *Brain : A Journal of Neurology, 139*(Pt 2), 578–587. https://doi.org/10.1093/brain/awv344

Cho, S. S., Ko, J. H., Pellecchia, G., Van Eimeren, T., Cilia, R., & Strafella, A. P. (2010). Continuous theta burst stimulation of right dorsolateral prefrontal cortex induces changes in impulsivity level. *Brain Stimulation, 3*(3), 170–176. https://doi.org/10.1016/j.brs.2009.10.002

Ciaramelli, E., De Luca, F., Kwan, D., Mok, J., Bianconi, F., Knyagnytska, V., Craver, C., Green, L., Myerson, J., & Rosenbaum, R. S. (2021). The role of ventromedial prefrontal cortex in reward valuation and future thinking during intertemporal choice. *eLife*, 10, e67387. https://doi.org/10.7554/eLife.67387.

Clithero, J. A., & Rangel, A. (2014). Informatic parcellation of the network involved in the computation of subjective value. *Social Cognitive and Affective Neuroscience, 9*(9), 1289–1302. https://doi.org/10.1093/scan/nst106

Coelho, S., Guerreiro, M., Chester, C., Silva, D., Maroco, J., Paglieri, F., & de Mendonça, A. (2016). Delay discounting in mild cognitive impairment. *Journal of Clinical and Experimental Neuropsychology, 1–11*. https://doi.org/10.1080/13803395.2016.1226269

Craver, C. F., Kwan, D., Steindam, C., & Rosenbaum, R. S. (2014). Individuals with episodic amnesia are not stuck in time. *Neuropsychologia, 57*, 191–195. https://doi.org/10.1016/j.neuropsychologia.2014.03.004

de Boer, N. S., Schluter, R. S., Daams, J. G., van der Werf, Y. D., Goudriaan, A. E., & van Holst, R. J. (2021). The effect of non-invasive brain stimulation on executive functioning in healthy controls: A systematic review and meta-analysis. *Neuroscience & Biobehavioral Reviews, 125*, 122–147. https://doi.org/10.1016/j.neubiorev.2021.01.013

Duff, N., Olsen, R., Walsh, Z., Salmon, K., Hunt, M., & Macaskill, A. (2024). EXPRESS: A fragile effect: The influence of episodic memory on delay discounting. *Quarterly Journal of Experimental Psychology, 2006*, 17470218241239289. https://doi.org/10.1177/17470218241239289

Eivazi, F., Hatami, J., Moradi, A., & Nazem-Zadeh, M.-R. (2022). Prediction of delay discounting in intertemporal decisions by future thinking: Accounting for fluency, contents, and functions of future thoughts. *Brain and Behavior, 12*(10), e2764. https://doi.org/10.1002/brb3.2764

El Haj, M., Boutoleau-Bretonnière, C., & Allain, P. (2020a). Memory of decisions: Relationship between decline of autobiographical memory and temporal discounting in Alzheimer's disease. *Journal of Clinical and Experimental Neuropsychology, 42*(4), 415–424. https://doi.org/10.1080/13803395.2020.1744527

El Haj, M., Boutoleau-Bretonnière, C., Moustafa, A., & Allain, P. (2020b). The discounted future: Relationship between temporal discounting and future thinking in Alzheimer's disease. *Applied Neuropsychology: Adult, 1–7*. https://doi.org/10.1080/23279095.2020.1764958

Fellows, L. K., & Farah, M. J. (2005). Dissociable elements of human foresight: A role for the ventromedial frontal lobes in framing the future, but not in discounting future rewards. *Neuropsychologia, 43*(8), 1214–1221. https://doi.org/10.1016/j.neuropsychologia.2004.07.018

Felton, J. W., Collado, A., Ingram, K. M., Doran, K., & Yi, R. (2019). Improvement of working memory is a mechanism for reductions in delay discounting among mid-age individuals in an urban medically underserved area. *Annals of Behavioral Medicine: A Publication of the Society of Behavioral Medicine, 53*(11), 988–998. https://doi.org/10.1093/abm/kaz010

Figner, B., Mackinlay, R. J., Wilkening, F., & Weber, E. U. (2009). Affective and deliberative processes in risky choice: Age differences in risk taking in the Columbia card task. *Journal of Experimental Psychology. Learning, Memory, and Cognition, 35*(3), 709–730. https://doi.org/10.1037/a0014983

Figner, B., Knoch, D., Johnson, E. J., Krosch, A. R., Lisanby, S. H., Fehr, E., & Weber, E. U. (2010). Lateral prefrontal cortex and self-control in intertemporal choice. *Nature Neuroscience, 13*(1), 538–539. https://doi.org/10.1038/nn.2516

Forbes, P. A. G., Nitschke, J. P., Hochmeister, N., Kalenscher, T., & Lamm, C. (2024). No effects of acute stress on monetary delay discounting: A systematic literature review and meta-analysis. *Neurobiology of Stress, 31*, 100653. https://doi.org/10.1016/j.ynstr.2024.100653

Franco-Watkins, A. M., Rickard, T. C., & Pashler, H. (2010). Taxing executive processes does not necessarily increase impulsive decision making. *Experimental Psychology, 57*(3), 193–201. https://doi.org/10.1027/1618-3169/a000024

Garzón, B., Kurth-Nelson, Z., Bäckman, L., Nyberg, L., & Guitart-Masip, M. (2023). Investigating associations of delay discounting with brain structure, working memory, and episodic memory. *Cerebral Cortex, 33*(5), 1669–1678. https://doi.org/10.1093/cercor/bhac164

Gelino, B. W., Schlitzer, R. D., Reed, D. D., & Strickland, J. C. (2024). A systematic review and meta-analysis of test-retest reliability and stability of delay and probability discounting. *Journal of the Experimental Analysis of Behavior, 121*(3), 358–372. https://doi.org/10.1002/jeab.910

Geng, Z., Wu, X., Wang, L., Zhou, S., Tian, Y., Wang, K., & Wei, L. (2020). Reduced delayed reward selection by Alzheimer's disease and mild cognitive impairment patients during intertemporal decision-making. *Journal of Clinical and Experimental Neuropsychology, 42*(3), 298–306. https://doi.org/10.1080/13803395.2020.1711873

Godefroy, V., Sezer, I., Bouzigues, A., Montembeault, M., Koban, L., Plassmann, H., & Migliaccio, R. (2023). Altered delay discounting in neurodegeneration: Insight into the underlying mechanisms and perspectives for clinical applications. *Neuroscience and Biobehavioral Reviews, 146*, 105048. https://doi.org/10.1016/j.neubiorev.2023.105048

Green, L., Myerson, J., & McFadden, E. (1997). Rate of temporal discounting decreases with amount of reward. *Memory & Cognition, 25*(5), 715–723. https://doi.org/10.3758/BF03211314

Guo, Y., Wu, H., Li, Z., Zhao, L., & Feng, T. (2022). Episodic future thinking predicts differences in delay discounting: The mediating role of hippocampal structure. *Frontiers in Psychology, 13*, 992245. https://doi.org/10.3389/fpsyg.2022.992245

Hassabis, D., Kumaran, D., Vann, S. D., Maguire, E., & a. (2007). Patients with hippocampal amnesia cannot imagine new experiences. *Proceedings of the National Academy of Sciences of the United States of America, 104*(5), 1726–1731. https://doi.org/10.1073/pnas.0610561104

He, Q., Chen, M., Chen, C., Xue, G., Feng, T., & Bechara, A. (2016). Anodal stimulation of the left DLPFC increases IGT scores and decreases delay discounting rate in healthy males. *Frontiers in Psychology, 7*, 1421. https://doi.org/10.3389/fpsyg.2016.01421

Hecht, D., Walsh, V., & Lavidor, M. (2013). Bi-frontal direct current stimulation affects delay discounting choices. *Cognitive Neuroscience, 4*(1), 7–11. https://doi.org/10.1080/17588928.2011.638139

Irish, M., & Piolino, P. (2016). Impaired capacity for prospection in the dementias—Theoretical and clinical implications. *The British Journal of Clinical Psychology, 55*(1), 49–68. https://doi.org/10.1111/BJC.12090

Irish, M., Addis, D. R., Hodges, J. R., Piguet, O., Addis, D. R., Cheng, T., Roberts, R. P., Schacter, D. L., Addis, D. R., Pan, L., Vu, M., Laiser, N., Schacter, D. L., Addis, D. R., Sacchetti, D. C., Ally, B. A., Budson, A. E., Schacter, D. L., Addis, D. R., et al. (2012). Considering the role of semantic memory in episodic future thinking: Evidence from semantic dementia. *Brain: A Journal of Neurology, 135*(Pt 7), 2178–2191. https://doi.org/10.1093/brain/aws119

Jimura, K., Chushak, M. S., Westbrook, A., & Braver, T. S. (2018). Intertemporal decision-making involves prefrontal control mechanisms associated with working memory. *Cerebral Cortex, 28*(4), 1105–1116. https://doi.org/10.1093/cercor/bhx015

Jurado, M. B., & Rosselli, M. (2007). The elusive nature of executive functions: A review of our current understanding. *Neuropsychology Review, 17*(3), 213–233. https://doi.org/10.1007/s11065-007-9040-z

Kable, J. W., & Glimcher, P. W. (2007). The neural correlates of subjective value during intertemporal choice. *Nature Neuroscience, 10*(12), 1625–1633. https://doi.org/10.1038/nn2007

Kable, J. W., & Glimcher, P. W. (2010). An "as soon as possible" effect in human intertemporal decision making: Behavioral evidence and neural mechanisms. *Journal of Neurophysiology, 103*(5), 2513–2531. https://doi.org/10.1152/jn.00177.2009

Kable, J. W., Caulfield, M. K., Falcone, M., McConnell, M., Bernardo, L., Parthasarathi, T., Cooper, N., Ashare, R., Audrain-McGovern, J., Hornik, R., Diefenbach, P., Lee, F. J., & Lerman, C. (2017). No effect of commercial cognitive training on brain activity, choice behavior, or cognitive performance. *The Journal of Neuroscience, 37*(31), 7390–7402. https://doi.org/10.1523/JNEUROSCI.2832-16.2017

Keeler, J. L., Peters-Gill, G., Treasure, J., Himmerich, H., Tchanturia, K., & Cardi, V. (2022). Difficulties in retrieving specific details of autobiographical memories and imagining positive future events in individuals with acute but not remitted anorexia nervosa. *Journal of Eating Disorders, 10*(1), 172. https://doi.org/10.1186/s40337-022-00684-w

Keidel, K., Rramani, Q., Weber, B., Murawski, C., & Ettinger, U. (2021). Individual differences in intertemporal choice. *Frontiers in Psychology, 12*. https://doi.org/10.3389/fpsyg.2021.643670
Kirby, K. N. (2009). One-year temporal stability of delay-discount rates. *Psychonomic Bulletin & Review, 16*(3), 457–462. https://doi.org/10.3758/PBR.16.3.457
Kirby, K. N., Petry, N. M., & Bickel, W. K. (1999). Heroin addicts have higher discount rates for delayed rewards than non-drug-using controls. *Journal of Experimental Psychology. General, 128*(1), 78–87.
Kirby, K. N., Winston, G. C., & Santiesteban, M. (2005). Impatience and grades: Delay-discount rates correlate negatively with college GPA. *Learning and Individual Differences, 15*(3), 213–222. https://doi.org/10.1016/j.lindif.2005.01.003
Kitamura, T., Macdonald, C. J., & Tonegawa, S. (2015). Entorhinal–hippocampal neuronal circuits bridge temporally discontiguous events. *Learning & Memory, 22*(9), 438–443. https://doi.org/10.1101/lm.038687.115
Kwan, D., Craver, C. F., Green, L., Myerson, J., Boyer, P., & Rosenbaum, R. S. (2012). Future decision-making without episodic mental time travel. *Hippocampus, 22*(6), 1215–1219. https://doi.org/10.1002/hipo.20981
Lambon Ralph, M. A., Ehsan, S., Baker, G. A., & Rogers, T. T. (2012). Semantic memory is impaired in patients with unilateral anterior temporal lobe resection for temporal lobe epilepsy. *Brain: A Journal of Neurology, 135*(Pt 1), 242–258. https://doi.org/10.1093/brain/awr325
Lebreton, M., Bertoux, M., Boutet, C., Lehericy, S., Dubois, B., Fossati, P., & Pessiglione, M. (2013). A critical role for the hippocampus in the valuation of imagined outcomes. *PLoS Biology, 11*(10), e1001684. https://doi.org/10.1371/journal.pbio.1001684
Lempert, K. M., Steinglass, J. E., Pinto, A., Kable, J. W., & Simpson, H. B. (2019). Can delay discounting deliver on the promise of RDoC? *Psychological Medicine, 49*(2), 190–199. https://doi.org/10.1017/S0033291718001770
Lempert, K. M., MacNear, K. A., Wolk, D. A., & Kable, J. W. (2020a). Links between autobiographical memory richness and temporal discounting in older adults. *Scientific Reports*. https://doi.org/10.1038/s41598-020-63373-1
Lempert, K. M., Mechanic-Hamilton, D. J., Xie, L., Wisse, L. E. M., de Flores, R., Wang, J., Das, S. R., Yushkevich, P. A., Wolk, D. A., & Kable, J. W. (2020b). Neural and behavioral correlates of episodic memory are associated with temporal discounting in older adults. *Neuropsychologia, 146*, 107549. https://doi.org/10.1016/j.neuropsychologia.2020.107549
Lempert, K. M., Parthasarathi, T., Linhares, S., Ruh, N., & Kable, J. W. (2024). Positive autobiographical memory recall does not influence temporal discounting: An internal meta-analysis of experimental studies. *Journal of Economic Psychology, 103*, 102730. https://doi.org/10.1016/j.joep.2024.102730
Lin, Y., & Feng, T. (2024). Lateralization of self-control over the dorsolateral prefrontal cortex in decision-making: A systematic review and meta-analytic evidence from noninvasive brain stimulation. *Cognitive, Affective, & Behavioral Neuroscience, 24*(1), 19–41. https://doi.org/10.3758/s13415-023-01148-7
Lindbergh, C. A., Puente, A. N., Gray, J. C., Mackillop, J., & Miller, L. S. (2014). Delay and probability discounting as candidate markers for dementia: An initial investigation. *Archives of Clinical Neuropsychology, 29*(7), 651–662. https://doi.org/10.1093/arclin/acu043
Liu, L., Feng, T., Chen, J., & Li, H. (2013). The value of emotion: How does episodic prospection modulate delay discounting? *PLoS One, 8*(11), e81717. https://doi.org/10.1371/journal.pone.0081717
Mariano, L. I., O'Callaghan, C., Guimarães, H. C., Gambogi, L. B., da Silva, T. B. L., Yassuda, M. S., Amaral, J. S., Caramelli, P., Hornberger, M., Teixeira, A. L., & de Souza, L. C. (2020). Disinhibition in frontotemporal dementia and Alzheimer's disease: A neuropsychological and behavioural investigation. *Journal of the International Neuropsychological Society: JINS, 26*(2), 163–171. https://doi.org/10.1017/S1355617719000973
Mazur, J. (1987). An adjusting procedure for studying delayed reinforcement. In M. L. Commons, J. E. Mazur, & J. A. Nevin (Eds.), *Quantitative analyses of behavior: Vol. 5: The effect of delay and of intervening events on reinforcement value* (pp. 55–73). Erlbaum.

McClure, S. M., Laibson, D. I., Loewenstein, G., & Cohen, J. D. (2004). Separate neural systems value immediate and delayed monetary rewards. *Science, 306*(5695), 503–507. https://doi.org/10.1126/science.1100907

McCue, R., McCormack, T., McElnay, J., Alto, A., & Feeney, A. (2019). The future and me: Imagining the future and the future self in adolescent decision making. *Cognitive Development, 50*, 142–156. https://doi.org/10.1016/j.cogdev.2019.04.001

Metcalfe, J., & Mischel, W. (1999). A hot/cool-system analysis of delay of gratification: Dynamics of willpower. *Psychological Review, 106*(1), 3–19.

Mok, J. N. Y., Kwan, D., Green, L., Myerson, J., Craver, C. F., & Rosenbaum, R. S. (2020). Is it time? Episodic imagining and the discounting of delayed and probabilistic rewards in young and older adults. *Cognition, 199*, 104222. https://doi.org/10.1016/j.cognition.2020.104222

Montchal, M. E., Reagh, Z. M., & Yassa, M. A. (2019). Precise temporal memories are supported by the lateral entorhinal cortex in humans. *Nature Neuroscience, 22*(2), 284–288. https://doi.org/10.1038/s41593-018-0303-1

Moro, A. S., Saccenti, D., Vergallito, A., Scaini, S., Malgaroli, A., Ferro, M., & Lamanna, J. (2023). Transcranial direct current stimulation (tDCS) over the orbitofrontal cortex reduces delay discounting. *Frontiers in Behavioral Neuroscience, 17*, 1239463. https://doi.org/10.3389/fnbeh.2023.1239463

Olschewski, S., Rieskamp, J., & Scheibehenne, B. (2018). Taxing cognitive capacities reduces choice consistency rather than preference: A model-based test. *Journal of Experimental Psychology: General, 147*(4), 462–484. https://doi.org/10.1037/xge0000403

Owens, M. M., Gray, J. C., Amlung, M. T., Oshri, A., Sweet, L. H., & MacKillop, J. (2017). Neuroanatomical foundations of delayed reward discounting decision making. *NeuroImage, 161*, 261–270. https://doi.org/10.1016/j.neuroimage.2017.08.045

Palombo, D. J., Keane, M. M., & Verfaellie, M. (2015). The medial temporal lobes are critical for reward-based decision making under conditions that promote episodic future thinking. *Hippocampus, 25*(3), 345–353. https://doi.org/10.1002/hipo.22376

Parthasarathi, T., McConnell, M. H., Luery, J., & Kable, J. W. (2017). The vivid present: Visualization abilities are associated with steep discounting of future rewards. *Frontiers in Psychology, 8*, 289. https://doi.org/10.3389/fpsyg.2017.00289

Patt, V. M., Hunsberger, R., Jones, D. A., & Verfaellie, M. (2023). The hippocampus contributes to temporal discounting when delays and rewards are experienced in the moment. *The Journal of Neuroscience: The Official Journal of the Society for Neuroscience, 43*(31), 5710–5722. https://doi.org/10.1523/JNEUROSCI.2250-22.2023

Peters, J., & Büchel, C. (2010). Episodic future thinking reduces reward delay discounting through an enhancement of prefrontal-mediotemporal interactions. *Neuron, 66*(1), 138–148. https://doi.org/10.1016/j.neuron.2010.03.026

Peters, J., & Büchel, C. (2011). The neural mechanisms of inter-temporal decision-making: Understanding variability. *Trends in Cognitive Sciences, 15*(5), 227–239. https://doi.org/10.1016/j.tics.2011.03.002

Peters, J., & D'Esposito, M. (2016). Effects of medial orbitofrontal cortex lesions on self-control in intertemporal choice. *Current Biology, 26*(19), 2625–2628. https://doi.org/10.1016/j.cub.2016.07.035

Pulcu, E., Trotter, P. D., Thomas, E. J., McFarquhar, M., Juhasz, G., Sahakian, B. J., Deakin, J. F. W., Zahn, R., Anderson, I. M., & Elliott, R. (2014). Temporal discounting in major depressive disorder. *Psychological Medicine, 44*(9), 1825–1834. https://doi.org/10.1017/S0033291713002584

Renoult, L., Davidson, P. S. R., Palombo, D. J., Moscovitch, M., & Levine, B. (2012). Personal semantics: At the crossroads of semantic and episodic memory. *Trends in Cognitive Sciences, 16*(11), 550–558. https://doi.org/10.1016/j.tics.2012.09.003

Renoult, L., Irish, M., Moscovitch, M., & Rugg, M. D. (2019). From knowing to remembering: The semantic–episodic distinction. *Trends in Cognitive Sciences, 23*(12), 1041–1057. https://doi.org/10.1016/J.TICS.2019.09.008

Reynolds, B. (2006). A review of delay-discounting research with humans: Relations to drug use and gambling. *Behavioural Pharmacology, 17*(8), 651–667. https://doi.org/10.1097/FBP.0b013e3280115f99
Rösch, S. A., Stramaccia, D. F., & Benoit, R. G. (2022). Promoting farsighted decisions via episodic future thinking: A meta-analysis. *Journal of Experimental Psychology. General, 151*(7), 1606–1635. https://doi.org/10.1037/xge0001148
Rung, J. M., & Madden, G. J. (2019). Demand characteristics in episodic future thinking II: The role of cues and cue content in changing delay discounting. *Experimental and Clinical Psychopharmacology, 27*(5), 482–495. https://doi.org/10.1037/pha0000260
Samuelson, P. A. (1937). A note on measurement of utility. *The Review of Economic Studies, 4*(2), 155. https://doi.org/10.2307/2967612
Sasse, L. K., Peters, J., & Brassen, S. (2017). Cognitive control modulates effects of episodic simulation on delay discounting in aging. *Frontiers in Aging Neuroscience, 9*, 58. https://doi.org/10.3389/fnagi.2017.00058
Schacter, D. L., Addis, D. R., & Buckner, R. L. (2007). Remembering the past to imagine the future: The prospective brain. *Nature Reviews Neuroscience, 8*(9), 657–661. https://doi.org/10.1038/nrn2213
Seinstra, M., Grzymek, K., & Kalenscher, T. (2015). Gender-specific differences in the relationship between autobiographical memory and intertemporal choice in older adults. *PLoS One, 10*(9), e0137061. https://doi.org/10.1371/journal.pone.0137061
Sekeres, M. J., Winocur, G., & Moscovitch, M. (2018). The hippocampus and related neocortical structures in memory transformation. *Neuroscience Letters, 680*, 39–53. https://doi.org/10.1016/j.neulet.2018.05.006
Senecal, N., Wang, T., Thompson, E., & Kable, J. W. (2012). Normative arguments from experts and peers reduce delay discounting. *Judgment and Decision making, 7*(5), 568–589.
Steinglass, J. E., Figner, B., Berkowitz, S., Simpson, H. B., Weber, E. U., & Walsh, B. T. (2012). Increased capacity to delay reward in anorexia nervosa. *Journal of the International Neuropsychological Society, 18*(4), 773–780. https://doi.org/10.1017/S1355617712000446
Steinglass, J. E., Lempert, K. M., Choo, T.-H., Kimeldorf, M. B., Wall, M., Walsh, B. T., Fyer, A. J., Schneier, F. R., & Simpson, H. B. (2017). Temporal discounting across three psychiatric disorders: Anorexia nervosa, obsessive compulsive disorder, and social anxiety disorder. *Depression and Anxiety, 34*(5), 463–470. https://doi.org/10.1002/da.22586
Strotz, R. H. (1956). Myopia and inconsistency in dynamic utility maximization. *Review of Economic Studies, 23*(3), 165–180.
Stuss, D. T., & Alexander, M. P. (2000). Executive functions and the frontal lobes: A conceptual view. *Psychological Research, 63*(3–4), 289–298. https://doi.org/10.1007/s004269900007
Sze, Y. Y., Stein, J. S., Bickel, W. K., Paluch, R. A., & Epstein, L. H. (2017). Bleak present, bright future: Online episodic future thinking, scarcity, delay discounting, and food demand. *Clinical Psychological Science, 5*(4), 683–697. https://doi.org/10.1177/2167702617696511
Tate, L. M. (2015). Temporal discounting rates and their relation to exercise behavior in older adults. *Physiology & Behavior, 152*(Pt A), 295–299. https://doi.org/10.1016/j.physbeh.2015.10.003
Tsao, A., Sugar, J., Lu, L., Wang, C., Knierim, J. J., Moser, M.-B., & Moser, E. I. (2018). Integrating time from experience in the lateral entorhinal cortex. *Nature, 561*(7721), 57–62. https://doi.org/10.1038/s41586-018-0459-6
Tulving, E. (1987). Multiple memory systems and consciousness. *Human Neurobiology, 6*(2), 67–80.
Wang, S., Kong, F., Zhou, M., Chen, T., Yang, X., Chen, G., & Gong, Q. (2017). Brain structure linking delay discounting and academic performance. *Human Brain Mapping, 38*(8), 3917–3926. https://doi.org/10.1002/hbm.23638
Weber, B. J., & Huettel, S. A. (2008). The neural substrates of probabilistic and intertemporal decision making. *Brain Research, 1234*, 104–115. https://doi.org/10.1016/j.brainres.2008.07.105

Wesley, M. J., & Bickel, W. K. (2014). Remember the future II: Meta-analyses and functional overlap of working memory and delay discounting. *Biological Psychiatry, 75*(6), 435–448. https://doi.org/10.1016/j.biopsych.2013.08.008

Zhao, X., Wang, Y., & Maes, J. H. R. (2022). The effect of working memory capacity and training on intertemporal decision making in children from low-socioeconomic-status families. *Journal of Experimental Child Psychology, 216*, 105347. https://doi.org/10.1016/j.jecp.2021.105347

The Neuroeconomics of Work: Computational and Neural Mechanisms of the Dynamics of Effort-Based Decisions

E. Scholey, S. Lugtmeijer, and M. A. Apps

Abstract The motivation to exert effort is fundamental to health, productivity, and well-being. From navigating life's challenges to completing everyday tasks, human behaviour is continually shaped by decisions about when and how much effort to invest. This chapter presents a neuroeconomic framework for understanding effort-based decision-making, integrating research from psychology, neuroscience, and behavioural economics. We outline how effort unfolds across distinct phases: identifying what options are available, choosing whether to act, sustaining effort during a task, and learning from the outcomes. We review key empirical findings and theoretical models conceptualising effort as a multidimensional, dynamic, and context-sensitive cost. At the neural level, we highlight the role of dopamine-driven fronto-striatal circuits in representing the subjective value of effort and explore how disruptions to these processes may contribute to motivational impairments in clinical and non-clinical populations. Finally, we identify future directions for interdisciplinary research into the mechanisms and dynamics of motivation, offering an integrative framework for understanding how and why we work.

1 Introduction

The motivation to exert effort is fundamental for health, wealth, and survival. Reduced motivation is linked to unemployment, poor educational outcomes, and poorer physical and mental health (Bautista et al., 2001; Clarke et al., 2010; Husain & Roiser, 2018; Le Heron et al., 2017; Robert et al., 2018; Vansteenkiste et al.,

Authors Selma Lugtmeijer and Matthew A. Apps have equally contributed to this chapter.

E. Scholey (✉) · S. Lugtmeijer (✉) · M. A. Apps (✉)
Centre for Human Brain Health, School of Psychology, University of Birmingham, Birmingham, UK

Institute for Mental Health, School of Psychology, University of Birmingham, Birmingham, UK
e-mail: exs165@bham.ac.uk; s.lugtmeijer@bham.ac.uk; m.a.j.apps@bham.ac.uk

D. V. Smith et al. (eds.), *Neuroeconomics: Core Topics and Current Directions*,
https://doi.org/10.1007/978-3-032-02925-6_17

2004). To understand the extensive impact of motivation, research across psychology, economics, psychiatry, neurology, and neuroscience is increasingly deploying neuroeconomics-inspired paradigms, to understand how we choose whether to exert effort or not (Salamone et al., 2007; Chong et al., 2015; Westbrook & Braver, 2015; Kool & Botvinick, 2018; Pessiglione et al., 2018; Contreras-Huerta et al., 2020). This has led to multiple insights into the computational and neural mechanisms underlying effort, its variation between healthy people, and how and why it can be impaired.

Here we outline the most recent work in this field, which highlights that there are multiple domains to motivation, various contexts which modulate our willingness to exert effort, and that choosing to exert effort or not dynamically waxes and wanes from moment to moment. We review and synthesise this work to move towards a neuroeconomic framework for motivation.

1.1 Effort-Based Decisions

The concept of effort-based decisions was first considered theoretically in a powerful proposal by Clark Hull (1943). He stated that if animals have a choice between two equally valuable rewards, such as two identical foods of equal size, their behaviour conforms to a 'law of less effort', almost ubiquitously choosing the option that is less difficult to obtain. This effort is not about the probability of succeeding at a task; even when there is no risk that an action will not lead to a reward, more difficult actions are avoided. Effort is therefore a subjective—or at least intrinsic—cost that an animal must pay in order to receive some rewards (Westbrook & Braver, 2015; Kool & Botvinick, 2018; Shenhav et al., 2013; Manohar et al., 2015). More effortful actions discount the value of rewards to a greater degree, such that more difficult tasks will only be pursued over easier ones when that higher exertion can bring a larger reward. In short, we are less likely to pursue food or money if it takes more effort to obtain them.

This theory has given rise to multiple experimental paradigms testing its tenets (Fig. 1a). Most tasks take a similar form: participants have to make a series of choices between a more difficult version of a task, such as higher grip force (Le Heron et al., 2017; Pessiglione et al., 2018), more complex arithmetic (Vassena et al., 2014, 2019; Matthews et al., 2023), or more switches between cognitive tasks (Kool et al., 2010; Kool & Botvinick, 2014; Apps et al., 2015; Chong et al., 2017) that require more effort but will gain more reward, or an easier version of the same task that requires less effort for less reward (Treadway et al., 2009). While choices vary considerably between people, the vast majority will consistently avoid high effort, only choosing it if the rewards for doing so are larger (Fig. 1b). Although decision-making is often used to probe motivation's effects on actions, it is clear that motivation plays a role in multiple phases of behaviour, from how we generate options, decide what to engage in, energise and exert the required control of behaviours, and how we learn from their consequences (Fig. 2) (Husain & Roiser,

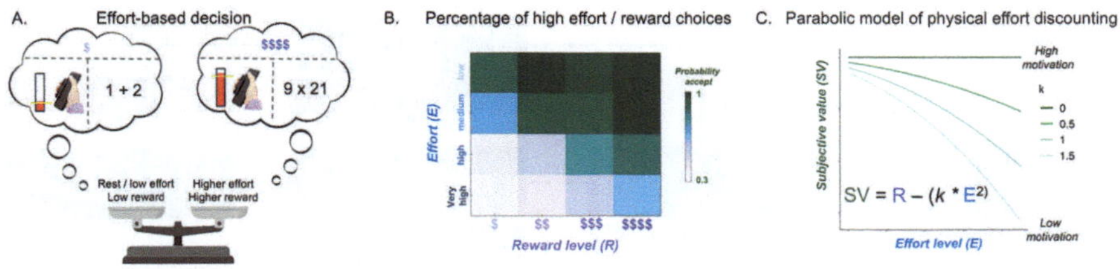

Fig. 1 Effort-based decisions and computational models of choices. (**a**) A typical effort-based decision-making task setup: Participants have to choose between a more difficult version of a task, e.g. higher physical effort (grip force) or a higher cognitive demand (harder maths puzzle), to gain more money, or an easier version of a task that requires less effort for less money. (**b**) Behavioural data from 31 healthy young adults performing a physical effort (grip force) task in which one option was to rest and gain a low reward, and the higher effort option varied in both the effort level and the reward level across trials. When the higher effort and reward option was to exert effort at the easiest level (1) and the reward at stake was high (4, top right), almost all participants consistently chose for the high option. In contrast, when the required effort was high (4) and the reward low (1, bottom left), participants often preferred to rest and gain a low reward. (**c**) An example of how a computational model of effort-based choice works. Choices depend on subjective value—the rewards being devalued or discounted by the costs of the effort. The model assumes that rewards (R) increase subjective value (SV), effort (E) decreases SV, and people discount the rewards by effort idiosyncratically—modelled with a discounting parameter (k). This schematic representation shows variable k changes how someone discounts one level of reward by different levels of effort, with higher values reflecting greater devaluation of rewards. Note that the function and corresponding figure are based on studies that provided evidence for a parabolic model of physical effort discounting

2018; Shenhav et al., 2013; Botvinick et al., 2004; Kerns et al., 2004; Ridderinkhof et al., 2004). Thus, across many phases during a task, our behaviours are only 'worth it' if the effort costs are low and rewards are high.

1.2 Computational Models of Effort-Based Choice

Using algorithms—computational models—to explain how people make decisions is a powerful neuroeconomic approach that can uncover underlying, but hidden, cognitive or neural mechanisms (Mars et al., 2012; Lockwood & Klein-Flugge, 2021). Models can precisely quantify variability between people and, when combined with brain imaging or neurophysiology, can provide a richer understanding of neural mechanisms. Computational models of 'effort-discounting' (Fig. 1c) have provided such benefits for understanding effort-based decisions (Pessiglione et al., 2018; Contreras-Huerta et al., 2020; Vogel et al., 2020). Most models assume that choices depend on a computation of subjective value (*SV*), where the level of difficulty of a task—the cost of effort (*E*)—is deducted from the magnitude and value of reward (*R*) on offer (Chong et al., 2017; Hartmann et al., 2013; Klein-Flügge et al., 2016). Thus, subjective value is higher as reward increases or as

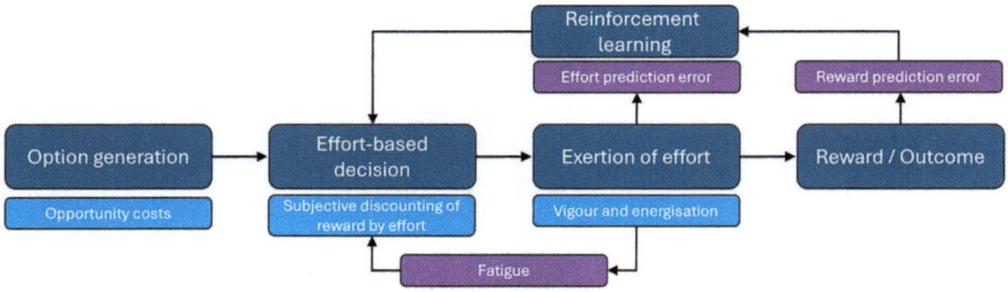

Fig. 2 Multiple phases in which motivation effects the allocation of effort. Motivation can be broken down into different phases. These phases comprise generating options, which is influenced by the opportunity costs of options in the environment; decisions of whether to initiate a behaviour, which involves evaluating, ascribing value and choosing; exertion and persistence, which involves energising the actions or cognitive control processes to a sufficient degree to succeed, receiving the reward, and learning from one's effortful actions. Throughout each, we must remain motivated in order that we continue to pursue behaviours and cognitive processes with the required vigour and control, meaning that a desire to avoid effort could impact behaviour in multiple ways during tasks. The value of exerting effort is then updated on a moment-to-moment basis, through effort prediction errors, reward prediction errors, as well as fatigue. All of these influence the subjective value of exerting effort in a dynamic manner during a task

effort decreases. Variability between people is accounted for through a discounting parameter (typically labelled 'k'), which is estimated across all the decisions made in the task, summarising the degree to which on average a person discounts reward by effort. A higher k reflects a person who devalues reward by effort to a greater degree and thus can be considered less motivated.

1.3 Brain Mechanisms of Effort

Studies using this neuroeconomic modelling approach have consistently found a network of anatomically connected fronto-striatal regions of the brain engaged during effort-based choice (Le Heron et al., 2017; Pessiglione et al., 2018; Vassena et al., 2014; Chong et al., 2017; Lopez-Gamundi et al., 2021; Forbes et al., 2023; Arulpragasam et al., 2018). The Blood Oxygenation Level Dependent (BOLD) response in dorsal portions of the anterior cingulate cortex (dACC), ventral tegmental area (VTA), ventral striatum (VS), anterior insula (AI), dorso-lateral prefrontal cortex (dlPFC), and frontal pole (FP) covaries with the subjective value of effort-discounted rewards during fMRI. Converging evidence from studies examining the effects of experimental and acquired lesions to these regions, as well as single-unit recording studies in monkeys, highlights this as a network that is important for deciding whether to exert effort (Le Heron et al., 2017). This network might be involved in motivation across multiple phases of behaviour (Fig. 3), with the dACC especially also implicated in the top-down control of cognitive processes during task behaviours (Botvinick et al., 1999, 2001, 2004; Kerns et al., 2004; MacDonald et al.,

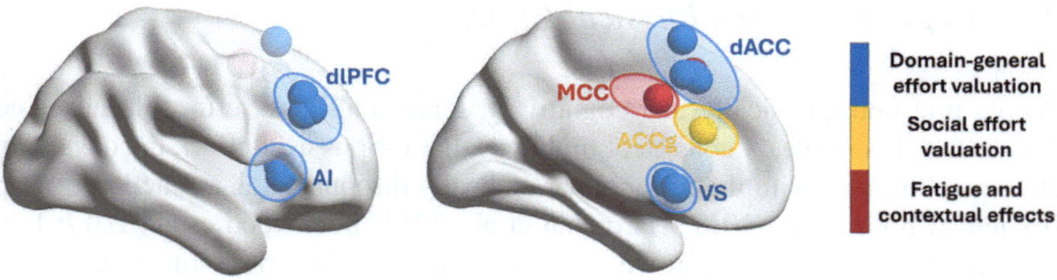

Fig. 3 The neural systems guiding effort-based decisions. Representation of the key regions engaged in deciding whether to exert effort, illustrated using spheres around Montreal Neurological Institute (MNI) standardized peak coordinates from fMRI studies (Vassena et al., 2014; Chong et al., 2017; Lopez-Gamundi et al., 2021; Lockwood et al., 2022; Müller et al., 2021; Inzlicht et al., 2018). Blue are regions in which the subjective value of exerting effort is processed in multiple studies, indicative of a domain-general network involved in deciding whether to exert effort in multiple contexts. Yellow are regions implicated specifically in deciding whether to exert effort to act prosocially and obtain other people rewards. Red are regions involved in processing contextual information, such as fatigue, that shifts the degree to which effort is discounted by rewards, and can switch effort to being valuable. Outline ovals represent approximate location of anatomical regions. Please note the dACC label corresponds to the typically large clusters that extend over cingulate sulcus and pre-SMA that are found in these studies. *dlPFC* dorso-lateral prefrontal cortex, *AI* anterior insula, *MCC* midcingulate cortex, *VS* ventral striatum; *dACC* dorsal anterior cingulate cortex. Visualised with the BrainNet Viewer (https://www.nitrc.org/projects/bnv/). Xia M, Wang J, He Y (2013) BrainNet Viewer: A Network Visualization Tool for Human Brain Connectomics. PLoS ONE 8 (7): e68910. doi:10.1371/journal.pone.0068910

2000; Shenhav et al., 2013; Frömer et al., 2021). Although it is clear that multiple neuromodulators may be involved in effort-based decision-making, most studies have only examined dopamine, finding that boosting dopamine increases people's willingness to choose to exert more effort (Chong et al., 2015, 2017; Westbrook & Braver, 2016; Le Heron et al., 2018; McGuigan et al., 2019; Westbrook et al., 2020).

Overall, this suggests there is a dopamine-driven fronto-striatal circuit (Fig. 3) that guides decisions, signalling the subjective value of effort-discounted rewards and motivating us to overcome the costs of effort.

2 The Domains of Motivation in Different Contexts

Motivation is multi-dimensional. We have to be motivated to exert different types of effort for different types of reward in different contexts. In the following sections, we outline the mechanisms underlying effort-based choice in different contexts.

2.1 *Cognitive Versus Physical Effort*

People find both physically and cognitively demanding tasks effortful and avoid those costs if possible. A consistent finding is that for physical effort, the devaluing effect is nonlinear, with the rewards parabolically discounted as a function of grip strength (Chong et al., 2017; Hartmann et al., 2013; Lockwood et al., 2017; Fig. 1c). This is less clear for cognitive effort, with some studies suggesting rewards are hyperbolically discounted when more difficult peripheral switches of attention are required to obtain them (Chong et al., 2017). However, this is likely to depend on the nature of the cognitive operations performed (Kool & Botvinick, 2018), with different cognitive processes having a different form to the cost function (Shenhav et al., 2017).

Intriguingly, despite differences in the computational nature of cognitive and physical effort costs, the same regions are implicated in signalling the subjective value of effort-discounted rewards, including the dACC, VS, and AI (Chong et al., 2017; Lopez-Gamundi et al., 2021) (Fig. 3). To date, only the dorsolateral amygdala has showed any difference, being engaged during cognitive but not physical effort-based choices (Chong et al., 2017; Hosking et al., 2015). Similarly, boosting dopamine has been shown to increase the motivation for both cognitive and physical effort, suggesting significant overlap in the neural mechanisms underlying how we motivate different forms of effort (Westbrook & Braver, 2016; McGuigan et al., 2019).

2.2 *Prosocial Effort: The Effort of Helping Others*

Prosocial behaviours—actions that benefit others—have been studied extensively in neuroeconomics (Fehr & Fischbacher, 2004). However, the field often overlooked that deciding to be prosocial usually requires us to choose whether to exert effort: for example, deciding whether to help a colleague at work or hold open the door for a stranger (Contreras-Huerta et al., 2020). By adapting typical effort-based decision tasks where one can obtain a reward for self, to include a prosocial condition where people must make a choice and exert effort, but to obtain a reward for a stranger, new insights into prosocial motivation have been revealed. Consistently, it has shown that people discount the value of rewards by both cognitive and physical effort to a significantly greater degree during prosocial compared to self-benefiting choices (Forbes et al., 2023; Lockwood et al., 2017, 2021; Depow et al., 2022; Gaule et al., 2024). Moreover, even when people make a choice to help, they exert less energy into high effort actions for others compared to themselves. The degree to which this is the case changes across the lifespan, and varies with psychological traits including empathy and apathy (Lockwood et al., 2017; Contreras-Huerta et al., 2022).

Research has revealed some of the neural mechanisms underlying self-benefitting and prosocial effort-based decisions overlap (Forbes et al., 2023; Lockwood et al.,

2022) in dACC and AI. However, this is not entirely the case. Damage to portions of the ventromedial prefrontal cortex selectively reduce the willingness to exert effort for others but not ourselves (Lockwood et al., 2024). Moreover, the VTA is engaged only when guiding decisions about whether to exert effort for self, and conversely a separate ACC sub-region, lying in the gyrus (ACCg) is engaged in motivating effort only when it is prosocial (Lockwood et al., 2022; Contreras-Huerta et al., 2023). Overall, this suggests that while there may be a domain-general system guiding effort-based decisions in multiple contexts, additional systems may be involved depending on the context in which an effort-based choice is made (Fig. 3).

3 The Dynamic Nature of Effort-Based Choices

The motivation to exert effort is not static, but instead fluctuates from moment-to-moment (Müller & Apps, 2019). Here, we discuss some of the processes underlying these fluctuations, and how computational modelling can capture how they influence effort-based decision-making.

3.1 Learning Whether to Exert Effort

How effortful an action is, and thus which actions to avoid, must often be gradually learned over time. Such 'effort learning' has recently begun to be investigated using tasks inspired by reinforcement learning theory—a hugely influential account for how humans, animals, and computers can learn to maximise rewards. Participants have to learn through trial and error how much cognitive or physical effort is required in order to successfully complete a task and obtain rewards (i.e. rather than maximise rewards, they learn to minimise effort costs). People indeed learn to avoid effort. Reinforcement learning algorithms, in which higher effort is treated as lower value, capture this process (Arulpragasam et al., 2018; Skvortsova et al., 2014; Hauser et al., 2017; Nagase et al., 2018), can capture this learning process. In these models, expectations about the effort required are updated through prediction errors (Rescorla & Wagner, 1972; Sutton & Barto, 1981) computed as the difference between expected effort cost and the actual effort required. These update future expectations about the effort required, allowing us to learn to avoid highly effortful behaviours.

Despite using the same computational rule for learning about effort and reward, and evidence that effort can modulate learning about rewards (Jarvis et al., 2022), studies have found dissociations in the brain areas that learn about them. Whilst reward prediction errors are associated with BOLD response in the VS and ventromedial prefrontal cortex (vmPFC), effort prediction errors correlate with activity in right AI and dorsal anterior cingulate cortex extending into pre-supplementary motor area (pre-SMA), the same areas that signal subjective value outlined above (Fig. 3). However, during decision-making, expected reward and effort are integrated in an overall value signal in VS (Skvortsova et al., 2014; Hauser et al., 2017; Nagase et al., 2018).

3.2 The Effects of Fatigue on Effort-Based Choice

Decisions of whether to exert effort can be significantly influenced by fatigue. Over time and with increased effort demands, fatigue causes a reduction in both task performance and the willingness to exert future cognitive or physical effort (Mackworth, 1964; Boksem & Tops, 2008; Marcora, 2009; Tanaka & Watanabe, 2012; Meyniel et al., 2013). However, these deficits can be alleviated through rest (Massar et al., 2018).

These findings have been integrated into computational models, explaining how fatigue and rest can dynamically influence the willingness to exert physical effort from moment to moment (Matthews et al., 2023; Meyniel et al., 2013, 2014; Blain et al., 2016, 2019; Müller et al., 2021). These models highlight that there are hidden states of fatigue: recoverable and unrecoverable. Both types increase with effort exertion; however, recoverable fatigue decreases with rest, whereas unrecoverable fatigue simply accumulates over time. These fatigue states influence the willingness to exert subsequent effort, by weighting the effort discounting parameter, k (Sect. 1.2). Higher levels of fatigue increase effort discounting, thus reducing the subjective value of exerting effort.

There is a wide network of brain regions involved in signalling and integrating cognitive and physical fatigue into effort-based decisions. There are two key findings from this research. One is that the VS appears to code subjective value in a dynamic manner, changing its response in line with how much fatigue is increasing the discounting of rewards by effort. However, the components of fatigue are processed in other regions, with the middle frontal gyrus signalling the unrecoverable aspects of fatigue (Blain et al., 2016, 2019; Müller et al., 2021), whilst the midcingulate cortex (MCC), lying posterior to the dACC, signals recoverable fatigue (Müller et al., 2021). This is consistent with the notion that whilst there is a core network of regions involved in effort-based decisions, the context a choice is made in is signalled in other regions that influence choice behaviour (Fig. 3).

3.3 The Opportunity Costs of Exerting Effort

The time we spend doing one effortful action is often at the expense of other activities we could be doing (Niv et al., 2006, 2007). This 'opportunity cost' imposed by the time doing an action is thought to be determined by the average rate at which you can obtain rewards from your actions in the environment. When average reward rates are high, there is a higher opportunity cost for moving slowly (the 'cost of sloth'), making it worthwhile to exert effort to move faster. Research shows these reward rates influence the 'vigour' of effortful movement, in terms of both speed and force production (Shadmehr et al., 2010; Guitart-Masip et al., 2011; Beierholm et al., 2013; Hird et al., 2022). Higher average reward rates also cause people to respond faster and make more errors when exerting cognitive effort (Otto & Daw, 2019). It has also been shown that average *effort* rates influence movement vigour (Sukumar et al., 2024) suggesting that both reward and effort influence the opportunity cost of acting in an environment.

How does the brain represent opportunity costs? Animal studies have found that dopamine in the mesolimbic pathway tracks average reward rates (Niv et al., 2007; Hamid et al., 2016), and indeed pharmacological manipulation of dopamine in humans changes how they make opportunity cost decisions (Constantino et al., 2017; Le Heron et al., 2020), as well as influencing movement vigour/force (Beierholm et al., 2013; Michely et al., 2020). The brain regions involved in tracking opportunity costs also implicate the domain-general circuit we have outlined, including the dACC and AI (Kolling et al., 2012; Wittmann et al., 2016; Khalighinejad et al., 2021).

4 Effort-Based Decisions Index Motivation Impairments in Clinical and Non-clinical Populations

Disruptions to motivation (e.g. apathy, anhedonia, and fatigue) are some of the most common symptoms in neurology and psychiatry, but are also present in a less severe form in the healthy population (Husain & Roiser, 2018). Effort-based decision tasks, combined with computational modelling, are beginning to offer insight into the underlying mechanisms of these motivational deficits (Le Heron et al., 2017; Contreras-Huerta et al., 2020; Treadway et al., 2009).

4.1 Individual Differences in Psychological Traits Are Linked to Effort-Based Decision-Making

Effort-based decision-making tasks are ideally set up for examining relationships with individual differences in psychological traits, as they can capture variability between participants, and measure this precisely with computational modelling.

Perhaps the most common finding is that the willingness to exert effort is associated with two key traits, apathy (defined as a reduction in goal-directed behaviour) and need for cognition (Lockwood et al., 2017; Contreras-Huerta et al., 2022; Cacioppo & Petty, 1982; Sandra & Otto, 2018; Jurgelis et al., 2021; Gheza et al., 2023; Mækelæ et al., 2023). The willingness to exert cognitive effort has also been associated with self-reported anxiety, cognitive function, behavioural activation, apathy, and self-efficacy (Bustamante et al., 2023). Moreover, the willingness to exert prosocial effort is associated with social dimensions of apathy, whereas effort to benefit oneself is associated with behavioural apathy (Lockwood et al., 2017; Contreras-Huerta et al., 2022). This suggests that effort sensitivity in different domains of behaviour may be transdiagnostic symptoms that are present in the healthy population.

Studies have shown age-related differences in effort-based decision-making (Lockwood et al., 2021; Westbrook et al., 2013; Crawford et al., 2023; Byrne & Anaraky, 2020; Ganesan & Steinbeis, 2021). Older adults, compared to younger adults, show increased discounting of rewards by cognitive effort (Westbrook et al., 2013; Crawford et al., 2023). However, the framing of the outcome, either exerting effort to attain a reward or avoid a loss, may influence age-related differences in effort discounting. Both younger and older adults are more willing to exert cognitive effort to avoid a loss than obtain a reward, with no age-related differences (Crawford et al., 2023). Yet, for physical effort, whilst older adults are again more willing to exert cognitive effort to avoid a loss, younger adults show the opposite pattern, being more willing to exert effort to obtain reward (Byrne & Anaraky, 2020). In young people, an awareness of effort is apparent even in childhood, although it does not appear to carry through into choice behaviour. For example, children are able to identify that higher levels of cognitive task difficulty are more effortful, but show no apparent aversion to choosing higher effort (Ganesan & Steinbeis, 2021). All of this highlights how variability in the willingness to exert effort is key to understanding motivation in daily life in healthy people.

4.2 *Effort and Impaired Motivation in Neurological and Psychiatric Disorders*

Clinical impairments to motivation are typically measured using questionnaires. Whilst useful for measuring symptom severity, these scales rely on patient insight and may not directly map onto cognitive processes. For instance, motivational impairments can be very different if they arise due to a loss of incentivisation by rewards, compared to a heightened sensitivity to the costs of effort. Computational approaches to motivational deficits can be used to discriminate between these and bridge the gap between clinical assessment and the underlying pathophysiology (e.g. lack of dopamine) (Pessiglione et al., 2018).

A heightened sensitivity to effort cost has been found in multiple conditions using this approach, including fronto-temporal dementia, Parkinson's disease (PD), depression, small vessel disease, and schizophrenia—all conditions associated with high levels of apathy, anhedonia, and fatigue (Husain & Roiser, 2018; Le Heron et al., 2017; Le Bouc et al., 2023).

Patients with Parkinson's disease have been repeatedly found to be less willing to invest physical effort for rewards compared to controls. However, when patients are on dopaminergic medication, they choose to invest more effort for rewards relative to off medication (Chong et al., 2015), an effect that extends to cognitive effort (McGuigan et al., 2019). Similarly, both dopaminergic and serotonergic medications increase the willingness to exert effort in depression (Pessiglione et al., 2018; Clery-Melin et al., 2011), and dopaminergic deficits are also linked to negative symptoms in schizophrenia (Hartmann et al., 2015). This evidence suggests that impairments to dopamine—alongside other neuromodulators—are a key component of severely impaired motivation and offer a potential avenue for future treatment. Thus, using effort-based decision-making paradigms suggests that augmenting both dopaminergic and serotonergic systems might be key for ameliorating impairments to motivation.

5 Future Directions

Since the conception of the law of least effort introduced by Hull, our understanding of the mechanisms of effort-based decisions has progressed. However, there are still many outstanding questions. Here, we offer four possible avenues for future research.

5.1 Neuromodulators

Whilst the dopaminergic cortico-striatal circuit is heavily implicated in effort-based decision-making (Salamone & Correa, 2024), there is not always a clear functional relationship between dopamine and effort (Walton & Bouret, 2019). Work has begun to create a unified account of dopamine that can account for its role in effort processing and its many other putative functions, as well as explaining how dopamine both is and isn't involved in effort processing in different studies (Collins & Frank, 2014; Berke, 2018; Westbrook et al., 2021).

As noted above, other neuromodulators beyond dopamine are also likely to be involved in motivating effortful exertion. Serotonin appears to be involved in effort discounting (Meyniel et al., 2016; Silveira et al., 2020; Michely et al., 2023; Nobis et al., 2023), encoding effort prediction errors (Scholl et al., 2017), and tracking opportunity costs (Khalighinejad et al., 2022). There is also evidence implicating other neuromodulators including GABA (Hu et al., 2022), noradrenaline (Varazzani et al., 2015; Borderies et al., 2020), and acetylcholine (Abdoust et al., 2024). One promising avenue may be using magnetic resonance spectroscopy (MRS) in humans to measure metabolites in the brain. For example, accumulation of glutamate could drive fatigue signals (Wiehler et al., 2022), and glutamate and other metabolites have been implicated in effort-based decisions (Strasser et al., 2020; Barakat et al., 2024). Techniques such as MRS, as well as further pharmacological manipulations in humans and animals, will be needed to disentangle the roles of different neuromodulators.

5.2 Oscillatory Mechanisms

Above we have highlighted many fMRI studies that provide insight into where in the brain effort is processed. However, there is a growing interest in understanding the temporal dynamics that underlie effort-based decisions, and in particular the neural oscillations that may be a key signature of how the brain processes information. Studies using EEG and MEG have shown that the influence of opportunity costs on cognitive effort has been linked to reduced theta band power (Lin et al., 2020), and beta band power in medial frontal cortex is also linked to cognitive control allocation and fatigue-related time-on-task effects in non-human primates (Wilson et al., 2016; Cavanagh & Frank, 2014). Intracranial EEG (iEEG) and invasive neurophysiological recordings, once only possible in animal models, are now possible in patients, which has opened the possibility of examining neural oscillations in deep structures in humans (Ricciardi et al., 2023). This has revealed that beta band power in the basal ganglia tracks moment-to-moment changes in effort in patients with Parkinson's disease (Hoy et al., in press). Advances in neurophysiological methods will likely further transform our understanding of effort processing.

5.3 Ecology

Neuroeconomics research is increasingly inspired by the animal ecology literature. This approach argues that cost-benefit decisions use brain systems that may be conserved across species, thus we can better understand human neuroeconomic decisions by using tasks and theories from ecology (Pearson et al., 2014; Mobbs et al., 2018). One example is the use of 'patch foraging' tasks, where the participant must choose between staying put and foraging for resources in one location ('patch'), or leaving to seek better patches elsewhere. These studies have told us much about reward decision-making during foraging (Le Heron et al., 2020; Hayden et al., 2011; Constantino & Daw, 2015; Eisenreich et al., 2019), and many studies are now also examining how effort costs are involved in these decisions (Sukumar et al., 2024; Toro-Serey et al., 2021). Ecological paradigms also allow a wider range of behavioural strategies to be expressed, which can help us understand individual differences (Bustamante et al., 2023). Foraging paradigms therefore offer a means for bringing together the tools of neuroscience, psychology, biology, and economics to better understand how we choose to exert effort.

5.4 The Effort Paradox

Most research in this chapter has treated effort as a cost we try to avoid. However, there is an 'effort paradox', in that sometimes we seek and value effort (Inzlicht et al., 2018). This paradox can be seen in many behaviours in everyday life, such as the desire to climb mountains or valuing effort to meet important deadlines. Recently, new paradigms are emerging that offer accounts for how effort can be both avoided and valued at different points in time (Lin et al., 2024). For example, when people are under pressure to exert effort to meet a deadline, neural correlates of this 'pressure' have been found in sub-regions of the putamen and mid-cingulate cortex (MCC) that lead to a switch in the value of effort from cost to benefit (Pisauro et al., 2024). Thus, the future of effort-based decision-making research may lie in understanding the situations and computations underlying when effort is not treated as a cost.

6 Conclusion

Every day we make decisions about whether to exert effort to obtain rewards, persist and motivate behaviours over extended periods of time, and adapt our decisions of whether to exert effort due to changes in our bodies, brains, and the external environment. Computational models have helped us identify the hidden neural mechanisms underlying multiple aspects of effort-based choice, including how people discount reward by effort, learn about effort over time, or how fatigue influences the motivation to exert effort. Research has identified a dopamine-driven fronto-striatal circuit that seems to play a central role in deciding whether to exert effort, controlling our behaviours when we exert ourselves, and updating our motivation over time. Disruptions to this circuit are linked to variability in motivation in healthy people and in disease. Yet there is much still to learn. Future advances will come from research offering new theories of motivation beyond the law of less effort, and integrating methods and knowledge across ecology, neuroscience, and economics. Such efforts may bring us closer to understanding why we sometimes think hard work isn't 'worth it'.

Acknowledgements This work was supported by a Biotechnology and Biological Sciences Research Council David Phillips Fellowship (BB/R010668/2; MAJA), a Jacobs Foundation Fellowship (MAJA), a Wellcome Trust Discovery Award (MAJA), and Medical Research Council IMPACT doctoral training scholarship (ES). We would like to thank all members of the Motivation and Social Neuroscience lab, past and present, who have supported research on effort processing.

References

Abdoust, M. E., Froböse, M. I., Schnitzler, A., Schreivogel, E., & Jocham, G. (2024). Distinct roles of dopamine and acetylcholine in delay- and effort-based decision making in humans. *bioRxiv*, 2023.11.20.566558. https://doi.org/10.1101/2023.11.20.566558

Apps, M. A. J., Grima, L. L., Manohar, S., & Husain, M. (2015). The role of cognitive effort in subjective reward devaluation and risky decision-making. *Scientific Reports, 5*, 16880. https://doi.org/10.1038/srep16880

Arulpragasam, A. R., Cooper, J. A., Nuutinen, M. R., & Treadway, M. T. (2018). Corticoinsular circuits encode subjective value expectation and violation for effortful goal-directed behavior. *Proceedings of the National Academy of Sciences, 115*, E5233–E5242. https://doi.org/10.1073/pnas.1800444115

Barakat, A., et al. (2024). A neurometabolic signature in the frontal cortex explains individual differences in effort-based decision-making. *bioRxiv*, 2024.01.23.576854. https://doi.org/10.1101/2024.01.23.576854

Bautista, L. M., Tinbergen, J., & Kacelnik, A. (2001). To walk or to fly? How birds choose among foraging modes. *Proceedings of the National Academy of Sciences of the United States of America, 98*, 1089–1094. https://doi.org/10.1073/pnas.98.3.1089

Beierholm, U., et al. (2013). Dopamine modulates reward-related vigor. *Neuropsychopharmacology, 38*, 1495–1503. https://doi.org/10.1038/npp.2013.48

Berke, J. D. (2018). What does dopamine mean? *Nature Neuroscience, 21*, 787–793. https://doi.org/10.1038/s41593-018-0152-y

Blain, B., Hollard, G., & Pessiglione, M. (2016). Neural mechanisms underlying the impact of daylong cognitive work on economic decisions. *Proceedings of the National Academy of Sciences of the United States of America, 113*, 6967–6972. https://doi.org/10.1073/pnas.1520527113

Blain, B., et al. (2019). Neuro-computational impact of physical training overload on economic decision-making. *Current Biology, 29*, 3289–3297. https://doi.org/10.1016/j.cub.2019.08.054

Boksem, M. A. S., & Tops, M. (2008). Mental fatigue: Costs and benefits. *Brain Research Reviews, 59*, 125–139. https://doi.org/10.1016/j.brainresrev.2008.07.001

Borderies, N., Bornert, P., Gilardeau, S., & Bouret, S. (2020). Pharmacological evidence for the implication of noradrenaline in effort. *PLoS Biology, 18*, e3000793. https://doi.org/10.1371/journal.pbio.3000793

Botvinick, M., Nystrom, L. E., Fissell, K., Carter, C. S., & Cohen, J. D. (1999). Conflict monitoring versus selection-for-action in anterior cingulate cortex. *Nature, 402*, 179–181. https://doi.org/10.1038/46035

Botvinick, M. M., Braver, T. S., Barch, D. M., Carter, C. S., & Cohen, J. D. (2001). Conflict monitoring and cognitive control. *Psychological Review, 108*, 624–652. https://doi.org/10.1037/0033-295X.108.3.624

Botvinick, M. M., Cohen, J. D., & Carter, C. S. (2004). Conflict monitoring and anterior cingulate cortex: An update. *Trends in Cognitive Sciences, 8*, 539–546. https://doi.org/10.1016/j.tics.2004.10.003

Bustamante, L. A., et al. (2023). Effort foraging task reveals positive correlation between individual differences in the cost of cognitive and physical effort in humans. *Proceedings of the National Academy of Sciences, 120*, e2221510120. https://doi.org/10.1073/pnas.2221510120

Byrne, K. A., & Anaraky, R. G. (2020). Strive to win or not to lose? Age-related differences in framing effects on effort-based decision-making. *The Journals of Gerontology: Series B, 75*(10), 2095–2105. https://doi.org/10.1093/geronb/gbz136

Cacioppo, J. T., & Petty, R. E. (1982). The need for cognition. *Journal of Personality and Social Psychology, 42*, 116. https://doi.org/10.1037/0022-3514.42.1.116

Cavanagh, J. F., & Frank, M. J. (2014). Frontal theta as a mechanism for cognitive control. *Trends in Cognitive Sciences, 18*, 414–421. https://doi.org/10.1016/j.tics.2014.04.012

Chong, T. T.-J., et al. (2015). Dopamine enhances willingness to exert effort for reward in Parkinson's disease. *Cortex*. https://doi.org/10.1016/j.cortex.2015.04.003

Chong, T. T.-J., et al. (2017). Neurocomputational mechanisms underlying subjective valuation of effort costs. *PLoS Biology, 15*, e1002598. https://doi.org/10.1371/journal.pbio.1002598

Clarke, D. E., Ko, J. Y., Lyketsos, C., Rebok, G. W., & Eaton, W. W. (2010). Apathy and cognitive and functional decline in community-dwelling older adults: Results from the Baltimore ECA longitudinal study. *International Psychogeriatrics, 22*, 819–829. https://doi.org/10.1017/S1041610209991402

Clery-Melin, M. L., et al. (2011). Why don't you try harder? An investigation of effort production in major depression. *PLoS One, 6*, e23178. https://doi.org/10.1371/journal.pone.0023178

Collins, A. G. E., & Frank, M. J. (2014). Opponent actor learning (OpAL): Modeling interactive effects of striatal dopamine on reinforcement learning and choice incentive. *Psychological Review, 121*, 337–366. https://doi.org/10.1037/a0037015

Constantino, S. M., & Daw, N. (2015). Learning the opportunity cost of time in a patch-foraging task. *Cognitive, Affective, & Behavioral Neuroscience, 15*, 837–853. https://doi.org/10.3758/s13415-015-0350-y

Constantino, S. M., et al. (2017). A neural mechanism for the opportunity cost of time. *bioRxiv, 173443*. https://doi.org/10.1101/173443

Contreras-Huerta, L. S., Pisauro, A., & Apps, M. A. (2020). Effort shapes social cognition and behaviour: A neuro-cognitive framework. *Neuroscience & Biobehavioral Reviews*, 10.31234/osf.io/mvwqz.

Contreras-Huerta, L. S., Lockwood, P. L., Bird, G., Apps, M. A., & Crockett, M. J. (2022). Prosocial behavior is associated with transdiagnostic markers of affective sensitivity in multiple domains. *Emotion, 22*, 820. https://doi.org/10.1037/emo0000813

Contreras-Huerta, L. S., et al. (2023). Neural representations of vicarious rewards are linked to interoception and prosocial behaviour. *NeuroImage, 119881*. https://doi.org/10.1016/j.neuroimage.2023.119881

Crawford, J. L., English, T., & Braver, T. S. (2023). Cognitive effort-based decision-making across experimental and daily life indices in younger and older adults. *The Journals of Gerontology: Series B, 78*(1), 40–50. https://doi.org/10.1093/geronb/gbac167

Depow, G. J., Lin, H., & Inzlicht, M. (2022). Cognitive effort for self, strangers, and charities. *Scientific Reports, 12*, 15009. https://doi.org/10.1038/s41598-022-19163-y

Eisenreich, B. R., Hayden, B. Y., & Zimmermann, J. (2019). Macaques are risk-averse in a freely moving foraging task. *Scientific Reports, 9*, 15091. https://doi.org/10.1038/s41598-019-51442-z

Fehr, E., & Fischbacher, U. (2004). Social norms and human cooperation. *Trends in Cognitive Sciences, 8*, 185–190. https://doi.org/10.1016/j.tics.2004.02.007

Forbes, P. A., et al. (2023). Acute stress reduces effortful prosocial behaviour. *eLife, 12*. https://doi.org/10.7554/eLife.87271

Frömer, R., Lin, H., Dean Wolf, C. K., Inzlicht, M., & Shenhav, A. (2021). Expectations of reward and efficacy guide cognitive control allocation. *Nature Communications, 12*, 1030. https://doi.org/10.1038/s41467-021-21315-z

Ganesan, K., & Steinbeis, N. (2021). Effort-related decision-making and its underlying processes during childhood. *Developmental Psychology, 57*, 1487–1496. https://doi.org/10.1037/dev0001228

Gaule, A., et al. (2024). Reduced prosocial motivation and effort in adolescents with conduct problems and callous-unemotional traits. *Journal of Child Psychology and Psychiatry*. https://doi.org/10.1111/jcpp.13945

Gheza, D., Kool, W., & Pourtois, G. (2023). Need for cognition moderates the relief of avoiding cognitive effort. *PLoS One, 18*, e0287954. https://doi.org/10.1371/journal.pone.0287954

Guitart-Masip, M., Beierholm, U. R., Dolan, R., Duzel, E., & Dayan, P. (2011). Vigor in the face of fluctuating rates of reward: An experimental examination. *Journal of Cognitive Neuroscience, 23*, 3933–3938. https://doi.org/10.1162/jocn_a_00090

Hamid, A. A., et al. (2016). Mesolimbic dopamine signals the value of work. *Nature Neuroscience, 19*, 117–126. https://doi.org/10.1038/nn.4173

Hartmann, M. N., Hager, O. M., Tobler, P. N., & Kaiser, S. (2013). Parabolic discounting of monetary rewards by physical effort. *Behavioural Processes, 100*, 192–196. https://doi.org/10.1016/j.beproc.2013.09.014

Hartmann, M. N., et al. (2015). Apathy but not diminished expression in schizophrenia is associated with discounting of monetary rewards by physical effort. *Schizophrenia Bulletin, 41*, 503–512. https://doi.org/10.1093/schbul/sbu102

Hauser, T. U., Eldar, E., & Dolan, R. J. (2017). Separate mesocortical and mesolimbic pathways encode effort and reward learning signals. *Proceedings of the National Academy of Sciences of the United States of America, 114*, E7395–E7404. https://doi.org/10.1073/pnas.1705643114

Hayden, B. Y., Pearson, J. M., & Platt, M. L. (2011). Neuronal basis of sequential foraging decisions in a patchy environment. *Nature Neuroscience, 14*, 933–939. https://doi.org/10.1038/nn.2856

Hird, E. J., et al. (2022). Dopamine and reward-related vigor in younger and older adults. *Neurobiology of Aging, 118*, 34–43. https://doi.org/10.1016/j.neurobiolaging.2022.06.003

Hosking, J. G., Cocker, P. J., & Winstanley, C. A. (2015). Prefrontal cortical Inactivations decrease willingness to expend cognitive effort on a rodent cost/benefit decision-making task. *Cerebral Cortex*. https://doi.org/10.1093/cercor/bhu321

Hoy, C. W., et al. (in press). Beta and theta oscillations track effort and previous reward in human basal ganglia and prefrontal cortex during decision making. *Proceedings of the National Academy of Sciences*. https://doi.org/10.1101/2023.12.05.570285

Hu, E. J., et al. (2022). Sensorimotor cortex GABA moderates the relationship between physical exertion and assessments of effort. *Journal of Neuroscience, 42*, 6121–6130. https://doi.org/10.1523/JNEUROSCI.2255-21.2022

Hull, C. (1943). *Principles of behavior*. Appleton-Century.

Husain, M., & Roiser, J. P. (2018). Neuroscience of apathy and anhedonia: a transdiagnostic approach. *Nature Reviews Neuroscience, 19*, 470–484. https://doi.org/10.1038/s41583-018-0029-9

Inzlicht, M., Shenhav, A., & Olivola, C. Y. (2018). The effort paradox: Effort is both costly and valued. *Trends in Cognitive Science, 22*, 337–349. https://doi.org/10.1016/j.tics.2018.01.007

Jarvis, H., et al. (2022). Effort reinforces learning. *Journal of Neuroscience, 42*, 7648–7658. https://doi.org/10.1523/JNEUROSCI.2223-21.2022

Jurgelis, M., et al. (2021). Heightened effort discounting is a common feature of both apathy and fatigue. *Scientific Reports, 11*, 1–12. https://doi.org/10.1038/s41598-021-01287-2

Kerns, J. G., et al. (2004). Anterior cingulate conflict monitoring and adjustments in control. *Science, 303*, 1023–1026. https://doi.org/10.1126/science.1089910

Khalighinejad, N., Garrett, N., Priestley, L., Lockwood, P., & Rushworth, M. F. S. (2021). A habenula-insular circuit encodes the willingness to act. *Nature Communications, 12*, 6329. https://doi.org/10.1038/s41467-021-26569-1

Khalighinejad, N., Manohar, S., Husain, M., & Rushworth, M. F. S. (2022). Complementary roles of serotonergic and cholinergic systems in decisions about when to act. *Current Biology, 32*, 1150–1162.e7. https://doi.org/10.1016/j.cub.2022.01.042

Klein-Flügge, M. C., Kennerley, S. W., Friston, K., & Bestmann, S. (2016). Neural signatures of value comparison in human cingulate cortex during decisions requiring an effort-reward trade-off. *Journal of Neuroscience, 36*, 10002–10015. https://doi.org/10.1523/JNEUROSCI.0292-16.2016

Kolling, N., Behrens, T. E. J., Mars, R. B., & Rushworth, M. F. S. (2012). Neural mechanisms of foraging. *Science, 336*, 95–98. https://doi.org/10.1126/science.1216930

Kool, W., & Botvinick, M. (2014). A labor/leisure tradeoff in cognitive control. *Journal of Experimental Psychology: General, 143*, 131–141. https://doi.org/10.1037/a0031048

Kool, W., & Botvinick, M. (2018). Mental labour. *Nature Human Behaviour, 2*, 899–908. https://doi.org/10.1038/s41562-018-0401-9

Kool, W., McGuire, J. T., Rosen, Z. B., & Botvinick, M. M. (2010). Decision making and the avoidance of cognitive demand. *Journal of Experimental Psychology: General, 139*, 665–682. https://doi.org/10.1037/a0020198

Le Bouc, R., et al. (2023). Effort avoidance as a core mechanism of apathy in frontotemporal dementia. *Brain, 146*, 712–726. https://doi.org/10.1093/brain/awac427

Le Heron, C., Apps, M. A. J., & Husain, M. (2017). The anatomy of apathy: a neurocognitive framework for amotivated behavior. *Neuropsychologia*. https://doi.org/10.1016/j.neuropsychologia.2017.07.003

Le Heron, C., et al. (2018). Distinct effects of apathy and dopamine on effort-based decision-making in Parkinson's disease. *Brain, 141*, 1455–1469. https://doi.org/10.1093/brain/awy110

Le Heron, C., et al. (2020). Dopamine modulates dynamic decision-making during foraging. *Journal of Neuroscience, 40*, 5273–5282. https://doi.org/10.1523/JNEUROSCI.2586-19.2020

Lin, H., Saunders, B., Friese, M., Evans, N. J., & Inzlicht, M. (2020). Strong effort manipulations reduce response caution: A preregistered reinvention of the ego-depletion paradigm. *Psychological Science, 31*, 531–547. https://doi.org/10.1177/0956797620904990

Lin, H., Westbrook, A., Fan, F., & Inzlicht, M. (2024). An experimental manipulation of the value of effort. *Nature Human Behaviour, 1–13*. https://doi.org/10.1038/s41562-024-01842-7

Lockwood, P., & Klein-Flugge, M. (2021). Computational modelling of social cognition and behaviour—a reinforcement learning primer. *Social Cognitive and Affective Neuroscience, 16*, 761–771. https://doi.org/10.1093/scan/nsaa040

Lockwood, P. L., et al. (2017). Prosocial apathy for helping others when effort is required. *Nature Human Behaviour, 1*, s41562-017–0131. https://doi.org/10.1038/s41562-017-0131

Lockwood, P. L., et al. (2021). Aging increases prosocial motivation for effort. *Psychological Science, 32*, 668–681. https://doi.org/10.1177/0956797620975781

Lockwood, P. L., et al. (2022). Distinct neural representations for prosocial and self-benefiting effort. *Current Biology, 32*, 4172–4185.e7. https://doi.org/10.1016/j.cub.2022.08.010

Lockwood, P. L., et al. (2024). Human ventromedial prefrontal cortex is necessary for prosocial motivation. *Nature Human Behaviour, 1–14*. https://doi.org/10.1038/s41562-024-01899-4

Lopez-Gamundi, P., et al. (2021). The neural basis of effort valuation: A meta-analysis of functional magnetic resonance imaging studies. *Neuroscience & Biobehavioral Reviews, 131*, 1275–1287. https://doi.org/10.1016/j.neubiorev.2021.10.024

MacDonald, A. W., Cohen, J. D., Stenger, V. A., & Carter, C. S. (2000). Dissociating the role of the dorsolateral prefrontal and anterior cingulate cortex in cognitive control. *Science, 288*, 1835–1838. https://doi.org/10.1126/science.288.5472.1835

Mackworth, J. F. (1964). Performance decrement in vigilance, threshold, and high-speed perceptual motor tasks. *Canadian Journal of Psychology/Revue canadienne de psychologie, 18*, 209–223. https://doi.org/10.1037/h0083302

Mækelæ, M. J., et al. (2023). Is it cognitive effort you measure? Comparing three task paradigms to the need for cognition scale. *PLoS One, 18*, e0290177. https://doi.org/10.1371/journal.pone.0290177

Manohar, S. G., et al. (2015). Reward pays the cost of noise reduction in motor and cognitive control. *Current Biology, 25*, 1707–1716. https://doi.org/10.1016/j.cub.2015.05.038

Marcora, S. (2009). Perception of effort during exercise is independent of afferent feedback from skeletal muscles, heart, and lungs. *Journal of Applied Physiology, 106*, 2060–2062. https://doi.org/10.1152/japplphysiol.90378.2008

Mars, R. B., Shea, N. J., Kolling, N., & Rushworth, M. F. (2012). Model-based analyses: Promises, pitfalls, and example applications to the study of cognitive control. *The Quarterly Journal of Experimental Psychology, 65*, 252–267. https://doi.org/10.1080/17470211003668272

Massar, S. A. A., Csatho, A., & Van der Linden, D. (2018). Quantifying the motivational effects of cognitive fatigue through effort-based decision making. *Frontiers in Psychology, 9*, 843. https://doi.org/10.3389/fpsyg.2018.00843

Matthews, J., et al. (2023). Computational mechanisms underlying the dynamics of physical and cognitive fatigue. *Cognition, 240*, 105603. https://doi.org/10.1016/j.cognition.2023.105603

McGuigan, S., et al. (2019). Dopamine restores cognitive motivation in Parkinson's disease. *Brain: A Journal of Neurology*. https://doi.org/10.1093/brain/awy341

Meyniel, F., Sergent, C., Rigoux, L., Daunizeau, J., & Pessiglione, M. (2013). Neurocomputational account of how the human brain decides when to have a break. *Proceedings of the National Academy of Sciences of the United States of America, 110*, 2641–2646. https://doi.org/10.1073/pnas.1211925110

Meyniel, F., Safra, L., & Pessiglione, M. (2014). How the brain decides when to work and when to rest: Dissociation of implicit-reactive from explicit-predictive computational processes. *PLoS Computational Biology, 10*, e1003584. https://doi.org/10.1371/journal.pcbi.1003584

Meyniel, F., et al. (2016). A specific role for serotonin in overcoming effort cost. *eLife, 5*, e17282. https://doi.org/10.7554/eLife.17282

Michely, J., et al. (2020). The role of dopamine in dynamic effort-reward integration. *Neuropsychopharmacology, 45*, 1448–1453. https://doi.org/10.1038/s41386-020-0669-0

Michely, J., Martin, I. M., Dolan, R. J., & Hauser, T. U. (2023). Boosting serotonin increases information gathering by reducing subjective cognitive costs. *Journal of Neuroscience, 43*, 5848–5855. https://doi.org/10.1523/JNEUROSCI.1416-22.2023

Mobbs, D., Trimmer, P. C., Blumstein, D. T., & Dayan, P. (2018). Foraging for foundations in decision neuroscience: Insights from ethology. *Nature Reviews Neuroscience, 19*, 419–427. https://doi.org/10.1038/s41583-018-0010-7

Müller, T., & Apps, M. A. J. (2019). Motivational fatigue: A neurocognitive framework for the impact of effortful exertion on subsequent motivation. *Neuropsychologia, 123*, 141–151. https://doi.org/10.1016/j.neuropsychologia.2018.04.030

Müller, T., Klein-Flügge, M. C., Manohar, S. G., Husain, M., & Apps, M. A. (2021). Neural and computational mechanisms of momentary fatigue and persistence in effort-based choice. *Nature Communications, 12*, 1–14. https://doi.org/10.1038/s41467-021-24927-7

Nagase, A. M., et al. (2018). Neural mechanisms for adaptive learned avoidance of mental effort. *Journal of Neuroscience, 38*, 2631–2651. https://doi.org/10.1523/JNEUROSCI.1995-17.2018

Niv, Y., Joel, D., & Dayan, P. (2006). A normative perspective on motivation. *Trends in Cognitive Sciences, 10*, 375–381. https://doi.org/10.1016/j.tics.2006.06.010

Niv, Y., Daw, N. D., Joel, D., & Dayan, P. (2007). Tonic dopamine: opportunity costs and the control of response vigor. *Psychopharmacology, 191*, 507–520. https://doi.org/10.1007/s00213-006-0502-4

Nobis, L., et al. (2023). Role of serotonin in modulation of decision-making in Parkinson's disease. *Journal of Psychopharmacology, 37*, 420–431. https://doi.org/10.1177/02698811221144636

Otto, A. R., & Daw, N. D. (2019). The opportunity cost of time modulates cognitive effort. *Neuropsychologia, 123*, 92–105. https://doi.org/10.1016/j.neuropsychologia.2018.05.006

Pearson, J. M., Watson, K. K., & Platt, M. L. (2014). Decision making: The neuroethological turn. *Neuron, 82*, 950–965. https://doi.org/10.1016/j.neuron.2014.04.037

Pessiglione, M., Vinckier, F., Bouret, S., Daunizeau, J., & Le Bouc, R. (2018). Why not try harder? Computational approach to motivation deficits in neuro-psychiatric diseases. *Brain, 141*, 629–650. https://doi.org/10.1093/brain/awx278

Pisauro, M. A., Pollicino, D., Fisher, L., & Apps, M. A. J. (2024). Neural and computational mechanisms of effort under the pressure of a deadline. *bioRxiv*, 2024.04.17.589910. https://doi.org/10.1101/2024.04.17.589910

Rescorla, R. A., & Wagner, A. R. (1972). Classical Conditioning II: Current Research and Theory. In W. F. Prokasy (Ed.), (pp. 64–99). Appleton-Century Crofts.

Ricciardi, L., Apps, M., & Little, S. (2023). Uncovering the neurophysiology of mood, motivation and behavioral symptoms in Parkinson's disease through intracranial recordings. *NPJ Parkinsons Disease, 9*, 1–16. https://doi.org/10.1038/s41531-023-00567-0

Ridderinkhof, K. R., Ullsperger, M., Crone, E. A., & Nieuwenhuiss, S. (2004). The role of the medial frontal cortex in cognitive control. *Science, 306*, 443–447. https://doi.org/10.1126/science.1100301

Robert, P., et al. (2018). Is it time to revise the diagnostic criteria for apathy in brain disorders? The 2018 international consensus group. *European Psychiatry, 54*, 71–76. https://doi.org/10.1016/j.eurpsy.2018.07.008

Salamone, J. D., & Correa, M. (2024). The neurobiology of activational aspects of motivation: Exertion of effort, effort-based decision making, and the role of dopamine. *Annual Review of Psychology, 75*, 1–32. https://doi.org/10.1146/annurev-psych-020223-012208

Salamone, J. D., Correa, M., Farrar, A., & Mingote, S. M. (2007). Effort-related functions of nucleus accumbens dopamine and associated forebrain circuits. *Psychopharmacology, 191*, 461–482. https://doi.org/10.1007/s00213-006-0668-9

Sandra, D. A., & Otto, A. R. (2018). Cognitive capacity limitations and need for cognition differentially predict reward-induced cognitive effort expenditure. *Cognition, 172*, 101–106. https://doi.org/10.1016/j.cognition.2017.12.004

Scholl, J., et al. (2017). Beyond negative valence: 2-week administration of a serotonergic antidepressant enhances both reward and effort learning signals. *PLoS Biology, 15*, e2000756. https://doi.org/10.1371/journal.pbio.2000756

Shadmehr, R., de Xivry, J. J. O., Xu-Wilson, M., & Shih, T. Y. (2010). Temporal discounting of reward and the cost of time in motor control. *Journal of Neuroscience, 30*, 10507–10516. https://doi.org/10.1523/JNEUROSCI.1343-10.2010

Shenhav, A., Botvinick, M. M., & Cohen, J. D. (2013). The expected value of control: An integrative theory of anterior cingulate cortex function. *Neuron, 79*, 217–240. https://doi.org/10.1016/j.neuron.2013.07.007

Shenhav, A., et al. (2017). Toward a rational and mechanistic account of mental effort. *Annual Review of Neuroscience, 40*, 99–124. https://doi.org/10.1146/annurev-neuro-072116-031526

Silveira, M. M., Wittekindt, S. N., Mortazavi, L., Hathaway, B. A., & Winstanley, C. A. (2020). Investigating serotonergic contributions to cognitive effort allocation, attention, and impulsive action in female rats. *Journal of Psychopharmacology, 34*, 452–466. https://doi.org/10.1177/0269881119896043

Skvortsova, V., Palminteri, S., & Pessiglione, M. (2014). Learning to minimize efforts versus maximizing rewards: Computational principles and neural correlates. *Journal of Neuroscience, 34*, 15621–15630. https://doi.org/10.1523/JNEUROSCI.1350-14.2014

Strasser, A., et al. (2020). Glutamine-to-glutamate ratio in the nucleus accumbens predicts effort-based motivated performance in humans. *Neuropsychopharmacology, 45*, 2048–2057. https://doi.org/10.1038/s41386-020-0760-6

Sukumar, S., Shadmehr, R., & Ahmed, A. A. (2024). Effects of reward and effort history on decision making and movement vigor during foraging. *Journal of Neurophysiology, 131*, 638–651. https://doi.org/10.1152/jn.00092.2023

Sutton, R. S., & Barto, A. G. (1981). Toward a modern theory of adaptive networks—Expectation and prediction. *Psychological Review, 88*, 135–170. https://doi.org/10.1037/0033-295X.88.2.135

Tanaka, M., & Watanabe, Y. (2012). Supraspinal regulation of physical fatigue. *Neuroscience & Biobehavioral Reviews, 36*, 727–734. https://doi.org/10.1016/j.neubiorev.2011.10.004

Toro-Serey, C., Kane, G. A., & McGuire, J. T. (2021). Choices favoring cognitive effort in a foraging environment decrease when multiple forms of effort and delay are interleaved. *Cognitive, Affective, & Behavioral Neuroscience, 1*, 1–24. https://doi.org/10.3758/s13415-021-00972-z

Treadway, M. T., Buckholtz, J. W., Schwartzman, A. N., Lambert, W. E., & Zald, D. H. (2009). Worth the 'EEfRT'? The effort expenditure for rewards task as an objective measure of motivation and anhedonia. *PLoS One, 4*, e6598. https://doi.org/10.1371/journal.pone.0006598

Vansteenkiste, M., Lens, W., De Witte, S., De Witte, H. A., & Deci, E. L. (2004). The 'why' and 'why not' of job search behaviour: Their relation to searching, unemployment experience, and well-being. *European Journal of Social Psychology, 34*, 345–363. https://doi.org/10.1002/ejsp.202

Varazzani, C., San-Galli, A., Gilardeau, S., & Bouret, S. (2015). Noradrenaline and dopamine neurons in the reward/effort trade-off: A direct electrophysiological comparison in behaving monkeys. *Journal of Neuroscience, 35*, 7866–7877. https://doi.org/10.1523/JNEUROSCI.0454-15.2015

Vassena, E., et al. (2014). Overlapping neural systems represent cognitive effort and reward anticipation. *PLoS One, 9*. https://doi.org/10.1371/journal.pone.0091008

Vassena, E., Deraeve, J., & Alexander, W. H. (2019). Task-specific prioritization of reward and effort information: Novel insights from behavior and computational modeling. *Cognitive, Affective, & Behavioral Neuroscience, 19*, 619–636. https://doi.org/10.3758/s13415-018-00685-w

Vogel, T. A., Savelson, Z. M., Otto, A. R., & Roy, M. (2020). Forced choices reveal a trade-off between cognitive effort and physical pain. *eLife, 9*, e59410. https://doi.org/10.7554/eLife.59410

Walton, M. E., & Bouret, S. (2019). What is the relationship between dopamine and effort? *Trends in Neurosciences, 42*, 79–91. https://doi.org/10.1016/j.tins.2018.10.001

Westbrook, A., & Braver, T. S. (2015). Cognitive effort: A neuroeconomic approach. *Cognitive, Affective, & Behavioral Neuroscience*. https://doi.org/10.3758/s13415-015-0334-y

Westbrook, A., & Braver, T. S. (2016). Dopamine does double duty in motivating cognitive effort. *Neuron, 89*, 695–710. https://doi.org/10.1016/j.neuron.2015.12.029

Westbrook, A., Kester, D., & Braver, T. S. (2013). What is the subjective cost of cognitive effort? Load, trait, and aging effects revealed by economic preference. *PLoS One, 8*, e68210. https://doi.org/10.1371/journal.pone.0068210

Westbrook, A., et al. (2020). Dopamine promotes cognitive effort by biasing the benefits versus costs of cognitive work. *Science, 367*, 1362–1366. https://doi.org/10.1126/science.aaz5891

Westbrook, A., Frank, M. J., & Cools, R. (2021). A mosaic of cost–benefit control over cortico-striatal circuitry. *Trends in Cognitive Sciences, 25*, 710–721. https://doi.org/10.1016/j.tics.2021.04.007

Wiehler, A., Branzoli, F., Adanyeguh, I., Mochel, F., & Pessiglione, M. (2022). A neuro-metabolic account of why daylong cognitive work alters the control of economic decisions. *Current Biology, 32*, 3564–3575.e5. https://doi.org/10.1016/j.cub.2022.07.010

Wilson, C. R. E., et al. (2016). Prefrontal markers and cognitive performance are dissociated during progressive dopamine lesion. *PLoS Biology, 14*, e1002576. https://doi.org/10.1371/journal.pbio.1002576

Wittmann, M. K., et al. (2016). Predictive decision making driven by multiple time-linked reward representations in the anterior cingulate cortex. *Nature Communications, 7*, 1–13. https://doi.org/10.1038/ncomms12327

Self-Control and Decision-Making

Contribution to 'Neuroeconomics: Core Topics and Current Directions'

Candace M. Raio

Abstract Using self-control during decision-making is critical to making choices that align with our broader goals. Nonetheless, conceptualizations of self-control differ dramatically across the literature, making it especially challenging to understand how self-control is implemented in choice behavior and related neural circuits. This chapter focuses on a subset of choice behavior in which individuals decide between options that satisfy distinct goals, specifically in the domain of dietary choice. The chapter first reviews neuroeconomic work that reveals a role for the ventral and lateral prefrontal cortex in self-control decisions, before highlighting studies that aim to parse how these brain regions selectively contribute to these decisions. It further examines factors that determine the success of self-control exertion and discusses different theories that explain the mechanistic role of self-control in decision-making. Finally, this chapter reviews how costs associated with exerting control play a crucial role in whether control is successfully used during decision-making and highlights ways in which neuroeconomic research can motivate future work to gain a more precise understanding of the neural and computational underpinnings of self-control decisions.

Throughout our lives, we encounter countless sources of temptation that can derail our higher-order goals or elicit behavior that does not conform to social norms. The use of "self-control"—broadly defined as the ability to resist these sources of temptation or override reflexive impulses in order to generate behavior that aligns with our goals—is essential to healthy functioning. Deficits in self-control have long been associated with a number of deleterious real-world outcomes, such as poorer academic performance (Duckworth et al., 2019), reduced financial savings (Moffitt et al., 2011; Benartzi & Thaler, 2013; Achtziger et al., 2015; Gathergood, 2012), worse physical (Moffitt et al., 2011; VanEpps et al., 2016) and mental (Duckworth

C. M. Raio (✉)
Department of Psychiatry, NYU Grossman School of Medicine, New York, NY, USA

Neuroscience Institute, NYU Grossman School of Medicine, New York, NY, USA
e-mail: candace.raio@nyulangone.org

D. V. Smith et al. (eds.), *Neuroeconomics: Core Topics and Current Directions*,
https://doi.org/10.1007/978-3-032-02925-6_18

et al., 2019; VanEpps et al., 2016) health, and higher rates of alcohol and substance use (Duckworth et al., 2019; De Ridder et al., 2012; Robson et al., 2020). Decades of empirical work have already established the importance of this regulatory capacity in the field of affective science, in which emotional states can be controlled and automatic reactions can be overridden in the service of a more adaptive emotional response (Gross, 2015). However, the field of decision science, and neuroeconomics in particular, now recognizes self-control as a central process that facilitates adaptive decision-making as well. Whether we are trying to meet a deadline at work, choose to eat healthier, avoid drugs and alcohol, or save money for the future, the ability to resist tempting rewards and choose in ways that do not interfere with these goals is critical for advantageous decision-making.

1 Conceptualizations of Self-Control in Neuroeconomics

The construct of self-control is among the most pervasively examined across the psychological, neuroscientific, and economic literature. While it is widely agreed upon that there is utility in using self-control to guide behavior and modulate decision-making, less clarity emerges when attempting to precisely define self-control. This is because across the psychology, neuroscience, and economic literature, notions of self-control differ in a domain-specific manner and can refer to a broad and diverse range of processes and behaviors. Self-control has been conceptualized as "willpower" (Metcalfe & Mischel, 1999), inhibitory control (Baumeister, 2014), the result of an impulsive (vs. deliberative) system (Kahneman, 2011), and sensitivity to time delay (Kable & Glimcher, 2007), and some have even questioned whether it can be distinguished from value-based choice at all (Berkman et al., 2017). More recent accounts have proposed that self-control might be better viewed as a construct describing a broad class of behavior comprising several cognitive components that both relate to each other but possibly stem from distinct cognitive or neural underpinnings (Kable, 2014; Inzlicht et al., 2021).

In this chapter, I will cover one subset of this class of choice behavior in which individuals must choose between choice options for whose outcomes satisfy distinct goals. This process can, theoretically, be characterized by goals that compete in temporal proximity, such as immediate versus delayed rewards (e.g., smoking a cigarette to relax in the moment or abstaining in order to promote long-term health), or extended to goals that conflict in the same temporal domain such as those that both have short- or long-term benefits (e.g., choosing to spend time exercising or working on a job assignment; Inzlicht et al., 2021). This chapter will not cover models of temporal discounting per se, which describe individuals' formal preferences for immediate versus delayed rewards, as this topic has motivated a large body of work across economics and psychology and is aptly covered in other chapters of this textbook (see Wagner et al. Chap. 14; Hare et al. Chap. 15; and Lempert et al., Chap. 16) However, it should be noted that features of temporal discounting play prominently in many types of self-control decisions since they generally involve choosing between immediate gratification versus long-term benefits.

2 Neuroeconomic Investigations into Self-Control Decisions

A classic self-control choice problem concerns decisions between consuming tempting, but unhealthy, foods versus healthier, but less tasty, foods that support broader goals to lose weight or maintain health. Over the last two decades, the field of neuroeconomics has capitalized on this pervasive decision problem as a broader model for self-control decisions by investigating the neural circuits underlying self-control behavior using dietary food choice paradigms coupled with functional Magnetic Resonance Imaging (fMRI). In these paradigms, choosers typically select between different food items that vary in taste and health level (Tusche & Hutcherson, 2018; Hare et al., 2009; Maier et al., 2015; Sullivan et al. 2015), whether to accept or reject a single food item depending on its taste and health attributes (Hare et al., 2011; Maier & Hare, 2023), and/or report how much they would pay for a particular food item (Cosme et al., 2019; Hutcherson et al., 2012). They also might be asked to make these decisions under different conditions, such as focusing on the health versus taste attributes of the food. As is standard now in these types of studies, after the task has been completed, a single trial is randomly selected and the food chosen on that trial is implemented. This feature of the task ensures that choices have real motivational meaning to participants because any trial may be selected, thus each decision is relevant to participants (Fig. 1).

Among the first neuroeconomic studies using this approach revealed a role for the dorsolateral and ventromedial prefrontal cortex (dlPFC and vmPFC, respectively) in representing different aspects of self-control decisions. Hare et al. (2009) found that the dlPFC, which plays a broader role in higher-order executive functions such as working memory (Wager & Smith, 2003; Arnsten, 2009), cognitive control (Nee et al., 2007; Wager et al., 2004), and cognitive forms of emotion regulation (Buhle et al., 2014), was more active when participants made choices consistent with their diets (i.e., used self-control) than when self-control attempts were unsuccessful, and that this was higher in participants categorized as "good" self-controllers (i.e., those that frequently made health-consistent choices). The vmPFC, which is widely thought to integrate value inputs and represent subjective value (SV; Bartra et al., 2013), reflected both "health" and "taste" ratings in "good" self-controllers but only taste ratings in "poor" self-controllers—pointing to a representation of prioritized attributes within goal-conflicting decision contexts in this brain region. In line with research stemming from the emotion regulation literature, the vmPFC and dlPFC showed significant functional connectivity during successful self-control trials, suggesting the dlPFC might modulate choice behavior either by enabling an adequate representation of higher-order goals (i.e., healthiness) or through a more canonical control deployment mechanism that enables goal-consistent choice (Fig. 2).

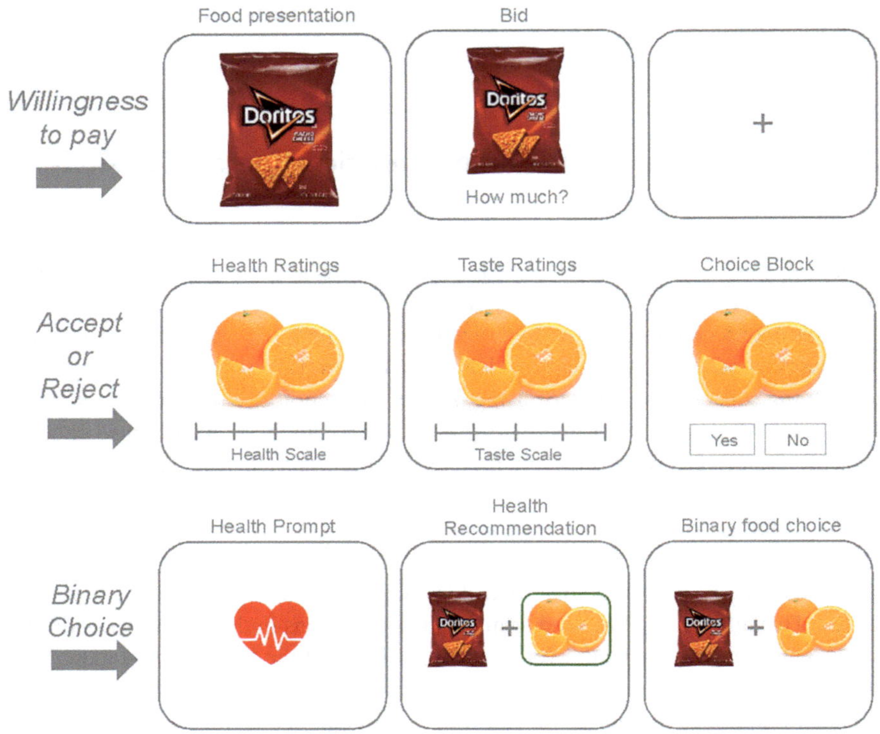

Fig. 1 Examples of self-control choice paradigms used in the neuroeconomic literature. Willingness-to-pay paradigms instruct participants to report how much they would pay for a given item to measure subjective value (top row). Accept/reject paradigms might ask participants to rate items on different features (e.g., health, taste) before reporting whether they would prefer to accept or reject the choice item in lieu of a default option (middle row). Participants may also repeat these choices, considering a health frame versus a taste frame. In binary choice paradigms, (bottom row) participants choose between two options that vary in attributes (e.g., health versus taste; smaller, immediate versus larger, delayed rewards)

Subsequent studies have extended the above findings using other analytic approaches and offer further support that the vmPFC, dlPFC, and related regions (e.g., inferior frontal gyrus, or IFG) serve as neural mechanisms underlying dietary self-control decisions (Hare et al., 2011; Hutcherson et al., 2012; Maier et al., 2015; Tusche & Hutcherson, 2018; Cosme et al., 2019). This has been confirmed not only via functional brain activity (i.e., BOLD signal change) during dietary choice tasks but also by work examining neuroanatomical (structural) differences in grey matter volume in these brain regions. For example, using voxel-based morphometry, Schmidt and colleagues pooled data across three dietary food choice studies and revealed that dlPFC and vmPFC volume was higher in those participants that chose healthier foods (i.e., demonstrated higher self-control) when prompted to focus on the health attributes of the foods. Brain volume in these regions also predicted out-

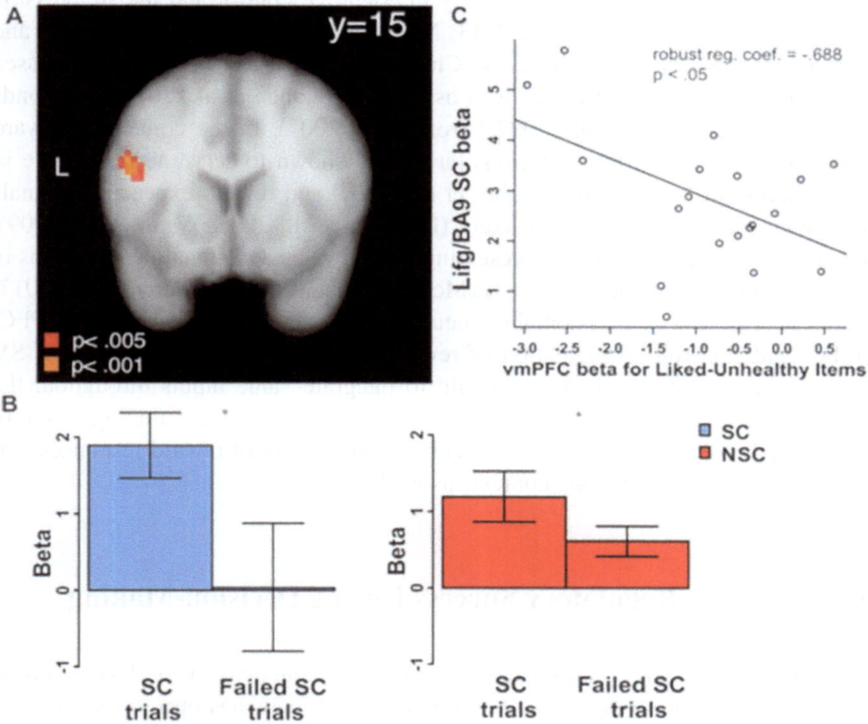

Fig. 2 (**a**) BOLD activity in left dlPFC was higher during trials where healthier choices were made in "good" self-controllers. (**b**) dlPFC activity was higher when healthier choices were made across both "good" and "poor" self-controllers (*$P < 0.05$, paired t-test). (**c**) In "good" self-controllers, left dlPFC (IFG/BA9) activity was negatively correlated with vmPFC activity during trials where tempting foods were not chosen. (Adapted with permission from Hare et al., 2009)

of-sample regulatory success during food choice in a task in which food craving was regulated while viewing palatable foods (Schmidt et al., 2018). In another study, the dlPFC and vmPFC were again found to be involved in self-control decisions using a distinct analytic approach in which patterns of brain activity across a specified cluster of voxels were tested for decoding accuracy (multi-voxel pattern analysis) of different choice attributes in both a dietary food choice and a prosocial decision-making task. Here, the dlPFC emerged as a core region that represented higher-order goal value and predicted regulatory success across the food and social choice tasks, pointing to this region as playing a domain-general role in representing distinct choice attributes that facilitate regulatory success during decision-making (Tusche & Hutcherson, 2018).

The finding that the dlPFC plays a role in self-control decisions is in accordance with the broader self-regulation literature, which has consistently demonstrated that the active and intentional regulation of rewards involves a network of prefrontal brain regions implicated more generally in cognitive control and the suppression of automatic responses (Duncan, 2013; Nee et al., 2007; Wager et al., 2004) and that include the dlPFC and vlPFC (McClure et al., 2004; Kober et al., 2010; Essex et al., 2012; Luo et al., 2012) as well as the IFG (Cohen et al., 2012; Simmonds et al., 2008; Chikazoe et al., 2007; Aron et al., 2007). In the context of reward responsivity, these prefrontal regions have been shown to serve a critical role in the reduction of both craving (Kober et al., 2010) and reward-expectancy signals in regions such as the ventral striatum (Delgado et al., 2008; Wang et al., 2009), which is widely implicated in representing reward value and signaling violations in expected rewards (prediction errors) (McClure et al., 2003; Diederen et al., 2017; Calderon et al., 2021). This domain of neuroeconomics research expands the vlPFC, dlPFC, and IFG role in regulation of reward to that of a value-modulator of SV signals in the vmPFC, which is thought to integrate value inputs throughout the brain and correlate with final choice outcomes. Collectively, these findings point to an essential role of the dlPFC in the deliberate inhibition of reward responses that generalize to decision-making contexts as well.

3 Increasing Regulatory Success During Decision-Making

A key objective of self-control decision work is to elucidate the neural mechanisms underlying self-control decisions, but a more recent focus has been on identifying factors that determine when self-control is better engaged in order to increase regulatory success. Neuroeconomic studies using dietary choice paradigms suggest a number of potential avenues. First, cognitive strategies commonly used to control affective responses have also been shown to increase goal-consistent choice. In one study (Hutcherson et al., 2012), choosers were instructed to use cognitive regulation strategies (i.e., “distancing” vs. “indulging”) to increase or decrease the SV of food rewards. Choice behavior revealed that food value increased or decreased accordingly with cognitive regulation efforts to alter food value. Additionally, vmPFC, dlPFC, and ventrolateral PFC (vlPFC) fluctuated depending on whether participants successfully up- or downregulated craving responses for food rewards. Specifically, when craving was successfully decreased, vmPFC responses to food images were more weakly related to food value, while dlPFC responses became strongly related, suggesting successful modulation of reward value was reflected in greater dlPFC engagement. When craving was upregulated (increased), this relation between dlPFC activity and food value diminished. This work points to the dynamic nature of these circuits during deliberate attempts to control reward responses. Further, more recent work has generalized the neural circuits implicated in affective control to that of self-control in decision contexts. Specifically, BOLD activity in the medial and lateral PFC during successful cognitive regulation of affective stimuli (positive vs. negative images) was associated with how successful these same choosers were at exercising self-control during a subsequent dietary choice task (Maier et al., 2015), suggesting that the neural circuits used to regulate affective responses generalize to decision contexts that require the flexible reappraisal of choice options to promote goal-consistent choice.

Negative affective states including stress and anxiety also play a substantial role in determining how successful self-control deployment is during decision-making. Stress exposure is widely known to be a prominent risk factor for self-control failure (Duckworth et al., 2013; Sinha 2008; Muraven, Tice & Baumeister, 1998), and mild to moderate stress has been shown to disrupt goal-directed decision control (Otto et al., 2013; Plessow et al., 2011; Plessow et al., 2012) and render the delay of gratification more difficult (Metcalfe & Mischel, 1999) by increasing the salience of temptation (Mather & Lighthall, 2012). Neuroscientific research demonstrates that the dlPFC and threat-relevant regions like the amygdala that share connectivity with the vmPFC are each particularly sensitive to the effects of stress exposure, with dlPFC function decreasing and amygdala reactivity increasing (Arnsten 2009; Hermans et al., 2014; van Oort et al., 2017). This points to stress as a potential factor that might interfere with self-control modulation of value during decision-making. Accordingly, stress prior to a dietary choice task led participants to weigh taste attributes more highly during food choice. Neurally, stress resulted in weaker functional connectivity between vmPFC and dlPFC, and the extent of this decrease was correlated with cortisol levels, a marker of neuroendocrine stress activity. Stress simultaneously *increased* connectivity patterns between the amygdala and vmPFC, which related to subjective reports of perceived stress (Maier et al., 2015), suggesting two distinct paths through which stress might simultaneously bias decisions away from goal-consistent choice (Fig. 3).

Finally, increases in goal-consistent choice have also been achieved through the attentional manipulations toward attributes that support goal-consistent choice. For example, explicitly providing cues signaling the healthy features of food items was associated with greater vmPFC representation of health attributes of foods, higher dlPFC and inferior frontal gyrus (IFG) activity, and healthier food choice more broadly (Hare et al., 2011), indicating that similar self-control networks are engaged when choosers are exogenously cued to consider broader health goals as when these goals are considered during spontaneous choice. In accordance with these findings, later work found that one reason why the deployment of self-control is so prone to failure, particularly under negative emotional affect (e.g., stress, anxiety, negative affect), may be because health attributes of these food items are processed more slowly and less automatically than the taste attributes associated with foods. In support of this, Sullivan et al. (2015) used mouse tracking data and found that choosers process health attributes slower than they do taste attributes, and that this latency difference explained substantial variability regarding whether self-control success was achieved, providing one explanation as to why explicit attentional signals toward health ratings appear to be so effective at increasing self-control decisions.

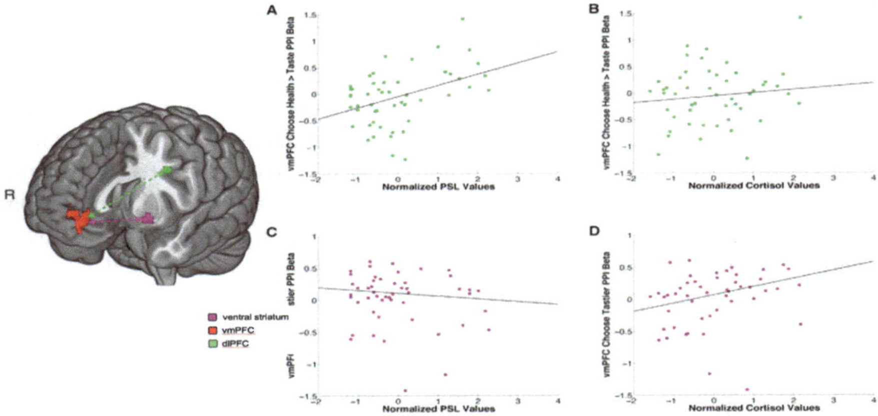

Fig. 3 Stress revealed two mechanistic paths to bias decisions away from goal-consistent (healthier) choice. vmPFC activity (red) tracked the SV of food. Magenta clusters represents ventral striatum voxels that showed both higher functional coupling with vmPFC during taste-driven choice in the stress group and those in which choices correlated more strongly with cortisol than perceived stress. Green clusters represents dlPFC voxels that showed both higher activity during self-controlled choices and those in which functional coupling during choice correlated more strongly with perceived stress than cortisol. (**a** and **b**). Scatterplots depict the strength of dlPFCs (green, **a** and **b**) and ventral striatum's (magenta, **c** and **d**) functional coupling with vmPFC for self-controlled (green) or tastier (magenta) choices against PSL (**a** and **c**) and cortisol levels (**b** and **d**), respectively. (Adapted with permission from Maier et al., 2015)

4 Different Accounts of Self-Control in Neuroeconomics

Collectively, this area of work has converged on a neural model of self-control decisions with the vmPFC representing an integrated SV signal and a circuit of prefrontal regions (vlPFC, dlPFC, and IFG) that modulate reward and value-related regions (VS, vmPFC) to collectively promote goal-consistent choice. Despite agreement regarding the involvement of these circuits, it should be noted that debate exists regarding the specific functional role these prefrontal regions play (Hutcherson et al., 2012; Berkman et al., 2017; Shenhav et al., 2016; Shenhav, 2017; Hutcherson & Tusche, 2022). For example, some accounts have argued that the dlPFC and related circuits modulate SV in the vmPFC and ventral striatum by enabling a representation of distinct choice attributes (e.g., health vs. taste) and what determines whether we make choices that align with our broader goals is how efficiently, and perhaps quickly (Sullivan et al., 2015), these attributes are represented and weighted when converging on a final choice (Tusche & Hutcherson, 2018; Hutcherson & Tusche, 2022). This account is supported by many of the neuroeconomic studies described above in which dlPFC shows preferentially higher activity to goal-relevant attributes or goal-consistent choice (Hare et al., 2009; 2011). Recent extensions of this account point to the role of the dlPFC in a later phase of the choice process (e.g., the "evidence accumulation" phase) rather than the initial assignment of value to different attributes (Hutcherson & Tusche, 2022). This would suggest that dlPFC activity does not necessarily modulate the initial valuation process itself but instead comes online when determining how these attributes align with the weight of a current goal. This would lead to the prediction that dlPFC is selectively involved in self-control choices *only* when conflict exists between the choice options *and* a chooser is considering the goal-inconsistent option. In support of this hypothesis, a recent study found evidence of higher dlPFC activity for goal-consistent choice when attributes of the choices were conflicting (e.g., healthier vs. tastier) *and* chooser's focus was on the more tempting choice (e.g., taste). Higher dlPFC activity was not observed when chooser's focus was on more "normative" goals, which instead yielded lower dlPFC activity (Hutcherson & Tusche, 2022), a hypothesis that tracked with and was predicted by computational modeling (i.e., drift-diffusion model).

The above account raises the question of what role "control" as represented by dlPFC activity plays in value-based decisions. It has been proposed that self-control is simply another feature of value-based choice, representing the cognitive effort it takes to choose in a goal-consistent manner as a weighting factor in an otherwise standard value-based decision (Berkman et al., 2017). Others have pointed out that this interpretation stands in contrast to the classic inhibitory role that lateral PFC (e.g., dlPFC, vlPFC, IFG) has been proposed to play in behavioral control and suggest that a more canonical control mechanism is likely at play during value-based decision-making (Shenhav, 2017). These accounts propose that the magnitude of control needed to choose in a goal-consistent manner can be calculated outside of standard value-based decision-making circuits, concentrated instead in regions such as the dorsal anterior cingulate cortex, dACC (Shenhav et al., 2016). After this calculation is made, control execution can then be deployed through lateral prefrontal regions such as the dlPFC. Therefore, this account points to distinct valuation systems for calculating the value of stimuli (such as rewards), and the value of exerting control over responses to those rewards. By positioning self-control decisions as emerging from the interaction between these two valuation systems, this account provides a unique means through which choosers can dynamically shift attention to different attributes (health vs. taste) that represent conflicting goals in standard self-control choice conflicts before arriving at a final SV representation (Shenhav, 2017).

5 Decision Costs Associated with Exercising Control

While much of the self-control literature has focused primarily on the benefits of using control to drive goal-directed behavior, the exertion of control also comes with costs. For example, the subjective experience of exerting control—whether in decision-making contexts or elsewhere—is universally known to be effortful and aversive. A growing body of theoretical and empirical frameworks from the cognitive and decision neuroscience literature suggests that this subjective perception of effort results from the exertion of control being registered as intrinsically "costly" (Westbrook & Braver, 2015; Kurzban et al., 2013; Kool & Botvinick, 2018; Boureau et al., 2015). These costs are believed to arise from the limited cognitive resources available to meet the demands of control and further point to these cognitive costs as a primary reason why self-control failures are so common. These frameworks predict that the use of self-control is ultimately determined by a cost-benefit decision that choosers calculate based on the perceived cognitive cost of using control relative to the perceived benefit. Thus, choosers should only engage in control if this calculus favors control being worth the cost of exertion (Westbrook & Braver, 2015; Kurzban et al., 2013; Kool & Botvinick, 2018). This value-based framework of control provides an alternative to psychological and lay theories about self-control failure (Baumeister et al., 1998) due to diminishing resources and repositions it instead as a rational decision process that dynamically updates whether expenditures of control continue to be worth the effort they require to meet task demands.

The notion that self-control exertion generates intrinsic cognitive costs and that these costs may play a seminal role in determining when self-control fails raises two important questions for the neuroeconomic literature. The first concerns whether we can measure the intrinsic costs of exerting self-control during decision-making to identify individual differences or cognitive states that are associated with lower self-control costs (and presumably, higher regulatory success). A growing body of cognitive neuroscience work has already demonstrated that humans tend to favor low cognitive effort over high cognitive effort when given the choice (Westbrook & Braver, 2015), that rewards requiring greater cognitive effort are often devalued (McGuire & Botvinick, 2010; Shenhav et al., 2013), and that offering motivational incentives can enhance performance in tasks requiring cognitive control (Kurzban et al., 2013). However, recent efforts in the specific domain of dietary choice have made some strides toward quantifying control costs, showing that choosers will in fact pay monetary costs to avoid using self-control and that negative affective states, such as exposure to stress, more than doubles what choosers will pay to avoid needing to exert control (Raio & Glimcher, 2021). Further supporting that these cognitive costs determine regulatory success, a recent paper used an analytical gradient framework that considers larger-scale changes across multiple brain networks to determine how neural network reconfigurations predict dietary choice. Successful regulators and those with a lower body mass index required a *lower* degree of neural reorganization to maintain goal-congruent choice, suggesting that goal-congruent choice may be more (or less) cognitively costly depending on the neural reorganization required to support regulatory success (Janet et al., 2024).

Acknowledging that self-control is cognitively costly and that these costs may in fact be measured, a second key question concerns whether there might be alternative decision strategies that choosers can adopt to circumvent these costs and assure regulatory success. Behavioral economics has long asserted that humans may seek to prospectively limit exposure to temptation to avoid self-control failures by adopting commitment devices (or "precommitment" strategies) that deliberately restrict access to temptation in future choice sets (Gul & Pesendorfer, 2001; Gul & Pesendorfer, 2004). Anyone that has deliberately not purchased junk food to

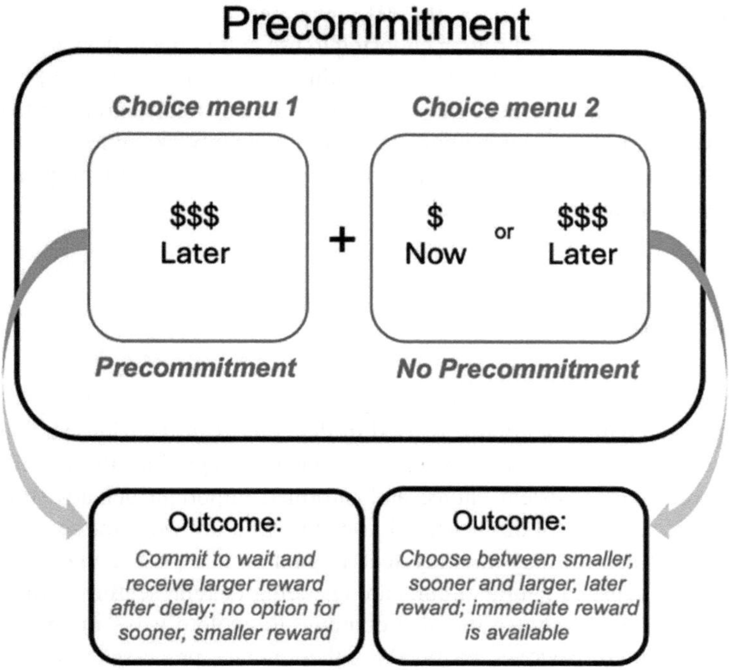

Fig. 4 Example of precommitment self-control task. Choosers decide initially between a choice menu in which only a larger, later reward appears, or a choice menu in which both the sooner, smaller, and larger, later reward are options. Precommitting to choice menu 1 removes the possibility of selecting the more immediately tempting reward, while choice menu 2 requires self-control to be exerted at the final choice phase

avoid having to resist eating it once home or turned off their Wi-Fi to avoid distractions on a work deadline has used a commitment device. These strategies resemble the concept of "situation modification" in emotion regulation theories (Gross, 2015) and "model-based" decision control strategies (Daw et al., 2006), both of which focus on planning actions in a proactive, goal-directed manner (Fig. 4). Although behavioral economists (Gul & Pesendorfer, 2001) have placed substantial focus on precommitment preferences as a model for self-control, the empirical literature on precommitment is sparse. However, those studies that have examined it report that when offered as a choice strategy, precommitment facilitates choices for larger, later rewards (Crockett et al., 2013; Soutschek et al., 2017, 2020, 2021; Studer et al., 2019) and that individuals will even incur costs to do this (Raio & Glimcher, 2021). Interestingly, precommitment appears to recruit a distinct network of prefrontal regions (e.g., anterior PFC; Crockett et al., 2013; Soutschek et al., 2017, 2021) than those that implement and execute control during choice (e.g., dlPFC). Neuroscientific investigations of precommitment have found that the frontopolar cortex, an anterior portion of the prefrontal cortex involved in planning, prospection, and anticipating future outcomes (Rushworth et al., 2011;

Boschin et al., 2015; Jimura et al., 2013; Fleming et al., 2010; Boorman et al., 2009), also plays a seminal role in choosing to use commitment devices. This work raises the intriguing possibility that evaluating self-control costs *prospectively* to determine if precommitment is necessary may rely on distinct neurocognitive choice mechanisms than those that support the more commonly examined choice phases reviewed in this chapter, enabling a potential intervention target to mitigate the need for costly self-control in the first place.

6 Future Directions

Neuroeconomic models of self-control provide a powerful framework for understanding how regulatory control is used in the context of decision-making. These models of self-control have revealed novel insights regarding the neural circuits that support goal-consistent choice and informed important debates surrounding the nature of self-control during decision-making. Future work for neuroeconomic models of self-control may not only seek to gain a more precise understanding of what these established neural circuits are encoding or representing during distinct stages of self-control choice but also continue to explore ways in which we can increase self-control success. By breaking these choice processes down to distinct choice stages, as recent work has done (Hutcherson & Tusche, 2022), we can start to capture more precision regarding when to intervene and increase regulatory success during decision-making.

Another promising avenue for neuroeconomic models of self-control is to expand the use of quantitative frameworks that view self-control as a cost-benefit decision. Such approaches have already been used in the standard cognitive effort and control literature (Shenhav et al., 2017; Botvinick & Braver, 2015). These approaches hold promise as they allow researchers to dissociate the perceived costs and benefits of exerting control rather than viewing control exertion as something that can fail or succeed in a binary fashion. Characterizing the heterogeneity in cost-benefit decisions to deploy control can also reveal targets for intervention, such as increasing motivational value to deploy control in some individuals or focusing instead on decreasing the perceived cost of exercising control in others.

Finally, the majority of neuroeconomic work has studied self-control decisions in the moment of encountering temptation, as opposed to earlier decision stages during which prospective strategies can circumvent the need for self-control in the first place. Future neuroeconomic work may seek to place a stronger focus on how choosers shape their environment to proactively ensure regulatory success, particularly in subsets of participants who are vulnerable to self-control failure (e.g., steep discounters, clinical disorders marked by self-control difficulties) or how the use of these strategies is adopted with experience and learning.

References

Achtziger, A., Hubert, M., Kenning, P., Raab, G., & Reisch, L. (2015). Debt out of control: The links between self-control, compulsive buying, and real debts. *Journal of Economic Psychology, 49*, 141–149.

Arnsten, A. F. T. (2009). Stress signalling pathways that impair prefrontal cortex structure and function. *Nature Reviews. Neuroscience, 10*(6), 410–422.

Aron, A. R., Behrens, T. E. J., Smith, S., Frank, M. J., & Poldrack, R. A. (2007). Triangulating a cognitive control network using diffusion-weighted magnetic resonance imaging (MRI) and functional MRI. *The Journal of Neuroscience, 27*, 3743–3752.

Badre, D., & D'Esposito, M. (2009). Is the rostro-caudal axis of the frontal lobe hierarchical? *Nature Reviews: Neuroscience, 10*(9), 659–669.

Bartra, O., McGuire, J. T., & Kable, J. W. (2013). The valuation system: A coordinate-based meta-analysis of BOLD fMRI experiments examining neural correlates of subjective value. *Neuroimage, 76*, 412–427.

Baumeister, R. F., Bratslavsky, E., Muraven, M., & Tice, D. M. (1998). Ego depletion: Is the active self a limited resource? *Journal of Personality and Social Psychology, 74*, 1252–1265.

Baumeister, R. F. (2014.) Self-regulation, ego depletion, and inhibition. *Neuropsychologia. 65*, 313–9.

Benartzi, S., & Thaler, R. H. (2013). Behavioral economics and the retirement savings crisis. *Science, 339*(6124), 1152–1153.

Berkman, E. T., Hutcherson, C. A., Livingston, J. L., Kahn, L. E., & Inzlicht, M. (2017). Self-control as value-based choice. *Current Directions in Psychological Science, 26*(5), 422–428.

Boorman, E. D., Behrens, T. E., Woolrich, M. W., & Rushworth, M. F. (2009). How green is the grass on the other side? Frontopolar cortex and the evidence in favor of alternative courses of action. *Neuron, 62*(5), 733–743.

Boschin, E. A., Piekema, C., & Buckley, M. J. (2015). Essential functions of primate frontopolar cortex. *Proceedings of the National Academy of Sciences, 112*(9), E1020–E1027.

Botvinick, M., & Braver, T. (2015). Motivation and cognitive control: From behavior to neural mechanism. *Annual Review of Psychology, 66*, 83–113.

Boureau, Y. L., Sokol-Hessner, P., & Daw, N. D. (2015). Deciding how to decide: Self-control and meta-decision making. *Trends in Cognitive Sciences, 19*(11), 700–710.

Buhle, J. T., Silvers, J. A., Wager, T. D., Lopez, R., Onyemekwu, C., Kober, H., Weber, J., & Ochsner, K. N. (2014). Cognitive reappraisal of emotion: A meta-analysis of human neuroimaging studies. *Cerebral Cortex, 24*, 2981–2990.

Calderon, C. B., De Loof, E., Ergo, K., Snoeck, A., Boehler, C. N., & Verguts, T. (2021). Signed reward prediction errors in the ventral striatum drive episodic memory. *The Journal of Neuroscience, 41*(8), 1716–1726.

Chikazoe, J., Konishi, S., Asari, T., Jimura, K., & Miyashita, Y. (2007). Activation of right inferior frontal gyrus during response inhibition across response modalities. *Journal of Cognitive Neuroscience, 19*, 69–80.

Cohen, J. R., Berkman, E. T., & Lieberman, M. D. (2012). Intentional and incidental self-control in ventrolateral PFC. In D. T. Stuss & R. T. Knight (Eds.), *Principles of frontal lobe functions* (pp. 417–440). Oxford University Press.

Cosme, D., Ludwig, R. M., & Berkman, E. T. (2019). Comparing two neurocognitive models of self-control during dietary decisions. *Social Cognitive and Affective Neuroscience, 14*(9), 957–966.

Crockett, M. J., Braams, B. R., Clark, L., Tobler, P. T., Robbins, T. W., & Kalenscher, T. (2013). Restricting temptations: Neural mechanisms of precommitment. *Neuron, 79*(2), 391–401.

Daw, N. D., O'Doherty, J. P., Dayan, P., Seymour, B., & Dolan, R. J. (2006). Cortical substrates for exploratory decisions in humans. *Nature, 441*(7095), 876–879.

De Ridder, D. T. D., et al. (2012). Taking stock of self-control: A meta-analysis of how trait self-control relates to a wide range of behaviors. *Personality and Social Psychology Review, 16*(1), 76–99.
Delgado, M. R., Gillis, M. M., & Phelps, E. A. (2008). Regulating the expectation of reward via cognitive strategies. *Nature Neuroscience, 11*, 880–881.
Diederen, K. M. J., Ziauddeen, H., Vestergaard, M. D., Spencer, T., Schultz, W., & Fletcher, P. C. (2017). Dopamine modulates adaptive prediction error coding in the human midbrain and striatum. *Journal of Neuroscience, 37*(7), 1708–1720.
Duckworth, A. L., Kim, B., & Tsukayama, E. (2013). Life stress impairs self-control in early adolescence. *Frontiers in Psychology, 3*, 608.
Duckworth, A. L., Taxer, J. L., Eskreis-Winkler, L., Galla, B. M., & Gross, J. J. (2019). Self-control and academic achievement. *Annual Review of Psychology, 70*(1), 373–399.
Duncan, J. (2013). The structure of cognition: Attentional episodes in mind and brain. *Neuron, 80*, 35–50.
Essex, B. G., Clinton, S. A., Wonderley, L. R., & Zald, D. H. (2012). The impact of the posterior parietal and dorsolateral prefrontal cortices on the optimization of long-term versus immediate value. *The Journal of Neuroscience, 32*, 15403–15413.
Fleming, S. M., Weil, R. S., Nagy, Z., Dolan, R. J., & Rees, G. (2010). Relating introspective accuracy to individual differences in brain structure. *Science, 329*(5998), 1541–1543.
Gathergood, J. (2012). Self-control, financial literacy and consumer over-indebtedness. *Journal of Economic Psychology, 33*(3), 590–602.
Gross, J. J. (2015). Emotion regulation: Current status and future prospects. *Psychological Inquiry, 26*(1), 1–26.
Gul, F., & Pesendorfer, W. (2001). Temptation and self-control. *Econometrica, 69*(6), 1403–1435.
Gul, F., & Pesendorfer, W. (2004). Self-control and the theory of consumption. *Econometrica, 72*(1), 119–158.
Hare, T. A. (2023). Goal-dependent hippocampal representations facilitate self-control. *The Journal of Neuroscience, 43*(46), 7822–7830.
Hare, T. A., Camerer, C. F., & Rangel, A. (2009). Self-control in decision-making involves modulation of the vmPFC valuation system. *Science, 324*, 646–648.
Hare, T. A., Malmaud, J., & Rangel, A. (2011). Focusing attention on the health aspects of foods changes value signals in vmPFC and improves dietary choice. *Science, 31*(30), 11077–11087.
Hermans, E. J., Henckens, M. J. A. G., Joels, M., & Fernandez, G. (2014). Dynamic adaptations of large-scale neural networks in response to acute stressors. *Trends in Neuroscience, 37*(6), 304–314.
Hutcherson, C. A., & Tusche, A. (2022). *Evidence accumulation, not 'self-control', explains dorsolateral prefrontal activation during normative choice eLife, 11*, e65661.
Hutcherson, C. A., Plassmann, H., Gross, J. J., & Rangel, A. (2012). Cognitive regulation during decision making shifts behavioral control between ventromedial and dorsolateral prefrontal value systems. *The Journal of Neuroscience, 32*(39), 13543–13554.
Inzlicht, M., Werner, K. M., Briskin, J. L., & Roberts, B. W. (2021). Integrating models of self-regulation. *Annual Review of Psychology, 72*, 319–345.
Janet, R., Smallwood, J., Hutcherson, C. A., Plassmann, H., Mckeowna, B., & Tusch, A. (2024). Body mass index–dependent shifts along large-scale gradients in human cortical organization explain dietary regulatory success. *PNAS, 121*(18), e2314224121.
Jimura, K., Chushak, M. S., & Braver, T. S. (2013). Impulsivity and self-control during intertemporal decision making linked to the neural dynamics of reward value representation. *The Journal of Neuroscience, 33*, 344–357.
Kable, J. W. (2014). Valuation, intertemporal choice, and self-control. In P. W. Glimcher & E. Fehr (Eds.), *Neuroeconomics* (2nd ed., pp. 173–192). Academic. ISBN 9780124160088.
Kable, J. W., & Glimcher, P. W. (2007). The neural correlates of subjective value during intertemporal choice. *Nature Neuroscience, 10*, 1625–1633.
Kahneman, D. (2011). *Thinking, fast and slow*. Anchor.

Kober, H., Mende-Siedlecki, P., Kross, E. F., Weber, J., Mischel, W., Hart, C. L., & Ochsner, K. N. (2010). Prefrontal-striatal pathway underlies cognitive regulation of craving. *PNAS, 107*, 14811–14816.

Kool, W., & Botvinick, M. (2018). Mental labour. *Nature Human Behavior, 2*, 899–908.

Kurzban, R., Duckworth, A., Kable, J. W., & Myers, J. (2013). An opportunity cost model of subjective effort and task performance. *Behavioral and Brain Science, 36*, 661–679.

Luo, S., Ainslie, G., Pollini, D., Giragosian, L., & Monterosso, J. R. (2012). Moderators of the association between brain activation and farsighted choice. *NeuroImage, 59*, 1469–1477.

Maier, S. U., & Hare, T. A. (2023). BOLD activity during emotion reappraisal positively correlates with dietary self-control success. *Social Cognitive and Affective Neuroscience, 18*(1), nsaa097.

Maier, S. U., Makawana, A. B., & Hare, T. (2015). Acute stress impairs self-control in goal-directed choice by altering multiple functional connections within the brain's decision circuits. *Neuron, 87*(3), 621–631.

Mather, M., & Lighthall, N. R. (2012). Both risk and reward are processed differently in decisions made under stress. *Current Directions in Psychological Science, 21*(2), 36–41.

McClure, S., Bern, G. S., & Montague, P. R. (2003). Temporal prediction errors in a passive learning task activate human striatum. *Neuron, 38*(2), 339–346.

McClure, S. M., Laibson, D. I., Loewenstein, G., & Cohen, J. D. (2004). Separate neural systems value immediate and delayed monetary rewards. *Science, 306*, 503–507.

McGuire, J. C., & Botvinick, M. (2010). Prefrontal cortex, cognitive control, and the registration of decision costs. *Proceedings of the National Academy of Sciences of the United States of America, 107*, 922–7926.

Metcalfe, J., & Mischel, W. (1999). A hot/cool-system analysis of delay of gratification: Dynamics of willpower. *Psychological Review, 106*(1), 3–19.

Moffitt, T. E., Arseneault, L., Belsky, D., Dickson, N., Hancox, R. J., Harrington, H., et al. (2011). A gradient of childhood self-control predicts health, wealth, and public safety. *Proceedings of the National Academy of Sciences, 108*(7), 2693–2698.

Muraven, M., Tice, D. M., & Baumeister, R. F. (1998). Self-control as limited resource: Regulatory depletion patterns. *Journal of Personality and Social Psychology, 74*, 774–789.

Nee, D. E., Wager, T. D., & Jonides. (2007). Interference resolution: Insights from a meta-analysis of neuroimaging tasks. *Cognitive, Affective, & Behavioral Neuroscience, 7*, 1–17.

Otto, A. R., Raio, C. M., Chiang, A., Phelps, E. A., & Daw, N. D. (2013). Working-memory capacity protects model-based learning from stress. *Proceedings of the National Academy of Sciences, 110*(52), 20941–20946.

Plessow, F., Fischer, R., Kirschbaum, C., & Goschke, T. (2011). Inflexibly focused under stress: Acute psychosocial stress increases shielding of action goals at the expense of reduced cognitive flexibility with increasing time lag to the stressor. *Journal of Cognitive Neuroscience, 23*, 3218–3227.

Plessow, F., Kiesel, A., & Kirschbaum, C. (2012). The stressed prefrontal cortex and goal-directed behaviour: Acute psychosocial stress impairs the flexible implementation of task goals. *Experimental Brain Research, 216*, 397–408.

Raio, C. M., & Glimcher, P. G. (2021). Quantifying the subjective cost of self-control in humans. *PNAS, 118*(35), e2018726118.

Robson, D. A., Allen, M. S., & Howard, S. J. (2020). Self-regulation in childhood as a predictor of future outcomes: A meta-analytic review. *Psychological Bulletin, 146*(4), 324.

Rushworth, M. F. S., Noonan, M. P., Boorman, E. D., Walton, M. E., & Behrens, T. E. (2011). Frontal cortex and reward-guided learning and decision-making. *Neuron, 70*, 1054–1069.

Schmidt, L., Tusche, A., Manoharan, N., Hutcherson, C., Hare, T., & Plassmann, H. (2018). Neuroanatomy of the vmPFC and dlPFC predicts individual differences in cognitive regulation during dietary self-control across regulation strategies. *Journal of Neuroscience, 38*(25), 5799–5806.

Shenhav, A. (2017). The perils of losing control: Why self-control is not just another value-based decision. *Psychological Inquiry, 28*(2–3), 148–152.

Shenhav, A., et al. (2013). The expected value of control: An integrative theory of anterior cingulate cortex function. *Neuron, 79*, 217–240.

Shenhav, A., Cohen, J. D., & Botvinick, M. M. (2016). Dorsal anterior cingulate cortex and the value of control. *Perspective Nature Neuroscience, 19*(10), 1286–1291.

Shenhav, A., Musslick, S., Lieder, F., Kool, W., Griffiths, T. L., Cohen, J. D., & Botvinick, M. M. (2017). Toward a rational and mechanistic account of mental effort. *Annual Review of Neuroscience, 40*, 99–124.

Simmonds, D. J., Pekar, J. J., & Mostofsky, S. H. (2008). Meta-analysis of go/no-go tasks demonstrating that fMRI activation associated with response inhibition is task-dependent. *Neuropsychologia, 46*, 224–232.

Sinha, R. (2008). Chronic stress, drug use, and vulnerability to addiction. *Ann N Y Acad Sci. 1141*, 105–30.

Soutschek, A., Ugazio, G., Crockett, M. J., Ruff, C. C., Kalenscher, T., & Tobler, P. N. (2017). Binding oneself to the mast: Stimulating frontopolar cortex enhances precommitment. *Social Cognitive and Affective Neuroscience, 12*(4), 635–642.

Soutschek, A., Moisa, M., Ruff, C. C., & Tobler, P. N. (2020). The right temporoparietal junction enables delay of gratification by allowing decision makers to focus on future events. *PLoS Biology, 18*(8), e3000800.

Soutschek, A., Moisa, M., Ruff, C. C., & Tobler, P. N. (2021). Frontopolar theta oscillations link metacognition with prospective decision making. *Nature Communications, 12*(1), 3943.

Studer, B., Koch, C., Knecht, S., & Kalenscher, T. (2019). Conquering the inner couch potato: Precommitment is an effective strategy to enhance motivation for effortful actions. *Philosophical Transactions of the Royal Society of London. Series B, Biological Sciences, 374*, 20180131.

Sullivan, N., Hutcherson, C., Harris, A., & Rangel, A. (2015). Dietary self-control is related to the speed with which attributes of healthfulness and tastiness are processed. *Psychological Science, 26*(2), 122–134.

Tusche, A., & Hutcherson, C. A. (2018). Cognitive regulation alters social and dietary choice by changing attribute representations in domain-general and domain-specific brain circuits. *eLife, 7*, e31185.

van Oort, J., Tendolkar, I., Hermans, E. J., Mulders, P. C., Beckmann, C. F., Schene, A. H., Fernández, G., & van Eijndhoven, P. F. (2017). How the brain connects in response to acute stress: A review at the human brain systems level. *Neuroscience and Biobehavioral Reviews, 83*, 281–297.

VanEpps, E. M., Downs, J. S., & Loewenstein, G. (2016). Advance ordering for healthier eating? Field experiments on the relationship between the meal order–consumption time delay and meal content. *Journal of Marketing Research, 53*(3), 369–380.

Wager, T. D., & Smith, E. E. (2003). Neuroimaging studies of working memory. *Cognitive, Affective, & Behavioral Neuroscience, 3*, 255–274.

Wager, T. D., Jonides, J., & Reading, S. (2004). Neuroimaging studies of shifting attention: A meta-analysis. *NeuroImage, 22*, 1679–1693.

Wang, G. J., et al. (2009). Evidence of gender differences in the ability to inhibit brain activation elicited by food stimulation. *Proceedings of the National Academy of Sciences USA, 106*, 1249–1254.

Westbrook, A., & Braver, T. S. (2015). Cognitive effort: A neuroeconomic approach. *Cognitive, Affective, & Behavioral Neuroscience, 15*, 395–415.

Part IV
Modulators of Decision-Making: Emotion and Social Context

Dominic S. Fareri, David V. Smith, and Patricia L. Lockwood

Our decisions are rarely made in isolated contexts. Rather, we make choices within complex environments that involve other people and evoke varied emotional states. As our world is becoming increasingly interconnected, it is critical to better understand how choice is shaped by emotion and social factors. This section explores myriad mechanisms through which choice is influenced by affect, feedback from others, and interpersonal relationship dynamics. By integrating insights from neuroscience, psychology, behavioral economics, and computational modeling, the chapters here consider the complex ways in which emotion and social context influence how we value choice options, exert and desire agency, and develop preferences.

A critical question in the literature surrounds the role of perceived control and emotion in shaping value-based choice. Agency over one's environment is often perceived as having rewarding and regulatory qualities. Yang and colleagues (this volume) detail how perceived control increases motivation, cognitive performance, and subjective well-being. Their chapter discusses work demonstrating that the reward value of control is represented within corticostriatal circuits and modulated by dopamine (Leotti & Delgado, 2014; Wang et al., 2020). They also note that these patterns have important implications for clinical conditions including depression and PTSD, which are characterized by disruptions in the ability to perceive control in the environment and view it as regulatory. The integration of these perspectives points to a broader framework in which control and emotion are deeply intertwined and interact, facilitating our assessment of value.

D. S. Fareri (✉)
Adelphi University, Garden City, NY, USA
e-mail: dfareri@adelphi.edu

D. V. Smith
Temple University, Philadelphia, PA, USA

P. L. Lockwood
University of Birmingham, Birmingham, UK

Affective responses to environmental stimuli may shape value by signaling emotional salience to our choice options. FeldmanHall and Teoh (this volume) review neuroimaging and computational modeling work demonstrating the ways in which internal emotional states help us accomplish this goal. Their synthesis of recent neuroimaging and modeling work converges on one possibility, that links between predictions and emotion are key. Multiple error learning signals (i.e., based on expected and experienced emotion arousal and valence) may interact to produce the value of a choice/outcome (Heffner et al., 2021). Building on this, Schmidt and Helion (this volume) examine how cognitive regulation of emotional arousal can interact with and alter our choices. This is particularly pertinent within the context of risk (e.g., Sokol-Hessner et al., 2009), wherein we are constantly weighing the value of potential positive and negative outcomes relative to some level of uncertainty. Schmidt and Helion present a framework suggesting that across multiple domains of risky decision-making, regulatory processes are engaged at various stages of choice and outcome evaluation, recruiting frontostriatal circuits supporting regulation and valuation processes (e.g., dorsolateral prefrontal cortex, medial prefrontal cortex). Together, these chapters suggest that emotion does not necessarily impede decision-making, but rather provides key signals that shape value computations.

Signals of emotion are often experienced and interpreted within social contexts, guided by social norms and expectations. Fareri and Dejoie (this volume) explore how social outcomes such as acceptance, rejection, and reciprocity from others act as reinforcement signals that reduce social uncertainty and facilitate relationship building (e.g., Brudner et al., 2023). These processes recruit putative reward-related and social cognitive neural circuits, and even extend to learning and sharing behavior within social media environments (e.g., Lindstrom et al., 2021; Scholz et al., 2017). Such experiences may lead to rejection, which neurocomputational studies have shown to impact learning of relational value (Babür et al., 2024) and the desire to restore control within one's environment (Dejoie et al., 2024). Interestingly, modeling approaches have also shown how interpersonal relationship dynamics can alter component processes supporting risky choice (i.e., risk attitudes, loss aversion; Fareri et al., 2022). In sum, this chapter highlights the importance of applying computational approaches to studies of social decision-making in real-world contexts noting the need for more ecologically valid paradigms.

Recent advances in social cognitive neuroscience have turned to naturalistic neuroimaging paradigms as one way to examine neural function in more ecologically valid ways. Chen and colleagues (this volume) review evidence regarding preference formation, highlighting the ventromedial prefrontal cortex, orbitofrontal cortex, and nucleus accumbens as key players in this process (van Holstein & Floresco, 2020; Fellows, 2007; Burton et al., 2015). Chen and colleagues discuss emerging work from their group using naturalistic movie viewing and a forced-choice preference task demonstrating that neural patterns during naturalistic movie viewing were significantly associated with neural patterns during preference decisions for the previously viewed movies. Their work suggests that preferences are instantiated by idiosyncratic neural patterns.

Taken together, the chapters in this section support the idea that decision-making is a complicated interplay of processes involving emotion, various social contexts, and perceived control. Arguing for the combination of behavioral, neural, and computational approaches, this section offers a rich, multidimensional account of how we arrive at varied types of choices within real-world contexts.

References

Babür, B. G., Leong, Y. C., Pan, C. X., & Hackel, L. M. (2024). Neural responses to social rejection reflect dissociable learning about relational value and reward. *Proceedings. National Academy of Sciences. United States of America, 121*, e2400022121.

Brudner, E. G., Fareri, D. S., Shehata, S. G., & Delgado, M. R. (2023). Social feedback promotes positive social sharing, trust and closeness. *Emotion, 23*, 1536–1548. https://doi.org/10.1037/emo0001182

Burton, A. C., Nakamura, K., & Roesch, M. R. (2015). From ventral-medial to dorsal-lateral striatum: Neural correlates of reward-guided decision-making. *Neurobiology of Learning and Memory, 117*, 51–59.

Chen, P. H. A., Chiu, T. P., Hsiao, P. A., & Chou, F. (this volume). From experiences to preferences: Neural insights through a naturalistic neuroimaging paradigm. In D. V. Smith, P. Lockwood, & D. S. Fareri (Eds.), *Neuroeconomics: Core topics and current directions*. Springer.

Dejoie, J. M., Ruiz, M., Brudner, E. G., & Fareri, D. S. (2024). Social rejection amplifies the value of choice. *Emotion, 25*(4), 929–939. https://doi.org/10.1037/emo0001456

Fareri, D. S., & Dejoie, J. M. (this volume). Modeling the social brain: Neurocomputational approaches to interpersonal learning and decision-making. In D. V. Smith, P. Lockwood, & D. S. Fareri (Eds.), *Neuroeconomics: Core topics and current directions*. Springer.

Fareri, D. S., Stasiak, J. E., & Sokol-Hessner, P. (2022). Choosing for others changes dissociable computational mechanisms underpinning risky decision-making. *Scientific Reports, 12*, 14361. https://doi.org/10.1038/s41598-022-18437-9

FeldmanHall, O., & Teoh, Y. Y. (this volume). Emotion and choice: The integral role of emotion in constructing value. In D. V. Smith, P. Lockwood, & D. S. Fareri (Eds.), *Neuroeconomics: Core topics and current directions*. Springer.

Fellows, L. K. (2007). The role of orbitofrontal cortex in decision making: A component process account. *Annals of the New York Academy of Sciences, 1121*, 421–430.

Heffner, J., Son, J. Y., & FeldmanHall, O. (2021). Emotion prediction errors guide socially adaptive behavior. *Nature Human Behavior, 5*(10), 1391–1401. https://doi.org/10.1038/s41562-021-01213-6

Leotti, L. A., & Delgado, M. R. (2014). The value of exercising control over monetary gains and losses. *Psychological Science, 25*(2), 596–604. https://doi.org/10.1177/0956797613514589

Lindström, B., Bellander, M., Schultner, D. T., Chang, A., Tobler, P. N., & Amodio, D. M. (2021). A computational reward learning account of social media engagement. *Nature Communications, 12*, 1311.

Schmidt, H., & Helion, C. A. (this volume). Integrated neural circuitry supporting emotion regulation and decision making. In D. V. Smith, P. Lockwood, & D. S. Fareri (Eds.), *Neuroeconomics: Core topics and current directions*. Springer.

Scholz, C., Baek, E. C., O'Donnell, M. B., Kim, H. S., Cappella, J. N., & Falk, E. B. (2017). A neural model of valuation and information virality. *National Academy of Sciences of the United States of America, 114*, 2881–2886. https://doi.org/10.1073/pnas.1615259114

Sokol-Hessner, P., Hsu, M., Curley, N. G., Delgado, M. R., Camerer, C. F., & Phelps, E. A. (2009). Thinking like a trader selectively reduces individuals' loss aversion.

National Academy of Sciences of the United States of America, 106(13), 5035–5040. https://doi.org/10.1073/pnas.0806761106

van Holstein, M., & Floresco, S. B. (2020). Dissociable roles for the ventral and dorsal medial prefrontal cortex in cue-guided risk/reward decision making. *Neuropsychopharmacology, 45*(4), 683–693.

Wang, K. S., Kashyap, M., & Delgado, M. R. (2020). The influence of contextual factors on the subjective value of control. *Emotion, 21*(4), 881–891. https://doi.org/10.1037/emo0000760

Yang, Y., Kwon, S., Derrell, A., & Delgado, M. R. (this volume). The value of perceived control. In D. V. Smith, P. Lockwood, & D. S. Fareri (Eds.), *Neuroeconomics: Core topics and current directions*. Springer.

The Value of Perceived Control

Yun-Yen Yang, Seh-Joo Kwon, Amanda Derrell, and Mauricio R. Delgado

Abstract This chapter explores the critical role of perceived control, defined as an individual's belief in their ability to influence outcomes, in shaping behavior, decision-making, and overall psychological well-being. Perceived control extends beyond mere objective capacity, profoundly impacting how individuals navigate daily challenges and complex life situations. The value of control is highlighted by the lengths to which people will go to preserve it, even at personal cost. The chapter underlines empirical evidence demonstrating that perceived control carries an intrinsic subjective value and enhances cognitive performance, emotional regulation, and resilience in the face of adversity, while its absence is strongly linked to psychiatric disorders, including anxiety and depression. Additionally, the chapter probes the neural underpinnings of perceived control in relation to motivation and reward systems, focusing in particular on cortical-striatal circuitry. Emphasizing the importance of understanding perceived control, how control varies across contexts such as task complexity, lifespan considerations, and socioeconomic conditions is also explored. These insights highlight the value of perceived control in promoting adaptive decision-making, psychological health, and resilience, offering a foundation for developing targeted interventions that enhance well-being across the lifespan.

1 Introduction

The belief in one's capacity to impact the environment shapes our perception of control. This perception, rather than tangible control, governs our interactions with our surroundings and informs our daily decisions. Such decisions can range from something as straightforward as selecting lottery numbers instead of accepting a computer's assignment, to more intricate situations such as navigating stress during

Y.-Y. Yang · S.-J. Kwon · A. Derrell · M. R. Delgado (✉)
Department of Psychology, Rutgers University, Newark, NJ, USA
e-mail: delgado@psychology.rutgers.edu

D. V. Smith et al. (eds.), *Neuroeconomics: Core Topics and Current Directions*,
https://doi.org/10.1007/978-3-032-02925-6_19

a pandemic, where a feeling of control can provide important coping mechanisms (Zheng et al., 2020). Indeed, this sense of control deeply influences our decision-making process, leading us to prioritize its preservation even at personal expense (Owens et al., 2014). For instance, when participating in a lottery, individuals often opt to pay more for a service that allows them to select their own numbers rather than accepting a randomly generated set, even though the probability of winning remains the same, highlighting the significance of perceived control and its intrinsic value, as individuals are willing to make financial sacrifices or pay more to maintain it. Decades of research have underscored the critical significance of perceived control in promoting overall well-being, and its pivotal role in upholding psychological and physical health by shaping individuals' coping mechanisms and cognitive performance (for review, see Leotti et al., 2010; Ly et al., 2019; Wallston et al., 1987). Conversely, disruptions in perceived control are fundamental features of various psychiatric disorders, including anxiety and depression (Gallagher et al., 2014; Liu et al., 2015).

Perceived control plays an important role in decision-making across various domains, impacting affective, motivational, and protective attributes (for review, see Wang et al., 2021). It promotes approach behavior and positive affect (Ly et al., 2019), drives individuals to seek competence (White, 1959), and helps resist stress in adverse situations (Maier & Seligman, 2016). In this chapter, we explore the many aspects of perceived control, looking back at its origins in early psychological theories to more recent insights from behavioral economics and neuroscience. We emphasize the dynamic nature of perceived control and how it is influenced by factors such as the availability of choices, task difficulty, and the relations between our actions and their outcomes. Next, we discuss the neural underpinnings of perceived control, concentrating on reward-processing-related brain areas such as the striatum and prefrontal cortex. Lastly, we emphasize how perceived control functions adaptively throughout the lifespan and how deficits in perceived control act as a cognitive shortcut that affects a spectrum of psychiatric disorders. This chapter highlights the critical role that perceived control plays in shaping our understanding of human motivation and decision-making processes.

2 The Structure of Control

The concept of perceived control was first introduced by Robert White in 1959 through the term "effectance motivation," which describes the feeling of efficacy to influence the environment (White, 1959). Subsequent theories aimed to explicate this construct, beginning with Rotter's theory on locus of control, which distinguished between internal and external beliefs regarding control over situations (Rotter, 1966). In a similar manner, Bandura's introduction of self-efficacy and response-efficacy further elucidated individuals' beliefs in their ability to perform actions and achieve desired outcomes (Bandura, 1978). Despite variations, these theories collectively interpret perceived control as the belief in one's capacity to

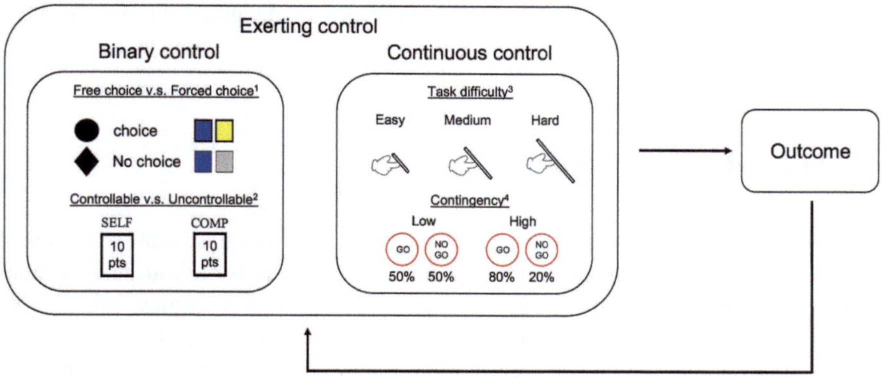

Fig. 1 Key elements contributing to perceived control. Perceived control can be examined in both binary and continuous states, as illustrated by two experimental paradigms for each. For binary control, participants are presented with opportunities to exert control. In one task (Leotti & Delgado, 2011), participants are either free to choose between a yellow key and a blue key (choice scenario) or are required to choose a computer-selected key (no-choice scenario). In another example (Wang & Delgado, 2019), participants choose between two options that vary in terms of the monetary value attached to them (i.e., willingness-to-pay) and they decide whether how much they are willing to pay to exert control (SELF) or allow the computer to make a choice (COMP). For continuous control, task parameters such as difficulty and reward contingency are manipulated. In the first task (Stevens et al., 2012), participants are asked to balance sticks of varying lengths using one finger. Shorter sticks, having a lower moment of inertia, are more challenging to balance, thereby reducing perceived control. A second task (Dorfman & Gershman, 2019) manipulates perceived control via reward contingency (low or high) using a go/nogo paradigm. These paradigms demonstrate that desirable outcomes following one's actions reinforce the preference for exerting control. Processes such as making free and controllable choices, reducing task difficulty, and increasing reward contingency enhance learning by boosting motivational drive through the outcomes achieved

exert influence over situations to attain rewards or avoid punishments. In experimental settings, this subjective perception of control has been manipulated in different ways (see Fig. 1): by allowing individuals to choose freely (e.g., Suzuki, 1997), by varying the difficulty of successfully executing behaviors (Stolz et al., 2020), or by providing behavioral contingencies that link actions to outcomes (e.g., Metcalfe et al., 2013; Tricomi et al., 2004). As such, perceived control can be operationally defined as the belief that performing a certain behavior will produce a desired effect. It exhibits (a) state-like characteristics, which fluctuate based on context and are influenced by action-outcome contingency and outcome valence and (b) trait-like characteristics, which are shaped by prior experiences of state-like perceived control (Liljeholm & O'doherty, 2012). The reward-based framework of perceived control highlights the reciprocal relationship between these state-like and trait-like aspects. Specifically, trait-like perceived control, informed by past experiences, influences how control is perceived in new situations, while state-like perceived control can alter these traits through mechanisms such as reinforcement learning. This interaction plays a critical role in decision-making, where the anticipation and experience of rewards reinforce the sense of control, thereby guiding future behavior (Ly et al., 2019).

3 The Value of Control

People use their decisions to exert control over their surroundings in daily life. These decisions range from basic perceptual choices made hundreds of thousands of times a day (e.g., where to focus attention in the visual field) to complex, life-altering decisions that occur maybe once in a lifetime (e.g., buying a house). Every voluntary action involves making choices, even when many are triggered by environmental signals and may not be fully conscious. Making choices is therefore naturally appealing as it is a potent way to communicate one's preferences and sense of value. Both humans and animals have a marked preference for scenarios that offer choices over those that do not, even when the option to choose does not enhance outcomes and requires greater effort (Bown et al., 2003; Leotti et al., 2010; Nussenbaum et al., 2024; Sullivan-Toole et al., 2017).

Human behavior studies have sought to quantify the intrinsic value of perceived control in experimental settings. Early research on the "illusion of control," where individuals believe they have control over outcomes despite lacking actual control, revealed that participants who selected their own lottery tickets valued them, on average, $1.96 higher and were less willing to sell them for $8.67 more than those who were assigned tickets (Langer, 1975). People also prefer making their own decisions rather than delegating them to others (Bartling et al., 2014). This preference underscores the intrinsic value placed on control, with individuals willing to forgo other rewards to enhance their sense of agency (Bobadilla-Suarez et al., 2017), such as sacrificing 8–15% of expected earnings for the sake of control (Owens et al., 2014). In some contexts, perception of control can increase the subjective value of an option by 30% (Wang & Delgado, 2019), with participants unwilling to relinquish control without substantial bonuses of at least 34% (Reis et al., 2023). Finally, participants are willing to bid more to learn outcomes of trials where they had control compared to random lotteries (Jiwa et al., 2021), suggesting a greater valuation of outcomes reflecting their own agency compared to unchosen actions. Taken together, these findings highlight the inherent value of perceived control by quantifying human behavior across various task conditions.

While behavioral evidence indicates that perceived control significantly impacts decision-making and that having control is both desirable and valuable, more recent investigations focusing on neuroimaging of the human brain have sought to address how perceived control is processed (see Fig. 2). Accordingly, the opportunity to exert control in the environment tends to recruit brain regions commonly observed in reward valuation studies focusing on corticostriatal circuits (Bartra et al., 2013). Specifically, this intrinsic value of choice opportunity is supported by neuroscientific evidence revealing heightened activity in regions such as the ventromedial prefrontal cortex (vmPFC) and ventral striatum when individuals anticipate making choices

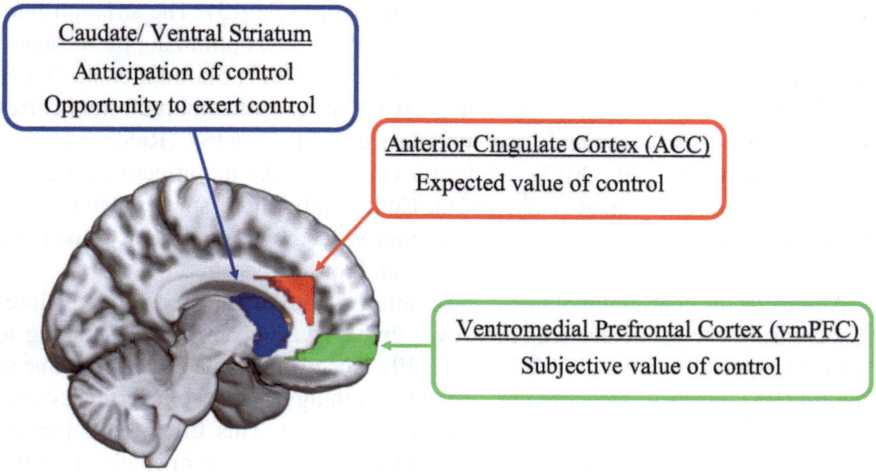

Fig. 2 Neural circuitry involved in the value of perceived control. Brain regions are depicted based on the Automated Anatomical Atlas (AAL; Rolls et al., 2020). Anticipation of control and the opportunity to exercise control engage areas typically associated with reward processing, including the caudate and ventral striatum. The intrinsic value of control, which pertains to a common currency for choice-inferred subjective values, is represented in regions such as the ventromedial prefrontal cortex (vmPFC). The expected value of control, which involves balancing the potential costs against anticipated rewards, is encoded in the anterior cingulate cortex (ACC)

compared to scenarios devoid of choice (Leotti & Delgado, 2011, 2014; Romaniuk et al., 2019; Wang & Delgado, 2019). This neural activation correlates positively with the perceived value of the choice opportunity (Fujiwara et al., 2013). Further, activity in the striatum associated with the opportunity for control can have indirect effects on cognitive performance, such as enhancing declarative memory (DuBrow et al., 2019; Murty et al., 2015; Ruiz et al., 2023; but also see Katzman & Hartley, 2020), or increasing decisions to persist in the pursuit of a goal following negative feedback (Bhanji & Delgado, 2014). Together, these findings underscore that the ability to make choices engages reward-processing neural circuits, which can have downstream effects on decision-making and other cognitive processes, emphasizing the profound intrinsic value of perceived control.

Perceived control extends beyond simple dichotomies, existing both as a binary state and along a continuum, reflecting the varying degrees to which individuals believe they can successfully execute actions (Bandura, 1978). A negative correlation between perceived task difficulty and perceived control has also been observed, where higher perceived difficulty diminishes perceived control by influencing performance expectations (Depasque Swanson & Tricomi, 2014; Stevens et al., 2012). That is, easier tasks enhance perceived control by increasing the likelihood of success, while more challenging tasks decrease it by raising the probability of failure. The prefrontal cortex and corticostriatal networks play a critical role in reward processing and goal-directed behavior by regulating perceived control and task complexity (Blain & Sharot, 2021; Shenhav et al., 2013). The striatum integrates feedback and expectations related to task difficulty, optimizing performance under varying levels of perceived control (Chiu et al., 2017; Skottnik et al., 2019). Additionally, the anterior cingulate cortex (ACC) and dorsolateral prefrontal cortex (dlPFC) evaluate task demands and modulate cognitive control (Ridderinkhof et al., 2004; Shenhav et al., 2013), while the vmPFC tracks the affective aspect of this intrinsic reward (Chew et al., 2021). Collectively, these studies demonstrate that perceived control exists on a continuum influenced by task difficulty, involving brain circuits related to goal-directed actions and reward systems.

Alongside the continuum of perceived control, adaptive control research explores how control is dynamically regulated over time, focusing on decision-making to optimize expected utility (Chiu & Egner, 2019). According to the Expected Value of Control (EVC) theory, decisions are made by weighing the costs of exerting control against the anticipated rewards (Shenhav et al., 2013). This theory incorporates perceived control into the analysis of effort allocation, with higher perceived control leading to increased effort (Blain & Sharot, 2021; Frömer et al., 2020; Shenhav et al., 2017). This overall expected value of control is assessed by the dorsal anterior cingulate cortex (dACC; Frömer et al., 2020; Shenhav et al., 2016). Following the computation of EVC, the dACC then activates the dlPFC to implement cognitive control.

4 Dopamine and Control

The rewarding aspect of perceived control underscores an intrinsic motivation to actively seek opportunities that enhance this sense of control (Gozli, 2019; Leotti et al., 2010). This inclination drives individuals to invest effort in behaviors aimed at gaining control aligning with theories of action-valence coupling (Boureau & Dayan, 2011; Cools et al., 2011; Panksepp & Green, 1998). According to these theories, dopamine-mediated projections to striatal regions, particularly the nucleus accumbens (NAcc), facilitate appetitive-driven actions, such as approach behaviors and active engagement, in pursuit of rewards. Every decision or action, regardless of its size, strengthens the sense of control. Importantly, the value of choice is informed by both the intrinsic motivation to engage in decision-making and the instrumental value of control (Cockburn et al., 2014; Devine et al., 2024). That is, while individuals show a propensity to favor situations where choice is available, this effect is boosted in situations where the informational value of future rewards is increased (Devine et al., 2024). This strengthening is closely linked to the concept of instrumental contingency, where actions are reinforced by desirable outcomes according to reinforcement learning principles (Cockburn et al., 2014; Dickinson & Balleine, 1994; Dorfman & Gershman, 2019; Liljeholm et al., 2011; Moscarello & Hartley, 2017; Shanks & Dickinson, 1991). Perceived control arises when specific actions reliably lead to desired outcomes, a phenomenon associated with learning instrumental action-outcome contingencies mediated by corticostriatal circuits and dopaminergic prediction error signals (Griffiths et al., 2014; Liljeholm et al., 2011; Ly et al., 2019; O'Doherty et al., 2004; Tricomi et al., 2004). Within the framework of action-outcome contingencies, it is evident that objective control is a significant predictor of perceived control. However, previous studies have identified divergences in scenarios involving outcome uncertainty and personal relevance (Reis et al., 2023; Romaniuk et al., 2023; Schwarz et al., 2022). A quadratic relation between contingency and perceived control reveals a tendency to overweight action-outcome contingencies under a sense of control (Schwarz et al., 2022). In other words, when people believe they have control over a situation, they are more likely to overestimate how much their actions actually influence the results.

Perceived control, particularly through action-outcome contingencies, acts as a motivational catalyst, with variations in dopaminergic transmission expected to positively correlate with increased approach behavior and response vigor (Berridge, 2012; Ly et al., 2019; Raab et al., 2024). Prior studies have supported this idea by demonstrating that individuals with a heightened sense of control exhibit elevated striatal dopaminergic levels (Balleine et al., 2021; Romaniuk et al., 2023; Vassena et al., 2019), and engage more frequently in control-seeking behaviors (Kayser et al., 2015). Collectively, these findings underscore the pivotal role of dopamine in explaining individual differences in behaviors related to seeking and exerting control, highlighting the significant intrinsic value of perceived control.

5 The Context of Perceived Control

With much of the existing literature mentioned above, there is a prevalent emphasis on the perception of control rather than the actual degree of control, which consistently proves pivotal in influencing outcomes. The main reason is that while there has been some exploration into manipulating actual control to understand its role, this area remains less prominent. For instance, studies utilizing driving simulations to manipulate objective control have demonstrated that the impact on stress levels varies depending on participants' perceived levels of control within a given environment (Agrigoroaei et al., 2013). Specifically, compared to participants who had a lower perception of control, a higher perception of control was associated with a larger stress response in the low controllability condition. Similarly, research into stressor controllability and conditioned fear responses has indicated that higher subjective confidence in controlling shocks, rather than objective learning performance, predicts reduced fear expression (Hartley et al., 2014). Further, under a certain level of objective controllability, lower subjectively perceived control levels correspond to greater acute stress responses (Liu et al., 2021). Taken together, these studies illustrate the importance of subjective perceptions of control in addition to objective controllability in shaping stress responses, particularly within different contexts.

While the significance of perceived control is widely emphasized, it is important to recognize that control is not always desirable in certain situations. Extensive literature in psychology and marketing has explored the determinants of choice deferral and its consequences (see reviews: Chernev et al., 2015; Misuraca et al., 2020). Factors such as increased decision task difficulty and greater choice set complexity can facilitate choice overload (Chernev et al., 2015; Scheibehenne et al., 2010), where too many choices can be counterproductive on effective decision-making and prompt individuals to defer control to others. In some cases, reliance on default rules may be preferable to active choosing, especially in contexts where individuals lack the information or motivation to make optimal choices (Sunstein, 2017), as the default leverages inertia, reduces cognitive costs, and signals that participation is beneficial (Mullainathan & Shafir, 2013; Sunstein, 2015). Experimental tasks with limited learning opportunities show that individuals tend to select tasks of intermediate difficulty to maximize learning (Baranes et al., 2014; Ten et al., 2021). Intermediate challenging tasks are also perceived as more engaging (Aston-Jones & Cohen, 2005; Sayali et al., 2022). Additionally, the addition of equally attractive options can increase the likelihood of choice deferral (Dhar & Simonson, 2003). Larger choice sets can be overwhelming, increasing cognitive conflict and leading to negative outcomes such as decision paralysis and reduced satisfaction with the chosen option (Chernev et al., 2015; Schwartz & Cheek, 2017). The contemporary "inverted U-shape" perspective on choice set size and satisfaction suggests that both too little choice (choice deprivation) and too much choice (choice overload) can be detrimental (Grant & Schwartz, 2011; Reutskaja et al., 2018). Further studies found that people perceived the sense of control differently according to contextual valence (e.g., gain or loss frames; Barlow, 2018; Herman & Tsakiris, 2020; Leotti & Delgado, 2014; Tanaka & Kawabata, 2020).

There are various reasons one may not prefer control, from avoiding responsibility and blame (Steffel et al., 2016) to minimizing the explicit confrontation of being out of control (Li et al., 2017). This line of studies suggests that when the act of choosing becomes burdensome, individuals are willing to relinquish control, and highlights limits in the desire for control. Importantly, however, even when deferring decisions, individuals are still exerting some form of control by opting out. Indeed, we often *choose* to defer our decisions based on expertise (e.g., retirement planning), effort (e.g., making a meal vs. take-out), or even negative feelings associated with regret or making the wrong choice in simple situations (e.g., picking the wine at a restaurant for a group of friends). These findings highlight how the subjective value of control is highly context-dependent and malleable (Wang et al., 2021). Ultimately, choosing to relinquish control to others or remain with a default represents a deliberate choice in itself, demonstrating that the act of decision deferral is a form of control tailored to context-specific needs.

6 Perceived Control Through the Lifespan

With many mechanisms of perceived control developing with age, one important consideration about perceived control is that it also varies across the lifespan. While the accuracy in detecting controllability improves over time, the preference for agency over choice begins in infancy. For instance, 4-month-old infants exhibit anger when their perceived control is restricted (Raab et al., 2022; Sullivan & Lewis, 2003). Early perceived control is crucial for the development of executive functions, as it underpins complex cognitive skills such as planning and goal-directed decision-making, and has been consistently linked to better school performance in children (Carlson, 2023; Skinner et al., 1998). During adolescence, teenagers show greater sensitivity to environmental controllability compared to children and adults, which affects instrumental learning and decision-making (Raab et al., 2024). While historically, increased risky behaviors in adolescence are thought to be a byproduct of a compromised control system, more recent studies argue that some adolescent risky behaviors may actually arise from *elevated* control and are thus strategic (Do et al., 2020). Together, adolescents' heightened sensitivity to perceived control and intentional use of control can make adolescents' decisions adaptive.

While the previous studies have demonstrated that perceived control continues to shift throughout adulthood such that it generally increases from early to mid-adulthood but declines from mid to late adulthood (albeit there are variations depending on the domain; Gatz & Karel, 1993; Lachman & Weaver, 1998; Robinson & Lachman, 2017; Specht et al., 2013), consistency in the preference for being in control across the lifespan has more recently been observed (Chantland et al., 2022). This persistent preference for control among older adults aligns with the notion that control is intrinsically rewarding (Bobadilla-Suarez et al., 2017; Owens et al., 2014; Stolz et al., 2020; Wang & Delgado, 2019; Yang & Delgado, 2025), rather than merely a means to achieve desired outcomes. Interestingly, older adults often prefer less autonomy and responsibility, seeking less information even when decisions affect them directly (Finucane et al., 2002; Mather, 2006; Reed et al., 2008). Despite feeling less in control, they might relinquish control due to deficits in reward learning (Chowdhury et al., 2013; Samanez-Larkin & Knutson, 2014, 2015) and reward anticipation (Dreher et al., 2008; Schott et al., 2007), which can reduce the frequency of obtaining rewards associated with exerting control (Ly et al., 2019). Over time, this reduction in reward frequency may diminish their inclination to engage in situations that offer opportunities to exercise control.

Perceived control continues to offer benefits in older age, such as reduced mortality risks, enhanced cognitive abilities, and a younger subjective age, highlighting its protective role (Bellingtier & Neupert, 2020; Infurna & Okun, 2015; Windsor et al., 2008). However, it is important to recognize that a high sense of control can be deleterious in certain circumstances, particularly when objective control is low (Robinson & Lachman, 2017). For example, individuals with a high sense of control but low health status may develop a potentially maladaptive "invincible" mindset (Chipperfield et al., 2016). This mindset, accompanied by denial of risk and absence of fear, can have detrimental consequences for health behaviors and approaches to seeking care (Chipperfield et al., 2017; Lachman et al., 2015). In fact, it has been suggested that a realistic assessment of one's control may be more beneficial than overly optimistic perceptions. Knowing when to relinquish control may be a form of wisdom, contributing to increased well-being (Chang et al., 2015; Lachman et al., 2011).

7 Transdiagnostic Implication for Psychopathologies

While much attention has been given to the value of perceived control that focuses on its affective and motivational attributes, its protective attribute is equally important. Pioneering research on "learned helplessness" demonstrated that perceived control can mitigate the negative effects of external stressors (Seligman et al., 1975). This protective effect indicates that controllability over an aversive stimulus reduces behavioral passivity, anxiety, and learning deficits (for review, see Maier & Seligman, 2016). Perceived stressor controllability has been shown to influence emotional, behavioral, and neurobiological responses. Emotionally, it is linked to blunted conditioned fear expression, improved fear recovery (Hartley et al., 2014), and mitigated feelings of helplessness (Meine et al., 2021). Behaviorally, it promotes increased effort exertion, persistence (Bhanji et al., 2016), more adaptive responses to future stressors (Cohodes et al., 2023), and enhanced performance, as well as reduced intensity and increased tolerability of painful stimuli (Beck et al., 2017; Bräscher et al., 2016). Neurobiologically, it is associated with reduced activation in key brain regions involved in threat processing, such as the bed nucleus of the stria terminalis and anterior insula (Limbachia et al., 2021). Additionally, it increases activation in the ventromedial prefrontal cortex, a region involved in encoding the subjective value of control (Wang & Delgado, 2019). Therefore, the ability to perceive control over the environment may provide an adaptive benefit by assisting in regulating emotions. Perceived control appears to function as an emotion regulation tool that can dampen stress reactivity and the associated negative emotions.

However, there are times when our perception of control becomes disturbed, leading to maladaptive behaviors that negatively impact our well-being. This is evident in various psychopathologies, where patients often report altered perceptions of control in response to specific environmental triggers or more broadly in their lives, resulting in detrimental behaviors. For example, individuals with depression often exhibit attentional bias, focusing more on negative stimuli (Kuehl et al., 2021). They also have a lower affective response to perceived control compared to healthy individuals (Chang et al., 2020), suggesting that they may assign a lower subjective value to control. Further studies support this finding, showing that greater depressive symptoms are negatively correlated with internal control and positively associated with the perception of external control (Inozu et al., 2012; Jasdi & Rahman, 2023; Rotter, 1966). Recent research also reports that depressed individuals, compared to healthy participants, exhibit less ventral striatal activation when anticipating the opportunity for control (Romaniuk et al., 2019) and show blunted reward-related neural responses to the receipt of rewards following the exertion of control (Chang et al., 2020). Therefore, it is important to recognize the potential of interventions that focus on enhancing aspects of perceived control to reduce depressive symptoms through longitudinal training (Ulleberg et al., 2021).

In the context of addiction, individuals often feel compelled to engage in substance use despite adverse consequences (O'Brien, 2011). Perceived control is a significant predictor of future behavior in treating substance abuse (Chavarria et al., 2012; Kadden & Litt, 2011). The inability to perceive control over drug-use behavior (Kalivas & Volkow, 2005) suggests that enhancing perceived control is critical for improving clinical outcomes (Kiburi et al., 2023). Strategies focused on increasing perceived control have been shown to reduce cue-induced cravings, and promote smoking cessation in nicotine-dependent individuals (Loeber et al., 2006; Ono et al., 2018; Schnoll et al., 2011). Similarly, individuals with post-traumatic stress disorder (PTSD) are characterized by diminished perceived control over emotional responses to stressors (Gallagher et al., 2014), and techniques such as recalling positive autobiographical memories have been effective in enhancing perceived control (Brown et al., 2016; Morina et al., 2018; Titcombe-Parekh et al., 2018). This high-perceived control intervention could possibly target the prefrontal cortex, which engages in top-down regulatory processes that focus on emotional regulation and executive function, mitigating specific neural changes associated with PTSD (Titcombe-Parekh et al., 2018).

Control is also proposed to play a critical role in obsessive-compulsive disorder (OCD), where individuals attempt to control their thoughts and rely on compulsive behaviors to manage external influences (Carr, 1974; Rachman & Silva, 2009). Empirical evidence suggests that the need to control thoughts, a sense of control, and beliefs about losing control are closely associated with OCD symptomatology (Gagné & Radomsky, 2020; Radomsky, 2022; Taylor et al., 2010). A mismatch between the desire for control and the actual sense of control tends to exacerbate obsessive-compulsive symptoms (Moulding et al., 2008, 2009). Furthermore, individuals with OCD have been found to exhibit lower external control, meaning they are less likely to perceive events as controlled by others or by chance, compared to nonclinical and anxiety disorder controls (Altın & Karanci, 2008; Inozu et al., 2012). Taken together, these findings suggest that impairments in perceived control are a common feature across psychopathologies, highlighting the importance of studying and understanding this psychological construct.

8 Conclusion and Future Directions

Our ability to perceive control in our environment profoundly impacts decision-making, well-being, and psychological health. This perception influences our daily choices and coping mechanisms during stressful events. Research highlights its intrinsic value, showing that individuals prioritize maintaining control, supported by corticostriatal circuit and dopaminergic system. Neuroscientific studies reveal that perceived control activates reward-processing brain areas, enhancing cognitive performance and emotional regulation, and that dopamine plays a critical role in linking control with motivation and persistence.

There is a growing body of research examining how perceived control varies across the lifespan, with different age groups exhibiting unique preferences and benefits (Lachman & Weaver, 1998; Robinson & Lachman, 2017). Yet, there is less research on how other factors can influence perceived control, such as differences across cultures, socioeconomic status (SES), or even individual trait differences. For instance, individuals from the United States report the highest levels of perceived control, while those from Japan report the lowest (Hornsey et al., 2019; Lachman, 2006). Interestingly, control beliefs are more important for the well-being of Americans than for the Japanese (Toyama, 2019; Toyama & Fuller, 2021). These cultural differences are likely related to variations in economic conditions, values, and religious beliefs (Lachman et al., 2011). With respect to SES, higher education and income are associated with greater perceived control (Mirowsky & Ross, 2007), whereas lower SES correlates with lower perceived control due to greater environmental constraints (Deci & Ryan, 1987; Lachman et al., 2011). Additionally, at the individual level, personality traits exhibit a modest yet consistent effect on perceived control, particularly in uncertain situations, with higher levels of openness and extraversion being positively correlated with perceived control (Schwarz et al., 2022).

Despite extensive research on the significance of perceived control for decision-making and overall well-being, knowledge on the role of social factors in shaping this perception is limited. Loneliness and social isolation can undermine perceived control, as individuals may feel inadequate in addressing their socioemotional needs (Cacioppo & Hawkley, 2009; Perlman, 2004). Yet, social network engagement and volunteering are linked to higher levels of perceived control (Curtis et al., 2016), with social activities, such as volunteering, enhancing perceived control by providing positive feedback and encouragement (Fuller-Iglesias et al., 2008; Shaw et al., 2004). Furthermore, research involving animal models suggests that social interactions can act as protective factors against substance abuse, with rats preferring social contact over drug administration when both options are available (Venniro et al., 2018). Future research may benefit from investigations of how various forms of social support and personal experiences influence perceived control in greater depth. By exploring these dynamics, we can develop more effective interventions to improve psychological health and resilience, thereby enhancing individuals' capacity to maintain a sense of control in diverse contexts.

References

Agrigoroaei, S., Polito, M., Lee, A., Kranz-Graham, E., Seeman, T., & Lachman, M. E. (2013). Cortisol response to challenge involving low controllability: The role of control beliefs and age. *Biological Psychology, 93*(1), 138–142. https://doi.org/10.1016/j.biopsycho.2013.01.003

Altın, M., & Karanci, A. N. (2008). How does locus of control and inflated sense of responsibility relate to obsessive-compulsive symptoms in Turkish adolescents? *Journal of Anxiety Disorders, 22*(8), 1303–1315. https://doi.org/10.1016/j.janxdis.2008.01.010

Aston-Jones, G., & Cohen, J. D. (2005). An integrative theory of locus coeruleus-norepinephrine function: Adaptive gain and optimal performance. *Annual Review of Neuroscience, 28*(1), 403–450. https://doi.org/10.1146/annurev.neuro.28.061604.135709

Balleine, B. W., Peak, J., Matamales, M., Bertran-Gonzalez, J., & Hart, G. (2021). The dorsomedial striatum: An optimal cellular environment for encoding and updating goal-directed learning. *Current Opinion in Behavioral Sciences, 41*, 38–44. https://doi.org/10.1016/j.cobeha.2021.03.004

Bandura, A. (1978). Self-efficacy: Toward a unifying theory of behavior change. *Advances in Behaviour Research and Therapy, 1*(4), 139–161. https://doi.org/10.1016/0146-6402(78)90002-4

Baranes, A. F., Oudeyer, P.-Y., & Gottlieb, J. (2014). The effects of task difficulty, novelty and the size of the search space on intrinsically motivated exploration. *Frontiers in Neuroscience, 8*, 317. https://doi.org/10.3389/fnins.2014.00317

Barlow, D. H. (2018). Disorders of emotion. In *The neurotic paradox* (Vol. 2, pp. 201–226). Routledge.

Bartling, B., Fehr, E., & Herz, H. (2014). The intrinsic value of decision rights: Intrinsic value of decision rights. *Econometrica, 82*(6), 2005–2039. https://doi.org/10.3982/ECTA11573

Bartra, O., McGuire, J. T., & Kable, J. W. (2013). The valuation system: A coordinate-based meta-analysis of BOLD fMRI experiments examining neural correlates of subjective value. *NeuroImage, 76*, 412–427. https://doi.org/10.1016/j.neuroimage.2013.02.063

Beck, B., Di Costa, S., & Haggard, P. (2017). Having control over the external world increases the implicit sense of agency. *Cognition, 162*, 54–60. https://doi.org/10.1016/j.cognition.2017.02.002

Bellingtier, J. A., & Neupert, S. D. (2020). Feeling young and in control: Daily control beliefs are associated with younger subjective ages. *The Journals of Gerontology: Series B, 75*(5), e13–e17. https://doi.org/10.1093/geronb/gbz015

Berridge, K. C. (2012). From prediction error to incentive salience: Mesolimbic computation of reward motivation: From prediction error to incentive salience. *European Journal of Neuroscience, 35*(7), 1124–1143. https://doi.org/10.1111/j.1460-9568.2012.07990.x

Bhanji, J. P., & Delgado, M. R. (2014). Perceived control influences neural responses to setbacks and promotes persistence. *Neuron, 83*(6), 1369–1375. https://doi.org/10.1016/j.neuron.2014.08.012

Bhanji, J. P., Kim, E. S., & Delgado, M. R. (2016). Perceived control alters the effect of acute stress on persistence. *Journal of Experimental Psychology: General, 145*(3), 356–365. https://doi.org/10.1037/xge0000137

Blain, B., & Sharot, T. (2021). Intrinsic reward: Potential cognitive and neural mechanisms. *Current Opinion in Behavioral Sciences, 39*, 113–118. https://doi.org/10.1016/j.cobeha.2021.03.008

Bobadilla-Suarez, S., Sunstein, C. R., & Sharot, T. (2017). The intrinsic value of choice: The propensity to under-delegate in the face of potential gains and losses. *Journal of Risk and Uncertainty, 54*(3), 187–202. https://doi.org/10.1007/s11166-017-9259-x

Boureau, Y.-L., & Dayan, P. (2011). Opponency revisited: Competition and cooperation between dopamine and serotonin. *Neuropsychopharmacology, 36*(1), 74–97. https://doi.org/10.1038/npp.2010.151

Bown, N. J., Read, D., & Summers, B. (2003). The lure of choice. *Journal of Behavioral Decision Making, 16*(4), 297–308. https://doi.org/10.1002/bdm.447

Bräscher, A.-K., Becker, S., Hoeppli, M.-E., & Schweinhardt, P. (2016). Different brain circuitries mediating controllable and uncontrollable pain. *The Journal of Neuroscience, 36*(18), 5013–5025. https://doi.org/10.1523/JNEUROSCI.1954-15.2016

Brown, A. D., Kouri, N. A., Rahman, N., Joscelyne, A., Bryant, R. A., & Marmar, C. R. (2016). Enhancing self-efficacy improves episodic future thinking and social-decision making in combat veterans with posttraumatic stress disorder. *Psychiatry Research, 242*, 19–25. https://doi.org/10.1016/j.psychres.2016.05.026

Cacioppo, J. T., & Hawkley, L. C. (2009). Perceived social isolation and cognition. *Trends in Cognitive Sciences, 13*(10), 447–454. https://doi.org/10.1016/j.tics.2009.06.005

Carlson, S. M. (2023). Let me choose: The role of choice in the development of executive function skills. *Current Directions in Psychological Science, 32*(3), 220–227. https://doi.org/10.1177/09637214231159052

Carr, A. T. (1974). Compulsive neurosis: A review of the literature. *Psychological Bulletin, 81*(5), 311–318. https://doi.org/10.1037/h0036473

Chang, W. C., Toh, Y., Fan, Q. Q., & Chen, A. S. H. (2015). Mind over body: Positive attitude and flexible control beliefs on positive aging. *BIOPHILIA, 2015*(3), 254–261. https://doi.org/10.14813/ibra.2015.254

Chang, Y., Wang, Y., Mei, S., Yi, W., & Zheng, Y. (2020). Blunted neural effects of perceived control on reward feedback in major depressive disorder. *Journal of Affective Disorders, 276*(9), 112–118. https://doi.org/10.1016/j.jad.2020.06.071

Chantland, E. C. M., Wang, K. S., Delgado, M. R., & Ravizza, S. M. (2022). Control preference persists with age. *Psychology and Aging, 37*(7), 843–847. https://doi.org/10.1037/pag0000708

Chavarria, J., Stevens, E. B., Jason, L. A., & Ferrari, J. R. (2012). The effects of self-regulation and self-efficacy on substance use abstinence. *Alcoholism Treatment Quarterly, 30*(4), 422–432. https://doi.org/10.1080/07347324.2012.718960

Chernev, A., Böckenholt, U., & Goodman, J. (2015). Choice overload: A conceptual review and meta-analysis. *Journal of Consumer Psychology, 25*(2), 333–358. https://doi.org/10.1016/j.jcps.2014.08.002

Chew, B., Blain, B., Dolan, R. J., & Rutledge, R. B. (2021). A neurocomputational model for intrinsic reward. *The Journal of Neuroscience, 41*(43), 8963–8971. https://doi.org/10.1523/JNEUROSCI.0858-20.2021

Chipperfield, J. G., Perry, R. P., Pekrun, R., Barchfeld, P., Lang, F. R., & Hamm, J. M. (2016). The paradoxical role of perceived control in late life health behavior. *PLoS One, 11*(3), e0148921. https://doi.org/10.1371/journal.pone.0148921

Chipperfield, J., Hamm, J., Perry, R., & Ruthig, J. (2017). Perspectives on studying perceived control in the twenty-first century. In *The happy mind: Cognitive contributions to well-being* (pp. 215–234). https://doi.org/10.1007/978-3-319-58763-9_12

Chiu, Y.-C., & Egner, T. (2019). Cortical and subcortical contributions to context-control learning. *Neuroscience & Biobehavioral Reviews, 99*, 33–41. https://doi.org/10.1016/j.neubiorev.2019.01.019

Chiu, Y.-C., Jiang, J., & Egner, T. (2017). The caudate nucleus mediates learning of stimulus–control state associations. *The Journal of Neuroscience, 37*(4), 1028–1038. https://doi.org/10.1523/JNEUROSCI.0778-16.2016

Chowdhury, R., Guitart-Masip, M., Lambert, C., Dayan, P., Huys, Q., Düzel, E., & Dolan, R. J. (2013). Dopamine restores reward prediction errors in old age. *Nature Neuroscience, 16*(5), 648–653. https://doi.org/10.1038/nn.3364

Cockburn, J., Collins, A. G. E., & Frank, M. J. (2014). A reinforcement learning mechanism responsible for the valuation of free choice. *Neuron, 83*(3), 551–557. https://doi.org/10.1016/j.neuron.2014.06.035

Cohodes, E. M., Odriozola, P., Mandell, J. D., Caballero, C., McCauley, S., Zacharek, S. J., Hodges, H. R., Haberman, J. T., Smith, M., Thomas, J., Meisner, O. C., Ellis, C. T., Hartley, C. A., & Gee, D. G. (2023). Neural effects of controllability as a key dimension of stress exposure. *Development and Psychopathology, 35*(1), 218–227. https://doi.org/10.1017/S0954579421001498

Cools, R., Nakamura, K., & Daw, N. D. (2011). Serotonin and dopamine: Unifying affective, activational, and decision functions. *Neuropsychopharmacology, 36*(1), 98–113. https://doi.org/10.1038/npp.2010.121

Curtis, R. G., Huxhold, O., & Windsor, T. D. (2016). Perceived control and social activity in midlife and older age: A reciprocal association? Findings from the German ageing survey. *The*

Journals of Gerontology Series B: Psychological Sciences and Social Sciences, 73(5), gbw070. https://doi.org/10.1093/geronb/gbw070

Deci, E. L., & Ryan, R. M. (1987). The support of autonomy and the control of behavior. *Journal of Personality and Social Psychology, 53*(6), 1024–1037. https://doi.org/10.1037/0022-3514.53.6.1024

Depasque Swanson, S., & Tricomi, E. (2014). Goals and task difficulty expectations modulate striatal responses to feedback. *Cognitive, Affective, & Behavioral Neuroscience, 14*(2), 610–620. https://doi.org/10.3758/s13415-014-0269-8

Devine, S., da Silva Castanheira, K., Fleming, S. M., & Otto, A. R. (2024). Distinguishing between intrinsic and instrumental sources of the value of choice. *Cognition, 245*, 105742. https://doi.org/10.1016/j.cognition.2024.105742

Dhar, R., & Simonson, I. (2003). The effect of forced choice on choice. *Journal of Marketing Research, 40*(2), 146–160. https://doi.org/10.1509/jmkr.40.2.146.19229

Dickinson, A., & Balleine, B. (1994). Motivational control of goal-directed action. *Animal Learning & Behavior, 22*(1), 1–18. https://doi.org/10.3758/BF03199951

Do, K. T., Sharp, P. B., & Telzer, E. H. (2020). Modernizing conceptions of valuation and cognitive-control deployment in adolescent risk taking. *Current Directions in Psychological Science, 29*(1), 102–109. https://doi.org/10.1177/0963721419887361

Dorfman, H. M., & Gershman, S. J. (2019). Controllability governs the balance between Pavlovian and instrumental action selection. *Nature Communications, 10*(1), 5826. https://doi.org/10.1038/s41467-019-13737-7

Dreher, J.-C., Meyer-Lindenberg, A., Kohn, P., & Berman, K. F. (2008). Age-related changes in midbrain dopaminergic regulation of the human reward system. *Proceedings of the National Academy of Sciences, 105*(39), 15106–15111. https://doi.org/10.1073/pnas.0802127105

DuBrow, S., Eberts, E. A., & Murty, V. P. (2019). A common mechanism underlying choice's influence on preference and memory. *Psychonomic Bulletin and Review, 26*(6), 1958–1966. https://doi.org/10.3758/s13423-019-01650-5

Finucane, M. L., Slovic, P., Hibbard, J. H., Peters, E., Mertz, C. K., & MacGregor, D. G. (2002). Aging and decision-making competence: An analysis of comprehension and consistency skills in older versus younger adults considering health-plan options. *Journal of Behavioral Decision Making, 15*(2), 141–164. https://doi.org/10.1002/bdm.407

Frömer, R., Lin, H., Dean Wolf, C. K., Inzlicht, M., & Shenhav, A. (2020). *When effort matters: Expectations of reward and efficacy guide cognitive control allocation*. https://doi.org/10.1101/2020.05.14.095935

Fujiwara, J., Usui, N., Park, S. Q., Williams, T., Iijima, T., Taira, M., Tsutsui, K.-I., & Tobler, P. N. (2013). Value of freedom to choose encoded by the human brain. *Journal of Neurophysiology, 110*(8), 1915–1929. https://doi.org/10.1152/jn.01057.2012

Fuller-Iglesias, H., Sellars, B., & Antonucci, T. C. (2008). Resilience in old age: Social relations as a protective factor. *Research in Human Development, 5*(3), 181–193. https://doi.org/10.1080/15427600802274043

Gagné, J.-P., & Radomsky, A. S. (2020). Beliefs about losing control, obsessions, and caution: An experimental investigation. *Behaviour Research and Therapy, 126*, 103574. https://doi.org/10.1016/j.brat.2020.103574

Gallagher, M. W., Bentley, K. H., & Barlow, D. H. (2014). Perceived control and vulnerability to anxiety disorders: A meta-analytic review. *Cognitive Therapy and Research, 38*(6), 571–584. https://doi.org/10.1007/s10608-014-9624-x

Gatz, M., & Karel, M. J. (1993). Individual change in perceived control over 20 years. *International Journal of Behavioral Development, 16*(2), 305–322. https://doi.org/10.1177/016502549301600211

Gozli, D. (2019). Free choice. In D. Gozli (Ed.), *Experimental psychology and human agency* (pp. 113–136). Springer. https://doi.org/10.1007/978-3-030-20422-8_6

Grant, A. M., & Schwartz, B. (2011). Too much of a good thing: The challenge and opportunity of the inverted U. *Perspectives on Psychological Science, 6*(1), 61–76. https://doi.org/10.1177/1745691610393523

Griffiths, K. R., Morris, R. W., & Balleine, B. W. (2014). Translational studies of goal-directed action as a framework for classifying deficits across psychiatric disorders. *Frontiers in Systems Neuroscience, 8*(MAY), 1–16. https://doi.org/10.3389/fnsys.2014.00101

Hartley, C. A., Gorun, A., Reddan, M. C., Ramirez, F., & Phelps, E. A. (2014). Stressor controllability modulates fear extinction in humans. *Neurobiology of Learning and Memory, 113*, 149–156. https://doi.org/10.1016/j.nlm.2013.12.003

Herman, A. M., & Tsakiris, M. (2020). Feeling in control: The role of cardiac timing in the sense of agency. *Affective Science, 1*(3), 155–171. https://doi.org/10.1007/s42761-020-00013-x

Hornsey, M. J., Greenaway, K. H., Harris, E. A., & Bain, P. G. (2019). Exploring cultural differences in the extent to which people perceive and desire control. *Personality and Social Psychology Bulletin, 45*(1), 81–92. https://doi.org/10.1177/0146167218780692

Infurna, F. J., & Okun, M. A. (2015). Antecedents and outcomes of level and rates of change in perceived control: The moderating role of age. *Developmental Psychology, 51*(10), 1420–1437. https://doi.org/10.1037/a0039530

Inozu, M., Yorulmaz, O., & Terzi, S. (2012). Locus of control in obsessive-compulsive (OC) and depression symptoms: The moderating effect of externality on obsessive-related control beliefs in OC symptoms. *Behaviour Change, 29*(3), 148–163. https://doi.org/10.1017/bec.2012.14

Jasdi, N. F. S., & Rahman, P. (2023). Symptoms of depression among health sciences students and its relation to locus of control. *Journal of Sustainability Science and Management, 18*(6), 70–81. https://doi.org/10.46754/jssm.2023.06.006

Jiwa, M., Cooper, P. S., Chong, T. T.-J., & Bode, S. (2021). Choosing increases the value of non-instrumental information. *Scientific Reports, 11*(1), 8780. https://doi.org/10.1038/s41598-021-88031-y

Kadden, R. M., & Litt, M. D. (2011). The role of self-efficacy in the treatment of substance use disorders. *Addictive Behaviors, 36*(12), 1120–1126. https://doi.org/10.1016/j.addbeh.2011.07.032

Kalivas, P. W., & Volkow, N. D. (2005). The neural basis of addiction: A pathology of motivation and choice. *American Journal of Psychiatry, 162*(8), 1403–1413. https://doi.org/10.1176/appi.ajp.162.8.1403

Katzman, P. L., & Hartley, C. A. (2020). The value of choice facilitates subsequent memory across development. *Cognition, 199*, 104239. https://doi.org/10.1016/j.cognition.2020.104239

Kayser, A. S., Mitchell, J. M., Weinstein, D., & Frank, M. J. (2015). Dopamine, locus of control, and the exploration-exploitation tradeoff. *Neuropsychopharmacology, 40*(2), 454–462. https://doi.org/10.1038/npp.2014.193

Kiburi, S. K., Kwobah, E. K., Paruk, S., & Chiliza, B. (2023). Feasibility, acceptability and preliminary efficacy of a cognitive behavior therapy text-message intervention among individuals with opioid use disorder in Kenya: A randomized feasibility trial. *BMC Digital Health, 1*(1), 14. https://doi.org/10.1186/s44247-023-00014-3

Kuehl, L. K., Deuter, C. E., Nowacki, J., Ueberrueck, L., Wingenfeld, K., & Otte, C. (2021). Attentional bias in individuals with depression and adverse childhood experiences: Influence of the noradrenergic system? *Psychopharmacology, 238*(12), 3519–3531. https://doi.org/10.1007/s00213-021-05969-7

Lachman, M. E. (2006). Aging-related declines adaptive beliefs and behaviors. *Current Directions in Psychological Science, 15*(6), 282–286.

Lachman, M. E., & Weaver, S. L. (1998). The sense of control as a moderator of social class differences in health and well-being. *Journal of Personality and Social Psychology, 74*(3), 763–773. https://doi.org/10.1037/0022-3514.74.3.763

Lachman, M. E., Neupert, S. D., & Agrigoroaei, S. (2011). Chapter 11—The relevance of control beliefs for health and aging. In K. W. Schaie & S. L. Willis (Eds.), *Handbook of the psychology of aging* (7th ed., pp. 175–190). Academic. https://doi.org/10.1016/B978-0-12-380882-0.00011-5

Lachman, M. E., Agrigoroaei, S., & Rickenbach, E. H. (2015). Making sense of control: change and consequences. In R. A. Scott & S. M. Kosslyn (Eds.), *Emerging trends in the social and behavioral sciences* (1st ed., pp. 1–16). Wiley. https://doi.org/10.1002/9781118900772.etrds0209

Langer, E. J. (1975). The illusion of control. *Journal of Personality and Social Psychology, 32*(2), 311–328. https://doi.org/10.1037/0022-3514.32.2.311

Leotti, L. A., & Delgado, M. R. (2011). The inherent reward of choice. *Psychological Science, 22*(10), 1310–1318. https://doi.org/10.1177/0956797611417005

Leotti, L. A., & Delgado, M. R. (2014). The value of exercising control over monetary gains and losses. *Psychological Science, 25*(2), 596–604. https://doi.org/10.1177/0956797613514589

Leotti, L. A., Iyengar, S. S., & Ochsner, K. N. (2010). Born to choose: The origins and value of the need for control. *Trends in Cognitive Sciences, 14*(10), 457–463. https://doi.org/10.1016/j.tics.2010.08.001

Li, X., Ye, Q., & Yang, G. (2017). The lack of dominance and choice deferral: Choosing to defer to cope with the feeling of being out of control. *The Journal of Social Psychology, 157*(6), 754–765. https://doi.org/10.1080/00224545.2017.1294141

Liljeholm, M., & O'doherty, J. P. (2012). Contributions of the striatum to learning, motivation, and performance: An associative account anatomical and functional delineations of the striatum. *Trends in Cognitive Sciences, 16*(9), 467–475. https://doi.org/10.1016/j.tics.2012.07.007.Contributions

Liljeholm, M., Tricomi, E., O'Doherty, J. P., & Balleine, B. W. (2011). Neural correlates of instrumental contingency learning: Differential effects of action-reward conjunction and disjunction. *Journal of Neuroscience, 31*(7), 2474–2480. https://doi.org/10.1523/JNEUROSCI.3354-10.2011

Limbachia, C., Morrow, K., Khibovska, A., Meyer, C., Padmala, S., & Pessoa, L. (2021). Controllability over stressor decreases responses in key threat-related brain areas. *Communications Biology, 4*(1), 42. https://doi.org/10.1038/s42003-020-01537-5

Liu, R. T., Kleiman, E. M., Nestor, B. A., & Cheek, S. M. (2015). The hopelessness theory of depression: A quarter-century in review. *Clinical Psychology: Science and Practice, 22*(4), 345–365. https://doi.org/10.1111/cpsp.12125

Liu, Q., Wu, J., Zhang, L., Sun, X., Guan, Q., & Yao, Z. (2021). The relationship between perceived control and hypothalamic–pituitary–adrenal axis reactivity to the trier social stress test in healthy young adults. *Frontiers in Psychology, 12*, 683914. https://doi.org/10.3389/fpsyg.2021.683914

Loeber, S., Croissant, B., Heinz, A., Mann, K., & Flor, H. (2006). Cue exposure in the treatment of alcohol dependence: Effects on drinking outcome, craving and self-efficacy. *British Journal of Clinical Psychology, 45*(4), 515–529. https://doi.org/10.1348/014466505X82586

Ly, V., Wang, K. S., Bhanji, J., & Delgado, M. R. (2019). A reward-based framework of perceived control. *Frontiers in Neuroscience, 13*(FEB), 1–11. https://doi.org/10.3389/fnins.2019.00065

Maier, S. F., & Seligman, M. E. P. (2016). Learned helplessness at fifty: Insights from neuroscience. *Psychological Review, 123*(4), 349–367. https://doi.org/10.1037/rev0000033

Mather, M. (2006). A review of decision-making processes: Weighing the risks and benefits of aging. In *When I'm 64*. National Academies Press (US). https://www.ncbi.nlm.nih.gov/sites/books/NBK83778/

Meine, L. E., Meier, J., Meyer, B., & Wessa, M. (2021). Don't stress, it's under control: Neural correlates of stressor controllability in humans. *NeuroImage, 245*, 118701. https://doi.org/10.1016/j.neuroimage.2021.118701

Metcalfe, J., Eich, T. S., & Miele, D. B. (2013). Metacognition of agency: Proximal action and distal outcome. *Experimental Brain Research, 229*(3), 485–496. https://doi.org/10.1007/s00221-012-3371-6

Mirowsky, J., & Ross, C. E. (2007). Life course trajectories of perceived control and their relationship to education. *American Journal of Sociology, 112*(5), 1339–1382. https://doi.org/10.1086/511800

Misuraca, R., Reutskaja, E., Fasolo, B., & Iyengar, S. (2020). How much choice is "good enough"? In R. Viale (Ed.), *Routledge handbook of bounded rationality* (1st ed., pp. 637–649). Routledge. https://doi.org/10.4324/9781315658353-51

Morina, N., Bryant, R. A., Doolan, E. L., Martin-Sölch, C., Plichta, M. M., Pfaltz, M. C., Schnyder, U., Schick, M., & Nickerson, A. (2018). The impact of enhancing perceived self-efficacy in torture survivors. *Depression and Anxiety, 35*(1), 58–64. https://doi.org/10.1002/da.22684

Moscarello, J. M., & Hartley, C. A. (2017). Agency and the calibration of motivated behavior. *Trends in Cognitive Sciences, 21*(10), 725–735. https://doi.org/10.1016/j.tics.2017.06.008

Moulding, R., Doron, G., Kyrios, M., & Nedeljkovic, M. (2008). Desire for control, sense of control and obsessive-compulsive checking: An extension to clinical samples. *Journal of Anxiety Disorders, 22*(8), 1472–1479. https://doi.org/10.1016/j.janxdis.2008.03.001

Moulding, R., Kyrios, M., Doron, G., & Nedeljkovic, M. (2009). Mediated and direct effects of general control beliefs on obsessive compulsive symptoms. *Canadian Journal of Behavioural Science/Revue Canadienne Des Sciences Du Comportement, 41*(2), 84–92. https://doi.org/10.1037/a0014840

Mullainathan, S., & Shafir, E. (2013). *Scarcity: Why having too little means so much*. Macmillan.

Murty, V. P., DuBrow, S., & Davachi, L. (2015). The simple act of choosing influences declarative memory. *Journal of Neuroscience, 35*(16), 6255–6264. https://doi.org/10.1523/JNEUROSCI.4181-14.2015

Nussenbaum, K., Katzman, P. L., Lu, H., Zorowitz, S., & Hartley, C. A. (2024). Sensitivity to the instrumental value of choice increases across development. *Psychological Science*. https://doi.org/10.1177/09567976241256961

O'Brien, C. (2011). Addiction and dependence in DSM-V. *Addiction, 106*(5), 866–867. https://doi.org/10.1111/j.1360-0443.2010.03144.x

O'Doherty, J., Dayan, P., Schultz, J., Deichmann, R., Friston, K., & Dolan, R. J. (2004). Dissociable roles of ventral and dorsal striatum in instrumental conditioning. *Science, 304*(5669), 452–454. https://doi.org/10.1126/science.1094285

Ono, M., Kochiyama, T., Fujino, J., Sozu, T., Kawada, R., Yokoyama, N., Sugihara, G., Murai, T., & Takahashi, H. (2018). Self-efficacy modulates the neural correlates of craving in male smokers and ex-smokers: An fMRI study. *Addiction Biology, 23*(5), 1179–1188. https://doi.org/10.1111/adb.12555

Owens, B. D., Grossman, Z., & Fackler, R. (2014). The control premium: A preference for payoff autonomy. *American Economic Association Stable, 6*(4), 138–161.

Panksepp, J., & Green, B. (1998). The periconscious substrates of consciousness: Affective states and the evolutionary origins of the self. *Journal of Consciousness Studies, 5*(5–6), 566–582.

Perlman, D. (2004). European and Canadian studies of loneliness among seniors. *Canadian Journal on Aging/La Revue Canadienne Du Vieillissement, 23*(2), 181–188. https://doi.org/10.1353/cja.2004.0025

Raab, H. A., Foord, C., Ligneul, R., & Hartley, C. A. (2022). Developmental shifts in computations used to detect environmental controllability. *PLoS Computational Biology, 18*(6), e1010120. https://doi.org/10.1371/journal.pcbi.1010120

Raab, H. A., Goldway, N., Foord, C., & Hartley, C. A. (2024). Adolescents flexibly adapt action selection based on controllability inferences. *Learning & Memory, 31*(3), a053901. https://doi.org/10.1101/lm.053901.123

Rachman, S., & de Silva, P. (2009). *Obsessive-compulsive disorder*. OUP Oxford.

Radomsky, A. S. (2022). The fear of losing control. *Journal of Behavior Therapy and Experimental Psychiatry, 77*, 101768. https://doi.org/10.1016/j.jbtep.2022.101768

Reed, A. E., Mikels, J. A., & Simon, K. I. (2008). Older adults prefer less choice than young adults. *Psychology and Aging, 23*(3), 671–675. https://doi.org/10.1037/a0012772

Reis, M., Pfister, R., & Schwarz, K. A. (2023). The value of control. *Journal of Behavioral Decision Making, 36*(4), e2325. https://doi.org/10.1002/bdm.2325

Reutskaja, E., Lindner, A., Nagel, R., Andersen, R. A., & Camerer, C. F. (2018). Choice overload reduces neural signatures of choice set value in dorsal striatum and anterior cingulate cortex. *Nature Human Behaviour, 2*(12), 925–935. https://doi.org/10.1038/s41562-018-0440-2

Ridderinkhof, K. R., Ullsperger, M., Crone, E. A., & Nieuwenhuis, S. (2004). The role of the medial frontal cortex in cognitive control. *Science, 306*(5695), 443–447. https://doi.org/10.1126/science.1100301

Robinson, S. A., & Lachman, M. E. (2017). Perceived control and aging: A mini-review and directions for future research. *Gerontology, 63*(5), 435–442. https://doi.org/10.1159/000468540

Rolls, E. T., Huang, C. C., Lin, C. P., Feng, J., & Joliot, M. (2020). Automated anatomical labelling atlas 3. *Neuroimage, 206*, 116189. https://doi.org/10.1016/j.neuroimage.2019.116189

Romaniuk, L., Sandu, A. L., Waiter, G. D., McNeil, C. J., Xueyi, S., Harris, M. A., Macfarlane, J. A., Lawrie, S. M., Deary, I. J., Murray, A. D., Delgado, M. R., Steele, J. D., McIntosh, A. M., & Whalley, H. C. (2019). The neurobiology of personal control during reward learning and its relationship to mood. *Biological Psychiatry: Cognitive Neuroscience and Neuroimaging, 4*(2), 190–199. https://doi.org/10.1016/j.bpsc.2018.09.015

Romaniuk, L., MacSweeney, N., Atkinson, K., Chan, S. W. Y., Barbu, M. C., Lawrie, S. M., & Whalley, H. C. (2023). Striatal correlates of Bayesian beliefs in self-efficacy in adolescents and their relation to mood and autonomy: A pilot study. *Cerebral Cortex Communications, 4*(4), tgad020. https://doi.org/10.1093/texcom/tgad020

Rotter, J. (1966). Generalized expectancies for internal versus external control of reinforcement. *Psychological Monographs: General and Applied, 80*(1), 1–28.

Ruiz, N. A., DuBrow, S., & Murty, V. P. (2023). Agency as a bridge to form associative memories. *Journal of Experimental Psychology: General, 152*(6), 1797–1813. https://doi.org/10.1037/xge0001356

Samanez-Larkin, G. R., & Knutson, B. (2014). Reward processing and risky decision making in the aging brain. In V. F. Reyna & V. Zayas (Eds.), *The neuroscience of risky decision making* (pp. 123–142). American Psychological Association. https://doi.org/10.1037/14322-006

Samanez-Larkin, G. R., & Knutson, B. (2015). Decision making in the ageing brain: Changes in affective and motivational circuits. *Nature Reviews Neuroscience, 16*(5), 278–289. https://doi.org/10.1038/nrn3917

Sayali, C., Rubin-McGregor, J., & Badre, D. (2022). *Policy abstraction as a predictor of cognitive effort*. https://doi.org/10.31234/osf.io/by7gk

Scheibehenne, B., Greifeneder, R., & Todd, P. M. (2010). Can there ever be too many options? A meta-analytic review of choice overload. *Journal of Consumer Research, 37*(3), 409–425. https://doi.org/10.1086/651235

Schnoll, R. A., Martinez, E., Tatum, K. L., Marcella, G., Bernath, A., Daron, F., & Reynolds, P. (2011). Increased self-efficacy to quit and perceived control over withdrawal symptoms predict smoking cessation following nicotine dependence treatment. *Addictive Behaviors, 36*(1–2), 122–127. https://doi.org/10.1016/j.addbeh.2010.08.024.Increased

Schott, B. H., Niehaus, L., Wittmann, B. C., Schutze, H., Seidenbecher, C. I., Heinze, H.-J., & Duzel, E. (2007). Ageing and early-stage Parkinson's disease affect separable neural mechanisms of mesolimbic reward processing. *Brain, 130*(9), 2412–2424. https://doi.org/10.1093/brain/awm147

Schwartz, B., & Cheek, N. N. (2017). Choice, freedom, and well-being: Considerations for public policy. *Behavioural Public Policy, 1*(1), 106–121. https://doi.org/10.1017/bpp.2016.4

Schwarz, K. A., Klaffehn, A. L., Hauke-Forman, N., Muth, F. V., & Pfister, R. (2022). Never run a changing system: Action-effect contingency shapes prospective agency. *Cognition, 229*, 105250. https://doi.org/10.1016/j.cognition.2022.105250

Seligman, M. E., Rosellini, R. A., & Kozak, M. J. (1975). Learned helplessness in the rat: Time course, immunization, and reversibility. *Journal of Comparative and Physiological Psychology, 88*(2), 542–547. https://doi.org/10.1037/h0076431

Shanks, D. R., & Dickinson, A. (1991). Instrumental judgment and performance under variations in action-outcome contingency and contiguity. *Memory & Cognition, 19*(4), 353–360. https://doi.org/10.3758/BF03197139

Shaw, B. A., Krause, N., Chatters, L. M., Connell, C. M., & Ingersoll-Dayton, B. (2004). Emotional support from parents early in life, aging, and health. *Psychology and Aging, 19*(1), 4–12. https://doi.org/10.1037/0882-7974.19.1.4

Shenhav, A., Botvinick, M. M., & Cohen, J. D. (2013). The expected value of control: An integrative theory of anterior cingulate cortex function. *Neuron, 79*(2), 217–240. https://doi.org/10.1016/j.neuron.2013.07.007

Shenhav, A., Cohen, J. D., & Botvinick, M. M. (2016). Dorsal anterior cingulate cortex and the value of control. *Nature Neuroscience, 19*(10), 1286–1291. https://doi.org/10.1038/nn.4384

Shenhav, A., Musslick, S., Lieder, F., Kool, W., Griffiths, T. L., Cohen, J. D., & Botvinick, M. M. (2017). Toward a rational and mechanistic account of mental effort. *Annual Review of Neuroscience, 40*(1), 99–124. https://doi.org/10.1146/annurev-neuro-072116-031526

Skinner, E. A., Zimmer-Gembeck, M. J., Connell, J. P., Eccles, J. S., & Wellborn, J. G. (1998). Individual differences and the development of perceived control. *Monographs of the Society for Research in Child Development, 63*(2/3), i–231. https://doi.org/10.2307/1166220

Skottnik, L., Sorger, B., Kamp, T., Linden, D., & Goebel, R. (2019). Success and failure of controlling the real-time functional magnetic resonance imaging neurofeedback signal are reflected in the striatum. *Brain and Behavior: A Cognitive Neuroscience Perspective, 9*(3), e01240. https://doi.org/10.1002/brb3.1240

Specht, J., Egloff, B., & Schmukle, S. C. (2013). Everything under control? The effects of age, gender, and education on trajectories of perceived control in a nationally representative German sample. *Developmental Psychology, 49*(2), 353–364. https://doi.org/10.1037/a0028243

Steffel, M., Williams, E. F., & Perrmann-Graham, J. (2016). Passing the buck: Delegating choices to others to avoid responsibility and blame. *Organizational Behavior and Human Decision Processes, 135*, 32–44. https://doi.org/10.1016/j.obhdp.2016.04.006

Stevens, D., Anderson, D. I., O'Dwyer, N. J., & Mark Williams, A. (2012). Does self-efficacy mediate transfer effects in the learning of easy and difficult motor skills? *Consciousness and Cognition, 21*(3), 1122–1128. https://doi.org/10.1016/j.concog.2012.03.014

Stolz, D. S., Müller-Pinzler, L., Krach, S., & Paulus, F. M. (2020). Internal control beliefs shape positive affect and associated neural dynamics during outcome valuation. *Nature Communications, 11*(1), 1–13. https://doi.org/10.1038/s41467-020-14800-4

Sullivan, M. W., & Lewis, M. (2003). Contextual determinants of anger and other negative expressions in young infants. *Developmental Psychology, 39*(4), 693–705. https://doi.org/10.1037/0012-1649.39.4.693

Sullivan-Toole, H., Richey, J. A., & Tricomi, E. (2017). Control and effort costs influence the motivational consequences of choice. *Frontiers in Psychology, 8*, 675. https://doi.org/10.3389/fpsyg.2017.00675

Sunstein, C. R. (2015). *Choosing not to choose: Understanding the value of choice*. Oxford University Press.

Sunstein, C. R. (2017). Default rules are better than active choosing (often). *Trends in Cognitive Sciences, 21*(8), 600–606. https://doi.org/10.1016/j.tics.2017.05.003

Suzuki, S. (1997). Effects of number of alternatives on choice in humans. *Behavioural Processes, 39*(2), 205–214. https://doi.org/10.1016/S0376-6357(96)00049-6

Tanaka, T., & Kawabata, H. (2020). Interface predictability changes betting behavior in computerized gambling. *Computers in Human Behavior, 110*, 106387. https://doi.org/10.1016/j.chb.2020.106387

Taylor, S., Coles, M. E., Abramowitz, J. S., Wu, K. D., Olatunji, B. O., Timpano, K. R., McKay, D., Kim, S.-K., Carmin, C., & Tolin, D. F. (2010). How are dysfunctional beliefs related to obsessive-compulsive symptoms? *Journal of Cognitive Psychotherapy, 24*(3), 165–176. https://doi.org/10.1891/0889-8391.24.3.165

Ten, A., Kaushik, P., Oudeyer, P.-Y., & Gottlieb, J. (2021). Humans monitor learning progress in curiosity-driven exploration. *Nature Communications, 12*(1), 5972. https://doi.org/10.1038/s41467-021-26196-w

Titcombe-Parekh, R. F., Chen, J., Rahman, N., Kouri, N., Qian, M., Li, M., Bryant, R. A., Marmar, C. R., & Brown, A. D. (2018). Neural circuitry changes associated with increasing self-efficacy in posttraumatic stress disorder. *Journal of Psychiatric Research, 104*, 58–64. https://doi.org/10.1016/j.jpsychires.2018.06.009

Toyama, M. (2019). *Longitudinal associations among personality, perceived control, and health for American and Japanese aging adults*. PhD North Dakota State University. https://www.proquest.com/docview/2312284114/abstract/4864C52B2C2748E8PQ/1

Toyama, M., & Fuller, H. R. (2021). Longitudinal associations between perceived control and health for American and Japanese aging adults. *The Gerontologist, 61*(6), 917–929. https://doi.org/10.1093/geront/gnaa135
Tricomi, E. M., Delgado, M. R., & Fiez, J. A. (2004). Modulation of caudate activity by action contingency. *Neuron, 41*(2), 281–292. https://doi.org/10.1016/S0896-6273(03)00848-1
Ulleberg, P., Berge, T., Lending, H. D., Skule, C., & Landrø, N. I. (2021). Is perceived control of depression related to therapy outcome for depression? A longitudinal study. *Journal of Psychiatric Research, 140*, 504–511. https://doi.org/10.1016/j.jpsychires.2021.06.025
Vassena, E., Van Opstal, F., Goethals, I., & Verguts, T. (2019). Striatal dopamine D2 binding correlates with locus of control: Preliminary evidence from [11C]raclopride positron emission tomography. *International Journal of Psychophysiology, 146*, 117–124. https://doi.org/10.1016/j.ijpsycho.2019.09.016
Venniro, M., Zhang, M., Caprioli, D., Hoots, J. K., Golden, S. A., Heins, C., Morales, M., Epstein, D. H., & Shaham, Y. (2018). Volitional social interaction prevents drug addiction in rat models. *Nature Neuroscience, 21*(11), 1520–1529. https://doi.org/10.1038/s41593-018-0246-6
Wallston, K. A., Wallston, B. S., Smith, S., & Dobbins, C. J. (1987). Perceived control and health. *Current Psychology, 6*(1), 5–25. https://doi.org/10.1007/BF02686633
Wang, K. S., & Delgado, M. R. (2019). Corticostriatal circuits encode the subjective value of perceived control. *Cerebral Cortex, 29*(12), 5049–5060. https://doi.org/10.1093/cercor/bhz045
Wang, K. S., Yang, Y.-Y., & Delgado, M. R. (2021). How perception of control shapes decision making. *Current Opinion in Behavioral Sciences, 41*, 85–91. https://doi.org/10.1016/j.cobeha.2021.04.003
White, R. W. (1959). Motivation reconsidered: The concept of competence. *Psychological Review, 66*(5), 297–333. https://doi.org/10.1037/h0040934
Windsor, T. D., Anstey, K. J., Butterworth, P., & Rodgers, B. (2008). Behavioral approach and behavioral inhibition as moderators of the association between negative life events and perceived control in midlife. *Personality and Individual Differences, 44*(5), 1080–1092. https://doi.org/10.1016/j.paid.2007.10.036
Yang, Y. Y., & Delgado, M. R. (2025). The integration of self-efficacy and response-efficacy in decision making. *Scientific reports, 15*(1), 1789
Zheng, Y., Wang, M., Zhou, S., & Xu, J. (2020). Functional heterogeneity of perceived control in feedback processing. *Social Cognitive and Affective Neuroscience, 15*(3), 329–336. https://doi.org/10.1093/scan/nsaa028

Integrated Neural Circuitry Supporting Emotion Regulation and Decision-Making

Helen Schmidt and Chelsea Helion

Abstract Recent research in psychology and neuroscience has begun to closely examine the relationship between emotion regulation and decision-making. Emotion regulation strategies allow people to change their thoughts and feelings during emotional situations and events. Successful regulation can be particularly important for a variety of decisions made in day-to-day life. This is especially true for decisions involving risk and reward, where highly rewarding and desirable outcomes are often paired with uncertainty in the likelihood of those outcomes. In this chapter, we aim to operationalize both emotion regulation and decision-making in the context of recent findings across psychology and neuroscience. We use an everyday, real-world example involving risk and reward to highlight the relationship between these processes behaviorally and neurocognitively. After presenting a framework, we layer in relevant findings in cognitive neuroscience to explore how the human brain supports these processes, both individually and in an integrated capacity. Finally, we highlight future research implications and open questions in the space.

1 Introduction

Hannah is considering buying her first house. She currently rents an apartment in a major metropolitan area in the United States and dreams of owning a home one day. Hannah is not the first of her friends to think about buying a home; at 36, she is the average age for first-time homebuyers in the United States (Brooks, 2023). Unfortunately for Hannah, in 2023, the housing market hit a breaking point. Mortgage rates soared to record levels and many would-be first-time homeowners were priced out of a prohibitively expensive and limited housing market (Friedman, 2024). Historically, homeownership has been seen as a worthwhile financial investment: in addition to signaling financial independence, it can also result in substantial

H. Schmidt (✉) · C. Helion
Department of Psychology & Neuroscience, Temple University, Philadelphia, PA, USA
e-mail: helen_schmidt@temple.edu

D. V. Smith et al. (eds.), *Neuroeconomics: Core Topics and Current Directions*,
https://doi.org/10.1007/978-3-032-02925-6_20

financial profit later in life (Goodman & Mayer, 2018). However, buying a house comes with a significant amount of risk, as there are many factors that are outside of one's control (e.g., one can set a firm budget but cannot control interest rates). In short, the outcome of Hannah's decision has the potential to significantly impact her life. These types of decisions can evoke strong negative emotions—including dread, fear, and anxiety—that often need to be regulated to maintain psychological well-being and achieve short- and long-term goals.

In this chapter, we will explore the relationship between decision-making and emotion regulation in the context of risky decision-making processes—those that have the potential for both rewards and losses—such as buying a home, switching jobs, or investing in stock options. We focus on risky decision-making because it has long-term and real-world implications for individual health and well-being (Rutledge et al., 2015). First, we will operationalize decision-making and emotion regulation and establish a framework in which to think about these two processes and their interaction. Second, we will briefly examine recent findings within this framework from research in psychology and cognitive neuroscience. Next, we will move through each stage of decision-making and demonstrate how a specific emotion regulation strategy, *cognitive reappraisal*, can guide thoughts and actions to potentially improve decision-making across a wide range of contexts. We will also underscore how the brain supports these processes and their interaction. Finally, we will conclude with unanswered questions in the space and implications for developmental and psychopathological populations.

2 What Do We Mean by Decision-Making?

Decision-making can be broadly defined as the cognitive process behind selecting an action or behavior, with the goal of achieving a desired outcome from a variety of possibilities. In this chapter, we focus specifically on risky decision-making, where possible outcomes include rewards and losses (Kahneman & Tversky, 1984). This type of decision-making has clear emotional relevance, where rewards are associated with positive emotions and losses are associated with negative emotions (Slovic et al., 2005). There is considerable uncertainty associated with risky decisions—decisions where there is a larger variety of possible outcomes or lesser clarity in the probabilities of desired outcomes. Uncertainty and risk tend to elicit negative emotions, providing an opportunity for emotion regulation (Flores-Torres et al., 2022; Heilman et al., 2010).

3 What Do We Mean by Emotion Regulation?

Emotion regulation refers to the act of exerting control over emotions with the goal of changing the experience, intensity, timing, or influence of these emotions (Gross,

2015). Most commonly, emotion regulation serves to decrease negative emotions (trying to calm down when feeling upset) or to increase positive emotions (giving oneself a pep talk before a public speech) (Gross, 2014). In this chapter, we will specifically examine emotion regulation in the context of the Process Model (Gross, 1998). This model posits that emotion regulation can have an impact at each stage of generating an emotional reaction to a given situation (Gross, 1998). There are five linear components to the Process Model: situation selection, situation modification, attention deployment, cognitive change, and response modulation. We will discuss how and when each component serves to impact an emotional response and provide an example for how each component may manifest in the context of decision-making.

Situation selection involves approaching or avoiding certain situations to facilitate or circumvent an emotional response to a situation. To return to our motivating example at the start of the chapter, Hannah may engage in situation selection by avoiding putting an offer in on a house that she's excited about in order to avoid the negative emotions that may arise if the offer was rejected. *Situation modification* involves actively trying to modify or change a situation to alter its emotional impact. In our example, this might mean that Hannah looks at multiple properties, to avoid getting too attached to the idea of one house. *Attention deployment* involves actively shifting attentional focus towards or away from an emotional situation or element to reduce its emotional impact. One common emotion regulation strategy that targets attention deployment is *distraction*, where someone may focus on a nonemotional aspect of a situation to reduce the overall emotional impact. By concentrating on nonemotional features, distraction can help amplify positive emotions by shifting attention toward or away from emotional elements. In our example, Hannah may go to the movies or plan a dinner with friends after putting in an offer, so that she doesn't ruminate on whether her offer will be accepted.

Cognitive change involves changing the evaluation of or meaning derived from an emotion-eliciting situation. An emotion regulation strategy that leverages cognitive change is *cognitive reappraisal*, wherein someone reframes a situation to alter its emotional impact. This might take the form of Hannah reminding herself that there are many houses in the area in which she could be happy, and that while a rejected offer would be disappointing, it could also represent an opportunity to find another great house elsewhere. Finally, *response modulation* involves directly influencing the physiological, experiential, or behavioral response to an emotion-eliciting situation. A response modulation strategy like *suppression* helps people inhibit expressive behaviors of an emotional reaction and potentially reduce their experience of those emotions. In our example, if Hannah's offer is rejected, she would avoid expressing disappointment or sadness to others, but would still feel it internally.

We posit that emotion regulation can serve as a tool that can be flexibly applied or manipulated as either an adaptive or maladaptive strategy for different types of decision-making scenarios. Consistent with this claim, findings from psychology and cognitive neuroscience suggest that unchecked emotional reactions when making risky decisions tend to be associated with worse outcomes (Martin

& Delgado, 2011). Under-regulation, or a tendency to leave emotional responses uncontrolled, is associated with risk-seeking behavior, while over-regulation, or regulating emotions to an excessive degree, is associated with risk-averse behavior. Thus, both under- and over-regulation are associated with suboptimal decision outcomes (Mitchell, 2011). Below, we outline potential contributions of emotion regulation at different stages of risky decision-making processes, which may be applicable across multiple contexts.

4 Focusing on the Intersection of Decision-Making and Emotion Regulation

Examining the neural and behavioral bases of risky decision-making through the lens of emotion regulation may yield new insights into both processes. We will also focus on risky decision-making because it has long-term and real-world implications for individual health and well-being (Rutledge et al., 2015). Below, we will outline six distinct components of the risky decision-making process: (1) identifying the problem, (2) evaluating possible outcomes, (3) taking decisive action, (4) anticipating the outcome, (5) dealing with the outcome, and (6) the post-decision experience. At each point, we identify the role that emotion regulation may play in this decision stage and identify neural circuitry that has been implicated in both processes. While we will identify contributions that different Process Model components may play throughout the decision-making process, we will largely focus on *cognitive reappraisal* given the extent to which its behavioral and neural bases have been characterized in the literature (Buhle et al., 2014; Braunstein et al., 2014; Denny et al., 2023; Webb et al., 2012).

Decision-Making as a Multi-stage Process As individuals navigate their day-to-day lives, they encounter situations that can be resolved or modified with decisive action in accordance with their momentary goals and values (Rangel et al., 2008). For decision-making to be initiated, an individual must first *identify the problem* (Doya, 2008). Once individuals have identified the problem, they can *evaluate possible outcomes* (Morris et al., 2021). This step involves weighing possible outcomes and comparing these outcomes to situational goals to identify an ideal outcome that aligns with these goals. Next, *decisive action* is taken (Sharpe et al., 2019). An individual decides to act to attempt to achieve the ideal outcome. In *anticipation of the outcome*, the individual wonders if the decision will achieve their goals and have the intended consequences (Lerner et al., 2015). Once an outcome becomes clear, the individual must *deal with the outcome* (Lerner et al., 2015). If the goals of the decisive action were not met, there may be a need to restart the decision-making process. He or she may also feel a range of emotions in response to the outcome, and sometimes individuals must navigate *post-decision experiences* like regret (Matarazzo et al., 2021). These types of decisions are common and made every day; importantly, however, at each stage of the decision-making process, there

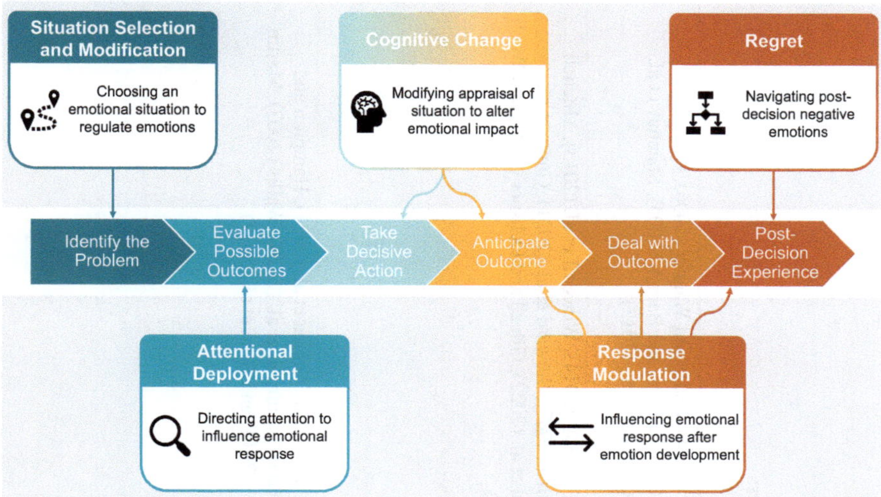

Fig. 1 Integrated decision-making and emotion regulation framework in the context of the Process Model

is an opportunity to modify an emotional drive, state, or response using emotion regulation to adjust goals, actions, and expected outcomes (Fig. 1).

5 Integration of Neural Circuitry Involved in Emotion Regulation and Decision-Making

Recent reviews in psychology and neuroscience have identified the central role of affect (Barrett & Bliss-Moreau, 2009) in both decision-making (Asutay & Västfjäll, 2022) and emotion regulation (Mitchell, 2011). By *affect*, we refer to an internal state characterized by a certain level of valence and arousal (Russell & Barrett, 1999; Russell, 2003). Consistent with this account, both processes involve the recruitment of similar prefrontal cortex regions implicated in executive control and the modulation of cortical and subcortical circuitry implicated in affective experience (Denny et al., 2023; Table 1). Further, the convergence of emotion dysregulation and deficits in effective decision-making (Grecucci & Sanfey, 2014) suggest a behavioral and neural relationship between these two processes. The majority of the studies in the literature have focused on cognitive reappraisal as a regulation strategy, and therefore in the remaining sections of this chapter, we will review neural circuitry involved in decision-making in the context of cognitive reappraisal (Fig. 2).

Table 1 Neural circuitry implicated in cognitive reappraisal at each stage of the decision-making process

Decision-making stage	Cognitive processes	Neural circuitry	Citations
Identify the problem	Conflict resolution Value assignment	dlPFC dmPFC vlPFC ACC	Oehrn et al. (2014), Badre and Wagner (2006), Dixon and Christoff (2014), Milham and Banich (2005), Shenhav et al. (2016) and Grecucci et al. (2013)
Evaluate possible outcomes	Value assignment Mentalization	vlPFC mPFC vmPFC dlPFC dmPFC OFC Amygdala VS TPJ PCC	Dixon and Christoff (2014), Winecoff et al. (2013), Mitchell (2011), Wallis, (2007), Steward et al. (2021), Gupta et al. (2011), Pool et al. (2022), Gilead and Ochsner (2021) and Webb et al. (2012)
Take decisive action	Action inhibition Motivational alignment	vlPFC dlPFC ACC VS OFC vmPFC Amygdala	Apšvalka et al. (2022), Tannou et al. (2021), Hampton and O'Doherty (2007), Filimon et al. (2020), Wallis (2007), Martin and Delgado (2011) and Doré et al. (2017)

Anticipate outcome	Navigate uncertainty Reward expectation	mPFC VS Amygdala BNST Insula	Mitchell (2011), Filimon et al. (2020), Pedersen et al. (2019), Hammack et al. (2021), Smith et al. (2014), Buhle et al. (2014), Banks et al. (2007), Liu et al. (2023) and Carlson and Mujica-Parodi (2010)
Deal with outcome and post-decision experience	Prediction errors Response inhibition Negative feedback	Amygdala vlPFC OFC VS Insula dlPFC dmPFC	Mitchell (2011), Gupta et al. (2011), Kim et al. (2018), Chua et al. (2009), Henderson and Norris (2013), Harmon-Jones et al. (2017), Bratec et al. (2015) and Jarcho et al. (2011)

While this table is not exhaustive and there may be other regions that are targeted in these processes, the listed regions play a prominent role in negative experiences and reward valuation, both of which are crucial elements to emotion regulation and decision-making

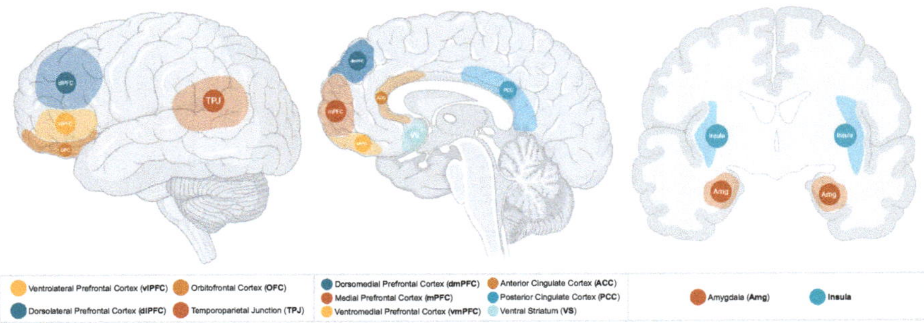

Fig. 2 Neural circuitry implicated in decision-making and cognitive reappraisal. Regions of interest (ROI) in lateral, sagittal, and coronal views (left to right) created using BioRender

5.1 Identifying the Problem

Let's return to the scenario of Hannah considering buying her first house. This decision represents a motivational conflict: she can either continue renting and prolong a familiar situation, or she can take a personal, emotional, and financial risk and buy her first home. She has identified her problem (choosing between a low-risk or high-risk outcome), and she can leverage cognitive reappraisal to regulate the emotional impacts that often come with making this kind of risky decision.

Neural Circuitry Implicated in Problem Identification The dorsolateral and dorsomedial prefrontal cortices (dlPFC, dmPFC) support conflict resolution, primarily by driving attention and cognitive control (Oehrn et al., 2014). The ventrolateral prefrontal cortex (vlPFC) may also facilitate flexible conflict resolution (Badre & Wagner, 2006) and helps individuals appropriately assign value to abstract and concrete concepts (Dixon & Christoff, 2014). The anterior cingulate cortex (ACC) and dorsal ACC are also heavily implicated in conflict monitoring and detection (Milham & Banich, 2005), helping to determine the expected value of motivation and control (Shenhav et al., 2016).

Modulation of Neural Activity via Emotion Regulation Prior research has found evidence that cognitive reappraisal can serve to modulate ACC activity during decision-making. In a study by Grecucci et al. (2013), researchers instructed participants to up- and downregulate (increase and decrease the interpretation of a negative event, respectively) in the context of an ultimatum game. Participants received a monetary offer from a computer player (either an equal or unequal split), reappraised the intentions and behavior of the player based on the regulation instruction, and decided whether to accept the offer. They found that decisions to accept an unfair offer when reappraising were associated with increased activity in the ACC, which is consistent with conflict resolution processes that may be involved in accepting an unfair offer by ignoring another's negative intentions. Taken together, this indicates that identifying the problem—whether it is accepting an unfair offer or taking a gamble on the housing market—recruits multiple neural regions in the frontal cortex involved in a diverse set of cognitive processes. This includes conflict detection, value assignment, and effort allocation. Furthermore, the neural and behavioral markers of these processes may be modulated by emotion regulation.

5.2 Evaluating Possible Outcomes

As Hannah explores different housing options, and tours properties around the city, she must evaluate her possible outcomes. She may have an emotional attachment to a given location or neighborhood, but the houses available in those areas may not align with her budget. She must consider if she wants to be near shops and restaurants, or if it's more important to be nearer to close friends. She tries to imagine what her life will look like if she lives in each home and considers whether it's the "right" place. This evaluative stage allows her to gather information before deciding, and she can again leverage an emotion regulation strategy like cognitive reappraisal to minimize her stress and improve her decision-related outcome.

Neural Circuitry Implicated in Outcome Evaluation When evaluating possible outcomes, many cortical and subcortical regions implicated in cognitive reappraisal are also involved in decision-making. These regions work together to not only support the evaluation of different outcomes and calculate probabilities of success (Kahneman & Tversky, 2000) but also support a number of cognitive processes. Notably, recent evidence highlights the role of affect as a weighted consideration when accumulating evidence and evaluating possible outcomes (Roberts et al., 2024). The ventrolateral prefrontal cortex (vlPFC) and medial prefrontal cortex (mPFC) assign value to different variables (Dixon & Christoff, 2014) while the dorsolateral and dorsomedial prefrontal cortices (dlPFC, dmPFC) support conflict resolution in addition to value assignment (Mitchell, 2011). The ventromedial prefrontal cortex (vmPFC) may also support encoding of emotional value (Winecoff et al., 2013), and evidence of an additive effect of value and emotion on vmPFC activation has also recently shown (Jaiswal et al., 2024). The orbitofrontal cortex (OFC) also processes reward and may aid in optimal solution consideration (Wallis, 2007), while the amygdala reduces negative affect associated with possible undesired outcomes (Steward et al., 2021; Gupta et al., 2011). Furthermore, the ventral striatum (VS) tracks outcome value, a critical component to evaluating reward in possible outcomes (Pool et al., 2022).

Modulation of Neural Activity via Emotion Regulation In addition to the valuation processes outlined above, mentalizing may also play a large role in the outcome evaluation process (Gilead & Ochsner, 2021). Each of her decision outcomes involves simulating her future self, likely recruiting mentalizing network regions such as temporoparietal junction (TPJ), mPFC, dmPFC, and posterior cingulate cortex (PCC). With regard to emotion regulation, perspective taking—a particularly

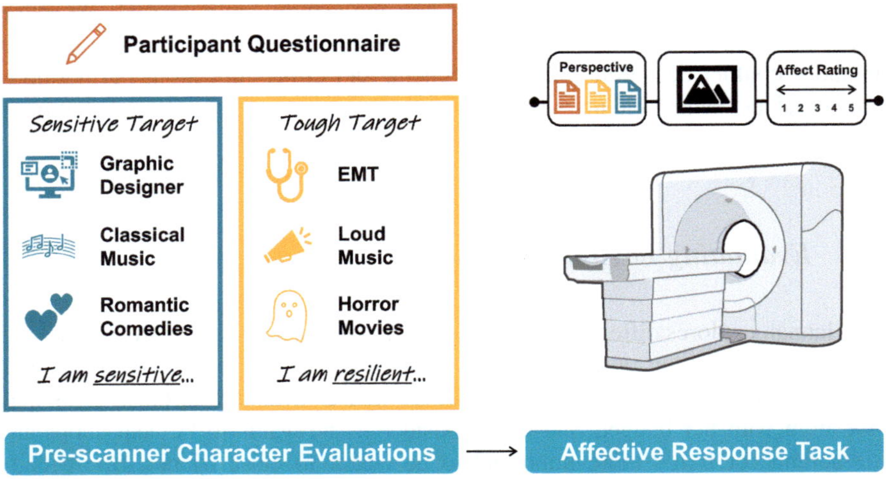

Fig. 3 Reproduction of experimental paradigm (Gilead et al., 2016) examining perspective-taking and emotional reactivity. Participants completed a questionnaire to report demographics and hobbies and viewed two character description questionnaires in the same format (pre-scanner character evaluations). In the scanner, participants were asked to view an affective image taking their own perspective or that of one of the evaluated characters

robust form of cognitive reappraisal (Webb et al., 2012)—modulates the recruitment of neural circuitry implicated in affective reactivity. In a study by Gilead et al. (2016), researchers had participants view affective images while taking their own perspective or that of another named person and rate their affective reactions to the images (Fig. 3). They found that perspective-taking was associated with activation in the mentalizing network and modulated affective processing. For Hannah, this type of reappraisal might take the form of imagining her future self in each home that she tours and picturing the positives and/or negatives associated with each property, thus shaping her emotional response to each outcome. Taken together, there is evidence that outcome evaluation recruits multiple neural regions across frontal and parietal cortices and subcortical regions. These regions support a range of cognitive processes, including value assignment, mentalizing, conflict resolution, and outcome optimization, all of which may be modulated by emotion regulation.

5.3 *Taking Decisive Action*

Although at the upper end of her budget, Hannah has found a house that checks every other box on her list—it's in a location near to both friends and restaurants and she can picture herself living here now and growing into it with a future family. She takes decisive action and puts an offer down on the house. Individual differences in risk-seeking or risk aversion may determine whether she feels nervous, energized, or exhausted by this process (Engelmann & Tamir, 2009). However, prior research indicates that she can once again use cognitive reappraisal to regulate her emotions.

Neural Circuitry Implicated in Taking Decisive Action Just before the point of taking decisive action, the vlPFC supports the opportunity for action inhibition (Apšvalka et al., 2022). It is rarely helpful to rush to action before weighing possible outcomes, and the vlPFC supports action inhibition and impulse control. The dlPFC and ACC also support a decision to act, such as Hannah putting an offer in on a house (Tannou et al., 2021; Hampton & O'Doherty, 2007). Decision enactment also involves a motivational perspective, to ensure that the action aligns with motivational goals and desired outcomes, recruiting the VS and OFC (Filimon et al., 2020; Wallis, 2007; Martin & Delgado, 2011).

Modulation of Neural Activity via Emotion Regulation Cognitive reappraisal consistently activates regions involved in cognitive control, like the dlPFC and vlPFC (Buhle et al., 2014). In addition to these more canonical regions involved in reappraisal, it is also possible that taking decisive action may involve a comparatively less-studied form of reappraisal, termed positivizing. This approach involves up-regulating the positive emotions associated with a stimulus (Doré et al., 2017) and may be particularly effective for decisions that involve elements of risk and reward. Prior neuroimaging research has found that engaging in positivizing is associated with increased function connectivity between the OFC, VS, vmPFC, and amygdala. Taken together, this indicates that taking decisive action recruits neural regions involved in action inhibition, increasing and decreasing emotional responses, and potentially up-regulating positive emotions.

5.4 Anticipating the Outcome

Now that Hannah has acted decisively and put in an offer, she must wait to hear if her offer is accepted. While waiting, she may feel anxious that the offer won't be accepted or worried that she overlooked something. In other words, Hannah experiences *uncertainty* following her decisive action while she anticipates the outcome. Like taking action, individual differences may account for variation in the emotional states and intensity elicited by this stage of the decision process. For example, individuals that have a lower tolerance for uncertainty (i.e., experience more negative affect when faced with unknowns) may experience more negative emotions during the anticipation period (Morriss et al., 2023), or may have more challenges emotionally disengaging from negative stimuli (Nanni-Zepeda et al., 2024), and thus may benefit more from the adoption of emotion regulation strategies like cognitive reappraisal.

Neural Circuitry Implicated in Outcome Anticipation After action is taken, and before the outcome is known, individuals must navigate the anticipation and uncertainty associated with waiting for the outcome. The mPFC supports outcome anticipation when expecting rewarding outcomes, which is often associated with motivational goals driving decisive action (Mitchell, 2011). The VS has also been implicated in information and reward anticipation (Filimon et al., 2020); subcortical regions like the amygdala and bed nucleus of the stria terminalis (BNST), a small subcortical structure connected to other limbic structures, are associated specifically with negative outcomes (Pedersen et al., 2019; Hammack et al., 2021). For example, Hannah is expecting her offer to be accepted (a rewarding outcome), but she must navigate a period of uncertainty as she waits (where both rewarding and disappointing outcomes are possible). Furthermore, although less well-studied, the insula may also help reduce uncertainty during this anticipatory period (Smith et al., 2014).

Modulation of Neural Activity via Emotion Regulation Emotion regulatory strategies like cognitive reappraisal may modulate neural activity in anticipation of negative outcomes in the amygdala and BNST (Pedersen et al., 2019; Hammack et al., 2021). The neural connectivity between the amygdala and mPFC are particularly central to emotion regulation processes (Buhle et al., 2014; Banks et al., 2007). Cognitive reappraisal may also modulate VS activity (Filimon et al., 2020; Liu et al., 2023), potentially preventing someone from getting too invested in an unlikely positive outcome (e.g., Hannah having an offer far below the listed price get accepted), and insular activity (Carlson & Mujica-Parodi, 2010), helping to reduce uncertainty and associated anxiety. Taken together, this indicates that outcome anticipation recruits neural regions involved in reward expectation and uncertainty reduction, and emotion regulation strategies to reduce negative affect associated with uncertainty may modulate the markers associated with these processes.

5.5 Dealing with the Outcome and Post-Decision Experience

After days of waiting, Hannah hears that her offer on the house was accepted! She feels relieved that her housing search is over, and she is excited about starting the new life she envisioned for herself. While this outcome is what Hannah wanted, others who bid on the same house may be simultaneously dealing with the outcome in a different way. Their housing search didn't conclude the way they wanted, and they may feel defeated to not have their offer accepted. These different outcomes provide different opportunities for emotion regulation.

Neural Circuitry Implicated in Post-Decision Emotional Outcomes While Hannah's ideal outcome became reality, other people may experience prediction errors between their expected and actual outcomes (Mitchell, 2011). Moreover, the context surrounding a decisive action may influence individual difficulties in accepting an outcome, as impulsivity during decisive action is associated with increased emotional reactivity and amygdala activation (Gupta et al., 2011). Therefore, outcomes do not always align with the ideal and may become tinged with regret over time. Post-decision regret may be supported by the vlPFC to encode negative feedback from a given decision and help an individual perhaps make a more effective decision in the future (Kim et al., 2018). Regret is also associated with greater activation in the OFC (Chua et al., 2009). Counterfactual thinking for negative outcomes, or comparing the reality of a given outcome to alternative possibilities, is also associated with mixed emotions and activation in the VS, insula, and dlPFC (Henderson & Norris, 2013).

Modulation of Neural Activity via Emotion Regulation Moreover, regret may require longer-term emotion regulation to prevent long-term emotional impacts (Reeck & LaBar, 2024). This may be especially true when individuals experience dissonance following a decision outcome, resulting in attitudinal adjustments to other possible outcomes (e.g., chosen outcome viewed as more positive and non-chosen outcomes seen as more negative) (Harmon-Jones et al., 2017). The dlPFC and dmPFC are thought to support this regulatory maintenance (Mitchell, 2011), and the vlPFC may additionally support regulation when predicted and actual outcomes differ (Bratec et al., 2015). Furthermore, decreased insula activity has been associated with subsequent decision-related attitude changes, perhaps to help individuals rationalize their decisions as they distance themselves from them (Jarcho et al., 2011). In this study by Jarcho et al. (2011), participants initially rated how much they liked hundreds of names and paintings. While in the fMRI scanner, they were shown a subset of similarly rated items (i.e., equally liked) and were asked to select preferences in two meaningful contexts: preferred names for future children and preferred paintings to hang in their home. Participants then rated their liking for all names and paintings again after exiting the scanner. Researchers found that attitudes towards selected preferences became more positive while attitudes towards rejected preferences became more negative. This attitude change was associated with decreased insula activity, which may suggest a neurocognitive mechanism for rationalization when choosing between equally attractive options. Taken together, this indicates that the post-decision experience recruits neural circuitry involved in prediction errors, feedback, and rationalization, and emotion regulation may modulate neural and behavioral markers associated with these processes.

6 Future Research and Considerations

While this chapter has focused on decision-making and emotion regulation in the context of healthy individuals, risk-related decision-making may be especially impacted in populations with mood or affective disorders (Lamba et al., 2020; Pawluk & Koerner, 2013; Sripada et al., 2009). This suggests that emotion regulation may play an important role in improving decision-making for these groups, and thus underscores the importance of exploring the neural and behavioral correlates at each stage of the decision-making process. Further research should continue to examine implications of emotion regulatory and risky decision-making success for clinical groups, such as those with eating disorders (Francesconi et al.,

2023), anxiety (Wake et al., 2020), and autism spectrum disorder (Sáez-Suanes et al., 2020).

Although the research summarized here has largely focused on young adult populations, prior research has consistently shown that emotion regulatory abilities vary substantially across the lifespan (Helion et al., 2019). Cognitive reappraisal doesn't emerge as an effective regulation strategy until adolescence, and individuals increase recruitment of dorsal and lateral prefrontal regions with age, suggesting that reappraisal is effective only once the neural circuitry is developed (Moreira et al., 2021; Silvers, 2022). Adolescents are also potentially particularly sensitive to reward, which may impact their ability to assess risk as they consider possible outcomes and take decisive action (Schreuders et al., 2018). This sensitivity may be especially important in social contexts, where adolescents are susceptible to social influence (Foulkes & Blakemore, 2016) and may act particularly impulsively (Smith et al., 2018).

While we discussed Gross's (1998) Process Model, we ultimately limited our discussion of decision-making and emotion regulation to focus on cognitive reappraisal, an emotion regulation strategy frequently examined in the neuroscience literature. However, there are alternative regulatory strategies that may or may not fit into the framework defined in this chapter, such as mindfulness (Jovanova et al., 2023), affect labeling (Torre & Lieberman, 2018), and interpersonal emotion regulation (Zaki & Williams, 2013). Other Process Model strategies have psychological processes and neural circuitry supporting them, but these are understudied relative to cognitive reappraisal (Webb et al., 2012). While the less-studied nature of some regulatory strategies can be explained by technical limitations (for example, studying situation selection is challenging in a functional neuroimaging experimental setting), others, like suppression and distraction, are alternative strategies to consider. Individual differences and situational factors may drive regulatory choices (Sheppes et al., 2011; Mitchell et al., 2025)—for some individuals, risky decision-making may be associated with high intensity negative affect, and they may opt for distraction, while those who consider it less affectively intense may instead select reappraisal.

Finally, additional research should continue to examine contexts in which emotion regulation is unhelpful for decision-making (Ford & Troy, 2019), an example we acknowledged but did not elaborate upon. While there is a general consensus in psychology and neuroscience that emotion regulation is helpful for decision-making (Heilman et al., 2010), there may be individual characteristics or situational factors that are associated with a beneficial role of high intensity, negative affect (e.g., social decision-making or interpersonal relationships) (Gross & John, 2003). Furthermore, for psychologists who study emotion regulation, the extent to which individual strategies may yield more or less optimal outcomes (i.e., what others have called a personalized science of emotion regulation) (Doré et al., 2016) remains unclear.

7 Conclusion

By entering the housing market and putting an offer down on her first house, Hannah experienced the full process of risky decision-making. While not all decisions involve such high risk, people make many decisions like these throughout their lifetimes. Likewise, affective states fluctuate depending on situational factors, interpersonal interactions, and internal motivations. Decision-making and emotion regulation processes are thus constantly working alongside each other in tandem, while overlapping neural circuitry in the brain works to support these cognitive processes and their associated behaviors.

References

Apšvalka, D., Ferreira, C. S., Schmitz, T. W., Rowe, J. B., & Anderson, M. C. (2022). Dynamic targeting enables domain-general inhibitory control over action and thought by the prefrontal cortex. *Nature Communications, 13*, 274. https://doi.org/10.1038/s41467-021-27926-w

Asutay, E., & Västfjäll, D. (2022). The continuous and changing impact of affect on risky decision-making. *Scientific Reports, 12*, 10613. https://doi.org/10.1038/s41598-022-14810-w

Badre, B., & Wagner, A. D. (2006). Computational and neurobiological mechanisms underlying cognitive flexibility. *PNAS, 103*(18), 7186–7191. https://doi.org/10.1073/pnas.0509550103

Banks, S. J., Eddy, K. T., Angstadt, M., Nathan, P. J., & Phan, K. L. (2007). Amygdala-frontal connectivity during emotion regulation. *Social Cognitive and Affective Neuroscience, 2*(4), 303–312. https://doi.org/10.1093/scan/nsm029

Barrett, L. F., & Bliss-Moreau, E. (2009). Chapter 4 affect as a psychological primative. *Advances in Experimental Social Psychology, 41*, 167–218. https://doi.org/10.1016/S0065-2601(08)00404-8

Bratec, S. M., Xie, X., Schmid, G., Doll, A., Schilbach, L., Zimmer, C., Wohlschläger, A., Riedl, V., & Sorg, C. (2015). Cognitive emotion regulation enhances aversive prediction error activity while reducing emotional responses. *NeuroImage, 123*, 138–148. https://doi.org/10.1016/j.neuroimage.2015.08.038

Braunstein, L. M., Herrera, S. J., & Delgado, M. R. (2014). Reappraisal and expected value modulate risk taking. *Cognition and Emotion, 28*(1), 172–181. https://doi.org/10.1080/02699931.2013.809330

Brooks, K. J. (2023, August 15). *It's taking Americans much longer in life to buy their first home*. CBS News. https://www.cbsnews.com/news/average-homebuyer-age-millennial-data-realtor/

Buhle, J. T., Silvers, J. A., Wager, T. D., Lopez, R., Onyemekwu, C., Kober, H., Weber, J., & Ochsner, K. N. (2014). Cognitive reappraisal of emotion: A meta-analysis of human neuroimaging studies. *Cerebral Cortex, 24*(11), 2981–2990. https://doi.org/10.1093/cercor/bht154

Carlson, J. M., & Mujica-Parodi, L. R. (2010). A disposition to reappraise decreases anterior insula reactivity during anxious anticipation. *Biological Psychiatry, 85*(3), 383–385. https://doi.org/10.1016/j.biopsycho.2010.08.010

Chua, H. F., Gonzalez, R., Taylor, S. F., Welsh, R. C., & Liberzon, I. (2009). Decision-related loss: Regret and disappointment. *NeuroImage, 47*(4), 2031–2040. https://doi.org/10.1016/j.neuroimage.2009.06.006

Denny, B. T., Jungles, M. L., Goodson, P. N., Dicker, E. E., Chavez, J., Jones, J. S., & Lopez, R. B. (2023). Unpacking reappraisal: A systematic review of fMRI studies of distancing and reinterpretation. *Social Cognitive and Affective Neuroscience, 18*(1), nsad050. https://doi.org/10.1093/scan/nsad050

Dixon, M. L., & Christoff, K. (2014). The lateral prefrontal cortex and complex value-based learning and decision making. *Neuroscience and Biobehavioral Reviews, 45*, 9–18. https://doi.org/10.1016/j.neubiorev.2014.04.011

Doré, B. P., Silvers, J. A., & Ochsner, K. N. (2016). Toward a personalized science of emotion regulation. *Social and Personality Psychology Compass, 10*(4), 171–187. https://doi.org/10.1111/spc3.12240

Doré, B. P., Boccagno, C., Burr, D., Hubbard, A., Long, K., Weber, J., Stern, Y., & Ochsner, K. N. (2017). Finding positive meaning in negative experiences engages ventral striatal and ventromedial prefrontal regions associated with reward valuation. *Journal of Cognitive Neuroscience, 29*(2), 235–244. https://doi.org/10.1162/jocn_a_01041

Doya, K. (2008). Modulators of decision making. *Nature Neuroscience, 11*, 410–416. https://doi.org/10.1038/nn2077

Engelmann, J. B., & Tamir, D. (2009). Individual differences in risk preference predict neural responses during financial decision-making. *Brain Research, 1290*, 28–51. https://doi.org/10.1016/j.brainres.2009.06.078

Filimon, F., Nelson, J. D., Sejnowski, T. J., Sereno, M. I., & Cottrell, G. W. (2020). The ventral striatum dissociates information expectation, reward anticipation, and reward receipt. *PNAS, 117*(26), 15200–15208. https://doi.org/10.1073/pnas.1911778117

Flores-Torres, J., Gómez-Pérez, L., & McRae, K. (2022). Positive cognitive reappraisal is beneficial for women's but not for men's IGT decision-making. *Motivation and Emotion, 46*(3), 350–365. https://doi.org/10.1007/s11031-022-09927-4

Ford, B. Q., & Troy, A. S. (2019). Reappraisal reconsidered: A closer look at the costs of an acclaimed emotion-regulation strategy. *Current Directions in Psychological Science, 28*(2), 195–203. https://doi.org/10.1177/0963721419827526

Foulkes, L., & Blakemore, S.-J. (2016). Is there heightened sensitivity to social reward in adolescence? *Current Opinion in Neurobiology, 40*, 81–85. https://doi.org/10.1016/j.conb.2016.06.016

Francesconi, M., Flouri, E., & Harrison, A. (2023). Decision-making difficulties mediate the association between poor emotion regulation and eating disorder symptoms in adolescence. *Psychological Medicine, 53*(8), 3701–3710. https://doi.org/10.1017/S003329172200037X

Friedman, N. (2024, January 19). *Home sales were the lowest in almost 30 years in 2023*. The Wall Street Journal. https://www.wsj.com/economy/housing/home-sales-likely-fell-to-15-year-low-in-2023-3da220e1

Gilead, M., & Ochsner, K. N. (2021). A guide to the neural bases of mentalizing. In M. Gilead & K. N. Ochsner (Eds.), *The neural basis of mentalizing* (pp. 3–16). Springer.

Gilead, M., Boccagno, C., Silverman, M., Hassin, R. R., Weber, J., & Ochsner, K. N. (2016). Self-regulation via neural simulation. *PNAS, 113*(36), 10037–10042. https://doi.org/10.1073/pnas.1600159113

Goodman, L. S., & Mayer, C. (2018). Homeownership and the American dream. *Journal of Economic Perspectives, 32*(1), 31–58. https://doi.org/10.1257/jep.32.1.31

Grecucci, A., & Sanfey, A. G. (2014). Emotion regulation and decision making. In J. J. Gross (Ed.), *Handbook of emotion regulation* (2nd ed., pp. 140–153). The Guilford Press.

Grecucci, A., Giorgetta, C., van't Wout, M, Bonini, N., & Sanfey, A. G. (2013). Reappraising the ultimatum: An fMRI study of emotion regulation and decision making. *Cerebral Cortex, 23*(2), 399–410. https://doi.org/10.1093/cercor/bhs028

Gross, J. J. (1998). The emerging field of emotion regulation: An integrative review. *Review of General Psychology, 2*(3), 271–299. https://psycnet.apa.org/doi/10.1037/1089-2680.2.3.271

Gross, J. J. (2014). Emotion regulation: Conceptual and empirical foundations. In J. J. Gross (Ed.), *Handbook of emotion regulation* (2nd ed., pp. 3–20). The Guilford Press.

Gross, J. J. (2015). Emotion regulation: Current status and future prospects. *Psychological Inquiry, 26*(1), 1–26. https://doi.org/10.1080/1047840X.2014.940781

Gross, J. J., & John, O. P. (2003). Individual differences in two emotion regulation processes: Implications for affect, relationships, and well-being. *Journal of Personality and Social Psychology, 85*(2), 348–362. https://doi.org/10.1037/0022-3514.85.2.348

Gupta, R., Koscik, T. R., Bechara, A., & Tranel, D. (2011). The amygdala and decision-making. *Neuropsychologia, 49*(4), 760–766. https://doi.org/10.1016/j.neuropsychologia.2010.09.029

Hammack, S. E., Braas, K. M., & May, V. (2021). Chemoarchitecture of the bed nucleus of the stria terminalis: Neurophenotypic diversity and function. *Handbook of Clinical Neurology, 179*, 385–402. https://doi.org/10.1016/B978-0-12-819975-6.00025-X

Hampton, A. N., & O'Doherty, J. P. (2007). Decoding the neural substrates of reward-related decision making with functional MRI. *PNAS, 104*(4), 1377–1382. https://doi.org/10.1073/pnas.0606297104

Harmon-Jones, E., Harmon-Jones, C., & Summerell, E. (2017). On the importance of both dimensional and discrete models of emotion. *Behavioral Science, 7*(4), 66. https://doi.org/10.3390/bs7040066

Heilman, R. M., Crişan, L. G., Houser, D., Miclea, M., & Miu, A. C. (2010). Emotion regulation and decision making under risk and uncertainty. *Emotion, 10*(2), 257–265. https://doi.org/10.1037/a0018489

Helion, C., Krueger, S. M., & Ochsner, K. N. (2019). Emotion regulation across the life span. *Handbook of Clinical Neurology, 163*, 257–280. https://doi.org/10.1016/B978-0-12-804281-6.00014-8

Henderson, S. E., & Norris, C. J. (2013). Counterfactual thinking and reward processing: An fMRI study of response to gamble outcomes. *NeuroImage, 64*, 582–589. https://doi.org/10.1016/j.neuroimage.2012.08.078

Jaiswal, S., Chakravarthula, L. N. C., & Padmala, S. (2024). Additive effects of monetary loss and positive emotion in the human brain. *eNeuro, 11*(4), ENEURO.0374-23.2024. https://doi.org/10.1523/ENEURO.0374-23.2024

Jarcho, J. M., Berkman, E. T., & Lieberman, M. D. (2011). The neural basis of rationalization: Cognitive dissonance reduction during decision-making. *Social Cognitive and Affective Neuroscience, 6*(4), 460–467. https://doi.org/10.1093/scan/nsq054

Jovanova, M., Cosme, D., Doré, B., Kang, Y., Stanoi, O., Cooper, N., Helion, C., Lomax, S., McGowan, A. L., Boyd, Z. M., Bassett, D. S., Mucha, P. J., Ochsner, K. N., Lydon-Staley, D. M., & Falk, E. B. (2023). Psychological distance intervention reminders reduce alcohol consumption frequency in daily life. *Scientific Reports, 13*, 12045. https://doi.org/10.1038/s41598-023-38478-y

Kahneman, D., & Tversky, A. (1984). Choices, values, and frames. *American Psychologist, 39*(4), 341–350. https://doi.org/10.1037/0003-066X.39.4.341

Kahneman, D., & Tversky, A. (Eds.). (2000). *Choices, values, and frames*. Cambridge University Press.

Kim, S., Hwang, S., & Lee, M. (2018). The benefits of negative yet informative feedback. *PLoS One, 13*(10), e0205183. https://doi.org/10.1371/journal.pone.0205183

Lamba, A., Frank, M. J., & FeldmanHall, O. (2020). Anxiety impedes adaptive social learning under uncertainty. *Psychological Science, 31*(5), 592–603. https://doi.org/10.1177/0956797620910993

Lerner, J. S., Li, Y., Valdesolo, P., & Kassam, K. S. (2015). Emotion and decision making. *Annual Review of Psychology, 66*, 799–823. https://doi.org/10.1146/annurev-psych-010213-115043

Liu, S., Li, S., Jiang, H., Zhang, Z., Gong, Y., Guo, X., & Liu, Z. (2023). The neural correlates underlying the regulation of anticipation on regret. *Behavioural Brain Research, 436*, 114075. https://doi.org/10.1016/j.bbr.2022.114075

Martin, L. N., & Delgado, M. R. (2011). The influence of emotion regulation on decision-making under risk. *Journal of Cognitive Neuroscience, 23*(9), 2569–2581. https://doi.org/10.1162/2Fjocn.2011.21618

Matarazzo, O., Abbamonte, L., Greco, C., Pizzini, B., & Nigro, G. (2021). Regret and other emotions related to decision-making: Antecedents, appraisals, and phenomenological aspects. *Frontiers in Psychology, 12*, 783248. https://doi.org/10.3389/fpsyg.2021.783248

Milham, M. P., & Banich, M. T. (2005). Anterior cingulate cortex: An fMRI analysis of conflict specificity and functional differentiation. *Human Brain Mapping, 25*(3), 328–335. https://doi.org/10.1002/hbm.20110

Mitchell, D. G. V. (2011). The nexus between decision making and emotion regulation: A review of convergent neurocognitive substrates. *Behavioural Brain Research, 217*(1), 215–231. https://doi.org/10.1016/j.bbr.2010.10.030
Mitchell, W. J., Stasiak, J., Martinez, S. A., Cliver, K. G., Gregory, D. F., Reisman, S., Schmidt, H., Murty, V. P., & Helion, C. (2025). Emotion regulation strategy use and forecasting in response to dynamic, multimodal stimuli. *Journal of Experimental Psychology. General, 154*(4), 1004–1024.
Moreira, J. F. G., McLaughlin, K. A., & Silvers, J. A. (2021). Characterizing the network architecture of emotion regulation neurodevelopment. *Cerebral Cortex, 31*(9), 4140–4150. https://doi.org/10.1093/cercor/bhab074
Morris, A., Phillips, J., Huang, K., & Cushman, F. (2021). Generating options and choosing between them depend on distinct forms of value representation. *Psychological Science, 32*(11), 1731–1746. https://doi.org/10.1177/09567976211005702
Morriss, J., Goh, K., Hirsch, C. R., & Dodd, H. F. (2023). Intolerance of uncertainty heightens negative emotional states and dampens positive emotional states. *Frontiers in Psychiatry, 14*, 1147970. https://doi.org/10.3389/fpsyt.2023.1147970
Nanni-Zepeda, M., DeGutis, J., Wu, C., Rothlein, D., Fan, Y., Grimm, S., Walter, M., Esterman, M., & Zuberer, A. (2024). Neural signatures of shared subjective affective engagement and disengagement during movie viewing. *Human Brain Mapping, 45*(4), e26622. https://doi.org/10.1002/hbm.26622
Oehrn, C. R., Hanslmayr, S., Fell, J., Deuker, L., Kremers, N. A., Do Lam, A. T., Elger, C. E., & Axmacher, N. (2014). Neural communication patterns underlying conflict detection, resolution, and adaptation. *Journal of Neuroscience, 34*(31), 10438–10452. https://doi.org/10.1523/JNEUROSCI.3099-13.2014
Pawluk, E. J., & Koerner, N. (2013). A preliminary investigation of impulsivity in generalized anxiety disorder. *Personality and Individual Differences, 54*(6), 732–737. https://psycnet.apa.org/doi/10.1016/j.paid.2012.11.027
Pedersen, W. S., Muftuler, L. T., & Larson, C. L. (2019). A high-resolution fMRI investigation of BNST and centromedial amygdala activity as a function of affective stimulus predictability, anticipation, and duration. *Social Cognitive and Affective Neuroscience, 14*(11), 1167–1177. https://doi.org/10.1093/scan/nsz095
Pool, E. R., Tord, D. M., Delplanque, S., Stussi, Y., Cereghetti, D., Vuilleumier, P., & Sander, D. (2022). Differential contributions of ventral striatum subregions to the motivational and hedonic components of the affective processing of reward. *Journal of Neuroscience, 42*(13), 2716–2728. https://doi.org/10.1523/JNEUROSCI.1124-21.2022
Rangel, A., Camerer, C., & Montague, P. R. (2008). A framework for studying the neurobiology of value-based decision making. *Nature Reviews Neuroscience, 9*, 545–556. https://doi.org/10.1038/nrn2357
Reeck, C., & LaBar, K. S. (2024). Reining in regret: Emotion regulation modulates regret in decision making. *Cognition and Emotion*, 1–8. https://doi.org/10.1080/02699931.2024.2357847
Roberts, I. D., HajiHosseini, A., & Hutcherson, C. A. (2024). How bad becomes good: A neurocomputational model of affect-informed choice. *Emotion, 24*(7), 1737–1752. https://doi.org/10.1037/emo0001347
Russell, J. A. (2003). Core affect and the psychological construction of emotion. *Psychological Review, 110*(1), 145–172. https://doi.org/10.1037/0033-295x.110.1.145
Russell, J. A., & Barrett, L. F. (1999). Core affect, prototypical emotional episodes, and other things called emotion: Dissecting the elephant. *Journal of Personality and Social Psychology, 76*(5), 805–819. https://doi.org/10.1037/0022-3514.76.5.805
Rutledge, R. B., Skandali, N., Dayan, P., & Dolan, R. J. (2015). Dopaminergic modulation of decision making and subjective well-being. *Journal of Neuroscience, 35*(27), 9811–9822. https://doi.org/10.1523/JNEUROSCI.0702-15.2015
Sáez-Suanes, G. P., García-Villamisar, D., del Pozo Armentia, A., & Dattilo, J. (2020). Emotional dysregulation and uncertainty intolerance as transdiagnostic mediators of anxiety in adults with

autism spectrum disorders and intellectual disability. *Research in Developmental Disabilities, 106*, 103784. https://doi.org/10.1016/j.ridd.2020.103784

Schreuders, E., Braams, B. R., Blankenstein, N. E., Peper, J. S., Güroğlu, B., & Crone, E. A. (2018). Contributions of reward sensitivity to ventral striatum activity across adolescence and early adulthood. *Child Development, 89*(3), 797–810. https://doi.org/10.1111/cdev.13056

Sharpe, M. J., Stalnaker, T., Schuck, N. W., Killcros, S., Schoenbaum, G., & Niv, Y. (2019). An integrated model of action selection: Distinct modes of cortical control of striatal decision making. *Annual Review of Psychology, 70*, 53–76. https://doi.org/10.1146/annurev-psych-010418-102824

Shenhav, A., Cohen, J., & Botvinick, M. (2016). Dorsal anterior cingulate cortex and the value of control. *Nature Neuroscience, 19*, 1286–1291. https://doi.org/10.1038/nn.4384

Sheppes, G., Scheibe, S., Suri, G., & Gross, J. J. (2011). Emotion-regulation choice. *Psychological Science, 22*(11), 1391–1396. https://doi.org/10.1177/0956797611418350

Silvers, J. A. (2022). Adolescence as a pivotal period for emotion regulation development. *Current Opinion in Psychology, 44*, 258–263. https://doi.org/10.1016/j.copsyc.2021.09.023

Slovic, P., Peters, E., Finucane, M. L., & Macgregor, D. G. (2005). Affect, risk, and decision making. *Health Psychology, 24*(4S), S35–S40. https://doi.org/10.1037/0278-6133.24.4.S35

Smith, A. R., Steinberg, L., & Chein, J. (2014). The role of the anterior insula in adolescent decision making. *Developmental Neuroscience, 36*(3–4), 196–209. https://doi.org/10.1159/000358918

Smith, A. R., Rosenbaum, G. M., Botdorf, M. A., Steinberg, L., & Chein, J. M. (2018). Peers influence adolescent reward processing, but not response inhibition. *Cognitive, Affective, & Behavioral Neuroscience, 18*, 284–295. https://doi.org/10.3758/s13415-018-0569-5

Sripada, C. S., Angstadt, M., Banks, S., Nathan, P. J., Liberzon, I., & Phan, K. L. (2009). Functional neuroimaging of mentalizing during the trust game in social anxiety disorder. *Neuroreport, 20*(11), 984–989. https://doi.org/10.1097/wnr.0b013e32832d0a67

Steward, T., Davey, C. G., Jamieson, A. J., Stephanou, K., Soriano-Mas, C., Felmingham, K. L., & Harrison, B. J. (2021). Dynamic neural interactions supporting the cognitive reappraisal of emotion. *Cerebral Cortex, 31*(2), 961–973. https://doi.org/10.1093/cercor/bhaa268

Tannou, T., Magnin, E., Comte, A., Aubry, R., & Joubert, S. (2021). Neural activation in risky decision-making tasks in healthy older adults: A meta-analysis of fMRI data. *Brain Sciences, 11*(8), 1043. https://doi.org/10.3390/brainsci11081043

Torre, J. B., & Lieberman, M. D. (2018). Putting feelings into words: Affect labeling as implicit emotion regulation. *Emotion Review, 10*(2), 116–124. https://doi.org/10.1177/1754073917742706

Wake, S., Wormwood, J., & Satpute, A. B. (2020). The influence of fear on risk taking: A meta-analysis. *Cognition and Emotion, 34*(6), 1143–1159. https://doi.org/10.1080/02699931.2020.1731428

Wallis, J. D. (2007). Orbitofrontal cortex and its contribution to decision-making. *Annual Review of Neuroscience, 30*, 31–56. https://doi.org/10.1146/annurev.neuro.30.051606.094334

Webb, T. L., Miles, E., & Sheeran, P. (2012). Dealing with feeling: A meta-analysis of the effectiveness of strategies derived from the process model of emotion regulation. *Psychological Bulletin, 138*(4), 775–808. https://doi.org/10.1037/a0027600

Winecoff, A., Clithero, J. A., Carter, R. M., Bergman, S. R., Wang, L., & Huettel, S. A. (2013). Ventromedial prefrontal cortex encodes emotional value. *Journal of Neuroscience, 33*(27), 11032–11039. https://doi.org/10.1523/JNEUROSCI.4317-12.2013

Zaki, J., & Williams, W. C. (2013). Interpersonal emotion regulation. *Emotion, 13*(5), 803–810. https://doi.org/10.1037/a0033839

Emotion and Choice: The Integral Role of Emotion in Constructing Value

Oriel FeldmanHall and Yi Yang Teoh

Abstract In the centuries-long history of decision-making research, emotion's role in choice has only been investigated relatively recently. Early theories of decision-making, which conceived of emotions as byproducts of choice, have evolved over time to emphasize emotion's integral role in choice. These theoretical advances have ushered in an explosion of empirical research on the modulatory influence of emotion in value-based decision-making. And yet, the field currently lacks a mechanistic understanding of where value comes from. While reinforcement learning models have been instrumental in revealing the computational principles by which people update the value of choice options based on feedback, these models shed little insight into the psychological mechanisms underpinning evaluative processes. We synthesize recent research on emotion and decision-making to propose how the field might close this gap: endogenous affective processes help transform choice options and decision outcomes into meaningful value signals. While the precise relation between affect and value—at a computational, algorithmic, and implementational level—remains to be described, such an endeavor will be facilitated by promising advancements in modeling and neuroimaging techniques.

1 Introduction

Researchers have been studying how humans make decisions for nearly 300 years. In 1738, with the Enlightenment period in full swing, Daniel Bernoulli introduced the concept of expected utility and risk aversion (Bernoulli, 1954). His theoretical treatise ushered in a new era of impassioned scientific discourse and research

O. FeldmanHall (✉)
Department of Cognitive and Psychological Sciences, Brown University, Providence, RI, USA

Carney Institute for Brain Sciences, Brown University, Providence, RI, USA
e-mail: Oriel@brown.edu

Y. Y. Teoh
Department of Cognitive and Psychological Sciences, Brown University, Providence, RI, USA

D. V. Smith et al. (eds.), *Neuroeconomics: Core Topics and Current Directions*,
https://doi.org/10.1007/978-3-032-02925-6_21

on human decision-making under uncertainty. How theorists and scientists have thought about human decision-making has steadily evolved over the centuries. In the early years, traditional economic models depicted decision-making as a purely rational process, driven by logic and the ability to calculate the utility of a choice. As it turns out, humans do not always have the capacity to calculate the utility associated with every option in a choice set. In fact, humans are quite constrained by their cognitive capacity. This idea, termed by Herbert Simon as "bounded rationality," argues that decision-making is a more intricate and complicated picture than originally thought (Simon, 1956, 1997). In the late 1970s, the belief that humans make purely rational calculations was finally put to bed when Prospect Theory (and other research spearheaded by Kahneman and Tversky) made its debut (Kahneman & Tversky, 1979). Human emotion, with its deep roots in human choice, began to come into focus.

The field's "affective awakening," which has been unfurling over the past 50 years, documents just how profoundly our emotions can shape our choices (Dukes et al., 2021). Some theories, largely those of the neoclassical variety, argue that emotions are mere byproducts of the decision-making process. Other theories place emotion as a more central figure (Frijda, 2001) that is critical for shaping how choice options become valued. More recent research argues that affect itself is an evaluation of the stimuli or people we encounter in the world, which helps to figure out what (or who) is valuable. In this chapter, we review in brief the different ways in which emotion is theorized to play a role in decision-making. We conclude by suggesting that affect may be the primary signal that transforms any external stimuli into something of value.

2 An Implicit Role of Emotion in Decision-Making

Neoclassical models of human decision-making, most famously Expected Utility Theory (EUT) (Samuelson, 1952; Von Neumann & Morgenstern, 1944), assume that choices are made to maximize utility. Utility is a numerical representation of an individual's preferences (or satisfaction) derived from a specific outcome, and it cannot be directly observed. Utility works by parametrizing rewards from the environment, where rewards typically refer to money or wealth. By mathematically quantifying "satisfaction," utility is meant to allow for a structured understanding of how much people *value* a specific outcome under uncertain conditions. Since its conception, utility has been evoked in a variety of forms (e.g., decision utility, experienced utility, anticipation utility, etc.), each of which has a different meaning. Regardless of which word precedes it, utility is deeply bound to the concept of value. It is argued that people make decisions based on the value of the available options, preferring more valuable options over less valuable alternatives. In short, utility indexes a person's subjective assessment of the advantages that an option provides relative to the alternatives (Levy & Glimcher, 2012; Rangel et al., 2008; Rangel & Hare, 2010).

EUT is a normative model describing how decisions *should* be made if your goal is to maximize utility; emotion is not explicitly acknowledged to play a role in shaping an outcome's value. The problem with EUT and related theories is that, in most real-world decision-making contexts, choices deviate from what is described in classic decision models. The era following neoclassicalism recognized that an individual's response to certain classes of events may be influenced by psychological or emotional factors, contributing to this deviation. Prospect Theory, for example, challenged traditional EUT by incorporating psychological insights into the differences between how people evaluate gains versus losses (Kahneman & Tversky, 1979), therefore arguing that the functional form of "maximizing utility" may not be as straightforward as originally presumed.

Fast forward to the dawn of the twenty-first century when Experienced Utility broke onto the scene. By revisiting ideas on pleasure and pain written more than two centuries before by Jeremy Bentham, Kahneman developed the idea that a pleasurable experience can be produced by an outcome of a decision (Kahneman et al., 1997). This idea took hold: Kahneman and other researchers conducted numerous studies exploring how people experience happiness, satisfaction, and other emotional states in real-time. The idea that a good decision is born out of maximizing the *experienced* utility of the chosen outcome (Kahneman et al., 1997), brought to the forefront that outcome value might be shaped, at least in part, by affective experiences (Berridge & O'Doherty, 2014).

To accommodate other psychological factors that might impact choice, decision models were adapted to allow probabilities to be weighted by their presumed emotional impact. For example, models can be tweaked to reflect that people assign higher subjective probabilities to emotionally charged events (e.g., being eaten by a shark) (Johnson & Tversky, 1983; Lieder et al., 2018), even if objective probabilities are lower for those events. However, even when emotions like happiness or regret are formally accounted for (Loomes & Sugden, 1982; Mellers, 2000; Mellers et al., 1999), emotion is often considered the byproduct of the outcome. For example, an individual might conjure up a counterfactual—a "what could have been" if another choice had been selected. They might feel regret at having made what seems to be the wrong choice, or perhaps feel profound joy at making what appears to be the right choice. In these scenarios, emotion is not considered a driver of choice, but is instead relegated to the aftermath of choice.

3 An Explicit Role for Emotion in Decision-Making

A surge of research on experienced utility gave way to the broader idea that decision-making models ought to include an explicit role for emotion and affect as part of the choice process. Perhaps emotional satisfaction or dissatisfaction does not only occur after a choice is made. Rather, emotion (i.e., conscious feelings or sensations accompanied by physiological changes) and affect (i.e., subjective feelings and bodily responses free of any implied cognitive structures, which give rise to emotions) comprise the very fabric of our daily lives (Birbaumer & Öhman, 1993; Clore & Ortony, 2013; Ortony et al., 1988) and therefore shape the choices we make. Work exploring this idea revealed that affect is indeed integral to the concept of value. In a seminal paper entitled "Risk as Feelings," Loewenstein et al. (2001) argued that rather than individuals making risk calculations in a purely cognitively rational manner, risky choices are deeply influenced by in-the-moment affective reactions to potential (not yet delivered) outcomes. In these cases, emotional responses serve as heuristics, commonly known as the Affect Heuristic, that can shape which choice is ultimately taken (i.e., gamble or refrain from gambling). In other words, emotions serve as shortcuts for making risky choices (where the goal is to maximize value given uncertain terms), and these emotional heuristics can lead to decisions that are driven more by one's affective state and less by objective, rational evaluation. A large swath of human behavior can be better explained by the theory that emotions are part and parcel of the process that biases our risky choices (Lucarelli et al., 2015; Mellers et al., 1999). The human risk-taking calculus depends on emotional states, not just on a cognitive assessment of risk (Lerner et al., 2015; De Martino et al., 2006).

The "risk as feelings" theory is rooted in the well-trodden framework of dual processes (Quartz, 2009). Dual-process models of decision-making argue that there are two systems at play: a fast, automatic affective system and a slower, deliberate cognitive one (Kahneman, 2011). Depending on your flavor of dual-process model (Evans & Stanovich, 2013), these two systems can either trade off against one another (like a seesaw) or work in concert. "Risk as feelings" argued for the former—the emotional system can essentially override the cognitive system when making decisions that involve risk. The rise of dual-process models calcified the notion that emotions play an essential role in decision-making, arguing that one of the main pathways driving choice is an emotional one. With the advent of neuroimaging, researchers discovered that decisions involving risk engaged regions like the medial prefrontal cortex and amygdala (FeldmanHall et al., 2019; Hsu et al., 2005; Kable & Glimcher, 2007; Levy et al., 2010), the same regions so often implicated in emotional experiences (Barrett & Bliss-Moreau, 2009; Barrett & Wager, 2006; LeDoux, 2000). This association helped to further solidify the link between emotion and decision-making under uncertainty (Levy & Glimcher, 2011; Louie & Glimcher, 2010). But these studies on risk and probability also revealed the involvement of other regions, such as the lateral prefrontal cortex, which is typically associated with the slower, more rational cognitive system (Fecteau et al., 2007; FeldmanHall et al., 2019; Tobler et al., 2009). That risky choices could engage such different regions of the brain suggested that perhaps the role of emotion in choice is more complicated than a simple heuristic, resulting in a knee-jerk reaction (Moors et al., 2013).

Further research began to test the boundary conditions of emotion's role during choice by leveraging the framework of goals and habits (FeldmanHall & Chang, 2018). Initially, goals were assumed to map onto the slow, cognitive system, and habits to the fast, emotional system outlined in dual-process models. However, despite the idea that goals are within the purview of the deliberate cognitive system, research began to reveal that even goals are shaped (and often usurped) by emotions (Tamir, 2016). For example, we may have the goal of losing weight because we feel negatively about not fitting into a pair of jeans. In this way, a negative emotion helped set the goal of losing weight. But it is also the case that emotions can lead us further away from achieving our goal, such as when we crave (and then consume) junk food—despite the goal of losing weight. The fact that emotions appear to drive *both* habits and goals (Lerner et al., 2015; Phelps et al., 2014) added to a steady stream of research illustrating how affect that is relevant to the decision at hand, known as *integral affect*, helps to shape the types of choices humans ultimately make.

In parallel, other research actively investigated the role of *incidental affect* on choice. Incidental affect refers to emotions that are unrelated to the decision or judgment at hand, such as when the stress we feel at work affects our interactions with our loved ones once we return home. It turns out that these unrelated emotional experiences can also have a profound influence on the choices we make (FeldmanHall et al., 2015; Garg et al., 2007; Morgado et al., 2015; Porcelli & Delgado, 2017). Anger and fear, for example, can have distinct (and opposing) effects on people's risk assessments and choices (Lerner & Keltner, 2001). While fear increases perceived risk, thereby promoting risk-averse choices, anger actually decreases perceived risk, thereby promoting risk-seeking choices. Longer-lasting emotional states, such as moods, have also been shown to impact choice, where, for example, positive moods can engender optimistic decision-making (Eldar et al., 2016).

The critical role of emotion in shaping choice is made none more apparent than in research on mood disorders (Dalgleish & Power, 1997), which highlights that affective dysfunction goes hand-in-hand with maladaptive patterns of decision-making (Admon & Pizzagalli, 2015; Fernandez & Johnson, 2016; Hammond et al., 2023; Houben & Kuppens, 2020; Pulcu & Browning, 2019; Russo & Nestler, 2013). Research shows that people with anxiety have difficulty learning when the outcomes of their choices change over time (Hammond et al., 2023), especially when the environment is highly uncertain (Beltzer et al., 2019; Lamba et al., 2020). These patterns seem to arise out of an asymmetric sensitivity to negative outcomes, which disproportionately biases how subsequent decisions are made (Aylward et al., 2019; Rabinak et al., 2017). For those with anxiety, a diminished ability to learn from aversive outcomes results in rigid choice policies (Lamba et al., 2024), illustrating how emotional hypersensitivity can hinder the proper updating of the mental models of the world. In other disorders of emotion, such as depression, experiencing blunted affect also leads to maladaptive mental models of the world. For example, those who experience anhedonia lack motivation to obtain valuable outcomes (Cooper et al., 2018; Huys et al., 2013). It is difficult to synthesize these seemingly disparate results (which span a variety of emotion disorders) into a compact, understandable model of how emotion helps orchestrate adaptive decision-making. Research on clinical populations continues to make inroads, and with that progress, a more granular, detailed picture of how exactly emotion and affect impact choice is taking shape

(Adolphs et al., 2019; Houben & Kuppens, 2020; O'Toole et al., 2020; Seah & Coifman, 2022).

Emotions, whether imperceptible or intensely consuming, are deeply influential for how we evaluate options, perceive risks, and ultimately make decisions (Lerner et al., 2015). But in this push to better understand emotion's role in choice, part of the equation has been overlooked. While we know that emotion can either demotivate or motivate people to make choices that maximize value, little is known about where exactly value comes from (Juechems & Summerfield, 2019).

4 Where Does Value Come From?

In most fields, whether it be behavioral economics, cognitive neuroscience, neuroeconomics, or psychology, the value of an option is rarely apparent (Khaw et al., 2015; Louie & Glimcher, 2010, 2012; Yamada et al., 2018). It must either be guessed or learned through one's own actions or by observing the behavior of others. To learn the value of an option (e.g., whether an exotic fruit is worth eating), a person has to assess whether an outcome was advantageous or disadvantageous (also known as rewards or punishments; in this case, whether the fruit tasted yummy or disgusting). Rewards and punishments are the fundamental principles underlying Reinforcement Learning (RL; Fig. 1a), which operationalizes learning as the updating of option-values based on errors between predicted and actual outcomes over time—prediction errors (Sutton & Barto, 1999). For instance, if a decision is more (or less) rewarding than expected, its value ought to be enhanced (or diminished) following a decision. A prediction error can either increase or decrease

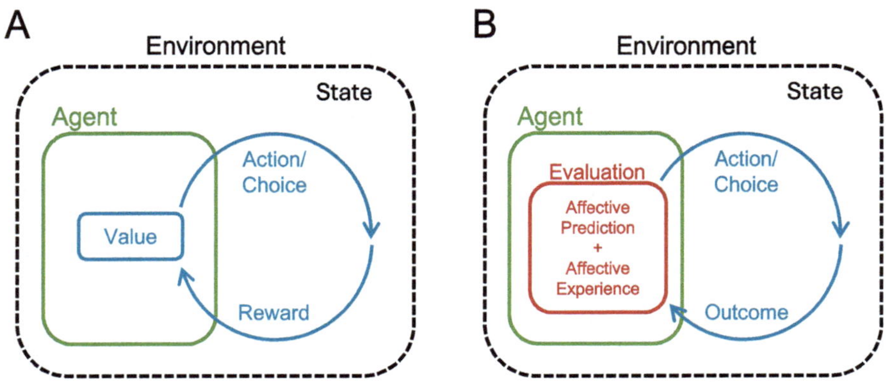

Fig. 1 Models of value-based decision-making. (**a**) Classic reinforcement learning models where agents' value-based choices and actions act directly on the environment to produce rewards/punishments as signals for value construction and learning. (**b**) Schematic of an affective evaluation model where agents' choice options and actions (which produce outcomes) are affectively assessed during decision-making in order to construct value

the likelihood that an agent makes the same choice (e.g., keep eating the same exotic fruit if delicious, or go back to mainstream bananas if yucky).

Although RL models and their variants formally describe the process by which value is updated (i.e., prediction errors), they shed little insight into where value comes from, often presupposing that outcomes in the external environment are self-evidently rewarding or punishing. In other words, the assumption is that it is always rewarding to receive more juice, more money, more smiles, more cookies, more love. Although the goal for RL is to simply maximize these rewards, these models are largely silent about more complex, multidimensional rewards that cannot be treated as a scalar value (FeldmanHall & Nassar, 2021). Put differently, in the real world, outcomes are yoked to complex "rewards" that must be *transformed* into psychologically meaningful percepts, and this must be done prior to constructing and updating the subjective value of our choices (Juechems & Summerfield, 2019). When applied to our example of people always seeking out *more* rewards, the truth is that people typically do not desire more cookies after an entire bag has already been consumed (Berridge & O'Doherty, 2014; Keramati & Gutkin, 2014).

Many stimuli out in the world are not inherently valuable, and there are situations in which stimuli require some transformation to become valuable. This suggests that the transformation of stimuli (or outcomes) to value is a vital missing piece of the equation. Affective scientists have been dancing around this idea for some time, and some in the field would argue that the transformation of stimuli, choice options, and outcomes into value is largely affective in nature (Barrett & Bliss-Moreau, 2009; Duncan & Barrett, 2007; Rolls, 2013). Rather than only occurring as a consequence of choice, affective responses serve as meaningful endogenous signals for the very *evaluation* of choice outcomes (Barrett & Bliss-Moreau, 2009; Berridge & Aldridge, 2008; Kahneman et al., 1997; Fig. 1b).

As an analogy, we might consider how the process of visual perception transforms light into objects and scenes that have meaning. Retinal receptors transduce light energy from the external environment into electrical signals that propagate via the optic nerve to the brain. These signals are iteratively processed in the occipital cortex to produce subjective perceptions of visual features, which include luminance, color, motion, depth, and texture (Lennie, 1998). While wavelength is objective, how each of us perceives a color is highly subjective. The binding and integration of these features further lead to the subjective perceptions of unified objects and scenes (Ashby et al., 1996; Treisman & Schmidt, 1982). Importantly, parsing a visual scene not only relies on the objective sensory input from external light but also prior knowledge, which is used to make predictions about configurations of features, objects, and scenes (Bar, 2004). Just as the eyes and brain work in concert to transform light into meaningful scenes, so too does the body provide important affective signals to help us make sense of the world (Jacobson, 2021). Valuable stimuli we encounter in the world come in various shapes and sizes: tangible primary sources such as food and water, intangible primary sources such as social feedback, and abstract secondary sources such as money. In each case, these stimuli reliably elicit physiological responses in the body and brain (Löw et al., 2008; Sander & Nummenmaa, 2021). Similar to the optic nerve, changes in autonomic arousal in the body are transmitted to the brain via the vagus nerve (Critchley & Garfinkel, 2017; Kreibig, 2010; Malezieux et al., 2023), where these physiological responses are processed by the insula and fed-forward to a distributed set of brain regions including the amygdala and prefrontal cortex to produce more complex emotion states (Barrett, 2017; Bradley & Lang, 2000; Mayer, 2011; Phan et al., 2004; Phillips et al., 1997; Seth & Friston, 2016). The result is that stimuli gain value as they are transformed through the affective prism we apply to the world.

Affective states are well-characterized by two dimensions that track the pleasantness (valence) and intensity (arousal) of the experience (Barrett & Bliss-Moreau, 2009; Russell, 1980). This two-dimensional architecture positions affect as a basic mechanism for evaluating a host of stimuli (Bach & Dayan, 2017) and for that reason has been implicated as a "common currency" for choice (Barrett & Bliss-Moreau, 2009; Levy & Glimcher, 2012; cf. Walasek & Brown, 2023). Indeed, choices resulting in outcomes that generate positive affect are "subjectively rewarding," and choices resulting in outcomes that generate negative affect are "subjectively punishing" (Berridge & Aldridge, 2008; Kahneman et al., 1997). Evidence for this theoretical perspective finds that a person's sensory awareness of their bodily signals—known as interoception (Barrett, 2017)—strongly modulates their sensitivity to rewarding outcomes, and thus, their ability to update the value of choice options during learning (Azzalini et al., 2021; Cabanac, 1971; Critchley & Garfinkel, 2017; Keramati & Gutkin, 2014; Pfeifer et al., 2017). This is true not only for concrete primary rewards, such as food, which directly affect our physiological states through consumption (Araujo et al., 2020), but also for more abstract secondary rewards, such as money (Hayes & Wedell, 2020).

Just as visual perception engages predictive processes to construct meaningful objects and scenes, affective states arise not only as reactive responses but also as *predictions* constructed from relevant prior knowledge (Barrett, 2017; Duncan & Barrett, 2007; Ellsworth, 2013; Emanuel & Eldar, 2023; Seth & Friston, 2016). In the field of visual perception, researchers have shown the critical role of predictive processing via work on perceptual illusions, where errors in prediction lead to optical illusions (Jovanovic et al., 2023). In a similar vein, research demonstrates how errors in affective forecasting systematically bias our evaluation of choice options. Evidence suggests that people overweight losses compared to gains during decision-making (i.e., loss aversion) because they erroneously expect losses to feel significantly worse than they actually do, leading to the persistent underestimation of option values (Kermer et al., 2006). This aligns with the broader idea that predictive processing specifically aims to anticipate the affective outcomes associated with future choice to facilitate decision-making (Barrett & Simmons, 2015; Marshall et al., 2019; Seth & Friston, 2016; Wilson & Gilbert, 2005).

However, persistent biases like those seen in loss aversion pose a quandary for classic RL (Baumeister et al., 2001; Gershman, 2015; Lefebvre et al., 2017; Sharot et al., 2011; Taylor, 1991). There are no assumptions in RL that suggest there should be asymmetrical learning depending on whether the update is from a reward or punishment. Yet, the literature finds patterns of asymmetrical updating, sometimes from positive prediction errors when people are optimistic (Lefebvre et al., 2017; Sharot et al., 2011), and other times from negative prediction errors when people are pessimistic (Baumeister et al., 2001; Gershman, 2015; Taylor, 1991). Why do these asymmetric updating patterns emerge? Theories of how we appraise our emotions offer a few clues (Scherer & Moors, 2019). Appraisals about the broader decision-making context can determine what affective states result from an outcome, in turn shaping how people learn from those outcomes. For example, people's beliefs about the causal structure of their environment determine whether positive or negative prediction errors are over-, under-, or equivalently weighted (Dorfman et al., 2019). To make it concrete, people might experience a monetary loss when interacting with another as more emotionally aversive if they believe the action to be intentional rather than accidental. This could lead to selectively updating only on the evaluation that there is malintent because of the uniquely negative affective state it generates.

Affective scientists have also long emphasized that these predictions shape not only the choices people make but also their subjective experiences of the outcomes of those choices (Houlihan et al., 2023; Otto & Eichstaedt, 2018; Verinis et al., 1968). The interaction between prediction and affect is strikingly illustrated in a recent study where students' responses were measured after receiving exam grades in the real world (Villano et al., 2020). Results found that students' emotional responses were driven by the gap between their expectations and reality, even more than the objective grade they received. This suggests that affective responses might be endogenous instantiations of "prediction errors" used for updating our models of the world (Houlihan et al., 2023; Huys & Renz, 2017). Put simply, prediction errors for value learning can be conceptualized as wholly affective in nature, capturing the deviation between affective forecasts and actual affective responses generated by the outcomes of choices.

A more recent test of this idea found that mathematically modeling affective prediction errors—by taking the difference between how one expects to feel and how one actually feels—not only predicts choice but does so better than classic reward prediction errors (in this case, formalized by money) (Heffner et al., 2021). Specifically, across a series of studies, Heffner et al. (2021) examined how different types of prediction errors are associated with punitive behavior in different economic games (Fig. 2). By decomposing affective predictions and affective experiences into their principal components of valence and arousal, the studies found that social decisions were primarily driven by valence prediction errors. Although classic reward prediction errors tracking the discrepancy between expected and actual monetary outcomes did predict people's punishment decisions, they were significantly less predictive than valence PEs. This suggests that affective experiences do play an outsized role in updating our models of the world. It has also been argued that affective error signals can do more than just evaluate the current options: Affective error signals appear to get integrated over time to produce longer-lasting moods that reflect global evaluations of the environment (Otto & Eichstaedt, 2018), which in turn can be used to make predictions about subsequent options (Eldar et al., 2016; Rutledge et al., 2014). This nascent but growing body of

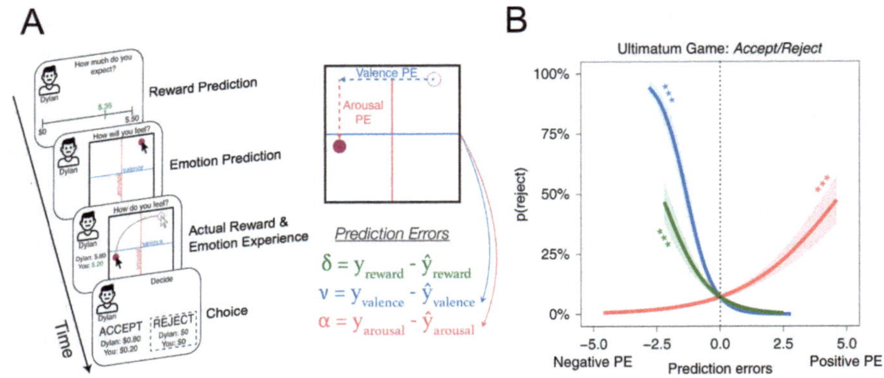

Fig. 2 Affective prediction errors drive value-based social decisions. (Reproduced in part with permission from Heffner et al. (2021, *Nature Human Behavior*)). (**a**) Participants completed the Ultimatum Game (UG) where they encountered unfair distributions of money and could choose whether to punish their partners. (**b**) The lines on each graph reflect the relationship between three kinds of prediction errors (PE) and the probability of punishment in the UG, where punishment penalizes participants' partner at a cost to the participant. The color of each line indicates the kind of PE predicting behavior: reward (green), arousal (red), and valence (blue). Negative values reflect negative PEs, indicating less money (reward), less pleasantness (valence), and less arousal than expected. Shaded areas reflect ±1 s.e. $^{***}P < 0.001$, $^{**}P < 0.01$, $^{*}P < 0.05$

work is beginning to form a sketch for the possible computations and algorithmic mechanisms of how affective states construct the value of a choice.

5 Future Research and Unanswered Questions

Over many decades, scientists, theorists, and philosophers have challenged the conventional notion of rationality (Helm, 2002), opening up new and exciting avenues for understanding human behavior (Frijda, 2001). While emotion and affect are now central to the story of human decision-making (Lerner et al., 2015), much is still unknown about how exactly (Scherer, 2009; Emanuel & Eldar, 2023) affect drives the construction of value across the levels of computation, algorithm, and biologically plausible implementation (Bach & Dayan, 2017). For example, is affect an entirely separate and distinct signal that shapes choice, or does it subsume other signals that are commonly believed to be *the* drivers of choice (i.e., most notably "reward," but also concepts like pleasure and hedonic value)? One possibility is that affect provides a separate signal that works in conjunction with a reward signal to shape value (Blain & Rutledge, 2020; Heffner et al., 2021; Smith et al., 2010). It remains an open question whether affect and reward signals are then integrated and mapped to a single scale for comparison (Levy & Glimcher, 2012), or distinctly encoded to produce value (Chib et al., 2009). Moreover, in order to understand how affect and emotion shape value, we must also understand what emotions represent about the environment (Bennett et al., 2022). While much remains to be learned about how the brain encodes these signals, new multivariate methods that enable the decoding of representational content (Kriegeskorte, 2011; Kriegeskorte & Bandettini, 2007; Kriegeskorte & Diedrichsen, 2019) have created tantalizing opportunities to probe the neural representations of affect, reward, and value and to explore whether putatively different signals (e.g., emotion, reward) are linked to distinct representational content. Adopting this level of analysis is challenging given the amorphous nature of emotion. However, it will enable scientists to finally shed light on what is happening inside the black box of affective evaluation.

References

Admon, R., & Pizzagalli, D. A. (2015). Dysfunctional reward processing in depression. *Current Opinion in Psychology, 4*, 114–118. https://doi.org/10.1016/j.copsyc.2014.12.011

Adolphs, R., Mlodinow, L., & Barrett, L. F. (2019). What is an emotion? *Current Biology, 29*(20), R1060–R1064. https://doi.org/10.1016/j.cub.2019.09.008

Ashby, F. G., Prinzmetal, W., Ivry, R., & Maddox, W. T. (1996). A formal theory of feature binding in object perception. *Psychological Review, 103*(1), 165–192. https://doi.org/10.1037/0033-295X.103.1.165

Aylward, J., Valton, V., Ahn, W.-Y., Bond, R. L., Dayan, P., Roiser, J. P., & Robinson, O. J. (2019). Altered learning under uncertainty in unmedicated mood and anxiety disorders. *Nature Human Behaviour, 3*(10), 1116–1123. https://doi.org/10.1038/s41562-019-0628-0

Azzalini, D., Buot, A., Palminteri, S., & Tallon-Baudry, C. (2021). Responses to heartbeats in ventromedial prefrontal cortex contribute to subjective preference-based decisions. *Journal of Neuroscience, 41*(23), 5102–5114. https://doi.org/10.1523/JNEUROSCI.1932-20.2021

Bach, D. R., & Dayan, P. (2017). Algorithms for survival: A comparative perspective on emotions. *Nature Reviews Neuroscience, 18*(5), 311–319.

Bar, M. (2004). Visual objects in context. *Nature Reviews Neuroscience, 5*(8), 617–629. https://doi.org/10.1038/nrn1476

Barrett, L. F. (2017). The theory of constructed emotion: An active inference account of interoception and categorization. *Social Cognitive and Affective Neuroscience, 12*(1), 1–23. https://doi.org/10.1093/scan/nsw154

Barrett, L. F., & Bliss-Moreau, E. (2009). Affect as a psychological primitive. In *Advances in experimental social psychology* (Vol. 41, pp. 167–218). Academic. https://doi.org/10.1016/S0065-2601(08)00404-8

Barrett, L. F., & Simmons, W. K. (2015). Interoceptive predictions in the brain. *Nature Reviews Neuroscience, 16*(7), 419–429. https://doi.org/10.1038/nrn3950

Barrett, L. F., & Wager, T. D. (2006). The structure of emotion: Evidence from neuroimaging studies. *Current Directions in Psychological Science, 15*(2), 79–83. https://doi.org/10.1111/j.0963-7214.2006.00411.x

Baumeister, R. F., Bratslavsky, E., Finkenauer, C., & Vohs, K. D. (2001). Bad is stronger than good. *Review of General Psychology, 5*(4), 323–370.

Beltzer, M. L., Adams, S., Beling, P. A., & Teachman, B. A. (2019). Social anxiety and dynamic social reinforcement learning in a volatile environment. *Clinical Psychological Science, 7*(6), 1372–1388. https://doi.org/10.1177/2167702619858425

Bennett, D., Davidson, G., & Niv, Y. (2022). A model of mood as integrated advantage. *Psychological Review, 129*(3), 513–541. https://doi.org/10.1037/rev0000294

Bernoulli, D. (1954). Exposition of a new theory on the measurement of risk. *Econometrica, 22*(1), 23–36. https://doi.org/10.2307/1909829

Berridge, K. C., & Aldridge, J. W. (2008). Decision utility, the brain, and pursuit of hedonic goals. *Social Cognition, 26*(5), 621–646.

Berridge, K. C., & O'Doherty, J. P. (2014). From experienced utility to decision utility. In P. W. Glimcher & E. Fehr (Eds.), *Neuroeconomics* (2nd ed., pp. 335–351). Academic. https://doi.org/10.1016/B978-0-12-416008-8.00018-8

Birbaumer, N., & Öhman, A. (1993). *The structure of emotion: Psychophysiological, cognitive, and clinical aspects*. Hogrefe & Huber. https://books.google.com/books?id=3Z1kQgAACAAJ

Blain, B., & Rutledge, R. B. (2020). Momentary subjective well-being depends on learning and not reward. *eLife, 9*, e57977. https://doi.org/10.7554/eLife.57977

Bradley, M. M., & Lang, P. J. (2000). Measuring emotion: Behavior, feeling, and physiology. In R. D. Lane & L. Nadel (Eds.), *Cognitive neuroscience of emotion* (pp. 242–276). Oxford University Press.

Cabanac, M. (1971). Physiological Role of Pleasure. *Science, 173*(4002), 1103–1107. https://doi.org/10.1126/science.173.4002.1103

Chib, V. S., Rangel, A., Shimojo, S., & O'Doherty, J. P. (2009). Evidence for a common representation of decision values for dissimilar goods in human ventromedial prefrontal cortex. *Journal of Neuroscience, 29*(39), 12315–12320. https://doi.org/10.1523/JNEUROSCI.2575-09.2009

Clore, G. L., & Ortony, A. (2013). Psychological construction in the OCC model of emotion. *Emotion Review, 5*(4), 335–343.

Cooper, J. A., Arulpragasam, A. R., & Treadway, M. T. (2018). Anhedonia in depression: Biological mechanisms and computational models. *Current Opinion in Behavioral Sciences, 22*, 128–135. https://doi.org/10.1016/j.cobeha.2018.01.024

Critchley, H. D., & Garfinkel, S. N. (2017). Interoception and emotion. *Current Opinion in Psychology, 17*, 7–14. https://doi.org/10.1016/j.copsyc.2017.04.020

Dalgleish, T., & Power, M. (1997). *Cognition and emotion: From order to disorder*. Psychology Press.

de Araujo, I. E., Schatzker, M., & Small, D. M. (2020). Rethinking food reward. *Annual Review of Psychology, 71*, 139–164. https://doi.org/10.1146/annurev-psych-122216-011643

De Martino, B., Kumaran, D., Seymour, B., & Dolan, R. J. (2006). Frames, biases, and rational decision-making in the human brain. *Science, 313*(5787), 684–687.

Dorfman, H. M., Bhui, R., Hughes, B. L., & Gershman, S. J. (2019). Causal inference about good and bad outcomes. *Psychological Science, 30*(4), 516–525. https://doi.org/10.1177/0956797619828724

Dukes, D., Abrams, K., Adolphs, R., Ahmed, M. E., Beatty, A., Berridge, K. C., Broomhall, S., Brosch, T., Campos, J. J., Clay, Z., Clément, F., Cunningham, W. A., Damasio, A., Damasio, H., D'Arms, J., Davidson, J. W., de Gelder, B., Deonna, J., de Sousa, R., et al. (2021). The rise of affectivism. *Nature Human Behaviour, 5*(7), 816–820. https://doi.org/10.1038/s41562-021-01130-8

Duncan, S., & Barrett, L. F. (2007). Affect is a form of cognition: A neurobiological analysis. *Cognition and Emotion, 21*(6), 1184–1211. https://doi.org/10.1080/02699930701437931

Eldar, E., Rutledge, R. B., Dolan, R. J., & Niv, Y. (2016). Mood as representation of momentum. *Trends in Cognitive Sciences, 20*(1), 15–24. https://doi.org/10.1016/j.tics.2015.07.010

Ellsworth, P. C. (2013). Appraisal theory: Old and new questions. *Emotion Review, 5*(2), 125–131. https://doi.org/10.1177/1754073912463617

Emanuel, A., & Eldar, E. (2023). Emotions as computations. *Neuroscience & Biobehavioral Reviews, 144*, 104977. https://doi.org/10.1016/j.neubiorev.2022.104977

Evans, J. St. B. T, & Stanovich, K. E. (2013). Dual-process theories of higher cognition: Advancing the debate. *Perspectives on Psychological Science, 8*(3), 223–241. https://doi.org/10.1177/1745691612460685

Fecteau, S., Pascual-Leone, A., Zald, D. H., Liguori, P., Théoret, H., Boggio, P. S., & Fregni, F. (2007). Activation of prefrontal cortex by transcranial direct current stimulation reduces appetite for risk during ambiguous decision making. *The Journal of Neuroscience, 27*(23), 6212. https://doi.org/10.1523/JNEUROSCI.0314-07.2007

FeldmanHall, O., & Chang, L. J. (2018). Social learning: Emotions aid in optimizing goal-directed social behavior. In R. Morris, A. Bornstein, & A. Shenhav (Eds.), *Goal-directed decision making* (pp. 309–330). Academic. https://doi.org/10.1016/B978-0-12-812098-9.00014-0

FeldmanHall, O., & Nassar, M. R. (2021). The computational challenge of social learning. *Trends in Cognitive Sciences, 25*(12), 1045–1057. https://doi.org/10.1016/j.tics.2021.09.002

FeldmanHall, O., Raio, C. M., Kubota, J. T., Seiler, M. G., & Phelps, E. A. (2015). The effects of social context and acute stress on decision making under uncertainty. *Psychological Science, 26*(12), 1918–1926.

FeldmanHall, O., Glimcher, P., Baker, A. L., NYU PROSPEC Collaboration, & Phelps, E. A. (2019). The functional roles of the amygdala and prefrontal cortex in processing uncertainty. *Journal of Cognitive Neuroscience, 31*(11), 1742–1754. https://doi.org/10.1162/jocn_a_01443

Fernandez, E., & Johnson, S. L. (2016). Anger in psychological disorders: Prevalence, presentation, etiology and prognostic implications. *Clinical Psychology Review, 46*, 124–135. https://doi.org/10.1016/j.cpr.2016.04.012

Frijda, N. H. (2001). The nature of pleasure. In *Unraveling the complexities of social life: A festschrift in honor of Robert B. Zajonc* (pp. 71–94). American Psychological Association. https://doi.org/10.1037/10387-006

Garg, N., Wansink, B., & Inman, J. J. (2007). The influence of incidental affect on consumers' food intake. *Journal of Marketing, 71*(1), 194–206.

Gershman, S. J. (2015). Do learning rates adapt to the distribution of rewards? *Psychonomic Bulletin & Review, 22*(5), 1320–1327. https://doi.org/10.3758/s13423-014-0790-3

Hammond, D., Xu, P., Ai, H., & Van Dam, N. T. (2023). Anxiety and depression related abnormalities in socio-affective learning. *Journal of Affective Disorders, 335*, 322–331. https://doi.org/10.1016/j.jad.2023.05.021

Hayes, W. M., & Wedell, D. H. (2020). Autonomic responses to choice outcomes: Links to task performance and reinforcement-learning parameters. *Biological Psychology, 156*, 107968. https://doi.org/10.1016/j.biopsycho.2020.107968

Heffner, J., Son, J.-Y., & FeldmanHall, O. (2021). Emotion prediction errors guide socially adaptive behaviour. *Nature Human Behaviour, 5*(10), 1391–1401. https://doi.org/10.1038/s41562-021-01213-6

Helm, B. W. (2002). Felt evaluations: A theory of pleasure and pain. *American Philosophical Quarterly, 39*(1), 13–30.

Houben, M., & Kuppens, P. (2020). Emotion dynamics and the association with depressive features and borderline personality disorder traits: Unique, specific, and prospective relationships. *Clinical Psychological Science, 8*(2), 226–239. https://doi.org/10.1177/2167702619871962

Houlihan, S. D., Kleiman-Weiner, M., Hewitt, L. B., Tenenbaum, J. B., & Saxe, R. (2023). Emotion prediction as computation over a generative theory of mind. *Philosophical Transactions of the Royal Society A, 381*(2251), 20220047.

Hsu, M., Bhatt, M., Adolphs, R., Tranel, D., & Camerer, C. F. (2005). Neural systems responding to degrees of uncertainty in human decision-making. *Science, 310*(5754), 1680–1683. https://doi.org/10.1126/science.1115327

Huys, Q. J. M., & Renz, D. (2017). A formal valuation framework for emotions and their control. *Biological Psychiatry, 82*(6), 413–420. https://doi.org/10.1016/j.biopsych.2017.07.003

Huys, Q. J. M., Pizzagalli, D. A., Bogdan, R., & Dayan, P. (2013). Mapping anhedonia onto reinforcement learning: A behavioural meta-analysis. *Biology of Mood & Anxiety Disorders, 3*(1), 12. https://doi.org/10.1186/2045-5380-3-12

Jacobson, H. (2021). The role of valence in perception: An ARTistic treatment. *The Philosophical Review, 130*(4), 481–531. https://doi.org/10.1215/00318108-9263939

Johnson, E. J., & Tversky, A. (1983). Affect, generalization, and the perception of risk. *Journal of Personality and Social Psychology, 45*(1), 20–31. https://doi.org/10.1037/0022-3514.45.1.20

Jovanovic, L., Trichanh, M., Martin, B., & Giersch, A. (2023). Strong perceptual consequences of low-level visual predictions: A new illusion. *Cognition, 230*, 105279. https://doi.org/10.1016/j.cognition.2022.105279

Juechems, K., & Summerfield, C. (2019). Where does value come from? *Trends in Cognitive Sciences, 23*(10), 836–850. https://doi.org/10.1016/j.tics.2019.07.012

Kable, J. W., & Glimcher, P. W. (2007). The neural correlates of subjective value during intertemporal choice. *Nature Neuroscience, 10*(12), 1625–1633. https://doi.org/10.1038/nn2007

Kahneman, D. (2011). *Thinking, fast and slow* (p. 499). Farrar, Straus and Giroux.

Kahneman, D., & Tversky, A. (1979). Prospect theory: An analysis of decision making under risk. *Econometrica, 47*, 263–292.

Kahneman, D., Wakker, P. P., & Sarin, R. (1997). Back to Bentham? Explorations of experienced utility. *The Quarterly Journal of Economics, 112*(2), 375–406.

Keramati, M., & Gutkin, B. (2014). Homeostatic reinforcement learning for integrating reward collection and physiological stability. *eLife, 3*, e04811. https://doi.org/10.7554/eLife.04811

Kermer, D. A., Driver-Linn, E., Wilson, T. D., & Gilbert, D. T. (2006). Loss aversion is an affective forecasting error. *Psychological Science, 17*(8), 649–653. https://doi.org/10.1111/j.1467-9280.2006.01760.x

Khaw, M. W., Grab, D. A., Livermore, M. A., Vossler, C. A., & Glimcher, P. W. (2015). The measurement of subjective value and its relation to contingent valuation and environmental public goods. *PLoS One, 10*(7), e0132842. https://doi.org/10.1371/journal.pone.0132842

Kreibig, S. D. (2010). Autonomic nervous system activity in emotion: A review. *Biological Psychology, 84*(3), 394–421. https://doi.org/10.1016/j.biopsycho.2010.03.010

Kriegeskorte, N. (2011). Pattern-information analysis: From stimulus decoding to computational-model testing. *NeuroImage, 56*(2), 411–421. https://doi.org/10.1016/j.neuroimage.2011.01.061

Kriegeskorte, N., & Bandettini, P. (2007). Analyzing for information, not activation, to exploit high-resolution fMRI. *NeuroImage, 38*(4), 649–662. https://doi.org/10.1016/j.neuroimage.2007.02.022

Kriegeskorte, N., & Diedrichsen, J. (2019). Peeling the onion of brain representations. *Annual Review of Neuroscience, 42*, 407–432. https://doi.org/10.1146/annurev-neuro-080317-061906

Lamba, A., Frank, M. J., & FeldmanHall, O. (2020). Anxiety impedes adaptive social learning under uncertainty. *Psychological Science*. https://doi.org/10.1177/0956797620910993

Lamba, A., Frank, M. J., & FeldmanHall, O. (2024). Keeping an eye out for change: Anxiety disrupts adaptive resolution of policy uncertainty. *Biological Psychiatry: Cognitive Neuroscience and Neuroimaging, 9*(11), 1188–1198. https://doi.org/10.1016/j.bpsc.2024.07.015

LeDoux, J. E. (2000). Emotion circuits in the brain. *Annual Review of Neuroscience, 23*, 155–184. https://doi.org/10.1146/annurev.neuro.23.1.155

Lefebvre, G., Lebreton, M., Meyniel, F., Bourgeois-Gironde, S., & Palminteri, S. (2017). Behavioural and neural characterization of optimistic reinforcement learning. *Nature Human Behaviour, 1*(4), 1–9. https://doi.org/10.1038/s41562-017-0067

Lennie, P. (1998). Single units and visual cortical organization. *Perception, 27*(8), 889–935. https://doi.org/10.1068/p270889

Lerner, J. S., & Keltner, D. (2001). Fear, anger, and risk. *Journal of Personality and Social Psychology, 81*, 146–159. https://doi.org/10.1037/0022-3514.81.1.146

Lerner, J. S., Li, Y., Valdesolo, P., & Kassam, K. S. (2015). Emotion and decision making. *Annual Review of Psychology, 66*(1), 799–823.

Levy, D. J., & Glimcher, P. W. (2011). Comparing apples and oranges: Using reward-specific and reward-general subjective value representation in the brain. *Journal of Neuroscience, 31*(41), 14693–14707. https://doi.org/10.1523/JNEUROSCI.2218-11.2011

Levy, D. J., & Glimcher, P. W. (2012). The root of all value: A neural common currency for choice. *Current Opinion in Neurobiology, 22*(6), 1027–1038. https://doi.org/10.1016/j.conb.2012.06.001

Levy, I., Snell, J., Nelson, A. J., Rustichini, A., & Glimcher, P. W. (2010). Neural representation of subjective value under risk and ambiguity. *Journal of Neurophysiology, 103*(2), 1036–1047. https://doi.org/10.1152/jn.00853.2009

Lieder, F., Griffiths, T. L., & Hsu, M. (2018). Overrepresentation of extreme events in decision making reflects rational use of cognitive resources. *Psychological Review, 125*(1), 1.

Loewenstein, G. F., Weber, E. U., Hsee, C. K., & Welch, N. (2001). Risk as feelings. *Psychological Bulletin, 127*(2), 267–286. https://doi.org/10.1037/0033-2909.127.2.267
Loomes, G., & Sugden, R. (1982). Regret theory: An alternative theory of rational choice under uncertainty. *The Economic Journal, 92*(368), 805–824. https://doi.org/10.2307/2232669
Louie, K., & Glimcher, P. W. (2010). Separating value from choice: Delay discounting activity in the lateral intraparietal area. *Journal of Neuroscience, 30*(16), 5498–5507. https://doi.org/10.1523/JNEUROSCI.5742-09.2010
Louie, K., & Glimcher, P. W. (2012). Efficient coding and the neural representation of value. *Annals of the New York Academy of Sciences, 1251*(1), 13–32. https://doi.org/10.1111/j.1749-6632.2012.06496.x
Löw, A., Lang, P. J., Smith, J. C., & Bradley, M. M. (2008). Both predator and prey: Emotional arousal in threat and reward. *Psychological Science, 19*(9), 865–873. https://doi.org/10.1111/j.1467-9280.2008.02170.x
Lucarelli, C., Uberti, P., Brighetti, G., & Maggi, M. (2015). Risky choices and emotion-based learning. *Journal of Economic Psychology, 49*, 59–73. https://doi.org/10.1016/j.joep.2015.04.004
Malezieux, M., Klein, A. S., & Gogolla, N. (2023). Neural circuits for emotion. *Annual Review of Neuroscience, 46*, 211–231. https://doi.org/10.1146/annurev-neuro-111020-103314
Marshall, A. C., Gentsch, A., Blum, A.-L., Broering, C., & Schütz-Bosbach, S. (2019). I feel what I do: Relating interoceptive processes and reward-related behavior. *NeuroImage, 191*, 315–324. https://doi.org/10.1016/j.neuroimage.2019.02.032
Mayer, E. A. (2011). Gut feelings: The emerging biology of gut–brain communication. *Nature Reviews Neuroscience, 12*(8), 453–466. https://doi.org/10.1038/nrn3071
Mellers, B. (2000). Choice and the relative pleasure of consequences. *Psychological Bulletin, 126*(6), 910–924. https://doi.org/10.1037/0033-2909.126.6.910
Mellers, B., Schwartz, A., & Ritov, I. (1999). Emotion-based choice. *Journal of Experimental Psychology: General, 128*(3), 332–345. https://doi.org/10.1037/0096-3445.128.3.332
Moors, A., Ellsworth, P. C., Scherer, K. R., & Frijda, N. H. (2013). Appraisal theories of emotion: State of the art and future development. *Emotion Review, 5*(2), 119–124. https://doi.org/10.1177/1754073912468165
Morgado, P., Sousa, N., & Cerqueira, J. J. (2015). The impact of stress in decision making in the context of uncertainty. *Journal of Neuroscience Research, 93*(6), 839–847.
O'Toole, M. S., Renna, M., Elkjær, E., Mikkelsen, M. B., & Mennin, D. S. (2020). A systematic review and meta-analysis of the association between complexity of emotion experience and behavioral adaptation. *Emotion Review, 12*(1), 23–38. https://doi.org/10.1177/1754073919876019
Ortony, A., Clore, G. L., & Collins, A. (1988). *The cognitive structure of emotions*. Cambridge University Press/Cambridge Core. https://doi.org/10.1017/CBO9780511571299
Otto, A. R., & Eichstaedt, J. C. (2018). Real-world unexpected outcomes predict city-level mood states and risk-taking behavior. *PLoS One, 13*(11), e0206923.
Pfeifer, G., Garfinkel, S. N., Gould van Praag, C. D., Sahota, K., Betka, S., & Critchley, H. D. (2017). Feedback from the heart: Emotional learning and memory is controlled by cardiac cycle, interoceptive accuracy and personality. *Biological Psychology, 126*, 19–29. https://doi.org/10.1016/j.biopsycho.2017.04.001
Phan, K. L., Wager, T. D., Taylor, S. F., & Liberzon, I. (2004). Functional neuroimaging studies of human emotions. *CNS Spectrums, 9*(4), 258–266. https://doi.org/10.1017/S1092852900009196
Phelps, E. A., Lempert, K. M., & Sokol-Hessner, P. (2014). Emotion and decision making: Multiple modulatory neural circuits. *Annual Review of Neuroscience, 37*(1), 263–287. https://doi.org/10.1146/annurev-neuro-071013-014119
Phillips, M. L., Young, A. W., Senior, C., Brammer, M., Andrew, C., Calder, A. J., Bullmore, E. T., Perrett, D. I., Rowland, D., Williams, S. C. R., Gray, J. A., & David, A. S. (1997). A specific neural substrate for perceiving facial expressions of disgust. *Nature, 389*(6650), 495–498. https://doi.org/10.1038/39051
Porcelli, A. J., & Delgado, M. R. (2017). Stress and decision making: Effects on valuation, learning, and risk-taking. *Current Opinion in Behavioral Sciences, 14*, 33–39.

Pulcu, E., & Browning, M. (2019). The misestimation of uncertainty in affective disorders. *Trends in Cognitive Sciences, 23*(10), 865–875. https://doi.org/10.1016/j.tics.2019.07.007

Quartz, S. R. (2009). Reason, emotion and decision-making: Risk and reward computation with feeling. *Trends in Cognitive Sciences, 13*(5), 209–215. https://doi.org/10.1016/j.tics.2009.02.003

Rabinak, C. A., Mori, S., Lyons, M., Milad, M. R., & Phan, K. L. (2017). Acquisition of CS-US contingencies during Pavlovian fear conditioning and extinction in social anxiety disorder and posttraumatic stress disorder. *Journal of Affective Disorders, 207*, 76–85. https://doi.org/10.1016/j.jad.2016.09.018

Rangel, A., & Hare, T. (2010). Neural computations associated with goal-directed choice. *Current Opinion in Neurobiology, 20*(2), 262–270. https://doi.org/10.1016/j.conb.2010.03.001

Rangel, A., Camerer, C., & Montague, P. R. (2008). A framework for studying the neurobiology of value-based decision making. *Nature Reviews Neuroscience, 9*(7), 545–556. https://doi.org/10.1038/nrn2357

Rolls, E. T. (2013). What are emotional states, and why do we have them? *Emotion Review, 5*(3), 241–247. https://doi.org/10.1177/1754073913477514

Russell, J. A. (1980). A circumplex model of affect. *Journal of Personality and Social Psychology, 39*(6), 1161–1178. https://doi.org/10.1037/h0077714

Russo, S. J., & Nestler, E. J. (2013). The brain reward circuitry in mood disorders. *Nature Reviews Neuroscience, 14*(9), 609–625. https://doi.org/10.1038/nrn3381

Rutledge, R. B., Skandali, N., Dayan, P., & Dolan, R. J. (2014). A computational and neural model of momentary subjective well-being. *Proceedings of the National Academy of Sciences, 111*(33), 12252–12257.

Samuelson, P. A. (1952). Probability, utility, and the independence axiom. *Econometrica, 20*(4), 670–678. https://doi.org/10.2307/1907649

Sander, D., & Nummenmaa, L. (2021). Reward and emotion: An affective neuroscience approach. *Current Opinion in Behavioral Sciences, 39*, 161–167. https://doi.org/10.1016/j.cobeha.2021.03.016

Scherer, K. R. (2009). Emotions are emergent processes: They require a dynamic computational architecture. *Philosophical Transactions of the Royal Society, B: Biological Sciences, 364*(1535), 3459–3474. https://doi.org/10.1098/rstb.2009.0141

Scherer, K. R., & Moors, A. (2019). The emotion process: Event appraisal and component differentiation. *Annual Review of Psychology, 70*(1), 719–745. https://doi.org/10.1146/annurev-psych-122216-011854

Seah, T. H. S., & Coifman, K. G. (2022). Emotion differentiation and behavioral dysregulation in clinical and nonclinical samples: A meta-analysis. *Emotion, 22*(7), 1686–1697. https://doi.org/10.1037/emo0000968

Seth, A. K., & Friston, K. J. (2016). Active interoceptive inference and the emotional brain. *Philosophical Transactions of the Royal Society, B: Biological Sciences, 371*(1708), 20160007. https://doi.org/10.1098/rstb.2016.0007

Sharot, T., Korn, C. W., & Dolan, R. J. (2011). How unrealistic optimism is maintained in the face of reality. *Nature Neuroscience, 14*(11), 1475–1479. https://doi.org/10.1038/nn.2949

Simon, H. A. (1956). Rational choice and the structure of the environment. *Psychological Review, 63*(2), 129.

Simon, H. A. (1997). *Models of bounded rationality: Empirically grounded economic reason* (Vol. 3). MIT Press.

Smith, D. V., Hayden, B. Y., Truong, T.-K., Song, A. W., Platt, M. L., & Huettel, S. A. (2010). Distinct value signals in anterior and posterior ventromedial prefrontal cortex. *Journal of Neuroscience, 30*(7), 2490–2495. https://doi.org/10.1523/JNEUROSCI.3319-09.2010

Sutton, R. S., & Barto, A. G. (1999). Reinforcement learning. *Journal of Cognitive Neuroscience, 11*(1), 126–134.

Tamir, M. (2016). Why do people regulate their emotions? A taxonomy of motives in emotion regulation. *Personality and Social Psychology Review, 20*(3), 199–222. https://doi.org/10.1177/1088868315586325

Taylor, S. E. (1991). Asymmetrical effects of positive and negative events: The mobilization-minimization hypothesis. *Psychological Bulletin, 110*(1), 67.

Tobler, P. N., Christopoulos, G. I., O'Doherty, J. P., Dolan, R. J., & Schultz, W. (2009). Risk-dependent reward value signal in human prefrontal cortex. *Proceedings of the National Academy of Sciences, 106*(17), 7185–7190. https://doi.org/10.1073/pnas.0809599106

Treisman, A., & Schmidt, H. (1982). Illusory conjunctions in the perception of objects. *Cognitive Psychology, 14*(1), 107–141. https://doi.org/10.1016/0010-0285(82)90006-8

Verinis, J. S., Brandsma, J. M., & Cofer, C. N. (1968). Discrepancy from expectation in relation to affect and motivation: Tests of McClelland's hypothesis. *Journal of Personality and Social Psychology, 9*(1), 47–58. https://doi.org/10.1037/h0025672

Villano, W., Otto, A., Ezie, C., Gillis, R., & Heller, A. (2020). Temporal dynamics of real-world emotion are more strongly linked to prediction error than outcome. *Journal of Experimental Psychology: General, 149*(9), 1755–1766. https://doi.org/10.1037/xge0000740

Von Neumann, J., & Morgenstern, O. (1944). *Theory of games and economic behavior*. (pp. xviii, 625. Princeton University Press.

Walasek, L., & Brown, G. D. A. (2023). Incomparability and incommensurability in choice: No common currency of value? *Perspectives on Psychological Science*, 17456916231192828. https://doi.org/10.1177/17456916231192828

Wilson, T. D., & Gilbert, D. T. (2005). Affective forecasting: Knowing what to want. *Current Directions in Psychological Science, 14*(3), 131–134. https://doi.org/10.1111/j.0963-7214.2005.00355.x

Yamada, H., Louie, K., Tymula, A., & Glimcher, P. W. (2018). Free choice shapes normalized value signals in medial orbitofrontal cortex. *Nature Communications, 9*(1), 162. https://doi.org/10.1038/s41467-017-02614-w

Modeling the Social Brain: Neurocomputational Approaches to Interpersonal Learning and Decision-Making

Dominic S. Fareri and Jordan M. Dejoie

Abstract Learning and decision-making are deeply embedded in our social world. Our choices occur in the presence of others, are shaped by social expectations, and can have significant impacts on both our and others' welfare. While these topics have been of interest to psychology and cognitive neuroscience for many years, recent efforts have incorporated model-based approaches to formalize the mechanisms through which our choices are influenced by social input. Combined with fMRI, our understanding of the role of neural circuits involved in reward-related behavior and social cognition in supporting such mechanisms has grown exponentially. In this chapter, we will first discuss the utility of incorporating computational approaches into studies of social learning and decision-making. Next, we will review recent evidence regarding how social outcomes can impact our choices and social learning. We will then discuss how social rejection can impact our desire for agency and social contact. We also discuss the importance of social sharing in real-time interactions and social media environments, and how this phenomenon fits into a framework of value-based choice that can shape social connection. Last, we will review recent work regarding how the value of social relationships shapes computational mechanisms underlying decisions to trust and take risks. We will conclude with future directions for the field, including merging computational approaches with individual differences, cutting-edge methodologies, and naturalistic paradigms to provide a more comprehensive picture of how the social world modulates choice and neural function.

1 Introduction

Our decisions are shaped by a combination of signals from internal states and external factors. For example, deciding where to go for lunch may depend on how hungry we are, our available options, and how much time we have. All these factors

D. S. Fareri (✉) · J. M. Dejoie
Gordon F. Derner School of Psychology, Adelphi University, Garden City, NY, USA
e-mail: dfareri@adelphi.edu

D. V. Smith et al. (eds.), *Neuroeconomics: Core Topics and Current Directions*,
https://doi.org/10.1007/978-3-032-02925-6_22

may contribute differential weight in helping us come to a final decision. Yet, we rarely act within a vacuum of isolation. Rather, our choices are frequently embedded within social environments and potentially shaped by myriad social factors (i.e., with whom will I go to eat lunch? What are their preferences? Do we like/want the same things?). Indeed, the contexts created by our social world influence many cognitive and emotional processes, particularly learning and decision-making. For example, the presence and influence of peers, authority figures, and abstract social norms can significantly alter how we perceive the value of different choice options. Further, social feedback in the form of praise or criticism of our performance, appearance, or preferences can shape our decisions by reinforcing specific behaviors or prompting course corrections. The social environment thus plays a critical role in guiding multiple facets of behavior, ultimately helping us to form relationships with others.

While the importance of the social world to decision-making has been a focus of study for some time now in psychology and cognitive neuroscience, efforts to formalize the mechanisms through which social context shapes choice have only recently begun to emerge. Incorporation of computational modeling approaches has become a powerful tool for quantifying and exploring social influences on decision-making. Such approaches allow for the formalized mapping of complex social inputs—such as the presence of competitors or partners, models of others' cooperation, or shared goals—onto processes that guide behavior and learning (Konovalov et al., 2018; Lockwood & Klein-Flügge, 2021; Ruff & Fehr, 2014; Vahed et al., 2024).

In conjunction with neuroimaging, these techniques have begun to elucidate how social computational processes are instantiated at the neural level (FeldmanHall & Nassar, 2021; Konovalov et al., 2018; Lockwood et al., 2020; Stanley & Adolphs, 2013). Robust literature across species has implicated corticostriatal circuits in reward-based decision-making, including dorsal and ventral medial prefrontal cortex (dmPFC, vmPFC), anterior cingulate cortex (ACC), the striatum, and the dopaminergic midbrain (Bartra et al., 2013; Haber & Knutson, 2010) (see also Schultz, this volume, Chapter "Reward Learning and Choice: From Theory to Neuronal Signals"; Terry & Huettel, this volume, Chapter "Resolving Uncertainty: Foundational Neuroeconomic Studies of Risk and Ambiguity"). Importantly, in conjunction with putative social cognitive circuits (e.g., temporoparietal junction and posterior cingulate/precuneus), these regions are also commonly implicated in studies of social decision-making and learning (see Fig. 1). Indeed, human neuroimaging research suggests that the engagement of both circuits is sensitive to a variety of social cues in our environment (e.g., making choices in the presence of known vs unknown others, pursuing shared goals, (dis)approval from others, evaluating and incorporating advice from others) (Behrens et al., 2008; Delgado et al., 2023; Diaconescu et al., 2020; Lockwood et al., 2018; Rudebeck et al., 2006). Yet, the nature of the precise computational processes that are being carried out by discrete regions has until recently been underexplored.

This chapter will explore how the computational mechanisms through which social environment shapes decision-making processes and the function of under-

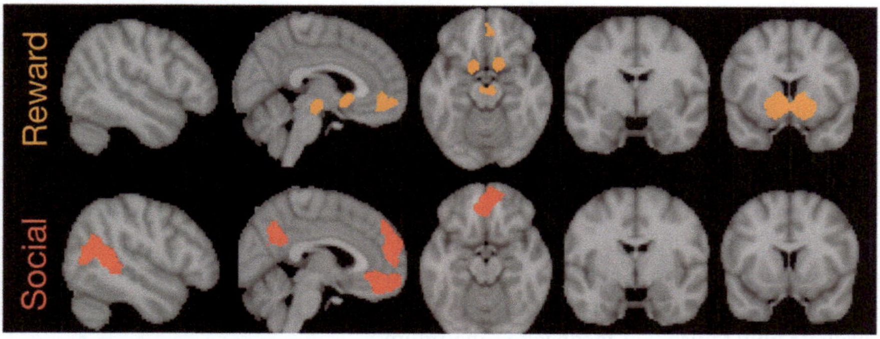

Fig. 1 Meta-analytic maps using the terms "reward" and "social" generated using Neurosynth. Maps show overlap particularly in medial prefrontal regions, indicating possible shared neural substrates for social and value-related information. (Adapted from Delgado et al., 2023)

lying neural architecture. We will first provide a brief overview of the utility of computational frameworks and model-based approaches for social decision-making. Next, we will discuss the multitude of ways in which feedback from others can shape behavior within social environments. We will then cover work highlighting how sharing of social information may be interpreted within a value-based decision-making space. Last, we will explore how model-based approaches can inform how specific social relationships modulate decision-making. We will conclude with considerations for future research in this space, with an eye toward combining computational approaches with other recent advances in social and decision neuroscience.

2 The Utility of Model-Based Approaches for Social Decision-Making

Traditional univariate approaches to data analysis have provided valuable insights into how individuals respond to social cues and what types of behaviors they exhibit within social environments. For example, we may be able to glean insight into differences in reactions to, or preferences for, cues exhibiting varied levels of trustworthiness or to finding out that someone likes us. However, such approaches do not often capture the complexity of real-world social interactions. Human decision-making is rarely linear or static; rather, it is shaped by dynamic social environments where feedback can be uncertain, delayed, or probabilistic, thus requiring online evaluation and adaptation to incoming social input. A key goal in navigating the social world is to reduce social uncertainty as a means to learn about others and adapt to changing social environments. Computational models have thus emerged as powerful tools to address these complexities, allowing researchers to quantify how individuals integrate social information and adapt behavior over time (FeldmanHall

& Shenhav, 2019; Vahed et al., 2024). The utility of these approaches lies in their ability to not only parameterize multiple component processes of learning and decision-making (i.e., prediction error, learning rate) but also operationalize psychological constructs (e.g., prior expectations, guilt aversion, loss aversion) within mathematical equations to understand the mechanisms that predict behavior (see also Schultz, this volume, Chapter "Reward Learning and Choice: From Theory to Neuronal Signals"; Wu & O'Doherty, this volume, Chapter "Neural Mechanisms of Reward, Decision-Making and Learning in a Social Context: Insights from Computational Neuroimaging"). These approaches are particularly important and useful in situations in which we receive feedback after a decision, needing to evaluate how to use that feedback in the future.

3 Social Outcomes Shape Learning and Experiences of Reward

Our actions within social environments are often met with a variety of types of feedback from others, including signals of social (dis)approval and reciprocity. Such feedback influences the impressions we form of others, how we feel about ourselves, and our willingness to share with others/interact with others. These outcomes are therefore crucial in our quest to reduce social uncertainty. Computational approaches centered on reinforcement learning frameworks (see Schultz, this volume, Chapter "Reward Learning and Choice: From Theory to Neuronal Signals") are frequently adapted to characterize social learning, using positive and negative social feedback (i.e., praise, criticism; reciprocity/violations of trust) to guide social expectations and inform choice.

Early work in this space focused on trust-based interactions as a model by which to characterize mechanisms of social learning (Fareri, 2019). Trust and cooperation are fundamental components of interpersonal interactions and societal systems that rely on managing risks, understanding others' expectations, and accounting for social norms. Choosing to trust requires dynamically updating initial appraisals of one's character based on patterns of experienced reciprocity (Chang et al., 2010). For example, in a trust game with fictional partners, reinforcement learning models demonstrated enhanced learning rates from trust game outcomes consistent with one's prior impressions of a partner (e.g., weighting reciprocity more heavily from someone we have a positive impression of), consistent with confirmation bias (Fareri et al., 2012a). Social prediction error signals in the ventral striatum were also observed here, further implicating this region in neurocomputational processes supporting social learning, consistent with a host of subsequent work (Bellucci et al., 2016; Fouragnan et al., 2013; Hackel et al., 2015; Jones et al., 2014; Lockwood et al., 2016; Stanley, 2016). We can also use the outcomes of social interactions to form impressions at the *group* level. For example, impressions formed about others' generosity based on interacting with them often generalize to group members, even when that information is no longer relevant. This process is suggestive of the formation of a social "habit" (Hackel et al., 2022), and the development of which may rely on model-free reinforcement learning mechanisms based on abstract social concepts (Hackel & Kalkstein, 2023).

While reciprocity is a key contributor to motivating trust decisions, higher order signals may also be involved. For example, accurate prediction of others' behaviors offers affirmation of one's strategy (i.e., policy) to trust (or distrust). Derivation of reward-learning signals based on policy affirmation is significantly predictive of one's likelihood to trust someone, and policy-based prediction errors emerge in the ventral striatum when we correctly predict the counterfactual behavior of another (Vanyukov et al., 2019). Policy-based prediction errors are also represented within the default mode network (Allen et al., 2024). Further, social norms are critical to the way in which we interpret and respond to others in social interactions, including inequity aversion (i.e., dislike of inequitable social outcomes) and guilt aversion (i.e., avoiding guilt associated with violating another's trust). vanBaar and colleagues constructed a "moral strategy model" to examine these potential competing motives, as well as greed, for motivating trust-based behavior using a novel Hidden Multiplier Trust Game. Here, one player (the investor) is unaware of the true multiplicative factor associated with their investment in the other player (the trustee) (van Baar et al., 2019). Findings indicated that people tend to exhibit robust individual differences in their motives for reciprocity, showing influence of both mechanisms across participants, each of which was associated with dissociable patterns of neural activity as determined by intersubject representational similarity analyses (IS-RSA). Guilt aversion was reflected in recruitment of the anterior insula, putamen, dmPFC, and dlPFC, whereas inequity aversion was reflected in recruitment of the anterior insula, vmPFC, and dACC. Interestingly, "moral opportunists"—people who vacillated between these two strategies—recruited associated guilt or inequity averse neural patterns respectively when the relevant strategy was used.

Complementing this framework, reciprocity may also be impacted by constructs including guilt and a feeling of obligation to others. Beneficiaries of help from others (i.e., offers to reduce pain) tend to choose to reciprocate such help by balancing self-interest with communal and obligation-driven concerns. A beneficiary may base decisions to accept help from others as a function of whether they think another person is acting genuinely or strategically; model-based analyses revealed that the degree to which this occurs is based on feelings of indebtedness, driven by guilt and obligation to others (Gao et al., 2024). Further, multivariate pattern analyses indicated that a neural utility model highlighting neural patterns associated with obligation significantly predicted reciprocity decisions. Specifically, communal concern was associated with activity in the ventromedial prefrontal cortex (vmPFC) and anterior insula, regions implicated in gratitude and guilt processing. Obligation, shaped by second-order beliefs about the benefactor's expectation for repayment, on the other hand, engaged the dorsomedial prefrontal cortex (dmPFC) and temporoparietal junction (TPJ). Taken together, the work reviewed in this section highlights a rich framework within which to characterize how social outcomes can shape interactions with others.

4 Learning from Social Rejection: Value, Beliefs, and Agency

Trust and reciprocity clearly provide useful information through which we can learn about the social environment. Yet, other types of social outcomes facilitate learning where we stand within our social circles and can significantly impact other cognitive processes. Acceptance and rejection from others can clue us into how others perceive us (i.e., relational value) (Izuma et al., 2008; Somerville et al., 2006, 2010), shaping our understanding of relational value separately from the potential reward value of a positive interaction. In a paradigm in which participants submitted responses about themselves to purported partners who would then rank participants relative to others (relational value), it was found that people form preferences for interaction partners based experienced relational value in addition to whether participants and partners were aligned in their preferences for each other (reward value) (Babür et al., 2024). Corresponding neuroimaging findings revealed dissociable neural mechanisms for updating both relational and reward value: updating relational value recruited regions typically implicated in processing social rejection (dACC, vACC, AI, and vlPFC), whereas updating reward value engaged the ventral striatum. Notably, participants' perceptions of relational value were biased by the frequency of rewarding outcomes, suggesting that acceptance and rejection can distort social judgments (Babür et al., 2024).

Negative social outcomes, including rejection and isolation, can have significant impacts beyond our perceptions of self and others. Experiences of rejection and isolation are associated with a host of negative physical and mental health outcomes (Delgado et al., 2023; Eisenberger, 2012; Eisenberger et al., 2016; Tomova et al., 2020). Yet, investigations into the consequences of rejection on other processes have only begun to emerge. Recent findings from our group have explored the impact of social rejection on the desire for agency (Dejoie et al., 2024). We offered participants the opportunity to engage in a lottery task or have the computer play the lottery for them after they had personal Instagram photos rated by purported anonymous peers who provided varying degrees of positive (likes) and negative (dislikes) feedback. Participants were significantly more likely to want to play the lottery themselves after interacting with peers who provided more negative feedback, suggesting a potential desire to regain control over one's environment; this pattern was modulated by self-reported affect (see Fig. 2). In a follow-up study (Dejoie et al., 2025), we examined whether experiences of rejection impact the desire for future social contact; instead of deciding whether to play a lottery, participants engaged in a willingness-to-pay task in which they indicated their preferences for social relative to nonsocial rewarding experiences after playing the same social media-inspired task. Here, we found that participants were less willing to spend more on social relative to nonsocial experiences after rejection (Fig. 3, top panel). Econometric modeling further revealed that individuals with lower social connectedness and higher sensitivity to negative social outcomes were more likely to choose the social option when it was priced higher than the nonsocial one (Fig. 3, middle and bottom panels). Taken together, these findings suggest that both social outcomes and

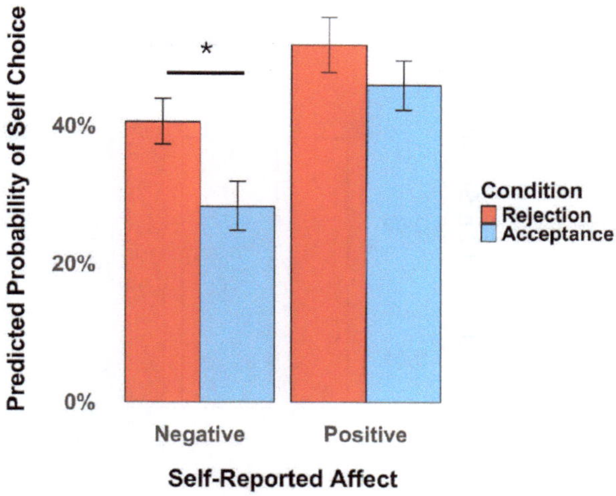

Fig. 2 Social rejection and affect shape the desire for agency. People who report negative affect after experiencing social rejection are significantly more likely to desire agency. (Figure adapted from Dejoie et al., 2024)

individual differences shape learning and decision-making, highlighting the value of computational approaches for capturing these processes.

5 Social Sharing Facilitates Interactions and Relationship Building

Being able to integrate social outcomes such as rejection or acceptance from others is thus a key mechanism through which we understand our place in the social world. This kind of feedback can be based on a multitude of inputs—physical appearance, past behavior, but a cornerstone of the human condition is the desire to share information and experiences with others. Sharing personal stories, thoughts, and experiences helps reveal common interests, preferences, and beliefs, fostering deeper connections and mutual understanding (Delgado et al., 2023).

Social sharing can be influenced by various factors, including the social context and norms governing what is considered appropriate to share. For instance, sharing personal achievements may be celebrated in one cultural or social group but perceived as boastful in another. Similarly, the impact of social sharing can vary depending on the relationship between the sharer and audience; sharing a personal story with a close friend may foster intimacy (Delgado et al., 2023). However, doing so with an acquaintance or someone who does not hold the same values, interests, or opinions may lead to friction, and/or misinterpretation, which could engender social rejection and maladaptive social behaviors, leading to isolation, loneliness, and poor health outcomes (Haihambo et al., 2025; Shen et al., 2025).

The dynamics of social sharing may be exacerbated in modern society, as sharing on social media may instantaneously result in positive or negative comments from others or simple (dis)approval in the form of a like or negative emoji. Such outcomes

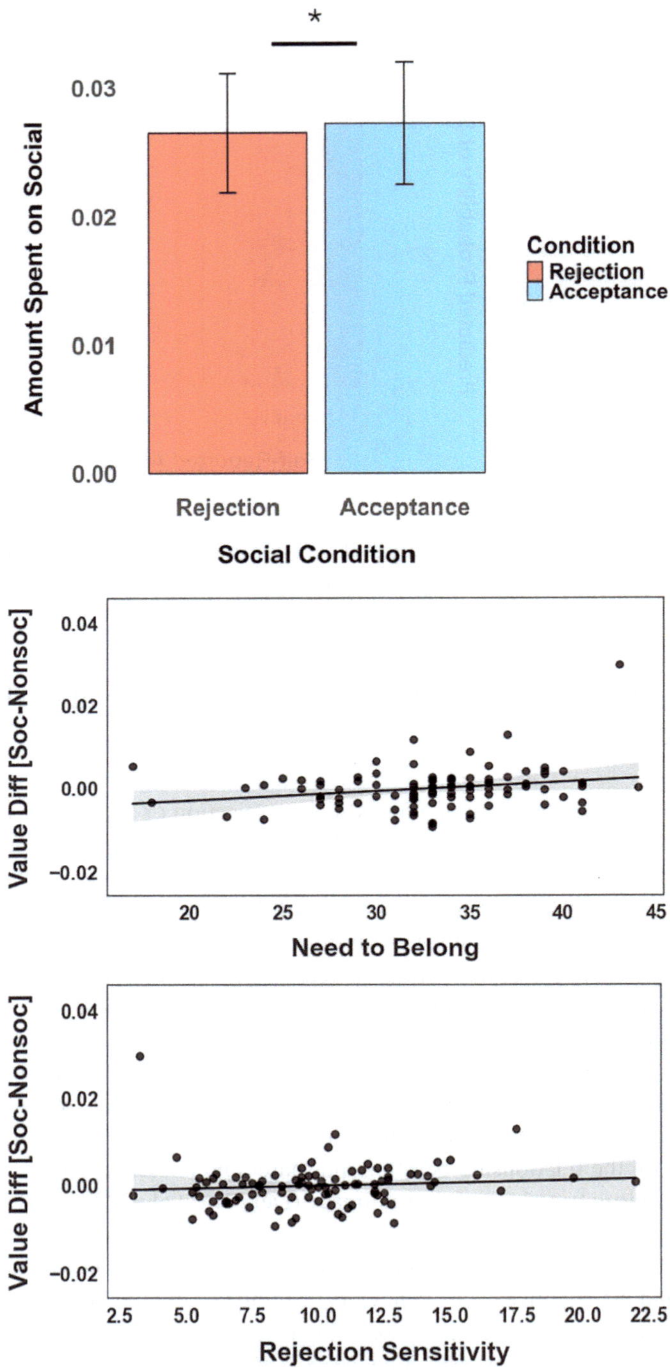

Fig. 3 Effects of social rejection on the willingness to pay for social experiences. People are less likely to pay for social experiences after social rejection ($p < .05$, top panel), but the degree to which they are is associated with the need to belong and rejection sensitivity (p's $< .05$, middle and bottom panels). (Adapted from Dejoie et al., 2025)

have the potential to be rewarding (Meshi et al., 2013) and can impact the likelihood of wanting to interact with specific individuals again. Sharing on social media may invoke value-based processes: neural systems supporting value-based choice (i.e., vmPFC, VS) might encode a general value-based "sharing" signal within a virality framework that integrates the value of the shared content, predictions of how it may be perceived, and how one's audience will perceive it in order to predict choice (Scholz et al., 2017, 2020).

Recent evidence has begun to support this framework. Using real-world Instagram data and a reinforcement learning approach, Lindstrom and colleagues demonstrated that posting behavior is sensitive to the frequency of received positive feedback and best predicted by a model that updates the value of posting as a function of the social reward gleaned from a "like" (Lindström et al., 2021). Importantly, these effects hold across social media platforms and were replicated in an experimental study mimicking a social media environment, suggesting a common value-based mechanism that broadly explains sharing on social media. Individual differences in socioemotional function may be a key modulator of social media sharing. In a novel social media-inspired experimental task, Brudner and colleagues found that people were more likely to want to share photos in the future with partners who provided more positive feedback on their photos and were more likely to invest with those partners in a subsequent trust game (Brudner et al., 2023). A loss-gain reinforcement learning model (Fareri et al., 2012a; Frank et al., 2007) revealed that this pattern was contingent upon differential weighting of positive relative to negative feedback on shared photos, which varied as a function of people's interpersonal emotion regulation abilities (Brudner et al., 2023). Those more likely to engage in social regulation of emotions were more likely to use positive social feedback to inform their impressions of others, supporting the idea that individual differences in traits and other sociocognitive abilities can affect the degree to which we are sensitive to incorporate feedback from others into subsequent behavior.

Social sharing can also come in the form of real-time dyadic or group interactions. Elegant work by Jolly and Chang investigated the role that gossip plays in learning about others and fostering closeness (Jolly & Chang, 2021). Using a multi-round public goods game, participants received different levels of information (complete, incomplete) about other players' contributions and were sometimes able to gossip about the group. Gossip about prior behavior predicted people's future contributions to the public pool and drove social connection within the group. Taken together, this line of research indicates that the ability to share information goes beyond influencing behavior on social media. Rather, social sharing shapes the ways in which we learn about both others *and* ourselves, and how our own actions might be interpreted by others in a manner that influences individual behavior, collective behavior, and group dynamics.

6 Mechanisms Through Which Social Relationships Shape Reward-Based Decision-Making

Sharing information and experiences most frequently occurs with those whom we have close interpersonal relationships. Yet, until recently, the nature of the ways in which interpersonal relationship dynamics may impact computational mechanisms underlying our choices and shared/collective experiences has remained an elusive question. Early studies on this topic indicated that positive experiences with close others are subjectively viewed as more rewarding relative to those who are less close or novel interaction partners, and this pattern was shown to be reflected in the strength of reward value signals in the ventral striatum and mPFC (Fareri et al., 2012b). Enhanced striatal reward signals also emerge when choosing to forgo monetary rewards to disclose personal information to a close other (Tamir & Mitchell, 2012). Building upon these findings, we have examined the neurocomputational mechanisms underlying motivations to trust close others. Using computational models implementing modified Rescorla-Wagner learning rules, we tested whether such decisions are driven by strong priors about close others relative to strangers (expectation model), or rather by the enhanced reward value of reciprocity from a friend (social value model). We found that a social value model best predicted participants' decisions to trust friends relative to strangers, and that computational parameters reflecting an enhanced value of reciprocity from a friend were associated with corticostriatal responses to reciprocity (Fareri et al., 2015). These findings were some of the first to demonstrate the neurocomputational mechanisms supporting trust in close relationships.

While choices to trust or collaborate are cornerstones of relationship building and maintenance, there are a variety of types of decisions that can involve other people and may be more consequential. Within the context of risky economic choice, evidence for how others factor into our decision calculus has been mixed (Polman & Wu, 2020). However, neurocomputational studies have begun to shed light on the nuances of how we might alter risk perception when our choices impact or involve other people. In these situations, early evidence indicated that people tend to conform (i.e., are "nudged") toward the risk preferences of others, as a function of how similar our risk preferences may be (Chung et al., 2015). Computationally, this was modeled as "other-conferred utility." a signal which tracks with vmPFC activation during risk evaluation. A common neural representation of subjective value for self and other when making choices that can impact others in the risk and intertemporal domains has also been found in the dmPFC (Piva et al., 2019), suggesting that multiple neural substrates contribute to our consideration of others in our choices.

A key remaining question pertains to how specific component processes underlying risk evaluation (i.e., risk attitudes (ρ), loss aversion (λ)) *change* in the context of others to support overall shifts in risk preferences in social contexts. Evidence from our group (Fareri et al., 2022) has shown that when people are tasked with making risky economic decisions, they become more conservative in their choices overall (i.e., choose a safe over a risky choice) when the consequences could impact others

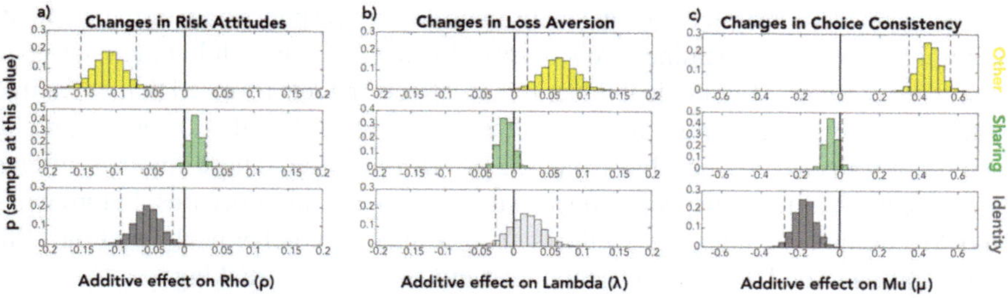

Fig. 4 Dissociable effects of social context on component processes of risky decision-making. Hierarchical Bayesian Modeling of risky choice revealed dissociable effects of social context on behavior. Analyses revealed consistent *changes* in risk attitudes (ρ), loss aversion (λ), and choice consistency (μ) when people evaluated risky decisions for others relative to themselves (top panel); no consistent effects on computational parameters underlying risky choice when participants were *sharing* the outcomes of risky choices with others relative to experiencing the outcomes only for themselves; and consistent effects of partner identity on risk attitudes and choice consistency when choices involved a close friend relative to a stranger. (Adapted from Fareri et al., 2022)

(friends, strangers). Using hierarchical Bayesian implementation of a prospect-theory model of risky economic choice, we saw that these conservative shifts were dependent upon nuanced changes in multiple component processes supporting risk evaluation: people became more risk averse, loss averse, and consistent (μ) when making choices that involved others relative to the self (Fig. 4). Interestingly, while we did not see any consistent effects of sharing outcomes with others on these parameters, we did find that the specific "other" affected by our decisions matters (consistent with related findings from other studies—(Guassi Moreira et al., 2018, 2021): risk aversion increased for choices that affected strangers, but decreased for choices that affected close friends. Taken together, these findings underscore the role of both relational and social contextual factors in shaping decision-making.

7 Conclusions: New Frontiers and Future Directions

A multitude of factors influence our choices within the social world. Here we have highlighted some of the computational mechanisms subserving (1) learning within social interactions; (2) ways in which valuation is shaped by social rejection; (3) social sharing; and (4) how relationship dynamics influence decision processes. By integrating social factors into neurocomputational frameworks, we are better able to account for the intricate social dynamics that characterize our daily lives. However, a number of interesting future directions remain ready to explore. For example, underlying many of the computational processes described in this chapter are likely changes in neurochemical processes. One recent study explored dopamine and serotonin dynamics in the substantia nigra pars reticulata during the ultimatum game in patients with Parkinson's disease (Batten et al., 2024).

Participants demonstrated distinct neuromodulatory responses to social context, with higher overall dopamine levels observed when playing with human relative to computer partners. Hierarchical linear mixed-effects models revealed that dopamine tracked trial-by-trial changes in offer value (akin to reward prediction errors), while serotonin reflected the absolute value of the current offer. These findings offer a novel neurochemical perspective on social decision-making processes. Future work should continue to integrate methods to modulate neurochemical function with computational modeling during social learning and decision-making.

Computational approaches can also benefit from the incorporation of trait-level personality differences in modulating social learning and choice. Psychopathy, for example, has been associated with an inverse association between guilt aversion and psychopathic traits: increased levels of such traits are associated with less aversion to guilt for acting in a non-trustworthy manner (Gong et al., 2019). Further, individual variation along dimensions of agreeableness, politeness, and compassion may impact the degree to which we follow policy-based strategies. People with higher levels of callous traits show reduced associations between policy signals and default mode network activation, whereas individuals higher on exploitative traits show enhanced policy signals in this network (Allen et al., 2024). Implementing metrics of personality traits and other (sub)clinical variation with computational approaches allows for a more holistic picture, capturing individual differences in computational processes underlying social choice.

Further, many studies to date have focused on relatively controlled laboratory experimental approaches, which may not fully capture the complexity of real-world social interactions. Future work should incorporate more naturalistic and ecologically valid environments, theories, and paradigms to track real-world behavior over time (Finn & Bandettini, 2021; Gonzalez & Rice, 2024; Parkinson et al., 2018; Tomova et al., 2020). For example, work implementing real-time strategic interactions in combination with model-based fMRI and brain stimulation techniques (repetitive TMS) has causally demonstrated that computational processes in the rTPJ support strategic social behavior (Hill et al., 2017). Recent advances have also emerged that can track in-the-moment changes in facial expressions (Cheong et al., 2023). One application of this kind of approach is to assess synchronization of facial expressions in dyads while collectively viewing emotional content (Cheong et al., 2020). Here increased synchronization of facial expressions of emotion across individuals tracks with increases in self-reported feelings of social closeness. Future work implementing such approaches with learning and decision-making paradigms and computational methods could better assess how physiological indices of emotion and social cognition are associated with relationship building. These could provide richer insights into how social decision-making unfolds in dynamic, real-life scenarios.

Social decision-making is deeply influenced by the context in which choices are made, shaped by feedback, relationships, and the broader social environment. Advances in computational modeling and neuroimaging have provided powerful tools to quantify these effects, revealing the dynamic interplay between individual attributes, social factors, and neural mechanisms. By refining computational and neural models and expanding their application to real-world contexts, future research can deepen our understanding of social decision-making. These efforts may lead to more effective strategies for enhancing social interactions, improving decision-making processes, and addressing complex societal challenges. Continued exploration of the intersection between social and computational sciences promises meaningful contributions to and implications for multiple fields, including psychology, behavioral economics, education, and public policy.

References

Allen, T. A., Hallquist, M. N., & Dombrovski, A. Y. (2024). Callousness, exploitativeness, and tracking of cooperation incentives in the human default network. *Proceedings of the National Academy of Sciences, 121*, e2307221121. https://doi.org/10.1073/pnas.2307221121

Babür, B. G., Leong, Y. C., Pan, C. X., & Hackel, L. M. (2024). Neural responses to social rejection reflect dissociable learning about relational value and reward. *Proceedings. National Academy of Sciences. United States of America, 121*, e2400022121. https://doi.org/10.1073/pnas.2400022121

Bartra, O., McGuire, J. T., & Kable, J. W. (2013). The valuation system: A coordinate-based meta-analysis of BOLD fMRI experiments examining neural correlates of subjective value. *NeuroImage, 76*, 412–427.

Batten, S. R., Bang, D., Kopell, B. H., Davis, A. N., Heflin, M., Fu, Q., Perl, O., Ziafat, K., Hashemi, A., Saez, I., Barbosa, L. S., Twomey, T., Lohrenz, T., White, J. P., Dayan, P., Charney, A. W., Figee, M., Mayberg, H. S., Kishida, K. T., Gu, X., & Montague, P. R. (2024). Dopamine and serotonin in human substantia nigra track social context and value signals during economic exchange. *Nature Human Behaviour, 8*, 718–728. https://doi.org/10.1038/s41562-024-01831-w

Behrens, T. E. J., Hunt, L. T., Woolrich, M. W., & Rushworth, M. F. S. (2008). Associative learning of social value. *Nature, 456*, 245–249.

Bellucci, G., Chernyak, S. V., Goodyear, K., Eickhoff, S. B., & Krueger, F. (2016). Neural signatures of trust in reciprocity: A coordinate-based meta-analysis. *Human Brain Mapping, 38*, 1233–1248.

Brudner, E. G., Fareri, D. S., Shehata, S. G., & Delgado, M. R. (2023). Social feedback promotes positive social sharing, trust, and closeness. *Emotion, 23*, 1536–1548. https://doi.org/10.1037/emo0001182

Chang, L. J., Doll, B. B., vant Wout, M, Frank, M. J., & Sanfey, A. G. (2010). Seeing is believing: Trustworthiness as a dynamic belief. *Cognitive Psychology, 61*, 87–105.

Cheong, J. H., Molani, Z., Sadhukha, S., & Chang, L. J. (2020). Synchronized affect in shared experiences strengthens social connection. *Communications Biology, 6*, 1099. https://doi.org/10.31234/osf.io/bd9wn

Cheong, J. H., Jolly, E., Xie, T., Byrne, S., Kenney, M., & Chang, L. J. (2023). Py-feat: Python facial expression analysis toolbox. *Affective Science, 4*, 781–796. https://doi.org/10.1007/s42761-023-00191-4

Chung, D., Christopoulos, G. I., King-Casas, B., Ball, S. B., & Chiu, P. H. (2015). Social signals of safety and risk confer utility and have asymmetric effects on observers' choices. *Nature Neuroscience, 18*, 912–916. https://doi.org/10.1038/nn.4022

Dejoie, J. M., Ruiz, M., Brudner, E. G., & Fareri, D. S. (2024). Social rejection amplifies the value of choice. *Emotion, 25*(4), 929–939. https://doi.org/10.1037/emo0001456

Dejoie, J. M., Ruiz, M., Gallardo, S., Sokol-Hessner, P., Smith, D. V., & Fareri D. S. (2025). The price of rejection: How social feedback influences spending on social experiences. *PsyArXiv*. Preprint. https://osf.io/preprints/psyarxiv/jpf7e_v1

Delgado, M. R., Fareri, D. S., & Chang, L. J. (2023). Characterizing the mechanisms of social connection. *Neuron, 111*, 3911–3925. https://doi.org/10.1016/j.neuron.2023.09.012

Diaconescu, A. O., Stecy, M., Kasper, L., Burke, C. J., Nagy, Z., Mathys, C., & Tobler, P. N. (2020). Neural arbitration between social and individual learning systems. *eLife, 9*, e54051. https://doi.org/10.7554/eLife.54051

Eisenberger, N. I. (2012). The pain of social disconnection: Examining the shared neural underpinnings of physical and social pain. *Nature Reviews Neuroscience, 13*, 421–434. https://doi.org/10.1038/nrn3231

Eisenberger, N. I., Moieni, M., Inagaki, T. K., Muscatell, K. A., & Irwin, M. R. (2016). In sickness and in health: The co-regulation of inflammation and social behavior. *Neuropsychopharmacology, 42*, 242–253.

Fareri, D. S. (2019). Neurobehavioral mechanisms supporting trust and reciprocity. *Frontiers in Human Neuroscience, 13*, 271. https://doi.org/10.3389/fnhum.2019.00271

Fareri, D. S., Chang, L. J., & Delgado, M. R. (2012a). Effects of direct social experience on trust decisions and neural reward circuitry. *Frontiers in Neuroscience, 6*, 148–117.

Fareri, D. S., Niznikiewicz, M. A., Lee, V. K., & Delgado, M. R. (2012b). Social network modulation of reward-related signals. *Journal of Neuroscience, 32*, 9045–9052.

Fareri, D. S., Chang, L. J., & Delgado, M. R. (2015). Computational substrates of social value in interpersonal collaboration. *The Journal of Neuroscience, 35*, 8170–8180. https://doi.org/10.1523/JNEUROSCI.4775-14.2015

Fareri, D. S., Stasiak, J. E., & Sokol-Hessner, P. (2022). Choosing for others changes dissociable computational mechanisms underpinning risky decision-making. *Scientific Reports, 12*, 14361. https://doi.org/10.1038/s41598-022-18437-9

FeldmanHall, O., & Nassar, M. R. (2021). The computational challenge of social learning. *Trends in Cognitive Sciences, 25*, 1045–1057. https://doi.org/10.1016/j.tics.2021.09.002

FeldmanHall, O., & Shenhav, A. (2019). Resolving uncertainty in a social world. *Nature Human Behaviour, 3*, 426–435. https://doi.org/10.1038/s41562-019-0590-x

Finn, E. S., & Bandettini, P. A. (2021). Movie-watching outperforms rest for functional connectivity-based prediction of behavior. *NeuroImage, 235*, 117963. https://doi.org/10.1016/j.neuroimage.2021.117963

Fouragnan, E., Chierchia, G., Greiner, S., Neveu, R., Avesani, P., & Coricelli, G. (2013). Reputational priors magnify striatal responses to violations of trust. *Journal of Neuroscience, 33*, 3602–3611.

Frank, M. J., Moustafa, A. A., Haughey, H. M., Curran, T., & Hutchison, K. E. (2007). Genetic triple dissociation reveals multiple roles for dopamine in reinforcement learning. *Proceedings of the National Academy of Sciences of the United States of America, 104*, 16311–16316.

Gao, X., Jolly, E., Yu, H., Liu, H., Zhou, X., & Chang, L. J. (2024). The psychological, computational, and neural foundations of indebtedness. *Nature Communications, 15*, 68. https://doi.org/10.1038/s41467-023-44286-9

Gong, X., Brazil, I. A., Chang, L. J., & Sanfey, A. G. (2019). Psychopathic traits are related to diminished guilt aversion and reduced trustworthiness during social decision-making. *Scientific Reports, 9*, 7307. https://doi.org/10.1038/s41598-019-43727-0

Gonzalez, M. Z., & Rice, M. A. (2024). Behavioural sciences need behavioural ecology. *Nature Human Behaviour, 8*, 1240–1242. https://doi.org/10.1038/s41562-024-01906-8

Guassi Moreira, J. F., Tashjian, S. M., Galván, A., & Silvers, J. A. (2018). Parents versus peers: Assessing the impact of social agents on decision making in young adults. *Psychological Science, 29*, 1526–1539. https://doi.org/10.1177/0956797618778497

Guassi Moreira, J. F., Tashjian, S. M., Galván, A., & Silvers, J. A. (2021). Computational and motivational mechanisms of human social decision making involving close others. *Journal of Experimental Social Psychology, 93*, 104086. https://doi.org/10.1016/j.jesp.2020.104086

Haber, S. N., & Knutson, B. (2010). The reward circuit: Linking primate anatomy and human imaging. *Neuropsychopharmacology, 35*, 4–26.

Hackel, L. M., & Kalkstein, D. A. (2023). Social concepts simplify complex reinforcement learning. *Psychological Science, 34*, 968–983. https://doi.org/10.1177/09567976231180587

Hackel, L. M., Doll, B. B., & Amodio, D. M. (2015). Instrumental learning of traits versus rewards: Dissociable neural correlates and effects on choice. *Nature Neuroscience, 18*, 1233–1235. https://doi.org/10.1038/nn.4080

Hackel, L. M., Kogon, D., Amodio, D. M., & Wood, W. (2022). Group value learned through interactions with members: A reinforcement learning account. *Journal of Experimental Social Psychology, 99*, 104267. https://doi.org/10.1016/j.jesp.2021.104267

Haihambo, N., Layiwola, D.-M., Blank, H., Hurlemann, R., & Scheele, D. (2025). Loneliness and social conformity: A predictive processing perspective. *Annals of the New York Academy of Sciences, 1547*, 5–17. https://doi.org/10.1111/nyas.15324

Hill, C. A., Suzuki, S., Polania, R., Moisa, M., O'Doherty, J. P., & Ruff, C. C. (2017). A causal account of the brain network computations underlying strategic social behavior. *Nature Neuroscience, 20*, 1142–1149.

Izuma, K., Saito, D. N., & Sadato, N. (2008). Processing of social and monetary rewards in the human striatum. *Neuron, 58*, 284–294.

Jolly, E., & Chang, L. J. (2021). Gossip drives vicarious learning and facilitates social connection. *Current Biology, 31*, 2539–2549.e6. https://doi.org/10.1016/j.cub.2021.03.090

Jones, R. M., Somerville, L. H., Li, J., Ruberry, E. J., Powers, A., Mehta, N., Dyke, J., & Casey, B. J. (2014). Adolescent-specific patterns of behavior and neural activity during social reinforcement learning. *Cognitive, Affective, & Behavioral Neuroscience, 14*, 683–697. https://doi.org/10.3758/s13415-014-0257-z

Konovalov, A., Hu, J., & Ruff, C. C. (2018). Neurocomputational approaches to social behavior. *Current Opinion in Psychology, Social Neuroscience, 24*, 41–47. https://doi.org/10.1016/j.copsyc.2018.04.009

Lindström, B., Bellander, M., Schultner, D. T., Chang, A., Tobler, P. N., & Amodio, D. M. (2021). A computational reward learning account of social media engagement. *Nature Communications, 12*(1), 1311.

Lockwood, P. L., & Klein-Flügge, M. C. (2021). Computational modelling of social cognition and behaviour—A reinforcement learning primer. *Social Cognitive and Affective Neuroscience, 16*, 761–771. https://doi.org/10.1093/scan/nsaa040

Lockwood, P. L., Apps, M. A. J., Valton, V., Viding, E., & Roiser, J. P. (2016). Neurocomputational mechanisms of prosocial learning and links to empathy. *Proceedings of the National Academy of Sciences of the United States of America, 113*, 9763–9768.

Lockwood, P. L., Wittmann, M. K., Apps, M. A. J., Klein-Flügge, M. C., Crockett, M. J., Humphreys, G. W., & Rushworth, M. F. S. (2018). Neural mechanisms for learning self and other ownership. *Nature Communications, 9*, 1–11. https://doi.org/10.1038/s41467-018-07231-9

Lockwood, P. L., Apps, M. A. J., & Chang, S. W. C. (2020). Is there a 'social' brain? Implementations and algorithms. *Trends in Cognitive Sciences, 24*, 802–813. https://doi.org/10.1016/j.tics.2020.06.011

Meshi, D., Morawetz, C., & Heekeren, H. R. (2013). Nucleus accumbens response to gains in reputation for the self relative to gains for others predicts social media use. *Frontiers in Human Neuroscience, 7*, 1–11.

Parkinson, C., Kleinbaum, A. M., & Wheatley, T. (2018). Similar neural responses predict friendship. *Nature Communications, 9*, 1–14.

Piva, M., Velnoskey, K., Jia, R., Nair, A., Levy, I., & Chang, S. W. (2019). The dorsomedial prefrontal cortex computes task-invariant relative subjective value for self and other. *eLife, 8*, e44939. https://doi.org/10.7554/eLife.44939

Polman, E., & Wu, K. (2020). Decision making for others involving risk: A review and meta-analysis. *Journal of Economic Psychology, Decision Making for Others, 77*, 102184. https://doi.org/10.1016/j.joep.2019.06.007

Rudebeck, P. H., Buckley, M. J., Walton, M. E., & Rushworth, M. F. S. (2006). A role for the macaque anterior cingulate gyrus in social valuation. *Science (New York, N.Y.), 313*, 1310–1312.

Ruff, C. C., & Fehr, E. (2014). The neurobiology of rewards and values in social decision making. *Nature Reviews Neuroscience, 15*(8), 549–562.

Scholz, C., Baek, E. C., O'Donnell, M. B., Kim, H. S., Cappella, J. N., & Falk, E. B. (2017). A neural model of valuation and information virality. *Proceedings of the National Academy of Sciences of the United States of America, 114*, 2881–2886. https://doi.org/10.1073/pnas.1615259114

Scholz, C., Jovanova, M., Baek, E. C., & Falk, E. B. (2020). Media content sharing as a value-based decision. *Current Opinion in Psychology, Privacy and Disclosure, Online and in Social Interactions, 31*, 83–88. https://doi.org/10.1016/j.copsyc.2019.08.004

Shen, C., Zhang, R., Yu, J., Sahakian, B. J., Cheng, W., & Feng, J. (2025). Plasma proteomic signatures of social isolation and loneliness associated with morbidity and mortality. *Nature Human Behaviour, 9*, 569–583. https://doi.org/10.1038/s41562-024-02078-1

Somerville, L. H., Heatherton, T. F., & Kelley, W. M. (2006). Anterior cingulate cortex responds differentially to expectancy violation and social rejection. *Nature Neuroscience, 9*, 1007–1008.

Somerville, L. H., Kelley, W. M., & Heatherton, T. F. (2010). Self-esteem modulates medial prefrontal cortical responses to evaluative social feedback. *Cerebral cortex (New York, NY: 1991), 20*, 3005–3013.

Stanley, D. A. (2016). Getting to know you: General and specific neural computations for learning about people. *Social Cognitive and Affective Neuroscience, 11*, 525–536. https://doi.org/10.1093/scan/nsv145

Stanley, D. A., & Adolphs, R. (2013). Toward a neural basis for social behavior. *Neuron, 80*, 816–826.

Tamir, D. I., & Mitchell, J. P. (2012). Disclosing information about the self is intrinsically rewarding. *Proceedings of the National Academy of Sciences of the United States of America, 109*, 8038–8043.

Tomova, L., Wang, K. L., Thompson, T., Matthews, G. A., Takahashi, A., Tye, K. M., & Saxe, R. (2020). Acute social isolation evokes midbrain craving responses similar to hunger. *Nature Neuroscience, 23*, 1597–1605. https://doi.org/10.1038/s41593-020-00742-z

Vahed, S., Galván, E. P., & Sanfey, A. G. (2024). Computational modeling of social decision-making. *Current Opinion in Psychology, 60*, 101884. https://doi.org/10.1016/j.copsyc.2024.101884

van Baar, J. M., Chang, L. J., & Sanfey, A. G. (2019). The computational and neural substrates of moral strategies in social decision-making. *Nature Communications, 10*(1), 1–14.

Vanyukov, P. M., Hallquist, M. N., Delgado, M., Szanto, K., & Dombrovski, A. Y. (2019). Neurocomputational mechanisms of adaptive learning in social exchanges. *Cognitive, Affective, & Behavioral Neuroscience, 19*, 985–997. https://doi.org/10.3758/s13415-019-00697-0

From Experiences to Preferences: Neural Insights Through a Naturalistic Neuroimaging Paradigm

Pin-Hao A. Chen, Tung-An P. Chiu, Po-Yuan A. Hsiao, and Feng-Chun B. Chou

Abstract In the digital age, recommendation systems personalize online content based on user preferences, enhancing satisfaction and loyalty. Understanding the factors that influence individual preferences is therefore crucial. While factors like repeated exposure, emotional states, social context, and memory can affect preferences, the neural mechanisms leading to divergent preferences from similar experiences remain underexplored. In our recent study, we investigated how subjective experiences are transformed into preferences, addressing the challenge of disentangling individual differences in subjective experiences from differences in transformation processes. To achieve this, we employed a naturalistic neuroimaging paradigm. Participants underwent a two-phase study: first, they watched diverse videos inside an fMRI scanner without knowledge of subsequent tasks, ensuring unbiased neural responses. In the second phase, they expressed preferences between video pairs through a binary forced-choice task and provided subjective appraisals for each video across various dimensions. Using machine learning methods, our findings showed that within-subject multivariate brain models could predict individuals' preferences with high accuracy. Further intersubject representational similarity analysis revealed a strong association between similarities in within-subject brain models and appraisal models. By combining naturalistic neuroimaging paradigms with advanced machine learning techniques, we offer a new approach to disentangling subjective experiences from decision-making behaviors. This research advances computational cognitive neuroscience and has practical implications for developing more effective personalized recommendation systems that account for individual uniqueness in preferences and choices.

P.-H. A. Chen (✉)
Department of Psychology, Neurobiology and Cognitive Science Center, Center for Artificial Intelligence and Advanced Robotics, Graduate Institute of Brain and Mind Sciences, National Taiwan University, Taipei, Taiwan
e-mail: andyphchen@ntu.edu.tw

T.-A. P. Chiu · P.-Y. A. Hsiao · F.-C. B. Chou
Department of Psychology, National Taiwan University, Taipei, Taiwan

D. V. Smith et al. (eds.), *Neuroeconomics: Core Topics and Current Directions*,
https://doi.org/10.1007/978-3-032-02925-6_23

This chapter reviews recent progress in investigating how subjective experiences are transformed into preferences, inspired by the recent development of personalized recommendation systems. Despite the recognized influence of factors, the neural transformations between subjective experiences and preferences remain underexplored. By utilizing a novel naturalistic neuroimaging paradigm and machine-learning methods, recent findings from our research team found that each brain engages distinct roles to encode and interpret experiential inputs, leading to divergent preference outcomes across individuals. Through reviewing these recent findings, this chapter aims to provide novel insights into the complexities of human decision-making from a computational cognitive neuroscience perspective.

1 Recommendation Systems: Transforming Digital Interactions into Personalized Preferences

In an age where digital experiences are tailored to individual preferences, recommendation systems have become an unseen force shaping our experiences on numerous online platforms. Designed to assist users by providing personalized suggestions based on their preferences and behaviors, these systems are integral to many online services, enhancing user experience by delivering customized recommendations. For example, Netflix uses its recommendation system for movie suggestions, Amazon employs it for product recommendations, and Spotify utilizes it for creating personalized playlists (Amatriain & Basilico, 2015; Belleflamme & Peitz, 2020). Over time, preferences captured in this manner often exhibit consistency across diverse contexts (Hardardottir, 2017; Liebe et al., 2016; Severin et al., 2001; Steenkamp & Maydeu-Olivares, 2015). This consistency indicates that preferences may represent a distinct dimension of personal characteristics, complementing traditional trait assessments. By employing sophisticated algorithms to analyze massive amounts of user data, individual preferences are effectively transformed into precise suggestions through recommendation systems.

Recommendation systems use advanced algorithms that analyze data from a vast user base. One of the most widely used methods is collaborative filtering (Bobadilla et al., 2009; Ekstrand et al., 2011), which is based on the idea that users who have shown similar preferences are likely to also agree on suggested new items. In general, this method can be categorized into two types: user-based and item-based collaborative filtering (Gong, 2010; Kim et al., 2010; Su & Khoshgoftaar, 2009; Wang et al., 2006). User-based collaborative filtering focuses on finding users with similar tastes. For example, if two users have rated several movies similarly, the system assumes they have similar preferences and recommends movies that one user liked to the other (Jia et al., 2015; Koohi & Kiani, 2016; Zhao & Shang, 2010). Item-based collaborative filtering, on the other hand, focuses on finding similarities between items. It examines the relationships between different items and recommends those similar to the ones the user has already shown interest in.

This approach is particularly useful in situations with many users and sparse data, as it leverages the similarities among items rather than relying on the similarities across users (Ajaegbu, 2021; Sarwar et al., 2001; Xue et al., 2019).

Implementing recommendation systems involves several key steps, including data collection, preprocessing, and the application of algorithms. Initially, data on user interactions, such as ratings, clicks, and purchase history, are collected. These data are then processed to create user-item matrices, which form the foundation of collaborative filtering algorithms. Once this matrix is established, the collaborative filtering algorithm is applied to generate recommendations (Bobadilla et al., 2009; Ekstrand et al., 2011; Schafer et al., 2007). For instance, in a movie streaming service, the platform collects data on users' viewing histories and ratings. This information is used to construct a user-item matrix, where each row represents a user, and each column represents a movie, with the entries indicating the ratings given by users (Asemi et al., 2022; Herlocker et al., 2004). Using user-based collaborative filtering, the recommendation system identifies users with similar viewing patterns and suggests movies that these similar users have liked. On the other hand, the recommendation system utilizes item-based collaborative filtering to analyze how all users have rated different movies to determine which movies are similar based on rating patterns. The system then recommends movies that are similar to those the user has rated highly.

Rather than relying on a single method, modern recommendation systems often employ a hybrid approach, combining collaborative filtering with other techniques, such as content-based filtering or context-aware recommendations, to improve the accuracy and relevance of suggestions. For example, while collaborative filtering might suggest movies based on user behavior, content-based filtering can refine these recommendations by considering factors such as genres, directors, and actors that the user prefers. This combination of methods helps overcome the limitations of each technique and enhances the overall performance of recommendation systems (Cui et al., 2020; Javari et al., 2014; Kim & Ahn, 2017; La et al., 2019; Shakirova, 2017; Symeonidis & Zioupos, 2017). Through the adoption and integration of these systems, many online platforms have greatly improved user engagement and satisfaction, leading to success in the competitive digital market.

In recent years, the success of recommendation systems has extended beyond e-commerce and streaming services to social media platforms like Instagram and Threads. These platforms not only collect and analyze user data to provide personalized content but also examine how individuals make decisions about their preferences. Understanding the factors that influence individuals' preferences is particularly crucial, as they play a significant role in our daily lives. Some research suggests that individuals tend to develop preferences for stimuli to which they are repeatedly exposed (Ballard et al., 2017; Mrkva & Van Boven, 2020; Zajonc, 2001). This phenomenon, known as the mere-exposure effect, indicates that familiarity can breed liking. Additionally, emotional states may significantly impact preference formation; positive emotions, for instance, can sway individuals toward certain choices (Ifcher & Zarghamee, 2011; Loewenstein et al., 2001; Yen & Chuang, 2008). Social context is another influential factor. Studies have demonstrated that peer opinions and cultural norms can shape individual preferences (Fershtman & Segal, 2018; Izuma & Adolphs, 2013; Robinson & Higgs, 2012), with social contexts enhancing how preferences are formed (Nakagawa et al., 2015). Memory also contributes to preference formation; repeated recall of a particular stimulus increases its likelihood of being selected when making choices (Bornstein et al., 2017; Pyo et al., 2022).

2 Neural Mechanisms Underlying Our Preference Decisions

To further explore how individuals decide on their preferences, researchers have turned to modeling methods (Ratcliff & McKoon, 2008). One prominent model is the drift-diffusion model (DDM), which describes how individuals accumulate evidence over time to make decisions. In a classic study, Hare et al. (2011) used the DDM to investigate how value signals are transformed into motor commands during simple choices in the brain. Their research demonstrated that value-based decision-making is a dynamic process wherein the brain continuously integrates value-related information until a threshold is reached, triggering a decision. Other studies have employed the DDM to examine the neural correlates of decision-making, revealing how brain regions such as the prefrontal cortex and striatum accumulate value-related information (Gold & Shadlen, 2007).

Building on the above findings, neuroscientists have utilized diverse experimental tasks to examine the neural mechanisms involved when individuals decide about their preferences. Among these experimental tasks, forced-choice tasks—where participants choose between two or more options—are the most commonly used. Additionally, rating tasks, where participants evaluate how much they like a particular stimulus, are also frequently employed. By utilizing these tasks, neuroscientists have identified specific brain regions involved in preference decision-making. For example, studies using forced-choice tasks have shown that the ventromedial prefrontal cortex (VMPFC) computes the value of different options for further decision-making. The VMPFC integrates information about stimuli and assigns value to them (Foo et al., 2014; Hare et al., 2010; van Holstein & Floresco, 2020). Findings from studies using rating tasks also reveal the involvement of the orbitofrontal cortex (OFC) when individuals make decisions about their preferences (Fellows, 2007; Keiflin et al., 2013; Kringelbach, 2005). Other neuroimaging studies have found that the dorsolateral prefrontal cortex (DLPFC) is involved in integrating value signals from other brain regions, especially when decisions require combining multiple sources of information or involve complex reasoning (Colombo et al., 2016; Foo et al., 2014; Hare et al., 2011). Furthermore, the nucleus accumbens (NAcc), a brain region involved in reward processing and reinforcement learning (Burton et al., 2015; Sugam et al., 2014), also plays a crucial role in deciding preferences. Together, these findings provide valuable insights into the neural mechanisms involved in how individuals make decisions about their preferences.

3 Taking a Naturalistic Neuroimaging Paradigm to Explore How Our Brains Transform Experiences into Preferences

However, despite these advances, understanding individual differences in preference decisions remains a challenge. While previous studies have provided insights into the neural mechanisms involved, we now recognize that the factors contributing to individual differences in preferences are multifaceted. One of the most common factors is the varying experiences and feelings that individuals derive from the same stimuli, such as movies or music (Chen et al., 2020; Dabas et al., 2024; Rentfrow et al., 2011; Vuoskoski & Eerola, 2011). Beyond these differing emotional responses, even when individuals share similar experiences, their preferences can diverge due to variations in the underlying mechanisms that transform these experiences into preferences. These mechanisms, often rooted in complex decision processes, determine how specific feelings translate into specific preferences (Garvert et al., 2015; Izuma & Adolphs, 2013). Traditionally, research has grappled with separating these two possible causes of preference divergence—differences in experience and differences in transformation processes—leading to a relative shortage of studies centered on the latter. For example, when participants are asked to choose between an apple and an orange, such experiments are not designed to differentiate between the experience of the stimuli and the transformation of that experience into a preference. This conflation makes it challenging to understand how these two distinct processes contribute to preference formation (Clairis & Pessiglione, 2022; Hare et al., 2011; Koban et al., 2023; Nitsch et al., 2024). Furthermore, participants often become aware of the purpose of these experiments, which can bias their decision-making responses (Rens et al., 2017; Si et al., 2021). As a result, there is a need for a new paradigm that provides a more detailed understanding of the neural mechanisms driving how individuals' experiences are transformed into their preferences.

Recently, neuroscientists have begun using naturalistic paradigms that mimic real-life environments and situations to enhance the ecological validity of their findings. These paradigms involve the use of stimuli such as movies, audio stories, or even virtual reality, presenting dynamic and multifaceted environments more representative of real-world experiences (Zhang et al., 2021). Nastase et al. (2020) similarly argued that naturalistic paradigms could provide more valid models of brain function and real-world behaviors, emphasizing the need to redesign experimental paradigms that reflect the complexity of real-life environments. These approaches offer several advantages, including capturing complex and dynamic interactions and eliciting more naturalistic responses. One of the main benefits is the ability to record neural responses to complex and unpredictable stimuli, leading to more robust and generalizable insights into real-world behaviors. For example, Finn et al. (2020) demonstrated how naturalistic paradigms can be used to study individual differences in brain activity during movie watching, revealing idiosyncratic neural patterns not easily captured in classical controlled experiments. Not only do naturalistic paradigms enhance the ecological validity of research findings, but they also allow the exploration of individual differences in complex behaviors, such as preference decision-making, that are often overlooked in highly controlled experimental settings (Grall & Finn, 2022). Thus, naturalistic neuroimaging paradigms have the potential to provide a deeper understanding of the neural mechanisms driving how experiences transform into preferences.

Recognizing the potential of naturalistic paradigms to investigate preference formation, we have begun designing studies that leverage these approaches. To investigate how neural mechanisms drive the transformation of experiences into preferences and what leads to individual differences in these decisions, Chou et al. (2024a) designed a novel two-phase study. This study comprised two phases: an in-scanner naturalistic viewing phase and a post-scanning task phase (Fig. 1). During the first phase, participants watched a series of videos on diverse topics inside an fMRI scanner (Chen et al., 2020; Chen & Qu, 2021; Chou et al., 2024b; Sonkusare et al., 2019; van Baar et al., 2021). Importantly, they were only instructed to watch the videos without any information about subsequent tasks, allowing

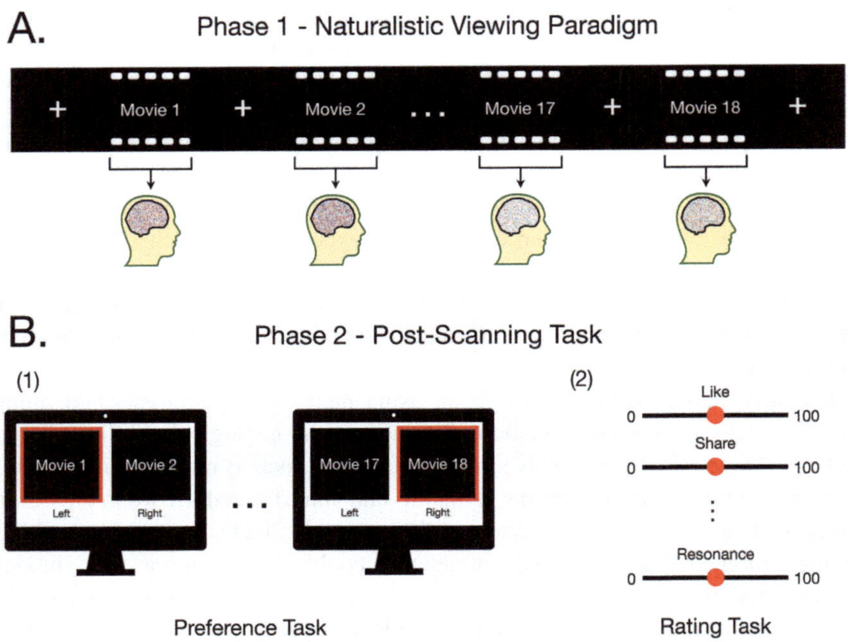

Fig. 1 Experimental procedure of the two-phase experiment. (**a**) During phase 1, participants completed the naturalistic viewing task in an fMRI scanner. Participants watched a series of videos, including scenes of nature, people interacting, and other varied content, separated by a fixed interval of fixation. (**b**) Phase 2 involved two behavioral tasks conducted outside the scanner. (1) In the Preference Task, participants were shown pairs of videos and were asked to choose which one they preferred by pressing the left or right button. (2) In the Rating Task, participants provided multidimensional ratings for each video, such as liking, sharing with others, and degrees of resonance, using a visual analog scale ranging from 0 to 100

researchers to capture neural responses to the stimuli without bias from anticipated evaluations (Fig. 1a). In the post-scanning phase, participants first completed a binary forced-choice task, choosing their preferred video from each pair they had previously watched (Fig. 1b(1)). This task assessed individual preferences through direct comparisons. Subsequently, they engaged in a rating task for each video, evaluating different dimensions, such as liking to capture their subjective appraisals (Fig. 1b(2)). This provided behavioral measures of their viewing experiences that complemented the neural data. This two-phase design effectively separated the processes contributing to preference divergence—differences in experience and differences in transformation processes—allowing researchers to investigate the mechanisms transforming subjective experiences into preferences (Dabas et al., 2024; Ma et al., 2023; Olsson et al., 2020).

Building upon this experimental design, we sought to identify the neural signatures associated with the transformation of experiences into preferences. To achieve this goal, we utilized whole-brain multivariate patterns from participants as they watched each video. Specifically, we computed the mean multivariate brain patterns across time for each video and used them as predictors of individuals' preferences (Koban et al., 2023). We employed Leave-One-Subject-Out (LOSO) cross-validation along with Support Vector Machine (SVM) algorithms (Suthaharan, 2016) to train a multivariate brain model that predicted preferences across all participants (Chang et al., 2022; Chen et al., 2023; Koban et al., 2023; Wager et al., 2013). This model demonstrated significant predictive performance with a balanced accuracy of 0.61, $p < .001$, indicating a robust relationship between brain response patterns during the viewing phase and subsequent preference decisions. Furthermore, we employed a similar approach to train individualized brain models for each participant to examine whether the relationship between brain response patterns and preference decisions was stronger within subjects than across subjects. We found that within-subject brain transformation models performed significantly better than across-subject models (i.e., within-subject balanced accuracy = 0.80; between-subject balanced accuracy = 0.61). These findings shed light on the crucial role of individual-specific neural mechanisms in the formation of preferences. These findings also reveal that variability in preference formation is not merely a consequence of differential experiences but is also profoundly shaped by the unique neural transformations these experiences undergo. The individualized nature of these neural processes suggests that each brain engages distinct roles to encode and interpret experiential inputs, leading to divergent preference outcomes across individuals. This study emphasizes the necessity of considering both the experiential content and the neural transformation processes when investigating the neural substrates of preference formation, thereby providing novel insights into the complexities of human decision-making from a computational cognitive neuroscience perspective.

Building upon the recognition of individualized neural processes, we further investigated whether these individual differences in neural transformation models were associated with differences in participants' appraisal models. We applied a similar approach to train a multivariate appraisal model that predicted preferences

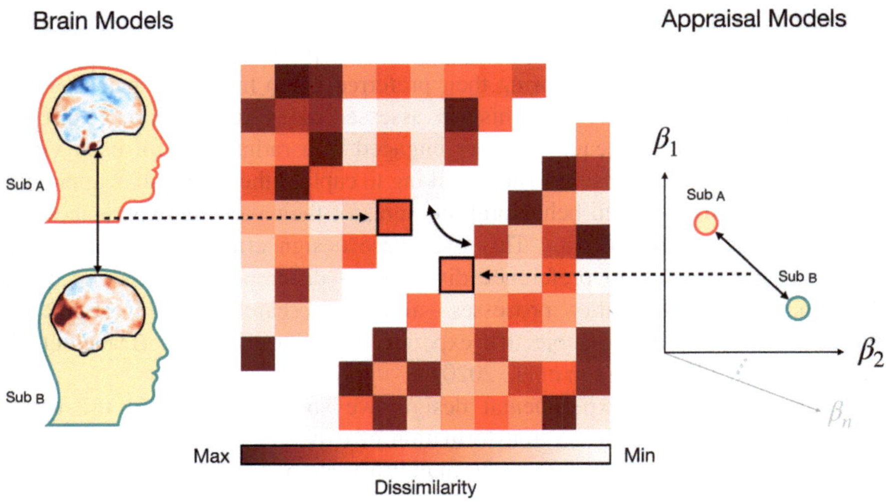

Fig. 2 Overview of the intersubject representational similarity analysis (ISRSA). We used ISRSA to examine whether intersubject similarities in the brain transformation models were associated with similarities in their appraisal models. Specifically, brain models from different subjects (e.g., Sub A and Sub B) are transformed into a dissimilarity matrix, where the color intensity represents the degree of dissimilarity between their brain patterns (left panel). Similarly, subjective appraisals are modeled in a similar way, producing a dissimilarity matrix based on their appraisals in the rating task at phase 2 (right panel). By correlating the dissimilarity matrices of brain models and subjective appraisals, ISRSA assesses how variations in brain patterns across individuals correspond to differences in their subjective experiences

within each individual. Then, we used intersubject representational similarity analysis (ISRSA) to test whether intersubject similarities in the brain transformation models were associated with similarities in their appraisal models (Fig. 2; Chen et al., 2020; Chen & Qu, 2021; Hsiao et al., 2024; van Baar et al., 2021). Our findings showed a strong association between similarities in the within-subject brain multivariate models and appraisal models, $r = 0.4$, $p < .001$, suggesting that these individual differences across different modalities are meaningful rather than mere noise (Dubois & Adolphs, 2016; Zadelaar et al., 2019). Interestingly, we also found that these individual differences in the brain models were not correlated with various dimensions of personality, implying that the transformations from subjective experiences to preferences are not simply driven by differences in existing personal traits, such as those measured by the Big Five Inventory or the Interpersonal Reactivity Index (Schäfer & Mehlhorn, 2017).

These results suggest that preferences might constitute a novel aspect of personal traits, independent of traditional personality measures. This idea aligns with the popularity of recommendation systems tailored to each individual. Researchers are just beginning to explore this possibility, challenging the classical view that personality traits alone account for individual differences in preference formation. This highlights the potential for individual differences in preferences to be considered an independent dimension of personality. Our study also provides an alternative approach for future research in neuroscience and psychology by disentangling different cognitive processes using an innovative paradigm that combines a naturalistic viewing phase with classical experimental tasks. This combination can help researchers uncover new insights into the complexities of subjective experiences and decision-making behaviors, which in turn can improve the efficiency of future recommendation systems (Soares & Viana, 2017). Furthermore, this hybrid approach can be extended to explore the mechanisms underlying other human behaviors, such as decision-making in social contexts or responses to marketing strategies (Bault & Rusconi, 2019). In summary, our innovative hybrid approach, utilizing a naturalistic viewing paradigm, offers substantial advantages, such as capturing neural responses to ecologically valid stimuli and reducing biases derived from task-based experimental designs. This new method provides an alternative way to explore the neural mechanisms of preference decisions (Clairis & Pessiglione, 2022; Williams et al., 2021).

4 Conclusion

In conclusion, our review provides substantial evidence for disentangling the complex processes involved in preference formation by distinguishing between subjective experiences and their subsequent transformation into preferences. Employing a naturalistic paradigm combined with advanced machine learning techniques, recent findings have revealed the idiosyncratic nature of these processes and found that individual differences in preferences constitute an independent dimension of personality. These insights not only deepen our understanding of the cognitive and neural mechanisms underpinning decision-making but also carry significant practical implications for developing more effective personalized recommendation systems. Given that preferences are deeply rooted in unique personal experiences, it is essential to incorporate authentic, everyday contexts into these systems. As digital interactions continue to play an increasingly prominent role in daily life, such interdisciplinary findings have the potential to refine both theoretical frameworks and practical solutions, leading to more engaging and adaptive online platforms. Ultimately, this perspective paves the way for advancements in neuroscience, psychology, and related fields by laying the groundwork for tools and models that more accurately capture the complexity and richness of human decision-making.

References

Ajaegbu, C. (2021). An optimized item-based collaborative filtering algorithm. *Journal of Ambient Intelligence and Humanized Computing, 12*(12), 10629–10636.

Amatriain, X., & Basilico, J. (2015). Recommender systems in industry: A Netflix case study. In F. Ricci, L. Rokach, & B. Shapira (Eds.), *Recommender systems handbook* (pp. 385–419). Springer US.

Asemi, A., Asemi, A., Ko, A., & Alibeigi, A. (2022). An integrated model for evaluation of big data challenges and analytical methods in recommender systems. *Journal of Big Data, 9*(1), 13.

Ballard, I. C., Hennigan, K., & McClure, S. M. (2017). Mere exposure: Preference change for novel drinks reflected in human ventral tegmental area. *Journal of Cognitive Neuroscience, 29*(5), 793–804.

Bault, N., & Rusconi, E. (2019). The art of influencing consumer choices: A reflection on recent advances in decision neuroscience. *Frontiers in Psychology, 10*, 3009.

Belleflamme, P., & Peitz, M. (2020). Ratings, reviews and recommendations. In *Handbook of cultural economics* (3rd ed., pp. 466–473). Edward Elgar Publishing.

Bobadilla, J., Serradilla, F., & Hernando, A. (2009). Collaborative filtering adapted to recommender systems of e-learning. *Knowledge-Based Systems, 22*(4), 261–265.

Bornstein, A. M., Khaw, M. W., Shohamy, D., & Daw, N. D. (2017). Reminders of past choices bias decisions for reward in humans. *Nature Communications, 8*, 15958.

Burton, A. C., Nakamura, K., & Roesch, M. R. (2015). From ventral-medial to dorsal-lateral striatum: Neural correlates of reward-guided decision-making. *Neurobiology of Learning and Memory, 117*, 51–59.

Chang, L. J., Li, X., Nguyen, K., Ranger, M., Begunova, Y., Chen, P.-H. A., Castrellon, J. J., Samanez-Larkin, G. R., Zald, D. H., Fareri, D. S., Delgado, M. R., & Tomova, L. (2022). A neural signature of reward. *bioRxiv*, 2022.08.23.504939. https://doi.org/10.1101/2022.08.23.504939

Chen, P.-H. A., & Qu, Y. (2021). Taking a computational cultural neuroscience approach to study parent-child similarities in diverse cultural contexts. *Frontiers in Human Neuroscience, 15*, 703999.

Chen, P.-H. A., Jolly, E., Cheong, J. H., & Chang, L. J. (2020). Intersubject representational similarity analysis reveals individual variations in affective experience when watching erotic movies. *NeuroImage, 216*, 116851. https://doi.org/10.1016/j.neuroimage.2020.116851

Chen, P.-H. A., Fareri, D., Güroğlu, B., Delgado, M. R., & Chang, L. J. (2023). Towards a neurometric-based construct validity of trust. *eLife*. https://doi.org/10.7554/eLife.90096.1

Chou, F.-C. B., Chiu, T.-A. P., Hsiao, P.-Y. A., & Chen, P.-H. A. (2024a). Unveiling the neural representations of preference with a naturalistic neuroimaging approach. In *Cognitive Neuroscience Society 2024a annual conference, Toronto, Canada*.

Chou, F.-C. B., Chen, Y.-C. C., Kuo, Y.-S. A., Hsiao, P.-Y. A., Lin, C. R., Huang, Y.-W. Y., Peng, W.-L., Pan, H.-J. S., & Chen, P.-H. A. (2024b). Facial insights: An exploration of similarity in spontaneous facial behaviors during pandemic lockdown. *PsyArXiv*. https://doi.org/10.31234/osf.io/46dvn

Clairis, N., & Pessiglione, M. (2022). Value, confidence, deliberation: A functional partition of the medial prefrontal cortex demonstrated across rating and choice tasks. *The Journal of Neuroscience, 42*(28), 5580–5592.

Colombo, B., Balzarotti, S., & Mazzucchelli, N. (2016). The influence of the dorsolateral prefrontal cortex on attentional behavior and decision making. A t-DCS study on emotionally vs. functionally designed objects. *Brain and Cognition, 104*, 7–14.

Cui, Z., Xu, X., Fei, X. U. E., Cai, X., Cao, Y., Zhang, W., & Chen, J. (2020). Personalized recommendation system based on collaborative filtering for IoT scenarios. *IEEE Transactions on Services Computing, 13*(4), 685–695.

Dabas, A., Bruckner, R., Schultz, H., & Benoit, R. G. (2024). Learning from imagined experiences via an endogenous prediction error. *bioRxiv*, 2024.06.24.600192. https://doi.org/10.1101/2024.06.24.600192

Dubois, J., & Adolphs, R. (2016). Building a science of individual differences from fMRI. *Trends in Cognitive Sciences, 20*(6), 425–443.

Ekstrand, M. D., Riedl, J. T., & Konstan, J. A. (2011). Collaborative filtering recommender systems. *Foundations and Trends® in Human–Computer Interaction, 4*(2), 81–173.

Fellows, L. K. (2007). The role of orbitofrontal cortex in decision making: A component process account. *Annals of the New York Academy of Sciences, 1121*, 421–430.

Fershtman, C., & Segal, U. (2018). Preferences and social influence. *American Economic Journal: Microeconomics, 10*(3), 124–142.

Finn, E. S., Glerean, E., Khojandi, A. Y., Nielson, D., Molfese, P. J., Handwerker, D. A., & Bandettini, P. A. (2020). Idiosynchrony: From shared responses to individual differences during naturalistic neuroimaging. *NeuroImage, 215*, 116828.

Foo, J. C., Haji, T., & Sakai, K. (2014). Prefrontal mechanisms in preference and non-preference-based judgments. *NeuroImage, 95*, 151–161.

Garvert, M. M., Moutoussis, M., Kurth-Nelson, Z., Behrens, T. E. J., & Dolan, R. J. (2015). Learning-induced plasticity in medial prefrontal cortex predicts preference malleability. *Neuron, 85*(2), 418–428.

Gold, J. I., & Shadlen, M. N. (2007). The neural basis of decision making. *Annual Review of Neuroscience, 30*, 535–574.

Gong, S. (2010). A collaborative filtering recommendation algorithm based on user clustering and item clustering. *Journal of Software Maintenance and Evolution: Research and Practice, 5*(7), 745–752.

Grall, C., & Finn, E. S. (2022). Leveraging the power of media to drive cognition: A media-informed approach to naturalistic neuroscience. *Social Cognitive and Affective Neuroscience, 17*(6), 598–608.

Hardardottir, H. (2017). Long term stability of time preferences and the role of the macroeconomic situation. *Journal of Economic Psychology, 60*, 21–36.

Hare, T. A., Camerer, C. F., Knoepfle, D. T., & Rangel, A. (2010). Value computations in ventral medial prefrontal cortex during charitable decision making incorporate input from regions involved in social cognition. *The Journal of Neuroscience, 30*(2), 583–590.

Hare, T. A., Schultz, W., Camerer, C. F., O'Doherty, J. P., & Rangel, A. (2011). Transformation of stimulus value signals into motor commands during simple choice. *Proceedings of the National Academy of Sciences of the United States of America, 108*(44), 18120–18125.

Herlocker, J. L., Konstan, J. A., Terveen, L. G., & Riedl, J. T. (2004). Evaluating collaborative filtering recommender systems. *ACM Transactions on Information and System Security, 22*(1), 5–53.

Hsiao, P.-Y. A., Kim, M. J., Chou, F.-C. B., & Chen, P.-H. A. (2024). Intersubject representational similarity analysis uncovers the impact of state anxiety on brain activation patterns in the human extrastriate cortex. *Brain Imaging and Behavior, 18*(2), 1–9.

Ifcher, J., & Zarghamee, H. (2011). Happiness and time preference: The effect of positive affect in a random-assignment experiment. *The American Economic Review, 101*(7), 3109–3129.

Izuma, K., & Adolphs, R. (2013). Social manipulation of preference in the human brain. *Neuron, 78*(3), 563–573.

Javari, A., Gharibshah, J., & Jalili, M. (2014). Recommender systems based on collaborative filtering and resource allocation. *Social Network Analysis and Mining, 4*(1), 234.

Jia, Z., Yang, Y., Gao, W., & Chen, X. (2015). User-based collaborative filtering for tourist attraction recommendations. In *2015 IEEE international conference on computational intelligence & communication technology* (pp. 22–25).

Keiflin, R., Reese, R. M., Woods, C. A., & Janak, P. H. (2013). The orbitofrontal cortex as part of a hierarchical neural system mediating choice between two good options. *The Journal of Neuroscience: The Official Journal of the Society for Neuroscience, 33*(40), 15989–15998.

Kim, K.-J., & Ahn, H. (2017). Recommender systems using cluster-indexing collaborative filtering and social data analytics. *International Journal of Production Research, 55*(17), 5037–5049.

Kim, H.-N., Ji, A.-T., Ha, I., & Jo, G.-S. (2010). Collaborative filtering based on collaborative tagging for enhancing the quality of recommendation. *Electronic Commerce Research and Applications, 9*(1), 73–83.

Koban, L., Lee, S., Schelski, D. S., Simon, M.-C., Lerman, C., Weber, B., Kable, J. W., & Plassmann, H. (2023). An fMRI-based brain marker of individual differences in delay discounting. *The Journal of Neuroscience: The Official Journal of the Society for Neuroscience, 43*(9), 1600–1613.

Koohi, H., & Kiani, K. (2016). User based collaborative filtering using fuzzy C-means. *Measurement, 91*, 134–139.
Kringelbach, M. L. (2005). The human orbitofrontal cortex: Linking reward to hedonic experience. *Nature Reviews. Neuroscience, 6*(9), 691–702.
La, A., Vo, P., & Vu, T. (2019). Adaptive collaborative filtering for recommender system. In *Graph-based representation and reasoning* (pp. 117–130). Springer.
Liebe, U., Hundeshagen, C., Beyer, H., & von Cramon-Taubadel, S. (2016). Context effects and the temporal stability of stated preferences. *Social Science Research, 60*, 135–147.
Loewenstein, G. F., Weber, E. U., Hsee, C. K., & Welch, N. (2001). Risk as feelings. *Psychological Bulletin, 127*(2), 267–286.
Ma, G.-F., Yang, X.-H., Tong, Y., & Zhou, Y. (2023). Graph neural networks for preference social recommendation. *PeerJ Computer Science, 9*, e1393.
Mrkva, K., & Van Boven, L. (2020). Salience theory of mere exposure: Relative exposure increases liking, extremity, and emotional intensity. *Journal of Personality and Social Psychology, 118*(6), 1118–1145.
Nakagawa, J., Takahashi, M., Okada, R., Matsushima, E., & Matsuda, T. (2015). Women's preference for a male acquaintance enhances social reward processing of material goods in the anterior cingulate cortex. *PLoS One, 10*(8), e0136168.
Nastase, S. A., Goldstein, A., & Hasson, U. (2020). Keep it real: Rethinking the primacy of experimental control in cognitive neuroscience. *NeuroImage, 222*, 117254.
Nitsch, A., Garvert, M. M., Bellmund, J. L. S., Schuck, N. W., & Doeller, C. F. (2024). Grid-like entorhinal representation of an abstract value space during prospective decision making. *Nature Communications, 15*(1), 1198.
Olsson, A., Knapska, E., & Lindström, B. (2020). The neural and computational systems of social learning. *Nature Reviews. Neuroscience, 21*(4), 197–212.
Pyo, T.-H., Lee, J. Y., & Park, H. M. (2022). The effects of consumer preference and peer influence on trial of an experience good. *JMR, Journal of Marketing Research, 59*(6), 1161–1178.
Ratcliff, R., & McKoon, G. (2008). The diffusion decision model: Theory and data for two-choice decision tasks. *Neural Computation, 20*(4), 873–922.
Rens, N., Bode, S., Burianová, H., & Cunnington, R. (2017). Proactive recruitment of frontoparietal and salience networks for voluntary decisions. *Frontiers in Human Neuroscience, 11*, 610.
Rentfrow, P. J., Goldberg, L. R., & Zilca, R. (2011). Listening, watching, and reading: The structure and correlates of entertainment preferences. *Journal of Personality, 79*(2), 223–258.
Robinson, E., & Higgs, S. (2012). Liking food less: The impact of social influence on food liking evaluations in female students. *PLoS One, 7*(11), e48858.
Sarwar, B., Karypis, G., Konstan, J., & Riedl, J. (2001). Item-based collaborative filtering recommendation algorithms. In *Proceedings of the 10th international conference on world wide web* (pp. 285–295).
Schäfer, T., & Mehlhorn, C. (2017). Can personality traits predict musical style preferences? A meta-analysis. *Personality and Individual Differences, 116*, 265–273.
Schafer, J. B., Frankowski, D., Herlocker, J., & Sen, S. (2007). Collaborative filtering recommender systems. In P. Brusilovsky, A. Kobsa, & W. Nejdl (Eds.), *The adaptive web: Methods and strategies of web personalization* (pp. 291–324). Springer.
Severin, V., Louviere, J. J., & Finn, A. (2001). The stability of retail shopping choices over time and across countries. *Journal of Retailing, 77*(2), 185–202.
Shakirova, E. (2017). Collaborative filtering for music recommender system. In *2017 IEEE conference of russian young researchers in electrical and electronic engineering (EIConRus)* (pp. 548–550).
Si, R., Rowe, J. B., & Zhang, J. (2021). Functional localization and categorization of intentional decisions in humans: A meta-analysis of brain imaging studies. *NeuroImage, 242*, 118468.
Soares, M., & Viana, P. (2017). The semantics of movie metadata: Enhancing user profiling for hybrid recommendation. In *Recent advances in information systems and technologies* (pp. 328–338).

Sonkusare, S., Breakspear, M., & Guo, C. (2019). Naturalistic stimuli in neuroscience: Critically acclaimed. *Trends in Cognitive Sciences, 23*(8), 699–714.
Steenkamp, J.-B. E. M., & Maydeu-Olivares, A. (2015). Stability and change in consumer traits: Evidence from a 12-year longitudinal study, 2002–2013. *JMR, Journal of Marketing Research, 52*(3), 287–308.
Su, X., & Khoshgoftaar, T. M. (2009). A survey of collaborative filtering techniques. *Advances in Artificial Intelligence, 2009*, 1–19.
Sugam, J. A., Saddoris, M. P., & Carelli, R. M. (2014). Nucleus accumbens neurons track behavioral preferences and reward outcomes during risky decision making. *Biological Psychiatry, 75*(10), 807–816.
Suthaharan, S. (2016). Support vector machine. In S. Suthaharan (Ed.), *Machine learning models and algorithms for big data classification: Thinking with examples for effective learning* (pp. 207–235). Springer US.
Symeonidis, P., & Zioupos, A. (2017). *Matrix and tensor factorization techniques for recommender systems*. Springer.
van Baar, J. M., Halpern, D. J., & FeldmanHall, O. (2021). Intolerance of uncertainty modulates brain-to-brain synchrony during politically polarized perception. *Proceedings of the National Academy of Sciences of the United States of America, 118*(20). https://doi.org/10.1073/pnas.2022491118
van Holstein, M., & Floresco, S. B. (2020). Dissociable roles for the ventral and dorsal medial prefrontal cortex in cue-guided risk/reward decision making. *Neuropsychopharmacology, 45*(4), 683–693.
Vuoskoski, J. K., & Eerola, T. (2011). The role of mood and personality in the perception of emotions represented by music. *Cortex; A Journal Devoted to the Study of the Nervous System and Behavior, 47*(9), 1099–1106.
Wager, T. D., Atlas, L. Y., Lindquist, M. A., Roy, M., Woo, C.-W., & Kross, E. (2013). An fMRI-based neurologic signature of physical pain. *The New England Journal of Medicine, 368*(15), 1388–1397.
Wang, J., de Vries, A. P., & Reinders, M. J. T. (2006). Unifying user-based and item-based collaborative filtering approaches by similarity fusion. In *Proceedings of the 29th annual international ACM SIGIR conference on research and development in information retrieval* (pp. 501–508).
Williams, T. B., Burke, C. J., Nebe, S., Preuschoff, K., Fehr, E., & Tobler, P. N. (2021). Testing models at the neural level reveals how the brain computes subjective value. *Proceedings of the National Academy of Sciences of the United States of America, 118*(43). https://doi.org/10.1073/pnas.2106237118
Xue, F., He, X., Wang, X., Xu, J., Liu, K., & Hong, R. (2019). Deep item-based collaborative filtering for top-N recommendation. *ACM Transactions on Information and System Security, 37*(3), 1–25.
Yen, H. R., & Chuang, S.-C. (2008). The effect of incidental affect on preference for the status quo. *Journal of the Academy of Marketing Science, 36*(4), 522–537.
Zadelaar, J. N., Weeda, W. D., Waldorp, L. J., Van Duijvenvoorde, A. C. K., Blankenstein, N. E., & Huizenga, H. M. (2019). Are individual differences quantitative or qualitative? An integrated behavioral and fMRI MIMIC approach. *NeuroImage, 202*, 116058.
Zajonc, R. B. (2001). Mere exposure: A gateway to the subliminal. *Current Directions in Psychological Science, 10*(6), 224–228.
Zhang, Y., Kim, J.-H., Brang, D., & Liu, Z. (2021). Naturalistic stimuli: A paradigm for multi-scale functional characterization of the human brain. *Current Opinion in Biomedical Engineering, 19*. https://doi.org/10.1016/j.cobme.2021.100298
Zhao, Z.-D., & Shang, M.-S. (2010). User-based collaborative-filtering recommendation algorithms on Hadoop. In *2010 third international conference on knowledge discovery and data mining* (pp. 478–481).

Part V
Social Preferences and Social Decision-Making

Patricia L. Lockwood, David V. Smith, and Dominic S. Fareri

Our decisions do not occur in a social vacuum. Instead, we often have to decide whether to help, harm, or learn from the consequences of our actions when they affect other people. Neuroeconomists have long highlighted how social preferences for fairness, trust, and altruism fundamentally shape how we interact with others (Fehr & Camerer, 2007). These preferences are exhibited when people make choices that result in a conflict between self and other preferences. More recently, researchers have extended these concepts to situations that are not financially costly for oneself to probe altruistic behaviours where they are perhaps purely other-regarding (Lockwood et al., 2020). This section considers several factors that influence social preferences and social decision-making in humans and in other animals. It considers helping or prosocial behaviours before revealing some of the influence and neural circuitry that underpins when we need to interact with another person strategically. The section then probes how other people's choices may, in fact, influence our own, before ending with an overview of how frameworks from ethology can provide new approaches to integrate findings from human and non-human animal research.

Prosociality, or decisions that benefit others, is essential for health and well-being across the lifespan (Lockwood et al., 2025) and even predicts labour market success (Kosee & Tincani, 2020). A key debate in neuroeconomics is the extent to which humans and other animals are willing to be prosocial or exhibit purely self-interest. Sandhu et al. (this volume) review the conceptualization and measurement

P. L. Lockwood (✉)
University of Birmingham, Birmingham, UK
e-mail: p.l.lockwood@bham.ac.uk

D. V. Smith
Temple University, Philadelphia, PA, USA

D. S. Fareri
Adelphi University, Garden City, NY, USA

of prosocial behaviour, the brain regions that may underpin it, and how stress, culture, age, and genetics modulate the willingness to be prosocial.

As well as frequently deciding whether to help others, we may find ourselves in situations where our goal is instead to compete or interact strategically for our own benefit. Neuroeconomists have created various games to elicit these behaviours. These games can be used with models to predict behaviour and estimate an equilibrium where no player has an incentive to change their decision. However, humans are often shown to deviate from the optimal behaviour, which may be due to different social preferences (Camerer, 2003). Konovalov (this volume) outlines current research examining the computational mechanisms and brain regions that are recruited during strategic decision-making and how these could be domain-specific or domain-general for social and non-social decision-making.

The next chapter reviews a different type of social decision-making, where observing someone else's choice changes our own. This social influence can have a strong effect on our economic preferences. Su and Lockwood (this volume) focus on how social influence changes the types of economic preference outlined in earlier chapters, such as risk-taking, temporal discounting, effort discounting, and moral preferences. The chapter highlights how such preferences are not fixed but fundamentally shaped by what others want.

Lastly, Shi et al. (this volume) consider complex social behaviours and the extent to which they can be measured in animal models. They put forward the idea of "comparative game theory" whereby similar paradigms can be used in humans and animals, with methodological and technical approaches tailored to what is possible in different species. They advocate for integrating naturalistic approaches such as wearable neuroimaging, which has recently shown promise (Fan et al., 2021).

In summary, this section covers the different types of social decisions and social preferences we encounter in everyday life. Whether we should help others, compete against them, or learn from their preferences. Again, themes from earlier in the book, including the importance of naturalistic paradigms and theories, and the central role that non-human animal research plays in understanding economic decision-making, are highlighted. Together, the section shows that social decisions are just as important, if not more, than those same decisions we make when they only affect ourselves. The chapters also commonly highlight the central debate regarding whether there are particular brain areas specialized for social behaviour or whether these brain systems are the same as for decision-making in general.

References

Camerer, C. (2003). *Behavioral game theory: Experiments in strategic interaction*. Princeton University Press.

Fan, S., Dal Monte, O., & Chang, S. W. (2021). Levels of naturalism in social neuroscience research. *IScience, 24*(7), 102702.

Fehr, E., & Camerer, C. F. (2007). Social neuroeconomics: The neural circuitry of social preferences. *Trends in Cognitive Sciences, 11*(10), 419–427.

Konovalov, A. (this volume). The neural mechanisms of strategic decision making. In D. V. Smith, P. L. Lockwood, & D. S. Fareri (Eds.), *Neuroeconomics: Core topics and current directions*. Springer. Preprint: https://doi.org/10.31234/osf.io/msyh5

Kosse, F., & Tincani, M. M. (2020). Prosociality predicts labor market success around the world. *Nature Communications, 11*, 5298.

Lockwood, P. L., Apps, M. A., & Chang, S. W. (2020). Is there a 'social' brain? Implementations and algorithms. *Trends in Cognitive Sciences, 24*(10), 802–813.

Lockwood, P. L., van den Bos, W., & Dreher, J. C. (2025). Moral learning and decision-making across the lifespan. *Annual Review of Psychology, 76*(1), 475–500.

Sandhu, T., Saju, S. R., Wang, R., & Tusche, A. (this volume). Prosociality and the brain: Understanding when and why we help, share, and cooperate. In D. V. Smith, P. L. Lockwood, & D. S. Fareri (Eds.), *Neuroeconomics: Core topics and current directions*. Springer. Preprint: https://doi.org/10.31234/osf.io/5gtcu_v1

Shi, W., Meisner, O. C., Jadi, M. P., Nandy, A. S., & Chang, S. W. C. (this volume). Comparative game theory: Bringing ethology back into decision neuroscience. In D. V. Smith, P. L. Lockwood, & D. S. Fareri (Eds.), *Neuroeconomics: Core topics and current directions*. Springer.

Su, Z., & Lockwood, P. L. (this volume). The neuroeconomics of social influence and contagion. In D. V. Smith, P. L. Lockwood, & D. S. Fareri (Eds.), *Neuroeconomics: Core topics and current directions*. Springer. Preprint: https://doi.org/10.31234/osf.io/tgmj3_v1

Prosociality and the Brain: Understanding When and Why We Help, Share, and Cooperate

Tavneen Sandhu, Sarah Saju, Ruien Wang, and Anita Tusche

Abstract Are humans naturally inclined to help others, or are we more focused on ourselves? How does the brain decide whether to put someone else's needs ahead of our own? Helping, sharing, and cooperating are key human behaviors that have shaped societies and allowed civilizations to thrive. These prosocial actions often involve giving up something—like time, effort, or resources—to benefit others. Despite the personal cost, prosocial actions play a crucial role in building relationships, improving well-being and health, and tackling global challenges like climate change. Prosocial behavior is a complex concept with many layers. Researchers from psychology, neuroscience, economics, and other fields use a range of tools and methods to study it. This chapter explores human prosociality in several ways. It looks at how researchers measure prosocial behavior, examines the brain regions involved in making prosocial decisions, and discusses the factors that influence when and why people act in prosocial ways. Specifically, it outlines how environmental influences like stress and culture, along with individual differences such as age and genetics, shape prosocial behaviors. Finally, the chapter explores exciting trends and future directions in research, helping us better understand what drives humans to care for one another.

1 Introduction

In 2018, Mamoudou Gassama risked his life to scale a Paris apartment building and rescue a 4-year-old dangling from a fifth-floor balcony. His heroic act went viral, raising a fundamental question: What compels people to help others, even at

T. Sandhu · A. Tusche (✉)
Department of Psychology, Queen's University, Kingston, ON, Canada

Centre for Neuroscience Studies, Queen's University, Kingston, ON, Canada
e-mail: anita.tusche@queensu.ca

S. Saju · R. Wang ·
Department of Psychology, Queen's University, Kingston, ON, Canada

D. V. Smith et al. (eds.), *Neuroeconomics: Core Topics and Current Directions*,
https://doi.org/10.1007/978-3-032-02925-6_24

great personal cost? While most prosocial behaviors—such as helping, sharing, and cooperating—are far less risky, they are essential for social cohesion.

Simply put, prosociality involves acts that benefit others, even when they require time, effort, or resources (Fehr & Fischbacher, 2003; Tusche & Bas, 2021; Pfattheicher et al., 2022). Prosocial behaviors influence how happy and healthy people are, how long they live (Post, 2014; Raposa et al., 2016; Hilbrand et al., 2017; Hui et al., 2020), and how socially connected or lonely they feel (Lanser & Eisenberger, 2022). On a larger scale, prosocial actions contribute to a society's economic success, such as how well its job market works (Kosse & Tincani, 2020; Cacioppo & Cacioppo, 2018). Human prosociality also plays a role in tackling global issues like sustainability and climate change (Klein et al., 2022; Hauser et al., 2014). Because of its importance, prosociality is studied in many fields, including psychology, neuroscience, evolutionary biology, sociology, behavioral economics, philosophy, and political science—all with unique focuses and contributions. For instance, neuroscientists contribute insights into the biological mechanisms underpinning prosociality. Economists bring expertise in modeling incentives and trade-offs that shape altruistic behavior. Psychologists offer a nuanced understanding of individual differences and contextual factors. Integrating these various perspectives allows for a more comprehensive understanding of human prosociality and its drivers.

This chapter focuses on understanding prosocial behavior by exploring four areas of research. First, we will examine how researchers study prosociality, using tools from psychology, neuroscience, and behavioral economics (disciplines that form the interdisciplinary field of neuroeconomics). We will then look at the brain regions involved in prosocial decision-making, including areas responsible for evaluating available choice options, exercising self-control, and understanding others' thoughts and feelings (like empathy). Next, we will discuss why people act prosocially in some situations but not others, considering how factors like stress, culture, age, and genetics influence behavior. Some factors make prosocial behavior flexible and situation-dependent, while others explain why some individuals are consistently more prosocial than others. Finally, we will highlight current trends and exciting future directions in cross-disciplinary research on prosociality.

2 Measuring Human Prosociality

How Do We Measure Human Prosociality? This question is complex because prosociality includes various behavioral phenomena like generosity, cooperation, and reciprocity, all of which can be measured in different ways. Think about a recent act of kindness, like holding the door for someone, donating to or volunteering at a charity, or even extraordinary acts like donating a kidney to a stranger. These examples, and many others (like Gassama's rescue of the child), fall under the umbrella of prosocial behavior. Different disciplines, such as psychology, economics, and neuroscience, focus on specific aspects of prosociality and use various tools to measure it (Böckler et al., 2016; Singer & Tusche, 2014).

Psychology commonly uses self-report questionnaires like the Prosocialness Scale (Caprara et al., 2005) and the Self-Report Altruism Scale (Rushton et al., 1981; Manzur & Olavarrieta, 2021) to measure prosocial tendencies. They are easy to administer, even to large groups, and provide standardized results. However, they can be biased since participants may give answers they think are socially acceptable. To capture actual behavior, early psychological studies often relied on "staged" real-world scenarios, such as helping an experimenter or a confederate with tasks like picking up dropped items, mailing a "lost" letter, or assisting with an unrelated task (Batson, 2011). Modern psychology now often uses behavioral tasks that offer better experimental control and allow repeated measurements. Examples include tasks involving charitable donations (Tusche et al., 2016; Schulreich et al., 2022, 2023; Hare et al., 2010), computer games (Leiberg et al., 2011), virtual reality simulations (Jin et al., 2024), or physical effort-based activities where participants exert effort on behalf of others. The latter exploit the fact that many prosocial acts in daily life require exerting effort, a cost that many people are averse to (Lockwood et al., 2017, 2022).

Economics often uses game-theoretical tasks to study social behaviors (Camerer, 2011). These simplified scenarios focus on key features of more complex real-world social interactions to study phenomena like generosity, cooperation, reciprocity, or trust. Excellent introductions to common economic games are available here (Thielmann et al., 2021; Van Dijk & De Dreu, 2021; Pisor et al., 2020). A well-known example is the dictator game (Forsythe et al., 1994; Kahneman et al., 1986), where one participant (the "dictator") decides how to split a resource, such as money (e.g., 10$), between themselves and another person. The recipient has no say, and the dictator's decision is final, similar to real-life situations like charitable giving or blood donation, where one individual fully controls a resource and must decide whether to share it. Decisions in these games are typically incentivized with real payoffs, ensuring participants take them seriously (Klein & Hilbig, 2019). Unlike some psychological behavioral paradigms, deception is rarely used in economic games (and economics more generally). Interestingly, while giving nothing maximizes personal benefit in the dictator game, participants usually share about 28% of the total resource (Engel, 2011). This challenges the traditional economic view of the *Homo Economicus*, the rational "economic man," who acts solely to maximize personal gain (Thaler, 2000). Instead, this generosity reflects concern for others or a desire to follow fairness norms (Guala & Mittone, 2010). However, behavior in the dictator game varies widely, with some giving nothing and others sharing generously (Engel, 2011), sparking research into what drives these individual differences. Economic games are now widely used in psychology and neuroscience, making them one of the most popular tools for studying prosociality.

Economic games like the dictator game can be played either once (a "one-shot" game) or over multiple rounds. In repeated games, participants often play several rounds with different partners, whose characteristics may vary to experimentally manipulate factors influencing social behavior. For example, people tend to be more generous when their partners are identifiable (e.g., via a picture), seen as deserving (e.g., a sick child), or belong to their ingroup rather than an outgroup—

a phenomenon known as "parochial altruism," which is closely linked to social discrimination and group conflicts (Lee & Feeley, 2016; Bas et al., 2025a, b, Pisor & Ross, 2024). Modified versions of the dictator game in repeated setups may also present participants with two proposed splits: one that benefits themselves at the partner's expense (selfish option) and another that benefits the partner at a cost to themselves (altruistic choice). These choices mimic real-world trade-offs in prosocial behavior, with the frequency of altruistic choices serving as a measure of altruism. Repeated rounds of economic tasks like the dictator game are particularly useful in neuroscientific studies, where they enable accurate brain measurements of noisy activity underlying prosocial behavior. They also support advanced analyses that require many trials, such as computational modeling, which estimates how factors like personal gain, concern for others, or fairness influence decisions (Tusche & Bas, 2021; Tusche & Hutcherson, 2018; Hutcherson & Tusche, 2022). When paired with neuroimaging, these models provide deeper insights into the brain mechanisms driving prosociality, making them a cornerstone of interdisciplinary research in neuroeconomics.

Integration of Assessments to Capture Distinct Facets of Prosociality Prosociality has been studied across various disciplines using different measurement tools, often independently. Recently, researchers have started integrating findings from these fields to better understand how the methods relate to one another and capture distinct aspects of human prosociality (Rhoads et al., 2021). For example, one influential study analyzed thousands of decisions from different economic games (Peysakhovich et al., 2014). The results showed that people's behavior in different cooperation games was positively correlated and consistent over time, suggesting a stable "cooperative phenotype." However, cooperation in these games did not align with prosocial acts aimed at enforcing social norms, such as altruistic punishment (i.e., punishing others for violating established norms through selfish or uncooperative behavior), suggesting that cooperation and social norm enforcement reflect distinct facets of prosocial behavior.

Another study combined 14 widely used measures from behavioral economics, neuroscience, and psychology to identify core aspects of prosociality (Böckler et al., 2016, 2018a, b). These measures included economic games, questionnaires, video games, and real-world actions like charitable donations. The study found that these tools capture three distinct dimensions of prosocial behavior: altruistically motivated actions (including cooperation), norm-driven behavior, and self-reported prosociality. Interestingly, self-reported prosociality did not always match actual altruistic or norm-driven behavior (Böckler et al., 2018a), emphasizing the importance of using multiple methods to fully understand prosociality. This multi-dimensional approach has also been useful in studying how prosocial behaviors can be influenced, for example, through mental training or episodic simulations (Böckler et al., 2018b; Campbell et al., 2022). Overall, this research highlights the value of diverse tools in capturing the full spectrum of prosocial behavior.

3 Neural Basis of Prosociality: The Altruistic Brain

Tools like economic games have inspired countless neuroimaging studies that reveal how the brain processes social motives and prosocial decisions. Reviews and meta-analyses provide detailed insights on this topic (Tusche & Bas, 2021; Bellucci et al., 2020; Lamm & Forbes, 2023; Wu & Hong, 2022; Luo, 2018; Bonandrini et al., 2025), including the neural basis of altruism, which we now examine. Prosociality broadly refers to behaviors benefiting others, regardless of motivation—such as enforcing social norms or boosting one's reputation. Altruism, a subset of prosociality, is driven by selfless concern for another's well-being, often at personal cost. Key brain regions involved in altruistic decision-making include those for value assessment, executive control, empathy, and mentalizing (Fig. 1). Let's explore how these processes shape our altruistic decisions.

Valuation Valuation plays a key role in social and altruistic decision-making and includes the ventral striatum (VS) and the ventromedial prefrontal cortex (VMPFC) (Bellucci et al., 2020; Pärnamets et al., 2020; Cutler & Campbell-Meiklejohn, 2019; Suzuki & O'Doherty, 2020; Ruff & Fehr, 2014). These brain regions help assess the "value" of different options, whether concrete (like money or food) or abstract (like social rewards) (Tusche & Hutcherson, 2018; Bartra et al., 2013; Clithero & Rangel, 2014). Damage to the VMPFC can impair prosocial behavior (Lockwood et al., 2024). Valuation-related brain areas compute and integrate choice-relevant factors—such as reward magnitude, probability, and timing—to enable complex

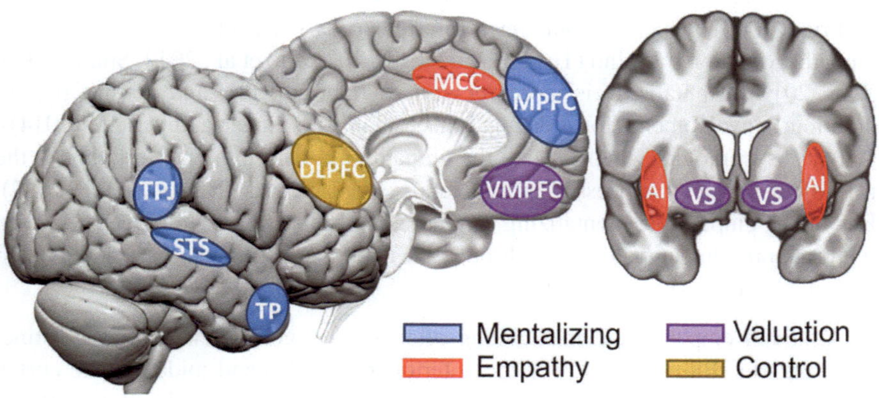

Fig. 1 Brain regions involved in altruistic decision-making. Colors reflect popular assumptions about each brain area's functional role in prosocial decision-making (note, though, that displayed brain regions are also involved in other mental processes). Highlighted brain regions include core brain areas involved in valuation (VMPFC, ventromedial prefrontal cortex; VS, ventral striatum), executive control (DLPFC, dorsolateral prefrontal cortex), empathy (AI, anterior insula; MCC, midcingulate cortex), and mentalizing (MPFC, medial prefrontal cortex; TPJ, temporoparietal junction; STS, superior temporal sulcus; TP, temporal pole). (Reprinted from Tusche and Bas (2021) CC BY 4.0)

decisions, like whether to prioritize personal benefit or act altruistically. These brain areas are also influenced by other brain networks, including those involved in attention, memory, and learning (Pärnamets et al., 2020; Gao et al., 2024). Additionally, broader social and emotional processes, such as empathy and mentalizing, provide inputs to valuation computations in prosocial contexts (Tusche & Bas, 2021).

Executive Control Executive control processes help us manage prosocial decisions by overriding selfish impulses, balancing short-term rewards with long-term benefits, and adapting behavior to social contexts or norms. By doing so, brain areas involved in executive control can enable us to prioritize fairness, empathy, or societal norms over instinctive self-interest. Key areas include the dorsolateral prefrontal cortex (DLPFC), which interacts with the brain's valuation network. The DLPFC has been suggested to serve different roles, such as inhibiting selfish tendencies, adjusting value signals during decision-making, and flexibly prioritizing goals or norms based on the situation or social expectations in a particular context (Tusche & Hutcherson, 2018; Hutcherson & Tusche, 2022; Hare et al., 2009; Wyss et al., 2024). Although the precise computational role of the DLPFC is debated, its importance in prosocial decision-making is well-established (Bellucci et al., 2020; Carlson & Crockett, 2018; Kwon et al., 2023).

Empathy and Mentalizing Understanding others' emotions and thoughts plays a crucial role in prosocial behaviors (Saulin et al., 2024). Here, we highlight two social processes that drive altruistic acts: empathy and mentalizing. Empathy refers to our ability to share someone else's emotions, like their happiness or distress (Singer & Tusche, 2014; Decety & Jackson, 2004). Mentalizing refers to our capacity to infer others' mental states, such as their beliefs, expectations, or goals (Frith & Frith, 2006), which is sometimes referred to as perspective-taking, cognitive empathy, or theory of mind (Tusche et al., 2016, 2023; Ho et al., 2022; Smith, 2006; Zaki, 2017). Mentalizing is more "cognitive" than empathy and doesn't require emotional involvement (for a detailed discussion, see Singer and Tusche (2014)). Classic psychological theories like the empathy-altruism hypothesis highlight the role of these social processes in motivating altruistic acts (Batson, 2011, 2022). Enhancing empathy or mentalizing can increase altruistic behavior (Batson, 2011), while suppressing empathy (e.g., through the dehumanization of others) is linked to antisocial actions (Bussey, 2020).

Although empathy and mentalizing often work together, they engage distinct brain regions. Empathy involves the anterior insula (AI) and midcingulate cortex (MCC), while mentalizing relies on areas like the medial prefrontal cortex (MPFC), temporoparietal junction (TPJ), posterior superior temporal sulcus (pSTS), and temporal poles (TP) (Schurz et al., 2021; Bzdok et al., 2012; Fan et al., 2011; Maliske et al., 2023). In complex naturalistic situations, empathy and mentalizing often co-activate and interact (Maliske et al., 2023). Numerous studies show that prosocial behaviors involve these processes and activate their underlying neural substrates (e.g., a meta-analysis of over 600 neuroimaging studies, see Bellucci et al. (2020)). For instance, a study found that activation in the anterior insula

(AI) reflects affective empathy for beneficiaries of charities, while activation in the temporoparietal junction (TPJ) reflects mentalizing regarding these people, with both predicting altruistic actions like charitable donations (Tusche et al., 2016). These findings suggest that individual differences in empathy and mentalizing can significantly shape prosocial decisions (Qiu et al., 2024).

Why Do We Care About Neural Underpinnings of Prosociality? Understanding the neural mechanisms driving prosocial behavior is important for several reasons. First, these insights can suggest targeted novel interventions to make people more generous or cooperative. For example, activation in empathy-related areas like the anterior insula (AI) prior to generous decisions suggests that differences in empathy may explain variation in prosociality across people and contexts, suggesting that cultivating social emotions boosts people's altruism (a notion confirmed in a large-scale intervention study (Böckler et al., 2018b)). Second, neuroscience can distinguish between related mental processes, such as empathy and mentalizing, which may appear similar in behavior but differ in underlying mechanisms (Tusche & Bas, 2021). Thus, it allows for improving theories of human prosociality. Third, neuroimaging captures various processes influencing social behavior, including value assessment and self-control, helping to explain how affective, cognitive, and motivational factors interact to shape social behaviors. The use of computational models has further advanced our understanding by revealing how these processes integrate in the brain and connect to psychological functions.

4 Moderating Factors

Prosocial behaviors can vary depending on individual and contextual factors. For example, a single night of sleep loss can reduce helping behavior the next day by impairing brain regions involved in mentalizing (Fig. 1). At the societal level, insufficient sleep predicts declines in daily acts of kindness and reduces charitable donations after daylight-saving time transitions, emphasizing the role of sleep in sustaining cooperative societies (Simon et al., 2022). Beyond sleep, other factors like age, genetics, culture, and stress also influence prosociality, as outlined in the following sections. Stress, in particular, has gained significant attention for its impact on prosocial behavior over the past decade, and we will explore its mechanisms in more depth. We will also explore how drivers of prosociality, like empathy and mentalizing, relate to these moderating factors.

4.1 Acute Stress

Stress is a common part of daily life, often occurring in social contexts. Does stress make us more or less prosocial? This question is crucial, as many situations where we encounter others in distress and need—like witnessing a child dangling from a balcony from our opening example—are inherently stressful. Research in psychology, neuroscience, and behavioral economics provides mixed findings on this relationship (Faber & Häusser, 2022; von Dawans et al., 2021; Buchanan & Preston, 2014). Below, we first summarize the mechanisms of acute stress, how it is measured, and why it might affect prosociality, before outlining its effects on prosociality.

4.1.1 Background on Acute Stress: Mechanisms, Measurement, and Theories

Mechanisms and Measurement When you speak in public, it's common to experience physical signs of acute stress, such as a racing heart, sweaty palms, or a dry mouth, all signs of the body preparing to respond to threats or challenges (Selye, 1950; O'Connor et al., 2021). The body undergoes changes in the nervous, cardiovascular, endocrine, and immune systems and triggers cognitive and emotional reactions (Schneiderman et al., 2005; Russell & Lightman, 2019). For instance, acute stress activates several systems, including the hypothalamic-pituitary-adrenal (HPA) axis and autonomic nervous system, releasing cortisol and noradrenaline (Schneiderman et al., 2005). Stress is commonly measured through self-reports, capturing subjective stress, and cortisol levels, reflecting the biological stress response (Crosswell & Lockwood, 2020). Other measures include alpha-amylase (a saliva enzyme), thermal infrared imaging (capturing facial heat changes), heart rate, or brain activity (Giannakakis et al., 2019; Dedovic et al., 2009; Engert et al., 2014). These varied measures complicate understanding stress's impact on prosocial behavior.

Inducing Stress in the Lab Common methods to study the effects of stress on prosocial behavior in controlled laboratory settings include the Trier Social Stress Test (TSST) and the Cold Pressor Test (CPT). The TSST involves public speaking (a mock job interview) and math tasks performed before a nonresponsive panel of judges (Fig. 2a). This social-evaluative threat—where others observe us and could negatively characterize our performance with little perceived controllability—reliably induces stress responses (Dickerson & Kemeny, 2004). TSST variants for children, group settings, virtual reality, or neuroimaging studies make it a versatile research tool (Von Dawans et al., 2011; Gunnar et al., 2021; Zimmer et al., 2019; Frisch et al., 2015; Spengler et al., 2020; Dedovic et al., 2005; Schiller et al., 2023). The CPT, where participants immerse their hand in cold water, induces physical stress (Lovallo, 1975) (Fig. 2b). To examine the effects of acute stress on (pro)social behaviors, researchers often compare stressed participants to a non-stressed control group.

Theoretical Models of Stress and Social Behavior Two main theories explain stress's impact on prosociality. The "(freeze-)fight-or-flight" model suggests stress increases self-serving, anti-social behavior to prioritize safety and resources

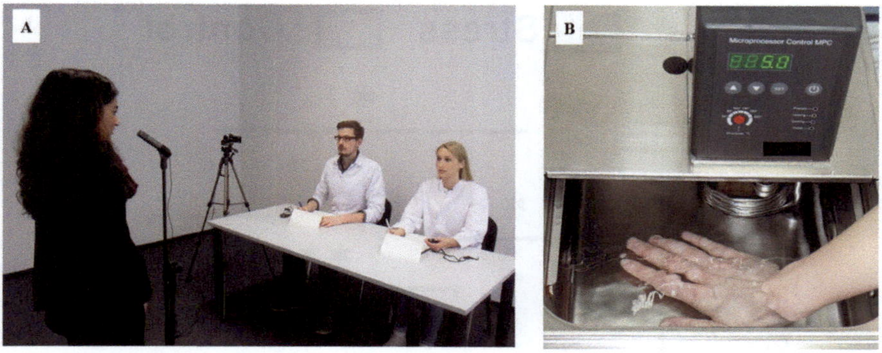

Fig. 2 Inducing stress in the laboratory. (**a**) The Trier Social Stress Test (TSST) involves a mock job interview and math test before a nonresponsive panel of judges. This paradigm evokes acute social stress by simulating real-world situations in which we are observed and evaluated by our social interaction partners. (**b**) The Cold Pressor Test (CPT) induces physical stress by having participants immerse one hand into a container of ice-cold water, typically around 0–5 °C (32–41 °F), for a set amount of time, often up to 2 min, or until they can no longer tolerate the discomfort. (Reprinted from Frisch et al. (2015) and Ommerborn et al. (2023) CC BY 4.0)

(Dedovic et al., 2009; Roelofs, 2017). In contrast, the "tend-and-befriend" model posits that stress enhances prosociality by encouraging social support and affiliation (Taylor, 2006). Notably, despite conflicting predictions, both models highlight the adaptive nature of stress responses shaped by social environments (von Dawans et al., 2021).

4.1.2 Understanding When and How Acute Stress Shapes Prosocial Behavior

Does Stress Make Us More or Less Prosocial? Research on acute stress and prosociality shows mixed results. Some studies find that stress increases generosity and helping (Buchanan & Preston, 2014; Hensel et al., 2022; von Dawans et al., 2012, 2019; Margittai et al., 2015; Ying et al., 2022; Takahashi et al., 2007; Zhen et al., 2021), such as shifting responses to injustice from punishing perpetrators to helping victims (Wang et al., 2024) (Fig. 3). Others report reduced prosocial behavior (Vinkers et al., 2013; Sollberger et al., 2016; Forbes et al., 2024), while some find no significant effect (Schulreich et al., 2022; Veszteg et al., 2021; Steinbeis et al., 2015; Azulay et al., 2022). These inconsistencies arise from differences in stress induction, measurement methods (of the stress responses and prosocial behavior), and the dynamic nature of stress, which may have different effects at different stages (Schwabe et al., 2022).

Moreover, individual and contextual factors such as personality (Ying et al., 2022; Sollberger et al., 2016; Singer et al., 2021) and prior life stress (Hensel et al., 2022) can influence whether stress promotes or inhibits prosociality. Consistently, a

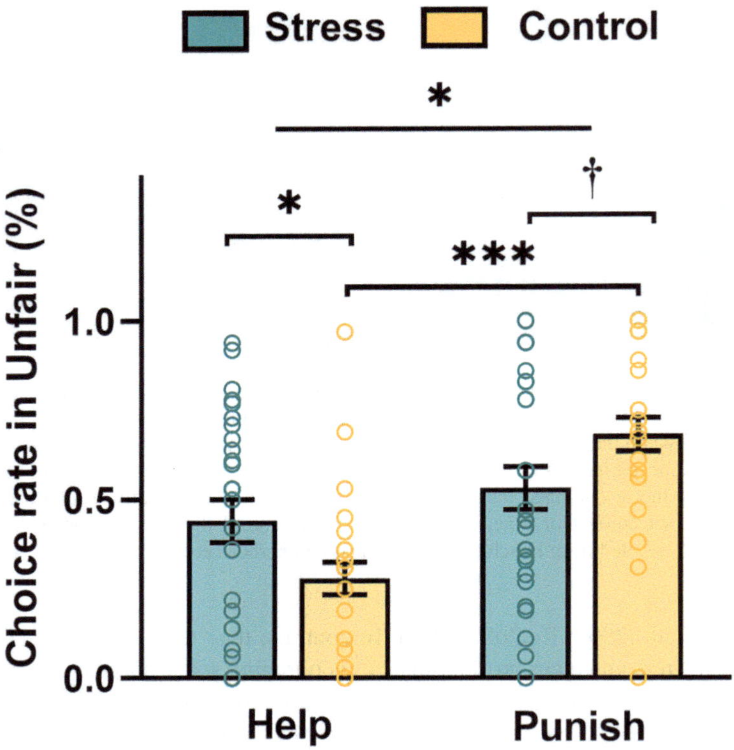

Fig. 3 People are often willing to interfere in the injustices of others by either punishing the transgressor or helping the victim, even if they are not involved in the unfair event. The figure illustrates that acute stress (green bars, induced by a cold pressor test, CPT) reduced the willingness to punish the violator while increasing the willingness to help victims, compared to a non-stressed control group (yellow bars). (Reprinted from Wang et al. (2024) CC BY 4.0)

recent meta-analysis that reviewed 23 studies (~2200 participants) found no consistent overall effect of stress on prosociality, instead emphasizing context dependence (Nitschke et al., 2022). For instance, stress appears to increase prosociality in simple decisions (e.g., binary choices like sharing or not) but not in complex ones. Factors such as stressor type (physical/social), sex, and timing may be less influential than previously thought. Social closeness plays a key role—stress tends to enhance generosity toward close others but not distant individuals (Margittai et al., 2015; Passarelli & Buchanan, 2020). In summary, stress does not have a straightforward effect on prosociality. Instead, its impact depends on contextual and individual factors, highlighting the complexity of social behavior under stress.

How Does Stress Shape Prosociality? To understand how stress can increase or decrease prosociality, let's take a closer look at the mental processes that drive our decisions to help others (as reviewed in Sect. 3 of this chapter), such as empathy, mentalizing, and self-control (von Dawans et al., 2021; Nitschke & Bartz, 2023).

Empathy often increases under acute stress, with participants who completed the stressful TSST (Fig. 2a) reporting greater empathy for emotionally charged situations (Zilioli et al., 2015; Wolf et al., 2015) (but see Buruck et al. (2014)). Neuroimaging studies confirm increased activity in empathy-related brain regions under stress, which correlates with greater generosity in economic games (Tomova et al., 2017). Trait empathy—a stable, enduring characteristic that reflects an individual's general capacity to empathize with others, often assessed in questionnaires—further moderates stress effects: highly empathetic individuals become more generous under stress, while those low in empathy show more selfish behavior (Azulay et al., 2022). This finding suggests stress amplifies intuitive, "gut-based" reactions, enhancing prosociality in empathetic individuals but reducing it in less empathetic ones (Yu, 2016).

The impact of stress on mentalizing is less clear. Higher cortisol levels after a stressful experience have been linked to reduced donations in participants with strong mentalizing abilities (Schulreich et al., 2022). This suggests that individuals who are otherwise good at perspective-taking become less attuned to others' needs under acute stress (for a similar argument regarding chronic stress, see Neff et al. (2021)). This effect may be mediated by executive control processes in the DLPFC (Fig. 1) (Schulreich et al., 2022). This finding is consistent with the notion that stress also weakens control processes (Maier et al., 2015); for reviews, see (Hermans et al., 2014; Shields et al., 2016; Starcke & Brand, 2012). Overall, current evidence suggests that stress may reduce cognitive processes like mentalizing or self-control while enhancing emotional processes like empathy.

4.2 Age

Age significantly influences prosocial behavior, with changes occurring throughout the lifespan. Much of the research in developmental psychology has focused on how prosociality emerges in childhood, demonstrating that even infants as young as 4 months old show a preference for prosocial actions over antisocial ones (Geraci & Surian, 2023; Kiley Hamlin et al., 2010; Geraci et al., 2025). These early tendencies suggest that certain aspects of morality are innate, providing a foundation for future social development (Woo et al., 2022). In recent years, researchers have turned their attention to how prosociality changes during adulthood, especially in light of an aging global population (Pollerhoff et al., 2025). Since many critical decisions in politics, economics, and society are made by older adults, understanding how prosocial tendencies evolve with age has become increasingly important (Mayr & Freund, 2020; Sparrow et al., 2021; Bailey et al., 2021; Bagaïni et al., 2023; Li et al., 2024).

Theories in lifespan development generally propose that prosociality increases as people age (Mayr & Freund, 2020; Pollerhoff et al., 2024; Freund & Blanchard-Fields, 2014). For instance, younger adults, often focused on personal growth and

Fig. 4 Older adults tended to make larger hypothetical donations to a national charity in most countries (27 out of 47; 57%). The colors represent β values from a standardized regression, where red indicates a positive association and blue indicates a negative association between age and charitable donations. These associations are visualized on a world map (left) and in greater detail for Europe (right). (Reprinted from Cutler et al. (2021) CC BY 4.0)

accumulating resources for survival and reproduction, prioritize self-interest. In contrast, middle-aged and older adults, who typically face shorter time horizons, tend to focus on intrinsic goals, such as contributing to others and acting on moral values. Other theories suggest that older adults are more prosocial because they have greater resources—such as time, money, or social capital—to share. Additional explanations highlight the role of shifting social roles, learned prosocial behaviors, cohort effects, or changes in how people weigh costs and benefits when making altruistic choices (within a value-based decision framework). Despite their differences, these theories all support the idea that prosocial behavior tends to increase with age throughout adulthood.

Recent findings strongly support an age-related increase in prosociality. A global study across 67 countries (46,576 individuals, ages 18–99) found that age positively predicted prosocial behavior (Fig. 4) (Cutler et al., 2021). Likewise, a meta-analysis of over 100,000 participants confirmed a small but significant linear effect, with older adults consistently more prosocial than younger ones across behavioral tasks (e.g., dictator game) and self-reports (Pollerhoff et al., 2024). Beyond financial generosity, older adults also engage more in effortful helping (Lockwood et al., 2021), suggesting prosociality grows across multiple dimensions. Notably, these increases begin in middle adulthood (ages 36–59), a stage often overlooked in developmental studies (Pollerhoff et al., 2024). These changes may stem from social, psychological, and biological factors, including age-related changes in the brain (Tusche & Bas, 2021; Hubbard et al., 2016) (also see the chapter by Tisdall & Seaman on age-related changes in decision circuits). Neglecting lifespan development throughout the whole lifespan risks missing key variations in prosocial behavior.

4.3 Genetics

Could genetic differences help explain why people vary so much in prosociality? Researchers commonly used twin studies and molecular genetic studies to explore this question.

Twin studies compare behaviors of identical twins (who share 100% of their genes) with fraternal twins (who share around 50%) to estimate the heritability of a phenotype, like observed prosociality. If identical twins are more similar in prosocial behaviors, this suggests a genetic influence, assuming their environments are similar. Twin studies estimate that genetics account for 10–69% of the variation in prosociality, depending on what facet of prosociality is measured (Cesarini et al., 2009; Rushton et al., 1986; Rushton, 2004; Knafo & Plomin, 2006; Israel et al., 2015; Koenig et al., 2007; Pedersen et al., 2015). Thus, prosociality is moderately heritable, comparable to other personality traits, across individuals with diverse cultural backgrounds (Knafo-Noam et al., 2018). Research also shows that genetic influences on prosociality increase with age, while shared environmental effects, such as parenting practices, decrease—from about 50% in toddlers to less than 10% by age seven (Knafo & Plomin, 2006). Among other reasons, this shift may occur because individuals genetically predisposed to prosociality might increasingly seek environments that align with their genetic tendencies (an example of so-called gene-environment correlations).

Interestingly, twin studies also highlight genetic contributions to empathy (48%) and—to a lesser degree—mentalizing (27%), both important drivers of prosocial behavior (see Fig. 1) (Knafo-Noam et al., 2018; Knafo et al., 2008a; Abramson et al., 2020). Mentalizing appears more influenced by environmental factors than empathy (Abramson et al., 2020), aligning with findings that brain development in childhood in parts of the social brain involved in mentalizing, like the TPJ (Fig. 1), is more sensitive to social experiences (van der Meulen et al., 2020).

Molecular genetic studies look for specific genes and polymorphisms linked to prosociality by examining the genetic similarity of unrelated people. Most molecular genetic studies on prosociality use a hypothesis-driven candidate gene approach that examines specific genes hypothesized to play a role based on prior biological knowledge. Genes related to neurotransmitter systems like serotonin and dopamine, as well as neuropeptides like oxytocin and vasopressin—involved in empathy, mentalizing, and social behaviors like cooperation—are implicated in the genetic basis of prosocial outcomes (for a detailed overview, see Conway and Slavich (2017)). For example, variations in the oxytocin receptor (OXTR) and vasopressin receptor (AVPR1A) genes are linked to differences in giving behavior in the dictator game (Knafo et al., 2008b; Israel et al., 2009). The latter is also linked to individual differences in the aversiveness of unfairness, another important driver of prosocial behaviors (Tanaka et al., 2023).

Genome-wide association studies (GWAS), another type of molecular genetic study, take a hypothesis-free approach, scanning the entire genome for links to prosociality. This method can uncover unknown genetic factors but requires very large samples, often tens of thousands of people. A recent study used GWAS to examine the genetic underpinnings of real-world helping behavior (Primes & Fieder, 2018), but they are still rare. Both approaches suggest that no single gene has a large effect on prosociality. Instead, it likely results from many small genetic effects working together (Knafo-Noam et al., 2018). Overall, these findings highlight the importance of considering genetic factors to fully understand the heterogeneity in prosociality across individuals.

4.4 *Culture*

Research on prosociality often relies on Western, educated, industrialized, rich, and democratic (WEIRD) populations (Rad et al., 2018; Henrich et al., 2010), which limits our understanding of global diversity and cross-cultural differences. Cultural values, social norms, and economic conditions in a society shape when and how prosociality is expressed.

Economic Conditions Seminal research used economic games like the dictator game in 15 small-scale societies to reveal significant cultural variability in altruistic and cooperative behaviors (Henrich et al., 2005). These findings demonstrated both the universal nature of prosociality—showing that individuals don't only act purely out of self-interest but care about others' well-being—and also the influence of economic and social environments (confirmed in a recent meta-analysis (Cochard et al., 2021)). For example, a study across 67 countries ($N = 50{,}396$) found that national income inequality correlates with citizens' prosocial intentions (Elbæk et al., 2023) (Fig. 5a), and a study across 76 countries showed that greatest wealth is associated with higher prosocially around the world (Vanags et al., 2025). Likewise, higher levels of market integration (meaning that communities are connected to and participate in a larger market system) and benefits individuals receive from cooperating with others in everyday life correlate with greater societal prosociality (Henrich et al., 2005). For example, the isolated Machiguenga of Peru tended toward self-interest, while the Lamalera whale hunters, with high market integration, exhibited more cooperation.

Social Norms Another cross-cultural study employed a binary dictator game in eight societies, ranging from Tanzanian hunter-gatherers to Westernized Europeans in Berlin. They showed that adult prosociality differs across societies (Fig. 5b, left panel) and aligns with societal norms judged as correct by community members (House et al., 2020). As children in these societies matured, their behavior and judgments increasingly mirrored adults', leading to distinct developmental trajectories across cultures (Fig. 5b, right panel). This highlights the role of culture-dependent norms in shaping both adult and child behaviors. Notably, middle childhood (ages 6–11) seems to be a particularly sensitive period for social learning and norm adoption (Henrich, 2016; Henrich & Muthukrishna, 2021).

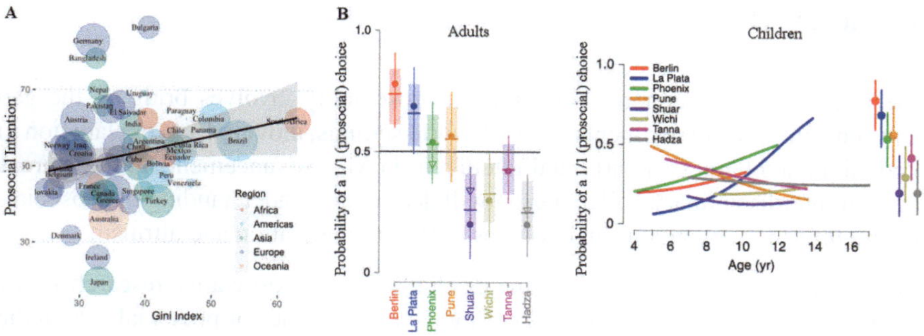

Fig. 5 (**a**) Individuals living in countries with high economic inequality (captured in higher values of the GINI index), where wealth or income is concentrated in the hands of a few, reported more prosocial intentions in the form of hypothetical donation intentions toward national and international charities. The bubble size represents the national sample size. (From Elbæk et al. (2023) CC BY 4.0). (**b**) In adults (*left*), prosocial allocations in a dictator game (a fair 1/1 split for either partner) varied significantly across eight populations. A mathematical model predicted cross-cultural differences in prosocial behavior, depicted by horizontal lines (regression estimates) and shaded bars (95% confidence intervals). In children (*right*), there were remarkable age-dependent differences in the predicted frequencies of the prosocial (fair) allocations in each population. The colored dots on the right indicate the adult averages for comparison in society. Populations sampled: Berlin (Germany; urban), La Plata (Argentina; urban), Phoenix (United States; urban), Pune (India; urban), Shuar (Ecuador; rural, small-scale horticulture and hunting), Wichí (Argentina; rural, sedentarized hunter-gatherers), Tanna (Vanuatu; rural, small-scale horticulture and hunting), and Hadza (Tanzania; rural, foraging and hunting). (Reprinted from House et al. (2020), Copyright 2025, with permission from Springer Nature)

Happiness Beyond determinants of prosociality, like economic conditions and social norms, cultural factors also influence the *effects* of prosociality. For example, altruistic acts can boost happiness and promote positive emotions across ages, sexes, and incomes (Aknin & Whillans, 2021; Curry et al., 2018; Aknin et al., 2022; Chen, 2024). Prosocial spending, one of the top six predictors of well-being globally in the World Happiness Report (Helliwell et al., 2021), underscores this universality. However, the happiness boost is stronger in individualistic than in collectivistic cultures—a classification that refers to the degree of integration among the members of a social group—emphasizing the need to consider cultural factors in understanding the consequences of prosocial behaviors (Weiss-Sidi & Riemer, 2023).

Overall, these findings highlight the need to expand studies to include non-WEIRD samples to form a more comprehensive understanding of prosociality, and how cultural norms and genetic factors (see Sect. 4.3) co-evolve to support large-scale cooperation in diverse environments (Henrich & Muthukrishna, 2021). In increasingly diverse societies, understanding cultural differences in prosociality is essential for fostering cooperation and social cohesion.

5 Future Directions

The future of neuroeconomic research on prosociality involves bridging the gap between controlled experiments and real-world settings, moving from explanation to prediction, and harnessing artificial intelligence (AI). Advancements in these areas through interdisciplinary collaboration will deepen our understanding of prosociality and inform interventions and policies to foster cooperation and altruism.

Enhancing Ecological Validity A major challenge in prosociality research is the reliance on controlled lab studies, ignoring the complexities of prosocial acts in the real world. Tools like virtual reality and portable neuroscience tools like functional near-infrared spectroscopy (fNIRS) and wearable electroencephalography (EEG) now allow studying brain activity in real-life contexts, such as group collaborations or volunteering. Combining these tools with ecological momentary assessments (EMAs)—short, real-time surveys on smartphones as people go about their day—can reveal how prosocial behaviors emerge in everyday situations (Wrzus & Neubauer, 2023), offering rich data on the interaction of individual and situational factors that shape prosociality in complex, dynamic settings outside the laboratory.

From Explanation to Prediction Research has focused mainly on describing decision-making processes and the factors influencing them rather than predicting behavior. While this has deepened our understanding of social, cognitive, and affective mechanisms and their biological foundations, it is still unclear if these insights can reliably forecast real-world prosocial behavior. Computational models, which quantify relationships between variables and can make predictions in new settings and populations, offer a promising solution (Tusche & Bas, 2021). These models could inform policies and interventions to address pressing societal challenges, such as promoting cooperative collective action on climate change or helping lonely individuals.

Artificial Intelligence AI tools like large language models (LLMs), such as ChatGPT, can analyze vast datasets, including social media posts, to study real-world prosociality. LLMs can also act as substitutes for humans in experiments, helping researchers explore prosocial behaviors. Additionally, AI-powered agents like social robots and companion chatbots, which are increasingly prevalent in schools or nursing homes, provide opportunities to study how human-AI behaviors compare to human-human prosociality (e.g., see (Akata et al., 2023). As these artificial agents become part of daily life, they may influence social norms, empathy, and mentalizing in human-AI interactions or lead to unintended effects, such as reduced sensitivity to human social cues, that drive prosocial behaviors (Adolphs & Tusche, 2017).

Conflict of Interest None.

References

Abramson, L., Uzefovsky, F., Toccaceli, V., & Knafo-Noam, A. (2020). The genetic and environmental origins of emotional and cognitive empathy: Review and meta-analyses of twin studies. *Neuroscience & Biobehavioral Reviews, 114*, 113–133.

Adolphs, R., & Tusche, A. (2017). From faces to prosocial behavior: Cues, tools, and mechanisms. *Current Directions in Psychological Science, 26*, 282–287.

Akata, E., et al. (2023). Playing repeated games with large language models. *arXiv preprint arXiv:2305.16867*.

Aknin, L. B., & Whillans, A. V. (2021). Helping and happiness: A review and guide for public policy. *Social Issues and Policy Review, 15*, 3–34.

Aknin, L. B., Dunn, E. W., & Whillans, A. V. (2022). The emotional rewards of prosocial spending are robust and replicable in large samples. *Current Directions in Psychological Science, 31*, 536–545.

Azulay, H., Guy, N., Pertzov, Y., & Israel, S. (2022). Empathy modulates the effect of stress reactivity on generous giving. *Frontiers in Neuroscience, 16*, 814789.

Bagaïni, A., et al. (2023). Aging and economic preferences: Cumulative meta-analyses of age differences in risk, time, social, and effort preferences. *The Journals of Gerontology: Series B, 78*, 1122–1135.

Bailey, P. E., Ebner, N. C., & Stine-Morrow, E. A. (2021). Introduction to the special issue on prosociality in adult development and aging: Advancing theory within a multilevel framework. *Psychology and Aging, 36*, 1.

Bartra, O., McGuire, J. T., & Kable, J. W. (2013). The valuation system: A coordinate-based meta-analysis of BOLD fMRI experiments examining neural correlates of subjective value. *NeuroImage, 76*, 412–427.

Bas, L. M. W., Xu, Y., Smallwood, J., & Tusche, A. (2025a). A Window into Social Preferences: Thought Patterns Reflect Generosity, Fairness, and Context. *bioRxiv*.

Bas, L. M., Roberts, I. D., Hutcherson, C. A., & Tusche, A. (2025b). A neurocomputational account of the link between social perception and social action. *eLife, 12*, RP92539.

Batson, C. D. (2011). *Altruism in humans*. Oxford University Press.

Batson, C. D. (2022). The empathy-altruism hypothesis: What it is and why it is important. In *Positive psychology: An international perspective* (pp. 12–32). https://doi.org/10.1002/9781119771418.ch2

Bellucci, G., Camilleri, J. A., Eickhoff, S. B., & Krueger, F. (2020). Neural signatures of prosocial behaviors. *Neuroscience & Biobehavioral Reviews, 118*, 186–195.

Bonandrini, R., Beveridge, F., & Steinbeis, N. (2025). A neuroimaging-based meta-analysis of prosocial behavior in childhood and adolescence. *Social Neuroscience, 20*(2), 132–142. https://doi.org/10.1080/17470919.2025.2541641

Böckler, A., Tusche, A., & Singer, T. (2016). The structure of human prosociality: Differentiating altruistically motivated, norm motivated, strategically motivated, and self-reported prosocial behavior. *Social Psychological and Personality Science, 7*, 530–541.

Böckler, A., Tusche, A., & Singer, T. (2018a). The structure of human prosociality revisited: Corrigendum and Addendum to Bockler, Tusche, and Singer (2016). *Social Psychological and Personality Science, 9*, 754–759.

Böckler, A., Tusche, A., Schmidt, P., & Singer, T. (2018b). Distinct mental trainings differentially affect altruistically motivated, norm motivated, and self-reported prosocial behaviour. *Scientific Reports-Uk, 8*(1), 13560.

Buchanan, T. W., & Preston, S. D. (2014). Stress leads to prosocial action in immediate need situations. *Frontiers in Behavioral Neuroscience, 8*, 5.

Buruck, G., Wendsche, J., Melzer, M., Strobel, A., & Dörfel, D. (2014). Acute psychosocial stress and emotion regulation skills modulate empathic reactions to pain in others. *Frontiers in Psychology, 5*, 517.

Bussey, K. (2020). Development of moral disengagement. In *The Oxford handbook of moral development* (pp. 306–326). Oxford University Press.
Bzdok, D., et al. (2012). Parsing the neural correlates of moral cognition: ALE meta-analysis on morality, theory of mind, and empathy. *Brain Structure and Function, 217*, 783–796.
Cacioppo, J. T., & Cacioppo, S. (2018). *Advances in experimental social psychology* (Vol. 58, pp. 127–197). Elsevier.
Camerer, C. F. (2011). *Behavioral game theory: Experiments in strategic interaction*. Princeton University Press.
Campbell, D., Tusche, A., & O'Connor, B. B. (2022). Imagination and the prosocial personality: Mapping the effect of episodic simulation on helping across prosocial traits. *Applied Cognitive Psychology, 36*, 685–698.
Caprara, G. V., Steca, P., Zelli, A., & Capanna, C. (2005). A new scale for measuring adults' prosocialness. *European Journal of Psychological Assessment, 21*, 77–89.
Carlson, R. W., & Crockett, M. J. (2018). The lateral prefrontal cortex and moral goal pursuit. *Current Opinion in Psychology, 24*, 77–82.
Cesarini, D., Dawes, C. T., Johannesson, M., Lichtenstein, P., & Wallace, B. (2009). Genetic variation in preferences for giving and risk taking*. *The Quarterly Journal of Economics, 124*, 809–842.
Chen, Y. (2024). Pro-sociality and happiness across national cultures: A hierarchical linear model. *Current Psychology, 43*(4), 3381–3394.
Clithero, J. A., & Rangel, A. (2014). Informatic parcellation of the network involved in the computation of subjective value. *Social Cognitive and Affective Neuroscience, 9*, 1289–1302.
Cochard, F., Le Gallo, J., Georgantzis, N., & Tisserand, J.-C. (2021). Social preferences across different populations: Meta-analyses on the ultimatum game and dictator game. *Journal of Behavioral and Experimental Economics, 90*, 101613.
Conway, C. C., & Slavich, G. M. (2017). Behavior genetics of prosocial behavior. In P. Gilbert (Ed.), *Compassion: Concepts, research and applications* (pp. 151–170). Routledge/Taylor & Francis Group.
Crosswell, A. D., & Lockwood, K. G. (2020). Best practices for stress measurement: How to measure psychological stress in health research. *Health Psychology Open, 7*, 2055102920933072.
Curry, O. S., et al. (2018). Happy to help? A systematic review and meta-analysis of the effects of performing acts of kindness on the well-being of the actor. *Journal of Experimental Social Psychology, 76*, 320–329.
Cutler, J., & Campbell-Meiklejohn, D. (2019). A comparative fMRI meta-analysis of altruistic and strategic decisions to give. *NeuroImage, 184*, 227–241.
Cutler, J., Nitschke, J. P., Lamm, C., & Lockwood, P. L. (2021). Older adults across the globe exhibit increased prosocial behavior but also greater in-group preferences. *Nature Aging, 1*, 880–888.
Decety, J., & Jackson, P. L. (2004). The functional architecture of human empathy. *Behavioral and Cognitive Neuroscience Reviews, 3*, 71–100.
Dedovic, K., et al. (2005). The Montreal Imaging Stress Task: Using functional imaging to investigate the effects of perceiving and processing psychosocial stress in the human brain. *Journal of Psychiatry and Neuroscience, 30*, 319–325.
Dedovic, K., Duchesne, A., Andrews, J., Engert, V., & Pruessner, J. C. (2009). The brain and the stress axis: The neural correlates of cortisol regulation in response to stress. *NeuroImage, 47*, 864–871.
Dickerson, S. S., & Kemeny, M. E. (2004). Acute stressors and cortisol responses: A theoretical integration and synthesis of laboratory research. *Psychological Bulletin, 130*, 355.
Elbæk, C. T., Mitkidis, P., Aarøe, L., & Otterbring, T. (2023). Subjective socioeconomic status and income inequality areassociated with self-reported morality across 67 countries. *Nature Communications, 14*, 5453.
Engel, C. (2011). Dictator games: A meta study. *Experimental Economics, 14*, 583–610.
Engert, V., et al. (2014). Exploring the use of thermal infrared imaging in human stress research. *PLoS One, 9*, e90782.

Faber, N. S., & Häusser, J. A. (2022). Why stress and hunger both increase and decrease prosocial behaviour. *Current Opinion in Psychology, 44*, 49–57.

Fan, Y., Duncan, N. W., De Greck, M., & Northoff, G. (2011). Is there a core neural network in empathy? An fMRI based quantitative meta-analysis. *Neuroscience & Biobehavioral Reviews, 35*, 903–911.

Fehr, E., & Fischbacher, U. (2003). The nature of human altruism. *Nature, 425*, 785–791.

Forbes, P. A. G., Aydogan, G., Braunstein, J., Todorova, B., Wagner, I. C., Lockwood, P. L., Apps, M. A. J., Ruff, C. C., & Lamm, C. (2024). Acute stress reduces effortful prosocial behaviour. *eLife, 12*, RP87271.

Forsythe, R., Horowitz, J. L., Savin, N. E., & Sefton, M. (1994). Fairness in simple bargaining experiments. *Games and Economic Behavior, 6*, 347–369.

Freund, A. M., & Blanchard-Fields, F. (2014). Age-related differences in altruism across adulthood: Making personal financial gain versus contributing to the public good. *Developmental Psychology, 50*, 1125.

Frisch, J. U., Häusser, J. A., & Mojzisch, A. (2015). The Trier Social Stress Test as a paradigm to study how people respond to threat in social interactions. *Frontiers in Psychology, 6*, 14.

Frith, C. D., & Frith, U. (2006). The neural basis of mentalizing. *Neuron, 50*, 531–534.

Gao, X., Jolly, E., Yu, H., Liu, H., Zhou, X., & Chang, L. J. (2024). The psychological, computational, and neural foundations of indebtedness. *Nature Communications, 15*(1), 68.

Geraci, A., & Surian, L. (2023). Intention-based evaluations of distributive actions by 4-month-olds. *Infant Behavior and Development, 70*, 101797.

Geraci, A., Surian, L., Tina, L. G., & Hamlin, J. K. (2025). Human newborns spontaneously attend to prosocial interactions. *Nature Communications, 16*(1), 6304.

Giannakakis, G., et al. (2019). Review on psychological stress detection using biosignals. *IEEE Transactions on Affective Computing, 13*, 440–460.

Guala, F., & Mittone, L. (2010). Paradigmatic experiments: The dictator game. *The Journal of Socio-Economics, 39*, 578–584.

Gunnar, M. R., et al. (2021). Validation of an online version of the Trier Social Stress Test in a study of adolescents. *Psychoneuroendocrinology, 125*, 105111.

Hare, T. A., Camerer, C. F., & Rangel, A. (2009). Self-control in decision-making involves modulation of the vmPFC valuation system. *Science, 324*, 646–648.

Hare, T. A., Camerer, C. F., Knoepfle, D. T., O'Doherty, J. P., & Rangel, A. (2010). Value computations in ventral medial prefrontal cortex during charitable decision making incorporate input from regions involved in social cognition. *The Journal of Neuroscience, 30*, 583–590.

Hauser, O. P., Rand, D. G., Peysakhovich, A., & Nowak, M. A. (2014). Cooperating with the future. *Nature, 511*, 220–223.

Helliwell, J. F., Layard, R., Sachs, J. D., & Neve, J.-E. D. (2021). *World happiness report 2021*. Sustainable Development Solutions Network.

Henrich, J. (2016). *The secret of our success: How culture is driving human evolution, domesticating our species, and making us smarter*. Princeton University Press.

Henrich, J., & Muthukrishna, M. (2021). The origins and psychology of human cooperation. *Annual Review of Psychology, 72*, 207–240.

Henrich, J., et al. (2005). "Economic man" in cross-cultural perspective: Behavioral experiments in 15 small-scale societies. *Behavioral and Brain Sciences, 28*, 795–815.

Henrich, J., Heine, S. J., & Norenzayan, A. (2010). The weirdest people in the world? *Behavioral and Brain Sciences, 33*, 61–83.

Hensel, L., Rohleder, N., & Niessen, C. (2022). Effects of psychosocial stress on prosociality: The moderating role of current life stress and thought control. *Stress, 25*, 235–245.

Hermans, E. J., Henckens, M. J., Joëls, M., & Fernández, G. (2014). Dynamic adaptation of large-scale brain networks in response to acute stressors. *Trends in Neurosciences, 37*, 304–314.

Hilbrand, S., Coall, D. A., Meyer, A. H., Gerstorf, D., & Hertwig, R. (2017). A prospective study of associations among helping, health, and longevity. *Social Science & Medicine, 187*, 109–117.

Ho, M. K., Saxe, R., & Cushman, F. (2022). Planning with theory of mind. *Trends in Cognitive Sciences, 26*, 959–971.

House, B. R., et al. (2020). Universal norm psychology leads to societal diversity in prosocial behaviour and development. *Nature Human Behaviour, 4*, 36–44.
Hubbard, J., Harbaugh, W. T., Srivastava, S., Degras, D., & Mayr, U. (2016). A general benevolence dimension that links neural, psychological, economic, and life-span data on altruistic tendencies. *Journal of Experimental Psychology: General, 145*, 1351.
Hui, B. P., Ng, J. C., Berzaghi, E., Cunningham-Amos, L. A., & Kogan, A. (2020). Rewards of kindness? A meta-analysis of the link between prosociality and well-being. *Psychological Bulletin, 146*, 1084.
Hutcherson, C. A., & Tusche, A. (2022). Evidence accumulation, not 'self-control', explains dorsolateral prefrontal activation during normative choice. *eLife, 11*, e65661.
Jin, K., Wu, J., Zhang, R., Zhang, S., Wu, X., Wu, T., et al. (2024). Observing heroic behavior and its influencing factors in immersive virtual environments. *Proceedings of the National Academy of Sciences, 121*(17), e2314590121.
Israel, S., et al. (2009). The oxytocin receptor (OXTR) contributes to prosocial fund allocations in the dictator game and the social value orientations task. *PLoS One, 4*, e5535.
Israel, S., Hasenfratz, L., & Knafo-Noam, A. (2015). The genetics of morality and prosociality. *Current Opinion in Psychology, 6*, 55–59.
Kahneman, D., Knetsch, J. L., & Thaler, R. H. (1986). Fairness and the assumptions of economics. *Journal of Business, 59*(4), S285–S300.
Kiley Hamlin, J., Wynn, K., & Bloom, P. (2010). Three-month-olds show a negativity bias in their social evaluations. *Developmental Science, 13*, 923–929.
Klein, S. A., & Hilbig, B. E. (2019). On the lack of real consequences in consumer choice research. *Experimental Psychology, 66*(1), 68–76.
Klein, S. A., Nockur, L., & Reese, G. (2022). Prosociality from the perspective of environmental psychology. *Current Opinion in Psychology, 44*, 182–187.
Knafo, A., & Plomin, R. (2006). Prosocial behavior from early to middle childhood: Genetic and environmental influences on stability and change. *Developmental Psychology, 42*, 771.
Knafo, A., Zahn-Waxler, C., Van Hulle, C., Robinson, J. L., & Rhee, S. H. (2008a). The developmental origins of a disposition toward empathy: Genetic and environmental contributions. *Emotion, 8*, 737.
Knafo, A., et al. (2008b). Individual differences in allocation of funds in the dictator game associated with length of the arginine vasopressin 1a receptor RS3 promoter region and correlation between RS3 length and hippocampal mRNA. *Genes, Brain and Behavior, 7*, 266–275.
Knafo-Noam, A., Vertsberger, D., & Israel, S. (2018). Genetic and environmental contributions to children's prosocial behavior: Brief review and new evidence from a reanalysis of experimental twin data. *Current Opinion in Psychology, 20*, 60–65.
Koenig, L. B., McGue, M., Krueger, R. F., & Bouchard, T. J., Jr. (2007). Religiousness, antisocial behavior, and altruism: Genetic and environmental mediation. *Journal of Personality, 75*, 265–290.
Kosse, F., & Tincani, M. M. (2020). Prosociality predicts labor market success around the world. *Nature Communications, 11*, 5298.
Kwon, S.-J., van Hoorn, J., Do, K. T., Burroughs, M., & Telzer, E. H. (2023). Neural representation of donating time and money. *The Journal of Neuroscience, 43*, 6297–6305.
Lamm, C., & Forbes, P. A. (2023). Neurobiology of prosociality. In *Handbook of prosociality: Development, mechanisms, promotion* (pp. 61–84). Cambridge University Press.
Lanser, I., & Eisenberger, N. I. (2022). Prosocial behavior reliably reduces loneliness: An investigation across two studies. *Emotion, 23*(6), 1781–1790.
Lee, S., & Feeley, T. H. (2016). The identifiable victim effect: A meta-analytic review. *Social Influence, 11*, 199–215.
Leiberg, S., Klimecki, O., & Singer, T. (2011). Short-term compassion training increases prosocial behavior in a newly developed prosocial game. *PLoS One, 6*, e17798.
Li, D., Cao, Y., Hui, B. P., & Shum, D. H. (2024). Are older adults more prosocial than younger adults? A systematic review and meta-analysis. *The Gerontologist, 64*, gnae082.

Lockwood, P. L., et al. (2017). Prosocial apathy for helping others when effort is required. *Nature Human Behaviour, 1*, 0131.
Lockwood, P. L., et al. (2021). Aging increases prosocial motivation for effort. *Psychological Science, 32*, 668–681.
Lockwood, P. L., et al. (2022). Distinct neural representations for prosocial and self-benefiting effort. *Current Biology, 32*, 4172–4185.e4177.
Lockwood, P. L., et al. (2024). Human ventromedial prefrontal cortex is necessary for prosocial motivation. *Nature Human Behaviour, 8*(7), 1403–1416.
Lovallo, W. (1975). The cold pressor test and autonomic function: A review and integration. *Psychophysiology, 12*, 268–282.
Luo, J. (2018). The neural basis of and a common neural circuitry in different types of pro-social behavior. *Frontiers in Psychology, 9*, 859.
Maier, S. U., Makwana, A. B., & Hare, T. A. (2015). Acute stress impairs self-control in goal-directed choice by altering multiple functional connections within the brain's decision circuits. *Neuron, 87*, 621–631.
Maliske, L. Z., Schurz, M., & Kanske, P. (2023). Interactions within the social brain: Co-activation and connectivity among networks enabling empathy and Theory of Mind. *Neuroscience & Biobehavioral Reviews, 147*, 105080.
Manzur, E., & Olavarrieta, S. (2021). The 9-SRA scale: A simplified 9-items version of the SRA scale to assess altruism. *Sustainability, 13*, 6999.
Margittai, Z., et al. (2015). A friend in need: Time-dependent effects of stress on social discounting in men. *Hormones and Behavior, 73*, 75–82.
Mayr, U., & Freund, A. M. (2020). Do we become more prosocial as we age, and if so, why? *Current Directions in Psychological Science, 29*, 248–254.
Neff, L. A., Nguyen, T. T., & Williamson, H. C. (2021). Too stressed to help? The effects of stress on noticing partner needs and enacting support. *Personality and Social Psychology Bulletin, 47*, 1565–1579.
Nitschke, J. P., & Bartz, J. A. (2023). The association between acute stress & empathy: A systematic literature review. *Neuroscience & Biobehavioral Reviews, 144*, 105003.
Nitschke, J. P., Forbes, P. A. G., & Lamm, C. (2022). Does stress make us more—or less—prosocial? A systematic review and meta-analysis of the effects of acute stress on prosocial behaviours using economic games. *Neuroscience & Biobehavioral Reviews, 142*, 104905.
O'Connor, D. B., Thayer, J. F., & Vedhara, K. (2021). Stress and health: A review of psychobiological processes. *Annual Review of Psychology, 72*, 663–688.
Ommerborn, M. A., et al. (2023). Effects on general pain perception and dental pulp sensibility in probable sleep bruxism subjects by experimentally induced pain in a pilot study. *Scientific Reports-Uk, 13*, 5836.
Pärnamets, P., Shuster, A., Reinero, D. A., & Van Bavel, J. J. (2020). A value-based framework for understanding cooperation. *Current Directions in Psychological Science, 29*, 227–234.
Passarelli, T. O., & Buchanan, T. W. (2020). How do stress and social closeness impact prosocial behavior? *Experimental Psychology, 67*(2), 123–131.
Pedersen, O. B., et al. (2015). The heritability of blood donation: A population-based nationwide twin study. *Transfusion, 55*, 2169–2174.
Peysakhovich, A., Nowak, M. A., & Rand, D. G. (2014). Humans display a 'cooperative phenotype' that is domain general and temporally stable. *Nature Communications, 5*, 4939.
Pfattheicher, S., Nielsen, Y. A., & Thielmann, I. (2022). Prosocial behavior and altruism: A review of concepts and definitions. *Current Opinion in Psychology, 44*, 124–129.
Pisor, A. C., & Ross, C. T. (2024). Parochial altruism: What it is and why it varies. *Evolution and Human Behavior, 45*, 2–12.
Pisor, A. C., Gervais, M. M., Purzycki, B. G., & Ross, C. T. (2020). Preferences and constraints: The value of economic games for studying human behaviour. *Royal Society Open Science, 7*, 192090.

Pollerhoff, L., Reindel, D. F., Kanske, P., Li, S.-C., & Reiter, A. M. F. (2024). Age differences in prosociality across the adult lifespan: A meta-analysis. *Neuroscience & Biobehavioral Reviews, 165*, 105843.

Pollerhoff, L., Saulin, A., Kurtz, M., Stietz, J., Peng, X. R., Hein, G., Tusche, A., Kanske, P., Shu-Chen, L. & Reiter, A. M. (2025). Adult age differences in the integration of values for self and other. *Scientific Reports, 15*(1), 12776.

Post, S. G. (2014). Altruism, happiness, and health: It's good to be good. In *An Exploration of the Health Benefits of Factors That Help Us to Thrive* (pp. 66–76).

Primes, G., & Fieder, M. (2018). Real-life helping behaviours in North America: A genome-wide association approach. *PLoS One, 13*, e0190950.

Qiu, X., Gao, M., Zhu, H., Li, W., & Jiang, R. (2024). Theory of mind, empathy, and prosocial behavior in children and adolescent: A meta-analysis. *Current Psychology, 43*(22), 19690–19707.

Rad, M. S., Martingano, A. J., & Ginges, J. (2018). Toward a psychology of Homo sapiens: Making psychological science more representative of the human population. *Proceedings of the National Academy of Sciences, 115*, 11401–11405.

Raposa, E. B., Laws, H. B., & Ansell, E. B. (2016). Prosocial behavior mitigates the negative effects of stress in everyday life. *Clinical Psychological Science, 4*, 691–698.

Rhoads, S. A., Cutler, J., & Marsh, A. A. (2021). A feature-based network analysis and fMRI meta-analysis reveal three distinct types of prosocial decisions. *Social Cognitive and Affective Neuroscience, 16*, 1214–1233.

Roelofs, K. (2017). Freeze for action: Neurobiological mechanisms in animal and human freezing. *Philosophical Transactions of the Royal Society, B: Biological Sciences, 372*, 20160206.

Ruff, C. C., & Fehr, E. (2014). The neurobiology of rewards and values in social decision making. *Nature Reviews Neuroscience, 15*, 549–562.

Rushton, J. P. (2004). Genetic and environmental contributions to pro-social attitudes: A twin study of social responsibility. *Proceedings of the Royal Society of London. Series B: Biological Sciences, 271*, 2583–2585.

Rushton, J. P., Chrisjohn, R. D., & Fekken, G. C. (1981). The altruistic personality and the self-report altruism scale. *Personality and Individual Differences, 2*, 293–302.

Rushton, J. P., Fulker, D. W., Neale, M. C., Nias, D. K., & Eysenck, H. J. (1986). Altruism and aggression: The heritability of individual differences. *Journal of Personality and Social Psychology, 50*, 1192.

Russell, G., & Lightman, S. (2019). The human stress response. *Nature Reviews Endocrinology, 15*, 525–534.

Saulin, A., Ma, Y., & Hein, G. (2024). Empathy incites a stable prosocial decision bias. *Cerebral Cortex, 34*(7), bhae272.

Schiller, B., et al. (2023). Social high performers under stress behave more prosocially and detect happy emotions better in a male sample. *Psychoneuroendocrinology, 156*, 106338.

Schneiderman, N., Ironson, G., & Siegel, S. D. (2005). Stress and health: Psychological, behavioral, and biological determinants. *Annual Review of Clinical Psychology, 1*, 607–628.

Schulreich, S., Tusche, A., Kanske, P., & Schwabe, L. (2022). Altruism under stress: Cortisol negatively predicts charitable giving and neural value representations depending on mentalizing capacity. *The Journal of Neuroscience, 42*, 3445–3460.

Schulreich, S., Tusche, A., Kanske, P., & Schwabe, L. (2023). Higher subjective socioeconomic status is linked to increased charitable giving and mentalizing-related neural value coding. *NeuroImage, 279*, 120315.

Schurz, M., et al. (2021). Toward a hierarchical model of social cognition: A neuroimaging meta-analysis and integrative review of empathy and theory of mind. *Psychological Bulletin, 147*, 293.

Schwabe, L., Hermans, E. J., Joëls, M., & Roozendaal, B. (2022). Mechanisms of memory under stress. *Neuron, 110*, 1450–1467.

Selye, H. (1950). Stress and the general adaptation syndrome. *British Medical Journal, 1*, 1383.

Shields, G. S., Sazma, M. A., & Yonelinas, A. P. (2016). The effects of acute stress on core executive functions: A meta-analysis and comparison with cortisol. *Neuroscience & Biobehavioral Reviews, 68*, 651–668.

Simon, E. B., Vallat, R., Rossi, A., & Walker, M. P. (2022). Sleep loss leads to the withdrawal of human helping across individuals, groups, and large-scale societies. *PLoS Biology, 20*, e3001733.

Singer, T., & Tusche, A. (2014). *Neuroeconomics* (pp. 513–532). Elsevier.

Singer, N., Sommer, M., Wüst, S., & Kudielka, B. M. (2021). Effects of gender and personality on everyday moral decision-making after acute stress exposure. *Psychoneuroendocrinology, 124*, 105084.

Smith, A. (2006). Cognitive empathy and emotional empathy in human behavior and evolution. *The Psychological Record, 56*, 3–21.

Sollberger, S., Bernauer, T., & Ehlert, U. (2016). Stress influences environmental donation behavior in men. *Psychoneuroendocrinology, 63*, 311–319.

Sparrow, E. P., Swirsky, L. T., Kudus, F., & Spaniol, J. (2021). Aging and altruism: A meta-analysis. *Psychology and Aging, 36*, 49.

Spengler, B., Hofer, J., Busch, H., Dzionsko, I., & Emslander, V. (2020). Implicit motives and children's salivary cortisol reactivity to an adapted version of the Trier Social Stress Test for Children (TSST-C). *Personality and Individual Differences, 162*, 110010.

Starcke, K., & Brand, M. (2012). Decision making under stress: A selective review. *Neuroscience & Biobehavioral Reviews, 36*, 1228–1248.

Steinbeis, N., Engert, V., Linz, R., & Singer, T. (2015). The effects of stress and affiliation on social decision-making: Investigating the tend-and-befriend pattern. *Psychoneuroendocrinology, 62*, 138–148.

Suzuki, S., & O'Doherty, J. P. (2020). Breaking human social decision making into multiple components and then putting them together again. *Cortex, 127*, 221–230.

Takahashi, T., Ikeda, K., & Hasegawa, T. (2007). Social evaluation-induced amylase elevation and economic decision-making in the dictator game in humans. *Neuroendocrinology Letters, 28*, 662–665.

Tanaka, H., et al. (2023). Association between arginine vasopressin receptor 1A (AVPR1A) polymorphism and inequity aversion. *Proceedings of the Royal Society B, 290*, 20230378.

Taylor, S. E. (2006). Tend and befriend: Biobehavioral bases of affiliation under stress. *Current Directions in Psychological Science, 15*, 273–277.

Thaler, R. H. (2000). From homo economicus to homo sapiens. *Journal of Economic Perspectives, 14*, 133–141.

Thielmann, I., Böhm, R., Ott, M., & Hilbig, B. E. (2021). Economic games: An introduction and guide for research. *Collabra: Psychology, 7*, 19004.

Tomova, L., et al. (2017). Increased neural responses to empathy for pain might explain how acute stress increases prosociality. *Social Cognitive and Affective Neuroscience, 12*, 401–408.

Tusche, A., & Bas, L. M. (2021). Neurocomputational models of altruistic decision-making and social motives: Advances, pitfalls, and future directions. *Wiley Interdisciplinary Reviews: Cognitive Science, 12*, e1571.

Tusche, A., & Hutcherson, C. (2018). Cognitive regulation alters social and dietary choice by changing attribute representations in domain-general and domain-specific brain circuits. *eLife, 7*, e31185.

Tusche, A., Böckler, A., Kanske, P., Trautwein, F. M., & Singer, T. (2016). Decoding the charitable brain: Empathy, perspective taking, and attention shifts differentially predict altruistic giving. *The Journal of Neuroscience, 36*, 4719–4732.

Tusche, A., Spunt, R. P., Paul, L. K., Tyszka, J. M., & Adolphs, R. (2023). Neural signatures of social inferences predict the number of real-life social contacts and autism severity. *Nature Communications, 14*, 4399.

Vanags, P., Cutler, J., Kosse, F., & Lockwood, P. L. (2025). Greater income and financial Well-being are associated with higher prosocial preferences and behaviors across 76 countries. *PNAS Nexus, 4*(2), pgae582. https://doi.org/10.1093/pnasnexus/pgae582

van der Meulen, M., et al. (2020). Genetic and environmental influences on structure of the social brain in childhood. *Developmental Cognitive Neuroscience, 44*, 100782.
Van Dijk, E., & De Dreu, C. K. (2021). Experimental games and social decision making. *Annual Review of Psychology, 72*, 415–438.
Veszteg, R. F., Yamakawa, K., Matsubayashi, T., & Ueda, M. (2021). Acute stress does not affect economic behavior in the experimental laboratory. *PLoS One, 16*, e0244881.
Vinkers, C. H., et al. (2013). Time-dependent changes in altruistic punishment following stress. *Psychoneuroendocrinology, 38*, 1467–1475.
Von Dawans, B., Kirschbaum, C., & Heinrichs, M. (2011). The Trier Social Stress Test for Groups (TSST-G): A new research tool for controlled simultaneous social stress exposure in a group format. *Psychoneuroendocrinology, 36*, 514–522.
von Dawans, B., Fischbacher, U., Kirschbaum, C., Fehr, E., & Heinrichs, M. (2012). The social dimension of stress reactivity: Acute stress increases prosocial behavior in humans. *Psychological Science, 23*, 651–660.
von Dawans, B., Ditzen, B., Trueg, A., Fischbacher, U., & Heinrichs, M. (2019). Effects of acute stress on social behavior in women. *Psychoneuroendocrinology, 99*, 137–144.
von Dawans, B., Strojny, J., & Domes, G. (2021). The effects of acute stress and stress hormones on social cognition and behavior: Current state of research and future directions. *Neuroscience & Biobehavioral Reviews, 121*, 75–88.
Wang, H., et al. (2024). Acute stress during witnessing injustice shifts third-party interventions from punishing the perpetrator to helping the victim. *PLoS Biology, 22*, e3002195.
Weiss-Sidi, M., & Riemer, H. (2023). Help others—Be happy? The effect of altruistic behavior on happiness across cultures. *Frontiers in Psychology, 14*, 1156661.
Wolf, O. T., et al. (2015). Enhanced emotional empathy after psychosocial stress in young healthy men. *Stress, 18*, 631–637.
Woo, B. M., Tan, E., & Hamlin, J. K. (2022). Human morality is based on an early-emerging moral core. *Annual Review of Developmental Psychology, 4*, 41–61.
Wrzus, C., & Neubauer, A. B. (2023). Ecological momentary assessment: A meta-analysis on designs, samples, and compliance across research fields. *Assessment, 30*, 825–846.
Wu, Y. E., & Hong, W. (2022). Neural basis of prosocial behavior. *Trends in Neurosciences, 45*, 749–762.
Wyss, A. M., Baumgartner, T., Guizar Rosales, E., Soutschek, A., & Knoch, D. (2024). Cathodal HD-tDCS above the left dorsolateral prefrontal cortex increases environmentally sustainable decision-making. *Frontiers in Human Neuroscience, 18*, 1395426.
Ying, L., Yan, Q., Shen, X., & Zhang, C. (2022). Social value orientation moderated the effect of acute stress on individuals' prosocial behaviors. *Frontiers in Psychology, 13*, 803184.
Yu, R. (2016). Stress potentiates decision biases: A stress induced deliberation-to-intuition (SIDI) model. *Neurobiology of Stress, 3*, 83–95.
Zaki, J. (2017). Moving beyond stereotypes of empathy. *Trends in Cognitive Sciences, 21*, 59–60.
Zhen, Z., et al. (2021). Acute psychosocial stress increases third-party helping but not punishing behavior. *Stress, 24*, 430–441.
Zilioli, S., Ponzi, D., Henry, A., & Maestripieri, D. (2015). Testosterone, cortisol and empathy: Evidence for the dual-hormone hypothesis. *Adaptive Human Behavior and Physiology, 1*, 421–433.
Zimmer, P., Buttlar, B., Halbeisen, G., Walther, E., & Domes, G. (2019). Virtually stressed? A refined virtual reality adaptation of the Trier Social Stress Test (TSST) induces robust endocrine responses. *Psychoneuroendocrinology, 101*, 186–192.

The Neural Mechanisms of Strategic Decision-Making

Arkady Konovalov

Abstract Strategic behavior emerges when two or more organisms interact. Our brains evolved to track others' behavior as part of the environment and build mental models of others' strategies and future actions. In this chapter, we review recent advances in the study of strategic decision-making in the human brain at the level of functional brain networks. We discuss general approaches to computational models of strategic behavior and evidence supporting their neural implementation. We then turn to specific issues, including neural representation of the social context, strategic uncertainty, and mechanisms such as belief-based learning, mentalization, and high-order strategic reasoning. We conclude with potential future directions and methodological developments.

1 Introduction

Strategic interactions are ubiquitous in animal and human groups. To survive, gain rewards, and maintain social status, organisms must learn to track and anticipate each other's actions. In most complex settings, this behavior requires mentalization, or theory of mind—the ability to form second-order beliefs or mental models of others' cognition (Frith & Frith, 2005).

A common example from daily life is bargaining at a street bazaar, where the buyer aims to secure a lower price while the seller seeks a higher one. Since neither party knows the other's true valuation, successful negotiation often involves multiple rounds of discussion and the development of a mental model of the other's beliefs. While this is an example of pure competition, strategic decision-making can also include cooperative elements and shared goals. For instance, when embarking on a new joint research project, it becomes crucial to assess and predict the reliability and trustworthiness of your collaborators.

A. Konovalov (✉)
School of Psychology, Centre for Human Brain Health (CHBH), University of Birmingham, Birmingham, UK
e-mail: a.konovalov@bham.ac.uk

D. V. Smith et al. (eds.), *Neuroeconomics: Core Topics and Current Directions*,
https://doi.org/10.1007/978-3-032-02925-6_25

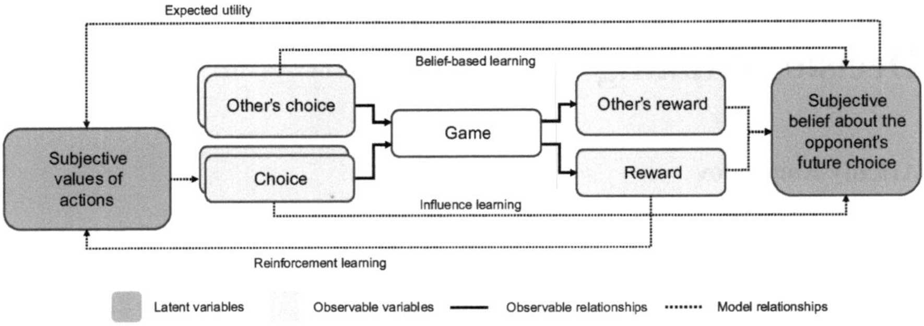

Fig. 1 A general model of dynamic strategic interactions in dyads. The decision-maker stores subjective values of actions, which guide their choice. The game structure transforms the choices of the two players (including the past history of choices) into rewards. The other's choices can affect the decision-maker's future beliefs through belief-based learning mechanisms. The decision-maker's own choice can affect future beliefs through influence learning (second-order beliefs). The decision-maker's rewards can affect the action values through reinforcement learning. Subjective beliefs determine subjective values of actions through expected utility calculations

Game theory formalizes strategic behavior, representing any interaction involving two or more agents as a set of players, their actions (strategies), probabilistic beliefs, and outcomes (rewards) (Maschler et al., 2020). Using a simple set of assumptions, game-theoretic models can predict the players' behavior, typically described as an *equilibrium,* a profile of strategies where no player has the incentive to deviate from the chosen course of action. Concepts such as the *Nash equilibrium* use frameworks such as expected utility and Bayesian learning to provide a useful benchmark for the case when the players behave optimally and rationally.

Empirical data, however, often contradict these predictions (Camerer, 2003). Humans tend to deviate from rational predictions and can be affected by various external factors such as interaction dynamics, social context, or game complexity. For instance, individuals learn and update their strategy as they play repeatedly, adapting to their opponent (Camerer & Ho, 1999; Erev & Roth, 1998) (Fig. 1). On the other hand, individual decisions can be affected by internal preferences, for instance, inequality aversion (Cox et al., 2017), where in some situations an individual might prefer receiving a fair split of money rather than taking the whole reward for herself.

Neural data provided general support for a dynamic learning account in strategic choices. Over the last 20 years, fMRI studies have identified a network often called the "social brain" (Adolphs, 2009) (Fig. 2). This network typically includes the right and left *temporoparietal junction* (TPJ), *dorsomedial prefrontal cortex* (dmPFC), *precuneus*, and *right* and *left temporal poles* (as well as the *superior temporal sulcus* (STS)). Some of these regions (particularly the TPJ) have been linked to *theory of mind* and show higher activity when a task requires mentalization (Carrington & Bailey, 2009; Coricelli, 2005; Koster-Hale & Saxe, 2013; Saxe, 2006a, b; Saxe & Kanwisher, 2003; Schaafsma et al., 2015; van Veluw & Chance, 2014).

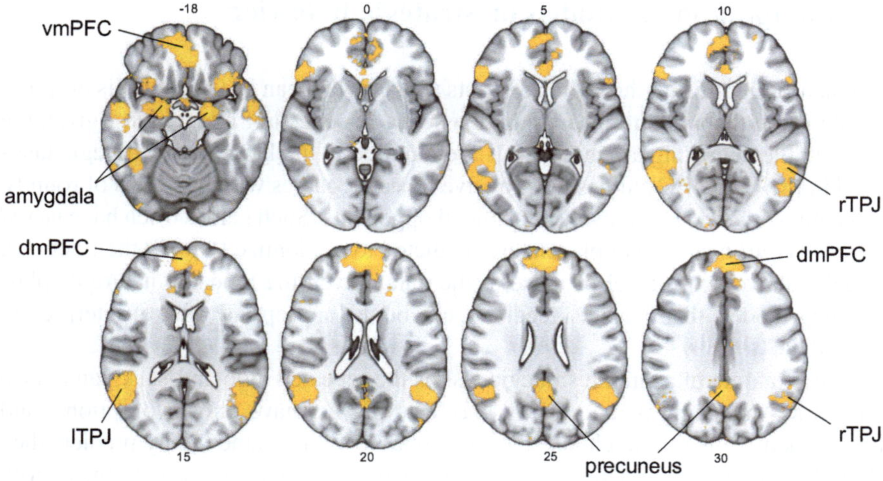

Fig. 2 Regions of the social brain network. An automated meta-analysis extracted from Neurosynth.org using the term "social" (1302 studies): the ventromedial prefrontal cortex (vmPFC), dorsomedial prefrontal cortex (dmPFC), amygdala, left and right temporoparietal junctions (lTPJ and rTPJ), and precuneus. The yellow color represents BOLD signal activations; the numbers represent axial MNI coordinates

Various theories offered interpretations of the specific functional role of the TPJ (Carter et al., 2012; Carter & Huettel, 2013; FeldmanHall et al., 2013; Morishima et al., 2012; Santiesteban et al., 2012; Tso et al., 2018; Young et al., 2010). Most importantly, TPJ seems to be often involved in strategic behavior (Hampton et al., 2008; Coricelli & Nagel, 2009; Hill et al., 2017; Ogawa & Kameda, 2019), including such examples as bluffing (Carter et al., 2012), trust games (Bhatt et al., 2010), consensus (Suzuki et al., 2015), and false advice (Behrens et al., 2008; Diaconescu et al., 2017).

Early research on theory of mind in the brain often used images and vignettes as task conditions (e.g., the Sally-Anne test (Korkiakangas et al., 2016)). In recent years, decision neuroscience and neuroeconomics (Dennison et al., 2022; Konovalov & Krajbich, 2019) have focused on the model-based accounts of the neural mechanisms of mentalization, using *strategic games* such as matching pennies or the Prisoner's dilemma as laboratory tasks (Camerer & Hare, 2014; Jiang et al., 2022; Wang & Kwan, 2023). Using well-specified games with monetary rewards and computational models can provide a more mechanistic interpretation of activity in the social brain, linking activity in certain areas to specific latent variables underlying the decision process (Charpentier & O'Doherty, 2018; Cutler & Campbell-Meiklejohn, 2019; Konovalov et al., 2018; Olsson et al., 2020; Rusch et al., 2020).

2 Computational Models of Strategic Behavior

The neuroscientific method often restricts the tasks that can be used for this purpose. While game theory can make predictions even for one-shot interactions, given the noise in the neural signal, many trials are required in the laboratory. Strategic tasks used in fMRI or EEG studies thus involve repeated games with hundreds of rounds, typically with only two players or artificial opponents. Such games often have *mixed strategy equilibria*, where players are predicted to randomize ("mix") their actions, which adds an extra level of noise to the data. Given the repeated nature of play, learning models that consider feedback can be the appropriate way of interpreting these data.

One example of a simple strategic task is the repeated hide-and-seek game, also known as "matching pennies" (Fig. 3a). The players have only two options, and one of them wins if both choose the same option. While game theory predicts they should randomize their actions, experiment participants pay attention to the history of play and the opponent's strategy (Fig. 3b, c).

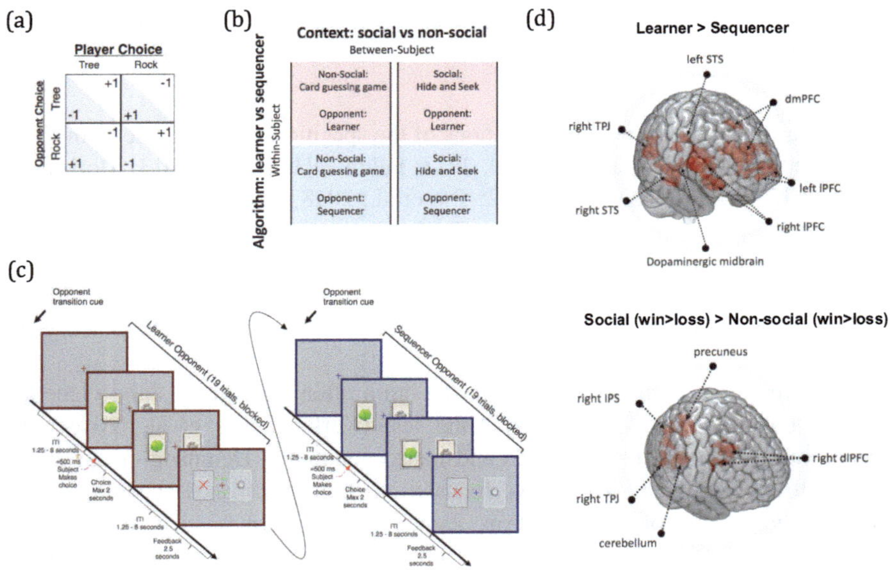

Fig. 3 Interaction between social context and strategic complexity in a hide-and-seek game. (**a**) The player and their opponent are choosing between two options: tree and rock. Their goal is to match their choice with the opponent, while the opponent is seeking to mismatch. (**b**) 2×2 manipulation of social context and the opponent strategy allows to disentangle the neural effects of context and algorithmic complexity. (**c**) The player is alternating between playing two computer opponents that are employing different strategies. (**d**) Playing against a reactive opponent (Learner) activates the social brain network, while social reward activates the TPJ, precuneus, and dorsolateral PFC. (See Konovalov et al. (2021) for details)

Computational models have a set of assumptions. Players are typically assumed to assign subjective values to their actions in each round (Fig. 1). Values determine choices, which are then transformed into rewards by the game task. Players receive feedback (their own and their opponent's rewards, as well as the opponent's choice) and use this feedback to guide their future strategy.

Feedback is hypothesized to affect players' strategic choices in multiple ways. Early behavioral economics studies showed that individuals use reinforcement learning in games by simply repeating rewarded actions (Erev & Roth, 1998; Roth & Erev, 1995). Under some conditions, such behavior can converge to the game-theoretic equilibrium (Camerer, 2003). Given the strong evidence for the neural mechanisms supporting reinforcement learning in humans and the use of prediction errors in non-social tasks (Botvinick et al., 2015), it is reasonable to expect these mechanisms to be shared by social and non-social choices (Lockwood et al., 2020). The value of actions is typically found to be encoded by the vmPFC (Bartra et al., 2013), while prediction errors correlate with activity in the striatum (Abler et al., 2006), which is also typically confirmed for strategic tasks.

However, simple reinforcement is often suboptimal in strategic situations. Another useful strategy is *fictitious play* or belief-based learning (Boylan & El-Gamal, 1993) (Fig. 1). The player can simply estimate the opponent's historical choice distribution (in other words, form *first-order beliefs*) and try to choose actions that are optimal given this distribution. There is converging neural evidence that such a tracking process might be performed by the regions of the dmPFC or ACC (anterior cingulate cortex) (Hampton et al., 2008; Zhu et al., 2012).

Mentalization allows for more complex strategies. If an agent knows that her actions are being tracked, she can form *second-order beliefs* and modify her strategy according to her opponent's reaction. One example of such behavior is the *level-k, cognitive hierarchy,* and *k-ToM* theories that assume that players have different levels of strategic sophistication (Camerer et al., 2004; d'Arc et al., 2020; Devaine et al., 2014; Nagel, 1995). An agent at level 0 would behave randomly or according to a simple heuristic, a level-1 agent optimally responds to that behavior, and a level-2 agent optimally responds to level 1 (in the level-k framework) or to a distribution of level 0 and level 1 (in the cognitive hierarchy framework); this reasoning process can reach higher levels.

In dyadic dynamic tasks, such behavior can be described using the *influence learning* model: if the agent is aware that her opponent is tracking the past history of her choices, she can adjust the strategy to influence the opponent's behavior (Fig. 1) (Hampton et al., 2008). Neural evidence suggests that increased levels of reasoning are linked to higher activity in the regions of the social brain (Coricelli & Nagel, 2009) and that in dynamic interactions, mentalization-driven prediction errors are linked to activity in the TPJ (Hampton et al., 2008). Disruption of activity in this region using transcranial magnetic stimulation (TMS) can lead to diminished use of high-order beliefs (Hill et al., 2017).

Despite the rapid progress in the field, many open questions remain. Some evidence suggests that beliefs and preferences are not independent and can co-evolve during strategic interactions (Rojek-Giffin et al., 2023). Neural activations during strategic tasks overlap with the regions of the social brain network (Konovalov et al., 2021), reward network (Zhu et al., 2012), default mode network (DMN) (Duan et al., 2012), and other areas with various functions. It is unclear whether these activations reflect domain-specific computations (related to the algorithmic demands of strategic decision-making) or domain-general (related to the mere presence of social context) (Lockwood et al., 2020; Rusch & Charpentier, 2021). It is unclear how these regions exactly perform the related computations and how different these activations are between non-strategic and strategic tasks (Parr et al., 2020). Finally, interactions between individual traits (such as social preferences) and strategic neural computations remain to be uncovered.

3 Social Context and Uncertainty

Two regions of the social brain most often linked to strategic behavior are the dmPFC and TPJ. Given that both areas are not uniquely linked to social behavior (Carter et al., 2013; Eickhoff et al., 2016), their precise computational roles are yet to be determined. One limiting factor is the ubiquitous presence of potential confounds in strategic tasks, including informational complexity, attentional demands, lack of control for social contexts, multiple linked processes (choice, uncertainty, valuation, preferences, just to name a few), and even varying visual presentations.

In early studies, activations in the social network areas were often linked to the presence of a social context or the task requiring mentalization. One recent study found that if the social context and computational demands are carefully controlled, higher activity in the dmPFC and TPJ is not simply modulated by social framing ("you are playing against a human opponent" versus "you are playing against a computer") but is instead driven by the strategy of the opponent, with more reactive partner eliciting stronger activations (Konovalov et al., 2021) (Fig. 3).

On the other hand, Reverberi et al. (2022) did not observe significant correlations of strategic complexity in the theory of mind networks but only registered them in the multiple demand network (MDN, which typically involves the dorsolateral prefrontal cortex, insula, and regions of the intraparietal sulcus) and only in subjects with higher cognitive skills. As this network can be modulated by activity in many non-social cognitive tasks, complexity remains an important potential confound for studies of strategic decision making.

Furthermore, converging evidence shows that basic decision processes (such as value encoding and comparison) are common between social and non-social decisions. Arabadzhiyska et al. (2022) used the combined EEG-MRI approach in a trust game to compare these two contexts and found that the temporal evidence integration process is very similar across the two contexts, reflected in the medial prefrontal cortex activity. However, the study also identified a degree of spatial dissociation, with social context engaging more posterior areas of the medial PFC and non-social context engaging more anterior areas.

Finally, uncertainty is also a common confound in strategic choices. Playing against an unpredictable opponent can be viewed as a lottery, introducing risk to decisions. Nagel et al. (2018) showed that the risk-processing network, which includes the insula, dmPFC, and parts of the parietal cortex, can also be activated during strategic choices. However, the study also manipulated the level of strategic thinking by changing the type of game the subjects played (a stag hunt game versus an entry game), showing that activity in this network also reflected the level of strategic complexity.

4 Tracking Others' Behavior

Converging evidence from several fields of neuroscience suggests that the dorso-medial prefrontal cortex (dmPFC) is the key region that encodes and stores social information, including other individuals' identities (Báez-Mendoza et al., 2021; Merritt et al., 2021). Using single neuron recordings in a non-strategic vignette task, Jamali et al. (2021) found that individual neurons can encode information about other people's beliefs and dissociate between the individual's and others' beliefs, predicting whether they are true or false.

These findings can be extended to dynamic interactions, proposing that the dmPFC can track and update beliefs about others' actions. In repeated games, the dmPFC's main functional role has been linked to updating first-order beliefs ("What is my opponent doing?") (Hampton et al., 2008; Hill et al., 2017). Applying inhibitory TMS to dmPFC can impair belief updating after negative prediction errors (Christian et al., 2024). However, simple encoding of action prediction errors could still be only a part of the full mechanism.

For instance, the dmPFC might encode not only the momentary updates in beliefs but also the general opponent tracking strategy or sensitivity to their actions. Konovalov et al. (2021) link the dmPFC activity to a computational model parameter that determines how sensitive the subject was to the opponent's actions in the preceding trial, across a whole block of trials played against that opponent.

Additionally, as discussed above, the dmPFC is also sensitive to strategic complexity and the number of reasoning steps required to find the optimal strategy. In many tasks, higher-order reasoning produces stronger dmPFC activations (Zhen & Yu, 2021), and functional connectivity between the dmPFC and TPJ reflects higher-order reasoning (Hill et al., 2017; Zhen & Yu, 2021).

5 Individual and Social Value

Another mechanism crucial for strategic decision-making is the dissociation between individual and social value. Many studies support the existence of some "common currency" encoding in the brain encompassing all types of decisions, involving the ventromedial prefrontal cortex (vmPFC) and basal ganglia (typically the ventral striatum) (Arabadzhiyska et al., 2022; Bartra et al., 2013; Ruff & Fehr, 2014). However, recent studies of strategic choices suggest a more complex mechanism of integration between individual and social rewards under various specific conditions of the game (group decision making, social hierarchy, or cooperative goals).

For instance, using a group decision-making task (a public goods game), Park et al. (2019) found a spatial dissociation between individual rewards and group utility. While the vmPFC encodes expected rewards for the individual, group utility is encoded in the lateral frontopolar cortex. These regions also change their connectivity with the ACC during strategy switching.

Additionally, the value-related activity in the PFC can reflect the social hierarchy. Using a non-strategic preference rating task, Kim et al. (2023) identified a spatial dissociation in the PFC indicating the social rank. The activity in the ventral mPFC increased with more alignment between the individual's preferences and a superior partner. In contrast, the rostral mPFC activity was modulated by goal-based conformity.

Finally, strategic decisions can involve both competitive and cooperative motives. Philippe et al. (2024) used artificial opponents that, unknowingly to the subjects, alternated between competitive and cooperative strategies in a matching pennies game. Subjects adapted to these changes and were able to arbitrate between competitive and cooperative learning strategies. The neural data showed that the vmPFC and striatum tracked the reliability of these strategies, reflecting learning above and beyond simple reinforcement.

Data from a lesion study suggest that the striatum might not even be necessary for successful strategic learning (Zhu et al., 2019). While patients with damage to the orbitofrontal cortex show worse learning in a strategic game, patients with lesions to the basal ganglia perform at the level of the control group and can anticipate the opponents' behavior. However, both groups are impaired when not given information about the strategic context (Zhu et al., 2019).

6 Mentalization and Higher-Order Reasoning

The temporoparietal junction (TPJ) is most often linked to attention reorientation and mentalization (Carter et al., 2013; Krall et al., 2015; Saxe & Kanwisher, 2003). In strategic tasks, its activity often correlates with second-order belief updates (Hampton et al., 2008; Hill et al., 2017; Konovalov et al., 2021), but its computational role must be more complex.

Given its cortical location, the TPJ is an easy target for brain stimulation interventions that elucidate some of the potential functions of this area. Disruption of activity in the TPJ tends to decrease reliance on mentalization and decreases the connectivity between the TPJ and dmPFC in the repeated inspection game (Hill et al., 2017). In a donation task involving moral values, disruption of the TPJ leads to a reduced impact of the conflict between moral and material values (Obeso et al., 2018). In the ultimatum game, inhibitory TMS administered to the TPJ changes the counterfactual evaluation of the offer (Ogawa et al., 2023). In the holdup game, disruption of TPJ activity using tDCS changes the subjects' perception of their own and others' benefits, while activation of the TPJ improves mentalization (Xiaoqin et al., 2023). Finally, activation of the TPJ in the patent race game increases reliance on belief-based learning (Zhang et al., 2024). Taken together, these results point to a causal role of the TPJ in strategic decisions, although it might be difficult to identify the common underlying computation given the diversity of tasks and choice contexts.

Most neural evidence suggests that the TPJ is primarily involved in instantaneous updating mechanisms rather than long-term strategy. Using a modified dictator game Park et al. (2021) suggest that the recruitment of the TPJ represents the integration of prediction in the face of conflicting evidence (e.g., violated expectations of generosity). Changes in feedback or strategy can modulate the functional connectivity of the TPJ. Konovalov et al. (2021) showed that connectivity between the TPJ and nucleus accumbens reflected losses against a reactive opponent strategy that requires mentalization. Using a bargaining game, Yang et al. (2022) found an increase in the interaction between the TPJ and the dorsolateral prefrontal cortex (dlPFC) when subjects engaged in strategic deception.

Recent evidence also suggests that the TPJ might play a role in *adaptive* mentalization. Using a dynamic cognitive hierarchy model, Buergi et al. (2024) link activity in the TPJ to the update not just on the opponent's actions but also on the opponent's level of reasoning (an individual who can achieve a high level of reasoning can also identify the level her opponent is playing at).

7 Future Directions

There are still many open questions in the study of strategic decision-making. Further clarification of the functional roles of the regions of the social networks and their interactions with the reward networks and the general task networks will most likely remain at the core of research on this topic.

As previously noted, it remains uncertain whether all or only some regions of the "social brain" are domain-specific in processing social interactions broadly and strategic thinking in particular (Lockwood et al., 2020; Rusch & Charpentier, 2021). Evidence suggests that disruptions in the TPJ may impair higher-order belief processing (Hill et al., 2017) while lesions in the prefrontal cortex can affect social behavior (Lockwood et al., 2024). However, further causal studies are necessary. Such research could incorporate brain stimulation or studies involving lesion patients, paired with tasks specifically designed to distinguish between social and non-social processing, as well as between strategic and non-strategic thinking (e.g., tasks requiring mentalization versus those based on simple observation).

One natural new direction is the use of hyperscanning (recording neural signals from several participants simultaneously), which still remains a relatively rare method in the study of strategic choice, given its cost and complexity (Liu et al., 2017; Xie et al., 2020; Zhang, 2018). One recent study suggested that neural synchrony between two interacting subjects requires mutual presence in the same space rather than simply predicting the other's choices (Kayhan et al., 2022).

The space of potential strategic tasks is rich and must expand beyond simple competitive zero-sum games that only require belief updating mechanisms. Tasks like trust games and social dilemma games also involve individual motivations such as guilt aversion, inequity aversion, and risk preferences, which can involve interaction with other neural substrates (van Baar et al., 2019). An interesting recent direction is the study of self-other mergence, or estimating own and others' abilities and preferences, with the dmPFC playing an important computational role (Wittmann et al., 2021).

In this chapter, we did not discuss hormones, which can potentially play an important role in strategic interactions, prosociality, and aggression. Recent evidence suggests that administering testosterone to male subjects can decrease strategic prosociality and the need to adapt to social expectations (Kutlikova et al., 2023), although the hormone's neural impact mechanism is yet to be determined.

Finally, MRI and EEG data are not the only window into the hidden mechanisms underlying strategic decisions. Other types of choice process data, such as response times and eye-tracking data, can also provide important insights into strategic decision-making (Cooper et al., 2019; Konovalov & Ruff, 2022). For instance, gaze and response times can inform others about a decision-maker's prosociality or hidden preferences and can thus be used strategically (Bavard et al., 2024; Fischbacher et al., 2022; Konovalov & Krajbich, 2023).

8 Conclusion

Research on strategic decision-making in the brain is an excellent example of an interdisciplinary collaboration between economics, psychology, and neuroscience. Further advancements in this topic will allow us to build better, neurally informed game-theoretic models and improve our understanding of the functional roles of specific brain regions. Identifying deficiencies in these functions and underlying computations can be useful in studying neurodiversity in social cognition, including autism and disorders such as psychopathy and apathy.

References

Abler, B., Walter, H., Erk, S., Kammerer, H., & Spitzer, M. (2006). Prediction error as a linear function of reward probability is coded in human nucleus accumbens. *NeuroImage, 31*(2), 790–795.

Adolphs, R. (2009). The social brain: Neural basis of social knowledge. *Annual Reviews of Psychology*. https://doi.org/10.1146/annurev.psych.60.110707.163514

Arabadzhiyska, D. H., Garrod, O. G. B., Fouragnan, E., Luca, E. D., Schyns, P. G., & Philiastides, M. G. (2022). A common neural account for social and nonsocial decisions. *Journal of Neuroscience, 42*(48), 9030–9044. https://doi.org/10.1523/JNEUROSCI.0375-22.2022

Báez-Mendoza, R., Mastrobattista, E. P., Wang, A. J., & Williams, Z. M. (2021). Social agent identity cells in the prefrontal cortex of interacting groups of primates. *Science, 374*(6566), eabb4149. https://doi.org/10.1126/science.abb4149

Bartra, O., McGuire, J. T., & Kable, J. W. (2013). The valuation system: A coordinate-based meta-analysis of BOLD fMRI experiments examining neural correlates of subjective value. *NeuroImage, 76*, 412–427. https://doi.org/10.1016/j.neuroimage.2013.02.063

Bavard, S., Stuchlý, E., Konovalov, A., & Gluth, S. (2024). Humans can infer social preferences from decision speed alone. *PLoS Biology, 22*(6), e3002686.

Behrens, T. E. J., Hunt, L. T., Woolrich, M. W., & Rushworth, M. F. S. (2008). Associative learning of social value. *Nature, 456*(7219), 245–249. https://doi.org/10.1038/nature07538

Bhatt, M. A., Lohrenz, T., Camerer, C. F., & Montague, P. R. (2010). Neural signatures of strategic types in a two-person bargaining game. *Proceedings of the National Academy of Sciences of the United States of America, 107*(46), 19720–19725. https://doi.org/10.1073/pnas.1009625107

Botvinick, M., Weinstein, A., Solway, A., & Barto, A. (2015). Reinforcement learning, efficient coding, and the statistics of natural tasks. *Current Opinion in Behavioral Sciences, 5*, 71–77. https://doi.org/10.1016/j.cobeha.2015.08.009

Boylan, R. T., & El-Gamal, M. A. (1993). Fictitious play: A statistical study of multiple economic experiments. *Games and Economic Behavior, 5*(2), 205–222.

Buergi, N., Aydogan, G., Konovalov, A., & Ruff, C. (2024). A neural fingerprint of adaptive mentalization. *OSF*. https://doi.org/10.31234/osf.io/npqy8

Camerer, C. (2003). *Behavioral game theory: Experiments in strategic interaction*. Princeton University Press.

Camerer, C. F., & Hare, T. A. (2014). The neural basis of strategic choice. In *Neuroeconomics* (pp. 479–492). Elsevier. http://linkinghub.elsevier.com/retrieve/pii/B9780124160088000255

Camerer, C., & Ho, T.-H. (1999). Experienced-weighted attraction learning in normal form games. *Econometrica, 67*(4), 827–874.

Camerer, C. F., Ho, T.-H., & Chong, J.-K. (2004). A cognitive hierarchy model of games*. *The Quarterly Journal of Economics, 119*(3), 861–898. https://doi.org/10.1162/0033553041502225

Carrington, S. J., & Bailey, A. J. (2009). Are there theory of mind regions in the brain? A review of the neuroimaging literature. *Human Brain Mapping, 30*(8), 2313–2335.

Carter, R. M., Bowling, D. L., Reeck, C., & Huettel, S. A. (2012). A distinct role of the temporal-parietal junction in predicting socially guided decisions. *Science, 337*(6090), 109–111. https://doi.org/10.1126/science.1219681

Carter, R. M., & Huettel, S. A. (2013). A nexus model of the temporal-parietal junction. *Trends in cognitive sciences, 17*(7), 328–336. https://doi.org/10.1016/j.tics.2013.05.007

Charpentier, C. J., & O'Doherty, J. P. (2018). The application of computational models to social neuroscience: Promises and pitfalls. *Social Neuroscience, 13*(6), 637–647. https://doi.org/10.1080/17470919.2018.1518834

Christian, P., Kaiser, J., Taylor, P. C., George, M., Schütz-Bosbach, S., & Soutschek, A. (2024). Belief updating during social interactions: Neural dynamics and causal role of dorsomedial prefrontal cortex. *Journal of Neuroscience, 44*(22). https://doi.org/10.1523/JNEUROSCI.1669-23.2024

Cooper, D. J., Krajbich, I., & Noussair, C. N. (2019). Choice-process data in experimental economics. *Journal of the Economic Science Association, 5*(1), 1–13. https://doi.org/10.1007/s40881-019-00075-z

Coricelli, G. (2005). Two-levels of mental states attribution: From automaticity to voluntariness. *Neuropsychologia, 43*(2), 294–300.

Coricelli, G., & Nagel, R. (2009). Neural correlates of depth of strategic reasoning in medial prefrontal cortex. *Proceedings of the National Academy of Sciences, 106*(23), 9163–9168.

Cox, C. A., Karam, A., & Murphy, R. J. (2017). Social preferences and cooperation in simple social dilemma games. *Journal of Behavioral and Experimental Economics, 69*, 1–3. https://doi.org/10.1016/j.socec.2017.05.002

Cutler, J., & Campbell-Meiklejohn, D. (2019). A comparative fMRI meta-analysis of altruistic and strategic decisions to give. *NeuroImage, 184*, 227–241. https://doi.org/10.1016/j.neuroimage.2018.09.009

d'Arc, B. F., Devaine, M., & Daunizeau, J. (2020). Social behavioural adaptation in autism. *PLoS Computational Biology, 16*(3), e1007700. https://doi.org/10.1371/journal.pcbi.1007700

Dennison, J. B., Sazhin, D., & Smith, D. V. (2022). Decision neuroscience and neuroeconomics: Recent progress and ongoing challenges. *WIREs Cognitive Science, 13*(3), e1589. https://doi.org/10.1002/wcs.1589

Devaine, M., Hollard, G., & Daunizeau, J. (2014). The social Bayesian brain: Does mentalizing make a difference when we learn? *PLoS Computational Biology, 10*(12), e1003992.

Diaconescu, A. O., Mathys, C., Weber, L. A. E., Kasper, L., Mauer, J., & Stephan, K. E. (2017). Hierarchical prediction errors in midbrain and septum during social learning. *Social Cognitive and Affective Neuroscience*. https://doi.org/10.1093/scan/nsw171

Duan, X., He, S., Liao, W., Liang, D., Qiu, L., Wei, L., Li, Y., Liu, C., Gong, Q., & Chen, H. (2012). Reduced caudate volume and enhanced striatal-DMN integration in chess experts. *NeuroImage, 60*(2), 1280–1286. https://doi.org/10.1016/j.neuroimage.2012.01.047

Eickhoff, S. B., Laird, A. R., Fox, P. T., Bzdok, D., & Hensel, L. (2016). Functional segregation of the human dorsomedial prefrontal cortex. *Cerebral Cortex, 26*(1), 304–321. https://doi.org/10.1093/cercor/bhu250

Erev, I., & Roth, A. E. (1998). Predicting how people play games: Reinforcement learning in experimental games with unique, mixed strategy equilibria. *American Economic Review, 88*, 848–881.

FeldmanHall, O., Mobbs, D., & Dalgleish, T. (2013). Deconstructing the brain's moral network: Dissociable functionality between the temporoparietal junction and ventro-medial prefrontal cortex. *Social Cognitive and Affective Neuroscience, 9*(3), 297–306.

Fischbacher, U., Hausfeld, J., & Renerte, B. (2022). Strategic incentives undermine gaze as a signal of prosocial motives. *Games and Economic Behavior, 136*, 63–91. https://doi.org/10.1016/j.geb.2022.07.006

Frith, C., & Frith, U. (2005). Theory of mind. *Current Biology, 15*(17), R644–R645.

Hampton, A. N., Bossaerts, P., & O'Doherty, J. P. (2008). Neural correlates of mentalizing-related computations during strategic interactions in humans. *Proceedings of the National Academy of Sciences of the United States of America, 105*(18), 6741–6746. https://doi.org/10.1073/pnas.0711099105

Hill, C. A., Suzuki, S., Polania, R., Moisa, M., O'doherty, J. P., & Ruff, C. C. (2017). A causal account of the brain network computations underlying strategic social behavior. *Nature Neuroscience*. https://doi.org/10.1038/nn.4602

Jamali, M., Grannan, B. L., Fedorenko, E., Saxe, R., Báez-Mendoza, R., & Williams, Z. M. (2021). Single-neuronal predictions of others' beliefs in humans. *Nature, 591*(7851), 610–614. https://doi.org/10.1038/s41586-021-03184-0

Jiang, Y., Wu, H.-T., Mi, Q., & Zhu, L. (2022). Neurocomputations of strategic behavior: From iterated to novel interactions. *WIREs Cognitive Science, 13*(4), e1598. https://doi.org/10.1002/wcs.1598

Kayhan, E., Nguyen, T., Matthes, D., Langeloh, M., Michel, C., Jiang, J., & Hoehl, S. (2022). Interpersonal neural synchrony when predicting others' actions during a game of rock-paper-scissors. *Scientific Reports, 12*(1), 12967. https://doi.org/10.1038/s41598-022-16956-z

Kim, D., Kim, J., & Kim, H. (2023). Distinctive roles of medial prefrontal cortex subregions in strategic conformity to social hierarchy. *Journal of Neuroscience, 43*(36), 6330–6341. https://doi.org/10.1523/JNEUROSCI.0549-23.2023

Konovalov, A., & Krajbich, I. (2019). Over a decade of neuroeconomics: What have we learned? *Organizational Research Methods, 22*(1), 148–173. https://doi.org/10.1177/1094428116644502

Konovalov, A., & Krajbich, I. (2023). Decision times reveal private information in strategic settings: Evidence from bargaining experiments. *The Economic Journal, 133*(656), 3007–3033. https://doi.org/10.1093/ej/uead055

Konovalov, A., & Ruff, C. C. (2022). Enhancing models of social and strategic decision making with process tracing and neural data. *Wiley Interdisciplinary Reviews: Cognitive Science, 13*(1), e1559.

Konovalov, A., Hu, J., & Ruff, C. C. (2018). Neurocomputational approaches to social behavior. *Current Opinion in Psychology, 24*, 41–47. https://doi.org/10.1016/j.copsyc.2018.04.009

Konovalov, A., Hill, C., Daunizeau, J., & Ruff, C. C. (2021). Dissecting functional contributions of the social brain to strategic behavior. *Neuron, 109*(20), 3323–3337.e5. https://doi.org/10.1016/j.neuron.2021.07.025

Korkiakangas, T., Dindar, K., Laitila, A., & Kärnä, E. (2016). The Sally–Anne test: An interactional analysis of a dyadic assessment. *International Journal of Language & Communication Disorders, 51*(6), 685–702. https://doi.org/10.1111/1460-6984.12240

Koster-Hale, J., & Saxe, R. (2013). Theory of mind: A neural prediction problem. *Neuron*. https://doi.org/10.1016/j.neuron.2013.08.020

Krall, S. C., Rottschy, C., Oberwelland, E., Bzdok, D., Fox, P. T., Eickhoff, S. B., Fink, G. R., & Konrad, K. (2015). The role of the right temporoparietal junction in attention and social interaction as revealed by ALE meta-analysis. *Brain Structure and Function, 220*(2), 587–604.

Kutlikova, H. H., Zhang, L., Eisenegger, C., van Honk, J., & Lamm, C. (2023). Testosterone eliminates strategic prosocial behavior through impacting choice consistency in healthy males. *Neuropsychopharmacology, 48*(10), 1541–1550. https://doi.org/10.1038/s41386-023-01570-y

Liu, T., Saito, G., Lin, C., & Saito, H. (2017). Inter-brain network underlying turn-based cooperation and competition: A hyperscanning study using near-infrared spectroscopy. *Scientific Reports, 7*(1), 8684. https://doi.org/10.1038/s41598-017-09226-w

Lockwood, P. L., Apps, M. A. J., & Chang, S. W. C. (2020). Is there a 'social' brain? Implementations and algorithms. *Trends in Cognitive Sciences, 24*(10), 802–813. https://doi.org/10.1016/j.tics.2020.06.011

Lockwood, P. L., Cutler, J., Drew, D., Abdurahman, A., Jeyaretna, D. S., Apps, M. A. J., Husain, M., & Manohar, S. G. (2024). Human ventromedial prefrontal cortex is necessary for prosocial motivation. *Nature Human Behaviour, 8*(7), 1403–1416. https://doi.org/10.1038/s41562-024-01899-4

Maschler, M., Zamir, S., & Solan, E. (2020). *Game theory*. Cambridge University Press.

Merritt, C. C., MacCormack, J. K., Stein, A. G., Lindquist, K. A., & Muscatell, K. A. (2021). The neural underpinnings of intergroup social cognition: An fMRI meta-analysis. *Social Cognitive and Affective Neuroscience, 16*(9), 903–914. https://doi.org/10.1093/scan/nsab034

Morishima, Y., Schunk, D., Bruhin, A., Ruff, C. C., & Fehr, E. (2012). Linking brain structure and activation in temporoparietal junction to explain the neurobiology of human altruism. *Neuron, 75*(1), 73–79. https://doi.org/10.1016/j.neuron.2012.05.021

Nagel, R. (1995). Unraveling in guessing games: An experimental study. *The American Economic Review, 85*(5), 1313–1326.

Nagel, R., Brovelli, A., Heinemann, F., & Coricelli, G. (2018). Neural mechanisms mediating degrees of strategic uncertainty. *Social Cognitive and Affective Neuroscience, 13*(1), 52–62. https://doi.org/10.1093/scan/nsx131

Obeso, I., Moisa, M., Ruff, C. C., & Dreher, J.-C. (2018). A causal role for right temporo-parietal junction in signaling moral conflict. *eLife, 7*, e40671. https://doi.org/10.7554/eLife.40671

Ogawa, A., & Kameda, T. (2019). Dissociable roles of left and right temporoparietal junction in strategic competitive interaction. *Social Cognitive and Affective Neuroscience*. https://doi.org/10.1093/scan/nsz082

Ogawa, A., Asano, S., Osada, T., Tanaka, M., Tochigi, R., Kamagata, K., Aoki, S., & Konishi, S. (2023). Role of right temporoparietal junction for counterfactual evaluation of partner's decision in ultimatum game. *Cerebral Cortex, 33*(6), 2947–2957. https://doi.org/10.1093/cercor/bhac252

Olsson, A., Knapska, E., & Lindström, B. (2020). The neural and computational systems of social learning. *Nature Reviews Neuroscience, 21*(4), Article 4. https://doi.org/10.1038/s41583-020-0276-4

Park, S. A., Sestito, M., Boorman, E. D., & Dreher, J.-C. (2019). Neural computations underlying strategic social decision-making in groups. *Nature Communications, 10*(1), 5287. https://doi.org/10.1038/s41467-019-12937-5

Park, B., Fareri, D., Delgado, M., & Young, L. (2021). The role of right temporoparietal junction in processing social prediction error across relationship contexts. *Social Cognitive and Affective Neuroscience, 16*(8), 772–781. https://doi.org/10.1093/scan/nsaa072

Parr, A. C., Coe, B. C., Munoz, D. P., & Dorris, M. C. (2020). A novel fMRI paradigm to dissociate the behavioral and neural components of mixed-strategy decision making from non-strategic decisions in humans. *European Journal of Neuroscience, 51*(9), 1914–1927. https://doi.org/10.1111/ejn.14586

Philippe, R., Janet, R., Khalvati, K., Rao, R. P. N., Lee, D., & Dreher, J.-C. (2024). Neurocomputational mechanisms involved in adaptation to fluctuating intentions of others. *Nature Communications, 15*(1), 3189. https://doi.org/10.1038/s41467-024-47491-2

Reverberi, C., Pischedda, D., Mantovani, M., Haynes, J.-D., & Rustichini, A. (2022). Strategic complexity and cognitive skills affect brain response in interactive decision-making. *Scientific Reports, 12*(1), Article 1. https://doi.org/10.1038/s41598-022-17951-0

Rojek-Giffin, M., Lebreton, M., Daunizeau, J., Fariña, A., Gross, J., & De Dreu, C. K. W. (2023). Learning rules of engagement for social exchange within and between groups. *Proceedings of the National Academy of Sciences, 120*(19), e2218443120. https://doi.org/10.1073/pnas.2218443120

Roth, A. E., & Erev, I. (1995). Learning in extensive-form games: Experimental data and simple dynamic models in the intermediate term. *Games and Economic Behavior, 8*(1), 164–212.

Ruff, C. C., & Fehr, E. (2014). The neurobiology of rewards and values in social decision making. *Nature Reviews Neuroscience, 15*(8), 549–562. https://doi.org/10.1038/nrn3776

Rusch, T., & Charpentier, C. J. (2021). Domain specificity versus process specificity: The "social brain" during strategic interaction. *Neuron, 109*(20), 3236–3238. https://doi.org/10.1016/j.neuron.2021.09.035

Rusch, T., Steixner-Kumar, S., Doshi, P., Spezio, M., & Gläscher, J. (2020). Theory of mind and decision science: Towards a typology of tasks and computational models. *Neuropsychologia, 146*, 107488. https://doi.org/10.1016/j.neuropsychologia.2020.107488

Santiesteban, I., Banissy, M. J., Catmur, C., & Bird, G. (2012). Enhancing social ability by stimulating right temporoparietal junction. *Current Biology, 22*(23), 2274–2277.

Saxe, R. (2006a). Uniquely human social cognition. *Current Opinion in Neurobiology, 16*(2), 235–239.

Saxe, R. (2006b). Why and how to study theory of mind with fMRI. *Brain Research, 1079*(1), 57–65.

Saxe, R., & Kanwisher, N. (2003). People thinking about thinking people: The role of the temporoparietal junction in "theory of mind". *NeuroImage, 19*(4), 1835–1842.

Schaafsma, S. M., Pfaff, D. W., Spunt, R. P., & Adolphs, R. (2015). Deconstructing and reconstructing theory of mind. *Trends in Cognitive Sciences, 19*(2), 65–72. https://doi.org/10.1016/j.tics.2014.11.007

Suzuki, S., Adachi, R., Bossaerts, P., Doherty, J. P. O., Suzuki, S., Adachi, R., Dunne, S., Bossaerts, P., & Doherty, J. P. O. (2015). Neural mechanisms underlying human consensus neural mechanisms underlying human consensus decision-making. *Neuron, 86*(2), 591–602. https://doi.org/10.1016/j.neuron.2015.03.019

Tso, I. F., Rutherford, S., Fang, Y., Angstadt, M., & Taylor, S. F. (2018). The "social brain" is highly sensitive to the mere presence of social information: An automated meta-analysis and an independent study. *PLoS One, 13*(5), e0196503.

van Baar, J. M., Chang, L. J., & Sanfey, A. G. (2019). The computational and neural substrates of moral strategies in social decision-making. *Nature Communications, 10*(1), 1483. https://doi.org/10.1038/s41467-019-09161-6

van Veluw, S. J., & Chance, S. A. (2014). Differentiating between self and others: An ALE meta-analysis of fMRI studies of self-recognition and theory of mind. *Brain Imaging and Behavior, 8*(1), 24–38.

Wang, H., & Kwan, A. C. (2023). Competitive and cooperative games for probing the neural basis of social decision-making in animals. *Neuroscience & Biobehavioral Reviews, 149*, 105158. https://doi.org/10.1016/j.neubiorev.2023.105158

Wittmann, M. K., Trudel, N., Trier, H. A., Klein-Flügge, M. C., Sel, A., Verhagen, L., & Rushworth, M. F. (2021). Causal manipulation of self-other mergence in the dorsomedial prefrontal cortex. *Neuron, 109*(14), 2353–2361.

Xiaoqin, W., Hao, S., Qin, L., Jiali, P., & Xiaomin, Z. (2023). A causal role for right temporoparietal junction in random pairing repeated holdup game. *Journal of Neuroscience, Psychology, and Economics, 16*(3), 155–169. https://doi.org/10.1037/npe0000175

Xie, H., Karipidis, I. I., Howell, A., Schreier, M., Sheau, K. E., Manchanda, M. K., Ayub, R., Glover, G. H., Jung, M., Reiss, A. L., & Saggar, M. (2020). Finding the neural correlates of collaboration using a three-person fMRI hyperscanning paradigm. *Proceedings of the National Academy of Sciences, 117*(37), 23066–23072. https://doi.org/10.1073/pnas.1917407117

Yang, R., Ma, Y., Pan, B. B., Bhatt, M. A., Lohrenz, T., Gu, H. G., ... & Luo, Q. (2022). Dynamic neural reconfiguration for distinct strategies during competitive social interactions. *NeuroImage, 263*, 119585.

Young, L., Camprodon, J. A., Hauser, M., Pascual-Leone, A., & Saxe, R. (2010). Disruption of the right temporoparietal junction with transcranial magnetic stimulation reduces the role of beliefs in moral judgments. *Proceedings of the National Academy of Sciences, 107*(15), 6753–6758.

Zhang, D. (2018). Computational EEG analysis for hyperscanning and social neuroscience. In C.-H. Im (Ed.), *Computational EEG analysis: Methods and applications* (pp. 215–228). Springer. https://doi.org/10.1007/978-981-13-0908-3_10

Zhang, H., Dong, Z., Xiong, G., & Wu, S. (2024). Modulating the activity of right temporoparietal junction increased belief-based learning in an economic game. *Journal of Neuroscience, Psychology, and Economics, 17*(1), 19–29. https://doi.org/10.1037/npe0000186

Zhen, S., & Yu, R. (2021). Neural correlates of recursive thinking during interpersonal strategic interactions. *Human Brain Mapping, 42*(7), 2128–2146. https://doi.org/10.1002/hbm.25355

Zhu, L., Mathewson, K. E., & Hsu, M. (2012). Dissociable neural representations of reinforcement and belief prediction errors underlie strategic learning. *Proceedings of the National Academy of Sciences, 109*(5), 1419–1424. https://doi.org/10.1073/pnas.1116783109

Zhu, L., Jiang, Y., Scabini, D., Knight, R. T., & Hsu, M. (2019). Patients with basal ganglia damage show preserved learning in an economic game. *Nature Communications, 10*(1), 802. https://doi.org/10.1038/s41467-019-08766-1

The Neuroeconomics of Social Influence and Contagion

Z. Su and P. L. Lockwood

Abstract People often adjust their behaviours and preferences to align with others, a phenomenon termed social influence or social contagion. Social influence can have a profound effect on our economic preferences, with important implications for financial decision-making and making rational choices. Here, we review recent findings in neuroeconomics, elucidating the psychological, computational and neural mechanisms that underpin social influence during value-based decision-making. We examine how others' decisions can shift our own economic preferences, regarding risk-taking, temporal discounting, effort discounting and moral preferences. Additionally, we discuss the lifespan similarities and differences in susceptibility to social influence. Finally, we cover studies revealing the neural basis of susceptibility to social influence and highlight key common brain areas across studies that could underpin these effects. We end by considering the implications of strong evidence of susceptibility to social influence on economic preferences, the challenges and opportunities of this influence and how future studies could harness it to create positive changes across the lifespan.

1 Introduction

Humans are ultra-social animals. They are deeply interdependent on one another and use social information to shape behaviour and cognition (Dunbar & Shultz, 2007; Tomasello, 2014). This interdependence often leads people to change their behaviour and preferences to align with others, a phenomenon known as social

Z. Su · P. L. Lockwood
Centre for Human Brain Health, School of Psychology, University of Birmingham, Birmingham, UK

Institute for Mental Health, School of Psychology, University of Birmingham, Birmingham, UK

Centre for Developmental Science, School of Psychology, University of Birmingham, Birmingham, UK
e-mail: zxs203@student.bham.ac.uk; p.l.lockwood@bham.ac.uk

D. V. Smith et al. (eds.), *Neuroeconomics: Core Topics and Current Directions*,
https://doi.org/10.1007/978-3-032-02925-6_26

influence (Cialdini & Goldstein, 2004) or social contagion (Wheeler, 1966). Understanding how and why people are susceptible to social influence is of paramount importance at both the individual and societal levels, for example, in the context of political decision-making and social cohesion (Andrews et al., 2020; Bavel et al., 2020; Miller & Prentice, 2016), where misinformation can contribute to negative outcomes. In the context of economic decision-making, knowing how others influence our own financial decisions is crucial for avoiding financial exploitation and making rational economic choices.

In this chapter, we outline the latest research on the psychological, computational and neural mechanisms underlying social influence within the framework of neuroeconomics. We describe how social influence drives shifts in preferences during value-based decision-making, a process that entails weighing rewards—such as monetary gains—against costs, including how long we need to wait, how uncertain an outcome is and how much effort is involved. We will also consider how these processes develop and differ across the lifespan and in health and disease.

1.1 Social Influence

Early work on social influence showed just how powerful other people's preferences can be on our own. In the pioneering studies of Sherif (1935) and Asch (1956), they demonstrated that knowing the preferences of a group can change our own perception. Two lines were shown to a group of people. One line was clearly bigger than the other. If the majority of members of the group intentionally misreported that the smallest line was the biggest, this was enough to change the report of an independent individual who was not a confederate of the experimenter.

Researchers propose that two distinct yet interconnected motivations underlie this form of social influence (Deutsch & Gerard, 1955; Toelch & Dolan, 2015). Firstly, people are motivated to align with others when conformity is itself rewarded—a process known as 'normative influence'. This influence helps people adapt to their social environment or fit within a social group, allowing them to demonstrate belonging and signal membership by following established group norms. In such cases, conformity is often rewarded with social approval, which strengthens or maintains social relationships and helps people avoid social punishment such as ostracism. Secondly, conformity may not always be directly rewarded, but its outcomes can be beneficial or advantageous, a process known as 'informational influence'. This type of influence helps people adopt the most appropriate behaviour in uncertain situations, especially when they believe others have better information or knowledge. In such cases, people view others as a valuable source of information, aiding them in making more accurate decisions.

A simple example makes this distinction clearer. Imagine you have just moved to a foreign country and are unfamiliar with the local customs. You go to a restaurant with your new local friends and notice several cups of water on the table. Instead of drinking the water, you observe everyone else using it to clean their cutlery.

Although surprised by this behaviour, you are guided by *informational influence* and follow suit, assuming it is the appropriate thing to do in this context. Later, during the dinner, everyone orders a local dish that you are not keen on trying. To avoid standing out and to fit in with the group, *normative influence* prompts you to order the same dish.

In addition to these common mechanisms driving social influence, people sometimes change their behaviour or preferences after simply learning what others value (Garvert et al., 2015; Moutoussis et al., 2016), even without experiencing group pressure (*absent normative influence*) or when there is no obvious benefit (*absent informational influence*). This type of non-rewarded social conformity is a compelling form of social influence and has been studied in terms of how it can change our economic preferences.

1.2 Social Influence on Economic Preferences

One way to test social influence within the field of neuroeconomics is to examine whether people's economic preferences can be shifted to be more aligned with one another. In the following sections, we will review the effect of social influence on different types of economic preferences: risk-taking attitude, temporal discounting, effort discounting and moral preference.

1.2.1 Risk-Taking Attitude

Economic risk can be defined as the variance of potential monetary outcomes, assuming the decision maker is aware of the objective probability distribution of these outcomes (Schonberg et al., 2011). Risk-taking attitudes reflect the level of risk people are willing to accept in exchange for a specific reward (Figner & Weber, 2011). People's risk-taking attitudes have substantial implications across various contexts, from substance abuse to high-risk stock investments (Dohmen et al., 2011; Platt & Huettel, 2008). Risk-taking attitudes are often assessed through monetary lottery tasks, which examine how people make decisions under risk (Rigoli et al., 2016; Sokol-Hessner et al., 2009; Tom et al., 2007). In such tasks, participants are asked to choose between two options with clearly stated probabilities and rewards: a secure option offering a guaranteed reward, and a risky option offering a potentially larger reward under uncertainty (Fig. 1a). For instance, would you prefer £40 for sure or £100 with a probability of 50%? A risk-neutral person would base their choice on the objectively calculated expected value of each option, leading them to select the latter. In contrast, a risk-averse person might prioritise the value of the secure option over the risky one, making them more likely to choose the guaranteed outcome. Conversely, a risk-seeking person might favour the option with a potentially higher reward, even if it also entails a greater potential loss.

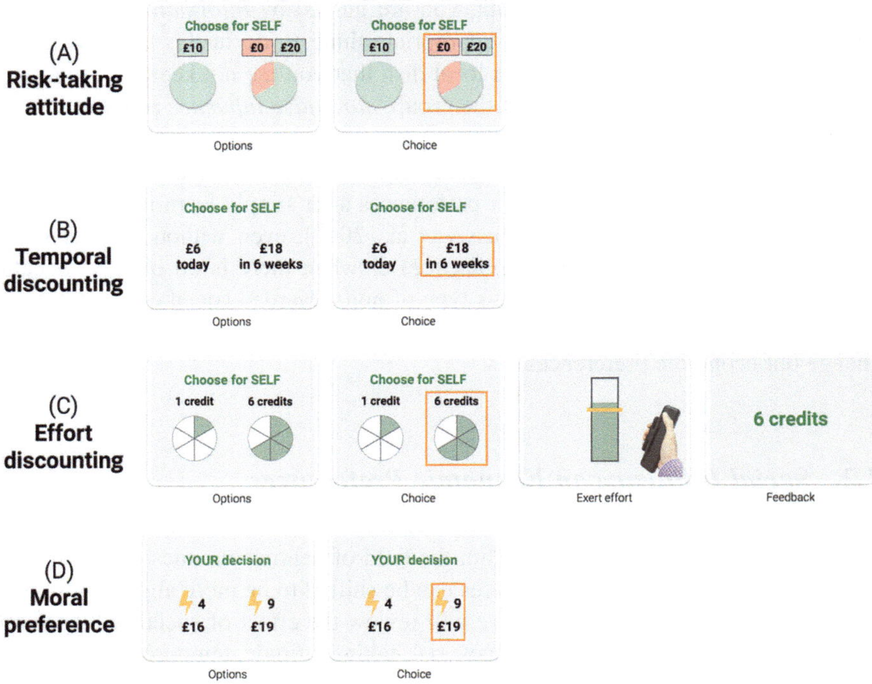

Fig. 1 Experimental tasks to measure different types of economic preferences. (**a**) Example of a monetary lottery task with a choice between two options with explicitly stated probabilities and rewards: a secure option providing a certain smaller reward and a risky alternative providing a potentially larger reward under uncertainty. Green full circles represent a guaranteed smaller reward, while the red and green 'spinner' represents a probabilistic outcome, offering either no reward or a larger reward (e.g. Devaine & Daunizeau, 2017). (**b**) Example of a temporal discounting task with a choice between an immediate smaller reward and a delayed larger reward (e.g. Garvert et al., 2015). (**c**) Example of a physical effort discounting task with a choice between a more effortful option with a higher reward and a less effortful option with a lower reward (e.g. Chong et al., 2017). (**d**) Example of a moral decision-making task involving a trade-off between self-interest and the welfare of others (e.g. Crockett et al., 2014). In this task, participants choose between a harmful option (high reward for themselves and high cost for others) and a helpful option (low reward for self and low cost for others). The lightning bolt signifies the number of painful electric shocks delivered to another person

Intriguingly, studies have demonstrated that our own attitudes towards risk can be modulated by observing other's risky behaviour. In one study, the authors used a social contagion-based risky gambling task to examine how risk-taking attitudes were influenced by others (Suzuki et al., 2016). The task included three types of trials: participants made their own choices between a risky gamble and a guaranteed $10 (self-trials), observed the choices of two other people (one who was more risk-averse and the other who was more risk-seeking; observe trials) and predicted the decisions that these people would make (predict trials). Some experimental blocks included only self-trials, while others required participants to alternate among all

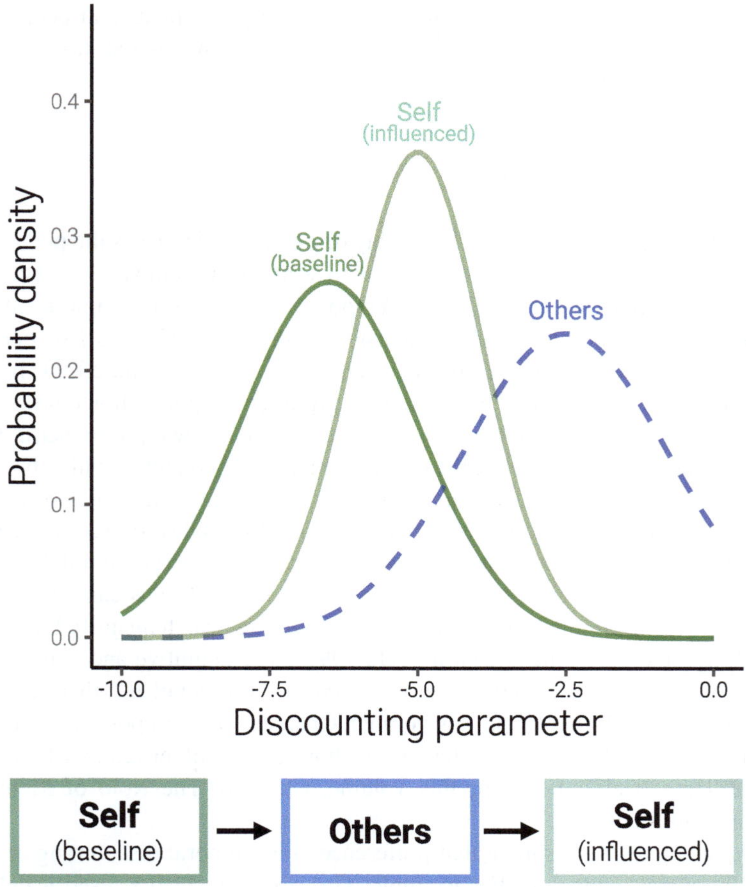

Fig. 2 Illustration of a shift in one's own economic preferences (self) following social influence (others). The participant's economic preferences are represented as normal probability distributions. After learning about the preference of others by observation, the participant's own preference shifts towards that of others, indicating an alignment of preferences driven by social influence

three trial types. They found that participants' behaviour showed a contagion effect, their risk-taking preferences shifted toward those of the people they observed and predicted (Fig. 2), regardless of whether others were more risk-averse or more risk-seeking. This finding replicated in an independent behavioural experiment. In another study with a similar design, participants were asked to choose between a secure lottery (90% winning a smaller reward) and a risky lottery (lower probability of getting a larger reward) before and after predicting other's choices (Devaine & Daunizeau, 2017). They again found a social contagion effect and that the extent of contagion was linked to the estimated differences between participants' own risk-taking preferences and those of others. Another study about social influence on gambling, conducted within the context of a group of three, also found participants' risk-taking preferences shifted to align with those of the majority in the group (Chung et al., 2015). Importantly, they also revealed that the extent of social influence was modulated by the similarity between participants and people in the group; that is, people were more influenced by a majority whose risk-taking attitudes were consistent with their own.

1.2.2 Temporal Discounting

Temporal discounting, also known as delay discounting, describes the phenomenon that a reward is valued less when one has to wait for it, compared to receiving it immediately (Frederick et al., 2002). Temporal discounting has significant implications for economic and financial decisions, influencing choices like opening a savings account or enrolling in a pension scheme (Angeletos et al., 2001). Temporal discounting preferences can be measured using inter-temporal choice tasks, where participants choose between two offers according to their own preferences: one is a smaller amount of reward delivered sooner (usually immediately), while the other is a larger amount of reward delivered after a longer delay (Fig. 1b). For example, people might need to choose between receiving £10 today or £30 next month. Typically, people favour smaller and sooner rewards over larger and delayed ones. People's preferences for immediate rewards over delayed ones can be effectively captured through computational models of discounting, which quantify how delayed rewards devalue over time (Rachlin et al., 1991). The cognitive and computational aspects of temporal discounting are discussed in more detail in other chapters of this volume (see Wagner et al., this volume, Chapter "Intertemporal Choice"; Hare et al., this volume, Chapter "Intertemporal Choices Are Influenced by Many Things Beyond Time Itself"; Lempert, this volume, Chapter "The Role of Memory in Temporal Discounting").

Several studies have shown that preferences for temporal discounting are influenced by social contagion. For example, Garvert et al. (2015) used a delegated inter-temporal choice task to demonstrate that temporal discounting preferences were susceptible to social influence. Participants first completed a series of above-mentioned inter-temporal choices for themselves, and then learned and predicted the choices of another partner (who was either more impulsive or more patient). Finally, in the last phase, they alternated between making inter-temporal choices for themselves and on behalf of others. They found that participants were able to quickly and accurately learn their partner's preferences. Moreover, even though these choices had no direct consequences for the participants, their temporal discounting preferences shifted in the direction of their partner's. Another study using the same experimental design with a much larger sample size replicated this result and further found that people were generally more susceptible to patient social influence compared to impulsive social influence (Moutoussis et al., 2016). In another study, with a modification where participants made inter-temporal choices only for themselves after observation, they also found that participants aligned their temporal discounting preferences with those of others, and the degree of these changes was associated with the estimated differences between participants' personal preferences and those of others (Devaine & Daunizeau, 2017). Another study on group-based social norms also found that participants' temporal preferences shifted significantly closer to the norm after learning about it (Apps & Ramnani, 2017). The social contagion of temporal discounting preferences is so robust that even people diagnosed with autistic spectrum conditions, characterised by social impairments, demonstrated shifts in their temporal discounting preferences after learning about others' preferences (Thomas et al., 2022).

1.2.3 Effort Discounting

Effort discounting refers to the phenomenon that the same reward is valued less if you have to exert effort, either physical or cognitive, to obtain it (Contreras-Huerta et al., 2020; Hull, 1943; Kool & Botvinick, 2018; Westbrook & Braver, 2015). Effort discounting preferences can be measured using effort-based decision-making paradigms, where participants are asked to make a series of choices between two options according to their preferences: one option offers a high reward but requires more effort, while the other offers a low reward with low effort (Fig. 1c). All things being equal, people tend to prefer options that require less effort, as effort is generally aversive (Chong et al., 2017; Hartmann et al., 2013; Klein-Flügge et al., 2015, 2016; Skvortsova et al., 2014; Westbrook & Braver, 2015).

Effort discounting preferences may also be transmitted through social influence. For example, in an attitude alignment study, researchers probed how observing others' effort discounting preferences influenced one's own (Devaine & Daunizeau, 2017). Participants were asked to make a series of effort-based choices, then to learn and predict someone else's choices (who was either more motivated or more apathetic), and finally to repeat the choice task themselves. They found that, based on both model-free and model-based analyses, participants adjusted their effort discounting preferences to align with others after learning about and predicting others' preferences. The extent of these changes was correlated with the estimated differences between participants' own preferences and those of others. In addition, a recent study on challenge-seeking behaviour, in which participants chose between solving more challenging or less challenging maths word problems requiring different levels of cognitive effort, found that participants became more inclined towards challenge-seeking after observing others' challenge-seeking decisions (Ogulmus et al., 2024). These studies suggest that effort discounting preferences, in both physical and cognitive domains, are influenced to some extent by the preferences of others.

1.2.4 Moral Preference

People sometimes make decisions that affect others and thus carry moral responsibility, such as offering help, refraining from causing harm or imposing punishment (Lockwood et al., 2024). Such social decisions play a crucial role in promoting social cohesion and strengthening communal bonds (Fehr & Camerer, 2007; Fehr & Fischbacher, 2003; Wu & Hong, 2022). These decisions often entail a trade-off between costs and benefits for oneself and others. Consequently, moral preferences can be assessed using tasks that trade-off self-interest and the welfare of others (Crockett et al., 2014; Lockwood et al., 2020). For example, people could be asked to choose between a harmful option (high reward for themselves and high cost for others) and a helpful option (low reward for self and low cost for others; Fig. 1d). Their choices reveal whether their moral preferences lean towards being more prosocial or more antisocial.

Intriguingly, moral economic preferences can also be influenced by other people (Lockwood et al., 2024). Yu et al. (2021) asked participants to make a series of private decisions that involved sacrificing some of their own money to reduce the intensity of painful electric shocks delivered to an anonymous stranger, before and after observing such decisions made by others (who were either more prosocial or more antisocial). They found observation of others led participants to adjust their moral preferences to more closely align with those of others, with this influence being stronger among participants who felt a greater sense of similarity to those they observed. Similar results were replicated in a study using an economic game that involved only monetary rewards and costs without pain (Panizza et al., 2021). Participants were asked to make decisions between a default offer and an alternative offer that provided a lower payoff to themselves but could either benefit or disadvantage a stranger compared to the default. The authors demonstrated that, after observing and predicting the choices of extremely prosocial or antisocial others, participants' moral preferences became polarised to align more closely with those of the observed others.

Together, all these studies suggest that people's economic preferences are not necessarily fixed. Instead, the preferences of others around us fundamentally shape our risk, time, effort and moral economic preferences. In all these instances, simply learning what others value can make us more risky or antisocial, with important implications for understanding decision-making in everyday life.

2 Lifespan Similarities and Differences in Social Influence

As is the case in many research studies, the majority, including those reviewed above, have focused on testing young adults around the ages of 18–40. However, it is not surprising that our susceptibility to social influence may change throughout our lifespan. Social influence requires self-other distinction, which likely emerges in infancy (Lockwood et al., 2024). The age at which economic preferences occur is challenging to specify, yet there is evidence that 5–6-year-olds can understand cost-benefit weightings of others' preferences (Jara-Ettinger et al., 2016). This is around the same age as the children who took part in the classical Bobo doll experiments by Albert Bandura, where observing an experimenter behaving aggressively towards a doll influenced similar aggressive behaviour in children who subsequently interacted with the same doll. These studies suggest that the propensity to confirm with others emerges early in our lives.

As we reach adolescence, several studies have documented how adolescents are more likely to engage in risk-seeking behaviour when in the presence of their peers. For example, epidemiological studies suggest adolescents are more likely to engage in illicit drug use or fast driving with peers present (Chein et al., 2011; Gardner & Steinberg, 2005; Pfeifer et al., 2011; Smith et al., 2014). However, this potential for peer influence can also result in positive outcomes, such as when others behave prosocially, i.e. helping or benefiting others (Armstrong-Carter et al., 2023; Do et al., 2019). There is some evidence that adolescents are particularly influenced by prosocial others compared to adults in their self report (Ahmed et al., 2020; Foulkes et al., 2018). Studies of economic choice behaviour, particularly risk-seeking, have been a focus of adolescent research given the possible real-world implications. These studies have integrated decision-making with computational modelling and shown stronger risk and temporal contagion in adolescents compared to adulthood (Reiter et al., 2019, 2021). Importantly, in these incentivised tasks, picking the risky or temporally impulsive option negatively impacts the bonus payment of the participant, making the decision an irrational economic choice. Together, this body of work suggests that risk and temporal preferences may be subject to particularly high levels of contagion in adolescents compared to adults.

In older adults, fewer studies have been conducted, yet initial evidence suggests that heightened susceptibility to types of economic social influence may re-emerge. A recent study tested a relatively large group of young (N = 76, aged 18–36) and older (N = 78, aged 60–80) adults (Su et al., 2025). They completed a social influence-based temporal discounting paradigm where they learnt the preferences of two other people (one more impulsive and one more patient) before making their choices again in different blocks. They found that despite similar accuracy in learning others' preferences, older adults were relatively more susceptible to learning from an impulsive other person. Another study using a collaborative delay discounting task found no age differences (Bixter & Rogers, 2019). However, participants' choices were not incentivised, as young adults received course credit and older adults received $30 regardless of their decisions. Consequently, their choices may not have reflected true preferences. Finally, a recent study using experience sampling showed that compared to young adults (aged 18–30 years), middle-aged (aged 31–59 years) and older adults (aged 60–80 years) were more likely to practise self-control when others were present enacting the desire, suggesting older people were less susceptible to social influence than young adults (Castrellon et al., 2024). The different patterns across these three studies highlight the importance of considering the domain of social influence studied and the age groups and incentives included. Longitudinal and mid-life samples with precise tasks and detailed computational models are needed to fully understand lifespan differences in social influence.

3 Neural Mechanisms

Previous studies have identified several brain regions implicated in social influence within group settings. A coordinate-based meta-analysis of functional neuroimaging studies suggested that the activation of ventral striatum (VS), posterior medial frontal cortex (pMFC)/dorsomedial prefrontal cortex (dmPFC) as well as anterior insula (aIC) is associated with people's deviation from group opinions, and furthermore, the deviation-related activation in the dmPFC predicts people's subsequent conformity to the group's majority opinion (Wu, 2016). To date, however, only a limited number of studies have investigated the neural mechanisms underlying social influence and contagion of economic preferences (Fig. 3).

3.1 Functional and Structural Neuroimaging

Initial evidence from functional neuroimaging studies indicates that the medial prefrontal cortex (mPFC), a pivotal region involved in both economic and social decision-making, might play a crucial role in social influence and contagion of economic preferences. In the above-reviewed study by Garvert et al. (2015), they combined the delegated inter-temporal choice task with a repetition suppression procedure to reveal a region in the mPFC where activity predicted subsequent susceptibility to social influence. The authors attributed the alignment of one's own temporal preferences with those of others to the adaptive learning plasticity of value representations within the mPFC. Another neuroimaging study of social influence of temporal discounting preferences found that dmPFC activation was associated with adherence to social norms, while ventromedial prefrontal cortex (vmPFC) activation was linked to both social conformity and economic decision-making (Apps & Ramnani, 2017).

Similarly, in the context of risk-taking preferences, Suzuki et al. (2016) found that the dmPFC, along with the dorsolateral prefrontal cortex (dlPFC) and inferior parietal lobule (IPL), was associated with preference updating when participants learnt about others' risk-taking preferences. In another study conducted in a group setting, the vmPFC was identified to encode the enhanced utility of options chosen by others, with its activation predicting subsequent conformity (Chung et al., 2015). The authors also found that participants showed increased response in the dorsal anterior cingulate cortex (dACC) and aIC when faced with incongruent choices made by others, but only when they subsequently chose to conform. Additionally, a study investigating the neurocomputational mechanisms of social influence on reinforcement learning found that there was increased functional coupling between the right temporoparietal junction (rTPJ) and the left putamen that predicted greater susceptibility to following the group consensus (Zhang & Gläscher, 2020).

Structural neuroimaging studies also support the central role of mPFC in social influence and contagion of economic preferences. A large longitudinal study

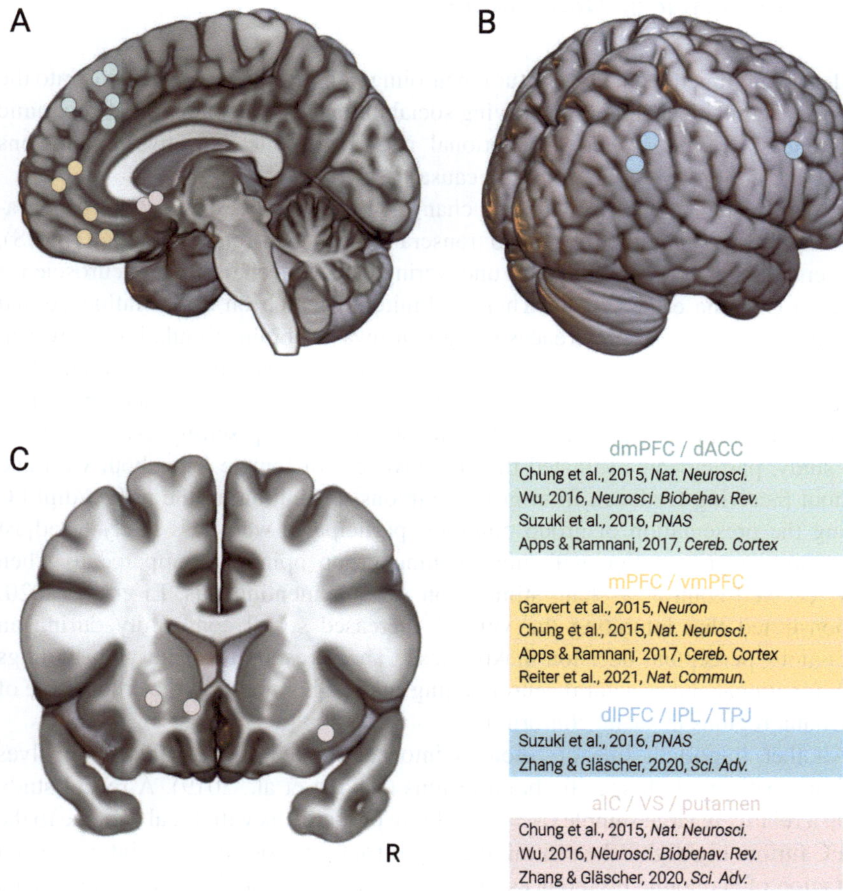

Fig. 3 Locations of neural responses during social influence and contagion of economic preferences from functional and structural neuroimaging studies. The marked locations correspond to the peak BOLD responses identified in MNI coordinates, displayed on MRI scans of the (**a**) medial plane, (**b**) lateral surface and (**c**) coronal plane. Abbreviations: *aIC* anterior insular cortex, *BOLD* blood-oxygen-level-dependent, *dACC* dorsal anterior cingulate cortex, *dlPFC* dorsolateral prefrontal cortex, *dmPFC* dorsomedial prefrontal cortex, *IPL* inferior parietal lobule, *MNI* Montreal Neurological Institute, *mPFC* medial prefrontal cortex, *TPJ* temporoparietal junction, *vmPFC* ventromedial prefrontal cortex, *VS* ventral striatum

integrating the delegated inter-temporal choice task, computational modelling and structural magnetic resonance imaging (MRI) techniques revealed that myelination within the mPFC could predict a reduction in participants' preference uncertainty (Reiter et al., 2021). This reduction, observed over a 1.5-year period in people aged 14–24, was associated with their decreased level of susceptibility to social influence.

3.2 Causal Neural Manipulation

Studies using functional and structural neuroimaging offer valuable insights into the potential neural mechanisms underlying social influence and contagion of economic preferences. However, their correlational nature prevents definitive conclusions about whether these brain regions are causally involved.

Experimental methods for inducing changes in neural processes, such as transcranial magnetic stimulation (TMS) and transcranial direct current stimulation (tDCS), have emerged as powerful tools for uncovering causal mechanisms in neuroscience research (Polanía et al., 2018). There is limited research on social influence and contagion of economic preferences using non-invasive brain stimulation. However, we can draw insights from studies examining social influence and contagion in other domains. For example, Klucharev et al. (2011) provided causal evidence for the role of the pMFC/dmPFC in social conformity within a group setting using TMS. In this study, participants evaluated the attractiveness of female faces both with and without receiving feedback from group opinions. By inhibiting the pMFC/dmPFC during the presentation of group opinions, participants were less likely to adjust their judgements to align with mismatching group opinions compared to when they received control or sham stimulation. Complementing this, Li et al. (2020) demonstrated that inhibiting the vmPFC increased social conformity during an ambiguous perceptual decision-making task. These results resonate with findings from functional and structural neuroimaging studies, highlighting the crucial role of prefrontal regions in social conformity.

Another important line of research into causal neural mechanisms involves studying patients with specific brain lesions (Vaidya et al., 2019). A recent study using a relatively large sample ($N = 33$) of rare participants with focal damage to the mPFC further highlights the central role of prefrontal regions in social influence and contagion of economic preferences (Su et al., 2025). This study, which integrated a delegated intertemporal choice task, Bayesian computational modelling and lesion-symptom mapping, revealed that the dmPFC plays a causal role in susceptibility to social influence, particularly when others were more impulsive in terms of temporal discounting preferences.

4 Future Directions

Other people fundamentally shape our preferences and decisions in everyday life. Studies of social influence have incorporated economic paradigms of learning and decision-making and shown that many preferences can be shifted to be more aligned with others. Future research is needed to thoroughly examine just how long these shifts in preference last as many studies have shown effects within the time frame of a task, yet it is uncertain whether people revert to their initial preferences in everyday life. Research has begun to document important differences across the lifespan in susceptibility to social influence, which are crucial to understand to prevent the negative consequences of maladaptive social influence. At the same time, we may be able to leverage people's susceptibility to social influence in positive ways, guiding them to make better financial decisions and decisions regarding health and well-being. New advances in neuroscience techniques like transcranial focused ultrasound may help reveal the precise causal mechanisms underlying social influence and how it can be changed. Finally, many of the above studies have focused on how we are influenced by individual other people with preferences quite different to ourselves, but the initial work on social influence highlighted just how powerful groups of people can be when they form a majority. Understanding how groups influence individuals or other groups is essential as we make decisions in a highly interconnected world.

References

Ahmed, S., Foulkes, L., Leung, J., Griffin, C., Sakhardande, A., Bennett, M., Dunning, D., Griffiths, K., Parker, J., & Kuyken, W. (2020). Susceptibility to prosocial and antisocial influence in adolescence. *Journal of Adolescence, 84*, 56–68.

Andrews, J. L., Foulkes, L., & Blakemore, S.-J. (2020). Peer influence in adolescence: Public-health implications for COVID-19. *Trends in Cognitive Sciences, 24*(8), 585–587. https://doi.org/10.1016/j.tics.2020.05.001

Angeletos, G.-M., Laibson, D., Repetto, A., Tobacman, J., & Weinberg, S. (2001). The hyperbolic consumption model: Calibration, simulation, and empirical evaluation. *Journal of Economic Perspectives, 15*(3), 47–68.

Apps, M. A. J., & Ramnani, N. (2017). Contributions of the medial prefrontal cortex to social influence in economic decision-making. *Cerebral Cortex, 27*(9), 4635–4648. https://doi.org/10.1093/cercor/bhx183

Armstrong-Carter, E., Do, K. T., Duell, N., Kwon, S.-J., Lindquist, K. A., Prinstein, M. J., & Telzer, E. H. (2023). Adolescents' perceptions of social risk and prosocial tendencies: Developmental change and individual differences. *Social Development, 32*(1), 188–203. https://doi.org/10.1111/sode.12630

Asch, S. E. (1956). Studies of independence and conformity: I. A minority of one against a unanimous majority. *Psychological Monographs: General and Applied, 70*(9), 1.

Bavel, J. J. V., Baicker, K., Boggio, P. S., Capraro, V., Cichocka, A., Cikara, M., Crockett, M. J., Crum, A. J., Douglas, K. M., Druckman, J. N., Drury, J., Dube, O., Ellemers, N., Finkel, E. J., Fowler, J. H., Gelfand, M., Han, S., Haslam, S. A., Jetten, J., et al. (2020). Using social and behavioural science to support COVID-19 pandemic response. *Nature Human Behaviour, 4*(5), 460–471. https://doi.org/10.1038/s41562-020-0884-z

Bixter, M. T., & Rogers, W. A. (2019). Age-related differences in delay discounting: Immediate reward, reward magnitude, and social influence. *Journal of Behavioral Decision Making, 32*(4), 471–484.

Castrellon, J. J., Zald, D. H., Samanez-Larkin, G. R., & Seaman, K. L. (2024). Adult age-related differences in susceptibility to social conformity pressures in self-control over daily desires. *Psychology and Aging, 39*(1), 102–112. https://doi.org/10.1037/pag0000790

Chein, J., Albert, D., O'Brien, L., Uckert, K., & Steinberg, L. (2011). Peers increase adolescent risk taking by enhancing activity in the brain's reward circuitry. *Developmental Science, 14*(2), F1–F10.

Chong, T. T.-J., Apps, M., Giehl, K., Sillence, A., Grima, L. L., & Husain, M. (2017). Neurocomputational mechanisms underlying subjective valuation of effort costs. *PLoS Biology, 15*(2), e1002598.

Chung, D., Christopoulos, G. I., King-Casas, B., Ball, S. B., & Chiu, P. H. (2015). Social signals of safety and risk confer utility and have asymmetric effects on observers' choices. *Nature Neuroscience, 18*(6), 912–916. https://doi.org/10.1038/nn.4022

Cialdini, R. B., & Goldstein, N. J. (2004). Social influence: Compliance and conformity. *Annual Review of Psychology, 55*, 591–621. https://doi.org/10.1146/annurev.psych.55.090902.142015

Contreras-Huerta, L. S., Pisauro, M. A., & Apps, M. A. (2020). Effort shapes social cognition and behaviour: A neuro-cognitive framework. *Neuroscience & Biobehavioral Reviews, 118*, 426–439.

Crockett, M. J., Kurth-Nelson, Z., Siegel, J. Z., Dayan, P., & Dolan, R. J. (2014). Harm to others outweighs harm to self in moral decision making. *Proceedings of the National Academy of Sciences, 111*(48), 17320–17325. https://doi.org/10.1073/pnas.1408988111

Deutsch, M., & Gerard, H. B. (1955). A study of normative and informational social influences upon individual judgment. *The Journal of Abnormal and Social Psychology, 51*(3), 629.

Devaine, M., & Daunizeau, J. (2017). Learning about and from others' prudence, impatience or laziness: The computational bases of attitude alignment. *PLoS Computational Biology, 13*(3), e1005422. https://doi.org/10.1371/journal.pcbi.1005422

Do, K. T., McCormick, E. M., & Telzer, E. H. (2019). The neural development of prosocial behavior from childhood to adolescence. *Social Cognitive and Affective Neuroscience, 14*(2), 129–139. https://doi.org/10.1093/scan/nsy117

Dohmen, T., Falk, A., Huffman, D., Sunde, U., Schupp, J., & Wagner, G. G. (2011). Individual risk attitudes: Measurement, determinants, and behavioral consequences. *Journal of the European Economic Association, 9*(3), 522–550.

Dunbar, R. I., & Shultz, S. (2007). Evolution in the social brain. *Science, 317*(5843), 1344–1347.

Fehr, E., & Camerer, C. F. (2007). Social neuroeconomics: The neural circuitry of social preferences. *Trends in Cognitive Sciences, 11*(10), 419–427.

Fehr, E., & Fischbacher, U. (2003). The nature of human altruism. *Nature, 425*(6960), Article 6960. https://doi.org/10.1038/nature02043

Figner, B., & Weber, E. U. (2011). Who takes risks when and why? Determinants of risk taking. *Current Directions in Psychological Science, 20*(4), 211–216.

Foulkes, L., Leung, J. T., Fuhrmann, D., Knoll, L. J., & Blakemore, S. (2018). Age differences in the prosocial influence effect. *Developmental Science, 21*(6), e12666. https://doi.org/10.1111/desc.12666

Frederick, S., Loewenstein, G., & O'Donoghue, T. (2002). Time discounting and time preference: A critical review. *Journal of Economic Literature, 40*(2), 351–401.

Gardner, M., & Steinberg, L. (2005). Peer influence on risk taking, risk preference, and risky decision making in adolescence and adulthood: An experimental study. *Developmental Psychology, 41*(4), 625.

Garvert, M. M., Moutoussis, M., Kurth-Nelson, Z., Behrens, T. E. J., & Dolan, R. J. (2015). Learning-induced plasticity in medial prefrontal cortex predicts preference malleability. *Neuron, 85*(2), 418–428. https://doi.org/10.1016/j.neuron.2014.12.033

Hartmann, M. N., Hager, O. M., Tobler, P. N., & Kaiser, S. (2013). Parabolic discounting of monetary rewards by physical effort. *Behavioural Processes, 100*, 192–196.

Hull, C. L. (1943). *Principles of behavior: An introduction to behavior theory*. Appleton-Century-Crofts.

Jara-Ettinger, J., Gweon, H., Schulz, L. E., & Tenenbaum, J. B. (2016). The naÏve utility calculus: Computational principles underlying commonsense psychology. *Trends in Cognitive Sciences, 20*(8), 589–604.

Klein-Flügge, M. C., Kennerley, S. W., Saraiva, A. C., Penny, W. D., & Bestmann, S. (2015). Behavioral modeling of human choices reveals dissociable effects of physical effort and temporal delay on reward devaluation. *PLoS Computationa l Biology, 11*(3), e1004116.

Klein-Flügge, M. C., Kennerley, S. W., Friston, K., & Bestmann, S. (2016). Neural signatures of value comparison in human cingulate cortex during decisions requiring an effort-reward trade-off. *Journal of Neuroscience, 36*(39), 10002–10015.

Klucharev, V., Munneke, M. A. M., Smidts, A., & Fernández, G. (2011). Downregulation of the posterior medial frontal cortex prevents social conformity. *The Journal of Neuroscience, 31*(33), 11934–11940. https://doi.org/10.1523/JNEUROSCI.1869-11.2011

Kool, W., & Botvinick, M. (2018). Mental labour. *Nature Human Behaviour, 2*(12), 899–908.

Li, Y., Wang, J., Ye, H., & Luo, J. (2020). Modulating the activity of vmPFC regulates informational social conformity: A tDCS study. *Frontiers in Psychology, 11*, 566977.

Lockwood, P. L., Klein-Flügge, M. C., Abdurahman, A., & Crockett, M. J. (2020). Model-free decision making is prioritized when learning to avoid harming others. *Proceedings of the National Academy of Sciences, 117*(44), 27719–27730. https://doi.org/10.1073/pnas.2010890117

Lockwood, P. L., van den Bos, W., & Dreher, J.-C. (2024). Moral learning and decision-making across the lifespan. *Annual Review of Psychology, 76*(1), 475–500.

Miller, D. T., & Prentice, D. A. (2016). Changing norms to change behavior. *Annual Review of Psychology, 67*(1), 339–361. https://doi.org/10.1146/annurev-psych-010814-015013

Moutoussis, M., Dolan, R. J., & Dayan, P. (2016). How people use social information to find out what to want in the paradigmatic case of inter-temporal preferences. *PLoS Computational Biology, 12*(7), e1004965. https://doi.org/10.1371/journal.pcbi.1004965

Ogulmus, C., Lee, Y., Chakrabarti, B., & Murayama, K. (2024). Social contagion of challenge-seeking behavior. *Journal of Experimental Psychology: General, 153*(10), 2573–2587. https://doi.org/10.1037/xge0001620

Panizza, F., Vostroknutov, A., & Coricelli, G. (2021). How conformity can lead to polarised social behaviour. *PLoS Computational Biology, 17*(10), e1009530. https://doi.org/10.1371/journal.pcbi.1009530

Pfeifer, J. H., Masten, C. L., Moore, W. E., Oswald, T. M., Mazziotta, J. C., Iacoboni, M., & Dapretto, M. (2011). Entering adolescence: Resistance to peer influence, risky behavior, and neural changes in emotion reactivity. *Neuron, 69*(5), 1029–1036.

Platt, M. L., & Huettel, S. A. (2008). Risky business: The neuroeconomics of decision making under uncertainty. *Nature Neuroscience, 11*(4), 398–403.

Polanía, R., Nitsche, M. A., & Ruff, C. C. (2018). Studying and modifying brain function with non-invasive brain stimulation. *Nature Neuroscience, 21*(2), 174–187.

Rachlin, H., Raineri, A., & Cross, D. (1991). Subjective probability and delay. *Journal of the Experimental Analysis of Behavior, 55*(2), 233–244. https://doi.org/10.1901/jeab.1991.55-233

Reiter, A. M. F., Suzuki, S., O'Doherty, J. P., Li, S.-C., & Eppinger, B. (2019). Risk contagion by peers affects learning and decision-making in adolescents. *Journal of Experimental Psychology: General, 148*(9), 1494. https://psycnet.apa.org/buy/2019-03040-001

Reiter, A. M. F., Moutoussis, M., Vanes, L., Kievit, R., Bullmore, E. T., Goodyer, I. M., Fonagy, P., Jones, P. B., NSPN Consortium, NSPN Consortium Representative, Bullmore, E., NSPN Principle Investigators, Bullmore, E., Dolan, R., Goodyer, I., Jones, P., NSPN Staff, Hauser, T., Neufeld, S., et al. (2021). Preference uncertainty accounts for developmental effects on susceptibility to peer influence in adolescence. *Nature Communications, 12*(1), 3823. https://doi.org/10.1038/s41467-021-23671-2

Rigoli, F., Rutledge, R. B., Dayan, P., & Dolan, R. J. (2016). The influence of contextual reward statistics on risk preference. *NeuroImage, 128*, 74–84. https://doi.org/10.1016/j.neuroimage.2015.12.016

Schonberg, T., Fox, C. R., & Poldrack, R. A. (2011). Mind the gap: Bridging economic and naturalistic risk-taking with cognitive neuroscience. *Trends in Cognitive Sciences, 15*(1), 11–19.

Sherif, M. (1935). A study of some social factors in perception. *Archives of Psychology (Columbia University), 187*, 60.

Skvortsova, V., Palminteri, S., & Pessiglione, M. (2014). Learning to minimize efforts versus maximizing rewards: Computational principles and neural correlates. *Journal of Neuroscience, 34*(47), 15621–15630.

Smith, A. R., Chein, J., & Steinberg, L. (2014). Peers increase adolescent risk taking even when the probabilities of negative outcomes are known. *Developmental Psychology, 50*(5), 1564–1568.

Sokol-Hessner, P., Hsu, M., Curley, N. G., Delgado, M. R., Camerer, C. F., & Phelps, E. A. (2009). Thinking like a trader selectively reduces individuals' loss aversion. *Proceedings of the National Academy of Sciences, 106*(13), 5035–5040. https://doi.org/10.1073/pnas.0806761106

Su, Z., Garvert, M. M., Zhang, L., Manohar, S. G., Vogel, T. A., Thomas, L., Balsters, J. H., Husain, M., Apps, M. A. J., & Lockwood, P. L. (2025). Older adults are relatively more susceptible to impulsive social influence than young adults . *Communications Psychology, 2*(1), 87. https://doi.org/10.1038/s44271-024-00134-0

Su, Z., Garvert, M. M., Zhang, L., Vogel, T. A., Cutler, J., Husain, M., Manohar, S. G., & Lockwood, P. L. (2025). Dorsomedial and ventromedial prefrontal cortex lesions differentially impact social influence and temporal discounting. *PLOS Biology, 23*(4), e3003079. https://doi.org/10.1371/journal.pbio.3003079

Suzuki, S., Jensen, E. L. S., Bossaerts, P., & O'Doherty, J. P. (2016). Behavioral contagion during learning about another agent's risk-preferences acts on the neural representation of decision-risk. *Proceedings of the National Academy of Sciences, 113*(14), 3755–3760. https://doi.org/10.1073/pnas.1600092113

Thomas, L., Lockwood, P., Garvert, M. M., & Balsters, J. (2022). Contagion of temporal discounting value preferences in neurotypical and autistic adults. *Journal of Autism and Developmental Disorders, 52*(3), 700–713. https://doi.org/10.1007/s10803-021-04962-5

Toelch, U., & Dolan, R. J. (2015). Informational and normative influences in conformity from a neurocomputational perspective. *Trends in Cognitive Sciences, 19*(10), 579–589. https://doi.org/10.1016/j.tics.2015.07.007

Tom, S. M., Fox, C. R., Trepel, C., & Poldrack, R. A. (2007). The neural basis of loss aversion in decision-making under risk. *Science, 315*(5811), 515–518. https://doi.org/10.1126/science.1134239

Tomasello, M. (2014). The ultra-social animal. *European Journal of Social Psychology, 44*(3), 187–194. https://doi.org/10.1002/ejsp.2015

Vaidya, A. R., Pujara, M. S., Petrides, M., Murray, E. A., & Fellows, L. K. (2019). Lesion studies in contemporary neuroscience. *Trends in Cognitive Sciences, 23*(8), 653–671.

Westbrook, A., & Braver, T. S. (2015). Cognitive effort: A neuroeconomic approach. *Cognitive, Affective, & Behavioral Neuroscience, 15*, 395–415.

Wheeler, L. (1966). Toward a theory of behavioral contagion. *Psychological Review, 73*(2), 179–192. https://doi.org/10.1037/h0023023

Wu, H. (2016). Neural signatures of social conformity: A coordinate-based activation likelihood estimation meta-analysis of functional brain imaging studies. *Neuroscience and Biobehavioral Reviews, 71*, 101–111.

Wu, Y. E., & Hong, W. (2022). Neural basis of prosocial behavior. *Trends in Neurosciences, 45*(10), 749–762.

Yu, H., Siegel, J. Z., Clithero, J. A., & Crockett, M. J. (2021). How peer influence shapes value computation in moral decision-making. *Cognition, 211*, 104641. https://doi.org/10.1016/j.cognition.2021.104641

Zhang, L., & Gläscher, J. (2020). A brain network supporting social influences in human decision-making. *Science Advances, 6*(34), eabb4159. https://doi.org/10.1126/sciadv.abb4159

Comparative Game Theory: Bringing Ethology Back into Social Decision Neuroscience

Weikang Shi, Olivia C. Meisner, Monika P. Jadi, Anirvan S. Nandy, and Steve W. C. Chang

Abstract Game theory has served as a very useful tool to study complex social behaviors. It provides the mathematical frameworks to formulate various social scenarios to explore behavioral patterns during the interaction. Studies with human subjects have shown that various factors including cognition and social traits influence performance and strategies, and neuroimaging results reveal that multiple brain regions, such as the social brain network and the reward circuits, play crucial roles in strategic interactions. On the other hand, various animal species also show the ability to participate in complex games, with different levels of similarity to human subjects. Animal models allow high-resolution access to neural information at spatial and temporal scales that are currently not possible with human subjects and are thus very suitable for studying the neural basis of strategic social interaction. With advancements in technology and data analysis methods, the field is witnessing

W. Shi
Wu Tsai Institute, Yale University, New Haven, CT, USA

Department of Psychology, Yale University, New Haven, CT, USA

Department of Neuroscience, Yale University School of Medicine, New Haven, CT, USA

O. C. Meisner
Department of Psychology, Yale University, New Haven, CT, USA

Department of Neuroscience, Yale University School of Medicine, New Haven, CT, USA

M. P. Jadi
Department of Neuroscience, Yale University School of Medicine, New Haven, CT, USA

Wu Tsai Institute, Yale University, New Haven, CT, USA

Department of Psychiatry, Yale University School of Medicine, New Haven, CT, USA

A. S. Nandy · S. W. C. Chang (✉)
Wu Tsai Institute, Yale University, New Haven, CT, USA

Department of Psychology, Yale University, New Haven, CT, USA

Department of Neuroscience, Yale University School of Medicine, New Haven, CT, USA

Kavli Institute for Neuroscience, Yale University School of Medicine, New Haven, CT, USA
e-mail: steve.chang@yale.edu

D. V. Smith et al. (eds.), *Neuroeconomics: Core Topics and Current Directions*,
https://doi.org/10.1007/978-3-032-02925-6_27

a surge of exciting new findings that reveal the neural mechanisms under more complex experimental settings. We therefore propose "comparative game theory" as a future direction for the field. We anticipate that by bringing ethology back to social neuroscience studies, we will take full advantage of animal models and understand brain function during strategic social interactions in their most naturalistic form.

1 Introduction

As a social species, our daily lives are characterized by a myriad of interactions aimed at achieving diverse goals. From negotiating with banks for favorable loans to enjoying competitive sports like soccer, wherein teams collaborate to attain victory, social dynamics surround us. To unravel and comprehend these intricate interactions, game theory emerges as a potent tool, employing mathematical frameworks to formulate and analyze social scenarios (Sanfey, 2007; Lee, 2008; Camerer, 2011). Originating from behavioral economics, game theory incorporates concepts such as utilities and payoffs to delineate people's strategies in various social settings (Camerer, 2011).

While early game theory portrayed individuals as self-interested players solely focused on maximizing personal rewards, it fell short in explaining the complexities of real-life social interactions. Contrary to the notion of humans as self-interested machines, our interactions are deeply rooted in complex cognitive, emotional, and social abilities, constrained by the structures of our brains (Kahneman, 2003, 2011; Kahneman & Tversky, 2012). Consequently, the collaboration between neuroscience, psychology, and economics has emerged in recent years to uncover the neurobiological foundations of social decision-making.

Undoubtedly, complex social dynamics are not exclusive to the human species. Many mammalian and non-mammalian species exhibit intricate social structures, employing diverse strategies in their interactions (Whitehead, 2008). Game theory has also proven instrumental in elucidating these phenomena (Smith, 1982; Trivers, 1985; Weibull, 1997; Dugatkin & Reeve, 2000). In the era of neuroeconomics, animal models have become invaluable tools for studying the neuroscience of complex behaviors (Glimcher & Fehr, 2013). Therefore, combining game theory and ethology provides a unique opportunity to explore the neural mechanisms behind complex social behaviors in diverse interacting scenarios.

This chapter endeavors to spotlight recent advancements deepening our understanding of the neural mechanisms governing complex social interactions. First, we will delve into how game theory serves as an apt framework for neuroscientists and neuroeconomists to explore human social decisions and their neural correlates. Subsequently, we will underscore the significance of using animal models to study the neural underpinnings of complex social decision-making in naturalistic behavioral contexts. Finally, we will conclude by proposing "comparative game theory" as a prospective avenue for the field, advocating for the integration of traditional ethology into neuroscience studies. This holistic approach, facilitated by technological and computational progress, allows the recording and manipulation of neural activities during naturalistic animal interactions, offering profound insights into the neurobiological and evolutionary foundations of complex social decisions.

2 Game Theory: The Tool to Understand Human Social Behaviors

Rooted in economics, game theory aims to understand the decision strategies in various games. Strategies are the patterns and rules of actions or decisions that subjects employ in response to the environments and partners during the games (Camerer, 2011). To demonstrate the game rules and illustrate the strategies, game theory commonly uses payoff matrices to indicate different action options and the outcomes associated with different action combinations among different players (Fig. 1, example payoff matrices for various games). Payoff matrices are useful tools to demonstrate which actions lead to the best outcomes, and therefore in theory choosing those actions by each player could be interpreted as strategies for playing such games. Game theory was first formally introduced by John von Neumann when he discovered the minimax theorem from the zero-sum games (Camerer, 2011), a family of competitive games in which one player's gain and other players' loss are completely balanced (Fig. 1a). Later efforts had started to focus on the mathematical properties of the payoff matrices. One of the most important discoveries in this regard is the Nash equilibrium. It describes the stability of the strategies applied among players: under certain scenarios, players cannot benefit from changing their current strategy if their partner maintains their current strategy (Camerer, 2011). One of the best examples of the Nash equilibrium is the prisoner's dilemma game (Fig. 1b). In this game, players can either defect or cooperate. Even though cooperation by both players has the highest payoff for both players, the Nash equilibrium is to both defect, because this is the pair of strategies that players from which could not deviate unilaterally because doing so would result in a worse outcome for them. John Nash and other researchers also expanded the application of game theory in broader scenarios such as cooperative games and multiplayer games. Researchers from other fields, such as computer science, political science and biology, also started to apply game theory to their own fields of study. With almost a century of development, game theory has become a very lively and interdisciplinary field today.

Game theory has attained such broad application because the core subject under study—games—is abstract and thus provides a quantitative framework for many scenarios. Real-life social interaction is one of the topics that benefit greatly from game theory. Namely, games themselves are forms of social interaction, and many other types of social interactions can be formulated as games, allowing them to be studied using game theory. Therefore, in recent years, game theory has drawn growing attention from psychologists and neuroscientists who are interested in understanding social behaviors. Early pioneers, such as Reinhard Selten, started to

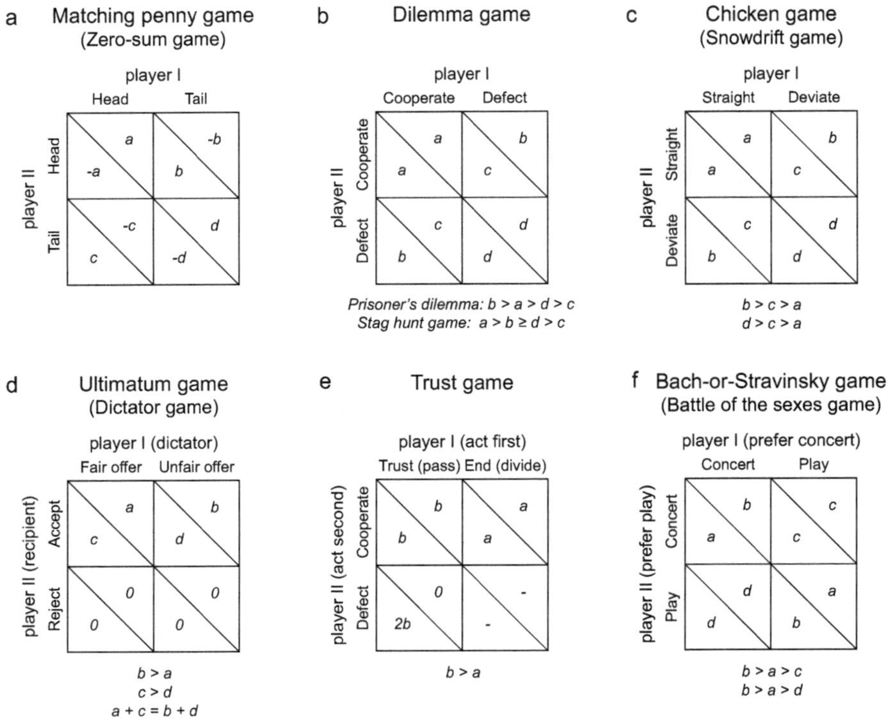

Fig. 1 Payoff matrices for various strategic games. a, b, c, and d indicate the outcome. (**a**) Matching penny game, a type of zero-sum game. (**b**) Dilemma game. If $b > a$, it is a prisoner's dilemma game. "Both defect" is the Nash equilibrium, because of $d > c$. If $a > b$; it is a stag hunt game, which is also referred to as "assurance" game. Stag hunt game has two Nash equilibria—"both stag" because of $a > b$, and "both hare" because of $d > c$. (**c**) Chicken game. It is also referred to as "snowdrift game." The relative values between b and d can vary and will lead to different strategies. (**d**) Ultimatum game. If player II does not have the "reject" option, it becomes a dictator game, which is also referred to as reward allocation game. (**e**) Trust game. If player I chooses to end the game, player II does not need to make a choice (the outcome is indicated as "–" in this case). (**f**) Bach-or-Stravinsky game. It is also referred to as "battle of the sexes" game. The relative values between c and d can vary

design laboratory paradigms and test game theory with experiments on human subjects (Friedman et al., 2004). With the experimental manipulations, researchers were able to examine the predictions of game theoretic models, and more importantly, the experimental game theory has moved the field into a new era of behavioral game theory. The groundbreaking studies done by Daniel Kahneman and Amos Tversky had shown that humans are not rational agents as economists had initially proposed. Human decision-making is limited and influenced by cognitive and psychological processes and thus contains a variety of biases and heuristics (Kahneman, 2003, 2011; Kahneman & Tversky, 2012). Such ideas have also expanded to the social domain and influenced game theory. Specifically, researchers have found that strategies that people use in the games are influenced by various factors including cognitive biases (Camerer, 2011; Holyoak & Morrison, 2012), emotions (Chang & Smith, 2015), memories (Camerer, 2011; Glimcher & Fehr, 2013), learnings (Fudenberg & Levine, 1998), social preferences (Fehr & Camerer, 2007), and communications (Camerer, 2011). In the era of behavioral game theory, the goal of the field is to understand how these cognitive and psychological factors affect strategic interactions (Camerer, 2011).

Game theory has thus become a major tool to study social behaviors. Psychologists and behavioral neuroscientists have been adapting and developing different versions of games to emphasize and focus on different factors that affect the strategies. For example, to examine the subject's social preference, researchers have designed games with different dilemmas (Fehr & Camerer, 2007). In the dictator game (Fig. 1d), in which one player, the dictator, allocates money between themself and another player, the recipient, the dictator does not always keep everything, indicating that the player is not purely self-interested and shows prosociality in making those decisions (Güth et al., 1982; Kahneman et al., 1986). The amount of money that the dictator allocates to the recipient also depends on several factors. When the dictator knows the identity of the recipient, they allocate more to the recipient (Eckel & Grossman, 1996; Dana et al., 2006). When comparing the one-shot games with the repeated games, researchers have found that the dictator allocates less to the recipient if they play with the same player multiple times compared with one-shot games (Engel, 2011; Brosig-Koch et al., 2017) and allocate more if the dictator and recipient alternate their roles in the games that encourage reciprocity (Grech & Nax, 2020). If the recipients also have the choice to reject the allocation and eliminate all of the reward, that is, the ultimatum game (Fig. 1d), the dictator tends to allocate more to avoid pure loss (Camerer, 2011; Engel, 2011). On the other hand, if the dictator knows that the recipient is not aware of them being in the game, they will allocate less to the recipient (Dana et al., 2006). These results indicate that in such reward allocation games, players' strategies balance self-interest to maximize their own reward and prosociality to benefit other players. In addition, the fact that the dictator would think about how the recipient understood the game rules and update their strategies accordingly indicates that theory of mind (ToM)—the ability to impute others' mental states (Frith & Frith, 2005)—plays an important role (Yoshida et al., 2008; Harré, 2022). Other example games involving ToM include the prisoner's dilemma game (Fig. 1b) (Press & Dyson, 2012) and the stag hunt game (Fig. 1b) (Yoshida et al., 2008). In these games, the choice of cooperating or defecting strategies depends on the depth of ToM reasoning (Press & Dyson, 2012).

With these fruitful results at hand, the field has started to focus on the neural basis of these strategic social interactions. In the next section, we will highlight the main findings that examined the neural activities of human subjects in various games.

3 Neural Basis of Game Theory

One of the first studies examining neural activity in strategic games was done by McCabe and colleagues (2001). In this study, the participants were playing a trust game (Fig. 1e) with either human or computer counterparts. In this two-player game with sequential actions, player one can first decide to either end the game and divide the reward with player two, or trust player two and pass down the reward. If player one chooses to trust player two, then the overall reward increases, and player two can decide to either divide the reward fairly or take the entire reward. Similar to other dilemma games, the trust game requires player one to have some levels of ToM when making the optimal decision. Functional neuroimaging from these participants showed that for the cooperative players, their neural activities of the prefrontal cortex were enhanced when cooperating with human counterparts, compared with computer counterparts. These results aligned with findings that the prefrontal cortex, especially the medial prefrontal cortex (mPFC), plays an important role in ToM (Frith & Frith, 2006). In addition to mPFC, temporoparietal junction (TPJ) is another brain region that has been found to play a major function in the ToM (Schurz et al., 2014) and during several strategic games (Gallagher et al., 2002; Kircher et al., 2009; Sripada et al., 2009; Speitel et al., 2019).

The social component of game theory is one of the main focuses when exploring its neural mechanism. In addition, under a neuroeconomics framework, game theory is also deeply rooted in reward systems. Therefore, many studies have focused on how the brain's reward circuitry is involved with game theory. Rilling and colleagues found that in a social dilemma game, the activities of the orbitofrontal cortex (OFC) and ventral striatum (VS), the major brain regions of the reward system, were enhanced when the players chose to mutually cooperate (Rilling et al., 2004). Such results implied that reciprocal altruism could be processed as a social reward in the brain, and thus activated the reward circuitry. Altruistic scenarios in other games are also shown to activate the reward system. Mesolimbic–striatal reward system, including the ventral tegmental area (VTA), and OFC were activated when human subjects made altruistic decisions in charity donation tasks (Moll et al., 2006; Harbaugh et al., 2007). In addition, fairness and punishment in the ultimatum games also modulated the neural activity of the reward system (Sanfey et al., 2003; Tabibnia et al., 2008), indicating that emotion, intertwined within the reward system, and cost-benefit calculations (Du & Chang, 2015) also contribute to the neural basis of game theory.

The mesolimbic system, especially the dopamine system, is well appreciated for its crucial role in reinforcement learning (Dayan & Balleine, 2002; Schultz, 2015). Indeed, bridging reinforcement learning with game theory has revealed great insights into the neural basis of strategic social interactions (Nowé et al., 2012). In repeated games, players could learn the payoff matrices and the expected actions of the counterpart via reinforcement learning, in order to choose their own optimal strategies (Nowé et al., 2012). In a repeated trust game (Fig. 1e), King-Casas and colleagues (2005; Tomlin et al., 2006) recorded the BOLD signals of both the investor (player one) and the trustee (player two) at the same time and found that caudate nucleus of both players encoded an anticipating reciprocal "trust" action of the trustee. They also found that there was a time lag in the peak of encoding between caudate nucleus and cingulate cortex, mirroring the "predictive error" signal found in the reinforcement learning tasks.

In conclusion, brain regions including the social brain network, the reward circuitry, and reinforcement learning system play crucial roles in the neural basis of game theory. Social prediction errors, predictions referenced to other individual's actions or outcomes, also seem to play a key role (Joiner et al., 2017). However, detailed neural mechanisms are often lacking from experiments done with human subjects due to methodological limitations. It is also challenging to study the causal function of brain regions implicated in strategic games with human subjects. Animal models can give us the access needed to explore neural activity with high resolution and also to probe the causal functions of these neural circuits. In the following sections, we will discuss the possibility of using animal models to study game theory, especially its neural basis, and highlight the benefits of using animal models.

4 Can Animals Play Strategic Games?

The application of game theory is not limited to explaining human behaviors. Other animals also have complicated interactions between individuals of the same species and across different species. Combined with evolution theory, John Maynard Smith and others proposed "evolutionary game theory" to understand animals' natural social interactions and decisions, such as cooperation, altruism, and competition, through the lens of natural selection (Smith, 1982; Weibull, 1997). It emphasizes how strategies evolved over time based on their success or fitness within a given environment. For example, the adapted models from game theory have been shown to explain the predator-prey interaction (Brown et al., 1999), mating behaviors (Roughgarden et al., 2006), as well as kin selection in altruistic behaviors (Ohtsuki, 2010). These fruitful results demonstrate that animals show complex strategic interactions ethologically and ecologically.

Such results provide the foundation to explore different species' capability in strategic social interactions and decisions. A large effort has been put into studying these questions in various experimental settings over the past several decades. Different species of nonhuman primates have been shown to participate in various strategic games. Chimpanzees (*Pan troglodytes*) have been shown to strategize when playing the ultimatum game (Fig. 1d) (Proctor et al., 2013), reward allocation games (Fig. 1d) (Melis et al., 2009), stag hunt games (Fig. 1b) (Bullinger et al., 2011), and competition games (Knofe et al., 2019). Other nonhuman primates also employed different strategies in complex social games, including bonobos (*Pan paniscus*) (Sánchez-Amaro et al., 2017; Sánchez-Amaro & Rossano, 2021), macaque species (*Macaca mulatta; Macaca fuscata; Macaca fascicularis*) (Reinhart et al., 2010; Brosnan et al., 2011, 2012; Kaigaishi et al., 2019; Sigmundson et al., 2021), capuchins (*Sapajus apella*) (Brosnan et al., 2006; Addessi et al., 2011; Smith et al., 2019), and common marmosets (*Callithrix jacchus*) (Burkart et al., 2007; Burkart & van Schaik, 2020; Martin et al., 2021; Sánchez-Amaro et al., 2021). In addition to nonhuman primates, many other animals are also able to strategize when interacting with conspecifics. For example, elephants (*Elephas maximus*) (Plotnik et al., 2011), wolves (*Canis lupus*) and dogs (*Canis familiaris*) (Marshall-Pescini et al., 2017), and different kinds of birds (Schwing et al., 2016; Heaney et al., 2017; Torres Ortiz et al., 2020) employ cooperative strategies in cooperation games, and wolves and dogs are able to coordinate to solve the social dilemma in coordination games (Bräuer et al., 2020).

While there is no doubt that many animals have the ability to strategize in sophisticated social interactions, it is particularly interesting and important to compare similar or dissimilar strategies across nonhuman species and with humans. In some cases, different species adapt very similar strategies. For example, capuchin monkeys, chimpanzees, and humans have been found to use very similar strategies in the stag hunt game (Fig. 1b) (Brosnan et al., 2011). On the other hand, different species can also adapt different strategies even when they are playing the same game, such as chimpanzees and human children in the same stag hunt games (Duguid et al., 2014). The difference in strategies employed by different species might be the result of their varying levels of prosociality inherent to the species' ethology. In a modified dictator game (Fig. 1d) designed for nonhuman primates (group service game) such that one animal has to pull a board to altruistically provide food to another animal, Burkart and colleagues compared performances among marmosets, capuchin monkeys, and macaques (Burkart & van Schaik, 2013). They found that marmosets had the highest and most stable pulling rates during the game, capuchin monkeys had the intermediate level, and macaques made no altruistic pullings. Importantly, these behavioral results are consistent with the breeding patterns of these species in the wild: marmosets breed cooperatively, macaques breed independently, and capuchin monkeys have an intermediate breeding system, indicating that their performance in the task could be a reflection of their natural prosociality, with marmoset having the highest prosociality and social tolerance (Burkart & van Schaik, 2013; Burkart & Finkenwirth, 2015).

Even within the same species, the prosocial traits of individuals also affect their strategies (Weinberg-Wolf & Chang, 2019). More prosocial marmoset individuals tend to offer more altruistic actions in the group service game and make more cooperative choices (Martin et al., 2021). Japanese macaques with higher social tolerance levels also performed better in the cooperative tasks compared with conspecifics with lower social tolerance (Kaigaishi et al., 2019). In fact, a wide range of social factors can influence prosocial traits and lead to different strategies. In a task where marmosets sat across from each other with a single lever between them, males and females demonstrated distinct strategies. In the competition condition, where the first marmoset to pull the lever received a larger reward, females tended to pull the lever faster (Sánchez-Amaro et al., 2021). However, in the snowdrift condition, where the first marmoset to pull received a smaller reward, females tended to wait longer before pulling, suggesting they were waiting for their partner to pull. In contrast, male marmosets did not differentiate between the two conditions, pulling the lever after the same amount of time in both scenarios (Sánchez-Amaro et al., 2021). Importantly, the different strategies played by different sexes are consistent with the food sharing behaviors of marmosets in the wild, where female breeders show the lowest levels of food sharing likely due to their energetic demands for reproducing (Guerreiro Martins et al., 2019). Dominance rank may also affect individuals' prosociality. For example, dominant macaques made more altruistic choices in a dictator game than the subordinate conspecifics (Massen et al., 2010; Chang et al., 2011).

Cognitive capacity may also lead different species to adopt different strategies (Melis & Raihani, 2023). Wang and Kwan compared the performance of macaques (Seo et al., 2009) and mice (Wang et al., 2022) playing matching penny games (Fig. 1a) (Wang & Kwan, 2023). They found that even though both species played against the computer with the identical algorithm, mice tended to adopt simpler strategies such as win-stay-lose-shift and had stronger choice biases, while macaques adopted their strategies based on the interactive outcome with their opponents. One might argue that the biggest challenge of cross-species comparisons is deciphering if behavioral differences are due to different training processes, instead of due to varied social or cognitive capabilities of different species. Researchers therefore need to carefully consider these alternatives.

Some studies have shown great insights into this challenging aspect of comparative game theory. Brosnan and colleagues carefully controlled the training procedures for capuchin monkeys, chimpanzees, and human subjects and found that with identical training, these species had very similar performance in the stag hunt game. Especially, when human subjects played the nonhuman primate friendly version, their performance was more similar to the nonhuman primates compared to typical human performance (Brosnan et al., 2011). In a more recent study, Moeller and colleagues compared humans' and macaques' performance in a coordination game "Bach-or-Stravinsky" (Fig. 1f, battle of the sexes game) in the same transparent touchscreen settings (Moeller et al., 2023). In this game, two players had opposite preferences for choosing one action over the other, and the optimal strategy was to dynamically take turns for each option to equalize the payoffs. Human dyads developed such coordinated behaviors, while macaque dyads converged to a simple and static strategy entailing strong biases. Interestingly, when paired and trained with human counterparts, macaques were able to learn and adopt the dynamic turn-taking strategy. In rodent models, researchers also showed that with careful control of the training process and the animal's motivation level, rats were able to use complex strategies in an iterated prisoner's dilemma game, similar to human subjects (Viana et al., 2010). Altogether, these investigations emphasize the importance of experimental and task designs in cross-species work. They showed the possibility that even with different prosocial and cognitive capacities from humans, animal subjects can develop and adopt human-like strategies in complex games with carefully considered training procedures.

Overall, across different species, some animals show similar strategies as humans, while some don't. The differences and similarities depend on the cognitive ability of different species, as well as their social traits. Comparing the strategies across species can help us understand our own social behaviors from an evolutionary perspective and also provide us with a chance to use animal models to study the neural basis of strategic social interactions in much greater detail. The next section will discuss some of the major findings on this aspect.

5 The Benefit of Animal Models in Game Theory

The biggest benefit of working with animal models is that they give us much more access to direct neural information at high spatial and temporal resolution that is very challenging or impossible to obtain from human subjects. Specifically, as discussed in the previous section, many species could display similar performance as human beings in various strategic games. As long as the context of the games is largely compatible with the animals' ethology, such similar performances can be leveraged to reveal the neural basis behind these strategic interactions that could translate to the human brain.

Over the years, macaque monkeys have been the most prevalent animal model to study the neural basis of decision strategies (Lee, 2008). In a modified dictator game involving pairs of macaques, an actor monkey had choices to allocate the reward to themselves, the partner monkey or neither of them (Fig. 2a). While the animals were performing the task, Chang and colleagues recorded the neural activities from the dictator animals' ACC and OFC (Chang et al., 2013). The results showed that the majority of OFC neurons encode the reward that was received by the "dictator" monkeys themselves. On the other hand, they found that many neurons in the ACC gyrus encode the reward outcome of the recipient, either exclusively or similarly to the reward received by the dictator. Finally, the majority of neurons in the ACC sulcus were found to encode the foregone reward (rewards that the dictator did not receive). These results indicated that neurons from different prefrontal cortical areas represent reward outcomes from the dictator game using different frames of reference—some neurons use self-centered frames of reference, whereas other neurons use either partner-centered or both-centered frames of reference. OFC and ACC sulcus were shown to be biased toward using a self-centered frame of reference, whereas ACC gyrus had a bias toward signaling the partner's reward outcome (i.e., partner- or both-centered). A subsequent study by the same team found that neurons in the BLA exhibited correlated value tuning between the reward of the dictator and the reward of the partner (Chang et al., 2015), indicating the role of BLA in shared value coding between self and other in social decision-making. When the researchers recorded the activity of ACC gyrus and BLA simultaneously during the same dictator game, they found that making prosocial choices or antisocial choices enhanced frequency-specific and direction-selective coordination of activity between BLA and ACC gyrus (Dal Monte et al., 2020), suggesting that specialized communication in the prefrontal-amygdala networks may guide social decision-making (Fig. 2a).

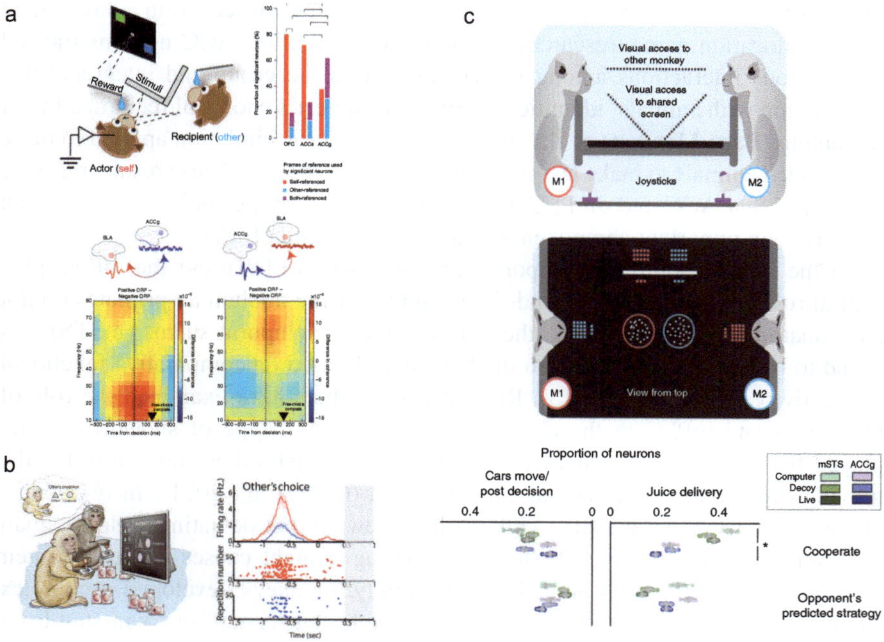

Fig. 2 Example studies that examine the neural mechanism while macaque monkeys participate in various games. (**a**) Macaques in a dictator game (Chang et al., 2013; Dal Monte et al., 2020). Top left panel demonstrates the experimental setup, in which animals are head fixed. Actor macaque decides to allocate the juice to self, other, both, or either. Top right panel summarizes the encoding properties of OFC, ACCs, and ACCg, with OFC and ACCs bias toward self-centered frame of reference and ACCg bias toward other-centered frame of reference. Bottom panel shows the frequency-specific and direction-selective coordination of activity between BLA and ACCg for the prosocial and antisocial choices, respectively. (**b**) Macaques in a prisoner's dilemma game (Haroush & Williams, 2015). Left panel demonstrates the experimental setup, in which animals are head fixed, but with free hand movement. Right panel highlights one example ACC neuron that predicts partners' choice to either cooperate (red) or defect (blue) that is yet unknown to the recorded animal. (**c**) Macaques in a chicken game (Ong et al., 2021). Top and middle panels show the experimental setup, in which animals are head fixed, but with free hand movement and free viewing of partner and shared screen. Bottom panel summarizes the percentage of neurons in mSTS and ACCg that encode animals' own strategies (cooperate or not) and partners' predicted strategies. Example studies like the one in (**c**) allow researchers to explore larger behavioral repertoires. (Panels (**a**) to (**c**) are adapted from published literature with permission)

ACC has also been investigated when pairs of macaques played an iterative version of the prisoner's dilemma game (Fig. 2b) (Haroush & Williams, 2015). In this study, researchers found that the activity of ACC neurons predicted partners' choice to either cooperate or defect that was yet unknown to the recorded animal. Further analysis revealed that ACC neurons kept track of the outcomes from the previous interactions in order to predict future choices (e.g., the likelihood of mutual cooperation was higher following mutual cooperation from the previous round), and disrupting the ACC activity led animals to make less mutually beneficial choices, likely due to a disruption in predicting partner's choices. Moreover, in a three-agent reward allocation game, researchers found that the dmPFC/ACC neurons tracked the previous interactions among individuals within the group and integrated this information with animals' identities to guide the subject's choice of reciprocating or retaliating (Báez-Mendoza et al., 2021). Notably, microstimulation applied to these neurons led animals to make less mutually beneficial choices. These results raise the possibility that ACC and dmPFC are key regions underlying social prediction and inference, an important component of the theory of mind (ToM).

In the human brain, the temporal-parietal junction (TPJ) and the mPFC play central roles in the theory of mind. Interestingly, while nonhuman primates do not have anatomically defined TPJ, the middle superior temporal sulcus (mSTS) was found to be a homologous area to the human TPJ based on comparative functional connectivity (Mars et al., 2013). Recently, researchers have examined the role of the mSTS and mPFC in the chicken game (Fig. 1c, a type of snowdrift game) played by macaques, in which two monkeys faced each other and controlled the moving direction of their virtual cars (Fig. 2c) (Ong et al., 2021). In this game, going straight without crashing yields a large reward, but deviating yields a small "chicken" reward. However, both animals going straight causes a crash of their card, no rewards for both animals. Behaviorally, monkeys developed a complex strategy that was more sophisticated than simple "tit-for-tat" or "win-stay-lose-shift" strategies by actively using and updating the payoffs for both themselves and the partners while taking into account the partners' predicted actions to guide their own actions. At the neural level, researchers found that mSTS carried key information about cooperative strategies.

Taken together, these studies in monkeys demonstrate the advantages of animal models in combining game theory with neuroscientific inquiries. Consistent with the human research, these studies also found the critical functions of OFC, mPFC areas such as ACC and dmPFC, mSTS, and BLA in strategic social interactions (Gangopadhyay et al., 2020), More importantly, several studies identified the specified and causal function of these brain regions and how the communication across these brain regions is involved in social decisions.

Recent efforts have also explored a larger variety of animal models. Jiang and colleagues recorded the OFC activity while rats were participating in a cooperative interaction, in which two rats need to poke their noses at the nose ports at the same time to receive a reward (Fig. 3a) (Jiang et al., 2021). They found that OFC neurons encoded various variables related to the strategies, including the payoffs, the outcomes, as well as the roles (initiator or follower) of the subjects and their partners. Such findings were consistent with the function of OFC in human studies (Rilling et al., 2002, 2004). Although rodents may not be capable of quite as complex strategies as humans and nonhuman primates, studying the neural mechanisms underlying social game processes in rodents can help us better understand more detailed physiological processes underlying mammalian social and interactive behaviors. The advantage of rodent models is the genetic tools they provide to study the neural mechanisms at cellular and subcellular levels, including the functions of different cell types and neuromodulators. Although these

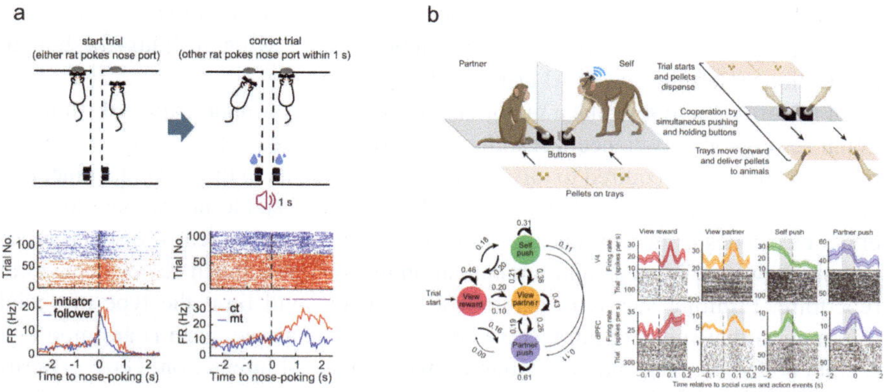

Fig. 3 Example studies that examine the neural mechanism in cooperative games with various animal species in freely moving conditions. (**a**) Rodents in a strategic cooperative interaction (Jiang et al., 2021). Top panel shows the experimental setup, in which animals are freely moving. Two rats are required to cooperatively poke their noses at the ports within 1 s to receive juice reward. Bottom panels show two example OFC neurons that encode the roles of the recorded animal (poke initiator or follower) and the trial types (ct: cooperative trial, or mt: missed trial). (**b**) Macaques in a strategic cooperative interaction (Franch et al., 2024). Top panel shows the experimental setup, in which animals are freely moving and neural activity is recorded with a wireless recording system. Bottom left panel shows the viewing and pushing behaviors and the transitions between behavioral epochs. Bottom right panel highlights example neurons from V4 and dlPFC and their activity aligned with different behavioral events. Compared with studies in Fig. 2, these experimental setups become more naturalistic, with no restraints on animals' actions. These two studies also highlight that animals across different species are able to do similar strategic interaction tasks. (Panels (**a**) and (**b**) are adapted from published literature with permission)

approaches have been underutilized in strategic social decision-making, in a recent study researchers found that the noradrenergic and cholinergic activity of the mouse frontal cortex encoded the key decision variables and played a causal role in the strategies when the animals were engaging in a matching penny game (Wang et al., 2024). Altogether, these results paved the way for future research to fully explore the opportunities of using rodent models to study the neural basis of game theory.

6 Future Direction: Game Theory in the Wild

In this chapter, we summarized the main concepts of game theory and the literature that examines the neural mechanisms behind strategic social interactions. We specifically reviewed that various animal species are capable of participating in strategic games, and their performances can be compared with those of human subjects. We also discussed and emphasized the advantage of introducing animal models when exploring the neural basis of social strategy. However, it is essential to carefully consider the key differences between animal models and humans—such as genetic backgrounds, brain anatomies, social structures, and cognitive capabilities—when interpreting results from animal studies and drawing insights for human applications (Hartung, 2008; Al Dahhan et al., 2019).

In addition, there remains a major challenge that is yet underexplored. In traditional settings, neuroscience research usually restrains the subjects' movements—human subjects are required to minimize their movements in the MRI machine, and animal subjects are head-restrained in order to record stable neural activity with either electrodes or microscopes, or monitor eye positions. In real-life social interactions, however, humans and nonhuman animals show a much richer behavioral repertoire. Limiting access to naturalistic behaviors will limit the type of social information that subjects can receive and send. For example, gaze is an important social cue for humans and nonhuman primates in social interaction. Gaze patterns have been found to influence people's strategies in cooperation and competition games (Jarick & Kingstone, 2015). People also strategize gaze behaviors to convey information to others in various games (Hausfeld et al., 2021). Nonhuman primates also show complex gaze patterns in naturalistic social interactions (Dal Monte et al., 2016, 2022; Fan et al., 2023) and actively use gaze to collect information while engaging in cooperation tasks (Fig. 3b) (Franch et al., 2024), "chicken" game (Fig. 2c) (Ong et al., 2021) and "Bach-or-Stravinsky" game (Moeller et al., 2023). In the "Bach-or-Stravinsky" game, if the subjects' view of the opponents was blocked, their strategies were changed and became less optimal (Moeller et al., 2023). In addition, the neural encodings between restrained and naturalistic conditions also appear to be different. For example, researchers found that PFC neurons in marmosets represented the same social information differently across head-restrained, freely moving, and communicational conditions, and the neural tunings can flexibly adapt to the dynamics of each condition (Jovanovic et al., 2022). Altogether, these results emphasized an important future direction for the field—to study the neural basis of game theory in its naturalistic form. Indeed, strategically balancing experimental control and ecological validity is increasingly becoming an important issue in social neuroscience research (Fan et al., 2021).

One way to explore this direction is to design experimental apparatus with more naturalistic settings (Fan et al., 2021; Miller et al., 2022). For human studies, researchers have begun to use more portable equipment to record neural activity while participating in games, such as functional near-infrared spectroscopy (fNIRS) (Zhang et al., 2023) and wearable electroencephalography (EEG) (Hafeez et al., 2021). Various wireless recording systems have been developed for rodents (Kim et al., 2020; Yang et al., 2022), bats (Zhang & Yartsev, 2019; Rose et al., 2021), marmosets (Roy & Wang, 2012; Walker et al., 2021; Jendritza et al., 2023; Meisner et al., 2024), and macaques (Schwarz et al., 2014; Franch et al., 2024; Shahidi et al., 2024; Testard et al., 2024) that allow researchers to explore the neural mechanisms when animals can move freely and behave more naturally than before (see Fig. 3b, an example study that recorded neural activities while macaques were performing a cooperative task in a freely moving manner).

Overall, a promising and important future direction for the field is to bring ethology back to the social decision neuroscience. The experimental designs could be naturalistic; however, they still require certain aspects of the social interactions to be abstracted and symbolized. Humans use game theory not only to strategize poker games but also in daily life, such as bargaining for goods. Animals never press buttons or pull levers in the wild; however, they can use complex strategies to forage food or claim territory in group settings. With the recent development of techniques to precisely and markerlessly track naturalistic behaviors of multiple individuals and to perform wireless neural recording, we are now witnessing a surge in computational neuroethology (Anderson & Perona, 2014; Datta et al., 2019; Mobbs et al., 2021; Cisek & Green, 2024). We anticipate that the study of strategic social interaction will follow this trend, and in the near future, we will see more exciting results revealing the neural mechanisms of our complex social behaviors in their naturalistic forms.

Acknowledgments This work was supported by the Wu Tsai Institute Postdoctoral Fellowship (W.S.), the National Science Foundation Graduate Research Fellowship (DGE2139841, O.C.M.), the National Institute of Mental Health (R21 MH126072, S.W.C.C., A.S.N., M.P.J.; R01 MH128190, S.W.C.C.; R01 MH120081, S.W.C.C.), and the Simons Foundation Autism Research Initiative (SFARI 875855, S.W.C.C., A.S.N., M.P.J.).

Conflict of Interest None.

References

Addessi, E., et al. (2011). How social context, token value, and time course affect token exchange in capuchin monkeys (*Cebus apella*). *International Journal of Primatology, 32*(1), 83–98.

Al Dahhan, N. Z., De Felice, F. G., & Munoz, D. P. (2019). Potentials and pitfalls of cross-translational models of cognitive impairment. *Frontiers in Behavioral Neuroscience, 13*, 48.

Anderson, D. J., & Perona, P. (2014). Toward a science of computational ethology. *Neuron, 84*(1), 18–31.

Báez-Mendoza, R., et al. (2021). Social agent identity cells in the prefrontal cortex of interacting groups of primates. *Science, 374*(6566), eabb4149.

Bräuer, J., Stenglein, K., & Amici, F. (2020). Dogs (*Canis familiaris*) and wolves (*Canis lupus*) coordinate with conspecifics in a social dilemma. *Journal of Comparative Psychology, 134*(2), 211–221.

Brosig-Koch, J., Riechmann, T., & Weimann, J. (2017). The dynamics of behavior in modified dictator games. *PLoS One, 12*(4), e0176199.

Brosnan, S. F., Freeman, C., & De Waal, F. B. M. (2006). Partner's behavior, not reward distribution, determines success in an unequal cooperative task in capuchin monkeys. *American Journal of Primatology, 68*(7), 713–724.

Brosnan, S. F., et al. (2011). Responses to the assurance game in monkeys, apes, and humans using equivalent procedures. *Proceedings of the National Academy of Sciences of the United States of America, 108*(8), 3442–3447.

Brosnan, S. F., Wilson, B. J., & Beran, M. J. (2012). Old World monkeys are more similar to humans than New World monkeys when playing a coordination game. *Proceedings. Biological Sciences/The Royal Society, 279*(1733), 1522–1530.

Brown, J. S., Laundré, J. W., & Gurung, M. (1999). The ecology of fear: Optimal foraging, game theory, and trophic interactions. *Journal of Mammalogy, 80*(2), 385–399.

Bullinger, A. F., et al. (2011). Coordination of chimpanzees (pan troglodytes) in a stag hunt game. *International Journal of Primatology, 32*(6), 1296–1310.

Burkart, J. M., & Finkenwirth, C. (2015). Marmosets as model species in neuroscience and evolutionary anthropology. *Neuroscience Research, 93*, 8–19.

Burkart, J. M., & van Schaik, C. (2013). Group service in macaques (*Macaca fuscata*), capuchins (*Cebus apella*) and marmosets (*Callithrix jacchus*): A comparative approach to identifying proactive prosocial motivations. *Journal of Comparative Psychology, 127*(2), 212–225.

Burkart, J. M., & van Schaik, C. P. (2020). Marmoset prosociality is intentional. *Animal Cognition, 23*(3), 581–594.

Burkart, J. M., et al. (2007). Other-regarding preferences in a non-human primate: Common marmosets provision food altruistically. *Proceedings of the National Academy of Sciences of the United States of America, 104*(50), 19762–19766.

Camerer, C. F. (2011). *Behavioral game theory: Experiments in strategic interaction*. Princeton University Press.

Chang, L. J., & Smith, A. (2015). Social emotions and psychological games. *Current Opinion in Behavioral Sciences, 5*, 133–140.

Chang, S. W. C., Winecoff, A. A., & Platt, M. L. (2011). Vicarious reinforcement in rhesus macaques (macaca mulatta). *Frontiers in Neuroscience, 5*, 27.

Chang, S. W. C., Gariépy, J.-F., & Platt, M. L. (2013). Neuronal reference frames for social decisions in primate frontal cortex. *Nature Neuroscience, 16*(2), 243–250.

Chang, S. W. C., et al. (2015). Neural mechanisms of social decision-making in the primate amygdala. *Proceedings of the National Academy of Sciences of the United States of America, 112*(52), 16012–16017.

Cisek, P., & Green, A. M. (2024). Toward a neuroscience of natural behavior. *Current Opinion in Neurobiology, 86*, 102859.

Dal Monte, O., et al. (2016). Live interaction distinctively shapes social gaze dynamics in rhesus macaques. *Journal of Neurophysiology, 116*(4), 1626–1643.

Dal Monte, O., et al. (2020). Specialized medial prefrontal–amygdala coordination in other-regarding decision preference. *Nature Neuroscience, 23*(4), 565–574.

Dal Monte, O., et al. (2022). Widespread implementations of interactive social gaze neurons in the primate prefrontal-amygdala networks. *Neuron, 110*(13), 2183–2197.e7.

Dana, J., Cain, D. M., & Dawes, R. M. (2006). What you don't know won't hurt me: Costly (but quiet) exit in dictator games. *Organizational Behavior and Human Decision Processes, 100*(2), 193–201.

Datta, S. R., et al. (2019). Computational neuroethology: A call to action. *Neuron, 104*(1), 11–24.

Dayan, P., & Balleine, B. W. (2002). Reward, motivation, and reinforcement learning. *Neuron, 36*(2), 285–298.

Du, E., & Chang, S. W. C. (2015). Neural components of altruistic punishment. *Frontiers in Neuroscience, 9*, 26.

Dugatkin, L. A., & Reeve, H. K. (2000). *Game theory and animal behavior*. Oxford University Press.

Duguid, S., et al. (2014). Coordination strategies of chimpanzees and human children in a Stag Hunt game. *Proceedings. Biological Sciences/The Royal Society, 281*(1796), 20141973.

Eckel, C. C., & Grossman, P. J. (1996). Altruism in anonymous dictator games. *Games and Economic Behavior, 16*(2), 181–191.

Engel, C. (2011). Dictator games: A meta study. *Experimental Economics, 14*(4), 583–610.

Fan, S., Dal Monte, O., & Chang, S. W. C. (2021). Levels of naturalism in social neuroscience research. *iScience, 24*(7), 102702.

Fan, S., et al. (2023). Closed-loop microstimulations of the orbitofrontal cortex during real-life gaze interaction enhance dynamic social attention. *bioRxiv: The Preprint Server for Biology* [Preprint]. Available at: https://doi.org/10.1101/2023.12.18.572176

Fehr, E., & Camerer, C. F. (2007). Social neuroeconomics: The neural circuitry of social preferences. *Trends in Cognitive Sciences, 11*(10), 419–427.

Franch, M., et al. (2024). Publisher correction: Visuo-frontal interactions during social learning in freely moving macaques. *Nature, 628*(8008), E3.
Friedman, D., Cassar, A., & Selten, R. (2004). *Economics lab: An intensive course in experimental economics*. Psychology Press.
Frith, C., & Frith, U. (2005). Theory of mind. *Current Biology: CB, 15*(17), R644–R646.
Frith, C. D., & Frith, U. (2006). The neural basis of mentalizing. *Neuron, 50*(4), 531–534.
Fudenberg, D., & Levine, D. K. (1998). *The theory of learning in games*. MIT Press.
Gallagher, H. L., et al. (2002). Imaging the intentional stance in a competitive game. *NeuroImage, 16*(3 Pt 1), 814–821.
Gangopadhyay, P., et al. (2020). Prefrontal–amygdala circuits in social decision-making. *Nature Neuroscience, 24*(1), 5–18.
Glimcher, P. W., & Fehr, E. (2013). *Neuroeconomics: Decision making and the brain*. Academic.
Grech, P. D., & Nax, H. H. (2020). Rational altruism? On preference estimation and dictator game experiments. *Games and Economic Behavior, 119*, 309–338.
Guerreiro Martins, E. M., et al. (2019). Food sharing patterns in three species of callitrichid monkeys (*Callithrix jacchus*, *Leontopithecus chrysomelas*, *Saguinus midas*): Individual and species differences. *Journal of Comparative Psychology, 133*(4), 474–487.
Güth, W., Schmittberger, R., & Schwarze, B. (1982). An experimental analysis of ultimatum bargaining. *Journal of Economic Behavior & Organization, 3*(4), 367–388.
Hafeez, T., et al. (2021). EEG in game user analysis: A framework for expertise classification during gameplay. *PLoS One, 16*(6), e0246913.
Harbaugh, W. T., Mayr, U., & Burghart, D. R. (2007). Neural responses to taxation and voluntary giving reveal motives for charitable donations. *Science, 316*(5831), 1622–1625.
Haroush, K., & Williams, Z. M. (2015). Neuronal prediction of opponent's behavior during cooperative social interchange in primates. *Cell, 160*(6), 1233–1245.
Harré, M. S. (2022). What can game theory tell us about an AI 'theory of mind'? *Games, 13*(3), 46.
Hartung, T. (2008). Thoughts on limitations of animal models. *Parkinsonism & Related Disorders, 14*(Suppl 2), S81–S83.
Hausfeld, J., von Hesler, K., & Goldlücke, S. (2021). Strategic gaze: An interactive eye-tracking study. *Experimental Economics, 24*(1), 177–205.
Heaney, M., Gray, R. D., & Taylor, A. H. (2017). Keas perform similarly to chimpanzees and elephants when solving collaborative tasks. *PLoS One, 12*(2), e0169799.
Holyoak, K. J., & Morrison, R. G. (2012). *The Oxford handbook of thinking and reasoning*. Oxford University Press.
Jarick, M., & Kingstone, A. (2015). The duality of gaze: Eyes extract and signal social information during sustained cooperative and competitive dyadic gaze. *Frontiers in Psychology, 6*, 1423.
Jendritza, P., Klein, F. J., & Fries, P. (2023). Multi-area recordings and optogenetics in the awake, behaving marmoset. *Nature Communications, 14*(1), 577.
Jiang, M., et al. (2021). Evolution and neural representation of mammalian cooperative behavior. *Cell Reports, 37*(7), 110029.
Joiner, J., et al. (2017). Social learning through prediction error in the brain. *npj Science of Learning, 2*, 8.
Jovanovic, V., et al. (2022). Behavioral context affects social signal representations within single primate prefrontal cortex neurons. *Neuron, 110*(8), 1318–1326.e4.
Kahneman, D. (2003). Maps of bounded rationality: Psychology for behavioral economics. *The American Economic Review, 93*(5), 1449–1475.
Kahneman, D. (2011). *Thinking, fast and slow*. Macmillan.
Kahneman, D., & Tversky, A. (2012). Prospect theory: An analysis of decision under risk. In *Handbook of the fundamentals of financial decision making* (World Scientific handbook in financial economics series) (pp. 99–127). World Scientific.
Kahneman, D., Knetsch, J. L., & Thaler, R. (1986). Fairness as a constraint on profit seeking: Entitlements in the market. *The American Economic Review, 76*(4), 728–741.

Kaigaishi, Y., Nakamichi, M., & Yamada, K. (2019). High but not low tolerance populations of Japanese macaques solve a novel cooperative task. *Primates: Journal of Primatology, 60*(5), 421–430.
Kim, J., et al. (2020). A bird's-eye view of brain activity in socially interacting mice through mobile edge computing (MEC). *Science Advances, 6*(49). Available at: https://doi.org/10.1126/sciadv.abb9841
King-Casas, B., et al. (2005). Getting to know you: Reputation and trust in a two-person economic exchange. *Science, 308*(5718), 78–83.
Kircher, T., et al. (2009). Online mentalising investigated with functional MRI. *Neuroscience Letters, 454*(3), 176–181.
Knofe, H., et al. (2019). Chimpanzees monopolize and children take turns in a limited resource problem. *Scientific Reports, 9*(1), 7597.
Lee, D. (2008). Game theory and neural basis of social decision making. *Nature Neuroscience, 11*(4), 404–409.
Mars, R. B., et al. (2013). Connectivity profiles reveal the relationship between brain areas for social cognition in human and monkey temporoparietal cortex. *Proceedings of the National Academy of Sciences of the United States of America, 110*(26), 10806–10811.
Marshall-Pescini, S., et al. (2017). Importance of a species' socioecology: Wolves outperform dogs in a conspecific cooperation task. *Proceedings of the National Academy of Sciences of the United States of America, 114*(44), 11793–11798.
Martin, J. S., et al. (2021). Prosociality, social tolerance and partner choice facilitate mutually beneficial cooperation in common marmosets, *Callithrix jacchus*. *Animal Behaviour, 173*, 115–136.
Massen, J. J. M., et al. (2010). Generous leaders and selfish underdogs: Pro-sociality in despotic macaques. *PLoS One, 5*(3), e9734.
McCabe, K., et al. (2001). A functional imaging study of cooperation in two-person reciprocal exchange. *Proceedings of the National Academy of Sciences of the United States of America, 98*(20), 11832–11835.
Meisner, O. C., et al. (2024). Development of a marmoset apparatus for automated pulling (MarmoAAP) to study cooperative behaviors. *bioRxiv: The Preprint Server for Biology* [Preprint]. Available at: https://doi.org/10.1101/2024.02.16.579531
Melis, A. P., & Raihani, N. J. (2023). The cognitive challenges of cooperation in human and nonhuman animals. *Nature Reviews Psychology, 2*(9), 523–536.
Melis, A. P., Hare, B., & Tomasello, M. (2009). Chimpanzees coordinate in a negotiation game. *Evolution and Human Behavior, 30*(6), 381–392.
Miller, C. T., et al. (2022). Natural behavior is the language of the brain. *Current Biology: CB, 32*(10), R482–R493.
Mobbs, D., et al. (2021). Promises and challenges of human computational ethology. *Neuron, 109*(14), 2224–2238.
Moeller, S., et al. (2023). Human and macaque pairs employ different coordination strategies in a transparent decision game. *eLife, 12*. Available at: https://doi.org/10.7554/eLife.81641
Moll, J., et al. (2006). Human fronto–mesolimbic networks guide decisions about charitable donation. *Proceedings of the National Academy of Sciences, 103*(42), 15623–15628.
Nowé, A., Vrancx, P., & De Hauwere, Y.-M. (2012). Game theory and multi-agent reinforcement learning. In M. Wiering & M. van Otterlo (Eds.), *Reinforcement learning: State-of-the-art* (pp. 441–470). Springer Berlin Heidelberg.
Ohtsuki, H. (2010). Evolutionary games in wright's island model: Kin selection meets evolutionary game theory. *Evolution, 64*(12), 3344–3353.
Ong, W. S., Madlon-Kay, S., & Platt, M. L. (2021). Neuronal correlates of strategic cooperation in monkeys. *Nature Neuroscience, 24*(1), 116–128.
Plotnik, J. M., et al. (2011). Elephants know when they need a helping trunk in a cooperative task. *Proceedings of the National Academy of Sciences of the United States of America, 108*(12), 5116–5121.

Press, W. H., & Dyson, F. J. (2012). Iterated prisoner's dilemma contains strategies that dominate any evolutionary opponent. *Proceedings of the National Academy of Sciences of the United States of America, 109*(26), 10409–10413.

Proctor, D., et al. (2013). Chimpanzees play the ultimatum game. *Proceedings of the National Academy of Sciences of the United States of America, 110*(6), 2070–2075.

Reinhart, C. J., et al. (2010). Targets and tactics of play fighting: Competitive versus cooperative styles of play in Japanese and Tonkean Macaques. *International Journal of Comparative Psychology/ISCP; Sponsored by the International Society for Comparative Psychology and the University of Calabria, 23*(2). Available at: https://doi.org/10.46867/ijcp.2010.23.02.05

Rilling, J., et al. (2002). A neural basis for social cooperation. *Neuron, 35*(2), 395–405.

Rilling, J. K., et al. (2004). Opposing BOLD responses to reciprocated and unreciprocated altruism in putative reward pathways. *Neuroreport, 15*(16), 2539–2543.

Rose, M. C., et al. (2021). Cortical representation of group social communication in bats. *Science, 374*(6566), eaba9584.

Roughgarden, J., Oishi, M., & Akçay, E. (2006). Reproductive social behavior: Cooperative games to replace sexual selection. *Science, 311*(5763), 965–969.

Roy, S., & Wang, X. (2012). Wireless multi-channel single unit recording in freely moving and vocalizing primates. *Journal of Neuroscience Methods, 203*(1), 28–40.

Sánchez-Amaro, A., & Rossano, F. (2021). Chimpanzees and bonobos use social leverage in an ultimatum game. *Proceedings. Biological Sciences/The Royal Society, 288*(1962), 20211937.

Sánchez-Amaro, A., et al. (2017). Chimpanzees, bonobos and children successfully coordinate in conflict situations. *Proceedings. Biological Sciences/The Royal Society, 284*(1856). Available at: https://doi.org/10.1098/rspb.2017.0259

Sánchez-Amaro, A., Burkart, J. M., & Rossano, F. (2021). Marmoset monkeys overcome dyadic social dilemmas while avoiding mutual defection. *Animal Behaviour, 173*, 93–104.

Sanfey, A. G. (2007). Social decision-making: Insights from game theory and neuroscience. *Science, 318*(5850), 598–602.

Sanfey, A. G., et al. (2003). The neural basis of economic decision-making in the ultimatum game. *Science, 300*(5626), 1755–1758.

Schultz, W. (2015). Neuronal reward and decision signals: From theories to data. *Physiological Reviews, 95*(3), 853–951.

Schurz, M., et al. (2014). Fractionating theory of mind: A meta-analysis of functional brain imaging studies. *Neuroscience and Biobehavioral Reviews, 42*, 9–34.

Schwarz, D. A., et al. (2014). Chronic, wireless recordings of large-scale brain activity in freely moving rhesus monkeys. *Nature Methods, 11*(6), 670–676.

Schwing, R., et al. (2016). Kea cooperate better with sharing affiliates. *Animal Cognition, 19*(6), 1093–1102.

Seo, H., Barraclough, D. J., & Lee, D. (2009). Lateral intraparietal cortex and reinforcement learning during a mixed-strategy game. *The Journal of Neuroscience: The Official Journal of the Society for Neuroscience, 29*(22), 7278–7289.

Shahidi, N., et al. (2024). Population coding of strategic variables during foraging in freely moving macaques. *Nature Neuroscience, 27*(4), 772–781.

Sigmundson, R., et al. (2021). Exploring the cognitive capacities of Japanese macaques in a cooperation game. *Animals: An Open Access Journal from MDPI, 11*(6). Available at: https://doi.org/10.3390/ani11061497

Smith, J. M. (1982). *Evolution and the theory of games*. Cambridge University Press.

Smith, M. F., et al. (2019). Capuchin monkeys (Sapajus [Cebus] apella) play Nash equilibria in dynamic games, but their decisions are likely not influenced by oxytocin. *American Journal of Primatology, 81*(4), e22973.

Speitel, C., Traut-Mattausch, E., & Jonas, E. (2019). Functions of the right DLPFC and right TPJ in proposers and responders in the ultimatum game. *Social Cognitive and Affective Neuroscience, 14*(3), 263–270.

Sripada, C. S., et al. (2009). Functional neuroimaging of mentalizing during the trust game in social anxiety disorder. *Neuroreport, 20*(11), 984–989.

Tabibnia, G., Satpute, A. B., & Lieberman, M. D. (2008). The sunny side of fairness: Preference for fairness activates reward circuitry (and disregarding unfairness activates self-control circuitry). *Psychological Science, 19*(4), 339–347.

Testard, C., et al. (2024). Neural signatures of natural behaviour in socializing macaques. *Nature, 628*(8007), 381–390.

Tomlin, D., et al. (2006). Agent-specific responses in the cingulate cortex during economic exchanges. *Science, 312*(5776), 1047–1050.

Torres Ortiz, S., et al. (2020). Problem-solving in a cooperative task in peach-fronted conures (*Eupsittula aurea*). *Animal Cognition, 23*(2), 265–275.

Trivers, R. (1985). *Social Evolution*. Benjamin/Cummings Publishing Company.

Viana, D. S., et al. (2010). Cognitive and motivational requirements for the emergence of cooperation in a rat social game. *PLoS One, 5*(1), e8483.

Walker, J. D., et al. (2021). Chronic wireless neural population recordings with common marmosets. *Cell Reports, 36*(2), 109379.

Wang, H., & Kwan, A. C. (2023). Competitive and cooperative games for probing the neural basis of social decision-making in animals. *Neuroscience and Biobehavioral Reviews, 149*, 105158.

Wang, H., et al. (2022). Pupil correlates of decision variables in mice playing a competitive mixed-strategy game. *eNeuro, 9*(2). Available at: https://doi.org/10.1523/ENEURO.0457-21.2022

Wang, H., et al. (2024). Frontal noradrenergic and cholinergic transients exhibit distinct spatiotemporal dynamics during competitive decision-making. *bioRxiv: The Preprint Server for Biology* [Preprint]. Available at: https://doi.org/10.1101/2024.01.23.576893

Weibull, J. W. (1997). *Evolutionary game theory*. MIT Press.

Weinberg-Wolf, H., & Chang, S. W. C. (2019). Differences in how macaques monitor others: Does serotonin play a central role? *Wiley Interdisciplinary Reviews: Cognitive Science, 10*(4), e1494.

Whitehead, H. (2008). *Analyzing animal societies: Quantitative methods for vertebrate social analysis*. University of Chicago Press.

Yang, Y., et al. (2022). Author correction: Wireless multilateral devices for optogenetic studies of individual and social behaviors. *Nature Neuroscience, 25*(10), 1394.

Yoshida, W., Dolan, R. J., & Friston, K. J. (2008). Game theory of mind. *PLoS Computational Biology, 4*(12), e1000254.

Zhang, W., & Yartsev, M. M. (2019). Correlated neural activity across the brains of socially interacting bats. *Cell, 178*(2), 413–428.e22.

Zhang, H., et al. (2023). Leader–follower behavioural coordination and neural synchronization during intergroup conflict. *Nature Human Behaviour, 7*(12), 2169–2181.

Part VI
Individual Differences in Neuroeconomics: Age, Sex and Gender, and Clinical Variation

David V. Smith, Dominic S. Fareri, and Patricia L. Lockwood

Although neuroeconomics often seeks to identify generalizable principles of value and choice, understanding how these processes vary across individuals is essential for building models that reflect the full range of human behavior. This section explores how cognitive, affective, and motivational systems differ across populations—whether due to aging, mental health, substance use, or sex and gender—and how those differences shape decision-making. The chapters also highlight tools such as model-based fMRI, lesion studies, and lifespan approaches that help quantify this variability. Together, they move the field toward more flexible, person-centered accounts of decision processes with translational relevance.

Aging provides a compelling window into individual variability. Lighthall and colleagues (this volume) synthesize research showing that while cognitive control tends to decline with age, brain regions often engaged during value-based decisions—including the ventromedial prefrontal cortex—typically exhibit more stable activation. This dissociation may explain how older adults preserve consistent preferences despite reductions in deliberative function, a pattern supported by findings on affective processing in aging (Samanez-Larkin & Knutson, 2015). Extending this foundation, Tisdall and Seaman (this volume) update the affect-integration-motivation (AIM) framework, illustrating how motivational shifts across the lifespan bias older adults toward emotionally meaningful outcomes. Their account aligns with socioemotional selectivity theory and links these shifts to changes in dopaminergic tone (Carstensen et al., 1999; Eppinger et al., 2011).

D. V. Smith (✉)
Temple University, Philadelphia, PA, USA
e-mail: david.v.smith@temple.edu

D. S. Fareri
Adelphi University, Garden City, NY, USA

P. L. Lockwood
University of Birmingham, Birmingham, UK

Building on this theme, Wyngaarden et al. (this volume) examine how such shifts in motivation and cognition may also increase susceptibility to decision biases—specifically, confirmation bias in financial contexts. Older adults, they argue, are more likely to favor belief-consistent information, especially in emotionally charged or ambiguous situations. This bias is linked to functional changes in socioaffective and internally directed cognition, including activity in the default mode and limbic networks. Their findings align with prior work showing age-related differences in neural responses to trust-related cues (Fareri et al., 2022) and with research linking exploitation vulnerability to changes in prefrontal and insular regions (Spreng et al., 2016; Han, 2022). Importantly, they also highlight early-stage evidence that noninvasive brain stimulation may offer a pathway to reduce such vulnerabilities.

Sex and gender offer additional axes of individual variability with widespread implications for neuroeconomic processes. Doren and Tobler (this volume) synthesize evidence that behavioral and neural differences linked to sex and gender appear across a range of domains, including risk taking, intertemporal choice, social preferences, and consumer behavior. They argue that many of these differences are not stable traits but are modulated by contextual and developmental factors, such as hormonal fluctuations or sociocultural expectations. In doing so, the chapter challenges the historical reliance on sex-averaged data in neuroimaging and advocates for the systematic inclusion of sex and gender as core variables in experimental design and computational modeling. This perspective echoes calls for more inclusive and mechanistic approaches to behavioral neuroscience (Cahill, 2006; Joel & Fausto-Sterling, 2016) and aligns with emerging evidence that biological and social variables interact to shape valuation and choice.

While aging and gender-related variability reveal how decision strategies adapt to normative changes, clinical populations show how these processes can break down. Rich and Levy (this volume) argue that dysfunction in conditions like depression, anxiety, and schizophrenia may reflect misalignments between domain-general and domain-specific value representations. This view is supported by evidence that psychopathology involves selective impairments in valuation, depending on context or task framing (Husain & Roiser, 2018; Halahakoon et al., 2020; Ng et al., 2019). Ramakrishnan et al. (this volume) focus on substance use disorders, showing how reinforcement learning models identify shifts from goal-directed to habitual behavior. Their synthesis draws on studies revealing altered prediction error signaling and impaired planning in both active users and individuals in remission (Sebold et al., 2014; Konova et al., 2020).

Across these chapters, a shared message emerges: individual variability is not just noise but a defining feature of decision-making. These insights raise critical questions: How do value computations evolve with age, biological sex, or illness? Which neural markers predict maladaptive patterns? And how can computational models be adapted to reflect diverse decision-making profiles? As neuroeconomics matures, addressing these questions will be essential for building more accurate, inclusive, and actionable models of choice.

References

Cahill, L. (2006). Why sex matters for neuroscience. *Nature Reviews Neuroscience, 7*(6), 477–484. https://doi.org/10.1038/nrn1909

Carstensen, L. L., Isaacowitz, D. M., & Charles, S. T. (1999). Taking time seriously: A theory of socioemotional selectivity. *American Psychologist, 54*(3), 165–181. https://doi.org/10.1037/0003-066X.54.3.165

Doren, A., & Tobler, P. N. (this volume). Sex and gender in neuroeconomics. In D. V. Smith, P. Lockwood, & D. S. Fareri (Eds.), *Neuroeconomics: Core topics and current directions*. Springer.

Eppinger, B., Hämmerer, D., & Li, S. C. (2011). Neuromodulation of reward-based learning and decision making in human aging. *Annals of the New York Academy of Sciences, 1235*(1), 1–17. https://doi.org/10.1111/j.1749-6632.2011.06230.x

Fareri, D. S., Hackett, K., Tepfer, L. J., Kelly, V., Henninger, N., Reeck, C., Giovannetti, T., & Smith, D. V. (2022). Age-related differences in ventral striatal and default mode network function during reciprocated trust. *NeuroImage, 251*, 119267. https://doi.org/10.1016/j.neuroimage.2022.119267

Halahakoon, D. C., Kieslich, K., O'Driscoll, C., et al. (2020). Reward-processing behavior in depressed participants relative to healthy volunteers: A systematic review and meta-analysis. *JAMA Psychiatry, 77*(12), 1286–1295. https://doi.org/10.1001/jamapsychiatry.2020.2139

Han, S. D. (2022). Vulnerability to financial exploitation in older age: A neuropsychology perspective. In A. Goudie (Ed.), *A fresh look at fraud* (pp. 94–111). Routledge. https://doi.org/10.4324/9781003017189-7

Husain, M., & Roiser, J. P. (2018). Neuroscience of apathy and anhedonia: A transdiagnostic approach. *Nature Reviews Neuroscience, 19*(8), 470–484. https://doi.org/10.1038/s41583-018-0029-9

Joel, D., & Fausto-Sterling, A. (2016). Beyond sex differences: New approaches for thinking about variation in brain structure and function. *Philosophical Transactions of the Royal Society, B: Biological Sciences, 371*(1688), 20150451. https://doi.org/10.1098/rstb.2015.0451

Konova, A. B., Lopez-Guzman, S., Urmanche, A., Ross, S., Louie, K., Rotrosen, J., & Glimcher, P. W. (2020). Computational markers of risky decision-making for identification of temporal windows of vulnerability to opioid use in a real-world clinical setting. *JAMA Psychiatry, 77*(4), 368–377. https://doi.org/10.1001/jamapsychiatry.2019.4013

Lighthall, N. R., Doheny, M. M., & Wang, S. (this volume). Decision neuroscience of aging: 20 years in. In D. V. Smith, P. Lockwood, & D. S. Fareri (Eds.), *Neuroeconomics: Core topics and current directions*. Springer.

Ng, T. H., Alloy, L. B., & Smith, D. V. (2019). Meta-analysis of reward processing in major depressive disorder reveals distinct abnormalities within the reward circuit. *Translational Psychiatry, 9*, 293. https://doi.org/10.1038/s41398-019-0644-x

Ramakrishnan, S. A., Muliya, M., Chakravarthy, V. S., & Parvaz, M. A. (this volume). Applications and evolution of neuroeconomics and reinforcement learning models in substance use disorders. In D. V. Smith, P. Lockwood, & D. S. Fareri (Eds.), *Neuroeconomics: Core topics and current directions*. Springer.

Rich, E. L., & Levy, B. J. (this volume). Domain-general and domain-specific values and their role in psychopathology. In D. V. Smith, P. Lockwood, & D. S. Fareri (Eds.), *Neuroeconomics: Core topics and current directions*. Springer.

Samanez-Larkin, G. R., & Knutson, B. (2015). Decision making in the ageing brain: Changes in affective and motivational circuits. *Nature Reviews Neuroscience, 16*(5), 278–289. https://doi.org/10.1038/nrn3917

Sebold, M., Deserno, L., Nebe, S., Schad, D. J., Garbusow, M., Hagele, C., et al. (2014). Model-based and model-free decisions in alcohol dependence. *Neuropsychobiology, 70*(2), 122–131. https://doi.org/10.1159/000362840

Spreng, R. N., Karlawish, J., & Marson, D. C. (2016). Cognitive, social, and neural determinants of diminished decision-making and financial exploitation risk in aging and dementia. *Journal of Elder Abuse & Neglect, 28*(4–5), 320–344. https://doi.org/10.1080/08946566.2016.1237918

Tisdall, L., & Seaman, K. L. (this volume). Decision making in the aging brain: An update of the evidence base for the 'affect-integration-motivation' framework. In D. V. Smith, P. Lockwood, & D. S. Fareri (Eds.), *Neuroeconomics: Core topics and current directions*. Springer. Preprint: https://doi.org/10.31234/osf.io/tnbky

Wyngaarden, J. B., III, Giovannetti, T., Krekelberg, B., & Smith, D. V. (this volume). Maladaptive decision making in older adults: Confirmation bias and financial exploitation. In D. V. Smith, P. Lockwood, & D. S. Fareri (Eds.), *Neuroeconomics: Core topics and current directions*. Springer. Preprint: https://doi.org/10.31234/osf.io/abnyd_v2

Decision Neuroscience of Aging: 20 Years In

Nichole R. Lighthall, Margaret M. Doheny, and Shensheng Wang

Abstract The present review presents the field of aging decision neuroscience, 20 years after the inception of this subfield. We begin with a historical overview of the area of study and highlight some current neural models of decision-making in aging. Typical changes to the brain that accompany aging are described, followed by a review of current lines of research that have advanced our understanding of how decision-making is impacted by advancing age. Four main areas of study are reviewed. First, we cover research on the dopamine system and feedback-based learning. Next, we review research on value integration and memory, followed by temporal discounting. We close out the literature review with discussion of studies in the area of social decision-making and its implications for understanding exploitation in aging. Finally, we end with conclusions and future directions that are likely to significantly advance the field over the next 20 years.

The subfield of decision neuroscience of aging began to emerge in the latter half of the 2010s, due in large part to National Institute on Aging-sponsored activities and initiatives. These initiatives recognized how growth in the aging population, coupled with the relevance of economic decision-making for health and well-being in late life, requires focused interdisciplinary research on the neural mechanisms of economic behavior (National Institute on Aging, 2006a, b). Interestingly, while the earliest publications and funding opportunities acknowledged that, in aging, decision-making is embedded in social contexts and economic decision-making has profound social consequences (Mather, 2006; Nielsen & Mather, 2011), decision neuroscience of aging and social affective neuroscience of aging have developed (mostly) in parallel to one another. As will be discussed later, these two subfields have increasingly merged over time. This merging has allowed the field to tackle some of the most complex and pressing problems highlighted at the inception of the subfield, including identification of mechanisms that contribute to scam

N. R. Lighthall (✉) · M. M. Doheny · S. Wang
Department of Psychology, University of Central Florida, Orlando, FL, USA
e-mail: nichole.lighthall@ucf.edu

D. V. Smith et al. (eds.), *Neuroeconomics: Core Topics and Current Directions*,
https://doi.org/10.1007/978-3-032-02925-6_28

vulnerability in aging and how to change institutions to account for unavoidable changes to decision-making with age (National Institute on Aging, 2006a).

Following the proposals from these early meetings of researchers at the intersection of neuroeconomics, social-affective science, and cognitive aging, decision neuroscience of aging has made substantial progress in understanding how the components of decision-making, and their neural mechanisms, change in aging (Lighthall, 2020 for review). An NIA-funded research network, the Scientific Research Network on Decision Neuroscience and Aging, also emerged and successfully fostered research in this area of study. This research growth led to a number of neural models of decision-making in aging. Some of the most influential and sustained models include those describing age-related changes to neurotransmitter systems and their impact on reward-based learning and decision-making in aging (particularly dopaminergic and serotonergic neuromodulators; e.g., Eppinger et al., 2011). And more recently, models that also integrated social-psychological perspectives on age-related changes to motivation (Carstensen et al., 1999; Mather & Carstensen, 2005) as well as cognitive perspectives on age-related changes to fluid and crystallized intelligence (Salthouse, 2004). Central among these models is the *Affect–Integration–Motivation* (AIM) framework (Samanez-Larkin & Knutson, 2015).

Under the *AIM* framework, affective neural components anticipate gains and losses (through dopaminergic projections to the nucleus accumbens and noradrenergic and dopaminergic projections to the anterior insula, respectively). These responses trigger valence/arousal outputs that are integrated with additional evaluative features through glutamatergic projections to the medial prefrontal cortex (mPFC). In the context of multiple-choice options and choices that involve long-term memory, integration can involve additional mPFC resources and the dorsolateral PFC (e.g., for choice comparison). Integrated choice value signals are projected to the motivational (action) component through glutamatergic projections to the dorsal striatum and supplementary motor area (for a detailed review of this framework and research that has tested its primary predictions, see Tisdall and Seaman in this volume). As described below, recent work on social decision-making has underscored the need for specific models of social decision-making in the aging brain. Answering this call is a recently developed extension of *AIM* for understanding social decision-making in aging, *CISDA* (*Changes in Integration for Social Decisions in Aging*; Frazier et al., 2019; Fig. 1). *CISDA* focuses on the choice value integration of decision processing as the most critical phase for the emergence of age differences in decision-making and identifies three age-variant component processes that impact choice value integration for social decisions: social-emotional processes, learning/memory systems, and theory of mind.

These reviews and theoretical papers capture much of the work conducted in the field of aging decision neuroscience to date (see also Hesse, 2015) and set the stage for current work in this area of study. The present review focuses on current pursuits and recent major advancements in aging decision neuroscience. While not exhaustive, this review provides a snapshot of ongoing research in this area of study approximately 20 years after its inception.

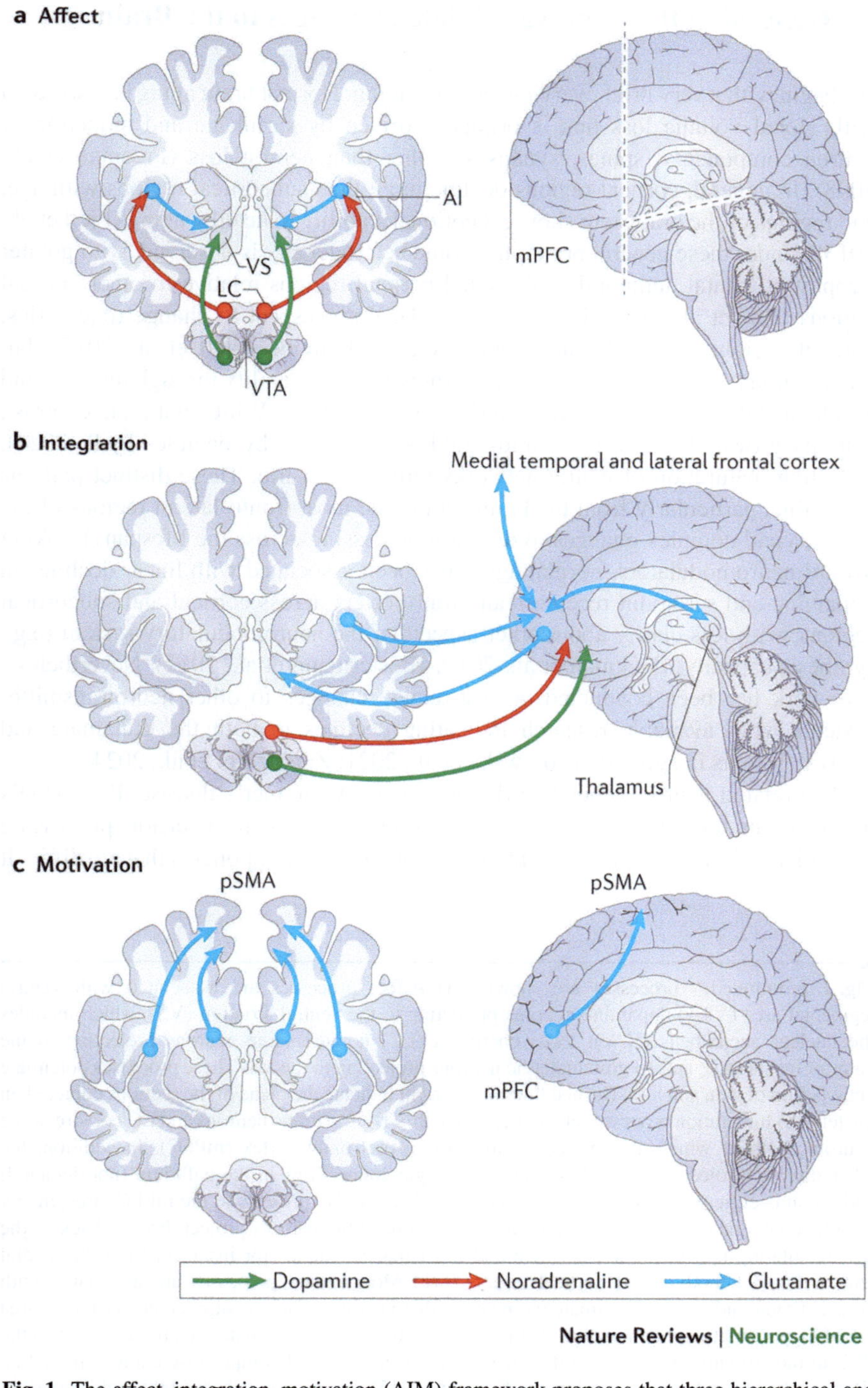

Fig. 1 The affect–integration–motivation (AIM) framework proposes that three hierarchical and sequential processes can occur before, and promote, choice. The brain regions involved in these

1 General Patterns of Age-Related Changes to the Brain

We begin with a very brief overview of brain aging. Normal brain aging is associated with global volume loss that is primarily driven by shrinkage and reduction of neuron components (spines, synapses) rather than neuron loss (Freeman et al., 2008). In general, cortical regions decline more than subcortical regions with age, and regions of the brainstem show exceptional stability in healthy aging (Fjell et al., 2013). Under these general principles, some heterogeneity is observed, with greater atrophy in frontal, temporal, and medial parietal regions relative to other cortical regions (Raz et al., 2005; Fjell et al., 2014). With respect to change trajectories, cortical regions exhibit linear atrophy across adulthood (Fjell et al., 2015), but other regions, such as the hippocampus, show relative stability through mid-life and accelerated decline in late adulthood (Fraser et al., 2015). White matter, in contrast, exhibits growth from early to mid-adulthood followed by decline (Fjell, 2024), reflecting maturation of neural networks through mid-life. These distinct patterns of decline, particularly from the brainstem to the most fronto-lateral regions of the cortex, yield complex changes to decision processing across the lifespan. Looking toward neuromodulators, normal aging has been associated with linear declines in dopamine and serotonin receptors and transporters across cortical and subcortical regions, with less clarity about other aspects of the neuromodulatory system (e.g., synthesis capacity; Bäckman et al., 2010; Klostermann et al., 2012). Nevertheless, less work has been conducted on age-related changes to other neurotransmitter systems, with available research indicating declines in both the glutamate and GABA systems in aging (Gasiorowska et al., 2021; Zuppichini et al., 2024).

Age-related neurodegenerative diseases (e.g., Alzheimer's disease, Parkinson's disease) can impact select regions and systems within the decision processing network preclinically, leading to impacts on processing components that are difficult

Fig. 1 (continued) processes are shown. (**a**) Affect processes are associated with ventral tegmental area (VTA) dopamine neurons projecting to the ventral striatum (VS), which includes the nucleus accumbens (NAc); locus coeruleus (LC) noradrenaline neurons projecting to the anterior insula (AI); and AI glutamatergic neurons projecting to the VS. These processes potentiate anticipation of gain and loss. Dashed lines on the right indicate the plane of the sections depicted on the left. (**b**) Integration processes are associated with VTA dopamine neurons and LC noradrenaline neurons, both of which also project to the medial prefrontal cortex (mPFC). In addition, the VS indirectly projects to the mPFC via GABAergic connections in the pallidum (not depicted) and glutamatergic projections from the thalamus. The AI also projects to the mPFC, presumably via glutamatergic connections. Finally, mPFC glutamatergic neurons project directly back to the VS, facilitating the integration of value and other relevant input (for instance, from the medial temporal and lateral frontal cortical regions). (**c**) Motivational processes are associated with dorsal striatal and insular glutamatergic neurons that project to the pre-supplementary motor area (pSMA), potentiating motor action. Healthy aging may degrade glutamatergic projections from the PFC to the striatum, which could diminish value integration and compromise choice optimality. (Republished from Samanez-Larkin and Knutson (2015). Permission conveyed through Copyright Clearance Center, Inc)

to differentiate from normal age-related changes. Over recent years, the field of aging decision neuroscience has increasingly included clinical and preclinical samples. This shift has provided a distinct view of decision-making in states of age-related neurogenerative disease compared with the healthy aging brain.

In the sections that follow, we highlight a few major lines of active research in the decision neuroscience of aging that have recently clarified trajectories of change to decision-making with age, challenged long-held beliefs in the field, and opened new areas of focus.

2 Dopamine System and Feedback-Based Learning

Dopamine plays a crucial role in reward processing (Schultz et al., 1997), feedback-based value learning, and goal-directed decision-making (Verharen et al., 2019). Previous research documents marked negative associations between age and dopamine receptors (D1 and D2) as well as transporters—most notably in frontal regions with relatively less decline in the striatum (Karrer et al., 2017). This decline forms the basis of the dopamine hypothesis of aging, which proposes that aging is associated with weakened dopaminergic error signals leading to downstream impacts on an array of cognitive functions (Nieuwenhuis et al., 2002; although the scope of functions under this hypothesis may be more limited than originally proposed, see Juarez et al., 2019; Karalija et al., 2024).

Moving beyond cross-sectional studies, more recent work has examined changes to the dopamine system longitudinally. Compared with estimates from cross-sectional studies, findings from the longitudinal studies indicate subtler changes to the dopamine system with age. One recent longitudinal positron emission tomography (PET) study over 5 years (Karalija et al., 2022) found that the decline in striatal dopamine D2-like receptors in older adults (aged 64–68) was roughly half of what past cross-sectional studies estimated (Karrer et al., 2017). Making the picture more complicated, the relationship between changes in dopamine function and decision-making is obscured by both inter- and intra-individual variability in dopamine function decline, as well as differences in study tasks.

Nevertheless, past and recent evidence consistently suggests that age-related changes to the dopamine system underlie changes to decision-making involving reward learning. For example, age-related dopamine reductions are associated with learning-based risky decisions (Green et al., 2023) but not reward-related decisions that do not involve learning (e.g., temporal discounting; Castrellon et al., 2019). Specifically, the former study found that lower dopamine receptor availability in the amygdala was negatively associated with flexibility in choice behavior during learning-based risky decisions when controlling for age (Green et al., 2023). This suggests that older adults with lower amygdala D2 receptor density are less likely to adjust their behavior after negative feedback, increasing their financial risk. In contrast to decline in receptor binding, dopamine synthesis capacity remains stable (Karrer et al., 2017), or even increases with age (Ciampa et al., 2022). The study yielding these latter findings further demonstrated a pattern of neural compensation wherein posterior parietal cortex atrophy over 13 years was associated with greater dopamine synthesis capacity among older adults. Further, the negative effects of posterior parietal cortex atrophy on cognition were reduced with higher levels of dopamine synthesis capacity, suggesting increased dopamine synthesis capacity may serve as a mechanism for cognitive resilience in aging even in the context of neural decline (Ciampa et al., 2022). Complementary findings have long been observed in studies of rodent models with age-related and experimentally lesioned central dopaminergic neurons. These studies demonstrated that decline/damage to dopamine-producing neurons upregulates dopamine synthesis enzymes (Zigmond et al., 1990). Further, in early Parkinson's disease, dopamine synthesis capacity remains stable while dopamine transporters decline (e.g., Nandhagopal et al., 2011). These findings lend additional support to the conclusion that dopamine synthesis capacity may represent a compensatory mechanism for cognitive resilience in (at least) the earlier stages of age-related neural decline and degenerative disease.

Mechanisms of accelerated reward learning decline in aging, on the other hand, include β-amyloid (Aβ) pathology. Work by Chen et al. (2023) using PET found that Aβ+ cognitively normal older adults showed poorer reward learning performance than both younger adults and Aβ- older adults. Using graph theoretical analysis, their study also demonstrated that Aβ+ older adults exhibited less neural network segregation than younger adults and Aβ- older adults. Thus, Alzheimer's disease (AD)-related brain pathology influences reward learning and reward-decision-making in aging, even among individuals who present as cognitively unimpaired.

3 Value Integration and Memory

It has been well established that the subjective value of choices depends on the individual weighting of outcome contingencies, including reward, effort, and outcome delays. Age may lead to shifts in the subjective value of outcome contingencies (e.g., how "long" does 1 year *feel*), leading to age differences in the weighting for choice features. However, recent work shows that when individual differences in the subjective value of choice features (e.g., outcome contingencies) are accounted for in a healthy adult lifespan sample, the resulting integrated subjective value signal in the medial prefrontal cortex is age-invariant and consistent across task contingencies (Seaman et al., 2018). Critically, as highlighted above, the age-invariance of subjective value mechanisms may be exclusive to decisions from description, rather than experience-dependent decisions. Importantly, age differences in decisions that depend on reinforcement learning may arise from age-related changes to the brain's reward network, but also from selective information processing and changes to episodic memory function.

With respect to selective processing, the "positivity effect" (Mather & Carstensen, 2005) under Socioemotional Selectivity Theory (Carstensen et al., 1999) proposes that age is associated with motivational shifts that lead older adults to preferentially attend to, and encode, positive over negative information, leading to more positively biased memories. However, this selection process requires additional neurocognitive resources, including prefrontal-dependent executive control functions that decline in normal aging (Kryla-Lighthall & Mather, 2009). Extending this line of research to value integration, older adults may similarly deploy selective encoding in value-based learning. This selective processing allows older adults to either meet socioemotional goals when prioritizing emotional goals (e.g., focusing on positive choice features) and/or compensate for age-related decline in learning and memory when prioritizing performance goals (e.g., focus "most important" information to manage cognitive load).

Toward performance goals, behavioral research on value-directed remembering paradigms has demonstrated that cognitively healthy older adults can selectively encode "high-value" information at the level of younger adults, but they do so at the expense of "low-value" information—forgetting more of this information compared with younger adults (Castel et al., 2002). Functional neuroimaging research examining value-directed remembering in aging showed that selectivity is accomplished in younger adults through increased activity in semantic processing regions during encoding of high-value information (Cohen et al., 2016). In the same study, however, older adults with greater value-directed memory *reduced engagement* of semantic processing regions during encoding of low-value information. More recent work has additionally demonstrated that selective encoding of high-value items in older adults is associated with greater integrity in white matter structures supporting semantic processing (left inferior fronto-occipital fasciculus; Hennessee et al., 2019). This structural neural correlate of value-directed remembering is notably distinct from that observed in younger adults (i.e., left uncinate fasciculus, Reggente et al., 2018). These findings indicate that, in aging, neural mechanisms that support selective processing during value learning are distinct in both activation patterns and underlying neural structures. Value-directed remembering strategies have also been examined in older adults with AD and behavioral-variant frontotemporal dementia (bvFTD; Wong et al., 2019). Relative to healthy, age-matched controls, individuals with AD could recall high-value items during learning trials and showed preferential recognition memory for high-value items but failed to show preferential recall. These results suggest that intentional selective encoding during value learning has potential utility in AD, but its efficacy is limited by patients' learning and memory impairments. Participants with bvFTD, however, could not successfully prioritize encoding of high-value items during learning due to inhibitory dysfunction. This suggests that prefrontal brain regions selectively impacted by bvFTD (Kipps et al., 2009) play a critical role in value-directed encoding and behavioral interventions to enhance value learning in bvFTD that depend on intentional selective encoding are unlikely to be successful.

Finally, it's important to note that information encountered during value learning may be preferentially encoded or retrieved in the absence of top-down strategies. Recent work by Noh et al. (2023) shows that decision-making can be simultaneously influenced by reinforcement learning and specific episodic experiences. In these value learning contexts, older age was associated with a decreasing influence of reinforcement learning and greater preservation of past choices. Additionally, greater memory precision (pattern separation in the representation of mnemonic stimuli) was associated with enhanced influence of memory sampling on choice, while poorer memory precision was associated with greater reliance on gist-based memory during memory sampling.

Collectively, these studies provide key insights into how age-related changes to different memory systems impact decision-making. They demonstrate how memory functions with greater preservation in aging (e.g., semantic memory), and those that can be intentionally leveraged in decision processing strategies (e.g., episodic encoding), may allow for age-related compensation during decision-making in aging. As one example, these recent studies suggest that interventions that enhance the *memorability* of critical, high-value information during value learning hold potential for improving value-based decision-making in healthy aging, and perhaps also AD.

4 Temporal Discounting

Temporal discounting is the phenomenon of devaluing future rewards relative to immediate or sooner rewards. Like other subjective value signals, valuation during intertemporal choice tasks has been associated with activation of the ventral striatum and vmPFC, with a potentially greater role for the medial PFC, lateral parietal cortex and subregions of the posterior cingulate cortex (PCC) for delay-specific value signals, and an expanded network that includes cognitive control regions (lateral PFC, ACC) as well as imagery/prospection regions (MTL; Peters & Büchel, 2011).

Many studies have been conducted on temporal discounting in aging, following the long-held assumption that age is associated with greater patience, and this would bear out in decreased temporal discounting with age. Critically, however, a recent meta-analysis of 37 cross-sectional studies (total $N = 104{,}737$) found no notable association between age and delay discounting across a range of designs and incentive types (Seaman et al., 2022) and a single study of >23,000 adults found no correlation between age and temporal discounting (Sanchez-Roige et al., 2018). Likewise, functional neuroimaging studies have not yielded clear age differences in the neural correlates of temporal discounting among healthy populations. Specifically, smaller initial studies suggested age-related differences in vmPFC subjective value signals during temporal discounting (Eppinger et al., 2012; Samanez-Larkin et al., 2011), but subsequent larger studies found no evidence of age difference in neural correlates (Seaman et al., 2018) or only under conditions of high decision conflict (e.g., age-related reduction of associated ventral striatum activation; Eppinger et al., 2018).

The picture of temporal discounting in the aging brain appears different, however, when considering populations with age-related cognitive impairments. Recent work by Lempert et al. (2020b) with a relatively large, diverse sample of cognitive healthy older adults ($n = 74$) and older adults with amnestic mild cognitive impairment

(MCI; $n = 26$) found that discounting of delayed rewards was associated with lower episodic memory and entorhinal cortical thickness (but was unrelated to age), and the observed effects on delay discounting were driven by older adults with MCI. This same research group found that, among cognitively healthy older adults, entorhinal cortical thickness was associated with individual differences in temporal discounting and recollection of specific perceptual details during autobiographical memory retrieval (Lempert et al., 2020a).

These findings highlight the role of medial temporal lobe structures in the expanded temporal discounting neural network across adulthood (Peters & Büchel, 2011) and align with prior work highlighting structural integrity of the entorhinal cortex as particularly important in determining temporal discounting even prior to older age (Owens et al., 2017). Notably, research with younger adults shows that the hippocampus supports the vividness of prospective thought during intertemporal decision-making (i.e., imagining the future; Lee et al., 2022). The fact that individual differences in hippocampal volume do not predict intertemporal decisions in aging may reflect the entorhinal cortex's vulnerability to early brain pathology in aging and AD (Braak & Braak, 1991), increasing its role in determining individual differences in delay discounting in late life. As decline in entorhinal cortex thickness among cognitively intact older adults has been linked to financial exploitation vulnerability (Fenton et al., 2024), future research should consider how changes to this region in normal aging and clinical/preclinical neurodegenerative disease may impact decisions about delayed outcomes in the context of scams and fraud. Additionally, a study conducted during the COVID-19 pandemic found that older adults showed increased perceived risk of COVID-19 transmission with personalized simulations involving social consequences (Sinclair et al., 2021). Future research should examine whether age-related changes to the brain, particularly in the entorhinal cortex, impact delay discounting for real-world decisions across both positive and negative delayed outcomes.

5 Social Decision-Making

Since the emergence of the subfield, decision neuroscience of aging research has shown significant growth in its focus on social decision-making in aging. A driving reason for this focus has been growing concerns over the exploitation of older adults by social partners or groups, with estimates for financial exploitation of seniors between 3.5% and 20% (Wood & Lichtenberg, 2017). However, other research indicates that older adults are more successful at social decision-making than younger adults due to components like life experience and wisdom (Lim & Yu, 2015). To clarify if and why age-related changes to social decision-making occur, the first framework for understanding changes to social decision-making in aging was developed—*Changes in Integration for Social in Aging* (*CISDA*; Frazier et al., 2019, Fig. 2). *CISDA* identifies the choice value integration phase as a processing stage where age-related differences in social decision-making are likely to arise.

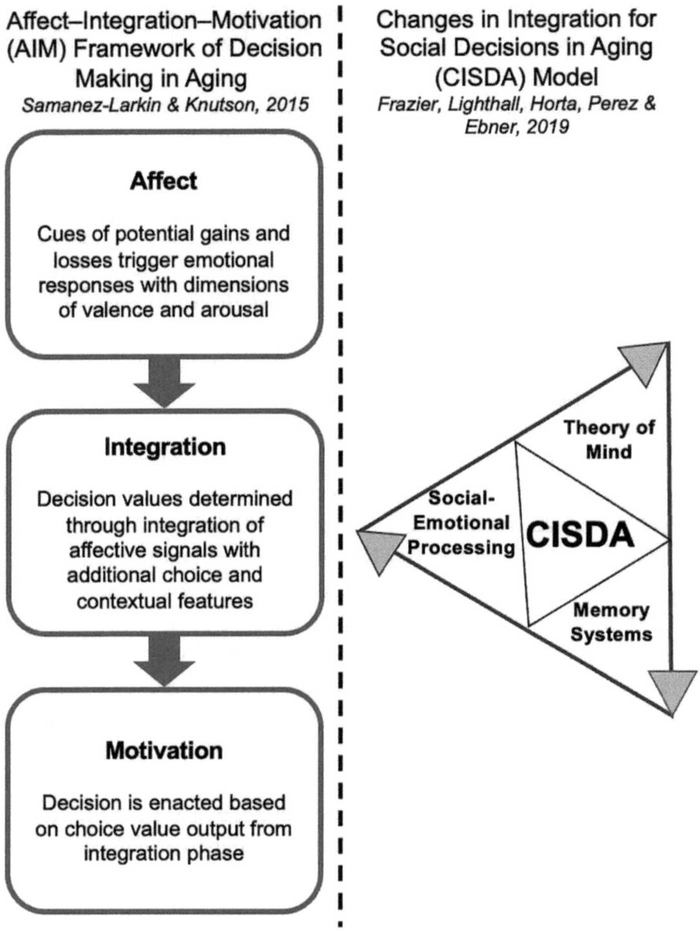

Fig. 2 Current models of decision-making in aging. The *Affect–Integration–Motivation* (*AIM*) framework is an evidence-based framework that provides a foundation for understanding the neural circuitry of motivated decision processing in aging (Samanez-Larkin & Knutson, 2015). This simplified view of the original process model delineates three hierarchical and sequential processes that change with age and their distinctive neural mechanisms: affect, integration, and motivation. The *Changes in Integration for Social Decisions in Aging* (*CISDA* Model provides an extension and specification to *AIM* and is a conceptual framework for understanding the social decision-making in aging (Frazier et al., 2019). CISDA identifies age-related changes to integration processes as a central mechanism for changes to social decision-making in aging. In *AIM* and *CISDA*, the integration phase of decision-making is understood as a neurocognitive process of applying mental functions to evaluate and combine different forms of choice-relevant signals/information to determine choice values (e.g., context-based interpretation of affective response, weighting possible outcomes with their probabilities). *CISDA* describes how three specific integrative processes, with different trajectories of change with age, may explain alterations to social decision-making in aging. These three integrative processes are social-emotional processes, learning/memory systems, and theory of mind

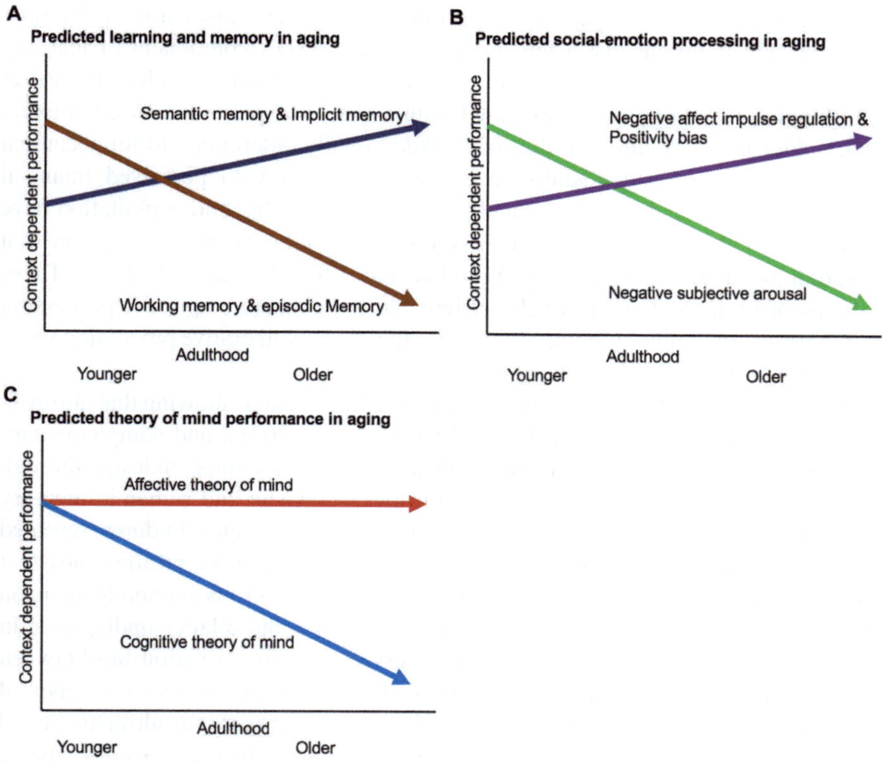

Fig. 3 CISDA proposes stability or change in decision-making performance as a function of aspects of learning and memory, social-emotional processing, and theory of mind. (**a**) Performance reliant on semantic or implicit memory is predicted to improve over adulthood, while performance reliant on working or episodic memory is predicted to decline. (**b**) Performance reliant on the ability to regulate negative affective impulses is predicted to improve across adulthood while performance reliant on negative subjective arousal is predicted to decline. (**c**) Performance reliant on affective theory of mind is predicted to remain stable across adulthood while performance reliant on cognitive theory of mind is predicted to decline. (Republished from Frazier et al. (2019). Permission conveyed through Copyright Clearance Center, Inc)

The framework further highlights three components of choice value integration that are common in social decision-making and have unique, complex trajectories of change with age due to brain aging and age-related motivational shifts: learning and memory, social-emotional processing, and theory of mind. Differences in these component change trajectories, depicted in Fig. 3, are due to differences in age-related patterns of decline in the components' underlying neural mechanisms, as well as adult developmental shifts in motivation.

Recent findings that support the CISDA model include research demonstrating that decline in the brain's episodic memory system is likely to impair social decision-making when optimal decisions rely on learning from specific past experiences, accurate recollection of relatively recent history, or simultaneous integration of multiple pieces of information to determine factors like social partner reliability. Here we will highlight just a few, relating to the CISDA component of learning and memory. First, as mentioned above, recent research has revealed that lower entorhinal cortex volume is associated with financial exploitation in cognitively intact older adults (Fenton et al., 2024). Additionally, alterations to hippocampal connectivity among older adults have been associated with perceived financial exploitation (Spreng et al., 2016), and financial vulnerability and exploitation have been associated with lower gray matter volumes and white matter integrity in medial temporal lobe regions impacted by AD (Han et al., 2016; Lamar et al., 2020). These patterns of brain decline are likely attributable to declarative episodic processing that is critical for deliberation, leading to a reduction in deliberative processing when evaluating scam- and fraud-related communication.

An interesting tie to these findings includes recent research showing that altruistic behavior generally increases with age (Sparrow et al., 2021), and complementary work finding that increased financial altruism in older adults, indexed through economic games, has been associated with lower episodic and semantic memory capacity (Weissberger et al., 2022). Importantly, these latter findings revealed that greater financial altruism was associated with a cognitive profile consistent with early Alzheimer's disease, indicating that overly generous economic decision behavior may be an early sign of the disease. Notably, these latter findings are in direct contrast to work on altruism that observed a positive relationship between altruism and cognitive performance when altruistic behavior was self-reported (Corrêa et al., 2019, 2020; James, 2011). Likewise, self-reported altruistic social activity is associated with greater gray matter volume in the posterior insula, middle cingulate gyrus, hippocampus, thalamus, superior temporal gyrus, anterior orbital gyrus, and middle occipital gyrus in a study of 8695 older adults (Liu et al., 2022). This mixed finding may reflect over-endorsement of altruistic social behaviors in studies that use self-report (e.g., social desirability effects). Another possibility is that practicing altruistic and other prosocial behaviors leads to positive health outcomes that benefit the brain (e.g., through positive social relationships, psychological well-being from sense of purpose) or the fact that those with lower cognition, including poor memory capacity, have a limited ability to exhibit some kinds of altruistic acts in the real world. Specifically, this study measured social altruistic behavior using two questions: Do you visit your friends' homes? Do you sometimes initiate conversations with younger people? It is sensible that decline in gray matter and cognition would limit one's ability to visit friends, and perhaps also initiate conversations with others. Thus, it is presently unclear whether the neurocognitive mechanism that drives exploitation vulnerability in normal aging and neurodegenerative disorders such as AD includes direct or indirect roles for learning and memory, or misplaced altruistic behaviors, indicating these will be important avenues for future research.

6 Future Directions for Research on Decision Neuroscience of Aging

In closing, 20 years after the conception of decision neuroscience of aging as a field of study, we find that many basic questions have been answered. We now understand how components of decision-making are impacted by brain aging, and we are beginning to clarify the ways in which these trajectories differ in states of age-related neurodegenerative disease. The area of study has multiple evidence-based, theoretical models that are helping to drive the field forward. And most importantly, we have expansions and translations from basic research toward more applied contexts that can provide solutions to real-world problems faced by older adults, their loved ones, and society. This review touches on a number of areas that require further research to advance the field.

First, the value of meta-analyses and studies with large samples and longitudinal methods is clearly demonstrated in this review. Age carries many stereotypes that can persist into research hypotheses over many empirical studies. Utilization of rigorous research approaches has already clarified that many aspects of decision-making undergo little to no change in normal aging. This brings us to the value of including clinical/pre-clinical populations in decision neuroscience of aging research. The inclusion of these populations has already shown that individuals with age-related neurodegenerative diseases may express alterations to decision-making that healthy older adults do not. These studies have also suggested that impaired decision-making might be one of the earliest signs of AD (Boyle et al., 2019; Stewart et al., 2019). With respect to new branches of research, understanding how older adults make economic decisions in modern technological contexts will be increasingly important. Cohort differences in familiarity with modern technologies persist in the "digital divide," indicating that age differences will be present in a number of new, technology-mediated decision contexts (Wang & Lighthall, 2025). As more of modern life takes place through digital devices, AI-mediation, and robotics, we must consider how these factors impact decision processing and behavior across the lifespan. One area of current research is how decision aids that use artificial intelligence can be most effectively and ethically utilized to support decision-making in aging (Chu et al., 2022). This frontier calls for the field to collaborate with additional disciplines, including engineering and computer science in order to make an impact over the next 20 years.

References

Bäckman, L., Lindenberger, U., Li, S.-C., & Nyberg, L. (2010). Linking cognitive aging to alterations in dopamine neurotransmitter functioning: recent data and future avenues. *Neuroscience and Biobehavioral Reviews, 34*, 670–677.

Boyle, P. A., Yu, L., Schneider, J. A., Wilson, R. S., & Bennett, D. A. (2019). Scam awareness related to incident Alzheimer dementia and mild cognitive impairment: A prospective cohort study. *Annals of Internal Medicine, 170*, 702–709.

Braak, H., & Braak, E. (1991). Neuropathological stageing of Alzheimer-related changes. *Acta Neuropathologica, 82*(4), 239–259.

Carstensen, L. L., Isaacowitz, D. M., & Charles, S. T. (1999). Taking time seriously: A theory of socioemotional selectivity. *American Psychologist, 54*(3), 165–181.

Castel, A. D., Benjamin, A. S., Craik, F. I. M., & Watkins, M. J. (2002). The effects of aging on selectivity and control in short-term recall. *Memory & Cognition, 30*(7), 1078–1085.

Castrellon, J. J., Seaman, K. L., Crawford, J. L., Young, J. S., Smith, C. T., Dang, L. C., et al. (2019). Individual differences in dopamine are associated with reward discounting in clinical groups but not in healthy adults. *Journal of Neuroscience, 39*(2), 321–332.

Chen, H. Y., Hsu, M., Jagust, W. J., & Berry, A. S. (2023). Graph theoretical analysis demonstrates associations among cortical network segregation, β-amyloid status, and reward-based learning in older age. *Alzheimer's & Dementia, 19*, e071996.

Chu, C. H., Nyrup, R., Leslie, K., Shi, J., Bianchi, A., Lyn, A., et al. (2022). Digital ageism: Challenges and opportunities in artificial intelligence for older adults. *The Gerontologist, 62*(7), 947–955.

Ciampa, C. J., Parent, J. H., Lapoint, M. R., Swinnerton, K. N., Taylor, M. M., Tennant, V. R., et al. (2022). Elevated dopamine synthesis as a mechanism of cognitive resilience in aging. *Cerebral Cortex, 32*(13), 2762–2772.

Cohen, M. S., Rissman, J., Suthana, N. A., Castel, A. D., & Knowlton, B. J. (2016). Effects of aging on value-directed moulation of semantic network activity during verbal learning. *NeuroImage, 125*, 1046–1062.

Corrêa, J. C., Ávila, M. P., Lucchetti, A. L., & Lucchetti, G. (2019). Altruistic behaviour, but not volunteering, has been associated with cognitive performance in community-dwelling older persons. *Psychogeriatrics, 19*(2), 117–125.

Corrêa, J. C., Ávila, M. P. W., Lucchetti, A. L. G., & Lucchetti, G. (2020). Altruism, volunteering and cognitive performance among older adults: A 2-year longitudinal study. *Journal of Geriatric Psychiatry and Neurology, 35*(1), 66–77.

Eppinger, B., Hämmerer, D., & Li, S. (2011). Neuromodulation of reward-based learning and decision making in human aging. *Annals of the New York Academy of Sciences., 1235*(1), 1–17.

Eppinger, B., Nystrom, L. E., & Cohen, J. D. (2012). Reduced sensitivity to immediate reward during decision-making in older than younger adults. *PLoS One, 7*(5), e36953.

Eppinger, B., Heekeren, H. R., & Li, S. C. (2018). Age differences in the neural mechanisms of intertemporal choice under subjective decision conflict. *Cerebral Cortex, 28*(11), 3764–3774.

Fenton L, Salminen LE, Lim AC, Weissberger GH, Nguyen AL, Axelrod J, Noriega-Makarskyy D, Yassine H, Mosqueda L, Han SD. (2024). Lower entorhinal cortex thickness is associated with greater financial exploitation vulnerability in cognitively unimpaired older adults. *Cerebral Cortex*. 34(9):bhae36.0

Fjell, A. M., McEvoy, L., Holland, D., Dale, A. M., & Walhovd, K. B. (2014). Alzheimer's disease neuroimaging initiative. What is normal in normal aging? Effects of aging, amyloid and Alzheimer's disease on the cerebral cortex and the hippocampus. *Progress in Neurobiology, 117*, 20–40.

Fjell, A. M. (2024). Aging brain from a lifespan perspective. In *Current topics in behavioral neurosciences*. Springer.

Fjell, A. M., Westlye, L. T., Grydeland, H., Amlien, I., Espeseth, T., Reinvang, I., Raz, N., et al. (2013). Critical ages in the life course of the adult brain: Nonlinear subcortical aging. *Neurobiology of Aging, 34*(10), 2239–2247.

Fjell, A. M., Grydeland, H., Krogsrud, S. K., Amlien, I., Rohani, D. A., Ferschmann, L., et al. (2015). Development and aging of cortical thickness correspond to genetic organization patterns. *Proceedings of the National Academy of Sciences, 112*(50), 15462–15467.

Fraser, M. A., Shaw, M. E., & Cherbuin, N. (2015). A systematic review and meta-analysis of longitudinal hippocampal atrophy in healthy human ageing. *NeuroImage, 112*, 364–374.

Frazier, I., Lighthall, N. R., Horta, M., Perez, E., & Ebner, N. C. (2019). CISDA: Changes in integration for social decisions in aging. *Wiley Interdisciplinary Reviews: Cognitive Science, 10*(3), e1490.

Freeman, S. H., Kandel, R., Cruz, L., Rozkalne, A., Newell, K., Frosch, M. P., et al. (2008). Preservation of neuronal number despite age-related cortical brain atrophy in elderly subjects without Alzheimer disease. *Journal of Neuropathology & Experimental Neurology, 67*(12), 1205–1212. https://doi.org/10.1007/7854_2024_476

Gasiorowska, A., Wydrych, M., Drapich, P., Zadrozny, M., Steczkowska, M., Niewiadomski, W., & Niewiadomska, G. (2021). The biology and pathobiology of glutamatergic, cholinergic, and dopaminergic signaling in the aging brain. *Frontiers in Aging Neuroscience, 13*, 654931.

Green, M. A., Crawford, J. L., Kuhnen, C. M., Samanez-Larkin, G. R., & Seaman, K. L. (2023). Multivariate associations between dopamine receptor availability and risky investment decision-making across adulthood. *Cerebral Cortex Communications, 4*(2), tgad008.

Han, D. S., Boyle, P. A., Yu, L., Arfanakis, K., James, B. D., Fleischman, D. A., & Bennett, D. A. (2016). Grey matter correlates of susceptibility to scams in community-dwelling older adults. *Brain Imaging and Behavior, 10*, 524–532.

Hennessee, J. P., Reggente, N., Cohen, M. S., Rissman, J., Castel, A. D., & Knowlton, B. J. (2019). White matter integrity in brain structures supporting semantic processing is associated with value-directed remembering in older adults. *Neuropsychologia, 129*, 246–254.

Hess, T. M., Strough, J., & Löckenhoff, C. (Eds.). (2015). *Aging and decision making: Empirical and applied perspectives*. Academic.

James, R. N. (2011). Charitable giving and cognitive ability. *International Journal of Nonprofit and Voluntary Sector Marketing, 16*(1), 70–83.

Juarez, E. J., et al. (2019). Reproducibility of the correlative triad among aging, dopamine receptor availability, and cognition. *Psychology and Aging, 34*(7), 921.

Karalija, N., Johansson, J., Papenberg, G., Wåhlin, A., Salami, A., Köhncke, Y., et al. (2022). Longitudinal dopamine D2 receptor changes and cerebrovascular health in aging. *Neurology, 99*(12), e1278–e1289.

Karalija, N., Papenberg, G., Johansson, J., Wåhlin, A., Salami, A., Andersson, M., et al. (2024). Longitudinal support for the correlative triad among aging, dopamine D2-like receptor loss, and memory decline. *Neurobiology of Aging, 136*, 125–132.

Karrer, T. M., Josef, A. K., Mata, R., Morris, E. D., & Samanez-Larkin, G. R. (2017). Reduced dopamine receptors and transporters but not synthesis capacity in normal aging adults: A meta-analysis. *Neurobiology of Aging, 57*, 36–46.

Kipps, C. M., Hodges, J. R., Fryer, T. D., & Nestor, P. J. (2009). Combined magnetic resonance imaging and positron emission tomography brain imaging in behavioural variant frontotemporal degeneration: Refining the clinical phenotype. *Brain, 132*(Pt 9), 2566–2578.

Klostermann, E. C., Braskie, M. N., Landau, S. M., O'Neil, J. P., & Jagust, W. J. (2012). Dopamine and frontostriatal networks in cognitive aging. *Neurobiology of Aging, 33*, 623.e15–623.e24.

Kryla-Lighthall, N., & Mather, M. (2009). The role of cognitive control in older adults' emotional well-being. In V. Bengtson, M. Silverstein, N. Putney, & D. Gans (Eds.), *Handbook of theories of aging* (pp. 323–344). Springer Press.

Lamar, M., Arfanakis, K., Yu, L., Zhang, S., Han, S. D., Fleischman, D. A., et al. (2020). White matter correlates of scam susceptibility in community-dwelling older adults. *Brain Imaging and Behavior, 14*, 1521–1530.

Lee, S., Parthasarathi, T., Cooper, N., Zauberman, G., Lerman, C., & Kable, J. W. (2022). A neural signature of the vividness of prospective thought is modulated by temporal proximity during intertemporal decision making. *Proceedings of the National Academy of Sciences, 119*(44), e2214072119.

Lempert, K. M., MacNear, K. A., Wolk, D. A., & Kable, J. W. (2020a). Links between autobiographical memory richness and temporal discounting in older adults. *Scientific Reports, 10*(1), 6431.

Lempert, K. M., Mechanic-Hamilton, D. J., Xie, L., Wisse, L. E., de Flores, R., Wang, J., et al. (2020b). Neural and behavioral correlates of episodic memory are associated with temporal discounting in older adults. *Neuropsychologia, 146*, 107549.

Lighthall, N. (2020). Neural mechanisms of decision-making in aging. *Wiley Interdisciplinary Reviews: Cognitive Science., 11*(1), e1519.

Lim, K. T., & Yu, R. (2015). Aging and wisdom: Age-related changes in economic and social decision making. *Frontiers in Aging Neuroscience, 7*. https://doi.org/10.3389/fnagi.2015.00120

Liu, Y., Zhang, Y., Thyreau, B., Tatewaki, Y., Matsudaira, I., Takano, Y., Hirabayashi, N., Furuta, Y., Hata, J., Ninomiya, T., Taki, Y., Ninomiya, T., Hata, J., Shibata, M., Honda, T., Ohara, T., Nakaji, S., Murashita, K., Mikami, T., et al. (2022). Altruistic social activity, depressive symptoms, and brain regional gray matter volume: Voxel-based morphometry analysis from 8,695 old adults. *The Journals of Gerontology: Series A, 77*(9), 1789–1797. https://doi.org/10.1093/gerona/glac093

Mather, M. (2006). A review of decision-making processes: Weighing the risks and benefits of aging. *When I'm, 64*(145), 145–173.

Mather, M., & Carstensen, L. L. (2005). Aging and motivated cognition: The positivity effect in attention and memory. *Trends in Cognitive Sciences, 9*(10), 496–502.

Nandhagopal, R., Kuramoto, L., Schulzer, M., Mak, E., Cragg, J., McKenzie, J., et al. (2011). Longitudinal evolution of compensatory changes in striatal dopamine processing in Parkinson's disease. *Brain, 134*(11), 3290–3298.

National Institute on Aging. (2006a). *Neuroeconomics of aging workshop*. https://www.nia.nih.gov/sites/default/files/d7/meeting-report_28.pdf. Date last accessed 15 Jan 2025.

National Institute on Aging. (2006b). *Neuroeconomics of aging RFA*. https://grants.nih.gov/grants/guide/rfa-files/rfa-ag-06-011.html. Date last accessed 15 Jan 2025.

Nielsen, L., & Mather, M. (2011). Emerging perspectives in social neuroscience and neuroeconomics of aging. *Social Cognitive and Affective Neuroscience, 6*(2), 149–164.

Nieuwenhuis, S., Ridderinkhof, K. R., Talsma, D., Coles, M. G., Holroyd, C. B., Kok, A., & Van der Molen, M. W. (2002). A computational account of altered error processing in older age: Dopamine and the error-related negativity. *Cognitive, Affective, & Behavioral Neuroscience, 2*, 19–36.

Noh, S. M., Singla, U. K., Bennett, I. J., & Bornstein, A. M. (2023). Memory precision and age differentially predict the use of decision-making strategies across the lifespan. *Scientific Reports, 13*(1), 17014.

Owens, M. M., Gray, J. C., Amlung, M. T., Oshri, A., Sweet, L. H., & MacKillop, J. (2017). Neuroanatomical foundations of delayed reward discounting decision making. *NeuroImage, 161*, 261–270.

Peters, J., & Büchel, C. (2011). The neural mechanisms of inter-temporal decision-making: Understanding variability. *Trends in Cognitive Sciences, 15*(5), 227–239.

Raz, N., Lindenberger, U., Rodrigue, K. M., Kennedy, K. M., Head, D., Williamson, A., et al. (2005). Regional brain changes in aging healthy adults: General trends, individual differences and modifiers. *Cerebral Cortex, 15*(11), 1676–1689.

Reggente, N., Cohen, M. S., Zheng, Z. S., Castel, A. D., Knowlton, B. J., & Rissman, J. (2018). Memory recall for high reward value items correlates with individual differences in white matter pathways associated with reward processing and Fronto-temporal communication. *Frontiers in Human Neuroscience, 12*, 241.

Salthouse, T. A. (2004). What and when of cognitive aging. *Current Directions in Psychological Science, 13*(4), 140–144.

Samanez-Larkin, G. R., & Knutson, B. (2015). Decision making in the ageing brain: Changes in affective and motivational circuits. *Nature Reviews Neuroscience, 16*(5), 278–289.

Samanez-Larkin, G. R., Mata, R., Radu, P. T., Ballard, I. C., Carstensen, L. L., & McClure, S. M. (2011). Age differences in striatal delay sensitivity during intertemporal choice in healthy adults. *Frontiers in Neuroscience, 5*, 126.

Sanchez-Roige, S., Fontanillas, P., Elson, S. L., 23 and Me Research Team, Pandit, A., Schmidt, E. M., et al. (2018). Genome-wide association study of delay discounting in 23,217 adult research participants of European ancestry. *Nature Neuroscience, 21*(1), 16–18.

Schultz, W., Dayan, P., & Montague, P. R. (1997). A neural substrate of prediction and reward. *Science, 275*(5306), 1593–1599.

Seaman, K. L., Brooks, N., Karrer, T. M., Castrellon, J. J., Perkins, S. F., Dang, L. C., et al. (2018). Subjective value representations during effort, probability and time discounting across adulthood. *Social Cognitive and Affective Neuroscience, 13*(5), 449–459.

Seaman, K. L., Abiodun, S. J., Fenn, Z., Samanez-Larkin, G. R., & Mata, R. (2022). Temporal discounting across adulthood: A systematic review and meta-analysis. *Psychology and Aging, 37*(1), 111–124.

Sinclair, A. H., Stanley, M. L., Hakimi, S., Cabeza, R., Adcock, R. A., & Samanez-Larkin, G. R. (2021). Imagining a personalized scenario selectively increases perceived risk of viral transmission for older adults. *Nature Aging, 1*(8), 677–683.

Sparrow, E. P., Swirsky, L. T., Kudus, F., & Spaniol, J. (2021). Aging and altruism: A meta-analysis. *Psychology and Aging, 36*(1), 49–56.

Spreng, R. N., Karlawish, J., & Marson, D. C. (2016). Cognitive, social, and neural determinants of diminished decision-making and financial exploitation risk in aging and dementia: A review and new model. *Journal of Elder Abuse & Neglect, 28*(4–5), 320–344.

Stewart, C. C., Yu, L., Wilson, R. S., Bennett, D. A., & Boyle, P. A. (2019). Healthcare and financial decision making and incident adverse cognitive outcomes among older adults. *Journal of the American Geriatrics Society, 67*, 1590–1595.

Tisdall, L., & Seaman, K. L. (2024, June 26). *Decision making in the aging brain: An update of the evidence base for the 'affect-integration-motivation' framework.* https://doi.org/10.31234/osf.io/tnbky

Tisdall, L., & Seaman, K. L. (this volume). Decision-making in the aging brain: An update of the evidence base for the "affect-integration-motivation" framework. In D. V. Smith, P. Lockwood, & D. S. Fareri (Eds.), *Neuroeconomics: Core topics and current directions*. Springer.

Verharen, J. P., Adan, R. A., & Vanderschuren, L. J. (2019). Differential contributions of striatal dopamine D1 and D2 receptors to component processes of value-based decision making. *Neuropsychopharmacology, 44*(13), 2195–2204.

Wang, S., & Lighthall, N. R. (2025, in press). *The role of age-related changes to memory in judgment and decision making involving artificial agents*. European Psychologist.

Weissberger, G. H., Samek, A., Mosqueda, L., Nguyen, A. L., Lim, A. C., Fenton, L., & Han, S. D. (2022). Increased financial altruism is associated with Alzheimer's disease neurocognitive profile in older adults. *Journal of Alzheimer's Disease, 88*(3), 995–1005.

Wong, S., Irish, M., Savage, G., Hodges, J. R., Piguet, O., & Hornberger, M. (2019). Strategic value-directed learning and memory in Alzheimer's disease and behavioural-variant frontotemporal dementia. *Journal of Neuropsychology, 13*(2), 328–353.

Wood, S., & Lichtenberg, P. A. (2017). Financial capacity and financial exploitation of older adults: Research findings, policy recommendations and clinical implications. *Clinical Gerontologist, 40*(1), 3–13.

Zigmond, M. J., Abercrombie, E. D., Berger, T. W., Grace, A. A., & Stricker, E. M. (1990). Compensations after lesions of central dopaminergic neurons: Some clinical and basic implications. *Trends in Neurosciences, 13*(7), 290–296.

Zuppichini, M. D., Hamlin, A. M., Zhou, Q., Kim, E., Rajagopal, S., Beltz, A. M., & Polk, T. A. (2024). GABA levels decline with age: A longitudinal study. *Imaging Neuroscience, 2*, 1–15.

Decision-Making in the Aging Brain: An Update of the Evidence Base for the "Affect-Integration-Motivation" Framework

Loreen Tisdall and Kendra L. Seaman

Abstract Shaped by accumulating experience, shifting goals, and neurobiological development, decision-making changes across the adult human lifespan. The Affect-Integration-Motivation (AIM) framework has served as a key model for understanding how affective and motivational processes, along with biological and neural processes, influence decision behavior. This chapter revisits and updates the AIM framework in light of recent neuroeconomic research focused on age-related change. Specifically, we summarize the current evidence on age-related differences in four critical aspects of decision-making under uncertainty, including (1) value learning, (2) risky choice, (3) temporal discounting, and (4) effort-based decision-making, addressing both behavioral patterns and their neural substrates. In addition, we highlight key gaps in the extant literature and argue that advancing the AIM framework requires integrating it with broader ecological models of decision-making and accounting for individual differences within age groups. This updated perspective aims to sharpen our understanding of how aging affects the motivational and affective foundations of decision behavior.

1 Introduction

Individuals' goals and motivations change across their life spans, which may affect decision-making as individuals perceive a dwindling number of opportunities to

---

The authors would like to thank Brian Knutson for his thoughtful comments and feedback on earlier drafts of this work.

L. Tisdall
Center for Cognitive and Decision Sciences, University of Basel, Basel, Switzerland

K. L. Seaman (✉)
Center for Vital Longevity, University of Texas at Dallas, Richardson, TX, USA

Department of Psychology, University of Texas at Dallas, Richardson, TX, USA
e-mail: kendra.seaman@utdallas.edu

D. V. Smith et al. (eds.), *Neuroeconomics: Core Topics and Current Directions*,
https://doi.org/10.1007/978-3-032-02925-6_29

compensate for negative consequences or other losses (Buchinger et al., 2022; Carstensen et al., 1999; Ravert et al., 2019). Moreover, with advancing age, changes are also occurring at the biological level, including brain structure and function (Bethlehem et al., 2022; Cabeza, 2002). This may bring about age differences in the cognitive and affective decision-making processes. To advance systematic neuroeconomic research, Samanez-Larkin and Knutson proposed the influential *Affect-Integration-Motivation (AIM) framework* 10 years ago, which described the brain circuits supporting the affective and motivational processes that influence decision-making (Samanez-Larkin & Knutson, 2015). Crucially, they explained how age-related changes to these circuits could impact older adults' decision-making. This chapter will serve as an update to this seminal paper. In the first part of the chapter, we briefly review the AIM framework before providing an updated literature review describing age-related changes in these associated brain circuits and how these changes impact decision-making with age. In the second part, we identify some main gaps in extant work and outline approaches to close them.

2 The Affect-Integration-Motivation Framework

Converging evidence suggests that multiple frontostriatal circuits support decision-making, and the AIM framework specifies how these brain circuits support different phases of the decision-making process through hierarchical, sequential processing (Fig. 1). First, anticipated gains and losses are encoded by regions of the brain that support affective processing. The ventral striatum (VS) and other regions innervated by dopaminergic projections from the ventral tegmental area (VTA) encode the anticipation of gains. In contrast, the anterior insula (AIns) encodes the anticipation of losses through noradrenergic connections with the locus coeruleus (LC). Next, these anticipated gain and loss signals, along with any other relevant information, are integrated into an expected value signal via indirect glutaminergic projections to the prefrontal cortex. This integration of potential gains and losses is thought to be supported by a core network of brain regions including the medial prefrontal cortex (MPFC) and posterior parietal cortex (PPC). Finally, this integrated signal provides motivation to act through motor circuits connecting the dorsal striatum and premotor cortex. Once an outcome has resolved, the actual experience of a gain or a loss is encoded in affective regions and the MPFC. Age-related changes in the structure and function of these brain networks could have implications for the decision-making abilities of older adults.

Since the AIM framework was published, there has been additional support for this framework and its role in supporting decision-making. For instance, diffusion-weighted imaging (DWI) studies have shown that white matter connections between these regions exist in humans and support decision-making across the lifespan (Han et al., 2018; Leong et al., 2016). Further, positron emission tomography studies have shown that the density of D2-like dopamine receptors are relatively preserved in the VS and MPFC across adulthood (Seaman et al., 2019) and that

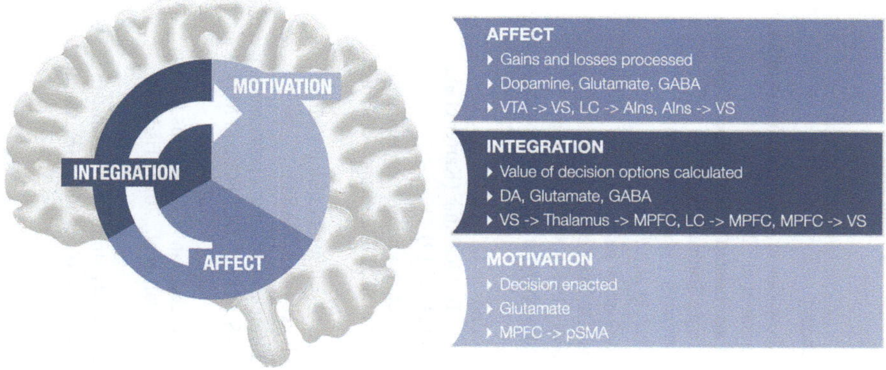

Fig. 1 Schematic of the AIM framework, outlining the sequential processes involved in decision-making: Affect, Integration, and Motivation. The diagram highlights the key neurotransmitter systems and associated brain regions relevant to each stage. *AIns* anterior insula, *GABA* gamma-aminobutyric acid, *LC* locus coeruleus, *MPFC* medial prefrontal cortex, *pSMA* pre-supplementary motor area, *VS* ventral striatum, *VTA* ventral tegmental area

individual differences in dopamine function are associated with decision-making (Castrellon et al., 2019). Similarly, functional magnetic resonance imaging (FMRI) and electroencephalography (EEG) studies have provided support for the integration of different types of information in the MPFC (Fernandes et al., 2022; Seaman et al., 2017, 2018). However, there have also been methodological developments, particularly in the study of effort-based decision-making, that have enabled a more nuanced understanding of these brain circuits to emerge. Specifically, FMRI studies have begun to parse these brain regions more finely, finding that distinct subregions of the VS (Suzuki et al., 2021) and MPFC (Clairis & Pessiglione, 2024; Yee et al., 2021) may encode different aspects of choice. However, as researchers have begun to use the AIM framework to make predictions, some weaknesses have also emerged. Notably, the temporal resolution of FMRI and the spatial resolution of EEG limit our ability to pull apart these distinct processing stages and may require specifically staged tasks to decompose these processes (Schonberg et al., 2010). Also, while useful, the AIM framework does not make specific predictions about how these brain circuits determine which action to pursue when posed with two (or more) options.

3 Changes in Decision-Making with Age

Following three of the main research topics covered in the original AIM framework review (Samanez-Larkin & Knutson, 2015), we outline recent work in value learning, risky choice, and temporal discounting below. We also introduce a new paragraph on effort-based decision-making, a growing research topic highly relevant for age differences in decision-making. This information is summarized in Table 1.

Table 1 Updated evidence for AIM framework claims of age-related changes in decision-making

Claim area	Original claim (2015)	Type of evidence	Updated evidence
Value learning	Age-related compromises in value integration account for deficits in value learning rather than age-related differences in gain anticipation	Behavioral	Some studies show slower value learning in older adults (e.g., Nassar et al., 2016) Others suggest age-related deficits depend on feedback type (e.g., Sojitra et al., 2018) Individual differences within older adults observed (Seaman et al., 2023)
		Neural	Attenuated anticipatory gain signals in VMPFC for older adults (De Boer et al., 2017) No age differences in reward prediction error-associated activity in some studies (Seaman et al., 2023) Age-related declines in D1 and D2 receptor availability may affect learning (Karrer et al., 2017; Seaman et al., 2019)
Risky decision-making	Age differences in risky choice are driven more by cognitive limitations than by differences in risk preference.	Behavioral	No robust age effects found in comprehensive reviews (Bagaïni et al., 2023; Frank & Seaman, 2023) Mixed results across different task paradigms (e.g., lottery choice tasks, BART)
		Neural	Age-related differences in neural processing of probability in VS, MPFC, DLPFC, and MTL (Su et al., 2018) Age-related differences in VS and AIns activity during BART (Tisdall et al., 2024; Tisdall & Mata, 2023) Increased connectivity strength in putamen functional network for older adults (Ren et al., 2023)

Temporal discounting	Older adults tend to discount less than younger adults, potentially due to reduced anticipation of immediate gains	Behavioral	Recent meta-analyses show small or negligible age differences in temporal discounting (Bagaïni et al., 2023; Seaman et al., 2022) Early consensus likely due to overestimation of effects in underpowered studies
		Neural	No age differences in neural subjective value representations across the adult lifespan (Seaman et al., 2018) Structural and functional connectivity in frontostriatal networks related to temporal discounting in older adults (Halfmann et al., 2015; Hänggi et al., 2016) Temporal discounting linked to entorhinal cortical thickness in older adults and those with cognitive impairment (Lempert et al., 2020a, b)
Effort-based decision-making	Not specifically addressed in the 2015 AIM framework	Behavioral	Recent meta-analysis shows significant increase in cognitive effort discounting with age (Bagaïni et al., 2023) Increased effort discounting linked to less engagement with demanding tasks in daily life (Crawford et al., 2023)
		Neural	Preserved representation of subjective value during effort discounting in MPFC and VS across lifespan, but age differences in medial temporal regions (Seaman et al., 2018) Effort discounting related to amyloid burden in healthy older adults and those with preclinical AD pathology (Aschenbrenner et al., 2023)

Note. The table was created using Claude (Anthropic, 2024), based on the authors' original writing

3.1 Changes in Value Learning

The value of a choice option is often not innate but must often be learned. If aging affects learning, decision-making that relies upon value learning over time may pose challenges, particularly if values must be updated to reflect a changing and dynamic environment. Research on age differences in value learning shows mixed results but generally indicates slower value learning in older adults (Eppinger et al., 2011). Such age-related differences in learning may stem from age-related differences in gain anticipation and value integration. Researchers often study reward prediction error, the difference between expected and actual reward outcomes, to investigate these possible mechanisms. In their review, Samanez-Larkin and Knutson (2015) concluded that age-related compromises in value integration can account for deficits in value learning rather than age-related differences in gain anticipation. For example, extant work demonstrated lower reward prediction error activity in VS and MPFC for older compared to younger adults but preserved VS activity for older versus younger in tasks assessing gain anticipation (Chowdhury et al., 2013; Samanez-Larkin et al., 2014). Results from DWI research yield evidence for diminished integrity of frontostriatal pathways. This also supports the notion of age effects on value integration since diminished frontostriatal integrity can fully statistically account for age-related differences in probabilistic reward learning (Samanez-Larkin et al., 2012).

3.1.1 Behavioral Differences

In uncertain situations, age differences in learning have been observed, yet the exact mechanisms are still debated and may depend on contextual features. For example, a comparison of younger and older adults completing a location-inference task based on previous outcomes (Nassar et al., 2016) suggests that learning deficits observed in healthy older adults may be driven by a diminished capacity to represent and use uncertain information to guide learning. In other words, instead of an overall impairment in the ability to learn from reward prediction errors, older adults may underestimate how much they need to learn in the first place. In line with previous work, learning rates in this study were found to be more variable for older individuals than for younger ones (Nassar et al., 2016).

Recent work (Sojitra et al., 2018) further supports the idea that learning declines with age but that the mechanisms of age-related deficits in learning may depend upon feedback. Specifically, learning from loss feedback (i.e., punishment) seems to take longer in older adults due to higher decision noise. In contrast, age-related deficits in learning from gain feedback (i.e., reward) seem to be due to age-related increases in the imbalance between gain and loss learning rates. However, in the context of learning in a social setting, recent computational work (Seaman et al., 2023) suggests that *both* older and younger adults learn differently from gain and loss feedback. Further, no systematic age differences existed in *how* older and younger adults learn from gain and loss feedback. Instead, they found individual differences within older adults: some older adults displayed learning, while others did not. Recent efforts to examine age differences in learning and potential cognitive mechanisms collectively point toward valence-specific learning deficits and substantial individual differences.

3.1.2 Neural Differences

Since the AIM framework was initially proposed, additional insights have been gained concerning neural substrates underlying age differences in value learning. For example, results from a probabilistic reward learning task (De Boer et al., 2017) suggest that aging is associated with attenuated anticipatory gain signals in the ventral MPFC (VMPFC), and that the VMPFC anticipatory gain signal predicts performance. In light of results suggesting age differences in social learning described above, along with the absence of age-related differences in reward prediction error-associated activity in the striatum, insular, and frontal regions (Seaman et al., 2023), these results suggest age differences in anticipatory gain signaling, not age differences in reward prediction error, drive age differences in value learning.

Age differences in dopaminergic function also constitute a critical element of the AIM framework, with work supporting differential rates of age-related change in dopaminergic function across cortical and subcortical regions. Indeed, the strength of the VMPFC anticipatory gain signal in the study by De Boer et al. (2017) was modulated by dopamine D1 receptor availability in the VS. Critically, D1 and D2 receptor activity is thought to subserve increased and decreased activity related to reward prediction errors, respectively, with D1 receptor activity leading to a reinforcement of the behaviors which led to better-than-expected outcomes, and D2 receptor activity leading to an attenuation of the behaviors which led to worse-than-expected outcomes. Regarding aging, meta-analytic work points toward a strong negative association between dopamine D2 receptor availability and age (Karrer et al., 2017), with additional empirical studies qualifying this effect as regionally heterogeneous and potentially nonlinear (Seaman et al., 2019). Together, these studies provide a window into potential mechanisms, with age-related declines in D1 and D2 receptor availability potentiating changes in learning based on differences in anticipatory signal and reward prediction errors.

3.2 Changes in Risky Decision-Making

Decisions are typically made under uncertainty, where outcomes are not guaranteed and may include potential losses alongside possible gains. Understanding what determines risky choices is essential in the face of increasing human life expectancy since older adults must make consequential decisions concerning their health, wealth, and security (Delaney et al., 2021; Duell et al., 2018; Shook et al., 2021; Wolfe et al., 2021). Based on their review of the literature, Samanez-Larkin and Knutson (2015) proposed that age differences in risky choice appear to be driven more by cognitive limitations than by differences in risk preference. Accordingly, there is greater age-related variability in the value integration signal, particularly for projections from MPFC to the VS (Leong et al., 2016), which can destabilize expected value estimates and subsequently influence risky choice, especially in tasks that require learning or updating value estimates. Furthermore, the AIM framework proposes that the expected value anticipation correlates with VS and MPFC activity. However, risk anticipation is linked with activity in the AIns (e.g., Kuhnen & Knutson, 2005).

3.2.1 Behavioral Differences

Research on age differences in risky decision-making using behavioral measures has yielded mixed results. For example, in a set of studies using a lottery choice task, older adults engaged in more risk-taking behaviors than their younger counterparts (Goh et al., 2016; Su et al., 2018). In contrast, in a lottery choice task that included gamble sets comprising a risky versus a certain option (O'Brien & Hess, 2020), older adults were more likely than younger adults to choose the sure option over the risky one. The studies mentioned above all assessed decisions from description, where magnitudes and probabilities are known, and thus do not involve learning or updating. Another set of studies requires individuals to acquire a representation of the proposed magnitudes and probabilities, which is known as decisions from experience, and age differences in these studies have been comparably heterogeneous. For example, studies using the Balloon Analogue Risk Task (BART) have found comparable levels of risk-taking between younger and older adults (Peng et al., 2020; Tisdall & Mata, 2023; Yu et al., 2016) but lower risk-taking in older compared to younger adults (Ren et al., 2023). Comprehensive efforts have been made to review and synthesize a growing body of research using behavioral measures of risky choice (Fig. 2a; Bagaïni et al., 2023; Frank & Seaman, 2023), with the results suggesting no robust age effects and no evidence for task-specific learning or memory demands moderating age effects on risky choice.

3.2.2 Neural Differences

Recent research into the neural substrates of age-related differences in risky choice has added further nuance to our current understanding. For example, an FMRI paradigm simulating lottery choices (Su et al., 2018) showed that although the magnitude of an option had a comparable positive effect on VS activity for younger and older adults, the probability was processed differently in these age groups. In younger but not older adults, probability parametrically modulated activation in VS, while in older but not younger adults, probability had a linear effect on activity in the

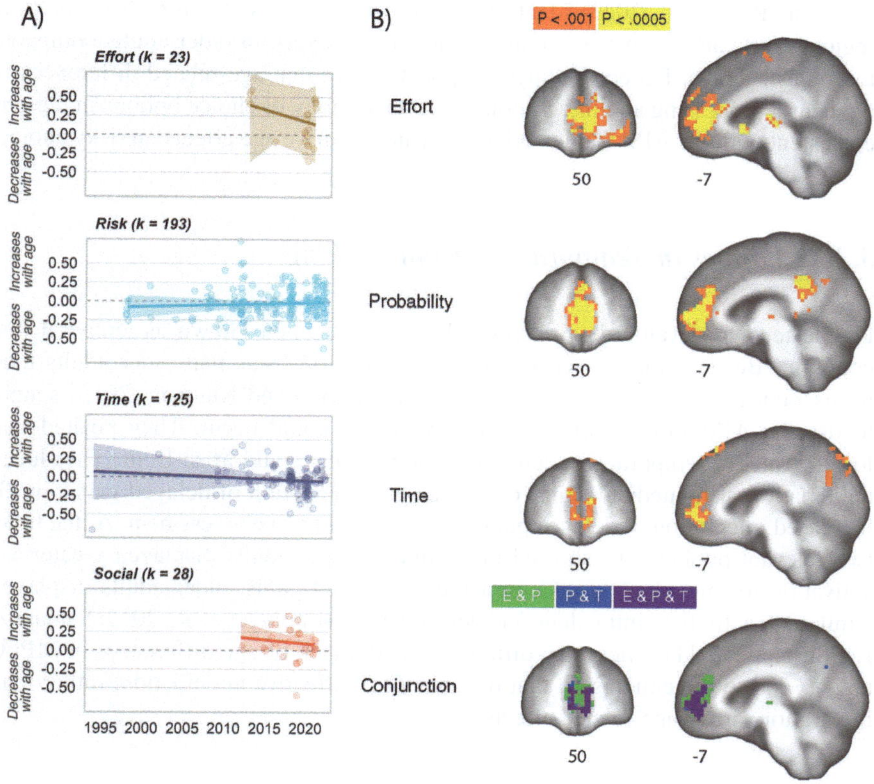

Fig. 2 Empirical evidence of constancy and change in decision-making with age. (**a**) Cumulative meta-analytic results of the effect of age on effort, risk, time, and social preferences. (Adapted from "Aging and Economic Preferences: Cumulative meta-analyses of age differences in risk, time, social, and effort preferences" by Bagaïni et al. (2023). Adapted with permission). (**b**) Neural representation of subjective value during effort, probability, and time tasks separately controlling for age (top three rows) and the conjunction of subjective value representation during effort and probability (green), probability and time (blue), and effort, probability, and time (purple) tasks (bottom row), controlling for age. (Adapted from "Subjective value representations during effort, probability and time discounting across adulthood," by Seaman et al. (2018). Adapted with permission). Please note that even though the components in **a** and **b** may be labeled with the same conceptual name, they are not necessarily operationalized in the same way

MPFC, dorsolateral prefrontal cortex (DLPFC), and medial temporal lobe. For the BART paradigm, recent work reported positive effects of age on VMPFC activity but negative effects on AIns activity (Tisdall et al., 2024; Tisdall & Mata, 2023) while also reporting a negative effect of age on VS activity, potentially representing reduced anticipatory gain signal in older adults. Going beyond individual studies, a meta-analysis of FMRI studies of decisions-from-experience paradigms in healthy older adults yielded clusters of common activation in bilateral DLPFC and left orbitofrontal cortex when decisions were made (Tannou et al., 2021). Further, recent research (Ren et al., 2023) identified increased connectivity strength in multiple regions associated with the putamen functional network in older adults compared to younger adults. Recent findings support the idea that age-related differences in the neural processing and integration of the uncertainty of choice options in regions consistent with the AIM framework may primarily drive age effects on risky choice.

3.3 Changes in Temporal Decision-Making

People tend to devalue delayed rewards, a phenomenon known as temporal discounting. Because the consensus in the field in 2015 was that older adults tend to discount less than younger adults, Samanez-Larkin and Knutson (2015) sought to use the AIM Framework to explain why this could occur. They posited that lower levels of temporal discounting in older adults could arise from (1) reduced anticipation of immediate gains, (2) increased anticipation of delayed gains, or (3) increased integration of delayed gains into an overall value assessment. At the time, two relevant publications showed that while younger adults displayed greater VS activation to immediate gains compared to delayed gains, older adults displayed similar VS activity to immediate and delayed gains (Eppinger et al., 2012; Samanez-Larkin et al., 2011). These age differences and a lack of age differences in MPFC activity led to the conclusion that differences in affective anticipation, rather than integration, drove age-related effects.

3.3.1 Behavioral Differences

Our understanding of age-related differences in temporal discounting has shifted dramatically since 2015. In terms of behavior, recent meta-analyses have shown that if there are age differences in temporal discounting, they are small ($r = -0.07$ to -0.04), not of practical importance, and unlikely to be detected using the typical sample sizes used in behavioral and neuroimaging research (Fig. 2a; Bagaïni et al., 2023; Seaman et al., 2022). Further, these reviews have shown that an erroneous early consensus was most likely due to a "winner's curse" in which seminal studies tended to be underpowered and overestimate effects of interest (Young et al., 2008). As described by Seaman et al. (2022) and then displayed by Bagaïni et al. (2023), the first study of temporal discounting in aging (Green et al., 1994) suggested a strong negative effect of age ($r = -0.72$) that was significantly greater than subsequent studies. Thus, the behavioral phenomenon Samanez-Larkin and Knutson attempted to explain in 2015 may not be robust.

3.3.2 Neural Differences

Our understanding of the neural mechanism underlying temporal discounting has also shifted dramatically over the last decade. While the neuroimaging studies cited in the original paper relied heavily on dual-processing models of decision-making (McClure et al., 2004), these models have largely fallen out of favor due to behavioral and neural evidence supporting single-system models (Kable & Glimcher, 2010). More recent studies have focused on single-system explanations that are also consistent with the AIM Framework. Regarding structure, two studies have shown that structural connectivity in frontostriatal networks is related to temporal discounting in older adults (Halfmann et al., 2015). Hänggi et al. (2016) also showed that functional connectivity within frontostriatal networks in older adults was related to temporal discounting. Others have shown that older adults who demonstrated impaired decision-making also showed diminished subjective value representations in the MPFC (Halfmann et al., 2015). However, these studies focused exclusively on older adults. When comparing behavior and neural subjective value representations across the adult lifespan ($N = 75$), Seaman et al. (2018) found no age differences (Fig. 2b). Collectively, these studies are consistent with the brain networks proposed by the AIM Framework, even if they do not support the specific age differences described originally.

Additional insights on temporal discounting have been gained by examining temporal discounting in samples that include participants with mild cognitive impairment, Alzheimer's disease (AD), and or other AD-related dementias. These studies have shown that neurological diseases that impact episodic memory show greater temporal discounting (Boutoleau-Bretonnière et al., 2023; Geng et al., 2020; Lempert et al., 2020a, b; Thoma et al., 2017). Further, two of these studies linked temporal discounting in older adults to entorhinal cortical thickness (Lempert et al., 2020a, b). This was also recently corroborated in a large-scale ($N = 1196$) study of younger adults from the Human Connectome Project (Garzón et al., 2023). These studies suggest that temporal discounting relies on the same neurocognitive machinery as memory, mainly autobiographical episodic memory. Although follow-up studies have contested this particular explanation (Garzón et al., 2023; Lempert et al., 2024), it remains possible that the neurocognitive systems involved in prospection, which rely on similar neural mechanisms as episodic memory and are impacted by neurodegenerative diseases, support intertemporal choice. Again, this is not inconsistent with the AIM Framework, which suggests information from memory systems would need to be integrated with other information to make decisions.

3.4 Changes in Effort-Based Decision-Making

People also tend to devalue rewards they must work for, a phenomenon known as effort discounting. In 2015, this was an emerging area of research, so Samanez-

Larkin and Knutson did not articulate hypotheses about the relationship between effort discounting and aging (although others have; see (Hess, 2014) for one example). Over the last decade, research in this area has exploded, with researchers developing new paradigms and models to measure this phenomenon. While there are still debates about what these different tasks are measuring (Toro-Serey et al., 2022) and how to model this change in choice behavior (Cuevas Rivera et al., 2020), some patterns have emerged. A recent meta-analysis suggests that effort discounting activates brain regions consistent with those proposed in the AIM Framework and regions like the Supplemental Motor Area (Varma et al., 2023). In terms of aging, a recent meta-analysis found a significant increase in cognitive effort discounting with age (Bagaïni et al., 2023), consistent with subjective reports of increased effort costs for cognitive tasks with age (Hess et al., 2016, 2022). Further, recent work has linked this effort discounting to real-world behavior, with those who display greater effort discounting in the lab also showing less engagement with demanding tasks in daily life activities (Crawford et al., 2023). Regarding brain circuits, one of the authors has shown preserved representation of subjective value during effort discounting in the MPFC and VS across the healthy adult lifespan (Fig. 2b), but age differences in medial temporal regions (Seaman et al., 2018). Another research group has shown that effort discounting is related to amyloid burden in healthy older adults and those with preclinical AD pathology (Aschenbrenner et al., 2023). Thus, although this is a rapidly growing research area, early evidence suggests that there are age effects in terms of *cognitive* effort and that this may be linked to age-related neurodegenerative diseases.

4 Current Gaps and Future Directions

Advancements in neuroeconomic methods and findings have provided valuable insights into how aging influences decision-making processes and how neural changes may support age-related differences. However, here, we discuss five broad areas worthy of further exploration to advance current knowledge and extend and further specify the AIM framework.

4.1 Domain Specificity

Extant work supports the notion of domain specificity in age effects on decision-making (Bagaïni et al., 2023; Liu et al., 2022). Yet, age differences and their neural substrates in highly relevant domains, including ethical, environmental, or social, remain underexplored (König, 2021). The Changes in Integration for Social Decisions in Aging (CISDA) model (Frazier et al., 2019) was proposed as an extension of the AIM framework to address this gap. CISDA emphasizes the unique challenges older adults face in decision-making with social components, such as when making decisions based on trust or for a partner, which require additional integration functions (i.e., for memory and theory of mind) to be added to the AIM framework. We see potential in following the domain-specific approach blueprinted by CISDA for other specific decision-making contexts, including in the domains of time, effort, and health.

4.2 *Operationalization of Uncertainty in Choice Paradigms*

To advance our understanding of the mechanisms underlying age differences in decision-making under uncertainty, we advocate for a more standardized and principled use of paradigms. For example, decision-making paradigms often represent different types of uncertainty (Frank & Seaman, 2023; Tisdall & Mata, 2023)—including *risk* (probabilities of each possible outcome are known or easily estimated, yet the actual outcome is unknown) and *ambiguity* (both the probability of each possible outcome and the actual outcome are unknown)—which elicit partly separable choice patterns and recruit distinct neural substrates. Research on age differences in decision-making under ambiguity versus risk has mainly focused on comparisons between children or adolescents and adults (Tymula et al., 2012; van den Bos & Hertwig, 2017). However, comprehensive efforts to quantify these differences across a broader age range, including older adults, are scarce and do not capture mechanisms at a neurobiological level (Gaubert et al., 2024; Liebherr et al., 2017). Furthermore, we propose that the much-needed development of a taxonomy that dissects decision-making paradigms into key decision components (e.g., gains, losses, time, probability, ambiguity), critical cognitive demands (e.g., memory, attention), as well as both integral and incidental affect would be a step in that direction. Building such a taxonomy will require researchers to align their work (e.g., via research consortia), but engaging with this challenge promises to harmonize research efforts and systematically map age differences in specific decision, cognitive, and affective processes as operationalized via a given decision paradigm.

4.3 *Synthesis*

While there have been efforts to synthesize extant research into the neural substrates of decision-making under uncertainty (Tannou et al., 2021; Wu et al., 2021), more synthesis is required that addresses (a) age differences in older adults as opposed to contrasts between children and adults (Wang et al., 2024) and (b) paradigm-specific processes (Wang et al., 2022) to get closer to the putative cognitive and affective mechanisms. Investing efforts into more (advanced) synthesis approaches would be helpful to extract common patterns from a literature that is heterogeneous and, in parts, fragmented, for example, through harmonization of individual data sets in individual participant data meta-analysis or the principled modeling of all available age comparisons using network meta-analysis.

4.4 Heterogeneity Across and Within Age Groups

Changes in individuals' goals and motivations across the life span may impact decision-making under uncertainty (Buchinger et al., 2022; Carstensen, 2006; Ravert et al., 2019), for instance, as a function of older adults' focus on maintenance and a (perceived) inability to recover from the potential losses of uncertain investment decisions, social interactions, or medical choices. By focusing on group differences and mean level change, rank order stability (Josef et al., 2016) and heterogeneity within older populations are often ignored. However, recent work (Rooks et al., 2020) suggests that different risk-taking profiles with separable neural structural connectivity may exist within older populations, with profile-specific structural connectivity differences being predictive of self-reported financial and health-related risk propensity. While requiring replication with a larger sample, this study (Rooks et al., 2020) could serve as a template for (a) comparing the kinds and composition of risk profiles in different age groups and investigating age differences in the associated connectomic profiles, but more importantly and (b) examining within-person change in longitudinal designs. Change occurs throughout the human life span, but while longitudinal designs are already mainstream for developmental questions in the first decades of life (Karcher & Barch, 2021; Poulton et al., 2015; Thompson et al., 2014), tracking change in older populations, while not absent (Bookheimer et al., 2019), has lagged. Understanding age effects on decision-making and its neural substrates will require tracking those changes.

4.5 Transferring Laboratory Insights to Real-World Settings

A final challenge will be to apply neuroeconomic research to everyday decision-making. We see three major opportunities to move beyond the lab to real life. First, objective and quantifiable measures are essential for evaluating actual decisions made by older adults. Tracking spending habits, health choices, living arrangements, or social interactions can provide valuable data related to real-life decision-making. However, researchers must find more efficient ways to identify and incorporate objective measures into research designs (Holtfreter et al., 2008; McCleskey & Gruda, 2021; Seaman et al., 2015; Wolfe et al., 2021). Second, cultural differences may impact age differences in decision-making (Anlló et al., 2024; Edelson & Reyna, 2021; Löckenhoff, 2018; Mata et al., 2016), for instance, as a function of culture modulating affect valuation (Tsai et al., 2006) or positivity bias and its neural substrates (Carstensen et al., 2000; Gong et al., 2020). Applying a cultural lens to the AIM framework may help overcome shortcomings in external validity that have capped the transfer of laboratory insights to real-world settings. Third, extant work has predominantly focused on description and explanation, but less on prediction. While the former has undoubtedly led to a deeper understanding of social, cognitive, and affective mechanisms of decision-making and their biological underpinnings, whether these insights are generalizable and precise enough to predict age-related change remains an open question. Bridging the gap between explanation and prediction represents a key milestone for scientific progress in translating laboratory insights into real-life solutions for helping aging decision-makers navigate an uncertain world.

5 Summary

Uncertainty presents an unavoidable feature of decision-making, challenging individuals to assess when to approach or avoid options in the context of individual motives and goals. As proposed a decade ago (Samanez-Larkin & Knutson, 2015), the AIM framework provides a tremendously helpful structure to understand the cascade of (neural) processes involved in translating sensory input into goal-oriented action and how aging might influence the different decision phases. As reviewed above, recent work has extended both our insights into the specific phases and their neural substrates and the AIM framework's reach to specific decision contexts with specific decision components. However, there are some gaps in the AIM framework with regard to decision components and contexts, such as effort-based decision-making and social decision-making. Moreover, while the AIM framework touches on key decision components like affect, integration, and motivation, considers their neural substrates, and aims to establish links to age differences observed in behavior (e.g., choices), all of these processes need to be viewed within the broader context of decision ecology. Specifically, integrating AIM with more general perspectives, for example, as proposed by Löckenhoff (2018), may allow us to account for factors such as environmental fit, compensatory strategies, and the adaptive role of motivation in decision-making across the lifespan. To fully harness the potential of the AIM framework, future research should continue fine-tuning the neural and behavioral insights provided by AIM while simultaneously exploring how these processes operate in diverse, real-world decision contexts.

References

Anlló, H., Bavard, S., Benmarrakchi, F., Bonagura, D., Cerrotti, F., Cicue, M., Gueguen, M., Guzmán, E. J., Kadieva, D., Kobayashi, M., Lukumon, G., Sartorio, M., Yang, J., Zinchenko, O., Bahrami, B., Silva Concha, J., Hertz, U., Konova, A. B., Li, J., et al. (2024). Comparing experience- and description-based economic preferences across 11 countries. *Nature Human Behaviour*. https://doi.org/10.1038/s41562-024-01894-9

Anthropic. (2024). *Claude (3.5 Sonnet)* [Large language model]. https://www.anthropic.com

Aschenbrenner, A. J., Crawford, J. L., Peelle, J. E., Fagan, A. M., Benzinger, T. L. S., Morris, J. C., Hassenstab, J., & Braver, T. S. (2023). Increased cognitive effort costs in healthy aging and preclinical Alzheimer's disease. *Psychology and Aging, 38*(5), 428–442. https://doi.org/10.1037/pag0000742

Bagaïni, A., Liu, Y., Bajrami, A., Son, G., Tisdall, L., & Mata, R. (2023). Aging and economic preferences: Cumulative meta-analyses of age differences in risk, time, social, and effort preferences. *The Journals of Gerontology: Series B, 78*(7), 1122–1135. https://doi.org/10.1093/geronb/gbad034

Bethlehem, R. A. I., Seidlitz, J., White, S. R., Vogel, J. W., Anderson, K. M., Adamson, C., Adler, S., Alexopoulos, G. S., Anagnostou, E., Areces-Gonzalez, A., Astle, D. E., Auyeung, B., Ayub, M., Bae, J., Ball, G., Baron-Cohen, S., Beare, R., Bedford, S. A., Benegal, V., et al. (2022). Brain charts for the human lifespan. *Nature, 604*(7906), 525–533. https://doi.org/10.1038/s41586-022-04554-y

Bookheimer, S. Y., Salat, D. H., Terpstra, M., Ances, B. M., Barch, D. M., Buckner, R. L., Burgess, G. C., Curtiss, S. W., Diaz-Santos, M., Elam, J. S., Fischl, B., Greve, D. N., Hagy, H. A., Harms, M. P., Hatch, O. M., Hedden, T., Hodge, C., Japardi, K. C., Kuhn, T. P., et al. (2019). The lifespan human connectome project in aging: An overview. *NeuroImage, 185*, 335–348. https://doi.org/10.1016/j.neuroimage.2018.10.009

Boutoleau-Bretonnière, C., Kapogiannis, D., & El Haj, M. (2023). The guaranteed euros: Probabilistic discounting in behavioural-variant frontotemporal dementia. *Journal of Neuropsychology*. https://doi.org/10.1111/jnp.12357

Buchinger, L., Richter, D., & Heckhausen, J. (2022). The development of life goals across the adult life span. *The Journals of Gerontology: Series B, 77*(5), 905–915. https://doi.org/10.1093/geronb/gbab154

Cabeza, R. (2002). Hemispheric asymmetry reduction in older adults: The HAROLD model. *Psychology and Aging, 17*(1), 85–100. https://doi.org/10.1037/0882-7974.17.1.85

Carstensen, L. L. (2006). The influence of a sense of time on human development. *Science, 312*(5782), 1913–1915. https://doi.org/10.1126/science.1127488

Carstensen, L. L., Isaacowitz, D. M., & Charles, S. T. (1999). Taking time seriously: A theory of socioemotional selectivity. *American Psychologist, 54*(3), 165–181. https://doi.org/10.1037/0003-066X.54.3.165

Carstensen, L. L., Pasupathi, M., Mayr, U., & Nesselroade, J. R. (2000). Emotional experience in everyday life across the adult life span. *Journal of Personality and Social Psychology, 79*(4), 644–655. https://doi.org/10.1037/0022-3514.79.4.644

Castrellon, J. J., Seaman, K. L., Crawford, J. L., Young, J. S., Smith, C. T., Dang, L. C., Hsu, M., Cowan, R. L., Zald, D. H., & Samanez-Larkin, G. R. (2019). Individual differences in dopamine are associated with reward discounting in clinical groups but not in healthy adults. *The Journal of Neuroscience, 39*(2), 321–332. https://doi.org/10.1523/JNEUROSCI.1984-18.2018

Chowdhury, R., Guitart-Masip, M., Lambert, C., Dayan, P., Huys, Q., Düzel, E., & Dolan, R. J. (2013). Dopamine restores reward prediction errors in old age. *Nature Neuroscience, 16*(5), 648–653. https://doi.org/10.1038/nn.3364

Clairis, N., & Pessiglione, M. (2024). Value estimation versus effort mobilization: A general dissociation between ventromedial and dorsomedial prefrontal cortex. *Journal of Neuroscience, 44*(17). https://doi.org/10.1523/JNEUROSCI.1176-23.2024

Crawford, J. L., English, T., & Braver, T. S. (2023). Cognitive effort-based decision-making across experimental and daily life indices in younger and older adults. *The Journals of Gerontology. Series B, Psychological Sciences and Social Sciences, 78*(1), 40–50. https://doi.org/10.1093/geronb/gbac167

Cuevas Rivera, D., Strobel, A., Goschke, T., & Kiebel, S. J. (2020). Modeling dynamic allocation of effort in a sequential task using discounting models. *Frontiers in Neuroscience, 14*. https://doi.org/10.3389/fnins.2020.00242

De Boer, L., Axelsson, J., Riklund, K., Nyberg, L., Dayan, P., Bäckman, L., & Guitart-Masip, M. (2017). Attenuation of dopamine-modulated prefrontal value signals underlies probabilistic reward learning deficits in old age. *eLife, 6*, e26424. https://doi.org/10.7554/eLife.26424

Delaney, R. K., Strough, J., Shook, N. J., Ford, C. G., & Lemaster, P. (2021). Don't risk it. Older adults perceive fewer future opportunities and avoid social risk taking. *The International Journal of Aging and Human Development, 92*(2), 139–157. https://doi.org/10.1177/0091415019900564

Duell, N., Steinberg, L., Icenogle, G., Chein, J., Chaudhary, N., Di Giunta, L., Dodge, K. A., Fanti, K. A., Lansford, J. E., Oburu, P., Pastorelli, C., Skinner, A. T., Sorbring, E., Tapanya, S., Uribe Tirado, L. M., Alampay, L. P., Al-Hassan, S. M., Takash, H. M. S., Bacchini, D., & Chang, L. (2018). Age patterns in risk taking across the world. *Journal of Youth and Adolescence, 47*(5), 1052–1072. https://doi.org/10.1007/s10964-017-0752-y

Edelson, S. M., & Reyna, V. F. (2021). How fuzzy-trace theory predicts development of risky decision making, with novel extensions to culture and reward sensitivity. *Developmental Review, 62*(September), 100986. https://doi.org/10.1016/j.dr.2021.100986

Eppinger, B., Hämmerer, D., & Li, S. C. (2011). Neuromodulation of reward-based learning and decision making in human aging. *Annals of the New York Academy of Sciences, 1235*(1), 1–17. https://doi.org/10.1111/j.1749-6632.2011.06230.x

Eppinger, B., Nystrom, L. E., & Cohen, J. D. (2012). Reduced sensitivity to immediate reward during decision-making in older than younger adults. *PLoS One, 7*, e36953. https://doi.org/10.1371/journal.pone.0036953

Fernandes, C., Macedo, I., Gonçalves, A. R., Pasion, R., Mata, R., Danese, G., Martins, I. P., Barbosa, F., & Marques-Teixeira, J. (2022). Neurophysiological examination of the affect–integration–motivation framework of decision-making in the aging brain: A registered report. *NeuroImage, 256*, 119189. https://doi.org/10.1016/j.neuroimage.2022.119189

Frank, C. C., & Seaman, K. L. (2023). Aging, uncertainty, and decision making—A review. *Cognitive, Affective, & Behavioral Neuroscience, 23*(3), 773–787. https://doi.org/10.3758/s13415-023-01064-w

Frazier, I., Lighthall, N. R., Horta, M., Perez, E., & Ebner, N. C. (2019). CISDA: Changes in integration for social decisions in aging. *Wiley Interdisciplinary Reviews: Cognitive Science, 10*(3), 1–15. https://doi.org/10.1002/wcs.1490

Garzón, B., Kurth-Nelson, Z., Bäckman, L., Nyberg, L., & Guitart-Masip, M. (2023). Investigating associations of delay discounting with brain structure, working memory, and episodic memory. *Cerebral Cortex, 33*(5), 1669–1678. https://doi.org/10.1093/cercor/bhac164

Gaubert, F., Borg, C., Roux, J.-C., & Chainay, H. (2024). Decision-making and ageing: Everyday life situations under risk and under ambiguity. *Quarterly Journal of Experimental Psychology, 77*(4), 747–766. https://doi.org/10.1177/17470218231182403

Geng, Z., Wu, X., Wang, L., Zhou, S., Tian, Y., Wang, K., & Wei, L. (2020). Reduced delayed reward selection by Alzheimer's disease and mild cognitive impairment patients during intertemporal decision-making. *Journal of Clinical and Experimental Neuropsychology, 42*(3), 298–306. https://doi.org/10.1080/13803395.2020.1711873

Goh, J. O. S., Su, Y.-S., Tang, Y.-J., McCarrey, A. C., Tereshchenko, A., Elkins, W., & Resnick, S. M. (2016). Frontal, striatal, and medial temporal sensitivity to value distinguishes risk-taking from risk-aversive older adults during decision making. *The Journal of Neuroscience, 36*(49), 12498–12509. https://doi.org/10.1523/JNEUROSCI.1386-16.2016

Gong, X., Fung, H. H., Zeng, G. Q., & Tse, C.-Y. (2020). Cultural relevance reduces the enhanced neural processing of positively valenced information in older adults. *The Journals of Gerontology: Series B, 75*(10), 2075–2085. https://doi.org/10.1093/geronb/gbz049

Green, L., Fry, A. F., & Myerson, J. (1994). Discounting of delayed rewards: A life-span comparison. *Psychological Science, 91*, 33–36. https://doi.org/10.1111/j.1467-9639.1991.tb00167.x

Halfmann, K., Hedgcock, W., Kable, J., & Denburg, N. L. (2015). Individual differences in the neural signature of subjective value among older adults. *Social Cognitive and Affective Neuroscience*, 1–10. https://doi.org/10.1093/scan/nsv078

Han, S. D., Arfanakis, K., Fleischman, D. A., Yu, L., Bennett, D. A., & Boyle, P. A. (2018). White matter correlates of temporal discounting in older adults. *Brain Structure and Function, 223*(8), 3653–3663. https://doi.org/10.1007/s00429-018-1712-3

Hänggi, J., Lohrey, C., Drobetz, R., Baetschmann, H., Forstmeier, S., Maercker, A., & Jäncke, L. (2016). Strength of structural and functional frontostriatal connectivity predicts self-control in the healthy elderly. *Frontiers in Aging Neuroscience, 8*. https://doi.org/10.3389/fnagi.2016.00307

Hess, T. M. (2014). Selective engagement of cognitive resources: Motivational influences on older adults' cognitive functioning. *Perspectives on Psychological Science, 9*(4), 388–407. https://doi.org/10.1177/1745691614527465

Hess, T. M., Smith, B. T., & Sharifian, N. (2016). Aging and effort expenditure: The impact of subjective perceptions of task demands. *Psychology and Aging, 31*(7), 653–660. https://doi.org/10.1037/pag0000127

Hess, T. M., Neupert, S. D., & Lothary, A. F. (2022). Aging attitudes and changes in the costs of cognitive engagement in older adults over 5 years. *Psychology and Aging, 37*(4), 456–468. https://doi.org/10.1037/pag0000685

Holtfreter, K., Reisig, M., & Pratt, T. (2008). Low self-control, routine activities, and fraud victimization. *Criminology, 46*(1), 189–220.

Josef, A. K., Richter, D., Samanez-Larkin, G. R., Wagner, G. G., Hertwig, R., & Mata, R. (2016). Stability and change in risk-taking propensity across the adult lifespan. *Journal of Personality and Social Psychology, 11*(3), 430–450. https://doi.org/10.1177/0963721412473755.Surging

Kable, J. W., & Glimcher, P. W. (2010). An 'as soon as possible' effect in human intertemporal decision making: Behavioral evidence and neural mechanisms. *Journal of Neurophysiology, 103*, 2513–2531. https://doi.org/10.1152/jn.00177.2009

Karcher, N. R., & Barch, D. M. (2021). The ABCD study: Understanding the development of risk for mental and physical health outcomes. *Neuropsychopharmacology, 46*(1), 131–142. https://doi.org/10.1038/s41386-020-0736-6

Karrer, T. M., Josef, A. K., Mata, R., Morris, E. D., & Samanez-Larkin, G. R. (2017). Reduced dopamine receptors and transporters but not synthesis capacity in normal aging adults: A meta-analysis. *Neurobiology of Aging, 57*, 36–46. https://doi.org/10.1016/j.neurobiolaging.2017.05.006

König, A. N. (2021). Domain-specific risk attitudes and aging—A systematic review. *Journal of Behavioral Decision Making, 34*(3), 359–378. https://doi.org/10.1002/bdm.2215

Kuhnen, C. M., & Knutson, B. (2005). The neural basis of financial risk taking. *Neuron, 47*(5), 763–770. https://doi.org/10.1016/j.neuron.2005.08.008

Lempert, K. M., MacNear, K. A., Wolk, D. A., & Kable, J. W. (2020a). Links between autobiographical memory richness and temporal discounting in older adults. *Scientific Reports, 10*(1), Article 1. https://doi.org/10.1038/s41598-020-63373-1

Lempert, K. M., Mechanic-Hamilton, D. J., Xie, L., Wisse, L. E. M., de Flores, R., Wang, J., Das, S. R., Yushkevich, P. A., Wolk, D. A., & Kable, J. W. (2020b). Neural and behavioral correlates of episodic memory are associated with temporal discounting in older adults. *Neuropsychologia, 146*. https://doi.org/10.1016/j.neuropsychologia.2020.107549

Lempert, K. M., Parthasarathi, T., Linhares, S., Ruh, N., & Kable, J. W. (2024). Positive autobiographical memory recall does not influence temporal discounting: An internal meta-analysis of experimental studies. *Journal of Economic Psychology, 103*, 102730. https://doi.org/10.1016/j.joep.2024.102730

Leong, J. K., Pestilli, F., Wu, C. C., Samanez-Larkin, G. R., & Knutson, B. (2016). White-matter tract connecting anterior insula to nucleus accumbens correlates with reduced preference for positively skewed gambles. *Neuron, 89*(1), 63–69. https://doi.org/10.1016/j.neuron.2015.12.015

Liebherr, M., Schiebener, J., Averbeck, H., & Brand, M. (2017). Decision making under ambiguity and objective risk in higher age—A review on cognitive and emotional contributions. *Frontiers in Psychology, 8*, 2128. https://doi.org/10.3389/fpsyg.2017.02128

Liu, Y., Bagaïni, A., Son, G., Kapoor, M., & Mata, R. (2022). Life-course trajectories of risk-taking propensity: A coordinated analysis of longitudinal studies. *The Journals of Gerontology: Series B*, gbac175. https://doi.org/10.1093/geronb/gbac175

Löckenhoff, C. E. (2018). Aging and decision-making: A conceptual framework for future research—A mini-review. *Gerontology, 64*(2), 140–148. https://doi.org/10.1159/000485247

Mata, R., Josef, A. K., & Hertwig, R. (2016). Propensity for risk taking across the life span and around the globe. *Psychological Science, 27*(2), 231–243. https://doi.org/10.1177/0956797615617811

McCleskey, J., & Gruda, D. (2021). Risk-taking, resilience, and state anxiety during the COVID-19 pandemic: A coming of (old) age story. *Personality and Individual Differences, 170*, 110485. https://doi.org/10.1016/j.paid.2020.110485

McClure, S. M., Laibson, D. I., Loewenstein, G., & Cohen, J. D. (2004). Separate neural systems value immediate and delayed monetary rewards. *Science, 306*, 503–507. https://doi.org/10.1126/science.1100907

Nassar, M. R., Bruckner, R., Gold, J. I., Li, S.-C., Heekeren, H. R., & Eppinger, B. (2016). Age differences in learning emerge from an insufficient representation of uncertainty in older adults. *Nature Communications, 7*(1), 11609. https://doi.org/10.1038/ncomms11609

O'Brien, E. L., & Hess, T. M. (2020). Differential focus on probability and losses between young and older adults in risky decision-making. *Aging, Neuropsychology, and Cognition, 27*(4), 532–552. https://doi.org/10.1080/13825585.2019.1642442

Peng, X., Lei, X., Xu, P., & Yu, J. (2020). The Age-related Neural Strategy Alterations in Decision Making Under Risk. *Neuroscience, 440*, 30–38. https://doi.org/10.1016/j.neuroscience.2020.05.017

Poulton, R., Moffitt, T. E., & Silva, P. A. (2015). The Dunedin multidisciplinary health and development study: Overview of the first 40 years, with an eye to the future. *Social Psychiatry and Psychiatric Epidemiology, 50*(5), 679–693. https://doi.org/10.1007/s00127-015-1048-8

Ravert, R. D., Murphy, L. M., & Donnellan, M. B. (2019). Valuing risk: Endorsed risk activities and motives across adulthood. *Journal of Adult Development, 26*(1), 1–11. https://doi.org/10.1007/s10804-018-9300-6

Ren, P., Hou, G., Ma, M., Zhuang, Y., Huang, J., Tan, M., Wu, D., Luo, G., Zhang, Z., & Rong, H. (2023). Enhanced putamen functional connectivity underlies altered risky decision-making in age-related cognitive decline. *Scientific Reports, 13*(1), 6619. https://doi.org/10.1038/s41598-023-33634-w

Rooks, B., Anthony, M., Chen, Q., Lin, Y., Baran, T., Zhang, Z., Lichtenberg, P. A., & Lin, F. (2020). A generic brain connectome map linked to different types of everyday decision-making in old age. *Brain Structure and Function, 225*(4), 1389–1400. https://doi.org/10.1007/s00429-019-02013-5

Samanez-Larkin, G. R., & Knutson, B. (2015). Decision making in the ageing brain: Changes in affective and motivational circuits. *Nature Reviews Neuroscience, 16*(May). https://doi.org/10.1038/nrn3917

Samanez-Larkin, G. R., Mata, R., Radu, P. T., Ballard, I. C., Carstensen, L. L., & McClure, S. M. (2011). Age differences in striatal delay sensitivity during intertemporal choice in healthy adults. *Frontiers in Neuroscience, 5*, 1–12. https://doi.org/10.3389/fnins.2011.00126

Samanez-Larkin, G. R., Levens, S. M., Perry, L. M., Dougherty, R. F., & Knutson, B. (2012). Frontostriatal White matter integrity mediates adult age differences in probabilistic reward learning. *Journal of Neuroscience, 32*(15), 5333–5337. https://doi.org/10.1523/JNEUROSCI.5756-11.2012

Samanez-Larkin, G. R., Worthy, D. A., Mata, R., McClure, S. M., & Knutson, B. (2014). Adult age differences in frontostriatal representation of prediction error but not reward outcome. *Cognitive, Affective, & Behavioral Neuroscience, 14*(2), 672–682. https://doi.org/10.3758/s13415-014-0297-4

Schonberg, T., Fox, C. R., & Poldrack, R. A. (2010). Mind the gap: Bridging economic and naturalistic risk-taking with cognitive neuroscience. *Trends in Cognitive Sciences, 15*, 11–19. https://doi.org/10.1016/j.tics.2010.10.002

Seaman, K. L., Stillman, C. M., Howard, D. V., & Howard, J. H. (2015). Risky decision-making is associated with residential choice in healthy older adults. *Frontiers in Psychology, 6*. https://doi.org/10.3389/fpsyg.2015.01192

Seaman, K. L., Leong, J. K., Wu, C. C., Knutson, B., & Samanez-Larkin, G. R. (2017). Individual differences in skewed financial risk-taking across the adult life span. *Cognitive, Affective, & Behavioral Neuroscience, 17*(6), 1232–1241. https://doi.org/10.3758/s13415-017-0545-5

Seaman, K. L., Brooks, N., Karrer, T. M., Castrellon, J. J., Perkins, S. F., Dang, L. C., Hsu, M., Zald, D. H., & Samanez-Larkin, G. R. (2018). Subjective value representations during effort, probability and time discounting across adulthood. *Social Cognitive and Affective Neuroscience, 13*(5), 449–459. https://doi.org/10.1093/scan/nsy021

Seaman, K. L., Smith, C. T., Juarez, E. J., Dang, L. C., Castrellon, J. J., Burgess, L. L., San Juan, M. D., Kundzicz, P. M., Cowan, R. L., Zald, D. H., & Samanez-Larkin, G. R. (2019). Differential regional decline in dopamine receptor availability across adulthood: Linear and nonlinear effects of age. *Human Brain Mapping, 40*(10), 3125–3138. https://doi.org/10.1002/hbm.24585

Seaman, K. L., Abiodun, S., Fenn, Z., Samanez-Larkin, G. R., & Mata, R. (2022). Temporal discounting across adulthood: A systematic review and meta-analysis. *Psychology and Aging, 37*(1), 111–124. https://doi.org/10.31234/osf.io/7ysxa

Seaman, K. L., Christensen, A. P., Senn, K. D., Cooper, J. A., & Cassidy, B. S. (2023). Age-related differences in the social associative learning of trust information. *Neurobiology of Aging, 125*, 32–40. https://doi.org/10.1016/j.neurobiolaging.2023.01.011

Shook, N. J., Delaney, R. K., Strough, J., Wilson, J. M., Sevi, B., & Altman, N. (2021). Playing it safe: Dispositional mindfulness partially accounts for age differences in health and safety risk-taking propensity. *Current Psychology, 40*(5), 2142–2152. https://doi.org/10.1007/s12144-019-0137-3

Sojitra, R. B., Lerner, I., Petok, J. R., & Gluck, M. A. (2018). Age affects reinforcement learning through dopamine-based learning imbalance and high decision noise—Not through Parkinsonian mechanisms. *Neurobiology of Aging, 68*, 102–113. https://doi.org/10.1016/j.neurobiolaging.2018.04.006

Su, Y.-S., Chen, J.-T., Tang, Y.-J., Yuan, S.-Y., McCarrey, A. C., & Goh, J. O. S. (2018). Age-related differences in striatal, medial temporal, and frontal involvement during value-based decision processing. *Neurobiology of Aging, 69*, 185–198. https://doi.org/10.1016/j.neurobiolaging.2018.05.019

Suzuki, S., Lawlor, V. M., Cooper, J. A., Arulpragasam, A. R., & Treadway, M. T. (2021). Distinct regions of the striatum underlying effort, movement initiation and effort discounting. *Nature Human Behaviour, 5*(3), 378–388. https://doi.org/10.1038/s41562-020-00972-y

Tannou, T., Magnin, E., Comte, A., Aubry, R., & Joubert, S. (2021). Neural activation in risky decision-making tasks in healthy older adults: A meta-analysis of fmri data. *Brain Sciences, 11*(8). https://doi.org/10.3390/brainsci11081043

Thoma, M. V., Maercker, A., & Forstmeier, S. (2017). Evidence for different trajectories of delay discounting in older adults with mild cognitive impairment and mild Alzheimer's disease. *The Journals of Gerontology: Series B, 72*(6), 956–965. https://doi.org/10.1093/geronb/gbw010

Thompson, P., Stein, J. L., Medland, S. E., Hibar, D. P., Vasquez, A. A., Renteria, M. E., Toro, R., Jahanshad, N., Schumann, G., Franke, B., Wright, M. J., Martin, N. G., Agartz, I., Alda, M., Alhusaini, S., Almasy, L., Almeida, J., Alpert, K., Andreasen, N. C., et al. (2014). The ENIGMA consortium: Large-scale collaborative analyses of neuroimaging and genetic data. *Brain Imaging and Behavior, 8*(2), 153–182. https://doi.org/10.1007/s11682-013-9269-5

Tisdall, L., & Mata, R. (2023). Age differences in the neural basis of decision-making under uncertainty. *Cognitive, Affective, & Behavioral Neuroscience, 23*(3), 788–808. https://doi.org/10.3758/s13415-022-01060-6

Tisdall, L., Frey, R., Wulff, D. U., Kellen, D., & Mata, R. (2024). Convergence of age differences in risk preference, impulsivity, and self-control: A multiverse analysis. *The Journals of Gerontology, Series B: Psychological Sciences and Social Sciences*, gbae092. https://doi.org/10.1093/geronb/gbae092

Toro-Serey, C., Kane, G. A., & McGuire, J. T. (2022). Choices favoring cognitive effort in a foraging environment decrease when multiple forms of effort and delay are interleaved. *Cognitive, Affective, & Behavioral Neuroscience, 22*(3), 509–532. https://doi.org/10.3758/s13415-021-00972-z

Tsai, J. L., Knutson, B., & Fung, H. H. (2006). Cultural variation in affect valuation. *Journal of Personality and Social Psychology, 90*(2), 288–307. https://doi.org/10.1037/0022-3514.90.2.288
Tymula, A., Rosenberg Belmaker, L. A., Roy, A. K., Ruderman, L., Manson, K., Glimcher, P. W., & Levy, I. (2012). Adolescents' risk-taking behavior is driven by tolerance to ambiguity. *Proceedings of the National Academy of Sciences of the United States of America, 109*(42), 17135–17140. https://doi.org/10.1073/pnas.1207144109
van den Bos, W., & Hertwig, R. (2017). Adolescents display distinctive tolerance to ambiguity and to uncertainty during risky decision making. *Scientific Reports, 7*, 1–11. https://doi.org/10.1038/srep40962
Varma, M. M., Zhen, S., & Yu, R. (2023). Not all discounts are created equal: Regional activity and brain networks in temporal and effort discounting. *NeuroImage, 280*, 120363. https://doi.org/10.1016/j.neuroimage.2023.120363
Wang, M., Zhang, S., Suo, T., Mao, T., Wang, F., Deng, Y., Eickhoff, S., Pan, Y., Jiang, C., & Rao, H. (2022). Risk-taking in the human brain: An activation likelihood estimation meta-analysis of the balloon analog risk task (BART). *Human Brain Mapping*, hbm.26041. https://doi.org/10.1002/hbm.26041
Wang, M., Deng, Y., Liu, Y., Suo, T., Guo, B., Eickhoff, S. B., Xu, J., & Rao, H. (2024). The common and distinct brain basis associated with adult and adolescent risk-taking behavior: Evidence from the neuroimaging meta-analysis. *Neuroscience & Biobehavioral Reviews, 160*, 105607. https://doi.org/10.1016/j.neubiorev.2024.105607
Wolfe, K., Sirota, M., & Clarke, A. D. F. (2021). Age differences in COVID-19 risk-taking, and the relationship with risk attitude and numerical ability. *Royal Society Open Science, 8*(9), 201445. https://doi.org/10.1098/rsos.201445
Wu, S., Sun, S., Camilleri, J. A., Eickhoff, S. B., & Yu, R. (2021). Better the devil you know than the devil you don't: Neural processing of risk and ambiguity. *NeuroImage, 236*, 118109. https://doi.org/10.1016/j.neuroimage.2021.118109
Yee, D. M., Crawford, J. L., Lamichhane, B., & Braver, T. S. (2021). Dorsal anterior cingulate cortex encodes the integrated incentive motivational value of cognitive task performance. *Journal of Neuroscience, 41*(16), 3707–3720. https://doi.org/10.1523/JNEUROSCI.2550-20.2021
Young, N. S., Ioannidis, J. P. A., & Al-Ubaydli, O. (2008). Why current publication practices may distort science. *PLoS Medicine, 5*(10), e201. https://doi.org/10.1371/journal.pmed.0050201
Yu, J., Mamerow, L., Lei, X., Fang, L., & Mata, R. (2016). Altered Value Coding in the Ventromedial Prefrontal Cortex in Healthy Older Adults. Frontiers in Aging Neuroscience, 8(August), 1–15. https://doi.org/10.3389/fnagi.2016.00210

Maladaptive Decision-Making in Older Adults: Confirmation Bias and Financial Exploitation

James B. Wyngaarden, Tania Giovannetti, Bart Krekelberg, and David V. Smith

Abstract Financial exploitation of older adults has become an increasingly urgent public health issue, with annual losses exceeding $10 billion in the United States. This exploitation often stems from age-related cognitive deficits, such as impairments in working memory and cognitive flexibility, which are linked to deterioration of the prefrontal cortex (PFC). A critical yet understudied component of age-related cognitive decline is confirmation bias—the tendency to interpret new information in a way that conforms to one's preexisting beliefs. As older adults focus more on present-oriented goals related to social and emotional satisfaction, their ability to recognize and respond to exploitative situations may decrease. This problem worsens with the progression to mild cognitive impairment (MCI) and Alzheimer's disease and related disorders, increasing vulnerability to fraud. Despite these links, little is known about how confirmation bias is associated with future cognitive decline and risk for fraud. Two main challenges impede progress: the lack of knowledge about the latent processes that shape confirmation bias in this population and the reliance on correlations between brain responses and behavior that do not clarify causal mechanisms. To address these issues, this chapter provides an overview of confirmation bias and its connection to aging. It then addresses adverse effects of confirmation bias before highlighting strengths of noninvasive brain stimulation as a method of investigating these relationships further.

1 Introduction

Maladaptive decision-making in older adults has emerged as a critical research area, particularly as cognitive decline often impacts optimal decision-making. Aging

J. B. Wyngaarden · T. Giovannetti · D. V. Smith (✉)
Temple University, Philadelphia, PA, USA
e-mail: david.v.smith@temple.edu

B. Krekelberg
Rutgers University—Newark, Newark, NJ, USA

D. V. Smith et al. (eds.), *Neuroeconomics: Core Topics and Current Directions*,
https://doi.org/10.1007/978-3-032-02925-6_30

alters brain function, affecting information processing, risk evaluation, and weighting of long-term versus short-term consequences, potentially reducing quality of life and increasing vulnerability to poor financial, health, and social outcomes. The AIM framework—focusing on Attention, Interpretation, and Memory—investigates these cognitive processes, highlighting how older adults allocate attention, interpret information, and rely on memory in decision-making (Samanez-Larkin & Knutson, 2015; Tisdall & Seaman, 2024). Social decision-making becomes more complex with age as cognitive and emotional shifts alter assessments of interactions and intentions. The CISDA model (Consideration of Intentions in Social Decision-Making Across Adulthood) outlines five key components: understanding context, identifying intentions, evaluating goals, regulating emotions, and adapting to feedback (Frazier et al., 2019). As aging affects each of these areas, decision-making in social contexts may lead to misjudgments, underscoring the need to understand how older adults navigate complex interactions.

A critical yet underexplored aspect of cognitive decline in aging is confirmation bias—the tendency to interpret new information in ways that align with preexisting beliefs (Nickerson, 1998). As older adults increasingly prioritize present-oriented goals tied to social and emotional satisfaction (Carstensen, 1995), their ability to detect exploitation may weaken. This vulnerability is further amplified as cognitive decline progresses, particularly in mild cognitive impairment (MCI) and Alzheimer's disease and related disorders (ADRD; Sun et al., 2020). Financial exploitation disproportionately affects older adults, contributing to more than $10 billion in annual losses in the United States (FTC, 2024). Many of these cases are linked to cognitive deficits stemming from the deterioration of the prefrontal cortex (PFC) and its connections (Rollwage et al., 2020; Yaple et al., 2019). Despite these associations, understanding of how confirmation bias contributes to fraud risk and cognitive decline remains limited.

This chapter explores the complex relationship between confirmation bias and aging, particularly how age-related cognitive and socioemotional changes shape decision-making processes (Fig. 1). In Sect. 2, we will delve into how confirmation bias develops with age, exploring both the socioemotional contributors and the cognitive and neural mechanisms underlying this bias. Section 3 examines the adverse effects of confirmation bias, with discussions on its potential adaptive functions, its role in financial exploitation, and its contribution to vulnerability to misinformation. We also ask whether confirmation bias may predict future cognitive decline. In Sect. 4, we review the potential of noninvasive brain stimulation techniques to modulate the cognitive processes related to confirmation bias, outlining promising avenues for future research and interventions. Finally, Sect. 5 offers conclusions and directions for further investigation into maladaptive decision-making in older adults.

2 Linking Confirmation Bias to Aging

While confirmation bias is well-documented in decision-making, questions remain regarding how socioemotional factors, cognitive processes, and neural mechanisms

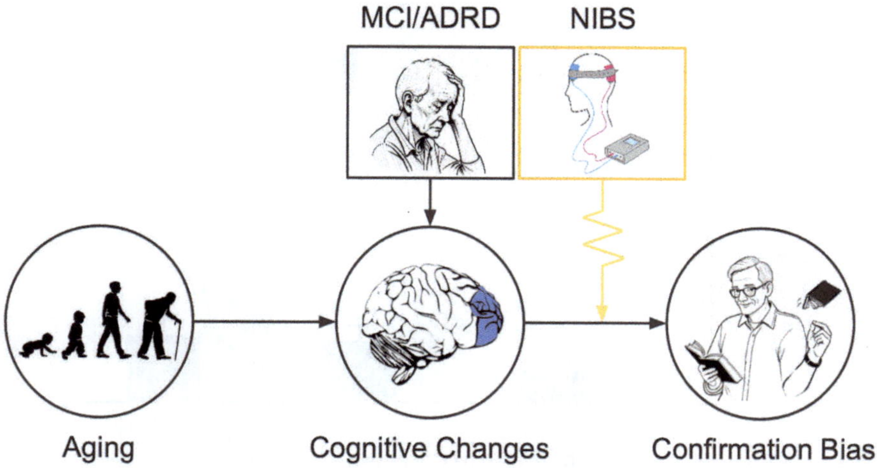

Fig. 1 Age-related cognitive changes lead to increased confirmation bias. Cognitive and socioemotional changes that occur in older adulthood contribute to confirmation bias, which can lead to negative downstream effects like heightened risk for financial exploitation. These changes are exacerbated by the onset of mild cognitive impairment (MCI) and Alzheimer's and related diseases (ADRD). The figure illustrates this cascade, showing how age-related prefrontal cortical changes enhance confirmation bias, which subsequently increases vulnerability to exploitation. The figure also highlights the potential of noninvasive brain stimulation (NIBS) to enhance responses from prefrontal cortical regions and reduce confirmation bias, mitigating risk. (Image components: Aging (Rebelo-Marques et al., 2018; CC BY 4.0); Cognitive Changes (Lundström, 2012; CC BY 3.0); NIBS (Park & Thompson, 2024; CC BY 4.0); MCI/ADRD & Confirmation Bias (DALL-E))

contribute to this bias. Despite its potential significance in explaining financial exploitation, confirmation bias has not been extensively studied in the context of aging. This section reviews socioemotional contributors and discusses hypothesized models of closely related processes underlying their neural mechanisms.

2.1 Confirmation Bias and Financial Exploitation

Financial exploitation takes many forms, including fraudulent schemes, scams, identity theft, theft, fund misuse, overcharging, coercion, and financial neglect (Peterson et al., 2014; Nguyen et al., 2021; Betz-Hamilton & Vincenti, 2018). Power of attorney abuse occurs when a trusted person misuses legal authority for personal gain (*United States Department of Justice*, 2023). Informal financial arrangements, like a child managing a parent's finances without legal documentation, can also lead to exploitation. Entitlement abuse occurs when family members use manipulation or threats to gain control of assets (Stum et al., 2017). Older adults acting as loan guarantors risk asset loss if debts are unpaid (British Columbia Law Institute, 2004). Exploitation ranges from stranger fraud to manipulation by close relatives,

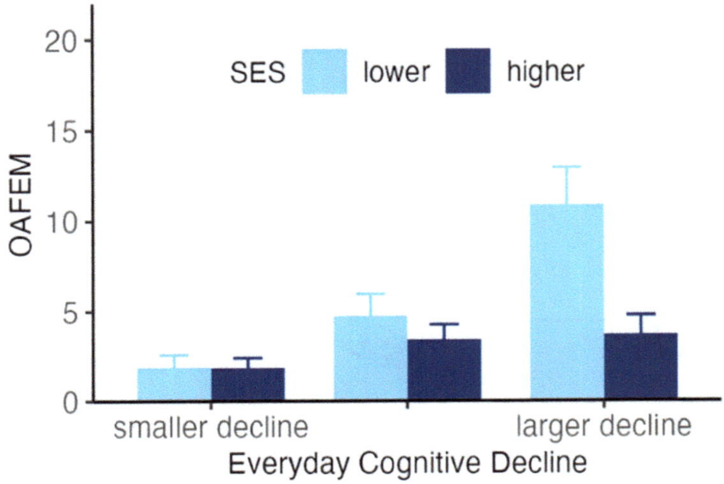

Fig. 2 Interaction between cognitive decline and socioeconomic status associated with risk for financial exploitation. Moderating effect of socioeconomic status on the association between everyday cognitive decline. Compared to individuals with higher socioeconomic status, those with lower socioeconomic status showed stronger positive associations between OAFEM and everyday cognitive decline. (This figure was adapted from Yang et al. (2025), CC BY 4.0)

with the risks often exacerbated by cognitive decline. Recent studies suggest that cognitively impaired older adults face heightened risk, particularly those from lower socioeconomic backgrounds (Fig. 2; Yang et al., 2025).

Confirmation bias heightens risk of financial exploitation by making individuals more susceptible to scams aligning with their preconceived beliefs (Cafferata & Tramontana, 2019). Scammers exploit this bias by presenting "confirming evidence" that reinforces existing beliefs (Park et al., 2010). For example, a scammer posing as a bank representative might exploit an older adult's fear of fraud, leading them to ignore inconsistencies and follow harmful instructions. From a neuroeconomic perspective, this represents a suboptimal valuation process where the perceived utility of acting on the scammer's instructions outweighs the actual risk, reflecting altered neural value computations in older adults.

The bias intensifies with emotionally charged issues, often playing on emotions like greed or fear (Charness & Dave, 2017; Horner et al., 2021). Such emotional manipulation, coupled with cognitive deficits in belief updating and control, impairs objective reevaluation of potential scams (Fenton et al., 2022; Spreng et al., 2016). Overall, confirmation bias increases vulnerability by making individuals more likely to accept confirmatory information, less able to evaluate contradictory evidence, and more prone to emotional manipulation. Neural mechanisms related to aging have also been linked to bias and financial decision-making. For example, older adults show blunted striatal responses but enhanced default-mode network (DMN) connectivity with the temporoparietal junction (TPJ) during reciprocated trust from friends, highlighting age-related differences in sensitivity to social closeness (Fareri et al., 2022). Cognitive decline, particularly with the onset of ADRD, exacerbates this bias (Fernandes et al., 2021).

2.2 *Socioemotional Contributors to Decision-Making and Aging*

Social and emotional factors significantly contribute to confirmation bias. According to socioemotional selectivity theory (SST), older adults become more selective in friendships, focusing more on maximizing positive emotions and viewing ingroup members more positively (Carstensen, 1995; Cassidy et al., 2020). SST's emphasis on emotional over knowledge goals may increase susceptibility to ingroup favoritism serving socioemotional needs (Vail & Juhl, 2015). Social identity and ingroup bias can widen social divides by leading individuals to favor information aligning with their social identities (Brewer, 1999; Steele et al., 2002). Cutler et al. (2021) demonstrated increased ingroup biases with age across a large global sample, highlighting the cross-cultural nature of this phenomenon. This dynamic creates "echo chambers" where people seek relationships and media reinforcing their beliefs (Terren & Borge-Bravo, 2021; Kroon et al., 2022). As aging progresses to MCI and ADRD, weakened regulatory control may amplify biases like ingroup bias (Cassidy et al., 2020). Thus, older adults' focus on ingroups may increase vulnerability to biases in selective attention and information interpretation biases. Older adults tend to focus more on positive information and memories—the "positivity effect"—making emotional experiences more predictable with age (Mather & Carstensen, 2005; Carstensen et al., 2011). They generally demonstrate better emotional self-regulation than younger adults (Gross et al., 1997). Emotionally charged issues amplify confirmation bias, as individuals seek and interpret evidence to protect their beliefs and self-esteem (Dunne, 2016).

Aging can also be associated with other undesirable emotional changes, including denial, guilt, loneliness, helplessness, suspicion, stubbornness, and anger when facing age-related changes (Loboprabhu et al., 2015; Rodda et al., 2011; Ryan & Patterson, 1987). Attempts to reduce uncertainty and maintain coherence in one's belief system also contribute to biased processing. When confronted with contradictory information, people may double down on existing beliefs to avoid the psychological discomfort of uncertainty (e.g., cognitive dissonance; Kruglanski et al., 2018). Taken together, desires for positive emotions, group belonging, and belief consistency can motivate selective information seeking, interpretation, and memory.

Age-related cognitive decline impairs decision-making by reducing objectivity in information processing, particularly when contradicting existing beliefs (Henderson et al., 2020). As cognitive resources decrease, older adults may rely more heavily on initial beliefs and trusted sources, exacerbating confirmation bias (Prat-Ortega & Rocha, 2018; Conway, 2000). For example, in a study on belief updating, older adults showed a larger update bias due to reduced updating of undesirable

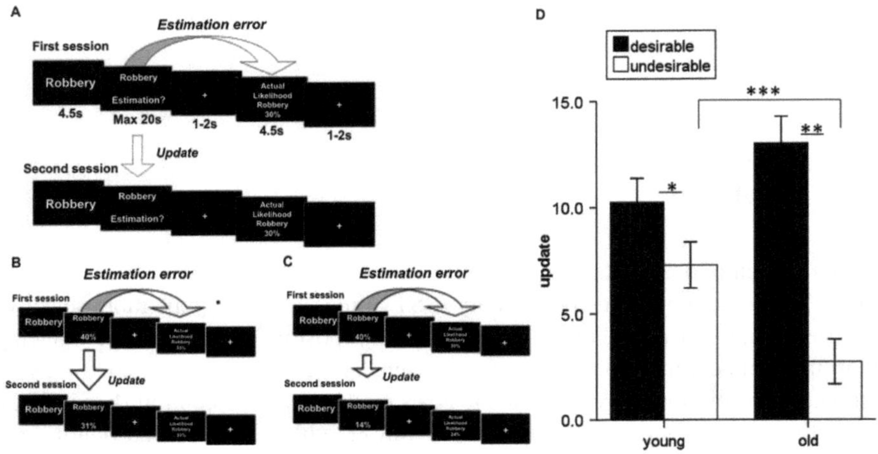

Fig. 3 Optimistic update bias increases in older age. Task design and age comparison of belief updating. (**a**) On each trial, participants were presented with 1 of 45 adverse life events and asked to estimate the likelihood of the event occurring to them in the next 5 years. They were then shown the average probability of that event happening to someone similar to them in the same sociocultural environment. Estimation error was calculated as the difference between participants' estimates and the provided average probability. The second session followed the same procedure. (**b**) For each event, belief updating was calculated as the difference between the first and second estimates. Trials were classified as "desirable" if the participants' initial estimate was higher than the provided probability (information better than expected, prompting optimistic adjustment). (**c**) Trials were classified as "undesirable" if the participants' initial estimate was lower than the provided probability (information worse than expected, prompting pessimistic adjustment). (**d**) Both young and older adults updated their beliefs more in response to desirable than undesirable information, but older adults showed a larger update bias due to reduced updating of undesirable information. (Adapted from Figs. 1 and 2 in Chowdhury et al. (2014),CC BY 3.0)

information relative to younger adults (Fig. 3; Chowdhury et al., 2014). Reduced perspective-taking and evaluation of complex information further contribute to biased decision-making (Fraga Dominguez et al., 2022). Consequently, lower cognitive complexity predicts biases like the fundamental attribution error (e.g., attributing actions of others to their character while attributing one's own actions to external factors), increasing their susceptibility to confirmation bias (Sui & Humphreys, 2017). These cognitive and neural mechanisms that contribute to confirmation bias in aging provide the foundation for understanding the adverse effects that manifest in daily decision-making, which we explore in the following section.

Other findings suggest that the selective accumulation of confirmatory evidence is driven primarily by confidence (Marcke et al., 2023). High confidence enhances the neural processing of consistent information while diminishing contradictory evidence, effectively gating information flow. When confidence is high, the neural representation of the initial decision is reinstated earlier, biasing subsequent evidence accumulation (Peters et al., 2017). While some studies show that older adults are better calibrated with respect to their confidence and their actual ability in specific tasks (e.g., vocabulary; Burns et al., 2016), others suggest a tendency toward overconfidence (Crawford & Stankov, 1996; Overhoff et al., 2022). The vmPFC has been identified as central to confirmation bias (Rollwage et al., 2020; Jiang, 2024; Kappes et al., 2020), as it tracks and represents confidence levels in decisions or beliefs (Gherman & Philiastides, 2018; Lebreton et al., 2015). Studies on the neural mechanisms of confidence have also implicated the parietal and frontal cortices during learning and perceptual decision-making, regions closely linked to accuracy and reaction times (Bounmy et al., 2023; Fetsch et al., 2014).

3 Adverse Effects of Confirmation Bias

Now that we have reviewed the socioemotional and cognitive processes related to confirmation bias, we will take a closer look at its impact on decision-making. Many cognitive biases, like the fundamental attribution error and positivity bias, increase with age as executive control and attentional resources decline (Murman, 2015; Sui & Humphreys, 2017; Carstensen & DeLiema, 2018). These biases can sometimes be adaptive, aiding in decision-making efficiency, maintaining coherent life narratives, and fostering social cohesion (Follett & Hess, 2002; Whittlestone, 2017; Bergerot et al., 2024). Despite these adaptive aspects of confirmation bias, it is crucial to recognize its detrimental effects on decision-making processes. The orientation toward confirmatory information and judgments can result in flawed reasoning and suboptimal decision-making. In this section, we will explore some of the advantageous outcomes of confirmation bias as a heuristic and how it can become maladaptive in the context of aging and financial exploitation. In the following sections, we outline some of the ways that these patterns contribute to maladaptive outcomes in older adults, including susceptibility to misinformation and risk for financial exploitation.

3.1 Age-Related Cognitive Bias Contributes to Vulnerability to Misinformation

Confirmation bias significantly contributes to vulnerability to misinformation, a risk that may be heightened in older adulthood (Ling, 2020). The refusal of scientific evidence has grown, notably around topics like smoking, evolution, vaccines, and climate change (McIntyre, 2018). This trend is often linked to biased interpretations shaped by group identity and preconceived beliefs. For instance, confirmation bias is particularly strong among climate change deniers, who are more likely to reject messages that contradict their views, thus intensifying polarization (Zhou & Shen, 2022). Similarly, older adults demonstrated confirmation bias in their evaluation of

COVID-19 public health information, which led to increased trust in misinformation that aligned with vaccine skepticism (Shiroma et al., 2023). Polarization fosters echo chambers, where individuals selectively engage with information that supports their beliefs while ignoring opposing views, further entrenching those beliefs (Modgil et al., 2021; Mao et al., 2021). Ingroup bias exacerbates this effect, increasing trust in misinformation when it comes from sources within one's own group (Knobloch-Westerwick et al., 2020; Nguyen et al., 2020). Older adults may be more susceptible to the illusory truth effect, where false statements seem truer after repeated exposure, due to cognitive decline and increased reliance on heuristic cues like repetition (Lyons, 2023). This, combined with difficulty detecting deception and the rise of generative AI capable of producing convincing fake content, poses significant risks for amplifying misinformation and targeting vulnerable populations like older adults (Brashier & Schacter, 2020; Xu et al., 2023; Shin et al., 2024; Monteith et al., 2024).

3.2 Confirmation Bias as a Predictor of Future Cognitive Decline?

Deficits in financial decision-making are indeed one of the first noticeable impairments to appear with cognitive decline, MCI, and Alzheimer's disease. For example, early signs of cognitive decline can include declining financial skills in patients with MCI in the year before developing Alzheimer's disease (Hsu & Willis, 2013). This early decline in financial abilities can have significant consequences for older adults' financial well-being and security. Interestingly, recent work suggests that sudden wealth loss (e.g., due to exploitation) has been linked to increased rates of subsequent cognitive decline (Cho et al., 2023). Given the link between financial abilities and cognitive health, identifying cognitive biases early could offer valuable insight into predicting an individual's risk for both financial exploitation and future cognitive decline.

Considering these and other negative effects associated with age-related cognitive decline, various approaches have been proposed for predicting the onset of MCI and ADRD, including specific cognitive tasks, as well as self-rated and informant-rated measures of everyday cognitive function (Rueda et al., 2015; Rabin et al., 2012). We proposed that more focused explorations of cognitive bias may offer another valuable dimension in predicting cognitive decline. By examining deviations from rational decision-making processes like confirmation bias, researchers can potentially identify subtle cognitive changes indicative of early cognitive impairment. Utilizing cognitive bias as a predictor may provide complementary insights into the cognitive processes affected by MCI or ADRD, offering a more comprehensive understanding of the onset and progression of cognitive impairment. This approach may enhance early detection efforts and facilitate interventions aimed at delaying cognitive decline and mitigating negative functional and financial consequences in at-risk populations.

Some research has already assessed the longitudinal influence of cognitive bias on aging. One study shows that "aging-related cognitions" are a significant predictor of health changes in later life (Wurm et al., 2007). Related findings have shown that age-related deficits in cognitive functioning may contribute to selection bias and attrition in longitudinal aging studies (Hernandez et al., 2024; Matthews et al., 2004), suggesting that biases could impact participation and retention in such studies and consequently contribute to health disparities. Although these findings do not directly assess cognitive biases as predictors of cognitive decline, they suggest that older adults who exhibit these biases may have reduced competency. However, further research is needed to explore this relationship in more depth.

3.3 Neuroeconomics of Confirmation Bias

Within the fields of decision neuroscience and neuroeconomics, several models have been proposed to explain confirmation bias. One model suggests that confirmation bias results from interactions between fast processes that form initial judgments and slower processes that evaluate subsequent evidence. The initial judgment biases the processing and integration of later evidence (Prat-Ortega & Rocha, 2018; Lange et al., 2021). This interplay between quick first impressions and deliberate decision-making is often linked to the orbitofrontal cortex (OFC) and anterior cingulate cortex (ACC), which track value representations over time (Herd et al., 2021). This perspective also involves executive control and higher-order cognitive processes like planning, inhibition, and cognitive flexibility, which help override automatic responses (Royall et al., 2002). Reduced executive control capacity hinders the ability to objectively consider contradictory evidence, making it difficult to override preexisting beliefs (Botvinick & Braver, 2015). Such reductions in executive function are well-documented in older adults. Aging diminishes the neural computational capacity needed for complex decision-making processes, such as deliberation, reasoning, and analysis (Type 2 processes; Mata, 2005), and this decline worsens with MCI and ADRD (Kirova et al., 2015; Belleville et al., 2017). Age-related declines in executive control may particularly impact responses to stimuli related to strangers versus the self or close friends, increasing bias toward familiar and positive stimuli (Sui & Humphreys, 2017).

Reduced cognitive flexibility has also been associated with several other key brain regions (Fig. 4). The orbitofrontal cortex (OFC) is involved in encoding expected outcomes, value representations, and reward learning (Mainen & Kepecs, 2009; Gallagher et al., 1999; Balleine et al., 2011). Abnormal OFC activity may contribute to biased integration of positive versus negative outcomes in line with prior beliefs. The anterior cingulate cortex (ACC) is also implicated in tracking expected outcomes during learning (Kosson et al., 2006) and emotion regulation (Stevens et al., 2011). Abnormalities in ACC function may disrupt accurate updating of expectations in the face of contradictory information, as well as the ability to regulate negative emotions in the context of inconsistency (Gasquoine, 2013).

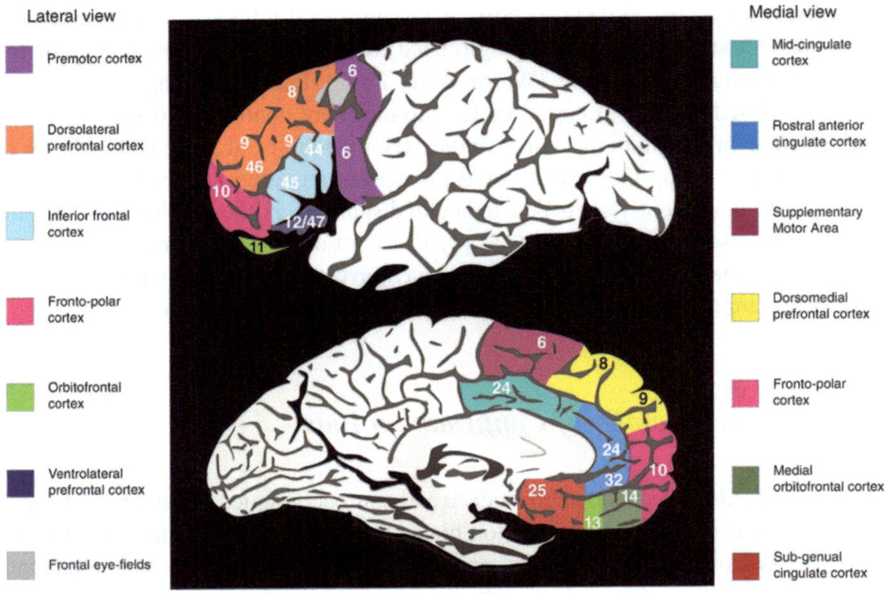

Fig. 4 Key regions of the prefrontal cortex, many of which are implicated in confirmation bias. The ventromedial prefrontal cortex (vmPFC), known for its role in decision-making and emotional regulation; the dorsolateral prefrontal cortex (dlPFC), involved in executive functions such as working memory and cognitive control; the orbitofrontal cortex (OFC), associated with reward processing and social decision-making; and the anterior cingulate cortex (ACC), involved in attentional control. Understanding the function of these regions provides insights into their contribution to confirmation bias and cognitive decline in older adulthood. (Reproduced from Friedman and Robbins (2022), CC BY 4.0)

Neural network frameworks have also been proposed for assessing cognitive biases (Korteling et al., 2018). Basic neural network properties like facilitation (i.e., easier activation after prior activation) and potentiation (strengthening of pathways from repeated activations) may make the brain more likely to process information compatible with existing beliefs/activations, which could underlie the tendency to favor confirmatory inputs. Biases may fundamentally arise from how biological neural networks inherently operate, via selectively processing inputs based on their compatibility with existing patterns of activation (for a more detailed argument, see Sprague et al., 2015). Slower, more fragmented transitions between different latent brain states during working memory encoding, maintenance, and retrieval mediate the relationship between older age and reduced cognitive flexibility (Lee et al., 2022). In these ways, impairment of the frontoparietal control network may also reduce the ability to override automatic, biased information processing (Korteling et al., 2018). Normal aging is associated with reorganization of functional brain networks. For example, stronger connections between default mode, executive control, and salience networks are associated with higher cognitive function in older adulthood; however, stronger connections of primary sensory networks with salience and basal ganglia networks were linked to lower function in aging Hughes et al., 2019). Taken together, these neuroeconomic models provide a framework for understanding how aging-related changes in neural structure and function can systematically bias decision-making processes, particularly through confirmation bias mechanisms. Having established the neural correlates of confirmation bias in aging (Fig. 4), we now turn to examining potential interventions that target these specific neural mechanisms through noninvasive brain stimulation techniques.

4 Modulating Decision-Making with Noninvasive Brain Stimulation

We have explored confirmation bias, its cognitive and neural underpinnings, and the adverse outcomes associated with it. Despite extensive research, direct evidence linking these concepts remains limited. Noninvasive brain stimulation (NIBS) techniques offer a unique advantage in studying cognitive processes like confirmation bias, especially in aging. Unlike correlational methods such as fMRI, NIBS allows for direct modulation of brain activity, enabling causal testing of cognitive functions (Zimerman & Hummel, 2010). For instance, transcranial alternating current stimulation (tACS) can explore the role of neural oscillations in cognition by entraining specific brain rhythms during tasks (Vosskuhl et al., 2018). NIBS also facilitates the study of network interactions by stimulating one node and observing the effects on broader neural connectivity (Sale et al., 2015). Additionally, it helps investigate how the roles of brain regions and networks in cognition evolve across the lifespan (Goldthorpe et al., 2020). Beyond research, NIBS holds potential for enhancing cognitive abilities, such as working memory, in both healthy and clinical populations by upregulating activity in key brain areas (Poppe et al., n.d.). This section reviews research employing NIBS to study aging, confirmation bias, and related cognitive processes.

4.1 Use of Noninvasive Brain Stimulation to Modulate Cognitive Processes Associated with Confirmation Bias and Aging

NIBS methods have been applied to study and potentially improve cognitive abilities (Poppe et al., n.d.). A major area of the research has centered on improving working memory, with examples using transcranial magnetic stimulation (TMS; Parkin et al., 2015), transcranial direct current stimulation (tDCS; Satorres et al., 2022), and transcranial alternating current stimulation (tACS; Nissim et al., 2023) being used to modulate neural activity. Much of this work has focused on applying anodal stimulation to the dorsolateral prefrontal cortex (dlPFC). While TMS has superior spatial specificity (e.g., Robertson & Pascual-Leone, 2003), tACS can modulate temporal patterns of stimulation very deliberately by entraining intrinsic oscillations through alternating currents (Helfrich et al., 2014; Krause & Cohen Kadosh, 2014). Exciting recent work has shown that tACS application using individualized theta rhythms can restore working memory in older adults (Reinhart & Nguyen, 2019).

NIBS techniques have also been used to enhance cognitive control abilities. For example, anodal tDCS (i.e., increases neural excitability) applied over the dlPFC has been shown to enhance abilities like inference control, selective attention, and proactive cognitive control in healthy individuals (Feeser et al., 2014). Other findings have shown specifically that anodal tDCS over the left dlPFC improved cognitive control (Dubreuil-Vall et al., 2021). Another study found that prefrontal tDCS enhanced proactive cognitive control as evidenced by changes in behavioral performance and gamma oscillations during a cognitive control task in individuals with schizophrenia (Boudewyn et al., 2020). TACS stimulation of frontal brain regions can impact theta oscillations, which are linked to cognitive control and decision-making (Wischnewski et al., 2023). Recent work has shown how tACS over frontal regions can modulate how people incorporate advice into the decision-making process (Wischnewski et al., 2021).

TMS and tDCS have also been used to assess and modulate the processing of social information in the brain. NIBS seems to be a promising approach for inducing social plasticity (modulation of neural substrates associated with social cognition) in developmental disorders like autism spectrum disorder (ASD), schizophrenia, and Williams syndrome (Boggio et al., 2015). NIBS techniques have been used to study implicit social cognition, which refers to the automatic processing of social information without conscious awareness or control. NIBS as a social cognitive intervention has shown benefits across multiple social cognitive and social behavioral skills by targeting and modulating activity in relevant brain regions (Marini et al., 2018). NIBS has also been used to study social decision-making processes (e.g., reducing rejection of unfair offers in an Ultimatum Game via dlPFC-TMS;; Fecteau, 2023). Other research suggests that the ability to modulate cerebro-cerebellar connectivity with noninvasive brain stimulation opens up possible clinical applications in social and affective processing in certain disorders Cattaneo et al., 2022).

4.2 Noninvasive Brain Stimulation and Reducing Bias

This research in working memory, cognitive control, decision-making, and social information processing can inform our understanding of confirmation bias through several mechanisms. First, improved cognitive control enables individuals to regulate their attention more effectively (Drigas & Karyotaki, 2017), allowing them to consider a broader range of information and perspectives rather than selectively attending to information that confirms preexisting beliefs. Second, enhanced working memory capacity may facilitate the retention and manipulation of contradictory information, making it easier to update beliefs in response to new evidence. Additionally, by increasing the efficiency of cognitive processes involved in reasoning and decision-making, such as inhibitory control and conflict resolution, noninvasive brain stimulation of the PFC may promote more objective evaluation of evidence and less reliance on biased heuristics. Overall, by bolstering cognitive control and working memory functions, NIBS targeted at the PFC could mitigate confirmation bias by facilitating more thorough and rational processing of information.

Other recent research has examined the efficacy of NIBS techniques in reducing bias during decision-making. One study shows that anodal vmPFC-tDCS reduced susceptibility to framing effects in a decision-making task when compared to cathodal (i.e., decreases neural excitability) stimulation (Winker et al., 2019). Other findings further support that anodal tDCS to the vmPFC increased rational decision-making by reducing cognitive biases like loss aversion (Kroker et al., 2022). Similar findings have shown that anodal vmPFC-tDCS led to more rational feedback evaluation by reducing biases in a gambling task (Kroker et al., 2023). These findings demonstrate the causal role of the vmPFC in evaluating rewards and tracking value during decision-making and feedback processing.

One important note is that the effects of NIBS on cognitive functions like memory in older adults need to be interpreted in the context of age-related structural and functional brain changes that could alter how stimulation impacts neural activity, although further research is needed to draw conclusions on this (e.g., Goldthorpe et al., 2020). In a related vein, abbreviated or accelerated stimulation protocols are sometimes used to make stimulation sessions more tolerable for older adults. For example, accelerated theta-burst TMS protocols that deliver stimulation in much shorter sessions can improve adherence in older patient groups (Blumberger et al., 2022). Overall, while NIBS techniques show promise for modulating cognitive processes in aging populations, further research is needed to optimize protocols specifically for older adults, taking into account both neural and practical considerations (Friedman & Robbins, 2022).

At-home tDCS and tACS are emerging as accessible and effective noninvasive brain stimulation techniques that can be self-administered by patients (Charvet et al., 2020; Riggs et al., 2018). These portable devices offer key advantages over traditional clinical treatments like TMS, which is likely to remain clinic-based due to its complexity and safety profile. By allowing patients to receive treatment at home, tDCS and tACS increase accessibility, reduce the burden of travel, and enhance adherence to treatment protocols. With appropriate training and remote supervision through telehealth, these devices can be used safely and effectively, incorporating built-in safeguards like dose control. At-home brain stimulation may also allow for more flexible treatment schedules, facilitating longer or more frequent sessions that would be impractical in a clinical setting. Ongoing research demonstrates the potential of at-home tDCS and tACS to treat a range of conditions, including depression, chronic pain, and Alzheimer's disease (Oh et al., 2023; Altomare et al., 2023). However, challenges remain, particularly in ensuring proper use and maintaining treatment fidelity without direct clinical oversight. As the technology and supporting infrastructure evolve, at-home brain stimulation has the potential to expand access to noninvasive therapies for diverse neurological and psychiatric conditions.

5 Conclusions and Future Directions

In conclusion, this chapter has shed light on the complex interplay between confirmation bias and aging, highlighting the socioemotional contributors and neural underpinnings of this phenomenon. Age-related changes in orientation toward emotionally rich and positive experiences, in combination with cognitive changes that reduce critical examination of one's existing beliefs and biases, all contribute to a heightened susceptibility to confirmation bias. These variables are enhanced as cognitive decline progresses to MCI and ADRD. Taken together, these changes leave older adults vulnerable to maladaptive outcomes, including financial exploitation and susceptibility to misinformation. While confirmation bias may serve as an adaptive process in certain contexts, its adverse effects, including susceptibility to financial exploitation and misinformation, underscore the importance of understanding and mitigating its impact, especially among older adults.

The findings presented in this chapter have significant clinical implications. Healthcare providers working with older adults should consider screening for confirmation bias as part of cognitive assessments, particularly when evaluating decision-making capacity. Financial advisors and elder law attorneys may benefit from recognizing signs of confirmation bias when advising clients. Additionally, educational interventions that explicitly address confirmation bias could be integrated into cognitive health programs for older adults, potentially serving as low-cost preventative measures against financial exploitation. As NIBS techniques become more refined, they may eventually transition from research tools to clinical interventions for at-risk individuals.

Moving forward, future research is needed to identify the causal mechanisms linking all these contributing variables. To this end, NIBS techniques emerge as a promising avenue for modulating related cognitive processes to identify key brain regions more explicitly and potentially as a method of reducing confirmation bias in later life. Specific research priorities should include (1) longitudinal studies examining confirmation bias as a predictor of financial exploitation vulnerability; (2) development of standardized assessment tools measuring confirmation bias in clinical settings; (3) optimization of NIBS protocols specifically for older populations, including home-based applications; and (4) investigation of combined cognitive training and NIBS approaches to enhance intervention efficacy. As global populations trend toward older age, and financial schemes, misinformation, and social interactions become more technological and complex, there is much to be gained from better understanding these relationships. By addressing these issues, future research can contribute to enhancing decision-making quality and promoting cognitive well-being in aging populations.

Acknowledgments This work was supported in part by a grant from the National Institute on Aging (R01-AG067011 to DVS and F31-AG085934 to JBW). The authors would like to thank Chelsea Helion feedback on an earlier version of this chapter.

References

Altomare, D., Benussi, A., Cantoni, V., Premi, E., Rivolta, J., Cupidi, C., Martorana, A., Santarnecchi, E., Padovani, A., Koch, G., & Borroni, B. (2023). Home-based transcranial alternating current stimulation (tACS) in Alzheimer's disease: Rationale and study design. *Alzheimer's Research & Therapy, 15*(1), 155. https://doi.org/10.1186/s13195-023-01297-4

Balleine, B. W., Leung, B. K., & Ostlund, S. B. (2011). The orbitofrontal cortex, predicted value, and choice. *Annals of the New York Academy of Sciences, 1239*(1), 43–50. https://doi.org/10.1111/j.1749-6632.2011.06270.x

Belleville, S., Fouquet, C., Hudon, C., Zomahoun, H. T. V., Croteau, J., & Consortium for the Early Identification of Alzheimer's Disease-Quebec. (2017). Neuropsychological measures that predict progression from mild cognitive impairment to Alzheimer's type dementia in older adults: A systematic review and meta-analysis. *Neuropsychology Review, 27*(4), 328–353. https://doi.org/10.1007/s11065-017-9361-5

Bergerot, C., Barfuss, W., & Romanczuk, P. (2024). Moderate confirmation bias enhances collective decision-making in reinforcement-learning agents. *bioRxiv*, 2023.11.21.568073. https://doi.org/10.1101/2023.11.21.568073

Betz-Hamilton, A. E., & Vincenti, V. B. (2018). Risk factors within families associated with elder financial exploitation by relatives with powers of attorney. *Journal of Family and Consumer Sciences, 110*(1), 19–27. https://doi.org/10.14307/JFCS110.1.19

Blumberger, D. M., Mulsant, B. H., Thorpe, K. E., McClintock, S. M., Konstantinou, G. N., Lee, H. H., Nestor, S. M., Noda, Y., Rajji, T. K., Trevizol, A. P., Vila-Rodriguez, F., Daskalakis, Z. J., & Downar, J. (2022). Effectiveness of standard sequential bilateral repetitive transcranial magnetic stimulation vs bilateral theta burst stimulation in older adults with depression: The FOUR-D randomized noninferiority clinical trial. *JAMA Psychiatry, 79*(11), 1065–1073. https://doi.org/10.1001/jamapsychiatry.2022.2862

Boggio, P. S., Asthana, M., Costa, T. L., Valasek, C. A., & Osório, A. A. (2015). Promoting social plasticity in developmental disorders with non-invasive brain stimulation techniques. *Frontiers in Neuroscience, 9*. https://doi.org/10.3389/fnins.2015.00294

Botvinick, M., & Braver, T. (2015). Motivation and cognitive control: From behavior to neural mechanism. *Annual Review of Psychology, 66*, 83–113. https://doi.org/10.1146/annurev-psych-010814-015044

Boudewyn, M. A., Scangos, K., Ranganath, C., & Carter, C. S. (2020). Using prefrontal transcranial direct current stimulation (tDCS) to enhance proactive cognitive control in schizophrenia. *Neuropsychopharmacology, 45*(11), 1877–1883. https://doi.org/10.1038/s41386-020-0750-8

Bounmy, T., Eger, E., & Meyniel, F. (2023). A characterization of the neural representation of confidence during probabilistic learning. *NeuroImage, 268*, 119849. https://doi.org/10.1016/j.neuroimage.2022.119849

Brashier, N. M., & Schacter, D. L. (2020). Aging in an era of fake news. *Current Directions in Psychological Science, 29*(3), 316–323. https://doi.org/10.1177/0963721420915872

Brewer, M. B. (1999). The psychology of prejudice: Ingroup love and outgroup hate? *Journal of Social Issues, 55*(3), 429–444. https://doi.org/10.1111/0022-4537.00126

Burns, K. M., Burns, N. R., & Ward, L. (2016). Confidence—More a personality or ability trait? It depends on how it is measured: A comparison of young and older adults. *Frontiers in Psychology, 7*, 518. https://doi.org/10.3389/fpsyg.2016.00518

Cafferata, A., & Tramontana, F. (2019). A financial market model with confirmation bias. *Structural Change and Economic Dynamics, 51*, 252–259. https://doi.org/10.1016/j.strueco.2019.08.004

Carstensen, L. L. (1995). Evidence for a life-span theory of socioemotional selectivity. *Current Directions in Psychological Science, 4*(5), 151–156.

Carstensen, L. L., & DeLiema, M. (2018). The positivity effect: A negativity bias in youth fades with age. *Current Opinion in Behavioral Sciences, 19*, 7–12. https://doi.org/10.1016/j.cobeha.2017.07.009

Carstensen, L. L., Turan, B., Scheibe, S., Ram, N., Ersner-Hershfield, H., Samanez-Larkin, G. R., Brooks, K. P., & Nesselroade, J. R. (2011). Emotional experience improves with age: Evidence based on over 10 years of experience sampling. *Psychology and Aging, 26*(1), 21–33. https://doi.org/10.1037/a0021285

Cassidy, B. S., Hughes, C., Lanie, S., & Krendl, A. C. (2020). Effects of executive ability on bias and ingroup perceptions in aging. *Psychology and Aging, 35*(2), 283–294. https://doi.org/10.1037/pag0000420

Cattaneo, Z., Ferrari, C., Ciricugno, A., Heleven, E., Schutter, D. J. L. G., Manto, M., & Van Overwalle, F. (2022). New horizons on non-invasive brain stimulation of the social and affective cerebellum. *The Cerebellum, 21*(3), 482–496. https://doi.org/10.1007/s12311-021-01300-4

Charness, G., & Dave, C. (2017). Confirmation bias with motivated beliefs. *Games and Economic Behavior, 104*, 1–23. https://doi.org/10.1016/j.geb.2017.02.015

Charvet, L. E., Shaw, M. T., Bikson, M., Woods, A. J., & Knotkova, H. (2020). Supervised transcranial direct current stimulation (tDCS) at home: A guide for clinical research and practice. *Brain Stimulation, 13*(3), 686–693. https://doi.org/10.1016/j.brs.2020.02.011

Cho, T.-C., Yu, X., Gross, A. L., Zhang, Y. S., Lee, J., Langa, K. M., & Kobayashi, L. C. (2023). Negative wealth shocks in later life and subsequent cognitive function in older adults in China, England, Mexico, and the USA, 2012-18: A population-based, cross-nationally harmonised, longitudinal study. *The Lancet. Healthy Longevity, 4*(9), e461–e469. https://doi.org/10.1016/S2666-7568(23)00113-7

Chowdhury, R., Sharot, T., Wolfe, T., Düzel, E., & Dolan, R. J. (2014). Optimistic update bias increases in older age. *Psychological Medicine, 44*(9), 2003–2012. https://doi.org/10.1017/S0033291713002602

Conway, M. (2000). Individual differences in attentional resources and social cognition: Elaboration and complexity in representations of others and self. In U. von Hecker, S. Dutke, & G. Sedek (Eds.), *Generative mental processes and cognitive resources: Integrative research on adaptation and control* (pp. 5–38). Springer Netherlands. https://doi.org/10.1007/978-94-011-4373-8_1

Crawford, J. D., & Stankov, L. (1996). Age differences in the realism of confidence judgements: A calibration study using tests of fluid and crystallized intelligence. *Learning and Individual Differences, 8*(2), 83–103. https://doi.org/10.1016/S1041-6080(96)90027-8

Cutler, J., Nitschke, J. P., Lamm, C., & Lockwood, P. L. (2021). Older adults across the globe exhibit increased prosocial behavior but also greater in-group preferences. *Nature Aging, 1*(10), 880–888.

Drigas, A., & Karyotaki, M. (2017). Attentional control and other executive functions. *International Journal of Emerging Technologies in Learning (iJET), 12*(03), Article 03. https://doi.org/10.3991/ijet.v12i03.6587

Dubreuil-Vall, L., Gomez-Bernal, F., Villegas, A. C., Cirillo, P., Surman, C., Ruffini, G., ... & Camprodon, J. A. (2021). Transcranial direct current stimulation to the left dorsolateral prefrontal cortex improves cognitive control in patients with attention-deficit/hyperactivity disorder: a randomized behavioral and neurophysiological study. Biological Psychiatry: *Cognitive neuroscience and neuroimaging, 6*(4), 439–448.

Dunne, R. (2016). Decision making is caused by information processing and emotion: A synthesis of two approaches to explain the phenomenon of confirmation bias. In *Irish conference on artificial intelligence and cognitive science*. https://api.semanticscholar.org/CorpusID:15518030

Elder Justice Initiative (EJI) | Elder Abuse and Elder Financial Exploitation Statutes | United States Department of Justice. (2023, October 17). https://www.justice.gov/elderjustice/prosecutors/statutes

Fareri, D. S., Hackett, K., Tepfer, L. J., Kelly, V., Henninger, N., Reeck, C., Giovannetti, T., & Smith, D. V. (2022). Age-related differences in ventral striatal and default mode network function during reciprocated trust. *NeuroImage, 256* 119267. https://doi.org/10.1016/j.neuroimage.2022.119267

Fecteau, S. (2023). Influencing human behavior with noninvasive brain stimulation: Direct human brain manipulation revisited. *The Neuroscientist, 29*(3), 317–331. https://doi.org/10.1177/10738584211067744

Feeser, M., Prehn, K., Kazzer, P., Mungee, A., & Bajbouj, M. (2014). Transcranial direct current stimulation enhances cognitive control during emotion regulation. *Brain Stimulation, 7*(1), 105–112. https://doi.org/10.1016/j.brs.2013.08.006

Federal Trade Commission. (2024). As nationwide fraud losses top $10 billion in 2023, FTC steps up efforts to protect the public. https://www.ftc.gov/news-events/news/press-releases/2024/02/nationwide-fraud-losses -top-10-billion-2023-ftc-steps-efforts-protect-public

Fenton, L., Weissberger, G. H., Boyle, P. A., Mosqueda, L., Yassine, H. N., Nguyen, A. L., Lim, A. C., & Han, S. D. (2022). Cognitive and neuroimaging correlates of financial exploitation vulnerability in older adults without dementia: Implications for early detection of Alzheimer's disease. *Neuroscience & Biobehavioral Reviews, 140*, 104773. https://doi.org/10.1016/j.neubiorev.2022.104773

Fernandes, C., Macedo, I., Barbosa, F., & Marques-Teixeira, J. (2021). Economic decision-making in the continuum between healthy aging and Alzheimer's disease: A systematic review of 20 years of research. *Neuroscience & Biobehavioral Reviews, 131*, 1243–1263. https://doi.org/10.1016/j.neubiorev.2021.10.030

Fetsch, C. R., Kiani, R., & Shadlen, M. N. (2014). Predicting the accuracy of a decision: A neural mechanism of confidence. *Cold Spring Harbor Symposia on Quantitative Biology, 79*, 185–197. https://doi.org/10.1101/sqb.2014.79.024893

Follett, K. J., & Hess, T. M. (2002). Aging, cognitive complexity, and the fundamental attribution error. *The Journals of Gerontology: Series B, 57*(4), P312–P323. https://doi.org/10.1093/geronb/57.4.P312

Fraga Dominguez, S., Ozguler, B., Storey, J. E., & Rogers, M. (2022). Elder abuse vulnerability and risk factors: Is financial abuse different from other subtypes? *Journal of Applied Gerontology, 41*(4), 928–939. https://doi.org/10.1177/07334648211036402

Frazier, I., Lighthall, N. R., Horta, M., Perez, E., & Ebner, N. C. (2019). CISDA: Changes in integration for social decisions in aging. *WIREs Cognitive Science, 10*(3), e1490. https://doi.org/10.1002/wcs.1490

Friedman, N. P., & Robbins, T. W. (2022). The role of prefrontal cortex in cognitive control and executive function. *Neuropsychopharmacology, 47*(1), Article 1. https://doi.org/10.1038/s41386-021-01132-0

Gallagher, M., McMahan, R. W., & Schoenbaum, G. (1999). Orbitofrontal cortex and representation of incentive value in associative learning. *Journal of Neuroscience, 19*(15), 6610–6614. https://doi.org/10.1523/JNEUROSCI.19-15-06610.1999

Gasquoine, P. G. (2013). Localization of function in anterior cingulate cortex: From psychosurgery to functional neuroimaging. *Neuroscience & Biobehavioral Reviews, 37*(3), 340–348. https://doi.org/10.1016/j.neubiorev.2013.01.002

Gherman, S., & Philiastides, M. G. (2018). Human VMPFC encodes early signatures of confidence in perceptual decisions. *eLife, 7*, e38293. https://doi.org/10.7554/eLife.38293

Goldthorpe, R. A., Rapley, J. M., & Violante, I. R. (2020). A systematic review of non-invasive brain stimulation applications to memory in healthy aging. *Frontiers in Neurology, 11*, 575075. https://doi.org/10.3389/fneur.2020.575075

Gross, J. J., Carstensen, L. L., Pasupathi, M., Tsai, J., Götestam Skorpen, C., & Hsu, A. Y. C. (1997). Emotion and aging: Experience, expression, and control. *Psychology and Aging, 12*(4), 590–599. https://doi.org/10.1037/0882-7974.12.4.590

Helfrich, R. F., Schneider, T. R., Rach, S., Trautmann-Lengsfeld, S. A., Engel, A. K., & Herrmann, C. S. (2014). Entrainment of brain oscillations by transcranial alternating current stimulation. *Current Biology, 24*(3), 333–339. https://doi.org/10.1016/j.cub.2013.12.041

Henderson, S. E., Lockhart, H. A., Davis, E. E., Emrich, S. M., & Campbell, K. L. (2020). Reduced attentional control in older adults leads to deficits in flexible prioritization of visual working memory. *Brain Sciences, 10*(8), 542. https://doi.org/10.3390/brainsci10080542

Herd, S., Krueger, K., Nair, A., Mollick, J., & O'Reilly, R. (2021). Neural mechanisms of human decision-making. *Cognitive, Affective, & Behavioral Neuroscience, 21*(1), 35–57. https://doi.org/10.3758/s13415-020-00842-0

Hernandez, R., Jin, H., Lee, P.-J., Schneider, S., Junghaenel, D. U., Stone, A. A., Meijer, E., Gao, H., Maupin, D., & Zelinski, E. M. (2024). Attrition from longitudinal ageing studies and performance across domains of cognitive functioning: An individual participant data meta-analysis. *BMJ Open, 14*(3), e079241. https://doi.org/10.1136/bmjopen-2023-079241

Horner, C. G., Galletta, D., Crawford, J., & Shirsat, A. (2021). Emotions: The unexplored fuel of fake news on social media. *Journal of Management Information Systems, 38*(4), 1039–1066. https://doi.org/10.1080/07421222.2021.1990610

Hsu, J. W., & Willis, R. (2013). Dementia risk and financial decision making by older households: The impact of information. *Journal of Human Capital, 2013*, 45. https://doi.org/10.2139/ssrn.2339225

Hughes, C., Cassidy, B. S., Faskowitz, J., Avena-Koenigsberger, A., Sporns, O., & Krendl, A. C. (2019). Age differences in specific neural connections within the default mode network underlie theory of mind. *NeuroImage, 191*, 269–277. https://doi.org/10.1016/j.neuroimage.2019.02.024

Institute, B. C. L. (2004). *Report on financial arrangements between older adults and family members: Loans and guarantees.* https://www.canlii.org/en/commentary/doc/2004CanLIIDocs210#!fragment//BQCwhgziBcwMYgK4DsDWszIQewE4BUBTADwBdoByCgSgBpltTCIBFRQ3AT0otokLC4EbDtyp8BQkAGU8pAELcASgFEAMioBqAQQByAYRW1SYAEbRS2ONWpA

Jiang, Y. (2024). A theory of the neural mechanisms underlying negative cognitive bias in major depression. *Frontiers in Psychiatry, 15*. https://doi.org/10.3389/fpsyt.2024.1348474

Kappes, A., Harvey, A. H., Lohrenz, T., Montague, P. R., & Sharot, T. (2020). Confirmation bias in the utilization of others' opinion strength. *Nature Neuroscience, 23*(1), 130–137. https://doi.org/10.1038/s41593-019-0549-2

Kirova, A.-M., Bays, R. B., & Lagalwar, S. (2015). Working memory and executive function decline across normal aging, mild cognitive impairment, and Alzheimer's disease. *BioMed Research International, 2015*, e748212. https://doi.org/10.1155/2015/748212

Knobloch-Westerwick, S., Mothes, C., & Polavin, N. (2020). Confirmation bias, ingroup bias, and negativity bias in selective exposure to political information. *Communication Research, 47*(1), 104–124. https://doi.org/10.1177/0093650217719596

Korteling, J. E., Brouwer, A.-M., & Toet, A. (2018). A neural network framework for cognitive bias. *Frontiers in Psychology, 9*, 1561. https://doi.org/10.3389/fpsyg.2018.01561

Kosson, D. S., Budhani, S., Nakic, M., Chen, G., Saad, Z. S., Vythilingam, M., Pine, D. S., & Blair, R. J. R. (2006). The role of the amygdala and rostral anterior cingulate in encoding expected outcomes during learning. *NeuroImage, 29*(4), 1161–1172. https://doi.org/10.1016/j.neuroimage.2005.07.060

Krause, B., & Cohen Kadosh, R. (2014). Not all brains are created equal: The relevance of individual differences in responsiveness to transcranial electrical stimulation. *Frontiers in Systems Neuroscience, 8*. https://doi.org/10.3389/fnsys.2014.00025

Kroker, T., Wyczesany, M., Rehbein, M. A., Roesmann, K., Wessing, I., & Junghöfer, M. (2022). Noninvasive stimulation of the ventromedial prefrontal cortex modulates rationality of human decision-making. *Scientific Reports, 12*, 20213. https://doi.org/10.1038/s41598-022-24526-6

Kroker, T., Wyczesany, M., Rehbein, M. A., Roesmann, K., Wessing, I., Wiegand, A., Bölte, J., & Junghöfer, M. (2023). Excitatory stimulation of the ventromedial prefrontal cortex reduces cognitive gambling biases via improved feedback learning. *Scientific Reports, 13*(1), 17984. https://doi.org/10.1038/s41598-023-43264-x

Kroon, A. C., Van Der Meer, T. G. L. A., & Pronk, T. (2022). Does information about bias attenuate selective exposure? The effects of implicit bias feedback on the selection of outgroup-rich news. *Human Communication Research, 48*(2), 346–373. https://doi.org/10.1093/hcr/hqac004

Kruglanski, A. W., Jasko, K., Milyavsky, M., Chernikova, M., Webber, D., Pierro, A., & Di Santo, D. (2018). Cognitive consistency theory in social psychology: A paradigm reconsidered. *Psychological Inquiry, 29*(2), 45–59. https://doi.org/10.1080/1047840X.2018.1480619

Lange, R. D., Chattoraj, A., Beck, J. M., Yates, J. L., & Haefner, R. M. (2021). A confirmation bias in perceptual decision-making due to hierarchical approximate inference. *PLoS Computational Biology, 17*(11), e1009517. https://doi.org/10.1371/journal.pcbi.1009517

Lebreton, M., Abitbol, R., Daunizeau, J., & Pessiglione, M. (2015). Automatic integration of confidence in the brain valuation signal. *Nature Neuroscience, 18*(8), 1159–1167. https://doi.org/10.1038/nn.4064

Lee, B., Cai, W., Young, C. B., Yuan, R., Ryman, S., Kim, J., Santini, V., Henderson, V. W., Poston, K. L., & Menon, V. (2022). Latent brain state dynamics and cognitive flexibility in older adults. *Progress in Neurobiology, 208*, 102180. https://doi.org/10.1016/j.pneurobio.2021.102180

Ling, R. (2020). Confirmation bias in the era of mobile news consumption: the social and psychological dimensions. *Digital Journalism, 8*(5), 596–604.

Loboprabhu, S., Molinari, V., & Asghar-Ali, A. A. (2015). Castaways: Addressing hostility and helplessness in severely lonely adults. *Journal of Psychiatric Practice, 21*(2), 93. https://doi.org/10.1097/01.pra.0000462602.94853.2f

Lundström, E. (2012). *Svenska: Prefrontala kortex av hjärnan* [Graphic]. Own work. https://commons.wikimedia.org/wiki/File:Prefrontal_cortex_of_the_brain.png

Lyons, B. A. (2023). Older Americans are more vulnerable to prior exposure effects in news evaluation. *Harvard Kennedy School Misinformation Review.*https://doi.org/10.37016/mr-2020-118

Mainen, Z. F., & Kepecs, A. (2009). Neural representation of behavioral outcomes in the orbitofrontal cortex. *Current Opinion in Neurobiology, 19*(1), 84–91. https://doi.org/10.1016/j.conb.2009.03.010

Mao, Y., Akyol, E., & Hovakimyan, N. (2021). Impact of confirmation bias on competitive information spread in social networks. *IEEE Transactions on Control of Network Systems, 8*(2), 816–827. https://doi.org/10.1109/TCNS.2021.3050117

Marcke, H. V., Denmat, P. L., Verguts, T., & Desender, K. (2023). Manipulating prior beliefs causally induces under- and overconfidence. *bioRxiv*, 2022.03.01.482511. https://doi.org/10.1101/2022.03.01.482511

Marini, M., Banaji, M. R., & Pascual-Leone, A. (2018). Studying implicit social cognition with noninvasive brain stimulation. *Trends in Cognitive Sciences, 22*(11), 1050–1066. https://doi.org/10.1016/j.tics.2018.07.014

Mata, R. (2005). A neurocomputational approach to decision making and aging. In W. Duch, J. Kacprzyk, E. Oja, & S. Zadrożny (Eds.), *Artificial neural networks: Formal models and their applications—ICANN 2005* (pp. 411–416). Springer. https://doi.org/10.1007/11550907_65

Mather, M., & Carstensen, L. L. (2005). Aging and motivated cognition: The positivity effect in attention and memory. *Trends in Cognitive Sciences, 9*(10), 496–502. https://doi.org/10.1016/j.tics.2005.08.005

Matthews, F. E., Chatfield, M., Freeman, C., McCracken, C., Brayne, C., & CFAS, M. (2004). Attrition and bias in the MRC cognitive function and ageing study: An epidemiological investigation. *BMC Public Health, 4*(1), 12. https://doi.org/10.1186/1471-2458-4-12

McIntyre, L. (2018). *Post-truth.* The MIT Press. https://doi.org/10.7551/mitpress/11483.001.0001

Modgil, S., Singh, R. K., Gupta, S., & Dennehy, D. (2021). A confirmation bias view on social media induced polarisation during Covid-19. *Information Systems Frontiers*, 1–25. https://doi.org/10.1007/s10796-021-10222-9

Monteith, S., Glenn, T., Geddes, J. R., Whybrow, P. C., Achtyes, E., & Bauer, M. (2024). Artificial intelligence and increasing misinformation. *The British Journal of Psychiatry, 224*(2), 33–35. https://doi.org/10.1192/bjp.2023.136

Murman, D. L. (2015). The impact of age on cognition. *Seminars in Hearing, 36*(3), 111–121. https://doi.org/10.1055/s-0035-1555115

Nguyen, V. X., Xiao, G., Zhou, J., Li, G., & Li, B. (2020). Bias in social interactions and emergence of extremism in complex social networks. *Chaos: An Interdisciplinary Journal of Nonlinear Science, 30*(10), 103110. https://doi.org/10.1063/5.0009943

Nguyen, A. L., Mosqueda, L., Windisch, N., Weissberger, G., Axelrod, J., & Han, S. D. (2021). Perceived types, causes, and consequences of financial exploitation: narratives from older adults. *The Journals of Gerontology: Series B, 76*(5), 996–1004. https://doi.org/10.1093/geronb/gbab010

Nickerson, R. S. (1998). Confirmation bias: A ubiquitous phenomenon in many guises. *Review of general psychology, 2*(2), 175–220.

Nissim, N. R., McAfee, D. C., Edwards, S., Prato, A., Lin, J. X., Lu, Z., Coslett, H. B., & Hamilton, R. H. (2023). Efficacy of transcranial alternating current stimulation in the enhancement of working memory performance in healthy adults: A systematic meta-analysis. *Neuromodulation: Journal of the International Neuromodulation Society, 26*(4), 728–737. https://doi.org/10.1016/j.neurom.2022.12.014

Oh, J., Jeon, S., Ha, T. H., Myung, W., Lee, S.-H., Ko, Y.-H., Kim, D. H., Lee, H.-Y., & Chae, J.-H. (2023). Effect of home-based self-administered transcranial direct stimulation in patients with mild to moderate major depressive disorder: A single-arm, multicentral trial. *Clinical Psychopharmacology and Neuroscience, 21*(2), 271–278. https://doi.org/10.9758/cpn.2023.21.2.271

Overhoff, H., Ko, Y. H., Fink, G. R., Stahl, J., Weiss, P. H., Bode, S., & Niessen, E. (2022). The relationship between response dynamics and the formation of confidence varies across the lifespan. *Frontiers in Aging Neuroscience, 14*. https://doi.org/10.3389/fnagi.2022.969074

Park, A. S., & Thompson, B. (2024). Non-invasive brain stimulation and vision rehabilitation: A clinical perspective. *Clinical and Experimental Optometry, 107*(6), 594–602.

Park, J., Konana, P., Gu, B., Kumar, A., & Raghunathan, R. (2010). *Confirmation bias, overconfidence, and investment performance: Evidence from stock message boards* (SSRN scholarly paper 1639470). https://doi.org/10.2139/ssrn.1639470

Parkin, B. L., Ekhtiari, H., & Walsh, V. F. (2015). Non-invasive human brain stimulation in cognitive neuroscience: A primer. *Neuron, 87*(5), 932–945. https://doi.org/10.1016/j.neuron.2015.07.032

Peters, M. A. K., Thesen, T., Ko, Y. D., Maniscalco, B., Carlson, C., Davidson, M., Doyle, W., Kuzniecky, R., Devinsky, O., Halgren, E., & Lau, H. (2017). Perceptual confidence neglects decision-incongruent evidence in the brain. *Nature Human Behaviour, 1*(7), 1–8. https://doi.org/10.1038/s41562-017-0139

Peterson, J. C., Burnes, D. P. R., Caccamise, P. L., Mason, A., Henderson, C. R., Wells, M. T., Berman, J., Cook, A. M., Shukoff, D., Brownell, P., Powell, M., Salamone, A., Pillemer, K. A., & Lachs, M. S. (2014). Financial exploitation of older adults: A population-based prevalence study. *Journal of General Internal Medicine, 29*(12), 1615–1623. https://doi.org/10.1007/s11606-014-2946-2

Poppe, A., Ritter, F. D. E., Bais, L., & Pustejovsky, J. E. (n.d.). *The efficacy of combining cognitive training and noninvasive brain stimulation: A transdiagnostic systematic review and meta-analysis.*

Prat-Ortega, G., & de la Rocha, J. (2018). Selective attention: A plausible mechanism underlying confirmation bias. *Current Biology, 28*(19), R1151–R1154. https://doi.org/10.1016/j.cub.2018.08.024

Rabin, L. A., Wang, C., Katz, M. J., Derby, C. A., Buschke, H., & Lipton, R. B. (2012). Predicting Alzheimer's disease: Neuropsychological tests, self-reports, and informant reports of cognitive difficulties. *Journal of the American Geriatrics Society, 60*(6), 1128–1134. https://doi.org/10.1111/j.1532-5415.2012.03956.x

Rebelo-Marques, A., De Sousa Lages, A., Andrade, R., Ribeiro, C. F., Mota-Pinto, A., Carrilho, F., & Espregueira-Mendes, J. (2018). Aging hallmarks: The benefits of physical exercise. *Frontiers in Endocrinology, 9*, 258. https://doi.org/10.3389/fendo.2018.00258

Reinhart, R. M. G., & Nguyen, J. A. (2019). Working memory revived in older adults by synchronizing rhythmic brain circuits. *Nature Neuroscience, 22*(5), Article 5. https://doi.org/10.1038/s41593-019-0371-x

Riggs, A., Patel, V., Paneri, B., Portenoy, R. K., Bikson, M., & Knotkova, H. (2018). At-home transcranial direct current stimulation (tDCS) with telehealth support for symptom control in chronically-ill patients with multiple symptoms. *Frontiers in Behavioral Neuroscience, 12*, 93. https://doi.org/10.3389/fnbeh.2018.00093

Robertson, E. M., & Pascual-Leone, A. (2003). Prefrontal cortex: Procedural sequence learning and awareness. *Current Biology, 13*(2), R65–R67. https://doi.org/10.1016/S0960-9822(02)01431-8

Rodda, J., Walker, Z., & Carter, J. (2011). Depression in older adults. *BMJ, 343*, d5219. https://doi.org/10.1136/bmj.d5219

Rollwage, M., Loosen, A., Hauser, T. U., Moran, R., Dolan, R. J., & Fleming, S. M. (2020). Confidence drives a neural confirmation bias. *Nature Communications, 11*(1), Article 1. https://doi.org/10.1038/s41467-020-16278-6

Royall, D. R., Lauterbach, E. C., Cummings, J. L., Reeve, A., Rummans, T. A., Kaufer, D. I., Curt LaFrance, W. Jr, & Coffey, C. E. (2002). Executive control function. *The Journal of Neuropsychiatry and Clinical Neurosciences, 14*(4), 377–405. https://doi.org/10.1176/jnp.14.4.377

Rueda, A. D., Lau, K. M., Saito, N., Harvey, D., Risacher, S. L., Aisen, P. S., Petersen, R. C., Saykin, A. J., Farias, S. T., & Initiative, A. D. N. (2015). Self-rated and informant-rated everyday function in comparison to objective markers of Alzheimer's disease. *Alzheimer's & Dementia, 11*(9), 1080–1089. https://doi.org/10.1016/j.jalz.2014.09.002

Ryan, M. C., & Patterson, J. (1987). Loneliness in the elderly. *Journal of Gerontological Nursing, 13*(5), 6–9. https://doi.org/10.3928/0098-9134-19870501-04

Sale, M. V., Mattingley, J. B., Zalesky, A., & Cocchi, L. (2015). Imaging human brain networks to improve the clinical efficacy of non-invasive brain stimulation. *Neuroscience & Biobehavioral Reviews, 57*, 187–198. https://doi.org/10.1016/j.neubiorev.2015.09.010

Samanez-Larkin, G. R., & Knutson, B. (2015). Decision making in the ageing brain: Changes in affective and motivational circuits. *Nature Reviews Neuroscience, 16*(5), Article 5. https://doi.org/10.1038/nrn3917

Satorres, E., Meléndez, J. C., Pitarque, A., Real, E., Abella, M., & Escudero, J. (2022). Enhancing immediate memory, potential learning, and working memory with transcranial direct current stimulation in healthy older adults. *International Journal of Environmental Research and Public Health, 19*(19), 12716. https://doi.org/10.3390/ijerph191912716

Shin, D., Koerber, A., & Lim, J. S. (2024). Impact of misinformation from generative AI on user information processing: How people understand misinformation from generative AI. *New Media & Society*, 14614448241234040. https://doi.org/10.1177/14614448241234040

Shiroma, K., Zimmerman, T., Xie, B., Fleischmann, K. R., Rich, K., Lee, M. K., Verma, N., & Jia, C. (2023). Older adults' trust and distrust in COVID-19 public health information: Qualitative critical incident study. *JMIR Aging, 6*, e42517. https://doi.org/10.2196/42517

Sprague, T. C., Saproo, S., & Serences, J. T. (2015). Visual attention mitigates information loss in small- and large-scale neural codes. *Trends in Cognitive Sciences, 19*(4), 215–226. https://doi.org/10.1016/j.tics.2015.02.005

Spreng, R. N., Karlawish, J., & Marson, D. C. (2016). Cognitive, social, and neural determinants of diminished decision-making and financial exploitation risk in aging and dementia: A review and new model. *Journal of Elder Abuse & Neglect, 28*(4–5), 320–344. https://doi.org/10.1080/08946566.2016.1237918

Steele, C. M., Spencer, S. J., & Aronson, J. (2002). Contending with group image: The psychology of stereotype and social identity threat. In *Advances in experimental social psychology* (Vol. 34, pp. 379–440). Academic. https://doi.org/10.1016/S0065-2601(02)80009-0

Stevens, F. L., Hurley, R. A., & Taber, K. H. (2011). Anterior cingulate cortex: unique role in cognition and emotion. *The Journal of neuropsychiatry and clinical neurosciences, 23*(2), 121–125.

Stum, M. S., Kilderia, R., & Peterson, M. (2017). When it's family: Examining entitlement as a risk factor for elder financial exploitation. *Innovation in Aging, 1*(Suppl 1), 366. https://doi.org/10.1093/geroni/igx004.1334

Sui, J., & Humphreys, G. W. (2017). Aging enhances cognitive biases to friends but not the self. *Psychonomic Bulletin & Review, 24*(6), 2021–2030. https://doi.org/10.3758/s13423-017-1264-1

Sun, T., Xie, T., Wang, J., Zhang, L., Tian, Y., Wang, K., Yu, X., & Wang, H. (2020). Decision-Making under ambiguity or risk in individuals with alzheimer's disease and mild cognitive impairment. *Frontiers in Psychiatry, 11*.

Terren, L. T. L., & Borge-Bravo, R. B. B. R. (2021). Echo chambers on social media: A systematic review of the literature. *Review of Communication Research, 9*. https://www.rcommunicationr.org/index.php/rcr/article/view/16

Tisdall, L., & Seaman, K. L. (2024). Decision making in the aging brain: An update of the evidence base for the 'affect-integration-motivation' framework. *OSF*. https://doi.org/10.31234/osf.io/tnbky

Vail, I. I. I., & Juhl, J. (2015). An appreciative view of the brighter side of terror management processes. *Social Sciences, 4*(4), Article 4. https://doi.org/10.3390/socsci4041020

Vosskuhl, J., Strüber, D., & Herrmann, C. S. (2018). Non-invasive brain stimulation: A paradigm shift in understanding brain oscillations. *Frontiers in Human Neuroscience, 12*. https://doi.org/10.3389/fnhum.2018.00211

Whittlestone, J. (2017). *The importance of making assumptions: Why confirmation is not necessarily a bias*. PhD, University of Warwick. http://webcat.warwick.ac.uk/record=b3110335~S15

Winker, C., Rehbein, M. A., Sabatinelli, D., Dohn, M., Maitzen, J., Roesmann, K., Wolters, C. H., Arolt, V., & Junghoefer, M. (2019). Noninvasive stimulation of the ventromedial prefrontal cortex indicates valence ambiguity in sad compared to happy and fearful face processing. *Frontiers in Behavioral Neuroscience, 13*. https://doi.org/10.3389/fnbeh.2019.00083

Wischnewski, M., Alekseichuk, I., & Schutter, D. J. L. G. (2021). Behavioral and electrocortical effects of transcranial alternating current stimulation during advice-guided decision-making. *Neuroimage: Reports, 1*(4), 100052. https://doi.org/10.1016/j.ynirp.2021.100052

Wischnewski, M., Alekseichuk, I., & Opitz, A. (2023). Neurocognitive, physiological, and biophysical effects of transcranial alternating current stimulation. *Trends in Cognitive Sciences, 27*(2), 189–205. https://doi.org/10.1016/j.tics.2022.11.013

Wurm, S., Tesch-Römer, C., & Tomasik, M. J. (2007). Longitudinal findings on aging-related cognitions, control beliefs, and health in later life. *The Journals of Gerontology: Series B, 62*(3), P156–P164. https://doi.org/10.1093/geronb/62.3.P156

Xu, D., Fan, S., & Kankanhalli, M. (2023). Combating Misinformation in the Era of Generative AI Models. Proceedings of the 31st ACM International conference on multimedia, 9291–9298. https://doi.org/10.1145/3581783.3612704

Yang, Y., Hackett, K., Katta, S., Ludwig, R. M., Jarcho, J., Giovannetti, T., Fareri, D. S., & Smith, D. V. (2025). Psychological, social, and health-related factors predict risk for financial exploitation. *Communications Psychology, 3*(1), 88. https://doi.org/10.1038/s44271-025-00266-x

Yaple, Z. A., Stevens, W. D., & Arsalidou, M. (2019). Meta-analyses of the n-back working memory task: fMRI evidence of age-related changes in prefrontal cortex involvement across the adult lifespan. *NeuroImage, 196* 16–31. https://doi.org/10.1016/j.neuroimage.2019.03.074

Zhou, Y., & Shen, L. (2022). Confirmation bias and the persistence of misinformation on climate change. *Communication Research, 49*(4), 500–523. https://doi.org/10.1177/00936502211028049

Zimerman, M., & Hummel, F. C. (2010). Non-invasive brain stimulation: Enhancing motor and cognitive functions in healthy old subjects. *Frontiers in Aging Neuroscience, 2*. https://doi.org/10.3389/fnagi.2010.00149

Sex/Gender in Neuroeconomics

Nick Doren and Philippe N. Tobler

Abstract Sex/gender is a primary constituent of individual differences in human behavior. In this chapter, we examine the role of sex/gender in economic decision-making, focusing on behavioral, neurobiological, and hormonal factors that underpin value-based choices. We report that two core regions of the brain valuation system, the striatum (male > female) and the ventromedial prefrontal cortex (female > male), show opposing sex/gender differences in gray matter volume, but their relations to function remain underexplored. Applying a five-step decision-making framework (representation, valuation, action selection, outcome evaluation, and learning), we explore how sex/gender differences and sex hormones impact each stage. We discuss how these differences or their absence find their expression in individual characteristics such as risk attitude, time preferences, prosociality, behavioral flexibility, or learning ability. For example, testosterone increases both temporal and social discounting and reduces connectivity between the temporoparietal junction and the striatum in men. Similarly, estrogen increases temporal discounting in women. Finally, we discuss the gradual passage from a binary conceptualization to a continuous representation of sex/gender and make a series of recommendations for future neuroeconomic studies.

1 Introduction

Individuals differ in their preferences, and neuroeconomics aims to characterize these differences in order to understand and predict value-based decisions (Kable & Levy, 2015). Sex constitutes a paradigmatic individual difference variable that can be investigated in various ways (e.g., discrete, continuous). Sex differences have

N. Doren · P. N. Tobler (✉)
Laboratory for Social and Neural Systems Research, Zurich Center for Neuroeconomics, Department of Economics, University of Zurich, Zurich, Switzerland

Zurich Center for Neuroscience, ETH Zurich and University of Zurich, Zurich, Switzerland
e-mail: phil.tobler@econ.uzh.ch

D. V. Smith et al. (eds.), *Neuroeconomics: Core Topics and Current Directions*,
https://doi.org/10.1007/978-3-032-02925-6_31

been reported separately by behavioral economics (e.g., women are more risk-averse than men (Eckel & Grossman, 2008)) and neuroscience (e.g., the male amygdala is larger than the female, controlling for total brain size (DeCasien et al., 2022)). However, these lines of research have rarely been considered by neuroeconomics. The reasons can be manifold. On the one hand, the power for investigating such questions binarily can become critically low with the often relatively small samples used in neuroeconomic studies. On the other hand, the politically sensitive nature of sex differences and the methodological intricacies (e.g., in how to correct for brain size, which can lead to apparent inconsistencies in sex-related findings (DeCasien et al., 2022)) may make neuroeconomists reluctant to investigate sex differences.

Outside neuroeconomics, however, large sample studies (Forde et al., 2020; Williams et al., 2021) have documented relatively stable neurobiological sex differences which in principle could relate to temporally stable sex differences in economic behavior (e.g., risk preferences (Frey et al., 2017)). To investigate the promise of this notion, we tested whether the regions showing neuroanatomical sex differences (Liu et al., 2020) intersect with the canonical network of regions processing subjective value (Bartra et al., 2013). We found overlap in core value regions, including the ventromedial prefrontal cortex and the ventral striatum (Fig. 1), suggesting that sex differences in economic behavior may indeed have a neural basis. Based on this, we discuss possible avenues for neuroeconomists interested in studying sex differences. We also focus on the role of sex hormones for economic behavior. While we consider biological aspects, we remain agnostic about the genetic or socially learned origin of sex differences. We review recent advancements in these various lines of research and combine them hypothetically with the hope of stimulating future neuroeconomic research.

2 Sex and/or Gender

Early work investigated sex as a binary variable that separated individuals into males and females. Although convenient in its dichotomy, such a separation has faced numerous challenges as it conflates and insufficiently resolves biological and socio-cultural individual differences. As a first step toward a paradigm change, later studies distinguished between biological sex (assessed with biomarkers such as reproductive organs, genes, and hormone levels) and sociocultural gender (typically assessed through self-report) (Torgrimson & Minson, 2005).

While sex and gender overlap, the match is not perfect (Fig. 2), in line with the notion that culture can diverge from biology. When asked to rate their own masculinity/femininity, men match their gender to their sex more frequently than women (Bittner & Goodyear-Grant, 2017). Given that socio-cultural experiences and environments as well as biological factors can influence neural structure and function, it becomes difficult to unequivocally separate the two variables. Moreover, it is often not possible to determine how the sex/gender of participants was assessed during recruitment. Therefore, we will use the term "sex/gender" (Kaiser et al.,

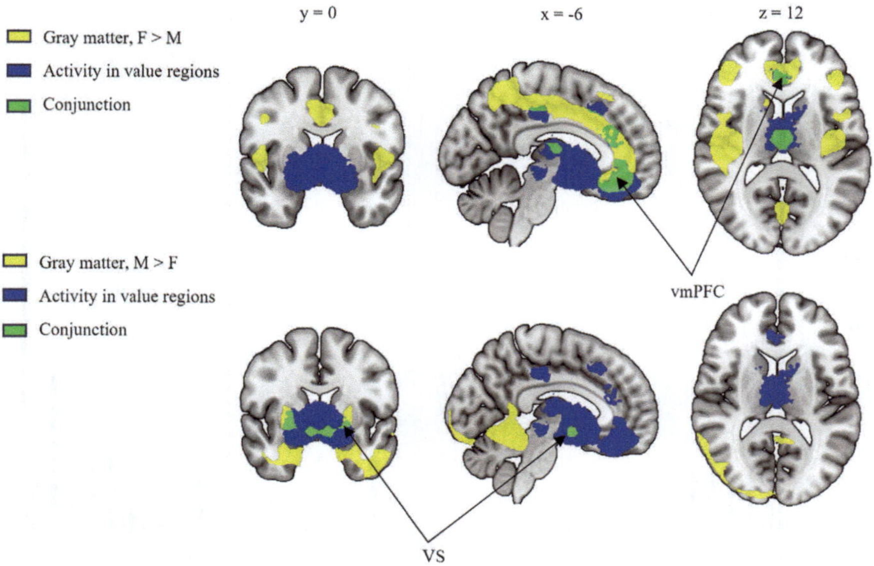

Fig. 1 Valuation regions show morphological sex differences. Overlap (green) between total brain volume-controlled sex differences in gray matter volume (yellow, top: female > male, bottom: male > female. Data extracted from Liu et al., 2020) and neural activity positively associated with subjective value (in blue; Data extracted from Bartra et al., 2013). Arrows indicate two regions commonly associated with valuation in neuroeconomics: ventral striatum (VS) and ventromedial prefrontal cortex (vmPFC)

2009; Jäncke, 2018) throughout but do not mean to imply that the two terms denote the same thing.

3 Value-Based Decision Framework

Neuroeconomic research aims to mechanistically explain value-based choice. A well-established framework decomposes value-based choice into five sub-processes: representation, valuation, action selection, outcome evaluation, and learning (Rangel et al., 2008) (Fig. 3a). In this review, we will discuss the role of sex/gender in neuroeconomics through the lens of these five steps.

3.1 Step 1: Representation

Value-based behavior and choice assign value to options (goods and actions). These options must be detected and identified by the decision-maker. Moreover, internal

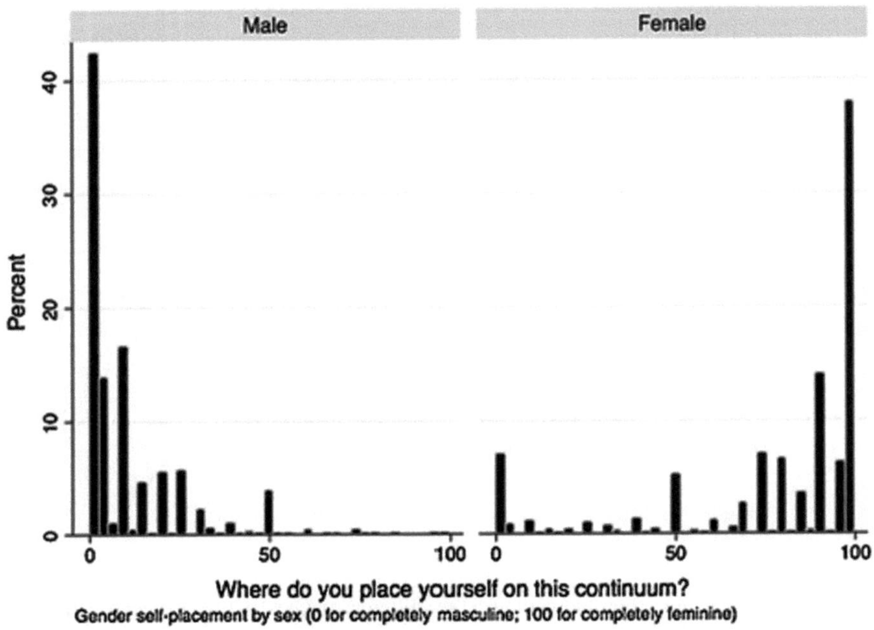

Fig. 2 Gender self-placement by sex. 6073 individuals self-placed their gender in a Web-based survey conducted during provincial elections in Canada over 2011 and 2012. Values of 0 and 100 represent 100% masculine and 100% feminine, respectively. The data suggest that using sex as a proxy for gender may be fully appropriate for about two fifths of the population. More men (77%) than women (68%) match their gender to their sex (i.e., place themselves between 0–10 for men and 90–100 for women). (Figure reprinted with permission from Bittner & Goodyear-Grant, 2017)

states (e.g., hunger, thirst) need to be registered. Sensory processing, attention, and interoception are the main processes of this stage. They facilitate the efficient allocation of cognitive resources by filtering out irrelevant information, but also can limit the space of considered options.

Noisy representation of numerical magnitudes has been suggested to (partially) explain individual differences in financial decision-making (Barretto-García et al., 2023). Moreover, sex/gender differences in number representation have been reported repeatedly. For example, during comparison of two-digit numbers, women tend to compare not only decades but also units, whereas men compare only decades, resulting in smaller compatibility effects. In other words, men showed smaller response time differences between compatible (both the decade and the unit of the larger number are larger) and incompatible (only the decade of the larger number is larger) trials (Huber et al., 2019). Sex/gender was the only variable that explained these differences even after accounting for other factors such as age or education.

Sex/gender differences have also been reported for non-symbolic number representations (e.g., clouds of dots), with men showing stronger activation than women

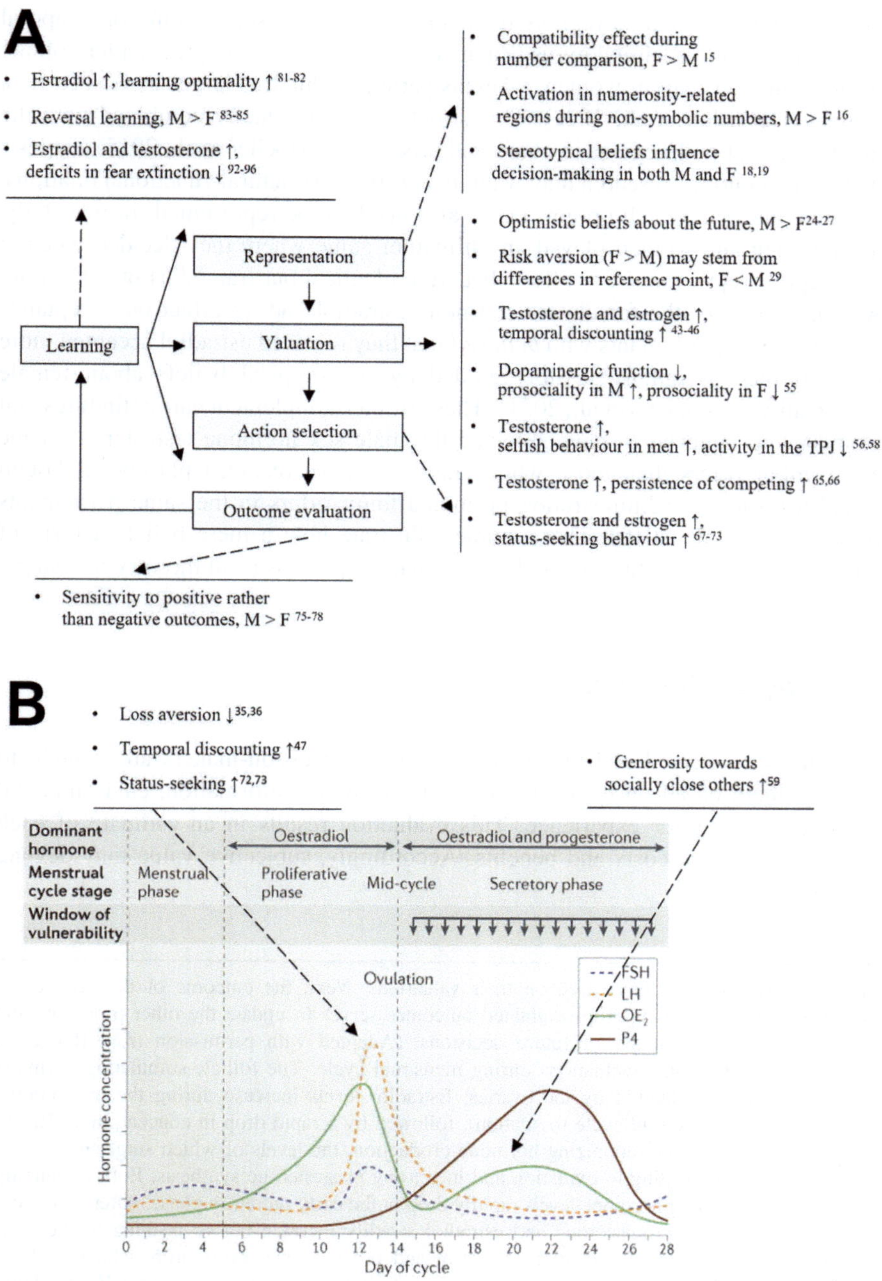

Fig. 3 Sex/gender differences and menstrual hormone fluctuations mapped to decision-making. **(a)** Five basic computational steps of value-based choice. Representation consists of detecting and identifying internal and external states as well as potential courses of action. Next, courses of actions and goods (options) receive a subjective value. Action selection picks one of the

in the numerosity-related regions including intraparietal sulcus, inferior temporal gyrus, and middle occipital gyrus (Zang et al., 2024). Temporal grey matter volume is larger in men than in women, whereas parietal volume is larger in women than in men (DeCasien et al., 2022). This structural heterogeneity combined with the involvement of both regions in numerical processing (Moeller et al., 2015) suggests that neuroeconomic research may want to investigate structural-functional relations.

The context in which options occur also needs to be represented. In one study, healthy men and women played an ultimatum game where they decided whether to accept or reject monetary offers that were identical but framed as fair or unfair. Administration of the female sex hormone estradiol had no effect on acceptance rate. However, participants who *believed* that they received estradiol accepted more offers framed as unfair, possibly guided by stereotypical beliefs about female prosociality (Coenjaerts et al., 2021). These results complement earlier findings that women who believed to have received the male sex hormone testosterone made more unfair offers than those who believed to have received placebo, although actual testosterone administration promoted fairer offers in the same participants (Eisenegger et al., 2010). These studies illustrate how a mere belief (a form of representation) can influence the decision-making process from the very first step.

3.2 *Step 2: Valuation*

After having detected and identified the options, decision-makers are thought to evaluate them based on their current needs, available information, environmental conditions, and prior experience. This evaluation results in an estimate of each option's subjective costs and benefits. Accordingly, subjective value may deviate

Fig. 3 (continued) options based on their valuations. Next, the outcome of the decision is evaluated. Finally, learning from evaluated outcomes serves to update the other processes and thereby improve the quality of future decisions. (Adapted with permission from Rangel et al., 2008). (**b**) Hormone fluctuations during menstrual cycle. The follicle-stimulating hormone induces estradiol production by the ovaries. Estradiol levels increase during the proliferative phase and peak before mid-cycle (ovulation), followed by a rapid drop in concentration. Rising estradiol levels stimulate luteinizing hormone production, the levels of which surge in the late-proliferative phase, leading to ovulation and increasing progesterone synthesis. Both luteinizing and follicle-stimulating hormone levels rapidly drop in the early secretory phase. After ovulation, the concentrations of progesterone and estradiol steadily increase before peaking in the mid-secretory phase. Without fertilization, estradiol and progesterone levels drop, which leads to endometrial shedding and the onset of menses. *FSH* follicle-stimulating hormone, *LH* luteinizing hormone, *OE2* estradiol, *P4* progesterone. (Figure adapted with permission from Wira et al., 2015). In (**a**), solid arrows represent the flow of the decision-making process. In (**a**) and (**b**), dashed arrows point to summaries of the main findings per step reported in the main text. ↑ and ↓ indicate increase and decrease, respectively. In (**a**), F and M indicate female and male, respectively

from objective value (Polanía et al., 2019). Valuation can be affected along multiple axes, including risk, time, and social factors (Serra, 2021).

3.2.1 Risk Preferences

A substantial literature portrayed women as more risk-averse than men (Eckel & Grossman, 2008). Harris and Jenkins (Eisenegger et al., 2010) explained the difference with women's greater perceived likelihood of negative outcomes and lesser expectation of enjoyment (Harris & Jenkins, 2006). Moreover, sex/gender differences in financial risk-taking may reflect differences in optimism bias (i.e., tendency to expect positive events even without them occurring) (Sharot, 2011). Indeed, men show more optimistic (but not necessarily realistic) beliefs about the future than women (Jacobsen et al., 2014; Bjuggren & Elert, 2019). Similarly, compared to women, men tend to overweigh positive information not only in risky (i.e., when the odds of choice outcomes are known) but also in uncertain (i.e., when the odds are at least partially unknown) situations (Karmarkar, 2023) and exhibit greater tolerance of uncertainty (Byrne et al., 2020). Thus, overly positive expectations contribute to sex/gender-related asymmetry in risk-taking.

Risk preferences are often viewed as stable primitives of choosers (Glimcher & Tymula, 2023). However, risk-taking can also be viewed as the malleable product of the environment that emerges from subjective payoff expectations. In line with this notion, one study suggested that risk-avoiding choices in women may stem from a lower reference point compared to men (Kettlewell et al., 2023). As the reference point is tied to payoff history, repeated experience of low payoffs may promote risk-averse behavior (by lowering the reference point and future payoff expectations). Thus, the authors argued that secular economic gender inequalities (e.g., gender wage gap) may contribute to risk-averse behavior exhibited by women. Furthermore, as risk aversion has been related to negative financial outcomes (Finke & Huston, 2003), experiencing low payoffs after risk-averse decisions could reinforce economic disadvantages of women.

Biologically, sex hormones appear to modulate economic risk-taking (Apicella et al., 2015). For instance, testosterone administration has been associated with promotion of risk-taking of non-professional traders in experimental asset markets (Cueva et al., 2015). The effect arose via testosterone increasing optimism about future price changes, in line with reports of amplified optimism bias in men (Jacobsen et al., 2014; Bjuggren & Elert, 2019; Karmarkar, 2023). Subsequent research showed a similar testosterone effect and increased tendency for market bubbles in professional traders (Nadler et al., 2018; Nofsinger et al., 2018).

By contrast, peri-ovulatory fluctuations in female sex hormones (Fig. 3b) have been consistently associated with reduced loss aversion (i.e., the tendency to weigh losses more than equivalent gains) (Lazzaro et al., 2016; Durante et al., 2020). However, the causal evidence linking estradiol to risky decision-making is mixed across species, highlighting the necessity for more diverse and better-powered study designs to capture estradiol's action in all its complexity (Islas-Preciado et al., 2020; Orsini et al., 2021; Ambrase et al., 2021).

From a neural perspective, the insula, a region commonly associated with risk aversion, has been shown to exhibit more activity during risky decision-making in women compared to men (Lee et al., 2009). These early findings were later replicated in adolescents (Korucuoglu et al., 2020). The insula has been associated with real-life stock trading (Häusler et al., 2018). By extension, neural and hormonal correlates of sex/gender differences in risky behavior may improve our understanding of such differences in real-life behavior.

3.2.2 Time Preferences

Women have been reported to be more patient than men (Dittrich & Leipold, 2014). This observation converges with that of endogenously or pharmacologically increased testosterone levels increasing temporal discounting (i.e., stronger preference for smaller sooner over larger later rewards) in men (Laube et al., 2020; Wu et al., 2020, 2022). Temporal discounting is also steeper mid-cycle (i.e., high estrogen levels) compared to other cycle phases (Lucas & Koff, 2017) and in those on hormonal contraception (Vincent et al., 2023), suggesting hormone-dependent regulation of intertemporal choices across sexes/genders.

Sex/gender-related differences in time preferences may also be underpinned by cultural stereotypes. For example, in one study, participants were asked to embody a virtual avatar whose gender either matched or differed from the self-reported gender of participants. The study found that women who embodied a male avatar became more impatient compared to women who embodied a female avatar (while men were not differentially affected by avatar sex/gender) (Bolt et al., 2021).

Collectively, the evidence points to hormonal and societal mechanisms explaining sex/gender differences in time preferences, although their exact neural underpinnings remain to be elucidated. Higher rates of discounting have been associated with increased ventral striatum (VS) activity and reduced ventromedial prefrontal cortex (vmPFC) activity (Frost & McNaughton, 2017). This finding somewhat mirrors the relatively larger VS volume and smaller vmPFC volume of men compared to women (Fig. 1). However, the links remain to be formally investigated. As higher temporal discounting has been associated with reduced likelihood of undergoing medical examinations (Keidel et al., 2021), better understanding of mechanisms behind sex/gender differences in intertemporal choice may also inform the sex/gender disparity in the use of preventive care (i.e., men undergoing fewer check-ups than women (Vaidya et al., 2012)).

3.2.3 Social Preferences

Women are typically more prosocial than men when it comes to donating and sharing (Bilén et al., 2021). One theoretical account proposed that because of differential socialization, sharing might carry more value for women, whereas selfishness would be preferable for men (Rand et al., 2016). This hypothesis is compatible with pharmacological downregulation of dopaminergic function increasing selfishness in women while promoting prosociality in men (Soutschek et al., 2017). Moreover, the same study found striatal activity during prosocial versus selfish choices to be higher in women compared to men, providing a neurobiological mechanism for sex/gender differences in prosocial behavior.

A mounting body of manipulation studies involves sex hormones in prosociality. For example, Wu and colleagues probed the generosity of male participants after receiving 150 mg of testosterone gel or placebo (Wu et al., 2019). Participants were asked to order the people in their social environment from 1 to 100 in terms of social distance (1 corresponding to most significant others, 50 to mere acquaintances, and 100 to complete strangers). Then, participants on each trial chose between a *generous* option providing an equal split for them and one of the ranked persons or a *selfish* option that provided a trial-specific monetary amount only for themselves. Unsurprisingly, the amount of money participants forewent to benefit others gradually decreased with social distance in both groups. Importantly, this decline was significantly steeper in the testosterone than the placebo group, suggesting that testosterone promotes selfish behavior.

A follow-up study replicated the effect of testosterone on social preferences. Moreover, it suggested that testosterone alters generous behavior by attenuating activity in the temporoparietal junction (TPJ; commonly associated with representing the mental states of others (Quesque & Brass, 2019)) parametrically related to amount foregone (Ou et al., 2021) (Fig. 4). Moreover, testosterone reduced functional connectivity between TPJ and regions associated with salience and value processing such as the insula and the striatum. Thus, testosterone appears to affect the encoding of prosocial value in activity and connectivity.

The same experimental task served to investigate the hormonal and behavioral converse of these findings in women. Increasing progesterone and decreasing estradiol levels across the menstrual cycle (Fig. 3b) account for increased generosity toward socially close others, but not toward strangers (Sellitto & Kalenscher, 2022). Taking these findings together suggests that it will be illuminating to study both women and men while manipulating the same substances and using neuroimaging. Given that the TPJ and striatum also show sex/gender differences in brain volume that are related to sex/gender differences in social contacts (Kiesow et al., 2020), a focus on these regions appears promising.

3.3 Step 3: Action Selection

After identifying and evaluating options, decision-makers are thought to compare alternatives based on their subjective values and eventually select one of them. Depending on the situation, the choice may require various degrees of behavioral

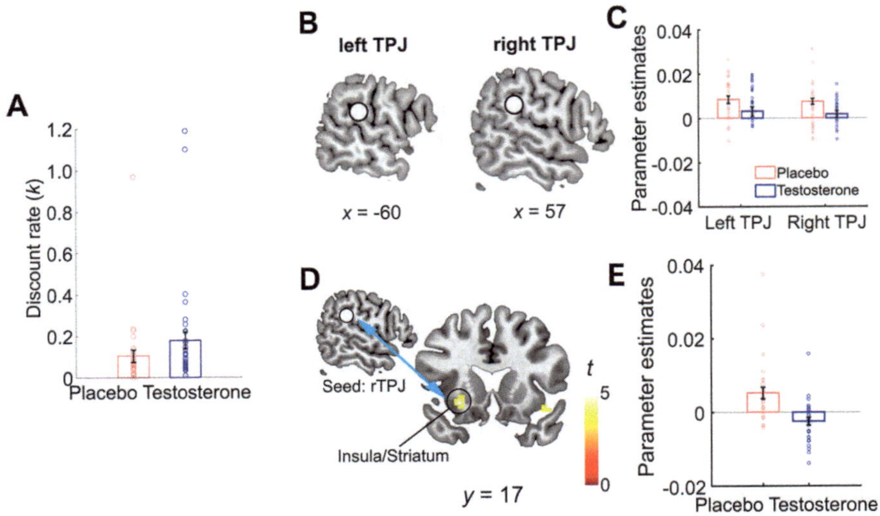

Fig. 4 Behavioral and neural effects of testosterone on generosity. (**a**) Testosterone increased the steepness of the social discounting function (i.e., decreased prosociality). (**b, c**) In bilateral temporoparietal junction (TPJ), activity associated with the degree of generosity was decreased in the testosterone compared to the placebo group. (**d, e**) Functional connectivity between TPJ and regions associated with reward and salience processing (insula and striatum) was decreased under testosterone. (Adapted with permission from Ou et al., 2021)

flexibility and cognitive control to override default behavior (e.g., a habit or a recently performed action).

Behaving flexibly facilitates exploration of new environments. In one animal study (Chen et al., 2021), male mice were more likely to perseverate in explorative behavior once started. By contrast, female mice learned more quickly during exploration and showed less noisy estimations of their values, relating reduced exploration to better learning of the environment. In agreement with this notion, women have been reported to adopt more cautious exploration under conditions of threat (Gagnon et al., 2016), although still little is known about sex/gender differences in human exploration (Mehlhorn et al., 2015).

Adaptively regulating flexibility of relevant behavioral patterns during action selection can be useful not only for survival in changing environments but also to outperform competitors and secure an advantage in a reward-maximization race. It is well established that testosterone is positively associated with competitiveness and desire to dominate in various domains (Stanton et al., 2021). For instance, testosterone administration enhances persistence of competing, even under conditions of high threat (Geniole et al., 2019) or of low chances to win (Kutlikova et al., 2021), thereby reinforcing the association between testosterone and the thrive for social dominance.

The status- and dominance-seeking effects of testosterone may come at a cost. For example, adolescent girls with higher endogenous testosterone levels make less profitable financial decisions by overbidding in an auction task to gain the status of the auction winner (Cardoos et al., 2017). Moreover, testosterone administration increases men's preference for status-enhancing (but costly) brands over brands of similar quality but lower status (Nave et al., 2018). Testosterone also eliminates the "audience effect," i.e., tendency to align one's behavior with others' expectations. Exogenous testosterone reduced prosociality in male participants who believed to be watched by female representatives of a non-governmental organization. This effect occurred even when prosocial choices had higher reward probability for participants than selfish ones, indicating a boosted desire to exhibit dominance at the cost of the overall gain (Kutlikova et al., 2023). These findings align with the notion that testosterone may promote risk-seeking to attain higher social status (Apicella et al., 2015). An administration study concurs and suggests that the effects of testosterone on risk-taking are specific to gaining social status rather than purely monetary rewards (Vermeer et al., 2020). Interestingly, testosterone-induced changes in risk-taking were found only in a group with a higher but not lower perceived social status, highlighting the importance of social components for sex/gender-related effects on economic decision-making (Fisk et al., 2017).

Not only testosterone but also other sex hormones are associated with competitive behavior in social status acquisition. For example, women with high estrogen levels select products that enhance physical appearance (Durante et al., 2011) and signal status (diamond rings and cars), compatible with a stronger desire to outperform attractive rival women compared to women at other stages of the ovulatory cycle (Durante et al., 2014). Moreover, the previously mentioned negative effect of ovulation on loss aversion reverses if status-enhancing products are at stake, suggesting a hormone-mediated overweighting of social dominance (Durante et al., 2020). However, tangible causal evidence from human studies is needed to disentangle individual hormone contributions amidst the complexity of fluctuations during the menstrual cycle (Arthur et al., 2022).

Taken together, these findings highlight the influence of sex hormones on action selection and can provide putative explanations for valuation-independent suboptimal behavior. Future studies may investigate sex/gender differences in human exploration and provide causal evidence for female sex hormones regulating action selection.

3.4 Step 4: Outcome Evaluation

After making a choice, decision-makers evaluate the experienced outcomes. In self-reports (Dhingra et al., 2021), women exhibit higher sensitivity to punishment than men, whereas men report higher sensitivity to reward. Moreover, in a monetary incentive delay task, men show stronger orbitofrontal activity to both gains and losses (relative to nil) than women (Dhingra et al., 2021). Following on from the stronger optimism bias in men, future research may want to test the hypothesis that compared to women, men show stronger neural differentiation in the weighting of gains and losses, using connectivity analyses with the common orbitofrontal seed.

An electrophysiological study found adolescent boys to exhibit stronger P3 responses (a neural indicator for the motivational significance of an incentive feedback) in reward than punishment trials (Greimel et al., 2018). These differences were significantly stronger in boys than girls, extending previous findings to a different developmental period. In the same vein, another electrophysiological study employed a simple guessing task where boys and girls had to choose which of two boxes had a reward inside. Boys were less likely to switch boxes after guessing wrongly and showed less feedback-related negativity (a neural indicator for comparison between expected and experienced outcomes) than girls, indicating an attenuated integration of negative feedback (Ding et al., 2017).

Sex/gender differences in outcome processing extend beyond monetary rewards. For instance, in a recent review, women exhibited greater neural activity in cognitive control-associated regions (e.g., cingulate cortex, inferior frontal gyrus) after unsuccessfully inhibiting their actions in stop signal and go/no-go tasks (typically used without any external reward). By contrast, men's neural responses were higher after successful inhibition (Weafer, 2020).

Testosterone administration can increase the affective ratings of both the obtained and the counterfactual monetary outcomes of choices (Wu et al., 2018). This evidence aligns with men exhibiting optimism bias during valuation, as reported above. Further understanding of sex/gender-related differences in outcome evaluation may facilitate recipient-specific tailoring of feedback provision in managerial (Burk & Wiese, 2021) and other settings.

3.5 Step 5: Learning

In the final stage of the decision-making process, the experienced outcome is compared to what has been expected, which may or may not generate prediction errors and updating of predictions. If the experience is better than expected, the value of the chosen alternative increases. Conversely, a negative prediction error reduces the value of the chosen alternative. Within the choice framework, learning allows building up predictions about the most likely outcomes of choice alternatives and thereby enables adaptive and efficient decisions.

The findings on sex/gender differences in learning are heterogeneous, which may partly reflect task-specificity. For example, two studies found diametrically opposite effects of estradiol administration on learning rates in men (Veselic et al., 2021; Joue et al., 2022). In one of these studies, participants performed a task with dynamically changing reward contingencies, such that high learning rates (i.e., relying on recent outcomes) were more suitable for reward maximization (Veselic et al., 2021). Conversely, lower learning rates (i.e., relying on previous expectations) were more optimal in the second study, where reward contingencies were static and changed slowly over time (Joue et al., 2022). While estradiol administration increased learning rates in the first study, it decreased them in the second, which can be reconciled by estradiol enhancing learning optimality in both. However, the two studies also used different doses (2 mg vs. 8 mg), complicating the comparison between them.

One paradigm commonly used in neuroeconomic research is reversal learning, where reward contingencies reverse from an acquisition phase to a reversal phase, requiring participants to learn new associations and disregard old ones. Across species, males appear to perform better than females in this task (LaClair et al., 2019; Fuss & Witte, 2019; Tascón et al., 2023). Endogenous testosterone levels correlate with reversal learning performance, suggesting that testosterone facilitates flexibility and adaptability to unexpected changes in the environment (Diekhof & Kraft, 2017).

Finally, it is important to consider also maladaptive learning. For instance, the inability to inhibit learned fear in response to stimuli that no longer predict threat features is seen in anxiety and trauma-related disorders. Cross-species studies have related increased activity in fear-processing brain networks, impaired fear extinction, and increased propensity to reinstate fear to low levels of estradiol and hormonal fluctuations during the menstrual cycle (Goldstein et al., 2010; Milad et al., 2010; Sartin-Tarm et al., 2020; Felmingham et al., 2021; Bierwirth et al., 2021). By extension, impairments in fear extinction can be counteracted by administration of estradiol (Zeidan et al., 2011; Wen et al., 2021). Testosterone administration appears to have similar effects (Frye et al., 2008; Maeng et al., 2017), which may be mediated by endogenous testosterone levels (Hutschemaekers et al., 2021). As fear-related disorders are more prevalent in women than men (Kessler et al., 2005), better understanding of sex/gender differences in learning mechanisms is paramount to provide tailored solutions to those concerned (Day & Stevenson, 2020).

4 Discussion

Neuroeconomics has made first steps into elucidating the biological mechanisms underpinning sex/gender differences in value-based decision-making. Emerging research has shown that these differences arise at behavioral, physiological, and neural levels. And yet, many questions remain open.

Although some studies linked specific brain regions to sex/gender-related differences in decision-making (Soutschek et al., 2017; Dhingra et al., 2021), the link is rarely directly investigated. Inspiration may come from large-scale studies documenting structural and functional sex/gender differences. For example, sex/gender differences are well-established for the amygdala (DeCasien et al., 2022; Ritchie et al., 2018). Given its association with loss aversion (De Martino et al., 2010), risk aversion (De Martino et al., 2006; Jung et al., 2018), and fear conditioning (Alvarez et al., 2008), the question arises whether the structural and functional differences underpin differences in decision-making.

Considering the mechanisms is often more informative than a simple determination of effects. Combining physiological and neural data with causal manipulations allows to infer such proximal mechanisms as exemplified here (Kettlewell et al., 2023; Soutschek et al., 2017; Ou et al., 2021). Moreover, authors sometimes interpret their sex/gender-specific findings with ultimate considerations through the lens of evolutionary psychology. For example, they may state that "*One reason that women are less politically ambitious than men is because, in the environments of our ancestors, there were fewer benefits for females in taking the risk to seek political power*" (Sweet-Cushman, 2016). However, one should keep in mind that such statements go beyond the data and are hard to falsify. Moreover, depiction of ancestral habits (e.g., labor division) as a driving force may create the impression that change is impossible, thereby reinforcing stereotypes and discrimination (Bachaud & Johns, 2023; Gonzalez & Rice, 2024). Thus, careful interpretation is paramount for successful translation of sex/gender-related findings from neuroeconomics into other scientific areas or policymaking.

Finally, leading neuroeconomists have suggested that future research in the field should focus on investigating a variety of decision-making dimensions (e.g., loss aversion or social preferences) and demonstrating real-life applications (Kable & Levy, 2015). The current review illustrates that the first suggestion has been at least partly addressed at the behavioral level. However, the translational potential remains largely untested and/or unrealized (Núñez et al., 2020). For example, social dominance-enhancing effects (Stanton et al., 2021; Geniole et al., 2019; Kutlikova et al., 2021, 2023) of testosterone may suggest therapeutic benefits for social anxiety disorder (Enter et al., 2016). Findings about testosterone-enhanced risk-taking resulting in financial bubbles and subsequent market crashes (Cueva et al., 2015; Nadler et al., 2018) may be leveraged to design efficient institutions where social and environmental factors are taken into account (e.g., by promoting diversity in investors). As the empowerment of sex/gender differences expression is one of the hallmarks of countries with higher economic stability and gender equality (Falk & Hermle, 2018), one should bear in mind that understanding these differences may promote socially and economically developed societies.

5 Future Directions

The focus on sex/gender is still emerging. Sociologist Robert Michels once said, "*Everything in the world is about sex, except sex. Sex is about power.*" (Seelig, 2018). This quote can be reinterpreted in the context of our review. One reason why research on sex/gender can be perilous is that a binary separation automatically halves the sample size and reduces statistical power. One way to address this issue is to use a continuous variable such as gender. Continuous measures of gender can indeed explain individual differences in brain structure and brain-behavior associations better than binary variables (Rauch & Eliot, 2022; Luckhoff et al.,

2024). Combined consideration of both variables may speak to the malleability of sex/gender differences in economic behavior.

Not only gender but also biological sex can be measured in a continuous fashion. One approach is to determine the level of sex hormones such as estradiol, progesterone, and testosterone and relate it to behavior (exemplified here (Coenjaerts et al., 2021; Ou et al., 2021; Kutlikova et al., 2023)). Another approach is to create a "brain-and-body sex score," combining multiple psychological (e.g., Big-5 traits) and physiological (e.g., weight, height, total brain volume, prenatal hormone exposure) variables into one continuous scale (Verweij et al., 2016; Vosberg et al., 2021). However, such measures are harder to interpret than simpler measures. A continuous approach can also be helpful for administration studies where hormone baselines may account for individual variability in drug effects and explain nonlinear relations with behavior. Thus, a continuous approach is feasible and potentially beneficial also for sex.

Better understanding sex/gender differences requires inclusive study designs. For decades, biological research has been male-centered, largely motivated by the argument that hormone changes during menstrual cycle would increase variation in the data and impede interpretation (Beery & Zucker, 2011). The outdatedness of this perspective is increasingly recognized (Will et al., 2017), with the aim of ending the underrepresentation of women in science. While some studies apply the original protocol (Ou et al., 2021) in women (Sellitto & Kalenscher, 2022), others include both sexes/genders in the same study (Joue et al., 2022). Furthermore, consideration of sex/gender differences concerns not only study participants but also the researchers themselves. Indeed, sex/gender-related experimenter bias can influence the outcomes of both animal and human studies (Chapman et al., 2018; Georgiou et al., 2022), highlighting the need to consider sex/gender at all levels of study design (Tannenbaum et al., 2019).

While the notion of gender fluidity and non-binarity gained attention in social debates, it has not significantly impacted research yet. Therefore, future studies may broaden their sex/gender-related inclusivity by actively recruiting gender fluid and non-binary participants. Understanding differences and similarities in decision-making across the whole spectrum of sex/gender-related identities will reduce oversimplification of research findings and provide valuable insights into emerging groups of economic actors (Coffman et al., 2024).

Some studies found no evidence for sex/gender differences in, or effects of sex hormones on, behavior (Lockwood et al., 2017; Nadler et al., 2019, 2021). First, this may of course be due to an objective absence of a difference. Second, null findings may partially arise from simultaneous but opposite effects of sex/gender on multiple variables involved in complex cognitive processes. This could result in a mutual cancellation of the effects in observable behavior. Moreover, it may also arise despite sex/gender-specific implementation of mental functions. For instance, meta-analyses found that while there were no overarching sex/gender differences in executive functions (Grissom & Reyes, 2019; Gaillard et al., 2021a), neuroimaging studies suggested that males and females engage different neural networks depending on the executive function task (Gaillard et al., 2021b). Finally, the presence/absence of sex/gender effects may be related to sensitivity of analytical approaches. In this regard, power calculations may be informative. However, it is noteworthy that even large-scale studies came to opposite conclusions regarding the presence of sex/gender differences in brain anatomy (Williams et al., 2021; Eliot et al., 2021), indicating that analytical approaches can play a crucial role in uncovering individual differences (DeCasien et al., 2022). Therefore, future studies need to experimentally isolate research variables and investigate them in multi-modality approaches while standardizing analytical frameworks. Moreover, quantification of the evidence in favor of the null hypothesis (absence of a sex/gender difference) with Bayes factors or equivalence tests would counteract publication bias and the impression that sex differences are more ubiquitous than they actually are in economic choice.

6 Conclusion

Neuroeconomics offers unique opportunities for researchers interested in decision-making and sex/gender differences. Insights from economic variables (e.g., risk preferences or delay discounting) relate to other domains such as sexual behavior (Johnson et al., 2021) or medical judgment (Saposnik et al., 2016) and can inform guidelines for real-life business and governmental policies (Robertson et al., 2017; Ryan, 2017). Future neuroeconomic studies will benefit from combining rigorously controlled study designs and standardized methods with diversity of studied cohorts. In light of the established sex/gender differences in core valuation regions (ventral striatum and ventromedial prefrontal cortex), this appears to be a worthwhile endeavor for future neuroeconomic research.

References

Alvarez, R. P., Biggs, A., Chen, G., Pine, D. S., & Grillon, C. (2008). Contextual fear conditioning in humans: Cortical-hippocampal and amygdala contributions. *The Journal of Neuroscience, 28*, 6211–6219.

Ambrase, A., Lewis, C. A., Barth, C., & Derntl, B. (2021). Influence of ovarian hormones on value-based decision-making systems: Contribution to sexual dimorphisms in mental disorders. *Frontiers in Neuroendocrinology, 60*, 1–18.

Apicella, C. L., Carré, J. M., & Dreber, A. (2015). Testosterone and economic risk taking: A review. *Adaptive Human Behavior and Physiology, 1*, 358–385.

Arthur, L. C., Casto, K. V., & Blake, K. R. (2022). Hormonal contraceptives as disruptors of competitive behavior: Theoretical framing and review. *Frontiers in Neuroendocrinology, 66*, 1–90.

Bachaud, L., & Johns, S. E. (2023). The use and misuse of evolutionary psychology in online manosphere communities: The case of female mating strategies. *Evolutionary Human Sciences, 5*, 1–15.

Barretto-García, M., et al. (2023). Individual risk attitudes arise from noise in neurocognitive magnitude representations. *Nature Human Behaviour, 7*, 1551–1567.
Bartra, O., McGuire, J. T., & Kable, J. W. (2013). The valuation system: A coordinate-based meta-analysis of BOLD fMRI experiments examining neural correlates of subjective value. *NeuroImage, 76*, 412–427.
Beery, A. K., & Zucker, I. (2011). Sex bias in neuroscience and biomedical research. *Neuroscience and Biobehavioral Reviews, 35*, 565–572.
Bierwirth, P., Sperl, M. F., Antov, M. I., & Stockhorst, U. (2021). Prefrontal theta oscillations are modulated by estradiol status during fear recall and extinction recall. *Biological Psychiatry: Cognitive Neuroscience and Neuroimaging, 6*, 1071–1080.
Bilén, D., Dreber, A., & Johannesson, M. (2021). Are women more generous than men? A meta-analysis. *Journal of the Economic Science Association, 7*, 1–18.
Bittner, A., & Goodyear-Grant, E. (2017). Sex isn't gender: Reforming concepts and measurements in the study of public opinion. *Political Behavior, 39*, 1019–1041.
Bjuggren, C. M., & Elert, N. (2019). Gender differences in optimism. *Applied Economics, 51*, 5160–5173.
Bolt, E., et al. (2021). Effects of a virtual gender swap on social and temporal decision-making. *Scientific Reports, 11*, 1–15.
Burk, C. L., & Wiese, B. S. (2021). How to alleviate the agony of providing negative feedback: Emotion regulation strategies affect hormonal stress responses to a managerial task. *Hormones and Behavior, 127*, 1–15.
Byrne, K. A., et al. (2020). Acute stress enhances tolerance of uncertainty during decision-making. *Cognition, 205*, 1–18.
Cardoos, S. L., et al. (2017). Social status strategy in early adolescent girls: Testosterone and value-based decision making. *Psychoneuroendocrinology, 81*, 14–21.
Chapman, C. D., Benedict, C., & Schiöth, H. B. (2018). Experimenter gender and replicability in science. *Science Advances, 4*, 1–7.
Chen, C. S., Knep, E., Han, A., Ebitz, R. B., & Grissom, N. M. (2021). Sex differences in learning from exploration. *eLife, 10*, 1–24.
Coenjaerts, M., et al. (2021). Sex differences in economic decision-making: Exogenous estradiol has opposing effects on fairness framing in women and men. *European Neuropsychopharmacology, 50*, 46–54.
Coffman, K. B., Coffman, L. C., & Ericson, K. M. (2024). *Non-binary gender economics*. National Bureau of Economic Research.
Cueva, C., et al. (2015). Cortisol and testosterone increase financial risk taking and may destabilize markets. *Scientific Reports, 5*, 1–16.
Day, H. L., & Stevenson, C. W. (2020). The neurobiological basis of sex differences in learned fear and its inhibition. *The European Journal of Neuroscience, 52*, 2466–2486.
De Martino, B., Kumaran, D., Seymour, B., & Dolan, R. J. (2006). Frames, biases, and rational decision-making in the human brain. *Science, 313*, 684–687.
De Martino, B., Camerer, C. F., & Adolphs, R. (2010). Amygdala damage eliminates monetary loss aversion. *Proceedings. National Academy of Sciences. United States of America, 107*, 3788–3792.
DeCasien, A. R., Guma, E., Liu, S., & Raznahan, A. (2022). Sex differences in the human brain: A roadmap for more careful analysis and interpretation of a biological reality. *Biology of Sex Differences, 13*, 13–43.
Dhingra, I., et al. (2021). Sex differences in neural responses to reward and the influences of individual reward and punishment sensitivity. *BMC Neuroscience, 22*, 1–14.
Diekhof, E. K., & Kraft, S. (2017). The association between endogenous testosterone level and behavioral flexibility in young men–evidence from stimulus-outcome reversal learning. *Hormones and Behavior, 89*, 193–200.
Ding, Y., et al. (2017). Gender differences in reward and punishment for monetary and social feedback in children: An ERP study. *PLoS One, 12*, 1–19.

Dittrich, M., & Leipold, K. (2014). Gender differences in time preferences. *Economics Letters, 122*, 413–415.
Durante, K. M., Griskevicius, V., Hill, S. E., Perilloux, C., & Li, N. P. (2011). Ovulation, female competition, and product choice: Hormonal influences on consumer behavior. *Journal of Consumer Research, 37*, 921–934.
Durante, K. M., Griskevicius, V., Cantu, S. M., & Simpson, J. A. (2014). Money, status, and the ovulatory cycle. *Journal of Marketing Research, 51*, 27–39.
Durante, K. M., Griskevicius, V., & Ulu, S. S. (2020). The effect of fertility on loss aversion. *Journal of Business Research, 120*, 599–607.
Eckel, C. C., & Grossman, P. J. (2008). Men, women and risk aversion: Experimental evidence. *Handbook of Experimental Economics Results, 1*, 1061–1073.
Eisenegger, C., Naef, M., Snozzi, R., Heinrichs, M., & Fehr, E. (2010). Prejudice and truth about the effect of testosterone on human bargaining behaviour. *Nature, 463*, 356–359.
Eliot, L., Ahmed, A., Khan, H., & Patel, J. (2021). Dump the "dimorphism": Comprehensive synthesis of human brain studies reveals few male-female differences beyond size. *Neuroscience and Biobehavioral Reviews, 125*, 667–697.
Enter, D., Terburg, D., Harrewijn, A., Spinhoven, P., & Roelofs, K. (2016). Single dose testosterone administration alleviates gaze avoidance in women with social anxiety disorder. *Psychoneuroendocrinology, 63*, 26–33.
Falk, A., & Hermle, J. (2018). Relationship of gender differences in preferences to economic development and gender equality. *Science, 362*, 1–7.
Felmingham, K. L., et al. (2021). Lower estradiol predicts increased reinstatement of fear in women. *Behaviour Research and Therapy, 142*, 1–7.
Finke, M. S., & Huston, S. J. (2003). The brighter side of financial risk: Financial risk tolerance and wealth. *Journal of Family and Economic Issues, 24*, 233–256.
Fisk, S. R., Miller, B. J., & Overton, J. (2017). Why social status matters for understanding the interrelationships between testosterone, economic risk-taking, and gender. *Sociology Compass, 11*, 1–18.
Forde, N. J., et al. (2020). Sex differences in variability of brain structure across the lifespan. *Cerebral Cortex, 30*, 5420–5430.
Frey, R., Pedroni, A., Mata, R., Rieskamp, J., & Hertwig, R. (2017). Risk preference shares the psychometric structure of major psychological traits. *Science Advances, 3*, 1–13.
Frost, R., & McNaughton, N. (2017). The neural basis of delay discounting: A review and preliminary model. *Neuroscience and Biobehavioral Reviews, 79*, 48–65.
Frye, C. A., Edinger, K., & Sumida, K. (2008). Androgen administration to aged male mice increases anti-anxiety behavior and enhances cognitive performance. *Neuropsychopharmacology, 33*, 1049–1061.
Fuss, T., & Witte, K. (2019). Sex differences in color discrimination and serial reversal learning in mollies and guppies. *Current Zoology, 65*, 323–332.
Gagnon, K. T., Cashdan, E. A., Stefanucci, J. K., & Creem-Regehr, S. H. (2016). Sex differences in exploration behavior and the relationship to harm avoidance. *Human Nature, 27*, 82–97.
Gaillard, A., Fehring, D. J., & Rossell, S. L. (2021a). A systematic review and meta-analysis of behavioural sex differences in executive control. *The European Journal of Neuroscience, 53*, 519–542.
Gaillard, A., Fehring, D. J., & Rossell, S. L. (2021b). Sex differences in executive control: A systematic review of functional neuroimaging studies. *The European Journal of Neuroscience, 53*, 2592–2611.
Geniole, S. N., et al. (2019). Testosterone reduces the threat premium in competitive resource division. *Proceedings of the Royal Society B, 286*, 1–8.
Georgiou, P., et al. (2022). Experimenters' sex modulates mouse behaviors and neural responses to ketamine via corticotropin releasing factor. *Nature Neuroscience, 25*, 1191–1200.
Glimcher, P. W., & Tymula, A. A. (2023). Expected subjective value theory (ESVT): A representation of decision under risk and certainty. *Journal of Economic Behavior and Organization, 207*, 110–128.

Goldstein, J. M., Jerram, M., Abbs, B., Whitfield-Gabrieli, S., & Makris, N. (2010). Sex differences in stress response circuitry activation dependent on female hormonal cycle. *The Journal of Neuroscience, 30*, 431–438.
Gonzalez, M. Z., & Rice, M. A. (2024). Behavioural sciences need behavioural ecology. *Nature Human Behaviour, 8*(7), 1240–1242.
Greimel, E., et al. (2018). Sex differences in the neural underpinnings of social and monetary incentive processing during adolescence. *Cognitive, Affective, & Behavioral Neuroscience, 18*, 296–312.
Grissom, N. M., & Reyes, T. M. (2019). Let's call the whole thing off: Evaluating gender and sex differences in executive function. *Neuropsychopharmacology, 44*, 86–96.
Harris, C. R., & Jenkins, M. (2006). Gender differences in risk assessment: Why do women take fewer risks than men? *Judgment and Decision making, 1*, 48–63.
Häusler, A. N., Kuhnen, C. M., Rudorf, S., & Weber, B. (2018). Preferences and beliefs about financial risk taking mediate the association between anterior insula activation and self-reported real-life stock trading. *Scientific Reports, 8*, 1–13.
Huber, S., Nuerk, H.-C., Reips, U.-D., & Soltanlou, M. (2019). Individual differences influence two-digit number processing, but not their analog magnitude processing: A large-scale online study. *Psychological Research, 83*, 1444–1464.
Hutschemaekers, M. H., de Kleine, R. A., Hendriks, G.-J., Kampman, M., & Roelofs, K. (2021). The enhancing effects of testosterone in exposure treatment for social anxiety disorder: A randomized proof-of-concept trial. *Translational Psychiatry, 11*, 1–7.
Islas-Preciado, D., et al. (2020). Risk-based decision making in rats: Modulation by sex and amphetamine. *Hormones and Behavior, 125*, 1–13.
Jacobsen, B., Lee, J. B., Marquering, W., & Zhang, C. Y. (2014). Gender differences in optimism and asset allocation. *Journal of Economic Behavior and Organization, 107*, 630–651.
Jäncke, L. (2018). Sex/gender differences in cognition, neurophysiology, and neuroanatomy. *F1000Research, 7*, 1–10.
Johnson, M. W., et al. (2021). Sexual discounting: A systematic review of discounting processes and sexual behavior. *Experimental and Clinical Psychopharmacology, 29*, 711–738.
Joue, G., et al. (2022). Sex differences and exogenous estrogen influence learning and brain responses to prediction errors. *Cerebral Cortex, 32*, 2022–2036.
Jung, W. H., Lee, S., Lerman, C., & Kable, J. W. (2018). Amygdala functional and structural connectivity predicts individual risk tolerance. *Neuron, 98*, 394–404.
Kable, J. W., & Levy, I. (2015). Neural markers of individual differences in decision-making. *Current Opinion in Behavioral Sciences, 5*, 100–107.
Kaiser, A., Haller, S., Schmitz, S., & Nitsch, C. (2009). On sex/gender related similarities and differences in fMRI language research. *Brain Research Reviews, 61*, 49–59.
Karmarkar, U. R. (2023). Gender differences in "optimistic" information processing in uncertain decisions. *Cognitive, Affective, & Behavioral Neuroscience, 23*, 827–837.
Keidel, K., Rramani, Q., Weber, B., Murawski, C., & Ettinger, U. (2021). Individual differences in intertemporal choice. *Frontiers in Psychology, 12*, 1–19.
Kessler, R. C., et al. (2005). The World Health Organization adult ADHD self-report scale (ASRS): A short screening scale for use in the general population. *Psychological Medicine, 35*, 245–256.
Kettlewell, N., Levy, J., Tymula, A., & Wang, X. (2023). *The gender reference point gap*. Paper No. 16138. https://hdl.handle.net/10419/272765
Kiesow, H., et al. (2020). 10,000 social brains: Sex differentiation in human brain anatomy. *Science Advances, 6*, 1–12.
Korucuoglu, O., et al. (2020). Adolescent decision-making under risk: Neural correlates and sex differences. *Cerebral Cortex, 30*, 2691–2707.
Kutlikova, H. H., et al. (2021). Not giving up: Testosterone promotes persistence against a stronger opponent. *Psychoneuroendocrinology, 128*, 1–9.
Kutlikova, H. H., Zhang, L., Eisenegger, C., van Honk, J., & Lamm, C. (2023). Testosterone eliminates strategic prosocial behavior through impacting choice consistency in healthy males. *Neuropsychopharmacology, 48*, 1541–1550.

LaClair, M., et al. (2019). Sex differences in cognitive flexibility and resting brain networks in middle-aged marmosets. *eNeuro, 6*, 1–19.
Laube, C., Lorenz, R., & van den Bos, W. (2020). Pubertal testosterone correlates with adolescent impatience and dorsal striatal activity. *Developmental Cognitive Neuroscience, 42*, 1–11.
Lazzaro, S. C., Rutledge, R. B., Burghart, D. R., & Glimcher, P. W. (2016). The impact of menstrual cycle phase on economic choice and rationality. *PLoS One, 11*, 1–15.
Lee, T. M., Chan, C. C., Leung, A. W., Fox, P. T., & Gao, J.-H. (2009). Sex-related differences in neural activity during risk taking: An fMRI study. *Cerebral Cortex, 19*, 1303–1312.
Liu, S., Seidlitz, J., Blumenthal, J. D., Clasen, L. S., & Raznahan, A. (2020). Integrative structural, functional, and transcriptomic analyses of sex-biased brain organization in humans. *Proceedings. National Academy of Sciences. United States of America, 117*, 1–10.
Lockwood, P. L., et al. (2017). Prosocial apathy for helping others when effort is required. *Nature Human Behaviour, 1*, 1–23.
Lucas, M., & Koff, E. (2017). Fertile women discount the future: Conception risk and impulsivity are independently associated with financial decisions. *Evolutionary Psychological Science, 3*, 261–269.
Luckhoff, H., Smit, R., Phahladira, L., Emsley, R., & Asmal, L. (2024). Sex versus gender associations with brain structure. *Journal of Clinical Neuroscience, 122*, 103–109.
Maeng, L., et al. (2017). Acute gonadotropin-releasing hormone agonist treatment enhances extinction memory in male rats. *Psychoneuroendocrinology, 82*, 164–172.
Mehlhorn, K., et al. (2015). Unpacking the exploration–exploitation tradeoff: A synthesis of human and animal literatures. *Decision, 2*, 191–215.
Milad, M. R., et al. (2010). The influence of gonadal hormones on conditioned fear extinction in healthy humans. *Neuroscience, 168*, 652–658.
Moeller, K., Willmes, K., & Klein, E. (2015). A review on functional and structural brain connectivity in numerical cognition. *Frontiers in Human Neuroscience, 9*, 1–14.
Nadler, A., Jiao, P., Johnson, C. J., Alexander, V., & Zak, P. J. (2018). The bull of wall street: Experimental analysis of testosterone and asset trading. *Management Science, 64*, 4032–4051.
Nadler, A., et al. (2019). Does testosterone impair men's cognitive empathy? Evidence from two large-scale randomized controlled trials. *Proceedings of the Royal Society B, 286*, 1–9.
Nadler, A., et al. (2021). *Does testosterone administration increase competitiveness, confidence and financial risk-taking in men?* PsyArXiv [Preprint] 2021 [cited 24 June 2024]. Available at https://psyarxiv.com/62af7/download?format=pdf
Nave, G., et al. (2018). Single-dose testosterone administration increases men's preference for status goods. *Nature Communications, 9*, 1–8.
Nofsinger, J. R., Patterson, F. M., & Shank, C. A. (2018). Decision-making, financial risk aversion, and behavioral biases: The role of testosterone and stress. *Economics and Human Biology, 29*, 1–16.
Núñez, F., Maraver, M. J., & Colzato, L. S. (2020). Sex hormones as cognitive enhancers? *Journal of Cognitive Enhancement, 4*, 228–233.
Orsini, C. A., et al. (2021). Regulation of risky decision making by gonadal hormones in males and females. *Neuropsychopharmacology, 46*, 603–613.
Ou, J., et al. (2021). Testosterone reduces generosity through cortical and subcortical mechanisms. *Proceedings. National Academy of Sciences. United States of America, 118*, 1–7.
Polanía, R., Woodford, M., & Ruff, C. C. (2019). Efficient coding of subjective value. *Nature Neuroscience, 22*, 134–142.
Quesque, F., & Brass, M. (2019). The role of the temporoparietal junction in self-other distinction. *Brain Topography, 32*, 943–955.
Rand, D. G., Brescoll, V. L., Everett, J. A., Capraro, V., & Barcelo, H. (2016). Social heuristics and social roles: Intuition favors altruism for women but not for men. *Journal of Experimental Psychology. General, 145*, 1–21.
Rangel, A., Camerer, C., & Montague, P. R. (2008). A framework for studying the neurobiology of value-based decision making. *Nature Reviews. Neuroscience, 9*, 545–556.

Rauch, J. M., & Eliot, L. (2022). Breaking the binary: Gender versus sex analysis in human brain imaging. *NeuroImage, 264*, 1–5.
Ritchie, S. J., et al. (2018). Sex differences in the adult human brain: Evidence from 5216 UK biobank participants. *Cerebral Cortex, 28*, 2959–2975.
Robertson, D. C., Voegtlin, C., & Maak, T. (2017). Business ethics: The promise of neuroscience. *Journal of Business Ethics, 144*, 679–697.
Ryan, L. V. (2017). Sex differences through a neuroscience lens: Implications for business ethics. *Journal of Business Ethics, 144*, 771–782.
Saposnik, G., Redelmeier, D., Ruff, C. C., & Tobler, P. N. (2016). Cognitive biases associated with medical decisions: A systematic review. *BMC Medical Informatics and Decision Making, 16*, 1–14.
Sartin-Tarm, A., Ross, M. C., Privatsky, A. A., & Cisler, J. M. (2020). Estradiol modulates neural and behavioral arousal in women with posttraumatic stress disorder during a fear learning and extinction task. *Biological Psychiatry: Cognitive Neuroscience and Neuroimaging, 5*, 1114–1122.
Seelig, B. J. (2018). *Constructing and deconstructing woman's power*. Routledge.
Sellitto, M., & Kalenscher, T. (2022). Variations in progesterone and estradiol across the menstrual cycle predict generosity toward socially close others. *Psychoneuroendocrinology, 140*, 1–10.
Serra, D. (2021). Decision-making: From neuroscience to neuroeconomics—An overview. *Theory and Decision, 91*, 1–80.
Sharot, T. (2011). The optimism bias. *Current Biology, 21*, 941–945.
Soutschek, A., et al. (2017). The dopaminergic reward system underpins gender differences in social preferences. *Nature Human Behaviour, 1*, 819–827.
Stanton, S. J., Welker, K. M., Bonin, P. L., Goldfarb, B., & Carré, J. M. (2021). The effect of testosterone on economic risk-taking: A multi-study, multi-method investigation. *Hormones and Behavior, 134*, 1–14.
Sweet-Cushman, J. (2016). Gender, risk assessment, and political ambition. *Politics and the Life Sciences, 35*, 1–17.
Tannenbaum, C., Ellis, R. P., Eyssel, F., Zou, J., & Schiebinger, L. (2019). Sex and gender analysis improves science and engineering. *Nature, 575*, 137–146.
Tascón, L., León, I., Fernández, R., & Cimadevilla, J. M. (2023). Reversal training discloses gender differences in a spatial memory task in humans. *Brain Sciences, 13*, 1–11.
Torgrimson, B. N., & Minson, C. T. (2005). Sex and gender: What is the difference? *Journal of Applied Physiology, 99*, 785–787.
Vaidya, V., Partha, G., & Karmakar, M. (2012). Gender differences in utilization of preventive care services in the United States. *Journal of Women's Health, 21*, 140–145.
Vermeer, A. L., Boksem, M., Gausterer, C., Eisenegger, C., & Lamm, C. (2020). *Testosterone increases risk-taking for status but not for money*. PsyArXiv [Preprint] 25 April 2020 [cited 24 June 2024]. Available at https://osf.io/preprints/psyarxiv/eu8jm/download
Verweij, K. J., Mosing, M. A., Ullén, F., & Madison, G. (2016). Individual differences in personality masculinity-femininity: Examining the effects of genes, environment, and prenatal hormone transfer. *Twin Research and Human Genetics, 19*, 87–96.
Veselic, S., et al. (2021). A causal role of estradiol in human reinforcement learning. *Hormones and Behavior, 134*, 1–10.
Vincent, B. T., Sztwiertnia, M., Koomen, R., & Warren, J. G. (2023). Discounting for money, food, and sex, over the menstrual cycle. *Evolutionary Psychological Science, 9*, 71–81.
Vosberg, D. E., et al. (2021). Sex continuum in the brain and body during adolescence and psychological traits. *Nature Human Behaviour, 5*, 265–272.
Weafer, J. (2020). Sex differences in neural correlates of inhibitory control. *Current Topics in Behavioral Neurosciences, 47*, 73–89.
Wen, Z., et al. (2021). Impact of exogenous estradiol on task-based and resting-state neural signature during and after fear extinction in healthy women. *Neuropsychopharmacology, 46*, 2278–2287.

Will, T. R., et al. (2017). Problems and progress regarding sex bias and omission in neuroscience research. *eNeuro, 4*, 1–10.
Williams, C. M., Peyre, H., Toro, R., & Ramus, F. (2021). Neuroanatomical norms in the UK Biobank: The impact of allometric scaling, sex, and age. *Human Brain Mapping, 42*, 4623–4642.
Wira, C. R., Rodriguez-Garcia, M., & Patel, M. V. (2015). The role of sex hormones in immune protection of the female reproductive tract. *Nature Reviews. Immunology, 15*, 217–230.
Wu, Y., et al. (2018). Single dose testosterone administration modulates emotional reactivity and counterfactual choice in healthy males. *Psychoneuroendocrinology, 90*, 127–133.
Wu, Y., et al. (2019). Testosterone administration increases social discounting in healthy males. *Psychoneuroendocrinology, 108*, 127–134.
Wu, Y., et al. (2020). Single dose testosterone administration increases impulsivity in the intertemporal choice task among healthy males. *Hormones and Behavior, 118*, 1–5.
Wu, Y., et al. (2022). Exogeneous testosterone increases sexual impulsivity in heterosexual men. *Psychoneuroendocrinology, 145*, 1–7.
Zang, Z., et al. (2024). Inter-individual, hemispheric and sex variability of brain activations during numerosity processing. *Brain Structure & Function, 229*, 459–475.
Zeidan, M. A., et al. (2011). Estradiol modulates medial prefrontal cortex and amygdala activity during fear extinction in women and female rats. *Biological Psychiatry, 70*, 920–927.

Domain-General and Domain-Specific Values and Their Role in Psychopathology

Alex Rich and Ifat Levy

Abstract This chapter explores the neural and behavioral mechanisms underlying decision-making across various contexts, highlighting the interplay between domain-general and domain-specific value representations. Drawing on evidence from neuroimaging and behavioral studies, we examine how the brain encodes value on a common scale to facilitate choices among qualitatively different options. Additionally, we discuss the clinical relevance of these findings, emphasizing disorder-specific value perturbations observed in psychiatric conditions such as binge eating disorder and post-traumatic stress disorder. The chapter also addresses methodological challenges in studying value representations and underscores the importance of standardized, replicable research designs. Finally, we highlight promising therapeutic interventions targeting disorder-specific valuation processes, illustrating their potential to improve treatment outcomes in psychiatric and neurodevelopmental disorders. This overview aims to bridge gaps in our current understanding and foster advancements in neuroeconomic research and clinical applications.

1 Introduction

How do we choose between fruit and ice cream for dessert? Many models of choice assume that we estimate the value of each option, integrating over multiple relevant

A. Rich
Department of Comparative Medicine, Yale School of Medicine, New Haven, CT, USA

Interdepartmental Neuroscience Program, Yale University, New Haven, CT, USA

I. Levy (✉)
Department of Comparative Medicine, Yale School of Medicine, New Haven, CT, USA

Interdepartmental Neuroscience Program, Yale University, New Haven, CT, USA

Departments of Neuroscience and Psychology and Wu-Tsai Institute, Yale University, New Haven, CT, USA
e-mail: ifat.levy@yale.edu

D. V. Smith et al. (eds.), *Neuroeconomics: Core Topics and Current Directions*,
https://doi.org/10.1007/978-3-032-02925-6_32

dimensions (e.g., taste, health, sweetness) and then compare the integrated values to yield choice (Kable & Glimcher, 2009). To facilitate such comparisons, the brain needs to represent values of highly different items on a common scale (Levy & Glimcher, 2012). However, suppose you finish dessert at the restaurant and head to a friend's house, where you are offered ice cream. If you have just consumed it, ice cream is likely to be devalued relative to your earlier representation. Such value updating requires some value representation that is specific to the ice cream (or, at least, to high-fat desserts). This suggests that both "common currency" and specific neural representations of value should exist.

Values are, by definition, subjective. Some people love ice cream and others do not; some take health considerations into account in their choices, and others choose solely based on taste (Hare et al., 2009). Subjective values also incorporate other properties of each available option, including the likelihood of receiving an outcome (Levy et al., 2010), the time of receipt (Kable & Glimcher, 2007), and the effort required to obtain it (Treadway et al., 2012). Individual decision-makers vary substantially in how these properties affect their valuation. For example, having to wait for the ice cream to be ready will reduce its value for an impatient decision-maker, but not for a patient one. As with the neural representation of value, decision characteristics may rely on a domain-general component, as well as on domain-specific mechanisms.

In this chapter, we will describe existing evidence for domain-general and domain-specific behavioral and neural representations, as well as gaps in our knowledge. We will also discuss the significant clinical relevance of recent advances in this area: Perturbations in valuation and decision-making have been observed across psychiatric populations, and these perturbations may be more general—encompassing all rewards, or even both rewards and punishments—or more specific to disorder-relevant content (e.g., encoding aberrations of food value among those with eating disorders). An exciting current direction in neuroeconomic research lies in understanding these varied value representations and their relevance to psychiatric conditions.

2 Decision-Making Across Domains

Humans (and animals) make decisions in a wide variety of contexts: from financial to medical decisions, and from deciding what to eat to deciding whom to spend time with. Self-report questionnaires suggest that individual decision characteristics, such as attitudes toward uncertainty, consist of both domain-general and domain-specific factors (Einav et al., 2012; Frey et al., 2017, 2023; Josef et al., 2016; Reeck et al., 2022). A similar picture begins to emerge from laboratory choice tasks. These tasks typically ask participants to choose between options that vary on one or more dimensions (Ben-Zion & Levy, 2025). For example, by making a series of choices between monetary lotteries that vary in the amount of money they offer, and in the probability of obtaining these amounts, participants reveal their "risk attitude."

Risk-averse individuals would prefer lotteries that offer small amounts at high probabilities, compared to those that offer higher amounts at lower probabilities; risk-seeking individuals would exhibit opposite preferences. Similarly, participants can make choices between options that offer different outcomes at different delays, or requiring different levels of effort, revealing their temporal and effort preferences, respectively. Laboratory experiments also assess learning—how participants adjust their value estimates and actions based on feedback. Many studies estimate such individual attitudes using monetary (or point) outcomes. The monetary domain is very useful because money is generally rewarding, and because there is typically no satiety for increasing amounts of money. As the significance of studying decision-making across domains is increasingly acknowledged, however, some studies have begun to compare individual attitudes across domains. At the behavioral level, these studies provide some evidence for consistent decision characteristics. For example, risk attitudes in the monetary domain were highly predictive of the same attitudes when choosing between probabilistic quantities of M&Ms. or water (Levy & Glimcher, 2011). Monetary risk attitudes were also predictive of choices between amounts of time spent with a social partner, or medication dosages (Seaman et al., 2016). Similarly, willingness to wait or exert effort for rewards was also consistent across domains (Seaman et al., 2016). At the same time, there is task-based evidence for a domain-specific component. For example, risk attitudes for monetary gains did not predict the same attitudes for monetary losses (Tymula et al., 2013). Similarly, individuals may have different rates of temporal discounting for different types of rewards (Estle et al., 2007), or rewards and punishments (Abdellaoui et al., 2018). Moreover, age-related changes in tolerance to uncertainty or temporal delay may be more pronounced in some domains compared to others (Josef et al., 2016; Seaman et al., 2016).

Consistent with behavioral findings, neural studies reveal several key areas of value representation (Fig. 1), which appear relevant in both domain-general and domain-specific contexts.

2.1 "Common Currency" Value Representations

Domain-general, or "common currency," representations of value may facilitate choices among qualitatively different options. A large number of neuroimaging studies support the existence of such representations, which encode value regardless of stimulus type, in regions of the prefrontal cortex (PFC), particularly the ventromedial prefrontal cortex (vmPFC), sometimes extending to the medial orbitofrontal cortex (mOFC), and the ventral striatum (VS), a structure of the basal ganglia (Bartra et al., 2013; Clithero & Rangel, 2014). These areas encode the value of immediate and future rewards (Kable & Glimcher, 2007), risky and ambiguous lotteries (Levy et al., 2010), and rewards of different categories (Levy & Glimcher, 2012), such as money, food, and consumer goods (Chib et al., 2009; Klein-Flügge et al., 2013; Levy & Glimcher, 2011). They integrate over taste and health (Hare et

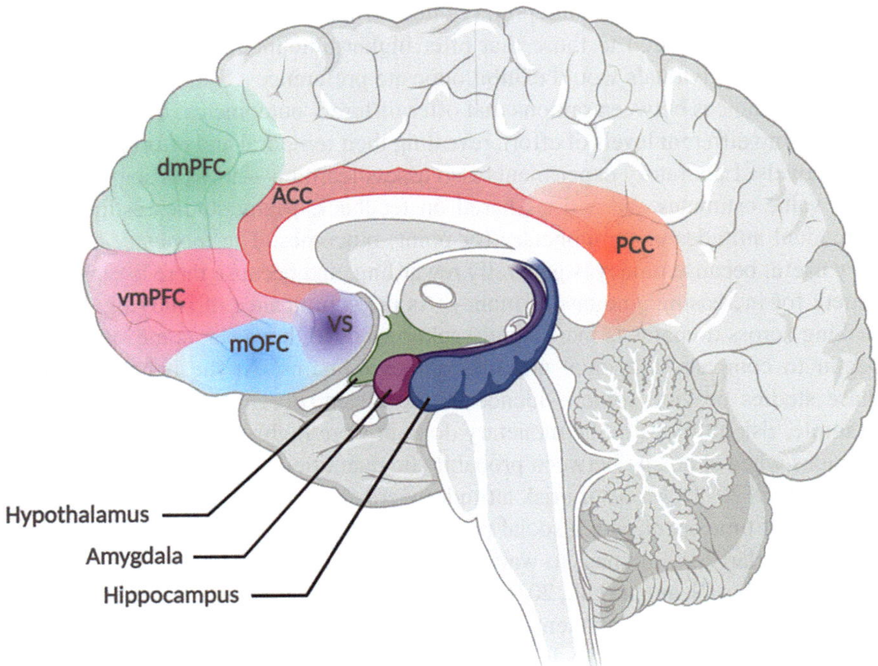

Fig. 1 Value representation regions of interest. *ACC* anterior cingulate cortex, *dmPFC* dorsomedial prefrontal cortex, *mOFC* medial orbitofrontal cortex, *PCC* posterior cingulate cortex, *VS* ventral striatum, *vmPFC* ventromedial prefrontal cortex

al., 2009) and various nutrients (Suzuki et al., 2017). There is some evidence that the same brain areas encode the value of punishments (Čeko et al., 2022; Tom et al., 2007), but this is less clear (Ben-Zion & Levy, 2025; Botvinik-Nezer et al., 2020).

Notably, much of the research documenting overlapping representations of value of different categories has relied on univariate analysis of functional magnetic resonance imaging (fMRI) and identification of regions of interest. While these analyses remain valuable, it is possible that regions of interest encompass distinct neural populations, each encoding value of a particular category. To partially overcome this issue, other studies have employed multivariate fMRI analysis techniques to uncover general and specific value encoding at greater resolution than that of regions of interest (Howard et al., 2015; McNamee et al., 2013). These studies revealed a relatively dorsal aspect of vmPFC, which continued to exhibit general value encoding across categories, providing stronger support for a domain-general value representation.

Value signals have also been consistently observed in several other brain areas, including more lateral regions of OFC (lOFC; Zhang et al., 2017) and subcortical regions. The amygdala responds to both rewards and punishments (Baxter & Murray, 2002; Feinstein et al., 2011; Hampton et al., 2007; Mustafa & Ali,

2019) and may have a role in salience processing (Kong & Zweifel, 2021). The hippocampus, implicated in memory encoding, also encodes value as part of the memory (Zeithamova et al., 2018), with over-representation of rewarded locations (Sato et al., 2020). Finally, the insula, primarily known for integrating bodily sensations with emotional states (Uddin et al., 2017), is also associated with value-based decision-making, especially in its anterior aspect (Loued-Khenissi et al., 2020). The extent to which value signals in these areas are domain-general or specific is not clear. More generally, the limited spatial and temporal resolution of fMRI highlights the significance of complementing research in healthy humans with electrophysiological recordings in animals and human patients. Indeed, single-unit recordings revealed value-related signals, in similar brain regions, in nonhuman primates (Padoa-Schioppa & Assad, 2006), and rodents (Burgos-Robles et al., 2017; Stalnaker et al., 2014). In humans with epilepsy, surgically implanted electrodes uncovered reward signals in single amygdala neurons (Jenison et al., 2011), and local field potentials in amygdala, OFC, and hippocampus reflected value-related signals (Manssuer et al., 2022). Overall, substantial research in humans and animals is consistent with value encoding, with support for domain-general representations in some of the areas.

2.2 Domain-Specific Value Representations

Domain-specific representations can facilitate value updating of specific options (e.g., ice cream) or categories (e.g., high-fat food) in response to changes in the environment, state, or available information. Some studies identifying "common currency" representations also reveal category-specific neural patterns. For example, food value representations have been identified in the hypothalamus and amygdala, while monetary value is processed in the posterior cingulate cortex (PCC; Jenison et al., 2011; Levy & Glimcher, 2011). Similarly, parts of the dorsomedial PFC and the anterior cingulate cortex (ACC) appear specialized to value encoding in social contexts (de la Vega et al., 2016; Ruff & Fehr, 2014), and insula activation has been found to mediate the relationship between affect and social decision biases (Harlé et al., 2012).

The multivariate fMRI studies that provided strong evidence for domain-general value representations also pointed to additional representations that were specific to distinct categories, in dmPFC, lOFC, hippocampus (Howard et al., 2015), and mOFC (McNamee et al., 2013). Consistent with this, several studies were able to extract information about identity from value-related areas (Howard & Kahnt, 2021; Zhang et al., 2017). Similarly, single-unit studies in animals also provide both value and identity information, for example, in rat OFC (Stalnaker et al., 2014), and reflect encoding of stimulus-specific values, for example, in mouse amygdala (Yang et al., 2023).

2.3 Summary: The Need for Both Domain-General and Domain-Specific Value Representations

Ultimately, current evidence suggests the involvement of both domain-specific and domain-general encoding, such that the best current models of decision-making are likely a combination of the two and the links between them (Čeko et al., 2022; Howard et al., 2015; McNamee et al., 2013; Ramsey & Ward, 2020; Spunt & Adolphs, 2017). A model of this nature requires a detailed understanding of general and specific choice-related representations, the connections between them, and their flexibility, which have been of increasing interest to the neuroeconomics literature. This distinction is of particular value for clinical research, as we discuss below.

3 Disorder Specificity

Variations in value-based decision-making and learning are common to psychiatric disorders (Homan et al., 2019; Jia et al., 2023; Levy & Schiller, 2020; Mukherjee et al., 2020; Powers et al., 2017; Pushkarskaya et al., 2015; Suthaharan et al., 2024). Some definitions of psychiatric disorders necessitate domain-specific behavioral manifestations. For example, a social anxiety disorder diagnosis requires fear, anxiety, or avoidance specific to social or performance situations, and binge eating disorder (BED) requires loss of control specific to eating (American Psychiatric Association, 2013). The cognitive processes underlying the observed behaviors, however, may be domain-specific or more general. An intriguing possibility is that some of the heterogeneity in symptom presentation may be explained by the extent of value-related computational aberration. For example, some individuals with BED may have a specific impairment in encoding the value of food or associating food rewards with the cues that predict them, while others may have a more general impairment in encoding reward value, or even in encoding both reward and punishment value.

Probing these questions can allow for the development of more effective therapies for psychiatric disorders: Whereas generalized aberrations may be better addressed through treatments such as cognitive behavioral therapy (CBT), which have proven effective across psychopathology (Ouimet et al., 2021; Ouimet & Ferguson, 2019), disorder-specific aberrations may require more targeted treatments. For example, trauma-focused therapies, such as prolonged exposure therapy and cognitive processing therapy, have emerged as recommended forms of treatment for post-traumatic stress disorder (PSTD; Watkins et al., 2018). Within and across diagnostic groups, significant differences are observed among individuals with respect to more holistic or specific variations in value encoding; accordingly, there is a need for treatment options for transdiagnostic psychopathology as well as targeted and individualized approaches.

Fig. 2 Disorder-specific value representations. Examples of this (depicted clockwise) include social stimuli in social anxiety disorder, substances in substance use disorders, trauma-specific stimuli in PTSD, phobic stimuli in specific phobia, money in gambling disorder, and food in eating disorders

As such, it is critical for neuroeconomic research to push current boundaries in investigations of the domain-general or domain-specific nature of value encoding to include "disorder specificity" (Fig. 2). This goes beyond the specificity of an item or category to probe the value encoding of disorder-specific stimuli, such as food in the context of BED or traumatic narratives in the context of PTSD. There is substantial behavioral and neural evidence that suggests such disorder-specific representations exist across a variety of conditions, including psychiatric, neurodevelopmental, and even metabolic disorders. Below we survey findings in some of these conditions. We focus on BED as a clear example where the behavioral deficits may be general or specific to food, but refer to other conditions, such as PTSD, substance-related and addictive disorders, and specific phobias.

3.1 *Disorder-Specific Value Representations: Behavioral Evidence*

Eating disorders exemplify a need for understanding general valuation across domains, as well as specific valuation of food and body. These disorders are often comorbid with a wide range of psychopathology and are associated with general differences in value-based learning and decision-making, as we will detail below; however, a diagnosis necessitates symptomatology that is food- or body-specific (American Psychiatric Association, 2013). To complicate things further, eating pathology is often associated with changes in body-weight status, which should also be taken into account, as obesity status has been repeatedly linked to changes in valuation (Coppin et al 2014; Dan et al., 2022; Zhang et al 2014). Binge eating disorder (BED), the psychiatric disorder most commonly co-occurring with obesity (Hudson et al., 2007), is one such example. In BED, behavioral evidence points to both general and food-specific differences in value-related processes. For example, in a working memory task with food and neutral stimuli, women with BED exhibited reduced learning compared to overweight controls regardless of stimulus type; the same participants, however, exhibited food-specific biases in a memory task manipulating recency (Svaldi et al., 2014b). This suggests that within the same individuals, there may be both general and food-specific alterations in value-based processes.

A similar approach was applied to studying response inhibition—the ability to suppress unwanted or inappropriate actions. A commonly used task for measuring response inhibition is the stop-signal task (Verbruggen et al., 2019). In this task, participants are required to perform an action (e.g. pressing a button) upon perceiving a cue (e.g., a visual stimulus). On a small number of trials, a “stop signal” (e.g., a visual or an auditory stimulus) appears sometime after the first cue, instructing participants to suppress the planned response. In individuals with BED, there are reports of increased difficulty in response inhibition with food cues, but not neutral cues (Svaldi et al., 2014a; Manasse et al., 2016a). Another study, however, reported deficits in women with BED performing the same task across food-specific stimuli, positive nonfood stimuli, and neutral stimuli, with no difference between stimulus types (Manasse et al., 2016b). A deficit independent of stimulus type was also observed with a slightly different measure, the go/no-go task, in which participants were instructed to respond to specified target stimuli and refrain from responding to other stimuli. Individuals with BED made more errors—compared to both obese individuals without BED and individuals with healthy weight—across food/body stimuli and neutral objects. More nuanced effects were observed when participants had to inhibit saccades elicited by peripheral stimuli. Individuals with BED had increased difficulty inhibiting saccades to both food and nonfood stimuli, but this difficulty was especially pronounced for food items (Schag et al., 2013). It is thus possible that deficits in behavioral inhibition may be specific to food in some individuals with BED, or under some conditions, but more general to rewarding stimuli, or even to any non-harmful stimuli, in other individuals or conditions.

Paradigms probing decision-making and attention have also been used to delineate valuation processes in BED (Colton et al., 2023). Some of these studies report general valuation differences across different contexts. For example, when making choices between delayed and immediate rewards, individuals with BED discounted the value of future food, money, and other consumable rewards (like massage time) more steeply than controls (Manwaring et al., 2011). Conversely, when attentional biases were more explicitly probed through a rapid serial visual presentation paradigm with food and neutral words, overweight women with BED showed more pronounced engagement with food stimuli over neutral stimuli than overweight and obese women without BED (Schmitz et al., 2015). Ultimately, findings are considerably mixed in BED concerning general and food-specific aberrations in valuation.

Literature encompassing a wide range of psychiatric conditions mirrors these behavioral findings in BED and obesity. Whereas food stimuli are specific to BED, disorder specificity in PTSD may include traumatic narratives; contamination-related stimuli (e.g., dirty surface) offer specificity to obsessive-compulsive disorder; and emotional stimuli (e.g., sad and happy faces) are of relevance to bipolar disorder. Valuation differences have been found in disorder-specific contexts across bipolar disorder (Vederman et al., 2012), borderline personality disorder (Kaiser et al., 2016), PTSD (Bardeen et al., 2016), depression (Lazarov et al., 2018), and obsessive-compulsive disorder (De Putter & Koster, 2017; Sizino da Victoria et al., 2012). It is important to note, however, that these findings do not imply that the impairment is specific to the disorder-related domain, as some of these studies do not include a control condition. Concerning addictive disorders, general valuation deficits have been observed (Hester et al., 2006; Lindheimer et al., 2020), in addition to drug- or alcohol-specific alterations (Franken et al., 2000; Hester et al., 2006; Hogarth & Hardy, 2018; MacKillop et al., 2010). Interestingly, preferential substance choice over other rewards has been shown across studies of alcohol, cocaine, and tobacco dependence in a show of converging support for the theory that drug dependence may be driven by a greater value assigned to drug rewards (Bickel et al., 2014; Chase et al., 2013; Hogarth & Hardy, 2018; Moeller et al., 2013). Similarly, alterations in both general and social-specific stimuli processing have been observed in neurodevelopmental disorders, such as autism spectrum disorder (Greenaway & Plaisted, 2005; Mo et al., 2019). Finally, recent studies have uncovered changes in decision-making under uncertainty in opiate use disorder (Konova et al., 2020) and PTSD (Jia et al., 2023) using a monetary choice task. These findings may be important not just for understanding the basic mechanisms of these disorders, but also for objective diagnosis, which does not rely on subjective self-report or clinician assessments. Overall, behavioral perturbations in valuation may thus be unique to the disorder-specific domain in some cases, but more general in others.

3.2 Disorder-Specific Value Representations: Neural Evidence

In line with behavioral findings, neuroimaging studies suggest variations in the neural representations of value that are both general and disorder-specific. For example, food images have been associated with greater activation of the OFC, ACC, and insula than neutral images in BED compared to overweight and healthy-weight controls (Schienle et al., 2009). Interestingly, this study also included a condition of disgust-inducing pictures; activation in response to these pictures did not differ among groups. Hyperactivation of value-related regions in response to food images seems to be a consistent finding (Balodis et al., 2015; Kober & Boswell, 2018). In future studies, it is important to include other reward categories to test whether these neural responses are specific to food or are reward-related more generally.

The increased value-related activation to food stimuli, together with some evidence of impaired value-based learning, led to a recent theory, positing that failures of behavioral control in BED may be linked to difficulties in updating food values (Voon, 2015; Waltmann et al., 2021). Similar perspectives have been of interest across eating disorders, as there is also evidence of general and disorder-specific valuation alterations in prefrontal and striatal regions. Interestingly, in anorexia nervosa, aberrant frontostriatal activity and connectivity have been observed in response to food stimuli, both with and without the presence of choice (Simon et al., 2019). Differentiating among eating disorder diagnoses remains an area of debate due to significant symptom overlap and individual differences (Burton & Abbott, 2017); as such, uncovering disorder-specific valuation alterations is highly relevant. For example, uncovering similarities and differences in food value encoding among BED, bulimia nervosa, and anorexia nervosa binge/purge subtype will aid in understanding and treating binge eating, a distressing symptom these disorders all share.

Beyond disordered eating, disorder-specific neural representations of value appear relevant in understanding various psychiatric disorders. Similar to eating disorders are substance use and alcohol use disorders, which may be linked to either substance-specific or general alterations in valuation processes. An early study in individuals with cocaine use disorder used a color-word Stroop task with both drug and neutral words (Goldstein et al., 2007). This study reported correlation between activation levels in parts of the valuation system (ACC and vmPFC) and the degree of negative valence attributed to drug words, as well as the number of errors committed for drug versus neutral cues. Since no other reward categories were included in the study, however, these neural findings may be more general rather than specific to drug rewards.

More recent studies revealed generalizable effects of drug use on neural responsiveness during learning. For example, striatal and OFC responsiveness to the received utility of a monetary lottery reward is diminished in people with cocaine addiction, which suggests that the computations critical to experiential learning may be disrupted in these individuals (Konova et al., 2023). This broad hyporesponsiveness is of notable contrast to major depressive disorder where, in response to nonspecific rewards (e.g., money, positive faces, preferred music), the VS and OFC exhibit diminished and elevated activity, respectively (Ng et al., 2019). Interestingly, in binge alcohol drinkers, the encoding of the expected reward value of nonalcoholic stimuli, predicting reward delivery, is blunted in prefrontal and amygdala-hippocampal regions (Tolomeo et al., 2023). Ultimately, findings of both drug- or alcohol-specific value encoding alterations and more general value-based learning deficits are consistent with theories of value updating as a key mechanism for disorder maintenance (Groman et al., 2022; Gueguen et al., 2021).

Anxiety-based disorders, such as specific phobias and PTSD, also lend themselves to similar investigations of their neural bases. There is evidence for phobia-specific aberrations in value encoding in some of the same brain regions implicated in addiction. For example, those with spider phobia had greater responses to spider than neutral stimuli in the amygdala, ACC, insula, and dorsomedial PFC. The amygdala was the only brain region in which this effect was observed even in the absence of direct attention (Straube et al., 2006), consistent with a phobia-specific role for the amygdala in automatic stimuli processing. Disorder-specific encoding in the amygdala, insula, hippocampus, and prefrontal cortex is also observed in PTSD: Exposure to trauma-specific narratives is accompanied by altered functional connectivity and blood flow in these regions, with greater PTSD severity linked to diminished connectivity changes (Cisler et al., 2014; Osuch et al., 2001). Like value-based learning deficits observed in addictive disorders, an impaired ability to update the value of fear-associated stimuli has been demonstrated in PTSD (Milad et al., 2009). This impaired recall of extinction memory was associated with deactivation of the hippocampus and vmPFC. Across a range of psychiatric disorders, it is evident that brain-based differences in valuation are often unique to disorder-specific contexts; moreover, difficulty in updating disorder-specific value representations may at least in part underlie disorder maintenance. A recent study, however, documented symptom-related activity in vmPFC and striatum of individuals with PTSD, in a monetary decision-making paradigm, completely unrelated to trauma (Jia et al. 2023). Interestingly, in that study, activity in the ventral striatum was qualitatively different in individuals with PTSD, compared to trauma-exposed controls.

3.3 Challenges and Considerations

While research on domain-specific and disorder-specific decision-making processes has both basic and clinical significance, it also faces several notable challenges. At the practical level, some types of outcomes cannot be actualized in the laboratory, for logistical or ethical reasons. For example, while researchers can induce a genuine rewarding experience with monetary rewards, they are much more limited in generating a sense of punishment or loss. A promising tool in overcoming this challenge is the use of gamified environments, including virtual reality, which are experienced as more similar to real life compared to traditional laboratory designs (Allen et al., 2024). Another challenge for clinical research is the interpretation of findings from control participants in these studies. For example, purging (or the ability to purge) can serve as a measure of food/body-related value in individuals with bulimia nervosa, but not in healthy controls. Similarly, there are challenges in identifying universally salient cues to compare with those that are domain- or disorder-specific. Many studies default to neutral cues, such as household items, or rely on monetary gains and losses. The monetary domain, however, like other domains, is multidimensional in ways that often intersect or challenge comparisons across domains. As an illustration, money is often considered a nonsocial reward, yet the value of money comes from society's agreement on its function, making it a "social construct" (Matyjek et al., 2020). It is possible that the social aspect of money, and indeed that of other categories such as trauma, contributes to findings of consistent behavior across social and nonsocial domains. Similarly, money is an instrument for obtaining other rewards, such as food, again complicating arguments for consistency across domains. Finally, as researchers aim to address these challenges, significant methodological heterogeneity has emerged. These differences in study designs, task characteristics, and stimuli make it difficult to make meaningful conclusions about disorder-specific processes (Antons et al., 2023). As neuroeconomic researchers increasingly aim to identify domain- and disorder-specific encoding, it is pertinent that the field engages in replication and standardization.

At the conceptual level, we also note that most of the findings described here—both those supporting the specificity of value representations and those pointing to "common currency" across domains—are correlational in nature and can be explained in ways that do not require value encoding. While activation patterns in several brain regions clearly reflect subjective value, there is much less evidence for a causal role of these value representations in decision-making (Hayden & Niv, 2021). Some evidence supports the presence of value signals even in the absence of choice (Lebreton et al., 2009; Levy et al., 2011; Tusche et al., 2010), but it is possible that decision-making relies primarily on direct learning of action policies rather than comparisons of stored values (Hayden & Niv, 2021). Value signals, for example in the OFC, might thus arise from "cognitive maps" of state space, rather than from value encoding (Behrens et al., 2018; Grueschow et al., 2015; Knudsen & Wallis, 2022; Schuck et al., 2016). This perspective implies that the impact of OFC and vmPFC lesions on decision-making (Bechara et al., 1994; Fellows & Farah, 2007; Yu et al., 2022) could be due to broader cognitive impairments rather than specific issues with value encoding. This discourse represents an evolving and exciting area of neuroeconomic research that directly supplements our understanding of domain-specific and domain-general representations.

4 Conclusion: Unanswered Questions and Future Directions

In the last 15 years, studies have increasingly included both domain-general and domain- or disorder-specific information to more thoroughly probe value encoding; however, many still tend to include only one or the other or compare domain/disorder-specific rewards to neutral stimuli which does not account for reward encoding generally. As the field aims to identify more tailored, evidence-based therapeutic targets, it is necessary to uncover what valuation processes are tailor-fit to decision-making aberrations observed in clinical practice. Recently, there have been promising advances in applying disorder-specific neuroeconomic findings to intervention development, such as food-specific inhibition, cue-approach, and attentional trainings for food devaluation in obesity (Itzkovitch et al., 2022; Stice et al., 2022; Yang et al., 2022). These trainings exhibit the potential to alter food-related preferences, which are a core difficulty of existing dietary interventions for obesity and overweight. By targeting value encoding processes that are disorder-specific, such as the updating of food values in obesity/BED, there is potential to alter decision-making behavior that contributes to psychiatric, metabolic, and neurodevelopmental pathology.

References

Abdellaoui, M., Gutierrez, C., & Kemel, E. (2018). Temporal discounting of gains and losses of time: An experimental investigation. *Journal of Risk and Uncertainty, 57*(1), 1–28. https://doi.org/10.1007/s11166-018-9287-1

Allen, K., Brändle, F., Botvinick, M., Fan, J. E., Gershman, S. J., Gopnik, A., Griffiths, T. L., Hartshorne, J. K., Hauser, T. U., Ho, M. K., de Leeuw, J. R., Ma, W. J., Murayama, K., Nelson, J. D., van Opheusden, B., Pouncy, T., Rafner, J., Rahwan, I., Rutledge, R. B., et al. (2024). Using games to understand the mind. *Nature Human Behaviour*, 1–9. https://doi.org/10.1038/s41562-024-01878-9

American Psychiatric Association, D. S. M. T. F., & American Psychiatric Association, D. S. (2013). *Diagnostic and statistical manual of mental disorders* (Vol. 5, Issue 5). American Psychiatric Association.

Antons, S., Müller, S. M., Neumann, P., Müller, A., & Steins-Loeber, S. (2023). Stimuli-specific inhibitory control in disorders due to addictive behaviours: A review of current evidence and discussion of methodological challenges. *Current Addiction Reports, 10*(4), 749–769. https://doi.org/10.1007/s40429-023-00517-z

Balodis, I. M., Grilo, C. M., & Potenza, M. N. (2015). Neurobiological features of binge eating disorder. *CNS Spectrums, 20*(6), 557–565. https://doi.org/10.1017/S1092852915000814

Bardeen, J. R., Tull, M. T., Daniel, T. A., Evenden, J., & Stevens, E. N. (2016). A preliminary investigation of the time course of attention bias variability in posttraumatic stress disorder: The moderating role of attentional control. *Behaviour Change, 33*(2), 94–111.

Bartra, O., McGuire, J. T., & Kable, J. W. (2013). The valuation system: A coordinate-based meta-analysis of BOLD fMRI experiments examining neural correlates of subjective value. *NeuroImage, 76*, 412–427. https://doi.org/10.1016/j.neuroimage.2013.02.063

Baxter, M. G., & Murray, E. A. (2002). The amygdala and reward. *Nature Reviews Neuroscience, 3*(7), Article 7. https://doi.org/10.1038/nrn875

Bechara, A., Damasio, A. R., Damasio, H., & Anderson, S. W. (1994). Insensitivity to future consequences following damage to human prefrontal cortex. *Cognition, 50*(1–3), 7–15.
Behrens, T. E., Muller, T. H., Whittington, J. C., Mark, S., Baram, A. B., Stachenfeld, K. L., & Kurth-Nelson, Z. (2018). What is a cognitive map? Organizing knowledge for flexible behavior. *Neuron, 100*(2), 490–509.
Ben-Zion, Z., & Levy, I. (2025). Representation of anticipated rewards and punishments in the human brain. *Annual Review of Psychology, 76*, 197–226. https://doi.org/10.1146/annurev-psych-022324-042614
Bickel, W. K., Johnson, M. W., Koffarnus, M. N., MacKillop, J., & Murphy, J. G. (2014). The behavioral economics of substance use disorders: Reinforcement pathologies and their repair. *Annual Review of Clinical Psychology, 10*, 641–677. https://doi.org/10.1146/annurev-clinpsy-032813-153724
Botvinik-Nezer, R., Holzmeister, F., Camerer, C. F., Dreber, A., Huber, J., Johannesson, M., Kirchler, M., Iwanir, R., Mumford, J. A., Adcock, R. A., Avesani, P., Baczkowski, B. M., Bajracharya, A., Bakst, L., Ball, S., Barilari, M., Bault, N., Beaton, D., Beitner, J., et al. (2020). Variability in the analysis of a single neuroimaging dataset by many teams. *Nature, 582*(7810), Article 7810. https://doi.org/10.1038/s41586-020-2314-9
Burgos-Robles, A., Kimchi, E. Y., Izadmehr, E. M., Porzenheim, M. J., Ramos-Guasp, W. A., Nieh, E. H., Felix-Ortiz, A. C., Namburi, P., Leppla, C. A., Presbrey, K. N., Anandalingam, K. K., Pagan-Rivera, P. A., Anahtar, M., Beyeler, A., & Tye, K. M. (2017). Amygdala inputs to prefrontal cortex guide behavior amid conflicting cues of reward and punishment. *Nature Neuroscience, 20*(6), Article 6. https://doi.org/10.1038/nn.4553
Burton, A. L., & Abbott, M. J. (2017). Conceptualising binge eating: A review of the theoretical and empirical literature. *Behaviour Change, 34*(3), 168–198.
Čeko, M., Kragel, P. A., Woo, C.-W., López-Solà, M., & Wager, T. D. (2022). Common and stimulus-type-specific brain representations of negative affect. *Nature Neuroscience, 25*(6), Article 6. https://doi.org/10.1038/s41593-022-01082-w
Chase, H. W., MacKillop, J., & Hogarth, L. (2013). Isolating behavioural economic indices of demand in relation to nicotine dependence. *Psychopharmacology, 226*(2), 371–380. https://doi.org/10.1007/s00213-012-2911-x
Chib, V. S., Rangel, A., Shimojo, S., & O'Doherty, J. P. (2009). Evidence for a common representation of decision values for dissimilar goods in human ventromedial prefrontal cortex. *Journal of Neuroscience, 29*(39), 12315–12320. https://doi.org/10.1523/JNEUROSCI.2575-09.2009
Cisler, J. M., Steele, J. S., Lenow, J. K., Smitherman, S., Everett, B., Messias, E., & Kilts, C. D. (2014). Functional reorganization of neural networks during repeated exposure to the traumatic memory in posttraumatic stress disorder: An exploratory fMRI study. *Journal of Psychiatric Research, 48*(1), 47–55.
Clithero, J. A., & Rangel, A. (2014). Informatic parcellation of the network involved in the computation of subjective value. *Social Cognitive and Affective Neuroscience, 9*(9), 1289–1302. https://doi.org/10.1093/scan/nst106
Colton, E., Wilson, K.-E., Chong, T. T.-J., & Verdejo-Garcia, A. (2023). Dysfunctional decision-making in binge-eating disorder: A meta-analysis and systematic review. *Neuroscience & Biobehavioral Reviews, 152*, 105250. https://doi.org/10.1016/j.neubiorev.2023.105250
Coppin, G., Nolan-Poupart, S., Jones-Gotman, M., & Small, D. M. (2014). Working memory and reward association learning impairments in obesity. *Neuropsychologia, 65*, 146–155.
Dan, O., Wertheimer, E. K., & Levy, I. (2022). A neuroeconomics approach to obesity. *Biological Psychiatry, 91*(10), 860–868. https://doi.org/10.1016/j.biopsych.2021.09.019
de la Vega, A., Chang, L. J., Banich, M. T., Wager, T. D., & Yarkoni, T. (2016). Large-scale meta-analysis of human medial frontal cortex reveals tripartite functional organization. *The Journal of Neuroscience, 36*(24), 6553–6562. https://doi.org/10.1523/JNEUROSCI.4402-15.2016
De Putter, L. M. S., & Koster, E. H. W. (2017). The effects of obsessive-compulsive symptoms and disorder-relevant stimuli on the dynamics of selective attention. *Journal of Obsessive-Compulsive and Related Disorders, 15*, 74–84. https://doi.org/10.1016/j.jocrd.2017.09.001

Einav, L., Finkelstein, A., Pascu, I., & Cullen, M. R. (2012). How general are risk preferences? Choices under uncertainty in different domains. *American Economic Review, 102*(6), 2606–2638. https://doi.org/10.1257/aer.102.6.2606

Estle, S. J., Green, L., Myerson, J., & Holt, D. D. (2007). Discounting of monetary and directly consumable rewards. *Psychological Science, 18*(1), 58–63. https://doi.org/10.1111/j.1467-9280.2007.01849.x

Feinstein, J. S., Adolphs, R., Damasio, A., & Tranel, D. (2011). The human amygdala and the induction and experience of fear. *Current Biology, 21*(1), 34–38. https://doi.org/10.1016/j.cub.2010.11.042

Fellows, L. K., & Farah, M. J. (2007). The role of ventromedial prefrontal cortex in decision making: Judgment under uncertainty or judgment per se? *Cerebral Cortex, 17*(11), 2669–2674. https://doi.org/10.1093/cercor/bhl176

Franken, I. H., Kroon, L. Y., Wiers, R. W., & Jansen, A. (2000). Selective cognitive processing of drug cues in heroin dependence. *Journal of Psychopharmacology, 14*(4), 395–400.

Frey, R., Pedroni, A., Mata, R., Rieskamp, J., & Hertwig, R. (2017). Risk preference shares the psychometric structure of major psychological traits. *Science Advances, 3*(10), e1701381. https://doi.org/10.1126/sciadv.1701381

Frey, R., Duncan, S. M., & Weber, E. U. (2023). Towards a typology of risk preference: Four risk profiles describe two-thirds of individuals in a large sample of the U.S. population. *Journal of Risk and Uncertainty, 66*(1), 1–17. https://doi.org/10.1007/s11166-022-09398-5

Goldstein, R. Z., Tomasi, D., Rajaram, S., Cottone, L. A., Zhang, L., Maloney, T., & Volkow, N. D. (2007). Role of the anterior cingulate and medial orbitofrontal cortex in processing drug cues in cocaine addiction. *Neuroscience, 144*(4), 1153–1159.

Greenaway, R., & Plaisted, K. (2005). Top-down attentional modulation in autistic spectrum disorders is stimulus-specific. *Psychological Science, 16*(12), 987–994. https://doi.org/10.1111/j.1467-9280.2005.01648.x

Groman, S. M., Thompson, S. L., Lee, D., & Taylor, J. R. (2022). Reinforcement learning detuned in addiction: Integrative and translational approaches. *Trends in Neurosciences, 45*(2), 96–105. https://doi.org/10.1016/j.tins.2021.11.007

Grueschow, M., Polania, R., Hare, T. A., & Ruff, C. C. (2015). Automatic versus choice-dependent value representations in the human brain. *Neuron, 85*(4), 874–885. https://doi.org/10.1016/j.neuron.2014.12.054

Gueguen, M. C., Schweitzer, E. M., & Konova, A. B. (2021). Computational theory-driven studies of reinforcement learning and decision-making in addiction: What have we learned? *Current Opinion in Behavioral Sciences, 38*, 40–48. https://doi.org/10.1016/j.cobeha.2020.08.007

Hampton, A. N., Adolphs, R., Tyszka, J. M., & O'Doherty, J. P. (2007). Contributions of the amygdala to reward expectancy and choice signals in human prefrontal cortex. *Neuron, 55*(4), 545–555. https://doi.org/10.1016/j.neuron.2007.07.022

Hare, T. A., Camerer, C. F., & Rangel, A. (2009). Self-control in decision-making involves modulation of the vmPFC valuation system. *Science, 324*(5927), 646–648. https://doi.org/10.1126/science.1168450

Harlé, K. M., Chang, L. J., van 't Wout, M, & Sanfey, A. G. (2012). The neural mechanisms of affect infusion in social economic decision-making: A mediating role of the anterior insula. *NeuroImage, 61*(1), 32–40. https://doi.org/10.1016/j.neuroimage.2012.02.027

Hayden, B. Y., & Niv, Y. (2021). The case against economic values in the orbitofrontal cortex (or anywhere else in the brain). *Behavioral Neuroscience, 135*(2), 192–201. https://doi.org/10.1037/bne0000448

Hester, R., Dixon, V., & Garavan, H. (2006). A consistent attentional bias for drug-related material in active cocaine users across word and picture versions of the emotional Stroop task. *Drug and Alcohol Dependence, 81*(3), 251–257.

Hogarth, L., & Hardy, L. (2018). Alcohol use disorder symptoms are associated with greater relative value ascribed to alcohol, but not greater discounting of costs imposed on alcohol. *Psychopharmacology, 235*, 2257–2266.

Homan, P., Levy, I., Feltham, E., Gordon, C., Hu, J., Li, J., Pietrzak, R. H., Southwick, S., Krystal, J. H., Harpaz-Rotem, I., & Schiller, D. (2019). Neural computations of threat in the aftermath of combat trauma. *Nature Neuroscience, 22*(3), 470–476. https://doi.org/10.1038/s41593-018-0315-x

Howard, J. D., & Kahnt, T. (2021). To be specific: The role of orbitofrontal cortex in signaling reward identity. *Behavioral Neuroscience, 135*(2), 210–217. https://doi.org/10.1037/bne0000455

Howard, J. D., Gottfried, J. A., Tobler, P. N., & Kahnt, T. (2015). Identity-specific coding of future rewards in the human orbitofrontal cortex. *Proceedings of the National Academy of Sciences of the United States of America, 112*(16), 5195–5200. https://doi.org/10.1073/pnas.1503550112

Hudson, J. I., Hiripi, E., Pope, H. G., & Kessler, R. C. (2007). The prevalence and correlates of eating disorders in the National Comorbidity Survey Replication. *Biological Psychiatry, 61*(3), 348–358. https://doi.org/10.1016/j.biopsych.2006.03.040

Itzkovitch, A., Bar Or, M., & Schonberg, T. (2022). Cue-approach training for food behavior. *Current Opinion in Behavioral Sciences, 47*, 101202. https://doi.org/10.1016/j.cobeha.2022.101202

Jenison, R. L., Rangel, A., Oya, H., Kawasaki, H., & Howard, M. A. (2011). Value encoding in single neurons in the human amygdala during decision making. *Journal of Neuroscience, 31*(1), 331–338. https://doi.org/10.1523/JNEUROSCI.4461-10.2011

Jia, R., Ruderman, L., Pietrzak, R. H., Gordon, C., Ehrlich, D., Horvath, M., Mirchandani, S., DeFontes, C., Southwick, S., Krystal, J. H., Harpaz-Rotem, I., & Levy, I. (2023). Neural valuation of rewards and punishments in posttraumatic stress disorder: A computational approach. *Translational Psychiatry, 13*(1), 101. https://doi.org/10.1038/s41398-023-02388-4

Josef, A. K., Richter, D., Samanez-Larkin, G. R., Wagner, G. G., Hertwig, R., & Mata, R. (2016). Stability and change in risk-taking propensity across the adult life span. *Journal of Personality and Social Psychology, 111*(3), 430–450. https://doi.org/10.1037/pspp0000090

Kable, J. W., & Glimcher, P. W. (2007). The neural correlates of subjective value during intertemporal choice. *Nature Neuroscience, 10*(12), 1625–1633. https://doi.org/10.1038/nn2007

Kable, J. W., & Glimcher, P. W. (2009). The neurobiology of decision: Consensus and controversy. *Neuron, 63*(6), 733–745. https://doi.org/10.1016/j.neuron.2009.09.003

Kaiser, D., Jacob, G. A., Domes, G., & Arntz, A. (2016). Attentional bias for emotional stimuli in borderline personality disorder: A meta-analysis. *Psychopathology, 49*(6), 383–396. https://doi.org/10.1159/000448624

Klein-Flügge, M. C., Barron, H. C., Brodersen, K. H., Dolan, R. J., & Behrens, T. E. J. (2013). Segregated encoding of reward-identity and stimulus-reward associations in human orbitofrontal cortex. *The Journal of Neuroscience, 33*(7), 3202–3211. https://doi.org/10.1523/JNEUROSCI.2532-12.2013

Knudsen, E. B., & Wallis, J. D. (2022). Taking stock of value in the orbitofrontal cortex. *Nature Reviews Neuroscience, 23*(7), Article 7. https://doi.org/10.1038/s41583-022-00589-2

Kober, H., & Boswell, R. G. (2018). Potential psychological & neural mechanisms in binge eating disorder: Implications for treatment. *Clinical Psychology Review, 60*, 32–44. https://doi.org/10.1016/j.cpr.2017.12.004

Kong, M.-S., & Zweifel, L. S. (2021). Central amygdala circuits in valence and salience processing. *Behavioural Brain Research, 410*, 113355. https://doi.org/10.1016/j.bbr.2021.113355

Konova, A. B., Lopez-Guzman, S., Urmanche, A., Ross, S., Louie, K., Rotrosen, J., & Glimcher, P. W. (2020). Computational markers of risky decision-making for identification of temporal windows of vulnerability to opioid use in a real-world clinical setting. *JAMA Psychiatry, 77*(4), 368–377. https://doi.org/10.1001/jamapsychiatry.2019.4013

Konova, A. B., Ceceli, A. O., Horga, G., Moeller, S. J., Alia-Klein, N., & Goldstein, R. Z. (2023). Reduced neural encoding of utility prediction errors in cocaine addiction. *Neuron, 111*(24), 4058–4070.e6. https://doi.org/10.1016/j.neuron.2023.09.015

Lazarov, A., Ben-Zion, Z., Shamai, D., Pine, D. S., & Bar-Haim, Y. (2018). Free viewing of sad and happy faces in depression: A potential target for attention bias modification. *Journal of Affective Disorders, 238*, 94–100.

Lebreton, M., Jorge, S., Michel, V., Thirion, B., & Pessiglione, M. (2009). An automatic valuation system in the human brain: Evidence from functional neuroimaging. *Neuron, 64*(3), 431–439. https://doi.org/10.1016/j.neuron.2009.09.040

Levy, D., & Glimcher, P. (2011). Comparing apples and oranges: Using reward-specific and reward-general subjective value representation in the brain. *Journal of Neuroscience, 31*(41), 14693–14707. https://doi.org/10.1523/JNEUROSCI.2218-11.2011

Levy, D., & Glimcher, P. (2012). The root of all value: A neural common currency for choice. *Current Opinion in Neurobiology, 22*(6), 1027–1038. https://doi.org/10.1016/j.conb.2012.06.001

Levy, I., & Schiller, D. (2020). Neural computations of threat. *Trends in Cognitive Sciences, 25*(2), 151–171. https://doi.org/10.1016/j.tics.2020.11.007

Levy, I., Snell, J., Nelson, A. J., Rustichini, A., & Glimcher, P. W. (2010). Neural representation of subjective value under risk and ambiguity. *Journal of Neurophysiology, 103*(2), 1036–1047.

Levy, I., Lazzaro, S. C., Rutledge, R. B., & Glimcher, P. W. (2011). Choice from non-choice: Predicting consumer preferences from blood oxygenation level-dependent signals obtained during passive viewing. *Journal of Neuroscience, 31*(1), 118–125.

Lindheimer, N., Nicolai, J., & Moshagen, M. (2020). General rather than specific: Cognitive deficits in suppressing task irrelevant stimuli are associated with buying-shopping disorder. *PLoS One, 15*(8), e0237093. https://doi.org/10.1371/journal.pone.0237093

Loued-Khenissi, L., Pfeuffer, A., Einhäuser, W., & Preuschoff, K. (2020). Anterior insula reflects surprise in value-based decision-making and perception. *NeuroImage, 210*, 116549. https://doi.org/10.1016/j.neuroimage.2020.116549

MacKillop, J., O'Hagen, S., Lisman, S. A., Murphy, J. G., Ray, L. A., Tidey, J. W., & Monti, P. M. (2010). Behavioral economic analysis of cue-elicited craving for alcohol. *Addiction, 105*(9), 1599–1607.

Manasse, S. M., Espel, H. M., Schumacher, L. M., Kerrigan, S. G., Zhang, F., Forman, E. M., & Juarascio, A. S. (2016a). Does impulsivity predict outcome in treatment for binge eating disorder? A multimodal investigation. *Appetite, 105*, 172–179. https://doi.org/10.1016/j.appet.2016.05.026

Manasse, S. M., Goldstein, S. P., Wyckoff, E., Forman, E. M., Juarascio, A. S., Butryn, M. L., Ruocco, A. C., & Nederkoorn, C. (2016b). Slowing down and taking a second look: Inhibitory deficits associated with binge eating are not food-specific. *Appetite, 96*, 555–559. https://doi.org/10.1016/j.appet.2015.10.025

Manssuer, L., Qiong, D., Wei, L., Yang, R., Zhang, C., Zhao, Y., Sun, B., Zhan, S., & Voon, V. (2022). Integrated amygdala, orbitofrontal and hippocampal contributions to reward and loss coding revealed with human intracranial EEG. *Journal of Neuroscience, 42*(13), 2756–2771. https://doi.org/10.1523/JNEUROSCI.1717-21.2022

Manwaring, J. L., Green, L., Myerson, J., Strube, M. J., & Wilfley, D. E. (2011). Discounting of various types of rewards by women with and without binge eating disorder: Evidence for general rather than specific differences. *The Psychological Record, 61*(4), 561–582. https://doi.org/10.1007/BF03395777

Matyjek, M., Meliss, S., Dziobek, I., & Murayama, K. (2020). A multidimensional view on social and non-social rewards. *Frontiers in Psychiatry, 11*, 818.

McNamee, D., Rangel, A., & O'Doherty, J. P. (2013). Category-dependent and category-independent goal-value codes in human ventromedial prefrontal cortex. *Nature Neuroscience, 16*(4), Article 4. https://doi.org/10.1038/nn.3337

Milad, M. R., Pitman, R. K., Ellis, C. B., Gold, A. L., Shin, L. M., Lasko, N. B., Zeidan, M. A., Handwerger, K., Orr, S. P., & Rauch, S. L. (2009). Neurobiological basis of failure to recall extinction memory in posttraumatic stress disorder. *Biological Psychiatry, 66*(12), 1075–1082. https://doi.org/10.1016/j.biopsych.2009.06.026

Mo, S., Liang, L., Bardikoff, N., & Sabbagh, M. A. (2019). Shifting visual attention to social and non-social stimuli in Autism Spectrum Disorders. *Research in Autism Spectrum Disorders, 65*, 56–64. https://doi.org/10.1016/j.rasd.2019.05.006

Moeller, S. J., Beebe-Wang, N., Woicik, P. A., Konova, A. B., Maloney, T., & Goldstein, R. Z. (2013). Choice to view cocaine images predicts concurrent and prospective drug use in cocaine addiction. *Drug and Alcohol Dependence, 130*(1), 178–185. https://doi.org/10.1016/j.drugalcdep.2012.11.001

Mukherjee, D., Filipowicz, A. L. S., Vo, K., Satterthwaite, T. D., & Kable, J. W. (2020). Reward and punishment reversal-learning in major depressive disorder. *Journal of Abnormal Psychology, 129*(8), 810–823. https://doi.org/10.1037/abn0000641

Mustafa, G., & Ali, N. (2019). Rewards, autonomous motivation and turnover intention: Results from a non-Western cultural context. *Cogent Business & Management, 6*(1), 1676090. https://doi.org/10.1080/23311975.2019.1676090

Ng, T. H., Alloy, L. B., & Smith, D. V. (2019). Meta-analysis of reward processing in major depressive disorder reveals distinct abnormalities within the reward circuit. *Translational Psychiatry, 9*(1), 1–10. https://doi.org/10.1038/s41398-019-0644-x

Osuch, E. A., Benson, B., Geraci, M., Podell, D., Herscovitch, P., McCann, U. D., & Post, R. M. (2001). Regional cerebral blood flow correlated with flashback intensity in patients with posttraumatic stress disorder. *Biological Psychiatry, 50*(4), 246–253.

Ouimet, A. J., & Ferguson, R. J. (2019). Innovations and advances in cognitive behavioral therapy: Insights from experimental psychopathology. *Journal of Experimental Psychopathology, 10*(3), 2043808719874966. https://doi.org/10.1177/2043808719874966

Ouimet, A. J., Dixon-Luinenburg, T., & Rooyakkers, M. (2021). Experimental psychopathology at the crossroads: Reflections on past, present, and future contributions to cognitive behavioural therapy. *International Journal of Cognitive Therapy, 14*(1), 133–159. https://doi.org/10.1007/s41811-020-00093-4

Padoa-Schioppa, C., & Assad, J. A. (2006). Neurons in the orbitofrontal cortex encode economic value. *Nature, 441*(7090), 223–226.

Powers, A. R., Mathys, C., & Corlett, P. R. (2017). Pavlovian conditioning–induced hallucinations result from overweighting of perceptual priors. *Science, 357*(6351), 596–600. https://doi.org/10.1126/science.aan3458

Pushkarskaya, H., Tolin, D., Ruderman, L., Kirshenbaum, A., Kelly, J. M., Pittenger, C., & Levy, I. (2015). Decision-making under uncertainty in obsessive–compulsive disorder. *Journal of Psychiatric Research, 69*, 166–173. https://doi.org/10.1016/j.jpsychires.2015.08.011

Ramsey, R., & Ward, R. (2020). Putting the nonsocial into social neuroscience: A role for domain-general priority maps during social interactions. *Perspectives on Psychological Science, 15*(4), 1076–1094. https://doi.org/10.1177/1745691620904972

Reeck, C., Mullette-Gillman, O. A., McLaurin, R. E., & Huettel, S. A. (2022). Beyond money: Risk preferences across both economic and non-economic contexts predict financial decisions. *PLoS One, 17*(12), e0279125. https://doi.org/10.1371/journal.pone.0279125

Ruff, C. C., & Fehr, E. (2014). The neurobiology of rewards and values in social decision making. *Nature Reviews Neuroscience, 15*(8), Article 8. https://doi.org/10.1038/nrn3776

Sato, M., Mizuta, K., Islam, T., Kawano, M., Sekine, Y., Takekawa, T., Gomez-Dominguez, D., Schmidt, A., Wolf, F., Kim, K., Yamakawa, H., Ohkura, M., Lee, M. G., Fukai, T., Nakai, J., & Hayashi, Y. (2020). Distinct mechanisms of over-representation of landmarks and rewards in the hippocampus. *Cell Reports, 32*(1), 107864. https://doi.org/10.1016/j.celrep.2020.107864

Schag, K., Teufel, M., Junne, F., Preissl, H., Hautzinger, M., Zipfel, S., & Giel, K. E. (2013). Impulsivity in binge eating disorder: Food cues elicit increased reward responses and disinhibition. *PLoS One, 8*(10), e76542. https://doi.org/10.1371/journal.pone.0076542

Schienle, A., Schäfer, A., Hermann, A., & Vaitl, D. (2009). Binge-eating disorder: Reward sensitivity and brain activation to images of food. *Biological Psychiatry, 65*(8), 654–661. https://doi.org/10.1016/j.biopsych.2008.09.028

Schmitz, F., Naumann, E., Biehl, S., & Svaldi, J. (2015). Gating of attention towards food stimuli in binge eating disorder. *Appetite, 95*, 368–374. https://doi.org/10.1016/j.appet.2015.07.023

Schuck, N. W., Cai, M. B., Wilson, R. C., & Niv, Y. (2016). Human orbitofrontal cortex represents a cognitive map of state space. *Neuron, 91*(6), 1402–1412.

Seaman, K. L., Gorlick, M. A., Vekaria, K. M., Hsu, M., Zald, D. H., & Samanez-Larkin, G. R. (2016). Adult age differences in decision making across domains: Increased discounting of social and health-related rewards. *Psychology and Aging, 31*(7), 737–746. https://doi.org/10.1037/pag0000131

Simon, J. J., Stopyra, M. A., & Friederich, H.-C. (2019). Neural processing of disorder-related stimuli in patients with anorexia nervosa: A narrative review of brain imaging studies. *Journal of Clinical Medicine, 8*(7), Article 7. https://doi.org/10.3390/jcm8071047

Sizino da Victoria, M., Nascimento, A. L., & Fontenelle, L. F. (2012). Symptom-specific attentional bias to threatening stimuli in obsessive-compulsive disorder. *Comprehensive Psychiatry, 53*(6), 783–788. https://doi.org/10.1016/j.comppsych.2011.12.005

Spunt, R. P., & Adolphs, R. (2017). A new look at domain specificity: Insights from social neuroscience. *Nature Reviews Neuroscience, 18*(9), 559–567. https://doi.org/10.1038/nrn.2017.76

Stalnaker, T. A., Cooch, N. K., McDannald, M. A., Liu, T. L., Wied, H., & Schoenbaum, G. (2014). Orbitofrontal neurons infer the value and identity of predicted outcomes. *Nature Communications, 5*(1), 3926.

Stice, E., Yokum, S., Gau, J., Veling, H., Lawrence, N., & Kemps, E. (2022). Efficacy of a food response and attention training treatment for obesity: A randomized placebo controlled trial. *Behaviour Research and Therapy, 158*, 104183. https://doi.org/10.1016/j.brat.2022.104183

Straube, T., Mentzel, H.-J., & Miltner, W. H. R. (2006). Neural mechanisms of automatic and direct processing of phobogenic stimuli in specific phobia. *Biological Psychiatry, 59*(2), 162–170. https://doi.org/10.1016/j.biopsych.2005.06.013

Suthaharan, P., Thompson, S. L., Rossi-Goldthorpe, R. A., Rudebeck, P. H., Walton, M. E., Chakraborty, S., Noonan, M. P., Costa, V. D., Murray, E. A., Mathys, C. D., Groman, S. M., Mitchell, A. S., Taylor, J. R., Corlett, P. R., & Chang, S. W. C. (2024). Lesions to the mediodorsal thalamus, but not orbitofrontal cortex, enhance volatility beliefs linked to paranoia. *Cell Reports, 43*(6), 114355. https://doi.org/10.1016/j.celrep.2024.114355

Suzuki, S., Cross, L., & O'Doherty, J. P. (2017). Elucidating the underlying components of food valuation in the human orbitofrontal cortex. *Nature Neuroscience, 20*(12), Article 12. https://doi.org/10.1038/s41593-017-0008-x

Svaldi, J., Naumann, E., Trentowska, M., & Schmitz, F. (2014a). General and food-specific inhibitory deficits in binge eating disorder. *International Journal of Eating Disorders, 47*(5), 534–542. https://doi.org/10.1002/eat.22260

Svaldi, J., Schmitz, F., Trentowska, M., Tuschen-Caffier, B., Berking, M., & Naumann, E. (2014b). Cognitive interference and a food-related memory bias in binge eating disorder. *Appetite, 72*, 28–36. https://doi.org/10.1016/j.appet.2013.09.014

Tolomeo, S., Baldacchino, A., & Steele, J. D. (2023). Blunted expected reward value signals in binge alcohol drinkers. *Journal of Neuroscience, 43*(31), 5685–5692.

Tom, S. M., Fox, C. R., Trepel, C., & Poldrack, R. A. (2007). The neural basis of loss aversion in decision-making under risk. *Science, 315*(5811), 515–518. https://doi.org/10.1126/science.1134239

Treadway, M. T., Buckholtz, J. W., Cowan, R. L., Woodward, N. D., Li, R., Ansari, M. S., & Zald, D. H. (2012). Dopaminergic mechanisms of individual differences in human effort-based decision-making. *Journal of Neuroscience, 32*(18), 6170–6176.

Tusche, A., Bode, S., & Haynes, J.-D. (2010). Neural responses to unattended products predict later consumer choices. *Journal of Neuroscience, 30*(23), 8024–8031.

Tymula, A., Rosenberg Belmaker, L. A., Ruderman, L., Glimcher, P. W., & Levy, I. (2013). Like cognitive function, decision making across the life span shows profound age-related changes. *Proceedings of the National Academy of Sciences, 110*(42), 17143–17148. https://doi.org/10.1073/pnas.1309909110

Uddin, L. Q., Nomi, J. S., Hebert-Seropian, B., Ghaziri, J., & Boucher, O. (2017). Structure and function of the human insula. *Journal of Clinical Neurophysiology, 34*(4), 300–306. https://doi.org/10.1097/WNP.0000000000000377

Vederman, A. C., Weisenbach, S. L., Rapport, L. J., Leon, H. M., Haase, B. D., Franti, L. M., Schallmo, M.-P., Saunders, E. F. H., Kamali, M. M., Zubieta, J.-K., Langenecker, S. A., & McInnis, M. G. (2012). Modality-specific alterations in the perception of emotional stimuli in bipolar disorder compared to healthy controls and major depressive disorder. *Cortex, 48*(8), 1027–1034. https://doi.org/10.1016/j.cortex.2011.03.017

Verbruggen, F., Aron, A. R., Band, G. P., Beste, C., Bissett, P. G., Brockett, A. T., Brown, J. W., Chamberlain, S. R., Chambers, C. D., Colonius, H., Colzato, L. S., Corneil, B. D., Coxon, J. P., Dupuis, A., Eagle, D. M., Garavan, H., Greenhouse, I., Heathcote, A., Huster, R. J., et al. (2019). A consensus guide to capturing the ability to inhibit actions and impulsive behaviors in the stop-signal task. *eLife, 8*, e46323. https://doi.org/10.7554/eLife.46323

Voon, V. (2015). Cognitive biases in binge eating disorder: The hijacking of decision making. *CNS Spectrums, 20*(6), 566–573. https://doi.org/10.1017/S1092852915000681

Waltmann, M., Herzog, N., Horstmann, A., & Deserno, L. (2021). Loss of control over eating: A systematic review of task based research into impulsive and compulsive processes in binge eating. *Neuroscience & Biobehavioral Reviews, 129*, 330–350. https://doi.org/10.1016/j.neubiorev.2021.07.016

Watkins, L. E., Sprang, K. R., & Rothbaum, B. O. (2018). Treating PTSD: A review of evidence-based psychotherapy interventions. *Frontiers in Behavioral Neuroscience, 12*, 258. https://doi.org/10.3389/fnbeh.2018.00258

Yang, Y., Qi, L., Morys, F., Wu, Q., & Chen, H. (2022). Food-specific inhibition training for food devaluation: A meta-analysis. *Nutrients, 14*(7), Article 7. https://doi.org/10.3390/nu14071363

Yang, T., Yu, K., Zhang, X., Xiao, X., Chen, X., Fu, Y., & Li, B. (2023). Plastic and stimulus-specific coding of salient events in the central amygdala. *Nature, 616*(7957), 510–519. https://doi.org/10.1038/s41586-023-05910-2

Yu, L. Q., Dana, J., & Kable, J. W. (2022). Individuals with ventromedial frontal damage display unstable but transitive preferences during decision making. *Nature Communications, 13*(1), Article 1. https://doi.org/10.1038/s41467-022-32511-w

Zeithamova, D., Gelman, B. D., Frank, L., & Preston, A. R. (2018). Abstract representation of prospective reward in the hippocampus. *Journal of Neuroscience, 38*(47), 10093–10101. https://doi.org/10.1523/JNEUROSCI.0719-18.2018

Zhang, Z., Manson, K. F., Schiller, D., & Levy, I. (2014). Impaired associative learning with food rewards in obese women. *Current Biology, 24*(15), 1731–1736. https://doi.org/10.1016/j.cub.2014.05.075

Zhang, Z., Fanning, J., Ehrlich, D. B., Chen, W., Lee, D., & Levy, I. (2017). Distributed neural representation of saliency controlled value and category during anticipation of rewards and punishments. *Nature Communications, 8*(1), Article 1. https://doi.org/10.1038/s41467-017-02080-4

Applications and Evolution of Neuroeconomics and Reinforcement Learning Models in Substance Use Disorders

Srinivasan A. Ramakrishnan, Madhuvanthi Muliya, V. Srinivasa Chakravarthy, and Muhammad A. Parvaz

Abstract This chapter examines the application of neuroeconomics and reinforcement learning (RL) models in understanding substance use disorders (SUDs). Neuroeconomics combines insights from neuroscience, psychology, and economics to investigate decision-making processes, particularly in contexts where maladaptive behaviors like addiction emerge. RL offers a computational framework to analyze how agents adapt their behaviors through interactions with the environment, focusing on reward processing, value computation, and prediction error signaling.

The chapter provides a detailed overview of key RL models used to study SUDs, including temporal difference RL, model-based and model-free approaches, and Bayesian frameworks. These models elucidate how decision-making processes are impaired in addiction, highlighting transitions from goal-directed to habitual behaviors, dysregulated reward systems, and altered risk-reward tradeoffs. RL-inspired frameworks also offer explanations for relapse and compulsive substance-seeking, emphasizing their utility in predicting addiction trajectories and informing interventions.

Looking ahead, future directions in RL modeling emphasize integrating advanced machine learning techniques, such as generative AI and hierarchical RL, to capture the complexity of addiction-related behaviors. Context-aware frameworks and multi-scale approaches combining RL with biophysical and theoretical models

S. A. Ramakrishnan
Department of Psychiatry, Icahn School of Medicine at Mount Sinai, New York, NY, USA

M. Muliya · V. S. Chakravarthy
Department of Biotechnology, Indian Institute of Technology, Madras, Chennai, Tamil Nadu, India

M. A. Parvaz (✉)
Department of Psychiatry, Icahn School of Medicine at Mount Sinai, New York, NY, USA

Department of Neuroscience, Icahn School of Medicine at Mount Sinai, New York, NY, USA

Department of Artificial Intelligence and Human Health, Icahn School of Medicine at Mount Sinai, New York, NY, USA
e-mail: muhammad.parvaz@mssm.edu

D. V. Smith et al. (eds.), *Neuroeconomics: Core Topics and Current Directions*,
https://doi.org/10.1007/978-3-032-02925-6_33

are discussed as avenues to bridge neural, behavioral, and environmental data. These innovations aim to refine predictive capabilities, personalize treatment strategies, and identify novel intervention points. By synthesizing diverse modeling approaches, this chapter underscores the transformative potential of RL in decoding the intricate mechanisms underlying SUDs and advancing addiction research.

1 Introduction

1.1 Neuroeconomics and Reinforcement Learning

Neuroeconomics emerged from conventional economic models to address the complexities of real-world decision-making. This process often involves trade-offs between costs and benefits, shaped by factors such as rewards, risks, and time constraints. These trade-offs are further influenced by the environment in which decisions are made. Neuroeconomics examines how the brain evaluates these factors, considering both constraints and environmental influences. By integrating insights from multiple disciplines, it provides a comprehensive framework for studying decision-making in both healthy and disordered states.

At the core of neuroeconomics is reinforcement learning (RL), which posits that decisions are shaped by experience-based learning. RL provides a dynamic framework for studying how choices are adapted over time through interactions with the environment. RL is often explained using an agent that takes actions, receives feedback in the form of rewards or penalties, and gradually learns to optimize cognitive resources for future decisions to maximize cumulative rewards. This process reflects fundamental behavioral principles, reinforcing actions that lead to positive outcomes while discouraging those associated with negative consequences. RL closely aligns with biological learning, encompassing both explicit learning (cognitive and reflective processes) and implicit learning (habits and conditioned responses) (Berry & Broadbent, 1987; Dayan et al., 2006; Yang & Li, 2012).

Computational models leveraging RL have been instrumental in dissecting these mechanisms, providing a structured approach to studying behavior. By simulating decision-making processes, RL models allow researchers to quantify value-based learning. These models serve as effective test beds, either as standalone frameworks or in combination with other computational approaches, to investigate the underlying psychophysiology of decision-making, its maladaptive aspects, and potential intervention strategies.

1.2 Computational Framework for the Reinforcement Learning Models

At its core, RL consists of two main components, an *agent* and an *environment*. The agent interacts with the environment by performing actions, for which it receives feedback in the form of rewards (positive, negative, or neutral). This feedback shapes the agent's *policy*, a mapping of its current state to the optimal action it should take. This process is repeated, allowing the agent to refine its strategy and optimize resource allocation over time. The primary objective of RL is to maximize cumulative rewards, enabling the agent to learn an optimal policy.

Building upon this framework, the computational approach to decision-making involves an *agent* interacting with its *environment*, learning from feedback, and refining its strategies over time. At each timestep, the agent selects an *action* (A_t) based on its current *state* (S_t) and its learned *policy* ($\pi(A_t|\, S_t)$), which maps states to actions. The environment then responds by transitioning the agent to a *new state* (S_{t+1}) and providing a *reward* (R_{t+1}) that indicates the desirability of the chosen action.

Through repeated interactions, the agent updates its *value function* ($V(S_t)$), which estimates the expected cumulative reward from each state, and its *Q-value function* ($Q(S_t, A_t)$), which evaluates specific state-action pairs. Learning is driven by *prediction error* (δ), the discrepancy between expected and received rewards, which helps refine future expectations and decision-making (Fig. 1 and Table 1).

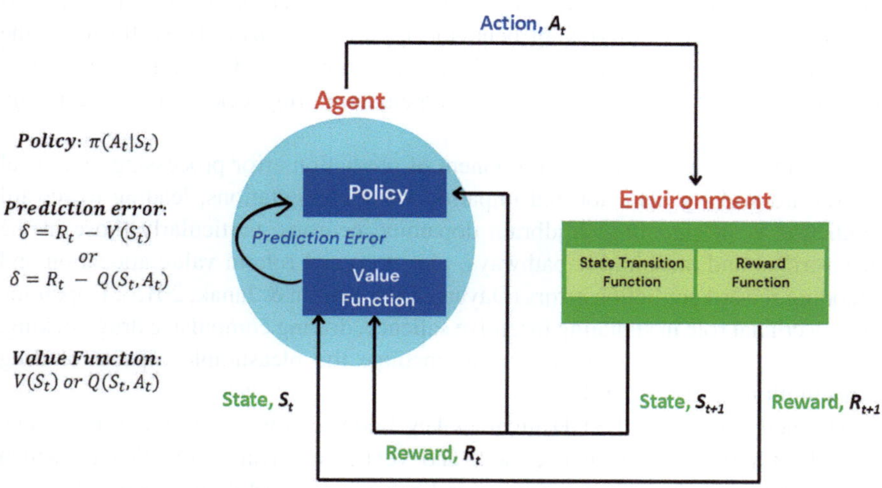

Fig. 1 The general RL loop—the conversation between the agent and the environment

Table 1 Model parameters of RL framework

Component	Definition
Agent	The decision-making entity that learns from interactions
Environment	The external system in which the agent operates and receives feedback
State (S_t)	The agent's current condition, which influences decision-making
Action (A_t)	A choice made by the agent that affects the environment
Reward (R_{t+1})	The feedback received for an action, representing its favorability
Policy ($\pi(A_t \| S_t)$),	The probability of deciding a particular action (A_t) given a specific state (S_t)
Value function ($V(S_t)$)	The expected long-term reward associated with a given state
Q-value ($Q(S_t, A_t)$)	The expected reward for taking a specific action in a given state
Prediction error (δ)	The difference between expected and received reward, guiding learning
State transition function	Defines how the system evolves from one state to another based on an action
Reward function	Specifies the reward associated with different state-action pairs

2 Aberrant Learning and Decision-Making in SUD

2.1 Neural Correlates and the Role of Dopamine

Substance use disorders (SUDs) are driven by complex neurobiological mechanisms of drug-seeking and drug-taking behavior. Figure 2 (Fonzo, 2018) illustrates the cyclical relationship between "liking" (hedonic impact) and "wanting" (incentive salience), both of which are critical in perpetuating drug-seeking and drug-taking behaviors.

Dopamine signaling, a key component of prediction error processing, is crucial for reward-seeking behavior and impaired value computations, leading to reward prediction error signaling. Midbrain dopamine neurons, particularly those in the mesostriatal and mesolimbic pathways, play a crucial role in value allocation and encoding reward prediction errors (Dayan, 2009; Keiflin & Janak, 2015). Dopamine plays a crucial role in signaling incentive salience, driving compulsive drug-seeking, while opioid and cannabinoid systems mediate the pleasurable aspects of drug consumption (Berridge, 2007).

The neurocircuitry of SUD involves key brain regions such as the amygdala, ventral striatum, orbitofrontal cortex, and ventral tegmental area (VTA), which regulate reward valuation, motivation, and behavioral adaptation (Koob & Volkow, 2016). A meta-analysis by Gowin et al. (2013) highlights deficits in these key neural substrates involved in risk computation, including the orbitofrontal cortex and the striatum, in individuals with SUDs. Increased drug consumption and associated behaviors have been linked to elevated striatal dopamine levels, potentially

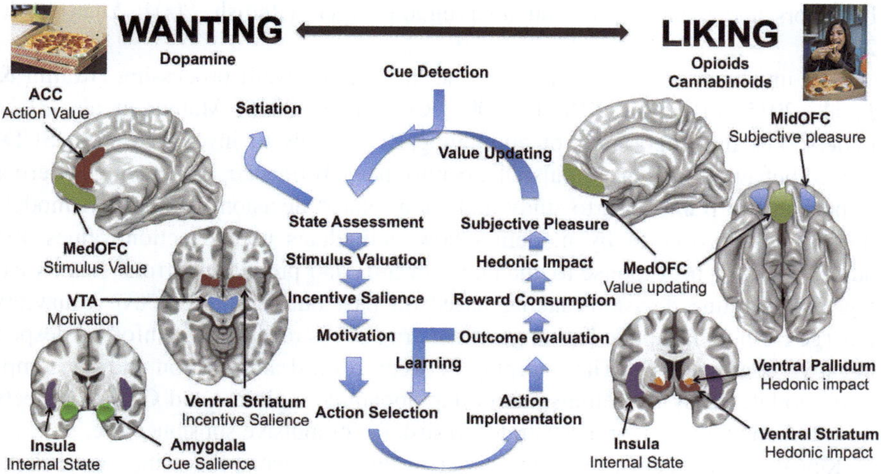

Fig. 2 Neuroanatomy and neurochemical substrates of aberrant decision-making and RL in SUD (Fonzo, 2018)

amplifying the value representation of the substance and reinforcing compulsive drug-seeking (Verdejo-Garcia et al., 2007).

From a neuroeconomic perspective, utility-based models suggest that neuromodulators, such as dopamine and serotonin, influence value and risk computations. The network model proposed by Balasubramani et al. (2015) posits that value and risk can be computed by striatal medium spiny neurons based on the dopamine receptors they express. In this model, value is conceptualized as expected reward, while risk is defined as the variance in reward. Additionally, the model explores serotonin's role in modulating impulsivity and risk-taking behavior. Supported by experimental evidence on dopamine-mediated plasticity in striatal value computation and serotonin's influence on risk preference, this framework generates predictions on risk behavior that can inform future translational research.

3 Reinforcement Learning Modeling for SUD

Computational models offer test beds to investigate the psychophysiology of SUD either as standalone frameworks or in combination with other computational approaches (Gueguen et al., 2021a; Groman et al., 2019, 2022). Prior research indicates that suboptimal decision strategies preceding substance use can serve as predictive markers for potential SUD development (Perry et al., 2008; Blair et al., 2018). By capturing the interaction between decision-making strategies and environmental outcomes, RL models allow researchers to simulate drug-seeking behaviors and identify potential intervention targets (Redish, 2004; Yuan et al., 2024).

The involvement of striatal dopamine in altered reward processing (Keiflin & Janak, 2015; Frank & O'Reilly, 2006; Cox et al., 2015; Mathar et al., 2017) underscores the importance of employing RL models to investigate core SUD-related behaviors such as impulsivity, persistence, abstinence, and craving, offering a mechanistic framework to study addiction-related decision-making. RL models mirror these processes by modeling how individuals update action values and adjust behavior in response to changing rewards and prediction errors. Just as RL algorithms refine decision-making based on past outcomes, addictive behaviors emerge from maladaptive learning, where drug use is repeatedly reinforced despite negative consequences. The interplay between reward anticipation and consumption, modulated by neurotransmitters like dopamine, opioids, and GABA, reflects how addiction alters neural substrates to sustain compulsive substance use.

Key RL components, including reward valuation, error processing, and behavioral updating, have shown to be impaired in SUD and contribute to poor decision-making strategies and maladaptive reward-based learning. This learning process encompasses both explicit learning (cognitive and reflective processes) and implicit learning (habits and conditioned responses) (Berry & Broadbent, 1987; Dayan et al., 2006; Yang & Li, 2012). By simulating value-based learning, these models quantify alterations in cognitive and habitual learning processes, shedding light on the mechanisms that sustain addiction.

Ultimately, RL provides a powerful framework for understanding addiction by linking behavioral manifestations to underlying neural mechanisms. In an RL framework (Fig. 1), the substance user acts as the agent, navigating an environment that includes access to substances, social influences, and emotional states. The user's actions choosing to use or abstain lead to consequences that shape future decision-making strategies.

At each time step, the agent (substance user) evaluates their *state* (e.g., stress, craving) and selects an *action* (e.g., drug use or abstinence) based on their learned *policy* ($\pi(A_t | S_t)$). The environment then responds by transitioning the user to a *new state* and providing a *reward*, reinforcing or discouraging the action taken. Over repeated interactions, the user updates their *value function*, which estimates the long-term benefits of different states, and refines their decision-making strategy. The *prediction error* (δ), the difference between expected and received rewards guides learning, shapes how future behaviors are adjusted. This RL-based modeling helps capture the transition from voluntary substance use to compulsive drug-seeking behavior, withdrawal, and relapse. Research indicates that individuals with SUD exhibit altered reward processing, leading to overvaluation of substance-related rewards and impaired learning from negative consequences (Schultz, 2011).

RL models have been instrumental in simulating these behavioral trajectories, identifying key decision points where interventions can be most effective (Everitt & Robbins, 2016; Redish et al., 2008). Various RL algorithms have been applied to model critical stages of addiction, including craving, persistence, withdrawal, and relapse (Dayan, 2009; Redish & Johnson, 2007; Gueguen et al., 2021b). This

Table 2 RL model parameters with SUD alignment

Component	Definition	Example in substance use
Agent	Decision-making entity	The substance user
Environment	External influences on decisions	Access to substances, social/emotional factors
State (S_t)	Current condition of the agent	Feeling stressed or experiencing cravings
Action (A_t)	Choice made by the agent	Using the substance or resisting temptation
Reward (R_{t+1})	Feedback-reinforcing behavior	Pleasure from drug use, regret from relapse
Policy ($\pi(A_t \| S_t)$)	The decision of particular action (A_t) in a specific state (S_t)	Influencers of substance use (access to the substance, socioeconomic conditions, emotions, etc.)
Value function ($V(S_t)$)	Expected long-term reward for a state	Anticipating relief from stress after drug use
Q-value ($Q(S_t, A_t)$)	Expected reward for an action in a state	Choosing between a small or large dose
Prediction error (δ)	Difference between expected and received reward	Expecting relief but feeling worse after use
State transition function	Determines how states evolve over time	Increased cravings after repeated use
Reward function	Assigns value to actions	Brain-reinforcing drug-seeking behavior

computational approach allows researchers to quantify how substance use alters learning and decision-making, bridging behavioral manifestations with underlying neurobiological mechanisms. Insights gained from RL models can help refine treatment strategies by targeting decision-making biases and maladaptive reinforcement patterns, offering a foundation for more personalized and effective interventions in SUD (Table 2).

3.1 Modeling Habitual and Goal-Directed Behaviors

RL suggests that deficits in decision-making in SUDs are driven either by reduced *model-based learning* (i.e., the ability to simulate and plan based on future outcomes (Lim et al., 2019; Garrett et al., 2023), and/or over-reliance on *model-free systems* (i.e., the dominance of habitual, reflexive decision-making (Groman et al., 2019). This imbalance may explain why individuals with SUD persist with substance use despite its adverse consequences.

Model-free RL methods, such as Q-learning and Actor-Critic algorithms, enable an agent to develop policies based on past actions and the resulting feedback from the environment. These methods rely on cached values of prior rewards, making them computationally efficient but inflexible (Watkins & Dayan, 1992; Sutton, 1988; Konda & Tsitsiklis, 1999). As such, model-free approaches are often likened to *habit-forming behaviors*, where actions become automatic and are no longer influenced by long-term outcomes.

In contrast, model-based RL methods involve constructing an internal model of the environment, including state transitions, actions, and associated rewards. Algorithms like DynaQ (Sutton, 1991) use this model to simulate outcomes and plan actions, while others, such as Dynamic Programming (Bellman, 1966; Bellman & Dreyfus, 1959) or Monte Carlo Tree Search (Coulom, 2007), simulate trajectories and backpropagate rewards to update value estimates. These methods reflect goal-directed behaviors, characterized by their flexibility and reliance on planning to achieve desired outcomes.

3.2 Fee (Meta) Parameters in RL

Meta parameters provide critical insights into decision-making processes and can be extended to explain the addictive behavior. The *learning rate* (α), also called *recency*, can characterize how fast an agent learns or integrates information for future decision-making, and *discount factor* (γ) defines how far rewards in the future are considered into current decisions (Lim et al., 2019; Avila Chauvet & Mejía Cruz, 2024).

Other parameters include *inverse temperature* (or reinforcement sensitivity, β), which regulates the agent's sensitivity to differences in value estimates during decision-making (Yuan et al., 2024; Lim et al., 2019). A higher inverse temperature reflects a preference for the most rewarding options, while a lower value indicates randomness in choices, often observed in individuals with SUDs, where impulsivity and impaired value differentiation dominate. The *eligibility trace decay* (λ) determines how far back in time an agent credits or leverages the past actions for current outcomes, with shorter traces potentially explaining the inability of individuals with SUD to integrate long-term consequences into their decisions (Kinley & Becker, 2024). *Choice stickiness* (or perseveration parameter, τ) refers to a tendency to repeat prior choices regardless of their outcomes (Lim et al., 2019; Kanen et al., 2019) aligning with habitual, model-free behavior seen in addiction. In contrast, *response stickiness* reflects a preference for repeating physical responses (e.g., pressing the same button) (Lim & Ersche, 2023), which may explain compulsive, automatic actions associated with substance use. The parameter, *extinction* (quantified by an extinction rate, α^{ext}), defines the capacity of the agent to learn from non-rewarding states (Lim et al., 2019). Together, these parameters capture the inflexibility in decision-making observed in SUD, where habitual actions and impaired sensitivity to changing reward structures perpetuate maladaptive behaviors. Modeling these parameters helps in pinpointing specific neural and behavioral deficits in SUD, offering targeted avenues for intervention.

(Note: While free parameter notation follows a convention, authors may choose to use different ones if necessary.)

3.3 Application of RL Modeling for SUD

Lim et al. (2019). investigated a tipping point in the transition between regulated or voluntary decision making to impulsive or automatic decision making. Using data from the Slips-of-Action task, the authors showed slower learning rates in individuals with cocaine use disorders, whereas perseverative responding or reinforcement sensitivity was comparable to those in healthy controls. The main model was consistent in demonstrating the weak impact of goal-directed behavior and showed no difference in the influence of habit system (demonstrated to be influenced by reinforcement sensitivity).

Groman et al. (2019) also showed evidence for the influence of model-based and model-free learning in assessing vulnerability to addiction and relapse in rodents. The authors found that higher vulnerability to methamphetamine self-administration in rats was correlated with model-free tendencies, whereas no such correlation was observed with model-based behavior. They attributed this finding to the rats' reduced ability to respond effectively to rewarding outcomes. Furthermore, the authors observed independent and distinct alterations in model-based and model-free learning in rats.

Redish and Johnson (2007) used model-based RL to suggest that the goal identification in goal-directed navigation would produce craving. The framework also models obsession by theorizing that exploration of alternative outcomes is blocked by seeking the path to the high outcome goal (craving) leading to a repeated search for the path (obsession). The proposed model can also explain how drug antagonists can reduce craving by possibly dampening the reward signal which would dampen value and potential goal recognition. The authors proposed an outcome-expectancy model of craving, a unified system of decision-making processing involving three sub-processes: a situation recognition system, a flexible planning system, and an inflexible habit system (Redish & Johnson, 2007).

Using dual RL framework (Daw et al., 2011) combined with a utility function (Balasubramani et al., 2014) to evaluate decision-making behaviors in alcohol users during a modified two-step learning task (Kool et al., 2016) under varying reward conditions, we showed that while alcohol nonusers demonstrated greater model-based control in high-reward conditions compared to low-reward conditions, this pattern was absent in alcohol users. Both groups exhibited reduced risk aversion in high-reward conditions, but alcohol users were significantly less risk-averse than nonusers. Additionally, greater model-based control in alcohol users was linked to a less risk-sensitive approach (Ramakrishnan et al., 2025).

By integrating RL models with utility-based functions, this research provides novel insights into the cognitive mechanisms underlying addiction. This combined modeling approach highlights the value of using complementary frameworks to disentangle decision-making processes and inform targeted strategies for restoring balance in cognitive control and risk sensitivity.

3.4 Bayesian Models

Bayesian models provide a probabilistic lens to examine decision-making in SUDs, emphasizing how individuals integrate prior experiences with new evidence to form beliefs and guide actions. In the context of SUDs, these models inform how substance users process uncertainty, assign subjective probabilities to outcomes, and adapt their behavior in environments with shifting rewards and risks. For instance, users may overemphasize prior experiences of substance-related rewards while undervaluing evidence of harm, leading to persistent substance use despite adverse consequences. These models have also been instrumental in studying relapse, framing it as a failure to recalibrate expectations about risk and reward after periods of abstinence. Yuan et al. (2024) examined relapse to alcohol through risk-taking propensity using the Balloon Analog Risk Task (BART). The model integrates prospect theory with belief updating in RL to capture the phenomenon of relapse as a heightened tendency to take risks, providing insight into how impaired decision-making processes contribute to relapse.

A key theory in SUD is Bechara's hypothesis, which states that while healthy individuals choose long-term advantageous alternatives, SUD users opt for higher short-term gains with larger punishment or penalties, resulting in long-term disadvantageous alternatives (Bechara et al., 1997). Chauvet and Cruz (2024) proposed an agent-based model derived from Bechara's hypothesis (Bechara et al., 1997) in the Iowa Gambling Task, where net gain is determined by subtracting losses from the gains which are weighted by their corresponding sensitivity parameters. The higher memory factor in the agent-based model was found to match with healthy participants' behavior in the task, whereas the model was unable to effectively replicate the substance users' behavior due to fluctuations in their behavior.

4 Non-RL Models for SUD

While RL provides a robust computational framework for understanding behavior, theoretical models delve into the tipping points or transitions in behavior under drug influence. These models, grounded in theoretical constructs, offer insights into the physiological and behavioral dynamics of drug use and self-administration.

4.1 Theoretical Models in SUD

Some of the earliest theoretical models in SUD research focus on identifying critical points that trigger behavioral transitions in response to drug consumption. One such example is the negative-feedback loop model or set-point model proposed by Ahmed and Koob (2005). This model views drug self-administration as a goal-directed behavior, predicting that drug users adjust their sensitivity to rewards to a "set-point" by increasing tonic dopamine signaling. This increase elevates basal dopamine levels, aligning the individual's reward sensitivity with their drug intake. The model further posits that drug demand depends on basal dopamine signaling and the sensitivity to reinforcing stimuli.

In contrast, the positive-feedback loop model, introduced by Norman and Tsibulsky (2006), suggests that self-administration of cocaine is not goal-directed. Instead, it occurs automatically when cocaine levels in the brain fall within a specific range, termed the "compulsion zone" bounded by the Priming threshold (i.e., the minimum drug level required to trigger a response), and the Satiety threshold (i.e., the drug level beyond which self-administration is inhibited). Within this zone, cocaine use is theorized to bypass considerations of consequences, compelling the user toward self-administration.

Panlilio et al. (2008) introduced the concept of an *interoceptive discriminative stimulus*, a threshold level of drug intake, above which reinforcement increases. They suggest that targeting this threshold could disrupt stable drug use, offering a potential intervention point.

5 Combining Theoretical and Computational Models

Integrative approaches and biophysical modeling approaches have opened new avenues for modeling the complex dynamics of substance use. For example, Gutkin et al. (2006). integrated a dynamic system model of nicotine's effects on dopamine with an RL model of action selection. This hybrid approach demonstrated how the pharmacological effects of drugs modulate reward signals and influence behavior.

Similarly, Qi et al. (2011). hypothesized addiction as a form of pathological learning. Their model used a system of Ordinary Differential Equations to simulate biochemical reactions and signal transduction processes, linking dopamine and glutamate signaling to synaptic plasticity in GABAergic medium spiny neurons. This framework emphasized the role of presynaptic dopamine metabolism, postsynaptic signal transduction, and AMPAR trafficking in addiction-related changes.

A recent review by Bonilla-Quintana and Rangamani (2022) extends biophysical modeling to study SUDs through the lens of dendritic spine dynamics and suggests that cocaine impairs long-term potentiation in medium spiny neuron spines, producing silent synapses and reverting the brain to a juvenile state. This state may underlie pathological drug-seeking behavior and increase relapse propensity.

Table 3 RL models and defining equations

Model	Nature	Key equations
Temporal difference RL (TDRL)	Predicts behavior based on reward prediction errors	$\delta = r + \gamma(V[s']) - V[s]$ or $\delta = r + \gamma \left(\max \sum_{a'} Q\left[s', a'\right]\right) - Q\left[s\right]\Big)$
Redish's TDRL model	Models escalating drug-seeking due to dopamine-driven prediction errors	$\delta = max\ \{\gamma^d[R(S_{t+1}) + V(S_{t+1})] - V(S_t) + D(S_t), D(S_t)\}$
Dezfouli's TDRL modification	Adjusts reward sensitivity over time to explain diminishing dopamine response	$\delta = \max\left\{\gamma^d\left[R\left(S_{t+1}\right) + V\left(S_{t+1}\right)\right] - V\left(S_t\right) + \left[D\left(S_t\right) - \kappa_t\right], \left[D\left(S_t\right) - \kappa_t\right]\right\} - \overline{R}_t$
Bayesian models	Probabilistic models integrating prior experiences and new evidence	No fixed equation, uses Bayesian updating
Bechara's hypothesis model	Model's preference for short-term high-reward despite long-term penalties	Net gain = Gains—Losses (weighted by sensitivity parameters)
Model-based and model-free RL	Explains transition from goal-directed to habitual behavior	$p\left(i, t\right) = \frac{e^{\beta V_t^i} + \tau C_t^i}{\sum_{k=1}^{n} e^{\beta V_t^k} + \tau C_t^k}$
Dual-system RL	Explores model-based vs. model-free influences on addiction	$Q_{\text{net}}(s_{A,\,t+1}, a_{j,\,t+1}) = \beta_{MB} Q_{MB}(s_{A,\,t}, a_j) + \beta_{MF} Q_{MF}(s_{A,\,t}, a_j)$
Redish and Johnson's craving model	Models craving as obsessive goal-seeking behavior	No fixed equation, uses outcome-expectancy framework

Thus, theoretical models provide insights into the mechanisms underlying drug self-administration and behavioral transitions, while integrative and biophysical approaches offer a more nuanced understanding of the physiological changes driving addiction. Together, these models complement RL frameworks, enriching our understanding of SUDs and paving the way for innovative interventions and diagnostic methodologies (Table 3).

6 Ethical and Practical Considerations

Ethical challenges in SUD modeling, such as ensuring data privacy and avoiding stigmatization, also present opportunities to build more robust frameworks. Ethical considerations for employing RL to model SUDs can be classified into deontological and consequential approaches. Deontological approach is based on evaluating future actions using constraints of a set of rules or principles, and consequential approach is based on considering purely the consequence of the action regardless of any rules. A comprehensive review (Vishwanath et al., 2024) elucidates three paradigms to ensure ethical RL modeling—a multi-objective paradigm: the incorporation of an ethical signal in addition to the reward signal, a constraint-based paradigm: a penalty imposed on the agent over unethical decisions, and a multi-agent paradigm: constraints imposing equitable resource distribution among agents.

7 Challenges and Opportunities

Advancements in the modeling of SUDs require both extending existing frameworks and leveraging new technologies. Current computational modeling approaches are limited in their scope as they exclusively model dopaminergic processes related to reward processing and risk and error detection. It is also important to note that while stimulants (e.g., cocaine and methamphetamine) directly increase synaptic dopamine (Kohno et al., 2022), opioids (e.g., heroin and fentanyl) indirectly influence dopamine neurons by inhibiting GABAergic interneurons in the VTA (Doyle & Mazei-Robison, 2021), and cannabis reduces neurotransmitter release via presynaptic inhibition (Augustin & Lovinger, 2022). Furthermore, current computational modeling approaches are also limited to modeling craving and reward liking and wanting (Johansen et al., 2024; Pool et al., 2016), whereas other addiction states, such as withdrawal that are associated with the alternative motivation of punishment avoidance (Lopez & Lerner, 2025; Vanderschuren et al., 2017), are often overlooked in computational modeling approaches. Addressing these conceptual and methodological challenges in RL modeling provides opportunities for refinement and translation, such as integrating RL with non-RL approaches, employing advanced machine learning techniques, and exploring the potential of generative AI and large language models (Fig. 3).

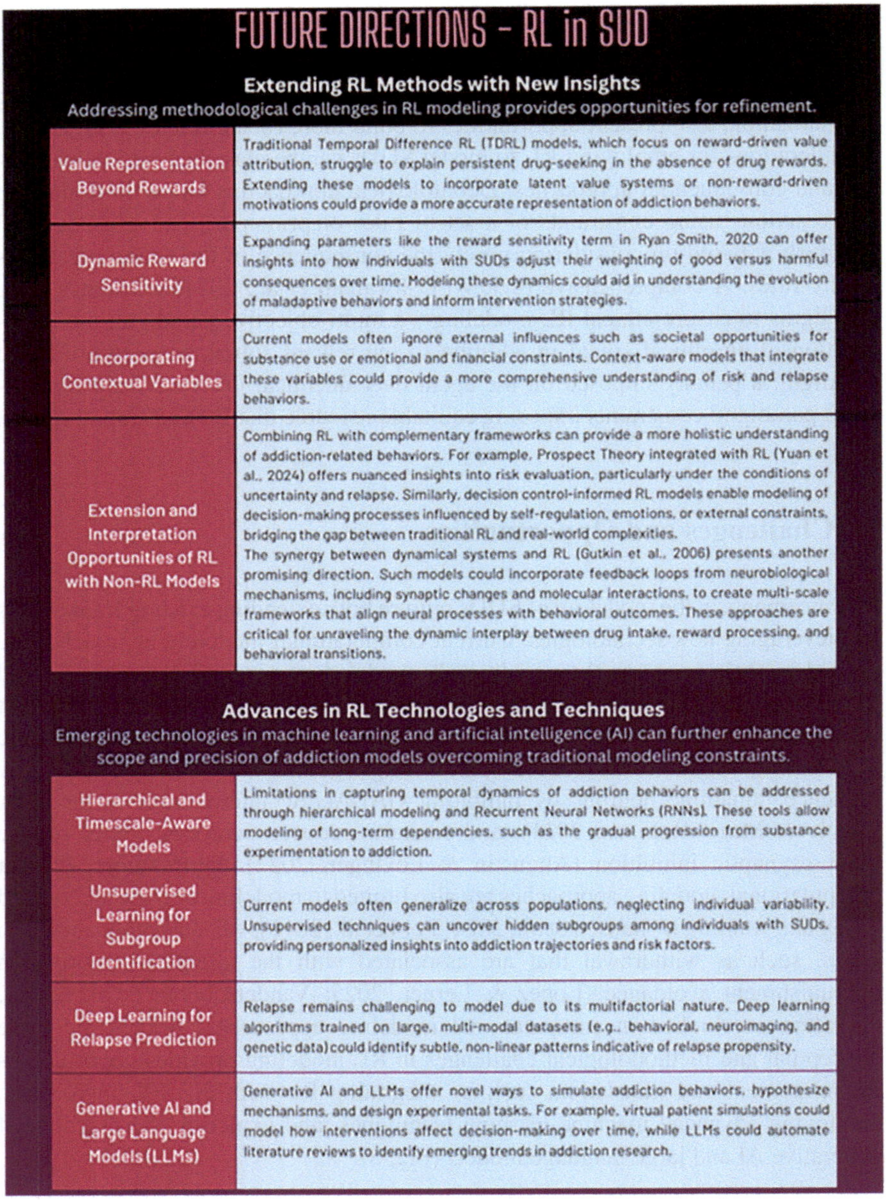

Fig. 3 Future directions elucidating potential RL modeling advancements and implementations to capture the whole behavior of SUD and translate them to diagnostic and intervention tools

References

Ahmed, S. H., & Koob, G. F. (2005). Transition to drug addiction: A negative reinforcement model based on an allostatic decrease in reward function. *Psychopharmacology, 180*, 473–490.

Augustin, S. M., & Lovinger, D. M. (2022). Synaptic changes induced by cannabinoid drugs and cannabis use disorder. *Neurobiology of Disease, 167*, 105670.

Avila Chauvet, L., & Mejía Cruz, D. (2024). Computational modeling of decision-making in substance abusers: Testing Bechara's hypotheses. *Frontiers in Psychology, 15*, 1281082.

Balasubramani, P. P., Chakravarthy, V. S., Ravindran, B., & Moustafa, A. A. (2014). An extended reinforcement learning model of basal ganglia to understand the contributions of serotonin and dopamine in risk-based decision making, reward prediction, and punishment learning. *Frontiers in Computational Neuroscience, 8*, 47. https://doi.org/10.3389/fncom.2014.00047

Balasubramani, P. P., Chakravarthy, V. S., Ravindran, B., & Moustafa, A. A. (2015). A network model of basal ganglia for understanding the roles of dopamine and serotonin in reward-punishment-risk based decision making. *Frontiers in Computational Neuroscience, 9*, 76. https://doi.org/10.3389/fncom.2015.00076

Bechara, A., Damasio, H., Tranel, D., & Damasio, A. R. (1997). Deciding advantageously before knowing the advantageous strategy. *Science, 275*, 1293–1295.

Bellman, R. (1966). Dynamic programming. *Science, 153*, 34–37.

Bellman, R., & Dreyfus, S. (1959). Functional approximations and dynamic programming. *Mathematical Tables and Other Aids to Computation, 13*, 247–251.

Berridge, K. C. (2007). The debate over dopamine's role in reward: The case for incentive salience. *Psychopharmacology, 191*, 391–431.

Berry, D. C., & Broadbent, D. E. (1987). The combination of explicit and implicit learning processes in task control. *Psychological Research, 49*, 7–15.

Blair, M. A., et al. (2018). Blunted frontostriatal blood oxygen level–dependent signals predict stimulant and marijuana use. *Biological Psychiatry: Cognitive Neuroscience and Neuroimaging, 3*, 947–958.

Bonilla-Quintana, M., & Rangamani, P. (2022). Can biophysical models of dendritic spines be used to explore synaptic changes associated with addiction? *Physical Biology, 19*, 041001.

Coulom, R. (2007). *International conference on computers and games* (pp. 72–83). Springer.

Cox, S. M., et al. (2015). Striatal D1 and D2 signaling differentially predict learning from positive and negative outcomes. *NeuroImage, 109*, 95–101.

Daw, N. D., Gershman, S. J., Seymour, B., Dayan, P., & Dolan, R. J. (2011). Model-based influences on humans' choices and striatal prediction errors. *Neuron, 69*, 1204–1215. https://doi.org/10.1016/j.neuron.2011.02.027

Dayan, P. (2009). Dopamine, reinforcement learning, and addiction. *Pharmacopsychiatry, 42*, S56–S65.

Dayan, P., Niv, Y., Seymour, B., & Daw, N. D. (2006). The misbehavior of value and the discipline of the will. *Neural Networks, 19*, 1153–1160.

Doyle, M. A., & Mazei-Robison, M. S. (2021). Opioid-induced molecular and cellular plasticity of ventral tegmental area dopamine neurons. *Cold Spring Harbor Perspectives in Medicine, 11*, a039362.

Everitt, B. J., & Robbins, T. W. (2016). Drug addiction: Updating actions to habits to compulsions ten years on. *Annual Review of Psychology, 67*, 23–50.

Fonzo, G. A. (2018). Diminished positive affect and traumatic stress: A biobehavioral review and commentary on trauma affective neuroscience. *Neurobiology of Stress, 9*, 214–230.

Frank, M. J., & O'Reilly, R. C. (2006). A mechanistic account of striatal dopamine function in human cognition: Psychopharmacological studies with cabergoline and haloperidol. *Behavioral Neuroscience, 120*, 497.

Garrett, N., Allan, S., & Daw, N. D. (2023). Model based control can give rise to devaluation insensitive choice. *Addiction Neuroscience, 6*, 100070.

Gowin, J. L., Mackey, S., & Paulus, M. P. (2013). Altered risk-related processing in substance users: Imbalance of pain and gain. *Drug and Alcohol Dependence, 132*, 13–21.
Groman, S. M., Massi, B., Mathias, S. R., Lee, D., & Taylor, J. R. (2019). Model-free and model-based influences in addiction-related behaviors. *Biological Psychiatry, 85*, 936–945. https://doi.org/10.1016/j.biopsych.2018.12.017
Groman, S. M., Thompson, S. L., Lee, D., & Taylor, J. R. (2022). Reinforcement learning detuned in addiction: Integrative and translational approaches. *Trends in Neurosciences, 45*, 96–105.
Gueguen, M. C., Schweitzer, E. M., & Konova, A. B. (2021a). Computational theory-driven studies of reinforcement learning and decision-making in addiction: What have we learned? *Current Opinion in Behavioral Sciences, 38*, 40–48.
Gueguen, M. C. M., Schweitzer, E. M., & Konova, A. B. (2021b). Computational theory-driven studies of reinforcement learning and decision-making in addiction: What have we learned? *Current Opinion in Behavioral Sciences, 38*, 40–48. https://doi.org/10.1016/j.cobeha.2020.08.007
Gutkin, B. S., Dehaene, S., & Changeux, J.-P. (2006). A neurocomputational hypothesis for nicotine addiction. *Proceedings of the National Academy of Sciences, 103*, 1106–1111.
Johansen, A. N., Acuff, S. F., & Strickland, J. C. (2024). Human laboratory models of reward in substance use disorder. *Pharmacology Biochemistry and Behavior, 241*, 173803.
Kanen, J. W., Ersche, K. D., Fineberg, N. A., Robbins, T. W., & Cardinal, R. N. (2019). Computational modelling reveals contrasting effects on reinforcement learning and cognitive flexibility in stimulant use disorder and obsessive-compulsive disorder: Remediating effects of dopaminergic D2/3 receptor agents. *Psychopharmacology, 236*(8), 2337–2358. https://doi.org/10.1007/s00213-019-05325-w
Keiflin, R., & Janak, P. H. (2015). Dopamine prediction errors in reward learning and addiction: From theory to neural circuitry. *Neuron, 88*, 247–263.
Kinley, I., & Becker, S. (2024). *Habits: Their definition, neurobiology, and role in addiction* (pp. 301–336). Springer.
Kohno, M., Dennis, L. E., McCready, H., & Hoffman, W. F. (2022). Dopamine dysfunction in stimulant use disorders: Mechanistic comparisons and implications for treatment. *Molecular Psychiatry, 27*, 220–229.
Konda, V., & Tsitsiklis, J. (1999). Actor-critic algorithms. *Advances in Neural Information Processing Systems, 12*. https://doi.org/10.1137/S0363012901385691
Koob, G. F., & Volkow, N. D. (2016). Neurobiology of addiction: A neurocircuitry analysis. *The Lancet Psychiatry, 3*, 760–773.
Kool, W., Cushman, F. A., & Gershman, S. J. (2016). When does model-based control pay off? *PLoS Computational Biology, 12*, e1005090. https://doi.org/10.1371/journal.pcbi.1005090
Lim, T. V., & Ersche, K. D. (2023). Theory-driven computational models of drug addiction in humans: Fruitful or futile? *Addiction Neuroscience, 5*, 100066.
Lim, T. V., et al. (2019). Impairments in reinforcement learning do not explain enhanced habit formation in cocaine use disorder. *Psychopharmacology, 236*, 2359–2371.
Lopez, G. C., & Lerner, T. N. (2025). How dopamine enables learning from aversion. *Current Opinion in Behavioral Sciences, 61*, 101476.
Mathar, D., et al. (2017). The role of dopamine in positive and negative prediction error utilization during incidental learning–insights from Positron Emission Tomography, Parkinson's disease and Huntington's disease. *Cortex, 90*, 149–162.
Norman, A. B., & Tsibulsky, V. L. (2006). The compulsion zone: A pharmacological theory of acquired cocaine self-administration. *Brain Research, 1116*, 143–152.
Panlilio, L. V., Thorndike, E. B., & Schindler, C. W. (2008). A stimulus-control account of regulated drug intake in rats. *Psychopharmacology, 196*, 441–450.
Perry, J. L., Nelson, S. E., & Carroll, M. E. (2008). Impulsive choice as a predictor of acquisition of IV cocaine self-administration and reinstatement of cocaine-seeking behavior in male and female rats. *Experimental and Clinical Psychopharmacology, 16*, 165.

Pool, E., Sennwald, V., Delplanque, S., Brosch, T., & Sander, D. (2016). Measuring wanting and liking from animals to humans: A systematic review. *Neuroscience & Biobehavioral Reviews, 63*, 124–142.

Qi, Z., Kikuchi, S., Tretter, F., & Voit, E. (2011). Effects of dopamine and glutamate on synaptic plasticity: A computational modeling approach for drug abuse as comorbidity in mood disorders. *Pharmacopsychiatry, 44*, S62–S75.

Ramakrishnan, S. A., et al. (2025). Impaired arbitration between reward-related decision-making strategies in alcohol users compared to alcohol non-users: A computational modeling study. *Digital Psychiatry and Neuroscience, 3*, 1.

Redish, A. D. (2004). Addiction as a computational process gone awry. *Science, 306*, 1944–1947.

Redish, A. D., & Johnson, A. (2007). A computational model of craving and obsession. *Annals of the New York Academy of Sciences, 1104*, 324–339.

Redish, A. D., Jensen, S., & Johnson, A. (2008). Addiction as vulnerabilities in the decision process. *Behavioral and Brain Sciences, 31*, 461–487.

Schultz, W. (2011). Potential vulnerabilities of neuronal reward, risk, and decision mechanisms to addictive drugs. *Neuron, 69*, 603–617. https://doi.org/10.1016/j.neuron.2011.02.014

Sutton, R. S. (1988). Learning to predict by the methods of temporal differences. *Machine Learning, 3*, 9–44.

Sutton, R. S. (1991). Dyna, an integrated architecture for learning, planning, and reacting. *ACM SIGART Bulletin, 2*, 160–163.

Vanderschuren, L. J., Minnaard, A. M., Smeets, J. A., & Lesscher, H. M. (2017). Punishment models of addictive behavior. *Current Opinion in Behavioral Sciences, 13*, 77–84.

Verdejo-Garcia, A., et al. (2007). The differential relationship between cocaine use and marijuana use on decision-making performance over repeat testing with the Iowa Gambling Task. *Drug and Alcohol Dependence, 90*, 2–11. https://doi.org/10.1016/j.drugalcdep.2007.02.004

Vishwanath, A., Dennis, L. A., & Slavkovik, M. (2024). Reinforcement learning and machine ethics: A systematic review. *arXiv preprint arXiv:2407.02425.*

Watkins, C. J., & Dayan, P. (1992). Q-learning. *Machine Learning, 8*, 279–292.

Yang, J., & Li, P. (2012). Brain networks of explicit and implicit learning. *PLoS One, 7*(8), e42993.

Yuan, W., et al. (2024). Computational markers of risky decision-making predict for relapse to alcohol. *European Archives of Psychiatry and Clinical Neuroscience, 274*, 353–362.

Part VII
From Neural Data to Societal Impact: Applications of Neuroeconomics

David V. Smith, Patricia L. Lockwood, and Dominic S. Fareri

Neural data are reshaping how we understand and influence decision-making—both at the level of the individual and across populations. By revealing how people attend to information, how they value outcomes, and how they resolve internal conflicts, neuroscience offers an unprecedented window into the cognitive and emotional machinery of economic life. This shift matters because many of the most stubborn problems in marketing, public policy, and behavioral economics—why people make inconsistent choices, why intentions don't predict actions, and why aggregate preferences shift in response to seemingly minor cues—cannot be resolved with behavioral data alone.

Recent work in consumer neuroscience demonstrates how brain-based measures can reveal preferences and mental states that are inaccessible through introspection or surveys. One chapter in this section illustrates how techniques such as multivariate pattern analysis and large-scale neural decoding can forecast real-world outcomes—such as crowdfunding success or music popularity—based solely on lab-acquired neural signals (Genevsky & Boksem, this volume). This forecasting capability reflects a broader trend across neuroscience, where models trained on vmPFC activity and other valuation-related signals have been shown to outperform self-report in anticipating behavioral outcomes at the population level (Yao & Wang, 2024; Javor et al., 2013).

In the domain of attention and visual processing, eye-tracking and neuroimaging are being used to understand how attention interacts with valuation mechanisms in real time. One contribution in this section examines how attentional drift diffusion

D. V. Smith (✉)
Temple University, Philadelphia, PA, USA
e-mail: david.v.smith@temple.edu

P. L. Lockwood
University of Birmingham, Birmingham, UK

D. S. Fareri
Adelphi University, Garden City, NY, USA

models capture the way attention not only reflects but actively shapes consumer preferences through gaze bias (Sullivan, this volume). These insights contribute to a growing literature that frames visual attention as a gatekeeper for valuation, implicating it in everything from product design to interface architecture (Krajbich et al., 2010; Shaw & Bagozzi, 2018).

A final chapter proposes a computational framework in which multiple decision systems—reflexive, Pavlovian, deliberative, and procedural—interact to guide behavior (Redish et al., this volume). This multi-system model has helped clarify longstanding puzzles in behavioral economics, such as the sunk cost fallacy and the endowment effect, by identifying the distinct neural substrates involved. These ideas align with findings that preference reversals can emerge when different systems (e.g., procedural vs. deliberative) are recruited depending on task structure and context (Sweis et al., 2018). The framework also supports recent policy innovations—such as contingency management in addiction treatment and memory cueing in disaster recovery—that harness specific neural processes to shift behavior in targeted, evidence-based ways (Pykett, 2013).

As neuroeconomic methods continue to develop, their relevance extends beyond individual-level explanation toward informing the design of environments, interventions, and policies that are grounded in empirically supported models of cognitive function. By integrating neuroscientific evidence with behavioral theory, neuroeconomics offers a framework for anticipating and shaping decision-making in applied contexts—ranging from consumer markets to public health. The chapters in this section collectively illustrate how neural data can contribute to more precise, mechanistically informed approaches to economic behavior, and how such approaches may ultimately enhance the effectiveness of institutional and commercial decision architectures.

References

Genevsky, A., & Boksem, M. (this volume). Applications in consumer neuroscience: Decoding and neuroforecasting. In D. V. Smith, P. Lockwood, & D. S. Fareri (Eds.), *Neuroeconomics: Core topics and current directions*. Springer. Preprint: https://doi.org/10.2139/ssrn.5001469

Javor, A., Koller, M., Lee, N., Chamberlain, L., & Ransmayr, G. (2013). Neuromarketing and consumer neuroscience: Contributions to neurology. *BMC Neurology, 13*(13). https://doi.org/10.1186/1471-2377-13-13

Krajbich, I., Armel, C., & Rangel, A. (2010). Visual fixations and the computation and comparison of value in simple choice. *Nature Neuroscience, 13*(10), 1292–1298. https://doi.org/10.1038/nn.2635

Pykett, J. (2013). Neurocapitalism and the new neuros: Using neuroeconomics, behavioural economics and picoeconomics for public policy. *Journal of Economic Geography, 13*(5), 845–869. https://doi.org/10.1093/jeg/lbs039

Redish, A. D., Chastain, H. S., Runge, C. F., Sweis, B. M., Allen, S. E., & Haldar, A. (this volume). Policy consequences of the new neuroeconomic framework. In D. V. Smith, P. Lockwood, & D. S. Fareri (Eds.), Neuroeconomics: Core topics and current directions. Springer. Preprint: https://doi.org/10.48550/arXiv.2409.07373.

Shaw, S. D., & Bagozzi, R. P. (2018). The neuropsychology of consumer behavior and marketing. *Consumer Psychology Review, 1*(2), 124–146. https://doi.org/10.1002/arcp.1006

Sullivan, N. J. (this volume). Attention dynamics: Antecedents to consumer choice. In D. V. Smith, P. Lockwood, & D. S. Fareri (Eds.), *Neuroeconomics: Core topics and current directions*. Springer.

Sweis, B. M., Abram, S. V., Schmidt, B. J., Seeland, K. D., MacDonald, A. W., Thomas, M. J., & Redish, A. D. (2018). Sensitivity to "sunk costs" in mice, rats, and humans. *Science, 361*(6398), 178–181. https://doi.org/10.1126/science.aar8644

Yao, X., & Wang, Y. (2024). Using neural data to forecast aggregate consumer behavior in neuromarketing: Theory, metrics, progress, and outlook. *Journal of Consumer Behaviour*. https://doi.org/10.1002/cb.2324

Applications in Consumer Neuroscience: Decoding and Neuroforecasting

Alexander Genevsky and Maarten Boksem

Abstract Consumer neuroscience is an emergent field of academic research based on the foundations and principles of neuroeconomics, while maintaining a focus on real-world applications. The promise of consumer neuroscience stems from the potential of using modern neuroimaging methods to provide relevant and actionable insights into the mental processes underlying consumer and market behavior. In this chapter, we provide a broad overview of the two areas of consumer neuroscience in which neuroeconomic perspectives and methods have shown the most promise. We first discuss the role of neuroimaging in the decoding of mental states and how these measures are informative for understanding consumers' responses, experiences, and associations. We then cover neuroforecasting—the use of neural data collected from laboratory samples to forecast market-level behavior. Finally, we close the chapter with an outlook for the future of the field.

1 Introduction

Over the past decade, the field of neuroeconomics has grown to include consumer neuroscience, an area of research focused on the application of neuroscientific methods and approaches to better understand consumer and market behavior. To this end, neuroimaging methods offer an important benefit over traditional behavioral methods by providing information about underlying mental processes without the need for participants to verbally articulate their thoughts, feelings, and preferences (Plassmann et al., 2015; Smidts et al., 2014). This represents a valuable opportunity for researchers and practitioners because many of the mental processes responsible for observed behavior occur below the level of conscious experience (Dijksterhuis, 2004; Winkielman & Berridge, 2004). Thus, research participants are often unable, or unwilling, to accurately report their true mental states and motivations (Nisbett

A. Genevsky (✉) · M. Boksem
Rotterdam School of Management, Erasmus University, Rotterdam, The Netherlands
e-mail: genevsky@rsm.nl

D. V. Smith et al. (eds.), *Neuroeconomics: Core Topics and Current Directions*,
https://doi.org/10.1007/978-3-032-02925-6_34

& Wilson, 1977; Wiles & Cornwell, 2012; Fisher, 1993). The hope is that by applying neuroscientific methodologies, we can gain insights that lead to a better understanding of consumers' mental processes, and subsequently to improved predictions of consumer behavior at the individual and aggregate levels.

From the outset, there has been enthusiasm that neuroscience would make significant impacts across a broad range of domains within consumer research (Plassmann et al., 2015). Indeed, valuable work in the field has applied neural data to marketing-relevant topics, including consumer valuation (Plassmann et al., 2008; Levy et al., 2011), branding (McClure et al., 2004; Yoon et al., 2006), product design and aesthetics (Reimann et al., 2010; der Laan et al., 2012), and the integration of price information (Karmarkar et al., 2015; Plassmann et al., 2007) (for reviews see Plassmann et al., 2015; Smidts et al., 2014; Clithero et al., 2023; Levallois et al., 2019; Karmarkar & Yoon, 2016; Ariely & Berns, 2010). However, to date, we find that the most promising contributions of consumer neuroscience to business practice fall into two categories: (1) the decoding of underlying mental states in response to marketing stimuli and the associated consumer decision-making processes and (2) the use of neural data to improve forecasts of market-level outcomes through neuroforecasting.

We begin with a discussion of the use of neural data to decode consumers' underlying mental states. This work has applied neuroeconomic approaches to explore consumers' emotional responses to marketing messages, the temporal experience of static and dynamic stimuli, and the assessment of consumer's perceptions of brand associations and image. Next, we consider the field of neuroforecasting, in which neural data are used to inform forecasts of real-world market-level consumer behavior. We discuss the origins of neuroforecasting, its unique experimental approach, existing empirical evidence, and theory building efforts.

2 Decoding Mental States from the Brain

2.1 Decoding Emotions

An extensive stream of literature in marketing has shown that emotions play an important role in how consumers respond to marketing stimuli (Vakratsas & Ambler, 1999; Bagozzi et al., 1999). For example, it has been shown that advertisements high in emotional content generate more online search behavior (Guitart & Stremersch, 2021), that positive emotions lead to more positive attitudes to ads, products, and brands (Burke & Edell, 1989; Derbaix, 1995; Pham et al., 2013; Holbrook & Batra, 1987; Pelsmacker & Geuens, 1999; Morris et al., 2002; Yeung & Wyer, 2004), and that positive emotions lead to higher purchase intention and increased sales (Morris et al., 2002; Binet & Field, 2009) (see Kranzbühler et al., 2020; Puccinelli et al., 2016 for meta-analyses).

However, measurement of these emotions is often difficult, in large part because individuals have limited access to their inner mental states (Winkielman & Berridge, 2004; Zajonc, 1980). Even when aware of their emotional state, people often find it difficult to accurately identify or put them into words (Nisbett & Wilson, 1977; Wiles & Cornwell, 2012; Fisher, 1993). In some cases, participants may willfully misreport their feelings for reasons of social desirability (Fisher, 1993; Arnold & Feldman, 1981; King & Bruner, 2000). In addition, emotional ratings are often reported retrospectively, thus relying on how well they can be retrieved from memory. Moreover, emotions are dynamic processes, fluctuating considerably in short periods of time. Often it is precisely these fluctuations, opaque in retrospective reports, that can be of great importance for understanding the experience of a stimulus (Do et al., 2008). For these reasons, it is highly beneficial to be able to measure emotions with high temporal resolution, and without requiring retrospective self-report responses.

This is where neuroscience insights and methods play an important role. Modern neuroimaging methods provide researchers with insight into affective processes by directly observing them in real-time from the consumer's brain with relatively high temporal resolution (Plassmann et al., 2015; Smidts et al., 2014). Importantly, emotions (as with other psychological processes) are considered distributed patterns of activations in a network of neural regions across the brain (Wager et al., 2015; Kragel & LaBar, 2016; Nummenmaa & Saarimäki, 2019). For these reasons, modern neuroscience makes extensive use of machine learning and pattern recognition techniques, referred to as multivariate pattern analysis (MVPA). This analytic approach is particularly suitable for analyzing distributed patterns of brain activation and how these patterns translate into psychological processes such as emotions (Norman et al., 2006; Kragel et al., 2018; Poldrack, 2011).

In short, MVPA statistical models specify how patterns of activity across the brain combine to predict the identity or intensity of a mental process. While traditional (univariate) approaches in neuroscience are mainly focused on identifying the function of brain areas, the MVPA approach is mainly interested in identifying mental processes. This approach aligns well with the goals of consumer neuroscience researchers. For example, in marketing, one is less concerned with understanding which specific brain area is involved in a certain psychological process. Rather, the main goal is to establish which mental processes are evoked by marketing stimuli irrespective of the underlying brain network.

Over the past 15 years, several papers have demonstrated the possibility of decoding emotions from multivariate brain measurements. Researchers in affective neuroscience have found that it is possible to decode the neural substrate of discrete emotions (e.g., happiness, sadness, fear, anger, and disgust) elicited by (audio)visual stimuli (Kassam et al., 2013; Saarimäki et al., 2016; Kragel et al., 2016; Eijlers et al., 2019). However, most work has been focused on investigating the neural representations of the underlying emotional dimensions of valence and arousal. A seminal study in this area (Baucom et al., 2012) showed that the pattern of brain activation evoked by emotional pictures could reliably predict the target emotion (see also Chang et al., 2015).

Building on these findings and applying them in a marketing context, Eijlers et al. (2020) first classified neural activity in response to emotional pictures and subsequently used these representations to estimate the level of valence and arousal in response to advertisements. In this study, participants viewed 100 images from the IAPS image database (Lench et al., 2011) and 150 print ads from various product categories and brands, while EEG activity was recorded. The researchers used these neural data to train a statistical model to differentiate high and low emotional valence and arousal as evoked by the IAPS images. Subsequently, this model was used to obtain a measure of (neural) valence and arousal evoked by an independent set of print ads. Finally, they estimated the relationship between the decoded levels of valence and arousal evoked by the print ads and external measures of ad effectiveness (as measured by notability and attitude toward the ad in an independent consumer panel). The results showed that the neural measure of arousal was positively associated with the notability of ads in the population at large, but was negatively associated with the attitude toward these ads.

Although static stimuli such as print ads are still relevant for marketing, the trend is clearly towards video and interactive media. Chan and colleagues (2020) tested whether affective neural representations evoked by static emotional stimuli (IAPS images) can be used to decode emotional experiences evoked by dynamic advertising stimuli (movie trailers). In this experiment, subjects were shown IAPS images and movie trailers in the MRI scanner. A support-vector machine (SVM) classifier was trained on the fMRI data obtained while viewing the IAPS images and then tested on the movie trailer fMRI data. The results showed that patterns of activation in the brain could be used to accurately predict both the overall emotional content of the trailers, as well as the time course of the changes in both valence and arousal across the trailer. These findings provide evidence that neural affective representations extracted from episodic affective states correspond with those of more naturalistic and dynamic emotional experiences. We will further discuss the link between affect and forecasting in section Neuroforecasting.

2.2 Decoding the Consumer Experience

While most of the work on neural decoding, particularly where it relates to consumer behavior, has focused on emotions, the approach has broader applications. A key tool in this area is a meta-analytic database called Neurosynth (Yarkoni et al., 2011), which to date mines over 14,000 peer-reviewed neuroscientific publications for brain activation associated with a large range of concepts (e.g., “emotion,” “attention,” “memory”). The database provides a statistical brain map representing the relationship between the concepts and the neural activity reported across regions of the brain. By testing the similarity between patterns of brain activation elicited by (marketing) stimuli and the association maps obtained from Neurosynth, it is possible to decode multiple underlying mental states simultaneously.

This is particularly relevant in consumer neuroscience because there is evidence from the marketing literature that mental states other than emotions are important for effective communication. For example, perceptual and linguistic responses (Shimojo et al., 2003; Schmälzle et al., 2015), memory, and attention (Kardes & Kalyanaram, 1992) have all been associated with consumer preferences. In addition, executive functions impacted by processing fluency, such as deliberation and inhibitory control, have been found to interact with brand preference. On the other hand, social cognition, encompassing psychological processes such as empathy and mentalizing, is closely related to the advertising literature on narrative transportation (van Laer et al., 2014).

Chan and colleagues (2023) leveraged the power of meta-analytic decoding to investigate which mental processes were most closely associated with advertising success, and at which points during ad presentation these processes were most relevant. In their study, participants viewed ads while undergoing scanning in the MRI. Neural activity evoked by the ads was decoded using Neurosynth to capture the dynamics of the mental processes evoked by the ads. The decoded mental states were then subsequently compared to a measure of attitude towards the ad. The results showed that, in addition to emotions, a broad range of mental states were strongly associated with self-reported attitudes. Interestingly, it was observed that the temporal dynamics of these processes differed substantially (see Fig. 1). That is, the emotional response was found to be particularly important early on in the ad (within the first 3 s), but then dropped off slightly, while mental processes associated with social cognition (i.e., mentalizing) became important later in the presentation but then remained important. On the other hand, processes related to perception and executive functions (or lack thereof) were important for the closing stages of the commercial, suggesting that making consumers engage in effortful cognitive processing while viewing an ad leads to lower attitudes, confirming the long-standing observation that information complexity negatively impacts comprehension.

This work is a good example of how consumer neuroscience can provide specific recommendations for marketing in practice. Specifically, it suggests that it is important for effective ads to evoke an early emotional response, aligning with recent behavioral work indicating that consumers make up their mind about engagement within the first 10–15 s (Chiong et al., 2024; Teixeira et al., 2012; Tong et al., 2020). In addition, these findings reveal the importance of eliciting processes associated with social cognition throughout the ad, echoing work showing that neural activity in the brain's social cognition system tracks message virality (Scholz et al., 2017) and that ad liking is associated with neural activity across participants at mentalizing regions in the brain (Chan et al., 2019).

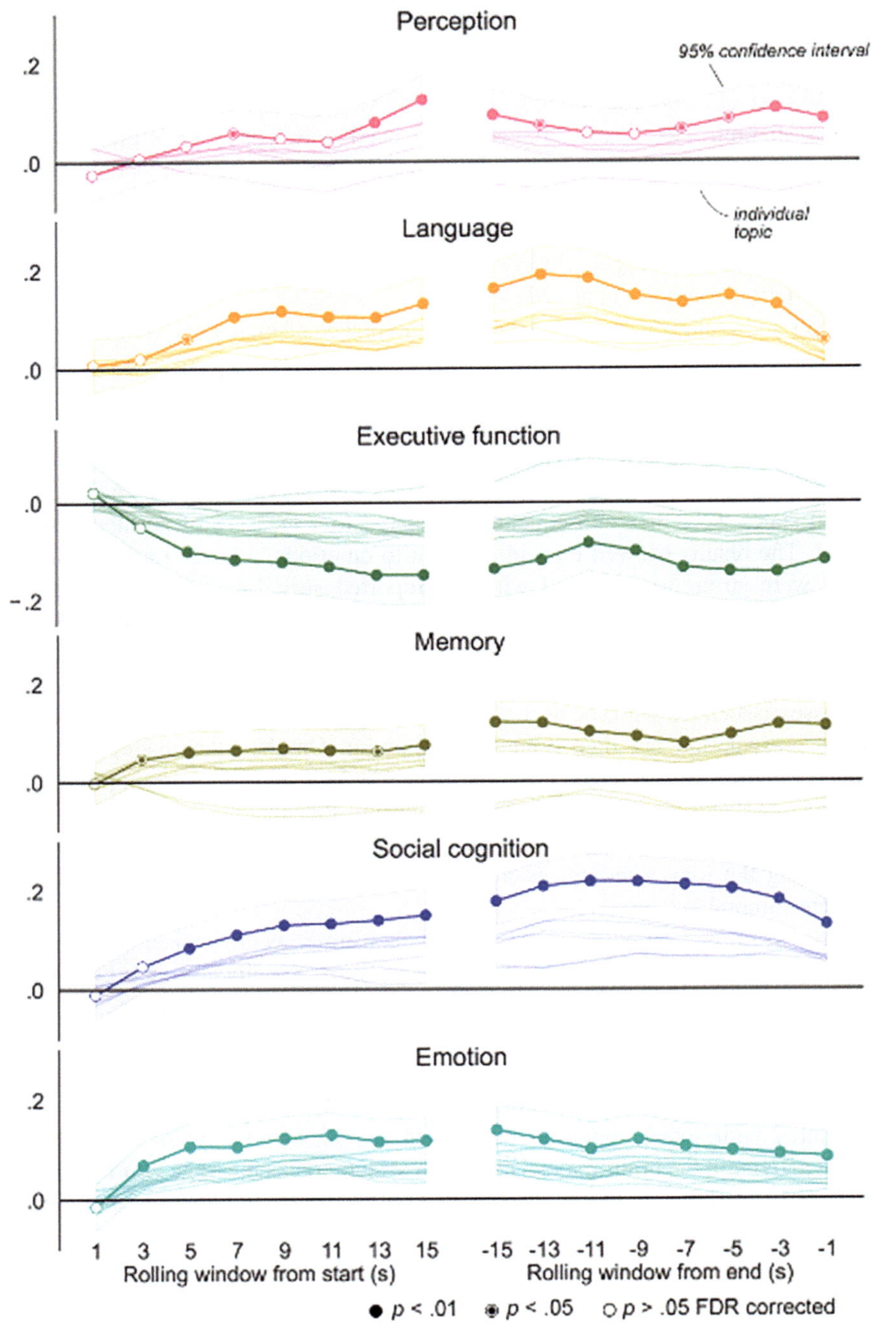

Fig. 1 Time-course of decoded mental processes predictive of self-reported attitude to advertisements. (Adapted from Chan et al., 2023)

2.3 Decoding (Brand) Associations

In addition to evoking mental processes through advertising and communications, a central function in marketing is to create strong and specific brand images—the mental associations of the brand in the minds of consumers (Keller, 1993). For example, a beer brand may try to create associations between the brand and the concepts of "young" and "party" that conjure up images of trendy young people at a party. A breakfast cereal, on the other hand, may try to associate the brand with the concept of "family" and an image of a loving family at the breakfast table. The extent to which these images and associations are elicited effectively and consistently in the minds of consumers is very difficult to measure. We know that most people are simply not very good at thinking about internal mental processes and often cannot express them accurately. An additional problem with asking people about brand associations is that it is difficult to determine whether the reported associations are actually retrieved from memory, or are constructed on the spot in response to the researchers' questions (Johar et al., 2006). Moreover, people are more inclined to offer associations with familiar and innocuous brands like McDonald's than those with more complex social connotations like Durex (Zaltman & Coulter, 1995). For these reasons, directly reading mental associations from the brains of consumers offers a more reliable way of measuring these brand images.

Findings from cognitive neuroscience support the idea that imagery and perception of specific visual content might be represented in the same cortical regions and in a similar representational format. In other words, models trained on fMRI data recorded during the visual perception of stimuli can decode which of these stimuli are then imagined at a later time (Lee et al., 2012; Cichy et al., 2012; Thirion et al., 2006). Indeed, the content of working memory—what people are thinking of at any moment without relying on direct visual input—can be decoded from the brain (Lewis-Peacock & Postle, 2008; Polyn et al., 2005). These findings suggest that it may indeed be possible to extract visual imagery from activity patterns within the brain, potentially allowing for brand images to be decoded from consumers' neural responses.

Chan et al. (2018) tested this hypothesis by asking participants to create a mental image that best fits a number of well-known brands (e.g., Apple, Disney, Heineken, Red Bull, Kellogg's, Microsoft, and Durex). They were then placed in an MRI scanner and asked to recall the mental images they had constructed previously while presented with the brand logos. In addition, participants viewed a large set of photos depicting different social situations: people working in an office, partying with friends, being intimate with a romantic partner, socializing with family, etc. In order to reveal the extent to which the association with the brand (the "brand image") resembles the images of the different social scenes, a classification model was trained on the brain activity evoked while looking at photos of these situations, and those evoked while thinking about the brands. Thus, a neural profile in terms of the similarity of each brand to the different social settings was created (see Fig. 2). It was observed, for example, that when participants thought about Heineken,

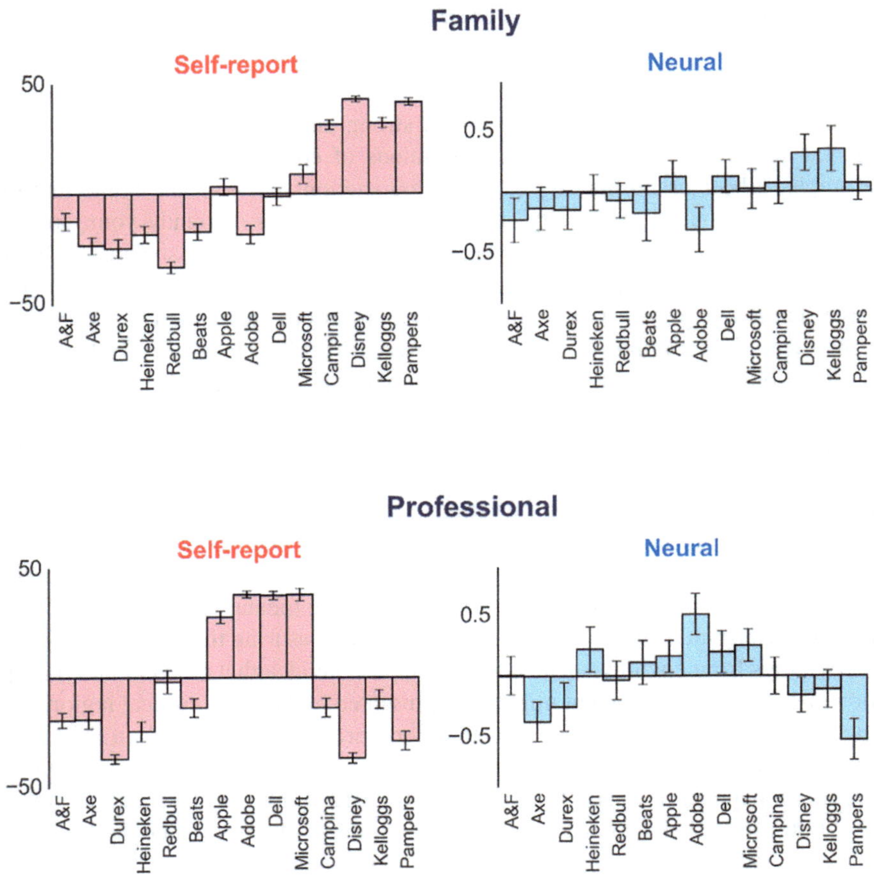

Fig. 2 Self-reported brand perceptions (pink) and neural context scores (blue) for the family context (top) and the professional context (bottom). Neural context scores represent the pattern similarity between the neural responses evoked by the brand imagery and the different social contexts. (Adapted from Chan et al., 2018)

their brain activity was similar to how the brain responds to "party" photos, but also that Heineken was fairly strongly associated with "work," but clearly considered less sexy than Red Bull. An additional insight from this study indicated that when the neural profiles of two brands are very similar, such as that of Apple and Beats, consumers reported that they would be more interested in co-branded product. This method could be a useful tool for marketers to predict which brand collaborations might be most successful.

A slightly different approach was taken by Chen et al. (2015). In this study, the researchers attempted to decode the personality traits associated with brands, rather than their brand images (Aaker, 1997). As in the previous study, participants viewed the logos of different brands in the MRI scanner and were asked to freely express the associations they had with these brands. A classifier was then trained to associate the patterns of brain activity acquired while participants viewed the brand logos with the personality characteristics as measured by a brand association scale (Aaker, 1997). The results showed that personality characteristics associated with a brand can be reliably decoded from patterns of brain activity.

2.4 *Interim Summary*

In this section, we have presented an overview of how contemporary methods from neuroscience provide a unique opportunity to capture measures of psychological processes, including emotions and mental associations, that are highly relevant for marketing. As reviewed above, there is growing evidence that mental states and mental associations can indeed be decoded from the brain and provide valuable insights for researchers and practitioners interested in understanding associations and mental states evoked in consumers by marketing stimuli.

3 Neuroforecasting

The decoding of mental states has proven valuable for exploring the processes underlying individual consumer decision-making. In practice, however, the value of understanding the motivations driving consumer behavior largely rests on its ability to inform managerial decision-making and improve predictions at the aggregate level. In this section, we address the application of neuroeconomic methods to the forecasting of real-world market-level outcomes—a recently established and quickly growing field of research called neuroforecasting.

Previous research in neuroeconomics, much of which has been covered in earlier chapters, has demonstrated that neural activity in specific brain areas can track decision-making processes and predict behavior, even before a conscious choice is made. This foundational work has paved the way for neuroforecasting, which aims to predict aggregate-level behavior, such as market or population trends, based on neural data collected in the laboratory. Unlike traditional methods that focus on individual choices, neuroforecasting uses data from a small sample to forecast the behavior of much larger groups—at the market, or even population levels.

In the following sections, we begin by addressing the key elements of neuroforecasting research and empirical design. We then review the foundational neuroforecasting studies, exploring their methodological and theoretical contributions and placing them within the broader context of consumer research. We do not delve into the fine details of each study, as there are a number of comprehensive reviews available for that purpose (Hakim & Levy, 2018; Genevsky & Yoon, 2022; Knutson & Genevsky, 2018; Yao & Wang, 2024). We end with a discussion of ongoing work aimed at understanding the mechanisms, optimization, and boundary conditions that define neuroforecasting in practice.

3.1 Empirical Approach

Neuroforecasting research, which aims to predict behaviors at the aggerate rather than individual level, necessitates a different empirical approach than utilized in traditional neuroeconomic studies. Typically, in psychology and neuroeconomics, data analysis focuses on individual-level predictions. In these contexts, data collected from an individual (neural and behavioral) are used to predict that individual's preferences and behavior. Thus, in these experimental designs, the independent and dependent variables are elicited from the same source. Alternatively, in neuroforecasting, the unit of analysis shifts from the individual (or sample) to the stimulus itself. The goal now is to understand and predict the broader impact of the stimulus at the aggregate level, rather than any individual response. The focus no longer lies on how differences in neural responses to a set of stimuli within an individual predict their relative preferences or choice, but rather, how shared responses across individuals to unique stimuli forecast their real-world, market-level outcomes (see Fig. 3).

An important consequence of this empirical approach is that it requires aggregate-level dependent variables that serve as the targets of our forecasts. Whereas traditional neuroeconomic experiments may rely on preference and choice data collected in the laboratory, neuroforecasting studies require access to market-level data. These data often represent a significant challenge for researchers. In many cases, market data are difficult to obtain and less controlled than traditional in-laboratory stimuli. Market outcome data collected by private firms, such as information about customers, sales figures, and online engagement metrics, are often highly proprietary and rarely shared outside of the company for competitive and legal reasons. Even when market data are available, one must carefully consider the numerous exogenous influences that impact their outcomes. For example, differences in promotional activities, communications, and availability might account for variance in real-world sales outcomes, but are presumably not associated with the neuropsychological processes driving choice that researchers are most interested in identifying and leveraging.

3.2 Primary Findings in the Field

Although the term "neuroforecasting" wasn't coined until 2018 (Knutson & Genevsky, 2018), the roots of the field can be traced back to a 2012 study by Berns and Moore on the forecasting of music preferences (Berns & Moore, 2012). The authors used data collected previously in a study on social influence in which

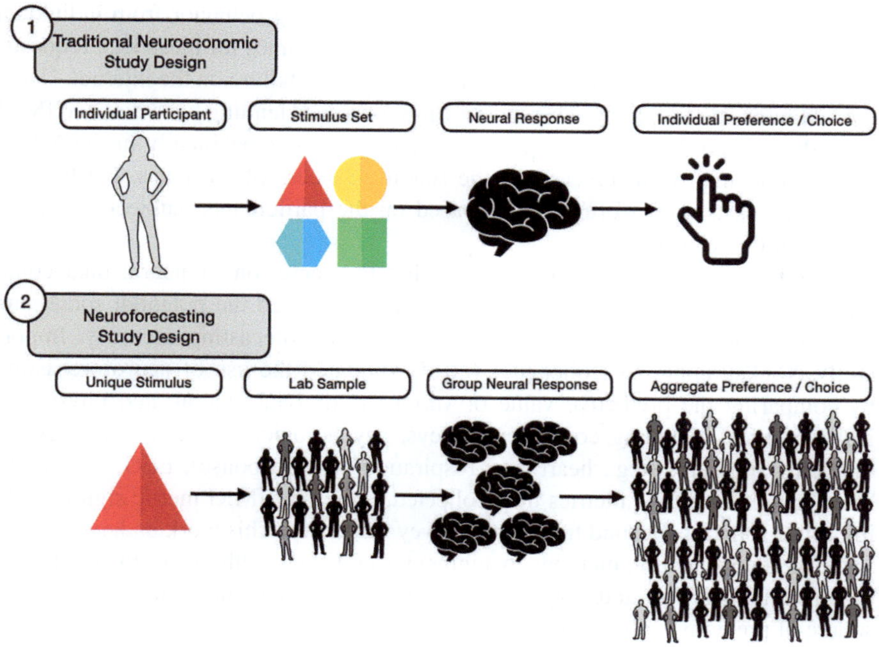

Fig. 3 The anatomy of a neuroforecasting study. (1) In traditional neuroeconomic studies, an individual is shown a set of stimuli or makes a series of decisions. Neural responses are recorded and used to predict the individual's preferences and choices regarding those stimuli. (2) In neuroforecasting experiments, stimuli are presented to a sample of laboratory participants. The sample's pooled neural responses (and behavioral metrics) are subsequently used in forecasts of aggregate outcomes at the market or population level

adolescents underwent fMRI while listening to unfamiliar songs. Years later, once market-level success of the songs became available, the neural and behavioral data were reanalyzed. The authors discovered that participants' self-reported preferences did not predict market success, but neural activity in the nucleus accumbens (NAcc) did. This work represents the first application of neural data to the forecasting of market-level outcomes.

This work was followed by focused efforts to explore the use of neural data for aggregate-level forecasts across a variety of domains. Falk et al. (2012) applied neuroforecasting in the public health domain, evaluating the effectiveness of neural responses to forecast the success of smoking cessation advertisements. They found that neural activity collected as participants viewed anti-smoking ads in the scanner predicted the volume of calls to smoking cessation hotlines better than self-reported effectiveness scores.

Genevsky and Knutson (2015) further expanded the application of neuroforecasting to prosocial behavior in a study of online microlending. In this study, participants in the fMRI scanner were presented with real microloan requests selected from a large online microloan platform. Initially, the goal of the study was to assess which features of loan requests most effectively elicited giving behavior from individual lenders. However, having access to the real-world outcomes for these loan requests, the authors went on to forecast their relative success or failure on the internet. Using the laboratory participants' self-report ratings, observed lending behavior, and fMRI data, they found that neural responses to loan requests (specifically in the NAcc) significantly improved forecasts of the funding success of the loans. Further, the neural forecasts outperformed those based on the participants' subjective ratings and donation behavior.

Together, these studies demonstrated that the inclusion of neural data could indeed improve aggregate-level forecasts and highlighted the potential social and economic impact of even modest improvements in forecasting accuracy. Importantly, Venkatraman and colleagues (2015) validated the use of neuroforecasting by comparing the predictive value of various traditional, neural, and physiological measures, including consumer surveys, psychological surveys, eye tracking, biometric measures (e.g., heart rate, respiration, skin response), EEG, and fMRI. They found that of the metrics they collected, only their fMRI metric significantly improved forecasts beyond traditional survey responses. This work underscores the value of neuroscientific methods to augment established methods for the prediction of consumer behavior and suggests that brain imaging data may represent a unique source of predictive information.

Genevsky et al. (2017) went on to apply neuroforecasting to predictions of crowdfunding campaigns. While in the scanner, participants were presented with crowdfunding projects from kickstarter.com and asked to choose which they would prefer to contribute to. Because these campaigns were newly listed, their final real-world outcomes were not known to the researchers at the time of data collection. Once the funding window for all of the projects on the website had closed, the authors assessed which laboratory measures, either behavioral or neural, best forecasted project success or failure. They found that while activity in several neural regions was associated with individual funding choices, only activity in the NAcc effectively predicted the real-world crowdfunding campaign success more accurately than behavioral measures.

Over the last 10 years, the applicability of neuroforecasting has been demonstrated across a diverse range of market-relevant domains. Kühn et al. (2016) tested neuroforecasting's application in retail marketing in a study on the impact of supermarket advertisements on chocolate sales, finding neural responses to ads and products predicted sales better than subjective ratings. In a pair of studies, Scholz et al. (2017). and Dore et al. (2019) showed that neural activity in reward-related brain regions predicted the sharing volume of *New York Times* articles better than article features or self-reported sharing intentions. Tong et al. (2020) then extended neuroforecasting to "attention markets" utilizing the engagement metrics of YouTube videos and demonstrating that neural activity, rather than self-reported viewing preferences, predicted real-world video engagement on the internet. Indeed, there is even emerging evidence that neural activity in the NAcc might hold predictive information regarding large-scale equity investment markets (Stallen et al., 2021; Van Brussel et al., 2024): In separate studies by Stallen et al.

(2021), Hakim & Levy (2018) and Van Brussel et al. (2024), activities in the NAcc of amateur and professional investors were better predictors of stock performance than investor choices or self-report measures.

Despite the early predominance of fMRI studies in the neuroforecasting literature, the relative affordability and accessibility of EGG has gained popularity in the field. In particular, the practical and logistical advantages of EEG have led to its widespread use in industry, particularly in private neuromarketing firms. Unlike fMRI studies, which typically emphasize spatially isolated neural activity in forecasting analyses, EEG studies often utilize alternative methods based on the increased temporal and spectral resolution of EEG data. For example, Boksem and Smidts (2015) applied spectral decomposition of participants' neural responses (the analysis of neural frequencies) to film trailers to predict the commercial success of movies. They found that particularly oscillations in the gamma range (>30 Hz) were predictive of the box office revenue for the movies, and that gamma improved these predictions beyond self-report measures. Guixeres and colleagues (2017) combined various EEG metrics to predict Super Bowl advertisement engagement, finding brain asymmetry in specific frequency ranges was the best predictor of YouTube views. Finally, Hakim et al. (2020) used EEG data to improve forecasting accuracy of consumer responses toward commercials.

Dmochowski et al. (2014) used Inter-Subject Correlations (ISC) to analyze the synchronization of neural activity across participants and found that correlated EEG activity while viewing television programs predicted real-world viewership and social media activity. Barnett and Cerf (2017) used a similar approach with movie trailers, finding that EEG data collected from viewers in theaters predicted box office success better than traditional ratings. Neural synchrony has also been found to contain predictive information regarding popularity on streaming music platforms (Leeuwis et al., 2021) and out-of-sample preferences for video stimuli (Chan et al., 2019).

3.3 *Understanding Neuroforecasting*

This body of research provides substantial evidence for the ability of neural data to improve forecasts of real-world market-level outcomes of interest to researchers and practitioners. However, a significant question remains regarding the mechanism that might explain the effectiveness of neurally derived predictions. How is it possible that neural data collected from a relatively small laboratory sample can significantly improve behavioral forecasts of market-level outcomes? Further, how can we understand results indicating that in some cases, forecasts based on the neural responses of laboratory respondents outperform forecasts based on the self-reported preference and observed behavior from those same individuals?

A first clue can be found in the Genevsky et al. (2017) neuroforecasting paper on crowdfunding (Genevsky et al., 2017). The authors find that neural activity in both primary affect (NAcc) and value integration (medial prefrontal cortex, MPFC) regions predicted individual funding decisions, but only activity in the NAcc scaled to effectively forecast aggregate outcomes. Subsequently, Knutson and Genevsky (2018) proposed that while various neural processes contribute to decision-making at the individual level, only a specific subset of these processes may remain relevant for predicting aggregate behavior. They argue that individual choice is a function of both primary affective responses, which are more widely shared across people, and deliberative processes which are more idiosyncratic to the individual. Thus, self-report and choice data collected at the individual level are a combination of both generalizable and idiosyncratic component processes. In neuroforecasting, where the aim is to forecast aggregate-level variables, the shared components of choice offer the most valuable and predictive source of information. On the other hand, while more idiosyncratic deliberative processes may account for the consistency and reliability in choice behavior within an individual, they represent unsystematic noise at the aggregate level.

Previous work in neuroeconomics provides some hypotheses regarding which components of the decision-making process might generalize best across individuals to subsequently account for aggregate-level behavior. In particular, the Affect-Integration-Motivation (AIM) framework (Samanez-Larkin et al., 2007) offers a useful framework through which to study neuroforecasting (see Fig. 4). The AIM framework suggests that choice stimuli initially evoke affective responses (either positive or negative) activating subcortical and cortical regions including the NAcc and anterior insula. This initial response is then further integrated through deliberative and reflective cognitive processes unique to the individual. Finally, the integrated representation generates a motivational state to approach or avoid the stimulus, which subsequently manifests as self-report and observable behavior. This framework suggests that initial affective responses to choice options (e.g., NAcc activity) may represent a more universally shared component of the decision-making process than later integrative neural responses (e.g., associated MPFC activity). Further, self-report and choice behavior then include both generalizable and idiosyncratic components of choice. As a result, forecasts based on only the more generalizable choice components indexed by primary affective responses in the brain may result in increased forecasting accuracy. Conversely, forecasts containing idiosyncratic components, including self-reported and choice, may result

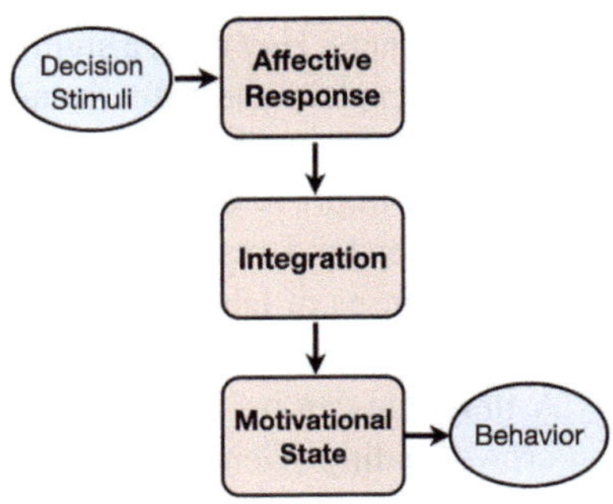

Fig. 4 The affect-integration-motivation (AIM) framework

in decreased accuracy due to the addition of non-generalizable noise into aggregate forecasts. In a test of this hypothesis, Genevsky and colleagues (2025) found that neural activity associated with early affective processing was more widely shared across individuals than later integrative or deliberative processes, and subsequently led to improved forecasts of aggregate behavior.

3.4 Open Questions and Future Directions in Neuroforecasting

In addition to the exploration of an underlying mechanism discussed above, many open questions in neuroforecasting revolve around issues of optimization and boundary conditions. Taking into account both theoretical and practical considerations, it becomes important that we understand when neural data may be most informative, and thus cost-effective to collect, and when it might not. For example, does neuroforecasting work equally well across contexts and choice domains? Based on the findings reviewed above, with a few notable exceptions, neural forecasts have been most broadly supported by generalizable neural activity associated with affective processes. This suggests that forecasts of consumer decisions in domains that heavily involve affective considerations, such as those related to hedonic and experiential products, might benefit most from neural data. However, it must be noted that in the existing neuroforecasting literature, studies of affectively oriented decisions are overrepresented, and thus may have led to findings specific to these domains. Whether or not neuroforecasting can be equally effective for forecasting outcomes in markets with more deliberative considerations is a topic of ongoing research.

There are also important open questions regarding the optimization of empirical approaches in neuroforecasting research (i.e., experimental designs and analysis methods). The initial set of neuroforecasting experiments reviewed above is almost exclusively characterized by efforts to forecast which stimuli within a set will perform better or worse at the aggregate level. Little work has explored how to optimize forecasting accuracy using different experimental designs, stimulus selection criteria, identification of critical stimulus features, and which tasks participants should perform while in the scanner. To date, empirical design choices have largely been based on existing paradigms from other neuroeconomic disciplines, not necessarily constructed to optimize the generalizable neural responses critical for neuroforecasting. Better understanding regarding how to conduct neuroforecasting research most efficiently and effectively represent a promising direction for future research and would offer substantial benefits for researchers and practitioners interested in addressing the scope of neuroforecasting's potential to inform real-world market decision-making.

Questions also remain regarding the relationship between the experimental sample and the aggregate market of interest. Does the effectiveness of neuroforecasting rely on the similarity between the sample and market? Using two independent neuroforecasting datasets, Genevsky and colleagues (2025) found that the accuracy of market forecasts derived from neural data was less impacted by the representativeness of the sample to the market than forecasts based on behavioral data. This suggests that neural activity capturing generalizable components of affect and value processing in the brain may represent a more universal index of preference across individuals compared to observed behavior and self-report measures.

Finally, while we have seen that multivariate pattern analysis has been pivotal in decoding consumer mental states, work on neuroforecasting has predominantly relied on univariate analysis approaches. However, in a recent study by Chan et al. (2023), a first step has been made to combine the decoding of mental states with neuroforecasting. The authors found that patterns of neural activity, particularly those associated with social-affective responses, improved the prediction of out-of-sample liking ratings for advertisements compared with traditional anatomically based neuroimaging measures and self-report measures. These findings align well with predictions derived from the AIM framework (see Fig. 4) that (social) affective processes may account for the ability of brain measures to forecast market-level behavior. The use of more nuanced approaches to neural analyses, as in this example, will become increasingly important as the field of neuroforecasting continues to grow and develop.

4 Conclusion

Over the past decade, the application of neuroeconomic approaches to business-relevant questions has propelled the development of the field of consumer neuroscience. This area of research is committed to applying the lessons and methods from the disciplines of neuroscience and neuroeconomics to better understand and predict consumer psychology and decision-making. While we remain optimistic about the potential for consumer neuroscience to contribute across a broad range of domains, reflection on the last 10 years of work indicates the greatest contributions have come from research on the decoding of consumer mental states and the forecasting of market-level outcomes.

The temporal resolution and real-time nature of neuroimaging methods have removed the limitations of black box models of consumer behavior, in which we only have access to stimulus inputs and resulting behavior. Using neural data related to psychological processes, researchers are able to decode emotions, consumer experiences, and brand associations. These methods have opened new avenues of research on consumer responses to marketing stimuli and their resulting decision-making process. In particular, multivariate analysis methods have proven particularly useful for the decoding and prediction of emotional and cognitive states. The ability to record brain responses in real time, unobtrusively, without requiring overt responses from participants that may interfere with ongoing mental processes, gives consumer neuroscience a unique advantage in providing insights into the experiences and cognitions underlying consumer behavior that are critical for the development of effective business strategies.

Finally, the evolving field of neuroforecasting has extended the applications of consumer neuroscience beyond the individual (or sample). A growing body of research has demonstrated the effectiveness of neural data to improve forecasts of real-world market-level outcomes across a wide range of product domains and consumer contexts. This work has shown that the brain responses of laboratory samples, primarily in affective and value-based neural circuits, hold hidden information capable of improving the accuracy of predictions of aggregate-level preference and behavioral outcomes. As the field continues to develop, ongoing work must focus on optimizing empirical designs, establishing boundary conditions, and exploring generalizability across product and consumer domains. In addition, neuroforecasting research on the generalizability of individual decision-making processes to aggregate-level behavior may contribute to the neuroeconomic literatures on affect and value-based decision-making.

In summary, the application of neuroeconomics in the form of consumer neuroscience has made significant contributions over the last decade. Still a nascent field, the continued integration of basic neuroscientific approaches to business-relevant research questions represents a promising direction. In particular, the use of neural metrics for the decoding of mental processes and the forecasting of market behavior offers the potential for improved communications, more effective intervention strategies, and better-informed strategic decision-making in business and public policy.

References

Aaker, J. L. (1997). Dimensions of brand personality. *Journal of Marketing Research, 34*, 347–356.

Ariely, D., & Berns, G. S. (2010). Neuromarketing: The hope and hype of neuroimaging in business. *Nature Reviews. Neuroscience, 11*, 284–292.

Arnold, H. J., & Feldman, D. C. (1981). Social desirability response bias in self-report choice situations. *Academy of Management Journal, 24*, 377–385.

Bagozzi, R. P., Gopinath, M., & Nyer, P. U. (1999). The role of emotions in marketing. *Journal of Academy of Market Science, 27*, 184.

Barnett, S. B., & Cerf, M. (2017). A ticket for your thoughts: Method for predicting movie trailer recall and future ticket sales using neural similarity among moviegoers. *Journal of Consumer Research, 44*(1), ucw083. https://doi.org/10.1093/jcr/ucw083

Baucom, L. B., Wedell, D. H., Wang, J., Blitzer, D. N., & Shinkareva, S. V. (2012). Decoding the neural representation of affective states. *NeuroImage, 59*, 718–727.

Berns, G. S., & Moore, S. E. (2012). A neural predictor of cultural popularity. *Journal of Consumer Psychology, 22*, 154–160.

Binet, L., & Field, P. (2009). Empirical generalizations about advertising campaign success. *Journal of Advertising Research, 49*, 130–133.

Boksem, M. A. S., & Smidts, A. (2015). Brain responses to movie trailers predict individual preferences for movies and their population-wide commercial success. *Journal of Marketing Research, 52*, 482–492.

Burke, M. C., & Edell, J. A. (1989). The impact of feelings on ad-based affect and cognition. *Journal of Marketing Research, 26*, 69–83.
Chan, H.-Y., Boksem, M. A. S., & Smidts, A. (2018). Neural profiling of brands: Mapping brand image in consumers' brains with visual templates. *Journal of Marketing Research, 55*, 600–615.
Chan, H.-Y., Smidts, A., Schoots, V. C., Dietvorst, R. C., & Boksem, M. A. S. (2019). Neural similarity at temporal lobe and cerebellum predicts out-of-sample preference and recall for video stimuli. *NeuroImage, 197*, 391–401.
Chan, H.-Y., Smidts, A., Schoots, V. C., Sanfey, A. G., & Boksem, M. A. S. (2020). Decoding dynamic affective responses to naturalistic videos with shared neural patterns. *NeuroImage, 216*, 116618.
Chan, H.-Y., et al. (2023). Neural signals of video advertisement liking: Insights into psychological processes and their temporal dynamics. *Journal of Marketing Research*. https://doi.org/10.1177/00222437231194319
Chang, L. J., Gianaros, P. J., Manuck, S. B., Krishnan, A., & Wager, T. D. (2015). A sensitive and specific neural signature for picture-induced negative affect. *PLoS Biology, 13*, e1002180.
Chen, Y.-P., Nelson, L. D., & Hsu, M. (2015). From "where" to "what": Distributed representations of brand associations in the human brain. *Journal of Marketing Research, 52*, 453–466.
Chiong, K. X., Shum, M., Webb, R., & Chen, R. (2024). Combining choice and response time data: A drift-diffusion model of mobile advertisements. *Management Science, 70*, 1238–1257.
Cichy, R. M., Heinzle, J., & Haynes, J.-D. (2012). Imagery and perception share cortical representations of content and location. *Cerebral Cortex, 22*, 372–380.
Clithero, J. A., Karmarkar, U. R., Nave, G., & Plassmann, H. (2023). Reconsidering the path for neural and physiological methods in consumer psychology. *Journal of Consumer Psychology*. https://doi.org/10.1002/jcpy.1397
der Laan, L. N. V., Ridder, D. T. D. D., Viergever, M. A., & Smeets, P. A. M. (2012). Appearance matters: Neural correlates of food choice and packaging aesthetics. *PLoS One, 7*, e41738.
Derbaix, C. M. (1995). The impact of affective reactions on attitudes toward the advertisement and the brand: A step toward ecological validity. *Journal of Marketing Research, 32*, 470–479.
Dijksterhuis, A. (2004). Think different: The merits of unconscious thought in preference development and decision making. *Journal of Personality and Social Psychology, 87*, 586–598.
Dmochowski, J. P., et al. (2014). Audience preferences are predicted by temporal reliability of neural processing. *Nature Communications, 5*, 4567.
Do, A. M., Rupert, A. V., & Wolford, G. (2008). Evaluations of pleasurable experiences: The peak-end rule. *Psychonomic Bulletin & Review, 15*, 96–98.
Doré, B. P., et al. (2019). Brain activity tracks population information sharing by capturing consensus judgments of value. *Cerebral Cortex, 29*, 3102–3110.
Eijlers, E., Smidts, A., & Boksem, M. A. S. (2019). Implicit measurement of emotional experience and its dynamics. *PLoS One, 14*, e0211496.
Eijlers, E., Boksem, M. A. S., & Smidts, A. (2020). Measuring neural arousal for advertisements and its relationship with advertising success. *Frontiers in Neuroscience (Switzerland), 14*, 736.
Falk, E. B., Berkman, E. T., & Lieberman, M. D. (2012). From neural responses to population behavior. *Psychological Science, 23*, 439–445.
Fisher, R. J. (1993). Social desirability bias and the validity of indirect questioning. *Journal of Consumer Research, 20*, 303.
Genevsky, A., & Knutson, B. (2015). Neural affective mechanisms predict market-level microlending. *Psychological Science, 26*, 1411–1422.
Genevsky, A., & Yoon, C. (2022). Neural basis of consumer decision making and neuroforecasting. In R. Kahle, T. M. Lowrey, & J. Huber (Eds.), *APA handbook of consumer psychology* (pp. 621–635). American Psychological Association. https://doi.org/10.1037/0000262-027
Genevsky, A., Yoon, C., & Knutson, B. (2017). When brain beats behavior: Neuroforecasting crowdfunding outcomes. *The Journal of Neuroscience, 37*, 8625–8634.
Genevsky, A., Tong, L., & Knutson, B. (2025). Neuroforecasting reveals generalizable components of choice. *PNAS Nexus, 4*(2), pgaf029.

Guitart, I. A., & Stremersch, S. (2021). The impact of informational and emotional television ad content on online search and sales. *Journal of Marketing Research, 58*, 299–320.

Guixeres, J., et al. (2017). Consumer neuroscience-based metrics predict recall, liking and viewing rates in online advertising. *Frontiers in Psychology, 8*, 1808.

Hakim, A., & Levy, D. J. (2018). A gateway to consumers' minds: Achievements, caveats, and prospects of electroencephalography-based prediction in neuromarketing. *Wiley Interdisciplinary Reviews: Cognitive Science, 10*, e1485.

Hakim, A., et al. (2020). Machines learn neuromarketing: Improving preference prediction from self-reports using multiple EEG measures and machine learning. *International Journal of Research in Marketing*. https://doi.org/10.1016/j.ijresmar.2020.10.005

Holbrook, M. B., & Batra, R. (1987). Assessing the role of emotions as mediators of consumer responses to advertising. *Journal of Consumer Research, 14*, 404.

Johar, G. V., Maheswaran, D., & Peracchio, L. A. (2006). Mapping the frontiers: Theoretical advances in consumer research on memory, affect, and persuasion. *Journal of Consumer Research, 33*, 139–149.

Kardes, F. R., & Kalyanaram, G. (1992). Order-of-entry effects on consumer memory and judgment: An information integration perspective. *Journal of Marketing Research, 29*, 343–357.

Karmarkar, U. R., & Yoon, C. (2016). Consumer neuroscience: Advances in understanding consumer psychology. *Current Opinion in Psychology, 10*, 160–165.

Karmarkar, U. R., Shiv, B., & Knutson, B. (2015). Cost conscious? The neural and behavioral impact of price primacy on decision making. *Journal of Marketing Research, 52*, 467–481.

Kassam, K. S., Markey, A. R., Cherkassky, V. L., Loewenstein, G., & Just, M. A. (2013). Identifying emotions on the basis of neural activation. *PLoS One, 8*, e66032.

Keller, K. L. (1993). Conceptualizing, measuring, and managing customer-based brand equity. *Journal of Marketing, 57*, 1–22.

King, M. F., & Bruner, G. C. (2000). Social desirability bias: A neglected aspect of validity testing. *Psychology & Marketing, 17*, 79–103.

Knutson, B., & Genevsky, A. (2018). Neuroforecasting aggregate choice. *Current Directions in Psychological Science, 27*, 110–115.

Kragel, P. A., & LaBar, K. S. (2016). Decoding the nature of emotion in the brain. *Trends in Cognitive Sciences, 20*, 444–455.

Kragel, P. A., Knodt, A. R., Hariri, A. R., & LaBar, K. S. (2016). Decoding spontaneous emotional states in the human brain. *PLoS Biology, 14*, e2000106.

Kragel, P. A., Koban, L., Barrett, L. F., & Wager, T. D. (2018). Representation, pattern information, and brain signatures: From neurons to neuroimaging. *Neuron, 99*, 257–273.

Kranzbühler, A.-M., Zerres, A., Kleijnen, M. H. P., & Verlegh, P. W. J. (2020). Beyond valence: A meta-analysis of discrete emotions in firm-customer encounters. *Journal of Academy of Market Science, 48*, 478–498.

Kühn, S., Strelow, E., & Gallinat, J. (2016). Multiple "buy buttons" in the brain: Forecasting chocolate sales at point-of-sale based on functional brain activation using fMRI. *NeuroImage, 136*, 122–128.

Lee, S.-H., Kravitz, D. J., & Baker, C. I. (2012). Disentangling visual imagery and perception of real-world objects. *NeuroImage, 59*, 4064–4073.

Leeuwis, N., Pistone, D., Flick, N., & van Bommel, T. (2021). A sound prediction: EEG-based neural synchrony predicts online music streams. *Frontiers in Psychology, 12*, 672980.

Lench, H. C., Flores, S. A., & Bench, S. W. (2011). Discrete emotions predict changes in cognition, judgment, experience, behavior, and physiology: A meta-analysis of experimental emotion elicitations. *Psychological Bulletin, 137*, 834–855.

Levallois, C., Smidts, A., & Wouters, P. (2019). The emergence of neuromarketing investigated through online public communications (2002–2008). *Business History, 63*, 1–40.

Levy, I., Lazzaro, S. C., Rutledge, R. B., & Glimcher, P. W. (2011). Choice from non-choice: Predicting consumer preferences from blood oxygenation level-dependent signals obtained during passive viewing. *The Journal of Neuroscience, 31*, 118–125.

Lewis-Peacock, J. A., & Postle, B. R. (2008). Temporary activation of long-term memory supports working memory. *The Journal of Neuroscience, 28*, 8765–8771.

McClure, S. M., et al. (2004). Neural correlates of behavioral preference for culturally familiar drinks. *Neuron, 44*, 379–387.

Morris, J. D., Woo, C., Geason, J. A., & Kim, J. (2002). The power of affect: Predicting intention. *Journal of Advertising Research, 42*, 7–17.

Nisbett, R. E., & Wilson, T. D. (1977). Telling more than we can know: Verbal reports on mental processes. *Psychological Review, 84*, 231–259.

Norman, K. A., Polyn, S. M., Detre, G. J., & Haxby, J. V. (2006). Beyond mind-reading: Multi-voxel pattern analysis of fMRI data. *Trends in Cognitive Sciences, 10*, 424–430.

Nummenmaa, L., & Saarimäki, H. (2019). Emotions as discrete patterns of systemic activity. *Neuroscience Letters, 693*, 3–8.

Pelsmacker, P. D., & Geuens, M. (1999). The advertising effectiveness of different levels of intensity of humour and warmth and the moderating role of top of mind awareness and degree of product use. *Journal of Marketing Communications, 5*, 113–129.

Pham, M. T., Geuens, M., & Pelsmacker, P. D. (2013). The influence of ad-evoked feelings on brand evaluations: Empirical generalizations from consumer responses to more than 1000 TV commercials. *International Journal of Research in Marketing, 30*, 383–394.

Plassmann, H., O'Doherty, J., & Rangel, A. (2007). Orbitofrontal cortex encodes willingness to pay in everyday economic transactions. *The Journal of Neuroscience, 27*, 9984–9988.

Plassmann, H., O'Doherty, J., Shiv, B., & Rangel, A. (2008). Marketing actions can modulate neural representations of experienced pleasantness. *Proceedings of the National Academy of Sciences of the United States of America, 105*, 1050–1054.

Plassmann, H., Venkatraman, V., Huettel, S., & Yoon, C. (2015). Consumer neuroscience: Applications, challenges, and possible solutions. *Journal of Marketing Research, 52*, 427–435.

Poldrack, R. A. (2011). Inferring mental states from neuroimaging data: From reverse inference to large-scale decoding. *Neuron, 72*, 692–697.

Polyn, S. M., Natu, V. S., Cohen, J. D., & Norman, K. A. (2005). Category-specific cortical activity precedes retrieval during memory search. *Science, 310*, 1963–1966.

Puccinelli, N. M., Grewal, D., Motyka, S., Andrzejewski, S. A. & Avnet, T. (2016). *Effects of affect on consumer behavior: a meta-analytic integration* (Marketing Science Institute Working Paper Series Report No. 16).

Reimann, M., Zaichkowsky, J., Neuhaus, C., Bender, T., & Weber, B. (2010). Aesthetic package design: A behavioral, neural, and psychological investigation. *Journal of Consumer Psychology, 20*, 431–441.

Saarimäki, H., et al. (2016). Discrete neural signatures of basic emotions. *Cerebral Cortex, 26*, 2563–2573.

Samanez-Larkin, G. R., et al. (2007). Anticipation of monetary gain but not loss in healthy older adults. *Nature Neuroscience, 10*, 787–791.

Schmälzle, R., Häcker, F. E. K., Honey, C. J., & Hasson, U. (2015). Engaged listeners: Shared neural processing of powerful political speeches. *Social Cognitive and Affective Neuroscience, 10*, 1137–1143.

Scholz, C., et al. (2017). A neural model of valuation and information virality. *Proceedings of the National Academy of Sciences of the United States of America, 114*, 2881–2886.

Shimojo, S., Simion, C., Shimojo, E., & Scheier, C. (2003). Gaze bias both reflects and influences preference. *Nature Neuroscience, 6*, 1317–1322.

Smidts, A., et al. (2014). Advancing consumer neuroscience. *Marketing Letters, 25*, 257–267.

Stallen, M., Borg, N., & Knutson, B. (2021). Brain activity foreshadows stock price dynamics. *The Journal of Neuroscience, 41*, JN-RM-1727-20.

Teixeira, T., Wedel, M., & Pieters, R. (2012). Emotion-induced engagement in internet video advertisements. *Journal of Marketing Research, 49*, 144–159.

Thirion, B., et al. (2006). Inverse retinotopy: Inferring the visual content of images from brain activation patterns. *NeuroImage, 33*, 1104–1116.

Tong, L. C., Acikalin, M. Y., Genevsky, A., Shiv, B., & Knutson, B. (2020). Brain activity forecasts video engagement in an internet attention market. *Proceedings of the National Academy of Sciences of the United States of America, 117*, 6936–6941.

Vakratsas, D., & Ambler, T. (1999). How advertising works: What do we really know? *The Journal of Marketing, 63*, 26–43.

van Brussel, L. D., Boksem, M. A. S., Dietvorst, R. C., & Smidts, A. (2024). Brain activity of professional investors signals future stock performance. *Proceedings of the National Academy of Sciences, 121*, e2307982121.

van Laer, T., de Ruyter, K., Visconti, L. M., & Wetzels, M. (2014). The extended transportation-imagery model: A meta-analysis of the antecedents and consequences of consumers' narrative transportation. *Journal of Consumer Research, 40*, 797–817.

Venkatraman, V., et al. (2015). Predicting advertising success beyond traditional measures: New insights from neurophysiological methods and market response modeling. *Journal of Marketing Research, 52*, 436–452.

Wager, T. D., et al. (2015). A Bayesian model of category-specific emotional brain responses. *PLoS Computational Biology, 11*, e1004066.

Wiles, J. A., & Cornwell, T. B. (2012). A review of methods utilized in measuring affect, feelings, and emotion in advertising. *Current Issues and Research in Advertising, 13*, 241–275.

Winkielman, P., & Berridge, K. C. (2004). Unconscious emotion. *Current Directions in Psychological Science, 13*, 120–123.

Yao, X., & Wang, Y. (2024). Using neural data to forecast aggregate consumer behavior in neuromarketing: Theory, metrics, progress, and outlook. *Journal of Consumer Behaviour, 23*(4), 2142–2159. https://doi.org/10.1002/cb.2324

Yarkoni, T., Poldrack, R. A., Nichols, T. E., Essen, D. C. V., & Wager, T. D. (2011). Large-scale automated synthesis of human functional neuroimaging data. *Nature Methods, 8*, 665–670.

Yeung, C. W. M., & Wyer, R. S. (2004). Affect, appraisal, and consumer judgment. *Journal of Consumer Research, 31*, 412–424.

Yoon, C., Gutchess, A. H., Feinberg, F., & Polk, T. A. (2006). A functional magnetic resonance imaging study of neural dissociations between brand and person judgments. *Journal of Consumer Research, 33*, 31–40.

Zajonc, R. B. (1980). Feeling and thinking: Preferences need no inferences. *The American Psychologist, 35*, 151–175.

Zaltman, G., & Coulter, R. H. (1995). Seeing the voice of the customer: Metaphor-based advertising research. *Journal of Advertising Research, 35*, 35–51.

Attention Dynamics: Antecedents to Consumer Choice

Nicolette J. Sullivan

Abstract This chapter reviews four decades of research on consumer attention and decision-making, focusing on how visual information influences preferences, attitudes, and choice. It traces methodological advances from early information display boards to modern eye tracking technologies, highlighting their role in capturing naturalistic attention patterns. The chapter explores visual marketing, examining how complexity, saliency, and repeated exposures shape consumer behavior. Studies show that attention allocation predicts preferences and choices, with concepts like gaze bias, visual equity, and the scanpath theory illustrating consistent patterns. The chapter concludes with valuation models, including the attentional drift diffusion model, which incorporates gaze data to explain how attention modulates evidence accumulation during choice. By integrating findings from consumer behavior, psychology, and neuroscience, the chapter details visual attention's role in consumer decision-making.

Researchers in business schools have been studying information processing to understand how we make decisions since the 1970s (Schulte-Mecklenbeck et al., 2017). There was an early acknowledgment that what people say and do often diverge (the "attitude-behavior discrepancy"; LaPiere, 1934), so techniques to assess the decision process without relying on self-reports were sought after. One of the first techniques, information display boards, featured cards labeled with attributes relevant to the choice that were turned over to reveal their content (Payne, 1976). This allowed researchers to observe the sequence of information acquisition and time spent considering each alternative and attribute. Mouse Lab computerized the classic information board; hovering one's mouse cursor over a box reveals its contents (Johnson et al., 1989; Payne et al., 1988). Mouse Lab Web (Willemsen & Johnson, 2011) and Flashlight (Schulte-Mecklenbeck et al., 2011) extend this to the

N. J. Sullivan (✉)
Department of Management, The London School of Economics and Political Science, London, UK
e-mail: N.Sullivan@lse.ac.uk

D. V. Smith et al. (eds.), *Neuroeconomics: Core Topics and Current Directions*,
https://doi.org/10.1007/978-3-032-02925-6_35

Internet, allowing researchers to use online platforms to increase sample sizes and reach more diverse demographics beyond the college campus lab.

One drawback of display board techniques is that participants must consciously choose to reveal hidden information. This active or explicit attention measure differs from organic or passive attention (Lohse & Johnson, 1996). For example, participants engage in a less extensive search when active attention measures are used (Van Raaij, 1977). Consequently, consumer researchers were quick to adopt eye tracking technology as a more unobtrusive method to track information search. Stationary eye trackers use infrared light to triangulate the eye's gaze location on a computer screen (Russo, 1978). Although participants' gaze must be calibrated in advance, stationary eye trackers better capture the attention process when consumers are not consciously monitoring their attention patterns. A recent advancement in eye tracking is the use of mobile eye tracking, which tracks gaze through wearable eyeglasses as participants move about the environment, and webcam eye tracking, which uses computer vision algorithms to detect the location of the eye and pupil through a participant's own webcam at home. Although this is not yet able to differentiate attention to highly detailed stimuli where small differences in gaze location matter, it enables less detailed studies and facilitates the use of more diverse participant populations. See Jenke and Sullivan (2024) for a review of and guidance on this new technology.

One feature of consumer research that distinguishes it from much of the research in psychology and neuroeconomics is the type of stimuli tested. Typically, neuroeconomics is concerned with response to objects of value themselves, such as foods, trinkets, or money. Consumer researchers also investigate response to the antecedents of choice: persuasion attempts such as advertisements. When studying the objects of choice themselves, consumer research also often departs from common methods in neuroeconomics by using multi-option (non-binary) choice contexts. For example, a planogram (Fig. 1a) is a schematic of a store's offerings and is a common choice context in consumer research. A choice matrix (Fig. 1b) is another commonly used stimulus, displaying options against their attributes in rows and columns. This chapter discusses the various approaches to understanding consumer attention within the field of marketing, along with the findings from four decades of work in this field. In particular, it will cover the field of visual marketing, which investigates attention to stimuli used in marketing practices such as advertisements, as well as visual search within such stimuli and among options. It will also cover research from consumer behavior, psychology, and neuroscience that contributes to our understanding of how attention influences consumers' preferences, valuations, and choices.

1 Visual Marketing

Individuals encounter many persuasion attempts in their daily lives, including print and digital ads, product packaging and displays, and influencer posts on social

A

B

	Marriot	Hilton	Best Western	Ritz Carlton	Super 8
Price per night	$190	$200	$125	$350	$85
Overall Rating (out of 5)	4.3	4.5	3.9	4.9	3.1
Location Rating (out of 5)	4	3	1	5	2
Service Rating (out of 5)	4	4	3	5	2
Style	Mid-range	Luxury	Mid-range	Luxury	Budget

Fig. 1 (**a**) A planogram (schematic) is a hypothetical layout of a store shelf on which various products are placed. Using planograms, experimenters vary the location of products on hypothetical images of store shelves to assess how location influences consumers' attention and choice. This information helps to inform firms about the relative value of different product positions in store and influences how much shops charge for placing products in different locations. Image designed by Freepik. (**b**) A choice matrix used to study attention to attributes and options during participant choice. This matrix displays a number of options and their attributes, allowing consumers to make trade-offs between attributes before choice, and is similar to ones seen across Internet product comparison websites. These computer displays, paired with eye tracking, allow researchers to observe the decision strategies used by consumers when making multi-attribute choices, such as comparing options within one attribute first (e.g., viewing all prices) before moving to the next attribute, or viewing all information about each option one at a time. Research has found that such patterns result in distinct types of choices among consumers

media. Given that these tactics likely have some, even if small, influence on choice, it is important to understand how such persuasion attempts succeed or fail at attracting attention and altering attitudes. Visual marketing is a subfield of marketing that investigates how the visual elements of marketing strategies influence consumer outcomes, such as brand awareness, attitudes, and purchase intent.

1.1 Attention Over Time

Consumers are typically exposed to the same persuasion attempt repeatedly, prompting researchers to investigate how responses change across multiple exposures. Noton and Stark (1971) introduced the scanpath theory, which proposed that subsequent exposures should generate the same pattern of fixations from one area of interest (AOI) to another (scanpaths) as seen in the first exposure. Pieters et al. (1999) tested this empirically using gaze data and a first-order Markov model (Eq. 1). Markov models are often used in consumer behavior eye tracking studies on stimuli such as Fig. 1a-b or advertisements to calculate switching probability p of a transition from one AOI (j) to another (k), where m is the frequency of saccades (rapid gaze shifts) between AOIs during exposure r.

$$p_{k|j}^{j} = \frac{m_{jk}^{r}}{\sum_{k=1}^{k} m_{jk}^{r}} \tag{1}$$

Observed saccade frequencies y_{ijkr} for individual i follow a negative binomial distribution (Eq. 2), where m^{r}_{jk} is a random variable from a gamma distribution with mean u^{r}_{jk} and dispersion v. This framework captures individual variance in attention patterns.

$$f\left(y_{ijkr}\right) = \left[\frac{\Gamma\left(y_{ijkr} + v\right)}{y_{ijkr}!\Gamma(v)}\right]\left(\frac{v}{u_{jk}^{r} + v}\right)\left(\frac{u_{jk}^{r}}{u_{jk}^{r} + v}\right)^{y_{ijkr}} \tag{2}$$

Using this model, Pieters et al. (1999) found that switching probabilities remained relatively stable across repeated exposures despite a decrease in the duration of fixations (times the eye stops to collect information). This finding has been replicated by others (e.g., Harris, 1993), but there has been concern in the literature about how repeated measurements influence consumer search and choice (Y. Li et al., 2022). Indeed, Liechty et al. (2003) used a hidden Markov model to test the theory that consumers' switch between two hidden attentional states across time during ad viewing: identifying objects and their meanings (local search) and locating objects in space (global search). Participants typically begin in the local state and terminate in the global state and transition between these states fewer than three times on average, with the most time spent in the local state.

1.2 Complexity and Saliency

Visual stimuli vary in complexity, and different types of complexity can affect their ability to influence individual choices. This is because individuals have limited cognitive resources, and the effectiveness of advertising depends on the availability of these resources relative to the complexity of the message itself (Peracchio & Meyers-Levy, 2005). Dense low-level, bottom-up perceptual features (color, luminance, and edges; Donderi, 2006; Huhmann, 2003) increase feature complexity, which can harm brand attitudes and memory due to the difficulty of processing the image. Such complexity has been found to increase attention to the ad overall but decrease attention to the brand itself (Pilelienė & Grigaliūnaitė, 2016). In contrast, design or structural complexity (Berlyne, 1958; Pieters et al., 2010), which arises from the irregularity, quantity, and dissimilarity of objects in a scene (Palmer, 1999), increases an ad's comprehensibility and attention to, and attitude toward, brands.

Another type of complexity is more conceptual, for example, original ads garnering more attention and brand recall (Pieters et al., 2002). Implicit or metaphorical ads, whose target product or brand requires processing to understand, receive more overall attention, fixations, and positive evaluations (Beard et al., 2024; Radach et al., 2003).

A product or ad's ability to attract attention is termed its visual equity, and its influence on choice has been found to be the inverse of its traditional memory-based brand equity (Chandon, 2002). Visual salience is a bottom-up feature of part of a visual scene that determines how much it stands out and attracts exogenous visual attention relative to other visual elements. This is determined, in part, by color, intensity, and line orientation. As a bottom-up feature of the visual system, visual salience can be predicted fairly accurately. Saliency maps are generated using known properties of the visual system to estimate how much elements of the scene "pop out" or attract attention (L. Itti et al., 1998). These maps have been shown to reliably predict gaze (Laurent Itti & Koch, 2001) and choice (Milosavljevic et al., 2012). Bottom-up attention is contrasted with top-down or endogenous attention, which is driven by the viewer's preferences and goals. A priority map combines both top-down and bottom-up factors to more accurately predict gaze than either alone (Fecteau & Munoz, 2006; White et al., 2017).

1.3 Attention to Text

Text is a prominent feature of nearly all persuasion attempts, as well as many decision contexts such as menus and product packaging. As a result, visual marketing has long been concerned with how individuals process text. Text processing comprises the vast majority of consumers' attention in ads (Q. Li et al., 2016), with one study showing that over 70% of the time was spent on text compared to images. Larger fonts received more initial gazes, followed by smaller text, and then images (Rayner et al., 2001). Fixations tend to remain segregated; within an ad, the probability of transitioning from image to text or vice versa was relatively low in one study, at around 20% (Rayner et al., 2008). Despite the dominance of attention to text, Wedel and Pieters (2000) find that only gaze to images, not text, promotes brand memory and recall.

Warning labels are a type of text-based messaging that is of increasing interest to policymakers and marketers alike. However, warnings on alcohol and tobacco products receive few, if any, fixations (Fischer et al., 1989; Thomsen & Fulton, 2007). Novel warnings modestly improved attention to labels (Krugman et al., 1994), and adding graphics further increased attention (Peterson et al., 2010). Text is not limited to product names and slogans; another type of text relevant to consumer behavior is attention to price, which operates very differently than reading text (Laurent & Vanhuele, 2023). Consumers spent a surprising amount of viewing time on the cents of a price, which influences how the price is encoded in memory and how price comparisons are made (Laurent & Vanhuele, 2023).

1.4 Dynamic Scenes

The majority of eye tracking studies, in neuroeconomics and elsewhere, investigate response to static images. This is partially a practical issue, as such stimuli are vastly easier to both design and analyze. However, many of the influence attempts we are exposed to feature moving images, such as TV, streaming, and social media ads. It is possible to create a dynamic saliency map of moving scenes that successfully predicts gaze location (Le Meur et al., 2007) with motion being an additional bottom-up feature driving saliency and gaze (Rumpf et al., 2020). While aligning gaze data with dynamic scenes can be cumbersome, this issue does not arise with other types of process data. For example, electroencephalogram (EEG) is a brain-based process tracing method that provides signals related to attention and stimulus valuation. A number of studies have demonstrated that EEG signals during movie or advertising clip viewing improve predictive validity of choice beyond behavioral measures, even out-of-sample (Dmochowski et al., 2014; Telpaz et al., 2015).

1.5 Heterogeneity

Differences between individuals have been a focus of study in visual marketing. Rosbergen et al. (1997) use a latent-class regression model of eye gaze data during naturalistic magazine viewing with ads of different types (e.g., pictorial, text) to identify three distinct consumer segments that were unobservable with demographic data alone. These segments differed in key outcome metrics such as brand attitude and recall. Other research paired mobile eye tracking with a real shopping environment and found that individuals differ significantly in their attention breadth—the size of the visual field attended to during shopping (Streicher et al., 2021). A shopper's attention breadth influenced their mindset, with wider breadths associated with more exploratory mindsets and more unplanned purchases.

2 Information Search

How consumers search for information relevant to their goals influences the choices they ultimately make. For example, first-sampled attributes and options are more likely to be chosen, even if they are of lower overall value (Russo et al., 1998; Russo et al., 2006; Sullivan & Huettel, 2021). Thus, studying this search process is central to our understanding of consumer behavior. Because search takes time and energy (i.e., is “costly”), consumers aim to find the optimal choice within the limits of their cognitive resources.

2.1 Gaze Bias

Studies have found a consistent pattern of gaze bias in choice contexts such as those illustrated in Fig. 1a. Specifically, the center of an ad or planogram receives the most attention (Atalay et al., 2012). This bias occurred at the outset of viewing, disappeared, and then re-emerged in the final seconds before choice is made. As illustrated throughout this chapter, attention predicted choice; central options were more likely to be chosen. Importantly, choice and central location independently predicted gaze location, with no interaction effect between them. However, other work has indicated that both the central *and* top locations receive most gaze (Chandon et al., 2009). Using mobile eye tracking in a store, gaze was found to be to the right and below eye level (M. Chen et al., 2021). Further work is needed to adjudicate between these accounts of gaze bias, which is likely to depend on context or consumer goals. For example, consumers frequently enter a retail setting with a target and an attention template to guide their search. An attention template is a mental representation of the search target that is stored in visual working memory (Olivers et al., 2011). A consumer may enter a grocery store with the goal of purchasing a specific product and having its packaging in mind. Gaze bias during search with templates—including negative templates for undesired objects (Berggren & Eimer, 2021)—may differ significantly and is currently understudied (Berggren & Eimer, 2018; Bravo & Farid, 2009).

2.2 Multi-attribute Search

When consumers have multiple attributes available during search, they may employ integrative search such as the weighted additive strategy, in which participants look at each attribute and then weigh and sum the attributes for each option (Payne et al., 1988). The Payne index (Payne, 1976) uses gaze data from multi-option, multi-attribute displays such as those in Fig. 1b to calculate the extent to which a participant gazes within an option, integrates all information for that option, then

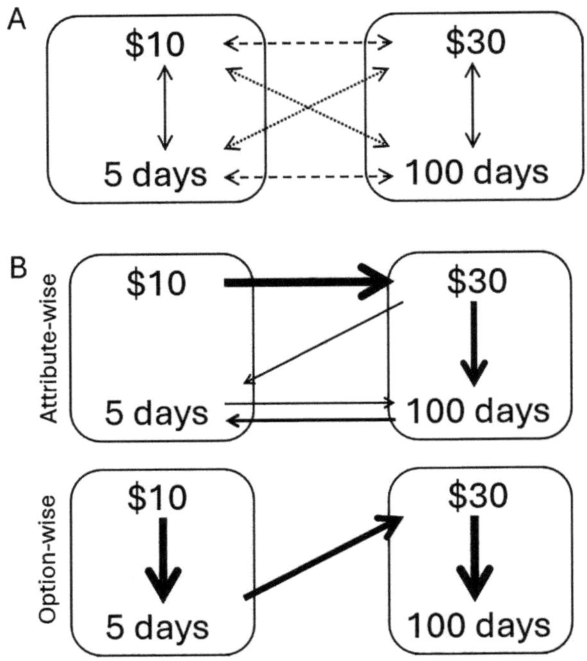

Fig. 2 (**a**) Payne Index. The Payne Index uses process data to quantify how much a participant engages in one of two common search patterns by calculating the number of transitions the eye makes between options and attributes. The two categories of search patterns used by researchers are attribute-wise (dashed lines) and option-wise (solid lines) transitions. A participant engaging in more attribute-wise search will make more transitions between attributes (for example, gazing back and forth between the monetary amounts, then between the days delay). Someone with a more option-wise search pattern will focus on one option first, inspecting both monetary amounts and days delay, before viewing information about the other option. Cross transitions between one option's attribute **a** and another option's attribute **b** (dotted lines) are not included in the Payne Index calculation. (**b**) Common transition patterns. Reeck et al. (2017) found two primary groups of participants, those who exhibited a primarily attribute-wise (top pattern) or option-wise pattern (bottom pattern). Here, more common transitions are visualized as wider arrows. This task used Mouse Lab Web (Willemsen & Johnson, 2011) rather than gaze data to track participants' transitions. Figure adapted from Reeck et al. (2017)

compares it to the next option, or makes comparisons between attributes, examining one attribute at a time across all options. The calculation (Eq. 3) contrasts the amount of within-option ("option-wise"; O; Fig. 2 solid lines) transitions against the number of across-attribute ("attribute-wise"; A; Fig. 2a dashed lines) transitions, relative to the total number of these transitions overall. Of note, this calculation excludes attribute-by-option transitions (Fig. 2a dotted lines).

$$PI = \frac{O - A}{O + A} \tag{3}$$

Figure 2a presents a choice between receiving a smaller sooner amount of money ($10 in 5 days) and waiting longer for a larger amount of money ($30 in 100 days). Some researchers argue that the Payne Index estimates whether an individual is engaging in a heuristic (e.g., "pick the higher amount" or "pick the shorter delay") or a more deliberative decision process, in which all attributes within an option are integrated into a single subjective value (SV) and then the option with the higher SV is selected. Reeck et al. (2017) tracked participants' transition patterns and applied a k-means clustering approach to sort participants into two groups based on their transition patterns. Two key patterns emerged (Fig. 2b) that remarkably resemble either attribute- or option-wise search patterns. Importantly, this emerged organically from their data without any model imposed on the transition patterns. When designing a study to use the Payne Index, it is important to note that the presentation format can bias participants toward one strategy, with more attributes inducing more attribute-wise search and vice versa (Bettman & Kakkar, 1977).

The Payne Index has been found to track participants' ability to delay gratification. Reeck et al. (2017) find that those who engage in more attribute-wise search exhibited more intertemporal patience than those who engage in more option-wise search. Amasino et al. (2019) confirm this, finding that attribute-wise searchers tend to use a heuristic that favors the larger amount, disregarding the delay. Other studies have found that adolescents engage in more option-wise strategies that shift to attribute-wise as they get older (Kwak et al., 2015). Importantly, it is not assumed that individuals stick to only one strategy throughout a task. Shi et al. (2013) use choice matrix data to develop a hierarchical hidden Markov model with three interconnected layers, the last of which captures strategy switching and is found to fit the data better than models that do not allow for strategy switching.

Conjoint Experiments (Green & Rao, 1971) are often used in the field of consumer behavior to understand how consumers make trade-offs between different attributes of a choice. In these experiments, options are presented in an option-by-attribute matrix such as in Fig. 1b, and both attributes and choice are fed into a logit model to estimate the part-worth utilities (relative importance) of each attribute or option, or even at each level of each attribute. Simulations are used to estimate market shares for the products tested. Eye tracking work has shown that, over time during a choice, individuals shift their attention toward more favored options and more important attributes, with these effects amplified after repeated choices (Meißner et al., 2016). Further research indicates participants neglect about half of attributes presented, and incorporating eye gaze data improves a latent class model, enabling better consumer segmentation (Yegoryan et al., 2020).

2.3 *Search Constraints*

In simple contexts such as those described above, individuals can easily view and consider all attributes. However, as mentioned earlier, most consumers do not have unlimited time and energy to make a choice. Time pressure, for example, induces different search patterns such as increased attribute-wise search (Pieters & Warlop, 1999). Additionally, in many search environments—especially online—consumers face an overwhelming array of options, making it necessary to use a stopping rule to decide whether to continue searching or select from already-seen options (Scholten et al., 2014). Eye tracking studies have shown that consumers stop searching if they find an option equal to or greater in value than their expectations or if they exceed the number of options they had planned to sample (Honka & Chintagunta, 2017; Reutskaja et al., 2011). Consumers may also prioritize viewing certain options or attributes over others (Joo, 2023). Such models are known as rational inattention models, in which consumers trade-off between the cost of acquiring additional information and the presumed benefits of that information (Bhui & Jiao, 2023; Sims, 2003).

Importantly, all of the above discuss external search, which involves acquiring information through external means such as eye gaze. However, a growing body of literature in consumer behavior highlights the importance of also understanding internal search, which involves sampling from an individual's memory for information about the choice (Johnson et al., 2007; Weber et al., 2007).

3 Valuation and Attention

The gaze cascade effect (Shimojo et al., 2003) proposed a bidirectional relationship between attention and preference, in which individuals gradually increase gaze frequency or duration to the option they will ultimately choose. According to this theory, greater underlying preference for an option drives more fixations to it (even nearly at the outset of viewing; Glaholt & Reingold, 2009), and, in turn, more fixations to an option also increase preference for it. Indeed, causally directing attention to options increases the likelihood of choosing them (F. Chen et al., 2024; Fisher, 2021), although this does not necessarily indicate a change in preference. This influence of attention on choice has been found to exist as early as the first quarter of a choice (Martinovici et al., 2023).

The influence of attention on choice is echoed by a computational model of increased popularity in psychology and neuroscience, the attentional drift diffusion model (aDDM; Krajbich et al., 2010; Krajbich et al., 2012). This builds on a traditional DDM, which assumes that individuals accumulate evidence in favor of each option until they have enough evidence for one and make the choice for that option. The aDDM (Eq. 4) posits that the value of the currently attended option (A) has full influence on weighted (δ) value accumulation (V), while unattended option(s) (U) are discounted by a certain amount Θ (that is, they do not have full influence on the accumulation of evidence). The aDDM does not assume whether attention drives preference, or vice versa, or both. This remains an open line of inquiry (Mormann & Russo, 2021).

$$V_t = V_{t-1} + \delta (A - \theta U) \quad (4)$$

The aDDM has been used in a number of ways to understand consumer attention. One well-known effect in consumer behavior is the positive influence of celebrity endorsements on brand value (Erdogan, 1999; Spry et al., 2011). However, until recently, it was unclear precisely how this effect worked. Using eye tracking, pupillometry, and aDDM, D'Ambrogio et al. (2023) found that ads with celebrities induced smaller pupil dilation, indicating less decision difficulty (Laeng et al., 2012). Moreover, the aDDM showed that participants were able to retain the unattended item's value in memory to a greater extent under the celebrity endorsement condition. The implications are significant; attention research clearly demonstrates that unattended options are at a disadvantage in choice. However, this disadvantage is substantially reduced for products with celebrity endorsements.

A valuable application of eye tracking data in consumer behavior research is to adjudicate between competing mechanisms underlying a behavioral effect. In consumer behavior literature, it is widely acknowledged that designating an option as the default increases its likelihood of being chosen (Dinner et al., 2011). Again, however, the precise mechanism behind this effect was unclear. Sullivan et al. (2024) addressed this gap by pairing the DDM and aDDM with a series of eye tracking and online studies to determine how defaults influence choice. They found that defaults affect choice primarily through two independent mechanisms: by increasing attention to default options and by enhancing their value during option comparison. This research provides a mechanistic explanation for a common decision-making nudge and quantifies its limitations.

Another example involves the influence of branding on choice. While it was previously known that branding affects choice—possibly functioning as an additional attribute relevant to the option—it was unclear whether brand information influences option evaluation directly or indirectly by altering the decision-making process. For example, branding might increase or decrease the impact of other attributes (e.g., price or quality). Using the aDDM, Yang et al. (2024) found that brands influence choice through *both* pathways. Their eye tracking data also revealed a unidirectional relationship: option value influenced attention to the brand, but not vice versa. This finding suggests that brands should focus on how product quality influences perceptions of their brand, rather than the reverse.

4 The Future of Attention in Consumer Research

It is convenient to assess attention using gaze, as eye tracking is a validated and well-established technique. However, our attention is increasingly being captured through auditory rather than visual channels due to the proliferation of audio streaming. This shift is also reflected in consumer choice, with shopping through voice-activated devices like Alexa and Siri becoming more common. Therefore, there is a need for work in consumer behavior on how these audio formats influence attention and choice. Other techniques, such as EEG, could be used to measure this type of attention. The field of auditory research has a rich literature that has shown increasing interest in auditory attention (Fritz et al., 2007), by developing models that could be adapted to study consumer attention (Kaya & Elhilali, 2017). It is worth noting that auditory attention, like visual attention, is influenced by both top-down and bottom-up factors (Driver, 2001; Kaya & Elhilali, 2014).

There are also several new methodological frontiers for attention research. One promising development is webcam eye tracking, which uses the participant's home webcam to triangulate gaze location on their computer screen. This has the promise to facilitate large sample sizes at more affordable prices and from more diverse participant pools worldwide. Although its current use is limited by low spatial accuracy and temporal resolution (Jenke & Sullivan, 2024), future advancements may enable more precise tracking, thus expanding its applicability in research. Another promising technology is the integration of virtual reality devices with eye tracking (Meißner et al., 2019). Virtual reality environments can provide a high degree of experimenter control while creating an illusion of reality for participants.

5 Conclusion

Four decades of research on consumer attention have shed light on how marketing actions influence consumer attitudes and choice and have revealed the mechanisms behind classic consumer effects. Heterogeneity among individuals emerges as a clear benefit of such methods, as idiosyncratic search leads to idiosyncratic choice. The use of eye tracking will continue to evolve, given the complexity of attention's role in consumer behavior and as the landscape of consumer choice, such as the increase in audio-based consumption, continues to advance. Integrating newer technologies like webcam-based eye tracking and virtual reality will offer deeper insights. Together, these insights underscore the critical importance of understanding attention in order to predict, influence, and optimize consumer decision-making.

References

Amasino, D. R., Sullivan, N. J., Kranton, R. E., & Huettel, S. A. (2019). Amount and time exert independent influences on intertemporal choice. *Nature Human Behaviour, 3*(4), 383–392.

Atalay, A. S., Bodur, H. O., & Rasolofoarison, D. (2012). Shining in the center: Central gaze cascade effect on product choice. *Journal of Consumer Research, 39*(4), 848–866.

Beard, E., Henninger, N. M., & Venkatraman, V. (2024). Making ads stick: Role of metaphors in improving advertising memory. *Journal of Advertising, 53*(1), 86–103. https://doi.org/10.1080/00913367.2022.2089302

Berggren, N., & Eimer, M. (2018). Object-based target templates guide attention during visual search. *Journal of Experimental Psychology: Human Perception and Performance, 44*(9), 1368.

Berggren, N., & Eimer, M. (2021). The guidance of attention by templates for rejection during visual search. *Attention, Perception, & Psychophysics, 83*, 38–57.

Berlyne, D. E. (1958). The influence of complexity and novelty in visual figures on orienting responses. *Journal of Experimental Psychology, 55*(3), 289.

Bettman, J. R., & Kakkar, P. (1977). Effects of information presentation format on consumer information acquisition strategies. *Journal of Consumer Research, 3*(4), 233–240.

Bhui, R., & Jiao, P. (2023). Attention constraints and learning in categories. *Management Science, 69*(9), 5394–5404.

Bravo, M. J., & Farid, H. (2009). The specificity of the search template. *Journal of Vision, 9*(1), 34–34.

Chandon, P. (2002). Do we know what we look at?: An eye-tracking study of visual attention and memory for Brands at the Point of purchase.

Chandon, P., Hutchinson, J. W., Bradlow, E. T., & Young, S. H. (2009). Does in-store marketing work? Effects of the number and position of shelf facings on brand attention and evaluation at the point of purchase. *Journal of Marketing, 73*(6), 1–17. https://doi.org/10.1509/jmkg.73.6.1

Chen, M., Burke, R. R., Hui, S. K., & Leykin, A. (2021). Understanding lateral and vertical biases in consumer attention: An in-store ambulatory eye-tracking study. *Journal of Marketing Research, 58*(6), 1120–1141. https://doi.org/10.1177/0022243721998375

Chen, F., Zheng, J., Wang, L., & Krajbich, I. (2024). Attribute latencies causally shape intertemporal decisions. *Nature Communications, 15*(1), 2948. https://doi.org/10.1038/s41467-024-46657-2

D'Ambrogio, S., Werksman, N., Platt, M. L., & Johnson, E. N. (2023). How celebrity status and gaze direction in ads drive visual attention to shape consumer decisions. *Psychology & Marketing, 40*(4), 723–734.

Dinner, I., Johnson, E. J., Goldstein, D. G., & Liu, K. (2011). Partitioning default effects: Why people choose not to choose. *Journal of Experimental Psychology. Applied, 17*(4), 332–341. https://doi.org/10.1037/a0024354

Dmochowski, J. P., Bezdek, M. A., Abelson, B. P., Johnson, J. S., Schumacher, E. H., & Parra, L. C. (2014). Audience preferences are predicted by temporal reliability of neural processing. *Nature Communications, 5*(1), 4567. https://doi.org/10.1038/ncomms5567

Donderi, D. C. (2006). Visual complexity: A review. *Psychological Bulletin, 132*(1), 73.

Driver, J. (2001). A selective review of selective attention research from the past century. *British Journal of Psychology, 92*(1), 53–78.

Erdogan, B. Z. (1999). Celebrity endorsement: A literature review. *Journal of Marketing Management, 15*(4), 291–314.

Fecteau, J. H., & Munoz, D. P. (2006). Salience, relevance, and firing: A priority map for target selection. *Trends in Cognitive Sciences, 10*(8), 382–390.

Fischer, P. M., Richards, J. W., Berman, E. J., & Krugman, D. M. (1989). Recall and eye tracking study of adolescents viewing tobacco advertisements. *JAMA, 261*(1), 84–89.

Fisher, G. (2021). A multiattribute attentional drift diffusion model. *Organizational Behavior and Human Decision Processes, 165*, 167–182. https://doi.org/10.1016/j.obhdp.2021.04.004

Fritz, J. B., Elhilali, M., David, S. V., & Shamma, S. A. (2007). Auditory attention—Focusing the searchlight on sound. *Current Opinion in Neurobiology, 17*(4), 437–455.

Glaholt, M. G., & Reingold, E. M. (2009). The time course of gaze bias in visual decision tasks. *Visual Cognition, 17*(8), 1228–1243. https://doi.org/10.1080/13506280802362962

Green, P. E., & Rao, V. R. (1971). Conjoint measurement-for quantifying judgmental data. *Journal of Marketing Research, 8*(3), 355–363.

Harris, C. (1993). On the reversibility of Markov scanning in free viewing. *Vision Search, 2*, 123–135.

Honka, E., & Chintagunta, P. (2017). Simultaneous or sequential? Search strategies in the US auto insurance industry. *Marketing Science, 36*(1), 21–42.

Huhmann, B. A. (2003). Visual complexity in banner ads: The role of color, photography, and animation. *Visual Communication Quarterly, 10*(3), 10–17.
Itti, L., & Koch, C. (2001). Computational modelling of visual attention. *Nature Reviews Neuroscience, 2*(3), 194–203. https://doi.org/10.1038/35058500
Itti, L., Koch, C., & Niebur, E. (1998). A model of saliency-based visual attention for rapid scene analysis. *IEEE Transactions on Pattern Analysis and Machine Intelligence, 20*(11), 1254–1259. https://doi.org/10.1109/34.730558
Jenke, L., & Sullivan, N. (2024). Attention and political choice: A Foundation for eye Tracking in political science.
Johnson, E. J., Payne, J. W., Schkade, D. A., & Bettman, J. R. (1989). *Monitoring information processing and decisions: The mouselab system*. DTIC Document.
Johnson, E. J., Häubl, G., & Keinan, A. (2007). Aspects of endowment: A query theory of value construction. *Journal of Experimental Psychology: Learning, Memory, and Cognition, 33*(3), 461.
Joo, J. (2023). Rational inattention as an empirical framework for discrete choice and consumer-welfare evaluation. *Journal of Marketing Research, 60*(2), 278–298.
Kaya, E. M., & Elhilali, M. (2014). Investigating bottom-up auditory attention. *Frontiers in Human Neuroscience, 8*, 327.
Kaya, E. M., & Elhilali, M. (2017). Modelling auditory attention. *Philosophical Transactions of the Royal Society B: Biological Sciences, 372*(1714), 20160101.
Krajbich, I., Armel, C., & Rangel, A. (2010). Visual fixations and the computation and comparison of value in simple choice. *Nature Neuroscience, 13*(10), 1292–1298.
Krajbich, I., Lu, D., Camerer, C., & Rangel, A. (2012). The attentional drift-diffusion model extends to simple purchasing decisions. *Frontiers in Psychology, 3*.
Krugman, D. M., Fox, R. J., Fletcher, J. E., & Rojas, T. H. (1994). Do adolescents attend to warnings in cigarette advertising? An eye-tracking approach. *Journal of Advertising Research, 34*(6), 39–53.
Kwak, Y., Payne, J. W., Cohen, A. L., & Huettel, S. A. (2015). The rational adolescent: Strategic information processing during decision making revealed by eye tracking. *Cognitive Development, 36*, 20–30. https://doi.org/10.1016/j.cogdev.2015.08.001
Laeng, B., Sirois, S., & Gredebäck, G. (2012). Pupillometry: A window to the preconscious? *Perspectives on Psychological Science, 7*(1), 18–27.
LaPiere, R. T. (1934). Attitudes vs. actions. *Social forces, 13*(2), 230–237.
Laurent, G., & Vanhuele, M. (2023). How do consumers read and encode a Price? *Journal of Consumer Research, 50*(3), 510–532. https://doi.org/10.1093/jcr/ucad005
Le Meur, O., Le Callet, P., & Barba, D. (2007). Predicting visual fixations on video based on low-level visual features. *Vision Research, 47*(19), 2483–2498. https://doi.org/10.1016/j.visres.2007.06.015
Li, Q., Huang, Z., & Christianson, K. (2016). Visual attention toward tourism photographs with text: An eye-tracking study. *Tourism Management, 54*, 243–258. https://doi.org/10.1016/j.tourman.2015.11.017
Li, Y., Krefeld-Schwalb, A., Wall, D. G., Johnson, E. J., Toubia, O., & Bartels, D. M. (2022). The more you ask, the less you get: When additional questions hurt external validity. *Journal of Marketing Research, 59*(5), 963–982.
Liechty, J., Pieters, R., & Wedel, M. (2003). Global and local covert visual attention: Evidence from a Bayesian hidden Markov model. *Psychometrika, 68*, 519–541.
Lohse, G. L., & Johnson, E. J. (1996). A comparison of two process tracing methods for choice tasks. *Organizational Behavior and Human Decision Processes, 68*(1), 28–43.
Martinovici, A., Pieters, R., & Erdem, T. (2023). Attention trajectories capture utility accumulation and predict brand choice. *Journal of Marketing Research, 60*(4), 625–645. https://doi.org/10.1177/00222437221141052
Meißner, M., Musalem, A., & Huber, J. (2016). Eye tracking reveals processes that enable conjoint choices to become increasingly efficient with practice. *Journal of Marketing Research, 53*(1), 1–17.

Meißner, M., Pfeiffer, J., Pfeiffer, T., & Oppewal, H. (2019). Combining virtual reality and mobile eye tracking to provide a naturalistic experimental environment for shopper research. *Journal of Business Research, 100*, 445–458.

Milosavljevic, M., Navalpakkam, V., Koch, C., & Rangel, A. (2012). Relative visual saliency differences induce sizable bias in consumer choice. *Journal of Consumer Psychology, 22*(1), 67–74. https://doi.org/10.1016/j.jcps.2011.10.002

Mormann, M., & Russo, J. E. (2021). Does attention increase the value of choice alternatives? *Trends in Cognitive Sciences, 25*(4), 305–315.

Noton, D., & Stark, L. (1971). Eye movements and visual perception. *Scientific American, 224*(6), 35–43.

Olivers, C. N., Peters, J., Houtkamp, R., & Roelfsema, P. R. (2011). Different states in visual working memory: When it guides attention and when it does not. *Trends in Cognitive Sciences, 15*(7), 327–334.

Palmer, S. E. (1999). *Vision science: Photons to phenomenology*. MIT press.

Payne, J. W. (1976). Task complexity and contingent processing in decision making: An information search and protocol analysis. *Organizational Behavior and Human Performance, 16*(2), 366–387.

Payne, J. W., Bettman, J. R., & Johnson, E. J. (1988). Adaptive strategy selection in decision making. *Journal of Experimental Psychology: Learning, Memory, and Cognition, 14*(3), 534.

Peracchio, L. A., & Meyers-Levy, J. (2005). Using stylistic properties of ad pictures to communicate with consumers. *Journal of Consumer Research, 32*(1), 29–40.

Peterson, E. B., Thomsen, S., Lindsay, G., & John, K. (2010). Adolescents' attention to traditional and graphic tobacco warning labels: An eye-tracking approach. *Journal of Drug Education, 40*(3), 227–244.

Pieters, R., & Warlop, L. (1999). Visual attention during brand choice: The impact of time pressure and task motivation. *International Journal of Research in Marketing, 16*(1), 1–16.

Pieters, R., Rosbergen, E., & Wedel, M. (1999). Visual attention to repeated print advertising: A test of scanpath theory. *Journal of Marketing Research, 36*(4), 424–438.

Pieters, R., Warlop, L., & Wedel, M. (2002). Breaking through the clutter: Benefits of advertisement originality and familiarity for brand attention and memory. *Management Science, 48*(6), 765–781.

Pieters, R., Wedel, M., & Batra, R. (2010). The stopping power of advertising: Measures and effects of visual complexity. *Journal of Marketing, 74*(5), 48–60.

Pilelienė, L., & Grigaliūnaitė, V. (2016). Influence of print advertising layout complexity on visual attention. *Eurasian Business Review, 6*(2), 237–251. https://doi.org/10.1007/s40821-015-0040-2

Radach, R., Lemmer, S., Vorstius, C., Heller, D., & Radach, K. (2003). Chapter 27 – Eye movements in the processing of print advertisements. In J. Hyönä, R. Radach, & H. Deubel (Eds.), *The mind's eye* (pp. 609–632). North-Holland.

Rayner, K., Rotello, C. M., Stewart, A. J., Keir, J., & Duffy, S. A. (2001). Integrating text and pictorial information: Eye movements when looking at print advertisements. *Journal of Experimental Psychology: Applied, 7*(3), 219.

Rayner, K., Miller, B., & Rotello, C. M. (2008). Eye movements when looking at print advertisements: The goal of the viewer matters. *Applied Cognitive Psychology: The Official Journal of the Society for Applied Research in Memory and Cognition, 22*(5), 697–707.

Reeck, C., Wall, D., & Johnson, E. J. (2017). Search predicts and changes patience in intertemporal choice. *Proceedings of the National Academy of Sciences, 114*(45), 11890–11895. https://doi.org/10.1073/pnas.1707040114

Reutskaja, E., Nagel, R., Camerer, C. F., & Rangel, A. (2011). Search dynamics in consumer choice under time pressure: An eye- tracking study. *The American Economic Review, 101*(2), 900–926. Retrieved from www.jstor.org/stable/29783694

Rosbergen, E., Pieters, R., & Wedel, M. (1997). Visual attention to advertising: A segment-level analysis. *Journal of Consumer Research, 24*(3), 305–314. https://doi.org/10.1086/209512

Rumpf, C., Boronczyk, F., & Breuer, C. (2020). Predicting consumer gaze hits: A simulation model of visual attention to dynamic marketing stimuli. *Journal of Business Research, 111*, 208–217. https://doi.org/10.1016/j.jbusres.2019.03.034

Russo, J. E. (1978). *Eye fixations can save the world: A critical evaluation and a comparison between eye fixations and other information processing methodologies*. ACR North American Advances.

Russo, J. E., Meloy, M. G., & Medvec, V. H. (1998). Predecisional distortion of product information. *Journal of Marketing Research, 35*(4), 438–452. https://doi.org/10.1177/002224379803500403

Russo, J. E., Carlson, K. A., & Meloy, M. G. (2006). Choosing an inferior alternative. *Psychological Science, 17*(10), 899–904. https://doi.org/10.1111/j.1467-9280.2006.01800.x

Scholten, M., Read, D., & Sanborn, A. (2014). Weighing outcomes by time or against time? Evaluation rules in intertemporal choice. *Cognitive Science, 38*(3), 399–438.

Schulte-Mecklenbeck, M., Murphy, R. O., & Hutzler, F. (2011). Flashlight–Recording information acquisition online. Computers in Human Behavior, 27(5), 1771–1782.

Schulte-Mecklenbeck, M., Johnson, J. G., Böckenholt, U., Goldstein, D. G., Russo, J. E., Sullivan, N. J., & Willemsen, M. C. (2017). Process-tracing methods in decision making: On growing up in the 70s. *Current Directions in Psychological Science, 26*(5), 442–450. https://doi.org/10.1177/0963721417708229

Shi, S. W., Wedel, M., & Pieters, F. (2013). Information acquisition during online decision making: A model-based exploration using eye-tracking data. *Management Science, 59*(5), 1009–1026.

Shimojo, S., Simion, C., Shimojo, E., & Scheier, C. (2003). Gaze bias both reflects and influences preference. *Nature Neuroscience, 6*(12), 1317–1322. http://www.nature.com/neuro/journal/v6/n12/suppinfo/nn1150_S1.html

Sims, C. A. (2003). Implications of rational inattention. *Journal of Monetary Economics, 50*(3), 665–690.

Spry, A., Pappu, R., & Bettina Cornwell, T. (2011). Celebrity endorsement, brand credibility and brand equity. *European Journal of Marketing, 45*(6), 882–909.

Streicher, M. C., Estes, Z., & Büttner, O. B. (2021). Exploratory shopping: Attention affects in-store exploration and unplanned purchasing. *Journal of Consumer Research, 48*(1), 51–76.

Sullivan, N. J., & Huettel, S. A. (2021). Healthful choices depend on the latency and rate of information accumulation. *Nature Human Behaviour, 5*(12), 1698–1706. https://doi.org/10.1038/s41562-021-01154-0

Sullivan, N. J., Doré, S., Breslav, A., Bachman, M., & Huettel, S. A. (2024). The Golden halo of defaults in simple choices. *Journal of Marketing Research.*

Telpaz, A., Webb, R., & Levy, D. J. (2015). Using EEG to predict consumers' future choices. *Journal of Marketing Research, 52*(4), 511–529.

Thomsen, S. R., & Fulton, K. (2007). Adolescents' attention to responsibility messages in magazine alcohol advertisements: An eye-tracking approach. *Journal of Adolescent Health, 41*(1), 27–34.

Van Raaij, W. F. (1977). Consumer information processing for different information structures and formats. *Advances in Consumer Research, 4*(1).

Weber, E. U., Johnson, E. J., Milch, K. F., Chang, H., Brodscholl, J. C., & Goldstein, D. G. (2007). Asymmetric discounting in intertemporal choice: A query-theory account. *Psychological Science, 18*(6), 516–523.

Wedel, M., & Pieters, R. (2000). Eye fixations on advertisements and memory for brands: A model and findings. *Marketing Science, 19*(4), 297–312.

White, B. J., Berg, D. J., Kan, J. Y., Marino, R. A., Itti, L., & Munoz, D. P. (2017). Superior colliculus neurons encode a visual saliency map during free viewing of natural dynamic video. *Nature Communications, 8*(1), 14263. https://doi.org/10.1038/ncomms14263

Willemsen, M. C., & Johnson, E. J. (2011). Visiting the decision factory: Observing cognition with MouselabWEB and other information acquisition methods. *A handbook of process tracing methods for decision research*, 21–42.

Yang, X., Retzler, C., Krajbich, I., Ratcliff, R., & Philiastides, M. G. (2024). Attention to brand labels affects, and is affected by, evaluations of product attractiveness. *Frontiers in Behavioral Economics, 2*. https://doi.org/10.3389/frbhe.2023.1274815

Yegoryan, N., Guhl, D., & Klapper, D. (2020). Inferring attribute non-attendance using eye tracking in choice-based conjoint analysis. *Journal of Business Research, 111*, 290–304. https://doi.org/10.1016/j.jbusres.2019.01.061

Policy Consequences of the New Neuroeconomic Framework

A. David Redish, Henri Scott Chastain, Carlisle Ford Runge, Brian M. Sweis, Scott E. Allen, and Antara Haldar

Abstract Current theories of decision-making suggest that the neural circuits in mammalian brains (including humans) computationally combine representations of the past (memory), present (perception), and future (agentic goals) to take actions that achieve the needs of the agent. How information is represented within those neural circuits changes what computations are available to that system, which changes how agents interact with their world to take those actions. We argue that the computational neuroscience of decision-making provides a new microeconomic framework (neuroeconomics) that offers new opportunities to construct policies that interact with those decision-making systems to improve outcomes. After laying out the computational processes underlying decision-making in mammalian brains, we present four applications of this logic with policy consequences: (1) precommitment to avoid falling into the trap of sunk costs, (2) media consequences for changes in housing prices after a disaster, (3) contingency management as a treatment for addiction, and (4) how social interactions underlie the success (and failure) of microfinance institutions.

A. D. Redish (✉)
Department of Neuroscience, University of Minnesota, Minneapolis, MN, USA
e-mail: redish@umn.edu

H. S. Chastain
Minnesota Center for Philosophy of Science, University of Minnesota, Minneapolis, MN, USA

C. F. Runge
Department of Applied Economics and Law School, University of Minnesota, St. Paul, MN, USA

B. M. Sweis
Department of Psychiatry, Department of Neuroscience, Icahn School of Medicine at Mount Sinai, New York, NY, USA

S. E. Allen
Department of Physics, Cornell University, Ithaca, NY, USA

A. Haldar
Faculty of Law, Judge Business School, University of Cambridge, Cambridge, UK

D. V. Smith et al. (eds.), *Neuroeconomics: Core Topics and Current Directions*,
https://doi.org/10.1007/978-3-032-02925-6_36

1 Introduction: Representation Matters

The promise of neuroeconomics is that neural processes will be key to understanding economic behavior and thus will have policy implications (Camerer et al., 2005; Glimcher et al., 2008; Rangel et al., 2008; Redish, 2022). Current neuroeconomic decision-making theories derive from computational neuroscience, which posits that neural circuits implement information processing (Marr, 1982; Churchland & Sejnowski, 1994; Dayan & Abbott, 2001). Information about the past (memory) is combined with information about the present (perception) and teleological information (agentic goals) to take actions that achieve the needs of the agent (Rangel et al., 2008; Redish, 2013; Mitchell, 2023).

One of the most important discoveries in computational science over the last century is that *how one represents information changes what one can do with it* (Cormen et al., 1990). A library that organizes fiction books by author but nonfiction books by subject makes it easy to find all of the nonfiction books about chaos theory quickly, but hard to find all of the nonfiction books by James Gleick. Similarly, it is easy to find all of the novels by Amy Tan quickly, but hard to find all of the fiction books about mother-daughter relationships. Because people tend to research nonfiction topics but authors have a fan base, this sorting structure aligned to different scenarios is more efficient than a single generic sorting would be. Analogously, animals (including humans) have multiple decision systems which represent information differently, to be better aligned to different situations they might find themselves in (Dickinson & Balleine, 1994; Niv et al., 2006; Gläscher et al., 2010; Kahneman, 2011; Redish, 2013).

All normative optimizations have process assumptions underneath them—the optimal actions to take are not independent of the processes that select them. The classical "rational model" of economics assumes that an agent follows a single-valued function over all possible outcomes, subject to an arbitrary budget constraint (Samuelson, 1938; Debreu, 1959; Mas-Colell et al., 1995). Implicit in this theory is the assumption that agents possess infinite resources of knowledge and abilities to find a global optimum. As pointed out by Simon (1972), sometimes it is better to take a pretty good action fast than to take the time to calculate the absolute best choice (say, if you are a zebra deciding which way to escape a lion bearing down on you). In this "bounded rationality" formulation, calculation time provides an additional cost to that utility function. Other models that have been proposed have memory or storage capacity limitations (Kahneman & Tversky, 1979, 2000; Shleifer, 2000; Akerlof, 2005; Gigerenzer & Brighton, 2009), but, taken to its full extent, this insight implies that representation matters—how data are stored and how information is transformed (through neural circuits) changes what the optimal actions are (Lieder & Griffiths, 2019). In this chapter, we argue that the neuroscience of decision-making provides a new framework within which to build economic models and theories. (See Levenstein et al. (2023) for a discussion of models, theories, and frameworks.)

Current theories of decision-making suggest that mammalian brains (including human ones) consist of multiple neural circuits that represent and process information about the past, present, and future in different ways that are optimized for different world conditions. The argument put forward in this chapter is that the computational neuroscience of decision-making provides a new microeconomic framework that offers (1) better descriptions of actual behavior, (2) new opportunities for policy that interact with those decision-making systems to improve outcomes, (3) better descriptions of the causes underlying the structures of group interactions, and (4) new understandings that can inform institutional policies because of their dependence on neural computations. In the following sections, we review the current neuropsychological theory of decision-making and then examine examples of these four consequences.

2 A Microeconomic Framework Derived from Neuroscience

Agents are evolved to take actions in an uncertain but probabilistically predictable world (Mitchell, 2023). As such, information (Shannon, 1948) about the state of the world can be used to drive behavior; neural firing patterns that differ as a function of different conditions in the world allow effectors within the agent (such as muscles) to act (Rieke et al., 1997). For example, the Mauthner cell in fish detects stimuli that increase in size on the retina (looming stimuli), and the fish swims away from that stimulus (likely a diving predator!) very fast (Diamond, 1971; Preuss et al., 2006). Importantly, the muscles in the fish are not responding to the predator, but to the firing of the Mauthner cell, which shares *mutual information* with the presence of a predator.

These neural circuits that transform information are computational processes with normative consequences. *Memory* in neural circuits, for example, arises through *content addressable* processes, such that successful retrieval is reactivation of a stored neural pattern (Hebb, 1949; Hertz et al., 1991). Small changes can arise with each retrieval, which can be mediated by the cues that produced that recollection (Kahana, 2020; Runge et al., 2023). These effects make memory unreliable (Loftus & Palmer, 1974; Wells & Loftus, 1984), but can also provide for behavioral flexibility (Mau et al., 2020; Sweis et al., 2021). *Perception* in neural circuits is fundamentally belief in a description of the world—we see trees and leaves, not green and brown pixels (Gibson, 1977; Rao & Ballard, 1999; Friston, 2005; Adams et al., 2013; Sterzer et al., 2018). Perception entails a process of categorization and generalization through parallel distributed processing that depends on experience and expertise (Grossberg, 1976; Hertz et al., 1991; McClelland & Rogers, 2003; Pettine et al., 2023). *Motivation* and *evaluation* are computational memory processes of their own (Balleine & Dickinson, 1991; Andermann & Lowell, 2017; Sharpe et al., 2021). For example, if one is ill after eating food, one learns to dislike that food, which is why you should eat weird-flavored foods before getting chemotherapy (Bernstein & Webster 1980). Finally, the *actions* themselves depend on neural representations of the physics of the body ("forward models" (Kawato, 1999)) and oscillatory processes that can step through those action sequences ("central pattern generators" (Grillner & Wallén, 1985)).

Computationally, neural circuits affect the solutions available to the system. For example, cortical circuits implement *divisive normalization* that ensures maximal efficiency within the range of actual stimuli being experienced (Carandini & Heeger, 2011). This is why we see white paper as white and black as black in both bright sunlight and a dim fluorescent-lit room, even though black paper in sunlight is actually thousands of times brighter than white paper in the fluorescent-lit room. Divisive normalization has been proposed to underlie range normalization in both normal economic behavior and after experience with extremes, such as after taking heroin (Webb et al., 2021; Gueguen et al., 2024). Similarly, because memory *pattern completion* processes depend on synaptic connections in distributed neural circuits that participate in multiple memories (Hebb, 1949; Hopfield, 1982; Hertz et al., 1991; Kahana, 2020), storing new memories can interfere with older memories unless those older memories are retrieved and restrengthened (Rumelhart et al., 1986; McClelland et al., 1995; Kahana, 2020). These retrieval/restrengthening processes can explain the effect of media on changes in housing prices seen after a disaster (Runge et al., 2023).

An important *open question* is the relationship between these computational processes and utility-based theories of value, given the complexities of the computation underlying the neural processes. In neuroscience, the single-valued utility function theory (Samuelson, 1938; Von Neumann & Morgenstern, 1944; Debreu, 1959; Mas-Colell et al., 1995) manifests as the hypothesis of *common currency*—that there exists a unitary neural representation of value without maintaining representations of the situation, goods obtained, or the information process leading to the decision. Although some data support a central value representation (Padoa-Schioppa & Assad, 2006, 2008; Kable & Glimcher, 2009; Levy & Glimcher, 2012), other data suggest incompatibilities with such a single-valued utility function (Berridge & Robinson, 2003; Rangel et al., 2008; Redish et al., 2008; Kurth-Nelson & Redish, 2010). In particular, how value is measured (willingness to pay, willingness to sell, as a choice between options) reliably produces preference orderings incompatible with a singular scale (Kahneman & Tversky, 2000; Lichtenstein & Slovic, 2006; Ahmed, 2010).

Current theories of action-selection identify four fundamentally different information processes to take actions (Redish, 2013). Although these go by different names in different neuropsychological literatures, most theories divide action-selection into these fundamental processes (Fig. 1).

Reflexes are simple stimulus-response pairs in which both the stimulus-detection and responses are genetically wired into the agent. Neurophysiologically, reflexes are primarily functions of spinal cord (Sherrington, 1906), in part because they evolved to respond quickly. Typical reflexes include things like throwing out your hands to catch yourself when you fall. Reflexes are usually ignored in economic models, but they can have important consequences on agentic control.

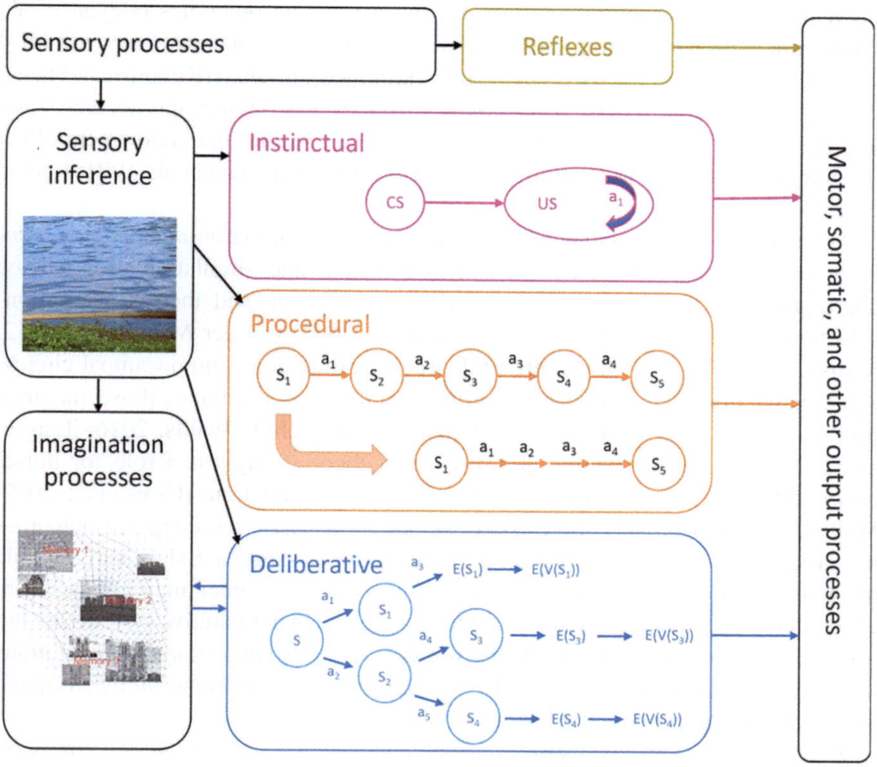

Fig. 1 The four decision-making systems. Each system uses different representations and different learning processes applied to different information representations to select goals. The *instinctual* system associates conditioned stimuli (CS) with unconditioned stimuli (US) that releases an action *a*. The *procedural* system transforms a series of situation-action pairs into an action chain. The *deliberative* system works with memory and imagination to search through outcomes. These depend on *sensory interference* processes that see water and grass, not green and blue. *Imagination* works through content-addressable processes. These all have consequences both for the development of preferences and for the expression of those preferences

Instinctual systems (sometimes called *Pavlovian actions*, or *stimulus-stimulus associations*) entail a limited repertoire of actions that one can learn when to release (Pavlov, 1927; Breland & Breland, 1961; LeDoux & Daw, 2018). In classical psychological theories, these systems transfer responses from an *unconditioned stimulus* to a *conditioned stimulus* (Domjan, 1998). They involve subcortical neural circuits of amygdala, hypothalamus, nucleus accumbens shell, and periaqueductal gray (McNally et al., 2011; LeDoux & Daw, 2018). They classically include fight or flight responses, but they also can be more complicated, such as mating dances. In humans, they include many prosocial behaviors (Redish, 2022), such as threatening and submissive postures, laughing with friends, a preference for fairness, and parochial altruism, in which one is cooperative (altruistic) to one's kith and kin and wary (xenophobic) towards strangers. Economically, instinctual systems are often modeled as providing additional utility to simple outcomes (Dayan et al., 2006), which have been proposed to underlie sign tracking (whereby the conditioned stimulus takes on characteristics of the unconditioned stimulus) (Robinson & Flagel, 2009; Colaizzi et al., 2020), a sensitivity to sunk costs (taking into account costs already spent) (Sweis et al., 2018a; Lind et al., 2023), and the endowment effect (excess value assigned to retaining what one owns) (Kahneman et al., 1991; Redish et al., 2016).

Deliberative systems (sometimes called *model-based*, *action-outcome*, or *planning systems*) entail a search through imagination and simulation of predicted consequences of one's actions, followed by evaluation, and then active action-selection (Niv et al., 2006; Buckner & Carroll, 2007; van der Meer et al., 2012; Redish, 2016). They involve prefrontal circuits setting goals, hippocampal circuits creating that imagined outcome, and orbitofrontal circuits evaluating those imagined possibilities (Stott & Redish, 2014; Redish 2016; Rich & Wallis, 2016; Tang et al., 2021; Knudsen & Wallis, 2022). Current theories suggest a role for dorsal prefrontal circuits in selecting actions based on those evaluations (Cisek et al., 2009; Wallis & Kennerley, 2011; Balewski et al., 2023). Because the imagined outcomes are processed serially (one at a time), deliberative systems are slow and variable in their execution (which options are examined in what order may change from encounter to encounter). It is important to recognize that deliberative systems are not economically rational in the classical sense as they depend on memory computations to determine what is searched and also on evaluation computations, which, as noted above, depend on experiential histories.

Procedural systems (sometimes called *model-free*, *situation-action*, or *habit systems*) entail a process of situation-recognition followed by release of a well-practiced action-chain (Rand et al., 1998; Jog et al., 1999; Dezfouli & Balleine, 2012; Smith & Graybiel, 2013; Mugan et al., 2024). The situation-recognition component depends on cortical sensory circuits, and the selection of the action-chain depends on dorsal striatal circuits. Some theories posit that procedural systems are economic in that they learn the value of taking an action in a situation and release those actions proportionally to that valuation (Samejima et al., 2005; Hikosaka et al., 2014), but these valuations are learned from past experiences, not derived from future expectations (Niv et al., 2006). Because of this, procedural systems are not sensitive to sudden changes in valuation and can drive perseverative motoric responses in over-learned conditions (Dickinson & Balleine, 1994).

An important *open question* is how actions are selected when these decision systems are in conflict. While the actual computations that underlie this deconfliction are controversial (Daw et al., 2005; Kable & Glimcher, 2009; McLaughlin et al., 2021), taking actions when the systems agree on the action to choose are easier and faster than when they are in conflict (Stroop, 1935; Guitart-Masip et al., 2012; Redish, 2016). It is also known that these systems interact in real behaviors. For example, deliberative tree-search is affected by Pavlovian evaluations that change under depression (Huys et al., 2012), and deliberative systems can train procedural systems over time (Klein, 1999). Moreover, while many tasks can be accomplished by multiple systems, changes in environmental conditions can encourage the use of one system over another. This important observation provides an access point to how institutional policies can be used to affect decision-making.

3 Examples Where the Computational Process Matters

In the second half of this chapter, we explore descriptive and prescriptive consequences of this new neuroeconomic framework. First, we address the sensitivity to sunk costs in both human and nonhuman animals, showing that a computational explanation better explains the sunk cost fallacy and precommitment to avoid it than a sociological explanation does. Second, we show how the computational processes of memory can explain housing prices in response to floods and other disasters and discuss how reminders can affect that memory process. While the observation that reminders improve memory is not novel, the computational process of memory explains how these reminders work. Third, we address how different scenarios impact preference ordering and suggest a novel explanation for contingency management as an addiction treatment, with novel policy consequences for improving contingency management. Finally, we discuss how computations underlying human prosociality affect the institution of microfinance, again with novel policy consequences. We have chosen these four examples to show the breadth and depth of this new neuroeconomic framework, from individual explanations for microeconomic observations (sunk costs) to policy systems that affect individuals (computational processes of memory, contingency management) to social policy and institutions (microfinance).

3.1 Sunk Costs and the Endowment Effect

Sunk costs are already spent and cannot be recovered. Because all futures include these same sunk costs, the optimal choice should be independent of those sunk costs. However, human choices often take sunk costs into account (Arkes & Blumer, 1985; McAfee et al., 2010; Sleesman et al., 2012). While many theories of the human sensitivity to sunk costs suggest a social cause (Staw, 1976; Arkes & Blumer, 1985; Kanodia et al., 1989; Staw & Ross, 1989), recent work has found that nonhuman animals are also sensitive to sunk costs (Wikenheiser et al., 2013; Sweis et al., 2018a; Redish et al., 2022). In temporal serial foraging tasks, subjects choose to wait out a delay to a goal or to leave and proceed to the next opportunity. This willingness to quit depends on the time already spent waiting (the sunk costs) (Sweis et al., 2018a). In humans, that sensitivity to sunk costs depends on attention paid to the counting down delay (Kazinka et al., 2021). In mice, that sensitivity depends on the neural circuits involved in the Instinctual/Pavlovian action selection system, particularly the nucleus accumbens shell (Sweis et al., 2018b). Consistent with this hypothesis, in a different experiment, stimulation of the amygdala inputs to

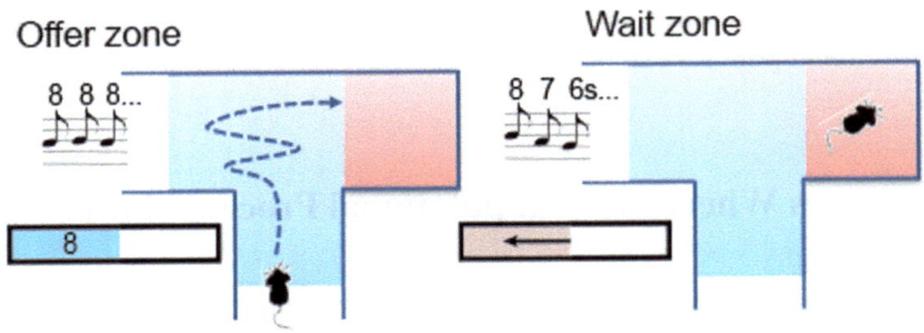

Fig. 2 In the offer zone, subjects are informed of the delay, but the delay does not count down; in the wait zone, the delay counts down. In the offer zone, subjects dither and deliberate about whether to enter the wait zone or not. Subjects show a sensitivity to sunk costs in the wait zone, but not the offer zone. Rodents hear a tone to inform them of the delay; humans get a download waitbar to inform them of the delay. All subjects learn to precommit in the offer zone to avoid the sunk cost fallacy in the wait zone. (See Sweis et al., 2018a, c; Kazinka et al., 2021; Huynh et al., 2021; Redish et al., 2022)

the nucleus accumbens shell increased sunk cost sensitivity in a conditioned place preference task (Lind et al., 2023).

Conditioned place preference is a behavioral phenomenon in which subjects prefer places where good things have happened to them over other places (Bardo et al., 1995; Tzschentke, 1998; Mueller & de Wit, 2011). Economically, conditioned place preference can be modeled as an increase in value of that place (Dayan et al., 2006), and may be analogous to the *endowment effect* seen in humans (Redish et al., 2016), in which people demand more money to sell an object than they are willing to pay for it (Kahneman et al., 1991). This hypothesis suggests that the endowment effect should depend on immediate, concrete experiences, because the instinctual system does not include a separate imagination component, and may be why concrete rewards are harder to forego (Bushong et al., 2010; Mischel, 2014).

In temporal serial foraging tasks with a separate "offer zone," where the delay does not count down, subjects do not include time spent in that offer zone in the sunk cost calculation (Sweis et al., 2018a). Mice, rats, and humans all learn to avoid wait-zone sunk costs, by making a separate decision in the offer zone (Sweis et al., 2018a, c; Kazinka et al., 2021; Huynh et al., 2021; Redish et al., 2022) (Fig. 2). This is a form of *precommitment*, in which one removes the opportunity to make a later decision that one thinks one will make "wrong" (Ainslie, 1992). Precommitment requires multiple valuation functions; it cannot arise if the valuation function is unitary (Kurth-Nelson & Redish, 2010). For example, one may recognize that if one goes to the party, one will drink (perhaps because of Pavlovian associations of alcohol and reward), thus one can deliberately decide not to go to the party. It is important to note that precommitment does not always entail deliberation overriding other systems. For example, learning the habit of driving home a different way to avoid the temptation of the casino is a process in which the procedural system provides the precommitment to avoid temptation (Elster, 1999; Schüll, 2012).

This neuroeconomic model not only provides explanations for the endowment effect and precommitment, but it specifically identifies the mechanisms through which the decision systems interact to produce these effects and suggests policies (precommitment) that can change behavior.

3.2 *Memory Implications*

The computational processes that underlie perception and memory also have economic consequences. Current theories suggest that memory depends on *pattern completion* in which Hebbian learning strengthening synapses that connect co-active neurons produce dynamic changes in neural firing such that a partial (or nearby) pattern will return to an originally stored pattern (Hebb, 1949; Hopfield, 1982; Hertz et al., 1991; Kahana, 2020). This recall process has consequences.

Both neurons and synapses participate in multiple memories, which implies that newly stored memories can *interfere* with older memories, moving the synapses away from the strengths needed to recall those older memories (Hasselmo, 1993; McClelland et al., 1995) (Fig. 3). When combined with a temporal context model that contains slowly changing representations of time (Howard et al., 2005), new

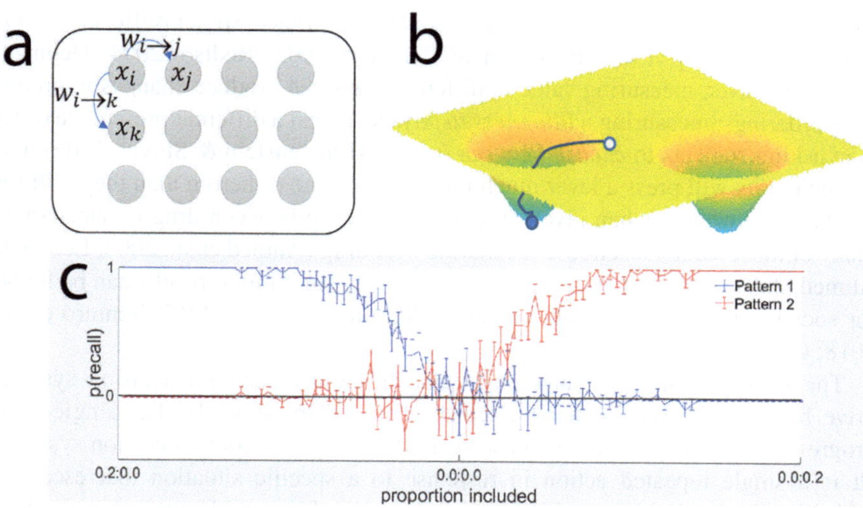

Fig. 3 Memory processes. (**a**) Current theories of human memory are based on attractor networks of parallel distributed processes, with the memories stored through the changing of synaptic weights between units. (**b**) Mathematically, memories can be understood as "valleys" in a memory space. A nearby representational pattern is transformed through those weights to find itself in the valley. This means that (**c**) the likelihood of falling into a memory is dependent on how deep the valley is (proportional to the strength of the weights) and can be reinstated (and strengthened) with reminders, such as from news media. (Modified after Runge et al., 2023)

experiences interfere with the recall of old memories. Runge et al. (2023) built a computational model of single event storage (such as a flood) and learned valuation (remember that valuation is its own memory process) and found that both assessed and remembered cost decayed with a set time course. Remembered cost decayed to a minimum level, while assessed cost eventually decayed down to baseline. Importantly, however, reminders could disrupt this decay, producing long-term memories of the disaster that did not fade away.

This memory model of disaster suggests that media reminders may be critical to this time course. Gallagher (2014) found that flood-induced declines in housing prices depended more on whether a household was in a given media market than on whether the house itself was damaged by the flood. Reminders of disasters (such as recognizing the anniversary of a disaster) keep the memory alive, while forgotten dangers (and a lack of reminders) make a mitigated disaster ignored (such as the Y2K bug, which was only not a disaster because of the expensive work done to prevent it, and has now been mostly forgotten) (Runge et al., 2023). Furthermore, those reminders can change the narrative, and even the memory itself (Loftus & Banaji, 1989; Sacchi et al., 2007).

3.3 Conflicting Decisions (Addiction and Addiction Treatment)

Addiction has been described as an overvaluation of drugs (Stigler & Becker, 1977; Becker & Murphy, 1988; Robinson & Berridge, 2003; Redish, 2004; Heyman, 2009). However, measuring value in different ways can produce changes in preference orderings: measuring *willingness to pay* can reveal a different preference order than asking subjects to *choose between options* (Lichtenstein & Slovic, 2006). For example, rats will press a lever much more for cocaine or heroin than they will for saccharin. However, when given an option to choose between drug or saccharine, many of those same rats choose saccharine over drug (Ahmed et al., 2002; Lenoir & Ahmed, 2007; Lenoir et al., 2007; Cantin et al., 2010). Similar results can be found for social rewards versus drugs of abuse (Martin & Iceberg, 2015; Venniro et al., 2018; Chow et al., 2022).

The neuroeconomic explanation is that different behavioral decision systems drive behavior in each scenario (Redish et al., 2016) (Fig. 4). The single-lever progressive-ratio experiment is most likely driven by procedural decision systems. (It is a single repeated action in response to a specific situation that escalates with experience. It is insensitive to outcome devaluation.) However, procedural decision systems are not well-suited to choosing between separate options. The two-lever choose-between scenario is much more likely driven by deliberative decision systems. Because these are two separate neural systems with separate evaluation processes, they can reveal incompatible preference orderings.

We can use this change in preference ordering to treat addiction.

Contingency management is one of the most successful treatments available for addiction (Higgins et al., 1991; Lussier et al., 2006; Petry, 2011). In *contingency*

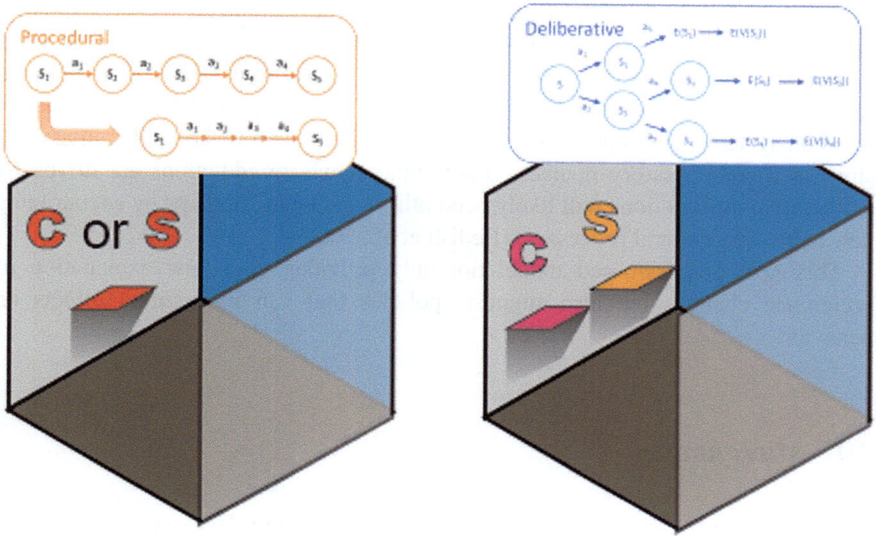

Fig. 4 When animals only have one option, and the experiment measures the number of lever presses the animal will make to receive a reward (cocaine or saccharine), this is measuring the *willingness to pay* and finds that the WTP(C) > WTP(S), allowing us to infer that C > S. In contrast, when animals have two options, and the experiment measures which option is chosen, this is measuring preferences through *choose between* processes and finds that S > C. These differences arise because the willingness-to-pay activates the procedural system, while choose between activates the deliberative system

management, a subject is provided a small (often monetary) reward for not using drugs in a recent time frame (typically a week). The original explanation for contingency management's success was economic—losing the reward was an opportunity cost that made the drug more expensive (Hursh, 1991; Higgins et al., 1991). However, contingency management works better than this economic hypothesis would predict (Regier & Redish, 2015; Davidson et al., 2024). An alternative explanation is that contingency management shifts the decision systems that the subject is using from either instinctual (drawn towards the drug) or procedural (willing to pay) systems to deliberative systems (choose between), much like the rats in the different experiments.

This proposed explanation for contingency management has important implications for implementation—variations that improve deliberative decision-making should improve contingency management. Deliberation depends on imagined expectations of outcome, and concrete outcomes are easier to imagine (Peters & Büchel, 2010), so more concrete outcomes should be more effective. They are (Petry et al., 2015). Deliberation depends on imagining outcomes, so improving one's ability to imagine future outcomes (through *episodic future thinking training*) should reduce drug use and relapse. It does (Snider et al., 2016; Stein et al., 2016). Deliberation depends on prefrontal cortical circuits interacting with hippocampus, accumbens core, and other structures. Improving those connections should reduce drug use and relapse. It does (Camchong et al., 2023).

Importantly, it is not that deliberative decisions are always better, but rather that, for those subjects whose addictions are driven by their procedural or instinctual decision systems, contingency management can help them make deliberative decisions in a compensatory manner. Other subjects whose addictions are driven by deliberative dysfunctions will likely need other treatments, perhaps by encouraging new habits (procedural processes) (Redish et al., 2008).

This new process-based model not only provides us a new explanation of preference changes but also suggests policies that can have direct effects on behavior.

3.4 Microfinance

Humans are social animals and have evolved systems and developed technologies that increase prosociality (Wilson, 2002, 2015; Fehr & Fischbacher, 2003; Bowles & Gintis, 2013; McCullough, 2020; Redish, 2022; Cushman, 2024). To achieve prosociality, people make themselves vulnerable to others, whether through risking loss or foregoing gain. These actions depend on how the decision systems interact with the external social structure (Allen et al., 2024). Instinctual trust will be based on kith and kin, on cooperation within one's community, but will include a wariness of strangers. Instinctual trust is unlikely to keep track of explicit transactions, but likely depends on regular social gatherings to create community. Deliberative trust depends on direct expectations of imagined outcomes and thus is often explicitly transactional. Procedural trust depends on the practiced regularity of behavior (as might occur in a sports team or a surgical team). Institutions that access all three systems will be more likely to foster prosocial behaviors.

As a concrete example, we can look at microfinance, which succeeded in providing loans to communities not normally reached by the formal banking sector (particularly in South Asia). We can compare two specific microfinance institutions: Grameen Bank and SKS India. Both institutions provided similar services to similar communities with similar historical and ethnic backgrounds, structured in similar ways (Haldar & Stiglitz, 2016). However, Grameen Bank has a strong track record of repayment (90% repayment over more than 25 million loans), while SKS India collapsed (facing default rates of around 90%). Grameen Bank leveraged "group lending," providing microloans ($500–$1000) to poor women (often in rural areas) organized into groups of 5–10 that met weekly with public repayment of loans (Bornstein, 1999; Yunus, 2007; Haldar & Stiglitz, 2013). The economic literature interpreted the microfinance success in terms of its joint liability structure (Stiglitz, 1990)—in theory at least, the Grameen Bank in Bangladesh provided loans to only some of the women in a group and would not provide additional loans until the first were repaid. When SKS Microfinance in India tried to replicate the joint liability model, however, it led to mass defaults.

The differences between these two banks suggest that the microfinance depended more on social pressure stemming from active peer-monitoring between borrowers and a robust relational contract between the bank and borrower than on the rational calculations of the joint liability. These strong community bonds incentivized the repayment of loans in the Grameen Bank but not SKS India. Compared to conventional formal banks, Grameen Bank deals were "informal" (requiring neither the provision of collateral by borrowers, nor formal contracts), which likely led to a more instinctual trust structure rather than transactional deliberative trust. Non-transactional deliberative trust was strengthened by the resilience of Grameen Bank in the face of the natural disasters of 1998. While SKS India ostensibly maintained this informal structure, it switched from a participatory model focused on poverty-reduction in the community to a for-profit model that made the lending relationship transactional rather than prosocial. Grameen Bank actively engaged instinctual trust with meaningful ritual and building of group social identity (weekly meetings, public repayment, reputation mechanisms, celebration of repayment within each group and socially punishing bad actors, and active peer monitoring that lowered information costs), until repayment became a matter of procedural trust. SKS India, on the other hand, relied on deliberative rather than instinctual trust, failing to create the community ties that activate the latter.

Experimental paradigms have found that when playing with the same people over and over again (i.e., living in a community), groups bifurcate, as more cooperators are identified, people become more willing to cooperate, while as more cheaters are identified, people become less willing to cooperate (Marwell & Ames, 1981; Fehr & Fischbacher, 2003). So observing that the others in your community are repaying the loans makes repaying them both the social norm of the community (instinctual), the thing one does (procedural), and the expectation that would be socially punished if not done (deliberative).

These neuroeconomic principles have direct consequences for how institutional policies change our behavior. Institutions that align with neuroeconomic principles that drive cooperation will drive people to cooperate, while misaligned institutions can have devastatingly negative consequences because they have made the wrong microeconomic assumptions. Importantly, while group lending was responsible for cooperative outcomes in the Grameen Bank, cooperation is not guaranteed by mechanically grouping individuals together; instead, groups need sufficient social salience to impact decisions: these nuances point to a need for new microeconomic models able to accommodate this socio- and neuroeconomic complexity.

4 Conclusions, Consequences, and Predictions

All normative optimizations contain process assumptions. Conversely, how data are stored and processed changes what the optimal choices will be. Neuroeconomics says that the computational processes implemented by neural systems that produce behavior provide us a new framework with which to understand microeconomics and suggest new models dependent on those computational processes.

In particular, the key prediction of this new framework is that the effects of behavioral manipulations will depend on which decision processes they access and how the behavioral manipulations interact with the underlying computational processes. Thus, for example, whether moral decisions are deontological or utilitarian in the classic trolley problem are driven by which decision process is activated (Greene et al., 2001; Moll et al., 2001); different framings are more likely to activate one decision process over the other, but it is the decision process that explains the behavior, not the framing (Costa et al., 2014). Some decision processes are sensitive to sunk costs, while others are not (Sweis et al., 2018a, b; Lind et al., 2023). A similar prediction can be made about loss aversion, the endowment effect, and other behavioral phenomena.

Furthermore, as we saw in the asset memory and contingency management examples, the specific computational processes underlying these behaviors have consequences as well. For example, making the second option more concrete makes it easier to imagine (Trope & Liberman, 2003), more likely to be pulled up in a search process (Kurth-Nelson et al., 2012), and thus slows down discounting rates (Peters & Büchel, 2010). Training someone to imagine future outcomes more deliberately can also slow down discounting rates (Radu et al., 2011) with consequences on drug demand curves (Snider et al., 2016; Stein et al., 2016). As another example, deliberative processes are based on sampling algorithms (Redish, 2016), and thus, like all sampling processes, they will be inconsistent, depending on which choices are sampled in which order (Stewart et al., 2006; Redish & Johnson, 2007); in contrast, procedural processes are based on well-practiced expertise and produce reliable, consistent action chains (Rand et al., 1998). This means that framing that pushes someone to be more deliberate in their choices (listing pros and cons) will likely produce more variable preferences than well-practiced procedural processes.

These consequences do not diminish with larger-scale policy questions, and thus, if one is going to attempt to create working institutions (such as microfinance lending institutions), one needs to take these computational processes into account. These computational processes have consequences both for our understanding of economic behavior and for the institutional policies that affect it. The neuroeconomic framework provides the entry point into these computational processes.

References

Adams, R. A., Stephan, K. E., Brown, H. R., et al. (2013). The computational anatomy of psychosis. *Frontiers in Psychiatry, 4*, 1–26.

Ahmed, S. H. (2010). Validation crisis in animal models of drug addiction: Beyond non-disordered drug use toward drug addiction. *Neuroscience and Biobehavioral Reviews, 35*, 172–184.

Ahmed, S. H., Kenny, P. J., Koob, G. F., & Markou, A. (2002). Neurobiological evidence for hedonic allostasis associated with escalating cocaine use. *Nature Neuroscience, 5*, 625–626.

Ainslie, G. (1992). *Picoeconomics*. Cambridge University Press.
Akerlof, G. A. (2005). *Explorations in pragmatic economics*. Oxford University Press.
Allen, S. E., Kizilcec, R. F., & Redish, A.D. (2024). A new model of trust based on neural information processing. *arXiv [econ.GN]*.
Andermann, M. L., & Lowell, B. B. (2017). Toward a wiring diagram understanding of appetite control. *Neuron, 95*, 757–778.
Arkes, H. R., & Blumer, C. (1985). The psychology of sunk cost. *Organizational Behavior and Human Decision Processes, 35*, 124–140.
Balewski, Z. Z., Elston, T. W., Knudsen, E. B., & Wallis, J. D. (2023). Value dynamics affect choice preparation during decision-making. *Nature Neuroscience, 26*, 1575–1583.
Balleine, B., & Dickinson, A. (1991). Instrumental performance following reinforcer devaluation depends upon incentive learning. *The Quarterly Journal of Experimental Psychology. B, 43B*, 279–296.
Bardo, M. T., Rowlett, J. K., & Harris, M. J. (1995). Conditioned place preference using opiate and stimulant drugs: A meta-analysis. *Neuroscience and Biobehavioral Reviews, 19*, 39–51.
Becker, G. S., & Murphy, K. M. (1988). A theory of rational addiction. *Journal of Political Economy, 96*, 675–700.
Bernstein, I. L., & Webster, M. M. (1980). Learned taste aversions in humans. *Physiology & Behavior, 25*, 363–366.
Berridge, K. C., & Robinson, T. E. (2003). Parsing reward. *Trends in Neurosciences, 26*, 507–513.
Bornstein, D. (1999). *The price of a dream: The story of the Grameen bank and the idea that is helping the poor to change their lives*. Simon & Schuster.
Bowles, S., & Gintis, H. (2013). *A cooperative species: Human reciprocity and its evolution*. Princeton University Press.
Breland, K., & Breland, M. (1961). The misbehavior of organisms. *The American Psychologist, 16*, 682–684.
Buckner, R. L., & Carroll, D. C. (2007). Self-projection and the brain. *Trends in Cognitive Sciences, 11*, 49–57.
Bushong, B., King, L. M., Camerer, C. F., & Rangel, A. (2010). Pavlovian processes in consumer choice: The physical presence of a good increases willingness-to-pay. *The American Economic Review, 100*, 1556–1571.
Camchong, J., Roediger, D., Fiecas, M., et al. (2023). Frontal tDCS reduces alcohol relapse rates by increasing connections from left dorsolateral prefrontal cortex to addiction networks. *Brain Stimulation, 16*, 1032–1040.
Camerer, C., Loewenstein, G., & Prelec, D. (2005). Neuroeconomics: How neuroscience can inform economics. *Journal of Economic Literature, 43*, 9–64.
Cantin, L., Lenoir, M., Augier, E., et al. (2010). Cocaine is low on the value ladder of rats: Possible evidence for resilience to addiction. *PLoS One, 5*, e11592.
Carandini, M., & Heeger, D. J. (2011). Normalization as a canonical neural computation. *Nature Reviews. Neuroscience, 13*, 51–62.
Chow, J. J., Beacher, N. J., Chabot, J. M., et al. (2022). Characterization of operant social interaction in rats: Effects of access duration, effort, peer familiarity, housing conditions, and choice between social interaction vs. food or remifentanil. *Psychopharmacology, 239*, 2093–2108.
Churchland, P. S., & Sejnowski, T. S. (1994). *The computational brain*. MIT Press.
Cisek, P., Puskas, G. A., & El-Murr, S. (2009). Decisions in changing conditions: The urgency-gating model. *The Journal of Neuroscience, 29*, 11560–11571.
Colaizzi, J. M., Flagel, S. B., Joyner, M. A., et al. (2020). Mapping sign-tracking and goal-tracking onto human behaviors. *Neuroscience and Biobehavioral Reviews, 111*, 84–94.
Cormen, T. H., Leiserson, C. E., & Rivest, R. L. (1990). *Introduction to algorithms*. MIT Press/McGraw-Hill.
Costa, A., Foucart, A., Hayakawa, S., et al. (2014). Your morals depend on language. *PLoS One*. 9(4): e94842.

Cushman, F. (2024). Computational social psychology. *Annual Review of Psychology, 75*, 625–652.
Davidson, R. M., Traxler, H. K., DeFulio, A., et al. (2024). Contingency management for monosubstance use disorders: Systematic review and assessment of predicted versus obtained effects. *Journal of Applied Behavior Analysis.*
Daw, N. D., Niv, Y., & Dayan, P. (2005). Uncertainty-based competition between prefrontal and dorsolateral striatal systems for behavioral control. *Nature Neuroscience, 8*, 1704–1711.
Dayan, P., & Abbott, L. F. (2001). *Theoretical neuroscience*. MIT Press.
Dayan, P., Niv, Y., Seymour, B., & Daw, N. D. (2006). The misbehavior of value and the discipline of the will. *Neural Networks, 19*, 1153–1160.
Debreu, G. (1959). *Theory of value: An axiomatic analysis of economic equilibrium.* Yale University Press.
Dezfouli, A., & Balleine, B. (2012). Habits, action sequences and reinforcement learning. *The European Journal of Neuroscience, 35*, 1036–1051.
Diamond, J. (1971). The Mauthner cell. In W. S. Hoar & D. J. Randall (Eds.), *Fish physiology* (pp. 265–346). Academic.
Dickinson, A., & Balleine, B. (1994). Motivational control of goal-directed action. *Animal Learning & Behavior, 22*, 1–18.
Domjan, M. (1998). *The principles of learning and behavior* (4th ed.). Brooks/Cole.
Elster, J. (1999). Gambling and addiction. In J. Elster & O.-J. Skog (Eds.), *Getting hooked* (pp. 208–234). Cambridge University Press.
Fehr, E., & Fischbacher, U. (2003). The nature of human altruism. *Nature, 425*, 785–791.
Friston, K. (2005). A theory of cortical responses. *Philosophical Transactions of the Royal Society of London. Series B, Biological Sciences, 360*, 815–836.
Gallagher, J. (2014). Learning about an infrequent event: Evidence from flood insurance take-up in the United States. *American Economic Journal of Applied Econometrics, 6*(3), 206–233.
Gibson, J. J. (1977). *The theory of affordances*. Hilldale, USA.
Gigerenzer, G., & Brighton, H. (2009). Homo Heuristicus: Why biased minds make better inferences. *Topics in Cognitive Science, 1*, 107–143.
Gläscher, J., Daw, N., Dayan, P., & O'Doherty, J. P. (2010). States versus rewards: Dissociable neural prediction error signals underlying model-based and model-free reinforcement learning. *Neuron, 66*, 585–595.
Glimcher, P. W., Camerer, C., & Poldrack, R. A. (Eds.). (2008). *Neuroeconomics: Decision making and the brain*. Academic.
Greene, J. D., Sommerville, R. B., Nystrom, L. E., et al. (2001). An fMRI investigation of emotional engagement and moral judgement. *Science, 293*, 2105–2108.
Grillner, S., & Wallén, P. (1985). Central pattern generators for locomotion, with special reference to vertebrates. *Annual Review of Neuroscience, 8*, 233–261.
Grossberg, S. (1976). Adaptive pattern classification and universal recoding: I. Parallel development and coding of neural feature detectors. *Biological Cybernetics, 23*, 121–134.
Gueguen, M. C. M., Anlló, H., Bonagura, D., et al. (2024). Recent opioid use impedes range adaptation in reinforcement learning in human addiction. *Biological Psychiatry, 95*, 974–984.
Guitart-Masip, M., Huys, Q. J. M., Fuentemilla, L., et al. (2012). Go and no-go learning in reward and punishment: Interactions between affect and effect. *NeuroImage, 62*, 154–166.
Haldar, A., & Stiglitz, J. E. (2013). Analyzing legal formality and informality: Lessons from the land titling and microfinance programs. In D. Kennedy & J. E. Stiglitz (Eds.), *Law and economics with Chinese characteristics: Institutions for promoting development in the twenty-first century* (pp. 112–147). Oxford University Press.
Haldar, A., & Stiglitz, J. E. (2016). Group lending, joint liability, and social capital: Insights from the Indian microfinance crisis. *Politics and Society, 44*, 459–497.
Hasselmo, M. E. (1993). Acetylcholine and learning in a cortical associative memory. *Neural Computation, 5*, 32–44.
Hebb, D. O. (1949). *The organization of behavior*. Wiley.

Hertz, J., Krogh, A., & Palmer, R. G. (1991). *Introduction to the theory of neural computation*. Addison-Wesley.

Heyman, G. (2009). *Addiction: A disorder of choice*. Harvard University Press.

Higgins, S. T., Delaney, D. D., Budney, A. J., et al. (1991). A behavioral approach to achieving initial cocaine abstinence. *The American Journal of Psychiatry, 148*, 1218–1224.

Hikosaka, O., Kim, H. F., Yasuda, M., & Yamamoto, S. (2014). Basal ganglia circuits for reward value-guided behavior. *Annual Review of Neuroscience, 37*, 289–306.

Hopfield, J. J. (1982). Neural networks and physical systems with emergent collective computational abilities. *Proceedings of the National Academy of Sciences of the United States of America, 79*, 2554–2558.

Howard, M. W., Fotedar, M. S., Datey, A. V., & Hasselmo, M. E. (2005). The temporal context model in spatial navigation and relational learning: Toward a common explanation of medial temporal lobe function across domains. *Psychological Review, 112*, 75–116.

Hursh, S. R. (1991). Behavioral economics of drug self-administration and drug abuse policy. *Journal of the Experimental Analysis of Behavior, 56*, 377–393.

Huynh, T., Alstatt, K., Abram, S. V., & Schmitzer-Torbert, N. (2021). Vicarious trial-and-error is enhanced during deliberation in human virtual navigation in a translational foraging task. *Frontiers in Behavioral Neuroscience, 15*, 586159.

Huys, Q. J. M., Eshel, N., O'Nions, E., et al. (2012). Bonsai trees in your head: How the Pavlovian system sculpts goal-directed choices by pruning decision trees. *PLoS Computational Biology, 8*, e1002410.

Jog, M. S., Kubota, Y., Connolly, C. I., et al. (1999). Building neural representations of habits. *Science, 286*, 1745–1749.

Kable, J. W., & Glimcher, P. W. (2009). The neurobiology of decision: Consensus and controversy. *Neuron, 63*, 733–745.

Kahana, M. J. (2020). Computational models of memory search. *Annual Review of Psychology, 71*, 107–138.

Kahneman, D. (2011). *Thinking, fast and slow*. Macmillan.

Kahneman, D., & Tversky, A. (1979). Prospect theory: An analysis of decision under risk. *Econometrica, 47*, 263–292.

Kahneman, D., & Tversky, A. (Eds.). (2000). *Choices, values, and frames*. Cambridge University Press.

Kahneman, D., Knetsch, J. L., & Thaler, R. H. (1991). The endowment effect, loss aversion, and status quo bias. *The Journal of Economic Perspectives, 5*, 193–206.

Kanodia, C., Bushman, R., & Dickhaut, J. (1989). Escalation errors and the sunk cost effect: An explanation based on reputation and information asymmetries. *Journal of Accounting Research, 27*, 59.

Kawato, M. (1999). Internal models for motor control and trajectory planning. *Current Opinion in Neurobiology, 9*, 718–727.

Kazinka, R., MacDonald, A. W., 3rd, & Redish, A. D. (2021). Sensitivity to sunk costs depends on attention to the delay. *Frontiers in Psychology, 12*, 604843.

Klein, G. (1999). *Sources of power: How people make decisions*. MIT Press.

Knudsen, E. B., & Wallis, J. D. (2022). Taking stock of value in the orbitofrontal cortex. *Nature Reviews Neuroscience, 23*, 428–438.

Kurth-Nelson, Z., & Redish, A. D. (2010). A reinforcement learning model of pre-commitment in decision making. *Frontiers in Behavioral Neuroscience, 4*, 184ff.

Kurth-Nelson, Z., Bickel, W. K., & Redish, A. D. (2012). A theoretical account of cognitive effects in delay discounting. *The European Journal of Neuroscience, 35*, 1052–1064.

LeDoux, J., & Daw, N. D. (2018). Surviving threats: Neural circuit and computational implications of a new taxonomy of defensive behaviour. *Nature Reviews Neuroscience, 19*, 269–282.

Lenoir, M., & Ahmed, S. H. (2007). Supply of a nondrug substitute reduces escalated heroin consumption. *Neuropsychopharmacology, 33*(9), 2272–2282.

Lenoir, M., Serre, F., Cantin, L., & Ahmed, S. H. (2007). Intense sweetness surpasses cocaine reward. *PLoS One, 2*, e698.

Levenstein, D., Alvarez, V. A., Amarasingham, A., et al. (2023). On the role of theory and modeling in neuroscience. *The Journal of Neuroscience, 43*, 1074–1088.

Levy, D. J., & Glimcher, P. W. (2012). The root of all value: A neural common currency for choice. *Current Opinion in Neurobiology, 22*, 1–12.

Lichtenstein, S., & Slovic, P. (Eds.). (2006). *The construction of preference*. Cambridge University Press.

Lieder, F., & Griffiths, T. L. (2019). Resource-rational analysis: Understanding human cognition as the optimal use of limited computational resources. *The Behavioral and Brain Sciences, 43*, e1.

Lind, E. B., Sweis, B. M., Asp, A. J., et al. (2023). A quadruple dissociation of reward-related behaviour in mice across excitatory inputs to the nucleus accumbens shell. *Communications Biology, 6*, 119.

Loftus, E. F., & Banaji, M. R. (1989). Memory modification and the role of the media. In *Suggestion and suggestibility* (pp. 279–293). Springer.

Loftus, E. F., & Palmer, J. C. (1974). Reconstruction of automobile destruction. *Journal of Verbal Learning and Verbal Behavior, 13*, 585–589.

Lussier, J. P., Heil, S. H., Mongeon, J. A., et al. (2006). A meta-analysis of voucher-based reinforcement therapy for substance use disorders. *Addiction, 101*, 192–203.

Marr, D. (1982). *Vision*. W. H. Freeman and Co.

Martin, L., & Iceberg, E. (2015). Quantifying social motivation in mice using operant conditioning. *Journal of Visualized Experiments, 102*, e53009.

Marwell, G., & Ames, R. E. (1981). Economists free ride, does anyone else?: Experiments on the provision of public goods, IV. *Journal of Public Economics, 15*, 295–310.

Mas-Colell, A., Whinston, M. D., & Green, J. R. (1995). *Microeconomic theory*. Oxford University Press.

Mau, W., Hasselmo, M. E., & Cai, D. J. (2020). The brain in motion: How ensemble fluidity drives memory-updating and flexibility. *eLife, 9*, e63550.

McAfee, R. P., Mialon, H. M., & Mialon, S. H. (2010). Do sunk costs matter? *Economic Inquiry, 48*, 323–336.

McClelland, J. L., & Rogers, T. T. (2003). The parallel distributed processing approach to semantic cognition. *Nature Reviews. Neuroscience, 4*, 310–322.

McClelland, J. L., McNaughton, B. L., & O'Reilly, R. C. (1995). Why there are complementary learning systems in the hippocampus and neocortex: Insights from the successes and failures of connectionist models of learning and memory. *Psychological Review, 102*, 419–457.

McCullough, M. E. (2020). *The kindness of strangers: How a selfish ape invented a new moral code*. Simon and Schuster.

McLaughlin, A. E., Diehl, G. W., & Redish, A. D. (2021). Potential roles of the rodent medial prefrontal cortex in conflict resolution between multiple decision-making systems. *International Review of Neurobiology, 158*, 249–281.

McNally, G. P., Johansen, J. P., & Blair, H. T. (2011). Placing prediction into the fear circuit. *Trends in Neurosciences, 34*(6), 283–292.

Mischel, W. (2014). *The marshmallow test: Mastering self-control*. Little, Brown, and Co.

Mitchell, K. J. (2023). *Free agents: How evolution gave us free will*. Princeton University Press.

Moll, J., Eslinger, P. J., & Oliveira-Souza, R. (2001). Frontopolar and anterior temporal cortex activation in a moral judgment task: Preliminary functional MRI results in normal subjects. *Arquivos de Neuro-Psiquiatria, 59*, 657–664.

Mueller, D., & de Wit, H. (2011). Conditioned place preference in rodents and humans. In J. Raber (Ed.), *Animal models of behavioral analysis* (pp. 133–152). Humana Press.

Mugan, U., Amemiya, S., Regier, P. S., & Redish, A. D. (2024). Navigation through the complex world—The neurophysiology of decision-making processes. In Y. Vandaele (Ed.), *Habits: Their definition, neurobiology, and role in addiction*. Springer.

Niv, Y., Joel, D., & Dayan, P. (2006). A normative perspective on motivation. *Trends in Cognitive Sciences, 10*, 375–381.

Padoa-Schioppa, C., & Assad, J. A. (2006). Neurons in the orbitofrontal cortex encode economic value. *Nature, 441*, 223–226.
Padoa-Schioppa, C., & Assad, J. A. (2008). The representation of economic value in the orbitofrontal cortex is invariant for changes of menu. *Nature Neuroscience, 11*, 95–102.
Pavlov, I. (1927). *Conditioned reflexes*. Oxford University Press.
Peters, J., & Büchel, C. (2010). Episodic future thinking reduces reward delay discounting through an enhancement of prefrontal-mediotemporal interactions. *Neuron, 66*, 138–148.
Petry, N. M. (2011). Contingency management: What it is and why psychiatrists should want to use it. *The Psychiatrist, 35*, 161–163.
Petry, N. M., Alessi, S. M., Barry, D., & Carroll, K. M. (2015). Standard magnitude prize reinforcers can be as efficacious as larger magnitude reinforcers in cocaine-dependent methadone patients. *Journal of Consulting and Clinical Psychology, 83*, 464–472.
Pettine, W. W., Raman, D. V., Redish, A. D., & Murray, J. D. (2023). Human generalization of internal representations through prototype learning with goal-directed attention. *Nature Human Behaviour, 7*, 442–463.
Preuss, T., Osei-Bonsu, P. E., Weiss, S. A., et al. (2006). Neural representation of object approach in a decision-making motor circuit. *The Journal of Neuroscience, 26*, 3454–3464.
Radu, P. T., Yi, R., Bickel, W. K., et al. (2011). A mechanism for reducing delay discounting by altering temporal attention. *Journal of the Experimental Analysis of Behavior, 96*, 363–385.
Rand, M. K., Hikosaka, O., Miyachi, S., et al. (1998). Characteristics of a long-term procedural skill in the monkey. *Experimental Brain Research, 118*, 293–297.
Rangel, A., Camerer, C., & Montague, P. R. (2008). A framework for studying the neurobiology of value-based decision making. *Nature Reviews Neuroscience, 9*, 545–556.
Rao, R. P. N., & Ballard, D. H. (1999). Predictive coding in the visual cortex: A functional interpretation of some extra-classical receptive-field effects. *Nature Neuroscience, 2*, 79–87.
Redish, A. D. (2004). The addiction process: A computational process gone awry. *Science, 306*(5703), 1944–1947.
Redish, A. D. (2013). *The mind within the brain: How we make decisions and how those decisions go wrong*. Oxford University Press.
Redish, A. D. (2016). Vicarious trial and error. *Nature Reviews Neuroscience, 17*, 147–159.
Redish, A. D. (2022). *Changing how we choose: The new science of morality*. MIT Press.
Redish, A. D., & Johnson, A. (2007). A computational model of craving and obsession. *Annals of the New York Academy of Sciences, 1104*, 324–339.
Redish, A. D., Jensen, S., & Johnson, A. (2008). A unified framework for addiction: Vulnerabilities in the decision process. *The Behavioral and Brain Sciences, 31*, 415–487.
Redish, A. D., Schultheiss, N. W., & Carter, E. C. (2016). The computational complexity of valuation and motivational forces in decision-making processes. *Current Topics in Behavioral Neurosciences, 27*, 313–333.
Redish, A. D., Abram, S. V., Cunningham, P. J., et al. (2022). Sunk cost sensitivity during change-of-mind decisions is informed by both the spent and remaining costs. *Communications Biology, 5*, 1337.
Regier, P. S., & Redish, A. D. (2015). Contingency management and deliberative decision-making processes. *Frontiers in Psychiatry, 6*, 0076.
Rich, E. L., & Wallis, J. D. (2016). Decoding subjective decisions from orbitofrontal cortex. *Nature Neuroscience, 19*, 973–980.
Rieke, F., Warland, D., de Ruyter van Steveninck, R, & Bialek, W. (1997). *Spikes*. MIT Press.
Robinson, T. E., & Berridge, K. C. (2003). Addiction. *Annual Reviews of Psychology, 54*, 25–53.
Robinson, T. E., & Flagel, S. B. (2009). Dissociating the predictive and incentive motivational properties of reward-related cues through the study of individual differences. *Biological Psychiatry, 65*, 869–873.
Rumelhart, D. E., McClelland, J. L., and the PDP Research Group. (1986). *Parallel distributed processing*. MIT Press.

Runge, C. F., Johnson, J. A., Nelson, E., & Redish, A. D. (2023). A neuroscience-based analysis of impacts of disaster memory on economic valuation. *Journal of Neuroscience, Psychology, and Economics, 16*, 24–49.
Sacchi, D. L. M., Agnoli, F., & Loftus, E. F. (2007). Changing history: Doctored photographs affect memory for past public events. *Applied Cognitive Psychology, 21*, 1005–1022.
Samejima, K., Ueda, Y., Doya, K., & Kimura, M. (2005). Representation of action-specific reward values in the striatum. *Science, 310*, 1337–1340.
Samuelson, P. A. (1938). A note on the pure theory of consumer's behaviour. *Economica, 5*, 61–71.
Schüll, N. D. (2012). *Addiction by design: Machine gambling in Las Vegas*. Princeton University Press.
Shannon, C. (1948). A mathematical theory of communication. *The Bell System Technical Journal, 27*, 379–423.
Sharpe, M. J., Batchelor, H. M., Mueller, L. E., et al. (2021). Past experience shapes the neural circuits recruited for future learning. *Nature Neuroscience, 24*, 391–400.
Sherrington, S. C. S. (1906). *The integrative action of the nervous system*. Yale University Press.
Shleifer, A. (2000). *Inefficient markets: An introduction to behavioral finance: An introduction to behavioral finance*. Oxford University Press.
Simon, H. A. (1972). Theories of bounded rationality. *Decision and Organization, 1*, 161–176.
Sleesman, D. J., Conlon, D. E., McNamara, G., & Miles, J. E. (2012). Cleaning up the big muddy: A meta-analytic review of the determinants of escalation of commitment. *AMJ, 55*, 541–562.
Smith, K. S., & Graybiel, A. M. (2013). A dual operator view of habitual behavior reflecting cortical and striatal dynamics. *Neuron, 79*, 361–374.
Snider, S. E., LaConte, S. M., & Bickel, W. K. (2016). Episodic future thinking: Expansion of the temporal window in individuals with alcohol dependence. *Alcoholism, Clinical and Experimental Research, 40*, 1558–1566.
Staw, B. (1976). Knee-deep in the big muddy: A study of escalating commitment to a chosen course of action. *Organizational Behavior and Human Performance, 16*, 27–44.
Staw, B. M., & Ross, J. (1989). Understanding behavior in escalation situations. *Science, 246*, 216–220.
Stein, J. S., Wilson, A. G., Koffarnus, M. N., et al. (2016). Unstuck in time: Episodic future thinking reduces delay discounting and cigarette smoking. *Psychopharmacology, 233*, 3771–3778.
Sterzer, P., Adams, R. A., Fletcher, P., et al. (2018). The predictive coding account of psychosis. *Biological Psychiatry, 84*, 634–643.
Stewart, N., Chater, N., & Brown, G. D. A. (2006). Decision by sampling. *Cognitive Psychology, 53*, 1–26.
Stigler, G. J., & Becker, G. S. (1977). De Gustibus Non Est Disputandum. *The American Economic Review, 67*, 76–90.
Stiglitz, J. E. (1990). Peer monitoring and credit markets. *World Bank Economic Review, 4*, 351–366.
Stott, J. J., & Redish, A. D. (2014). A functional difference in information processing between orbitofrontal cortex and ventral striatum during decision-making behavior. *Philosophical Transactions of the Royal Society of London. Series B, Biological Sciences, 369*(1655), 20130472.
Stroop, J. R. (1935). Studies of interference in serial verbal reactions. *Journal of Experimental Psychology, 18*, 643–662.
Sweis, B. M., Abram, S. V., Schmidt, B. J., et al. (2018a). Sensitivity to "sunk costs" in mice, rats, and humans. *Science, 361*, 178–181.
Sweis, B. M., Larson, E. B., Redish, A. D., & Thomas, M. J. (2018b). Altering gain of the infralimbic-to-accumbens shell circuit alters economically dissociable decision-making algorithms. *Proceedings of the National Academy of Sciences of the United States of America, 115*, E6347–E6355.
Sweis, B. M., Thomas, M. J., & Redish, A. D. (2018c). Mice learn to avoid regret. *PLoS Biology, 16*, e2005853.

Sweis, B. M., Mau, W., Rabinowitz, S., & Cai, D. J. (2021). Dynamic and heterogeneous neural ensembles contribute to a memory engram. *Current Opinion in Neurobiology, 67*, 199–206.
Tang, W., Shin, J. D., & Jadhav, S. P. (2021). Multiple time-scales of decision-making in the hippocampus and prefrontal cortex. *eLife, 10*.
Trope, Y., & Liberman, N. (2003). Temporal construal. *Psychological Review, 110*, 403–421.
Tzschentke, T. M. (1998). Measuring reward with the conditioned place preference paradigm: A comprehensive review of drug effects, recent progress and new issues. *Progress in Neurobiology, 56*, 613–672.
van der Meer, M. A. A., Kurth-Nelson, Z., & Redish, A. D. (2012). Information processing in decision-making systems. *The Neuroscientist, 18*, 342–359.
Venniro, M., Zhang, M., Caprioli, D., et al. (2018). Volitional social interaction prevents drug addiction in rat models. *Nature Neuroscience, 21*, 1520–1529.
Von Neumann, J., & Morgenstern, O. (1944). *Theory of games and economic behavior*. Wiley.
Wallis, J. D., & Kennerley, S. W. (2011). Contrasting reward signals in the orbitofrontal cortex and anterior cingulate cortex. *Annals of the New York Academy of Sciences, 1239*, 33–42.
Webb, R., Glimcher, P. W., & Louie, K. (2021). The normalization of consumer valuations: Context-dependent preferences from neurobiological constraints. *Management Science, 67*, 93–125.
Wells, G. L., & Loftus, E. F. (Eds.). (1984). *Eyewitness testimony*. Cambridge University Press.
Wikenheiser, A., Stephens, D. W., & Redish, A. D. (2013). Subjective costs drive overly-patient foraging strategies in rats on an intertemporal foraging task. *Proceedings of the National Academy of Sciences of the United States of America, 110*, 8308–8313.
Wilson, D. S. (2002). *Darwin's cathedral: Evolution, religion, and the nature of society*. University of Chicago Press.
Wilson, D. S. (2015). *Does altruism exist?: Culture, genes, and the welfare of others*. Yale University Press.
Yunus, M. (2007). *Banker to the poor: Micro-lending and the battle against world poverty*. Perseus Books.

Index

D. V. Smith et al. (eds.), *Neuroeconomics: Core Topics and Current Directions*,
https://doi.org/10.1007/978-3-032-02925-6

MIX
Papier aus verantwortungsvollen Quellen
Paper from responsible sources
FSC® C105338

If you have any concerns about our products,
you can contact us on
ProductSafety@springernature.com

In case Publisher is established outside the EU,
the EU authorized representative is:
Springer Nature Customer Service Center GmbH
Europaplatz 3, 69115 Heidelberg, Germany

Printed by Libri Plureos GmbH
in Hamburg, Germany